THE REWIRING BRAIN

THE REWIRING BRAIN

A Computational Approach to Structural Plasticity in the Adult Brain

Edited by

ARJEN VAN OOYEN
VU University Amsterdam, Amsterdam, The Netherlands

MARKUS BUTZ-OSTENDORF
Biomax Informatics AG, Planegg, Germany

ELSEVIER

ACADEMIC PRESS
An imprint of Elsevier

Academic Press is an imprint of Elsevier
125 London Wall, London EC2Y 5AS, United Kingdom
525 B Street, Suite 1800, San Diego, CA 92101-4495, United States
50 Hampshire Street, 5th Floor, Cambridge, MA 02139, United States
The Boulevard, Langford Lane, Kidlington, Oxford OX5 1GB, United Kingdom

British Library Cataloguing-in-Publication Data
A catalogue record for this book is available from the British Library

Library of Congress Cataloging-in-Publication Data
A catalog record for this book is available from the Library of Congress

ISBN: 978-0-12-803784-3

Publisher: Nikki Levy
Acquisition Editor: Melanie Tucker
Editorial Project Manager: Kristi Anderson
Production Project Manager: Chris Wortley
Cover Designer: Matthew Limbert

Typeset by MPS Limited, Chennai, India

Contents

II

HOMEOSTATIC STRUCTURAL PLASTICITY

5. Network Formation Through Activity-Dependent Neurite Outgrowth: A Review of a Simple Model of Homeostatic Structural Plasticity

ARJEN VAN OOYEN

6. Clustered Arrangement of Inhibitory Neurons Can Lead to Oscillatory Dynamics in a Model of Activity-Dependent Structural Plasticity

ROSANNA C. BARNARD, ISTVAN Z. KISS, SIMON F. FARMER AND LUC BERTHOUZE

7. A Detailed Model of Homeostatic Structural Plasticity Based on Dendritic Spine and Axonal Bouton Dynamics

MARKUS BUTZ-OSTENDORF AND ARJEN VAN OOYEN

8. Critical Periods Emerge from Homeostatic Structural Plasticity in a Full-Scale Model of the Developing Cortical Column

SEBASTIAN RINKE, MIKAËL NAVEAU, FELIX WOLF AND MARKUS BUTZ-OSTENDORF

9. Lesion-Induced Dendritic Remodeling as a New Mechanism of Homeostatic Structural Plasticity in the Adult Brain

STEFFEN PLATSCHEK, HERMANN CUNTZ, THOMAS DELLER AND PETER JEDLICKA

III

STRUCTURAL PLASTICITY AND CONNECTIVITY

10. The Role of Structural Plasticity in Producing Nonrandom Neural Connectivity

PAUL MILLER

11. Structural Plasticity and the Generation of Bidirectional Connectivity

DANIEL MINER, FELIX Z. HOFFMANN, FLORENCE KLEBERG AND JOCHEN TRIESCH

12. Spike Timing-Dependent Structural Plasticity of Multicontact Synaptic Connections

MORITZ DEGER

13. Selection of Synaptic Connections by Wiring Plasticity for Robust Learning by Synaptic Weight Plasticity

NAOKI HIRATANI AND TOMOKI FUKAI

IV

STRUCTURAL PLASTICITY AND LEARNING AND MEMORY

14. Within a Spine's Reach

WILL DEBELLO AND KAREN ZITO

15. Impact of Structural Plasticity on Memory Capacity

PANAGIOTIS BOZELOS AND PANAYIOTA POIRAZI

16. Long-Term Information Storage by the Interaction of Synaptic and Structural Plasticity

MICHAEL FAUTH, FLORENTIN WÖRGÖTTER AND CHRISTIAN TETZLAFF

17. Impact of Structural Plasticity on Memory Formation and Decline

ANDREAS KNOBLAUCH

V

NEUROGENESIS-RELATED STRUCTURAL PLASTICITY

18. Adult Neurogenesis and Synaptic Rewiring in the Hippocampal Dentate Gyrus

ARJEN VAN OOYEN, GERTRAUD TEUCHERT-NOODT, KEREN GRAFEN AND MARKUS BUTZ-OSTENDORF

19. Modifications in Network Structure and Excitability May Drive Differential Activity-Dependent Integration of Granule Cells into Dentate Gyrus Circuits During Normal and Pathological Adult Neurogenesis

QUINTON SKILLING, JAMES ROACH, ALISON L. ALTHAUS, GEOFFREY G. MURPHY, LEONARD SANDER AND MICHAL ZOCHOWSKI

20. Computational Perspectives on Adult Neurogenesis

KRISTOFOR D. CARLSON, FRED ROTHGANGER AND JAMES B. AIMONE

21. Restricted Boltzmann Machine Models of Hippocampal Coding and Neurogenesis

RORY FINNEGAN, MARK SHAW AND SUZANNA BECKER

VI
STRUCTURAL PLASTICITY AND PATHOLOGY

22. Modeling the Impact of Lesions in the Brain

SOL LIM, FRANCES HUTCHINGS AND MARCUS KAISER

23. Network Models of Epilepsy-Related Pathological Structural and Functional Alterations in the Dentate Gyrus

IVAN RAIKOV, MARK PLITT AND IVAN SOLTESZ

24. Computational Models of Stroke Recovery

VITTORIO SANGUINETI

25. Neural Plasticity in Human Brain Connectivity: The Effects of Deep Brain Stimulation

TIM J. VAN HARTEVELT, HENRIQUE M. FERNANDES, ANGUS B.A. STEVNER, GUSTAVO DECO AND MORTEN L. KRINGELBACH

List of Contributors

James B. Aimone Sandia National Laboratories, Albuquerque, NM, United States

Alison L. Althaus University of Michigan, Ann Arbor, MI, United States

Rosanna C. Barnard University of Sussex, Falmer, United Kingdom

Suzanna Becker McMaster University, Hamilton, ON, Canada

Luc Berthouze University of Sussex, Falmer, United Kingdom

Panagiotis Bozelos Foundation for Research and Technology-Hellas (FORTH), Crete, Greece; Democritus University of Thrace, Alexandroupolis, Greece

Craig E. Brown University of Victoria, Victoria, BC, Canada; University of British Columbia, Vancouver, BC, Canada

Markus Butz-Ostendorf Biomax Informatics AG, Planegg, Germany

Kristofor D. Carlson Sandia National Laboratories, Albuquerque, NM, United States

Hermann Cuntz Goethe University Frankfurt, Frankfurt, Germany; Ernst Strüngmann Institute (ESI) for Neuroscience in Cooperation with Max Planck Society, Frankfurt, Germany; Frankfurt Institute for Advanced Studies, Frankfurt, Germany

Will DeBello University of California, Davis, CA, United States

Gustavo Deco Pompeu Fabra University, Barcelona, Spain

Moritz Deger University of Cologne, Cologne, Germany

Thomas Deller Goethe University Frankfurt, Frankfurt, Germany

Simon F. Farmer Institute of Neurology, London, United Kingdom

Michael Fauth University of Göttingen, Göttingen, Germany

Henrique M. Fernandes University of Oxford, Oxford, United Kingdom; Aarhus University, Aarhus, Denmark

Rory Finnegan McMaster University, Hamilton, ON, Canada

Tomoki Fukai RIKEN Brain Science Institute, Wako, Japan; Japan Science and Technology Agency, Kawaguchi, Japan

Shaoyu Ge Stony Brook University, Stony Brook, NY, United States

Kimberly Gerrow University of Victoria, Victoria, BC, Canada

Keren Grafen Office of Biological Medicine, Bielefeld, Germany

Naoki Hiratani RIKEN Brain Science Institute, Wako, Japan; University College London, London, United Kingdom

Felix Z. Hoffmann Frankfurt Institute for Advanced Studies, Frankfurt am Main, Germany; Max Planck Institute for Brain Research, Frankfurt am Main, Germany

Anthony Holtmaat University of Geneva, Geneva, Switzerland; Chair Alain Rossier

Frances Hutchings Newcastle University, Newcastle upon Tyne, United Kingdom

Peter Jedlicka Goethe University Frankfurt, Frankfurt, Germany

Marcus Kaiser Newcastle University, Newcastle upon Tyne, United Kingdom

Vassilis Kehayas University of Geneva, Geneva, Switzerland; Lemanic Neuroscience Doctoral School, Geneva, Switzerland

Gregory W. Kirschen Stony Brook University, Stony Brook, NY, United States

Istvan Z. Kiss University of Sussex, Falmer, United Kingdom

Florence Kleberg Frankfurt Institute for Advanced Studies, Frankfurt am Main, Germany

Andreas Knoblauch University of California, Davis, CA, United States

Morten L. Kringelbach University of Oxford, Oxford, United Kingdom; Aarhus University, Aarhus, Denmark

Sol Lim Newcastle University, Newcastle upon Tyne, United Kingdom; Indiana University, Bloomington, IN, United States

Paul Miller Brandeis University, Waltham, MA, United States

Daniel Miner Frankfurt Institute for Advanced Studies, Frankfurt am Main, Germany

Geoffrey G. Murphy University of Michigan, Ann Arbor, MI, United States; University of Michigan Medical School, Ann Arbor, MI, United States

Mikaël Naveau Jülich Aachen Research Alliance, Forschungszentrum Jülich, Jülich, Germany; Caen Normandy University, Caen, France

Steffen Platschek Goethe University Frankfurt, Frankfurt, Germany

Mark Plitt Stanford University, Stanford, CA, United States

Panayiota Poirazi Foundation for Research and Technology-Hellas (FORTH), Crete, Greece

Ivan Raikov Stanford University, Stanford, CA, United States

Sebastian Rinke Technische Universität Darmstadt, Darmstadt, Germany

James Roach University of Michigan, Ann Arbor, MI, United States

Fred Rothganger Sandia National Laboratories, Albuquerque, NM, United States

Kurt A. Sailor Pasteur Institute, Paris, France

Leonard Sander University of Michigan, Ann Arbor, MI, United States

Vittorio Sanguineti University of Genoa, Genova, Italy

Mark Shaw University of Toronto, Toronto, ON, Canada

Quinton Skilling University of Michigan, Ann Arbor, MI, United States

Ivan Soltesz Stanford University, Stanford, CA, United States

Angus B.A. Stevner University of Oxford, Oxford, United Kingdom; Aarhus University, Aarhus, Denmark

Christian Tetzlaff Max Planck Institute for Dynamics and Self-Organization, Göttingen, Germany

Gertraud Teuchert-Noodt University of Bielefeld, Bielefeld, Germany

Jochen Triesch Frankfurt Institute for Advanced Studies, Frankfurt am Main, Germany

Tim J. van Hartevelt University of Oxford, Oxford, United Kingdom; Aarhus University, Aarhus, Denmark

Arjen van Ooyen VU University Amsterdam, Amsterdam, The Netherlands

Felix Wolf Technische Universität Darmstadt, Darmstadt, Germany

Florentin Wörgötter University of Göttingen, Göttingen, Germany

Karen Zito University of California, Davis, CA, United States

Michal Zochowski University of Michigan, Ann Arbor, MI, United States

Editorial

1 INTRODUCTION

Current thinking about how the brain learns from experience and encodes memories focuses on changes in synaptic strength (synaptic plasticity), while neuronal morphology and the pattern of synaptic connections are usually considered fixed. However, the adult brain is not as hard-wired as traditionally thought. By extending or withdrawing axons and dendrites, or changing the number of axonal boutons and dendritic spines (the pre- and postsynaptic parts of synapses), neurons can establish new synaptic connections or break existing ones (structural plasticity).[1] Structural changes and rewiring of connections accompany memory formation[2] and are triggered by adult neurogenesis,[3] brain and peripheral lesions,[4–6] and neurodegeneration.[7] Structural plasticity is known since the late 1960s, but its guiding principles and relevance for brain function and pathology remain poorly understood.

Exploring structural plasticity can be greatly assisted by mathematical and computational modeling.[8] In general, building models forces one to be precise and self-consistent, potentially exposing inconsistencies, hidden assumptions and missing pieces of experimental data.[9] Importantly, models enable one to deduce the implications of cellular and molecular processes for higher-level system dynamics. Models can test the plausibility of hypotheses, generate new hypotheses and make testable predictions, thereby guiding further experimental research. Moreover, models can help explain and integrate existing experimental data.[10]

Despite the evidence for structural plasticity in the brain, most traditional neural network models have fixed neuronal morphologies and a static connectivity pattern, with plasticity merely arising from changes in the strength of existing synapses (synaptic plasticity). In this volume, we for the first time bring together modeling studies that investigate the implications of structural plasticity for brain function and pathology. The book opens with a section on the experimental background of structural plasticity, followed by sections on models of homeostatic structural plasticity, models exploring the implications of structural plasticity for connectivity and learning and memory, models of neurogenesis-related structural plasticity, and models of structural plasticity in neurological disorders.

2 EXPERIMENTAL BACKGROUND

Chapter 1, Structural Plasticity and Cortical Connectivity, reviews the current state of knowledge of structural plasticity in the adult brain. Structural plasticity occurs at multiple levels, ranging from changes in axonal boutons and dendritic spines to large-scale rearrangements of axonal and dendritic trees. The chapter describes structural plasticity following enriched experience and sensory deprivation or stimulation, outlines the implications of structural plasticity for cognition and cortical connectivity, and discusses the activity-dependent and -independent processes that control structural plasticity.

In the hippocampus and the olfactory bulb, new neurons are continuously generated and integrated into pre-existing networks throughout life. Chapter 2, Structural Plasticity Induced by Adult Neurogenesis, reviews the morphological and electrophysiological evidence for neurogenesis-induced synaptic rewiring, highlighting the progress that has been made in understanding the advantages that adult-generated cells confer to brain plasticity and function.

After stroke, neurons that survive must rebuild and establish new synaptic connections. Chapter 3, Structural Neural Plasticity During Stroke Recovery, provides an up-to-date perspective on structural plasticity in the stroke-affected brain, reviewing experimental data indicating that axonal sprouting, dendritic spine formation and synaptogenesis surge in the weeks following stroke, causing rewiring of connections and partial recovery of lost functions.

As shown in Chapter 2, Structural Plasticity Induced by Adult Neurogenesis, neurogenesis-induced rewiring occurs in a manner that maintains a homeostatic balance of electrical activity. Likewise, Chapter 3, Structural Neural Plasticity During Stroke Recovery, discusses the possibility that structural changes following stroke may represent a homeostatic response to restore normal levels of activity. Chapter 4, Is Lesion-Induced Synaptic Rewiring Driven by Activity Homeostasis?, further elaborates on the role of activity homeostasis in synaptic rewiring, arguing that after loss of input, neurons strive to bring their activity back to a desired level by making new synaptic connections or breaking existing ones. The chapter reviews in vitro and in vivo experiments, including stroke and sensory deprivation studies, that suggest that activity homeostasis may drive the spine dynamics and axonal and dendritic remodeling that underlie the restructuring of synaptic connectivity following lesions.

3 HOMEOSTATIC STRUCTURAL PLASTICITY

Chapter 5, Network Formation Through Activity-Dependent Neurite Outgrowth: A Review of a Simple Model of Homeostatic Structural Plasticity, reviews one of the first models of (homeostatic) structural plasticity, showing how a wide range of phenomena can emerge from activity-dependent neurite outgrowth. In the model, each neuron grows out when its electrical activity is below a homeostatic set-point and retracts when it is above the set-point. As a result, the neurons self-organize, via a transient phase of high connectivity, into a synaptically connected network. At equilibrium the network has critical connectivity and characteristic differences in cell size, as well as the ability to self-repair following cell loss (stroke) and to reorganize after insertion of new cells (neurogenesis).

Chapter 6, Clustered Arrangement of Inhibitory Neurons Can Lead to Oscillatory Dynamics in a Model of Activity-Dependent Structural Plasticity, uses the same model to demonstrate that not only the ratio of inhibitory to excitatory cells but also the spatial arrangement of cells determines network activity and structure, with spatial clustering of inhibitory cells leading to complex oscillations in activity and connectivity.

Chapter 7, A Detailed Model of Homeostatic Structural Plasticity Based on Dendritic Spine and Axonal Bouton Dynamics, presents a detailed model of homeostatic structural plasticity based on dendritic spine and axonal bouton dynamics. In the model, spine and bouton numbers are governed by bell-shaped growth curves, with an optimal neuronal activity level for the creation of new spines and boutons, a homeostatic set-point above which, and a minimum level below which, spines and boutons are deleted. The model serves as proof-of-concept that brain repair after lesions can be brought about by homeostatic structural plasticity, producing functional and structural changes like those observed experimentally after lesions, such as spine turnover, axonal spouting, activity restoration, remapping of inputs, and alterations in network topology.

Chapter 8, Critical Periods Emerge From Homeostatic Structural Plasticity in a Full-Scale Model of the Developing Cortical Column, describes how the model of structural plasticity (MSP) introduced in the previous chapter is implemented in NEST, a well-known simulator for large-scale spiking neural networks. Using NEST simulations the authors show that critical periods emerge from homeostatic structural plasticity in a full-scale model of the developing cortical column. The timing of these periods was found to be set by the developing inhibition. Furthermore, the chapter presents a new algorithm for optimally parallelizing MSP that makes use of the Barnes-Hut algorithm from particle physics.

Chapter 9, Lesion-Induced Dendritic Remodeling as a New Mechanism of Homeostatic Structural Plasticity in the Adult Brain, proposes that in addition to the activity-dependent spine and bouton dynamics explored in the previous two chapters, another form of homeostatic structural plasticity based on remodeling of dendritic morphology may be operating after lesions. Using realistic compartmental models of healthy and denervated cells, the authors show that denervation-induced dendritic retraction boosts neuronal excitability, precisely compensating the denervation-associated loss of synapses. This dendritic plasticity is hypothesized to act in synergy with structural plasticity of the axon initial segment.

The next two sections explore further forms of structural plasticity, the interactions between structural plasticity and synaptic plasticity, and the implications of structural plasticity for connectivity and learning and memory.

4 STRUCTURAL PLASTICITY AND CONNECTIVITY

Chapter 10, The Role of Structural Plasticity in Producing Nonrandom Neural Connectivity, studies the impact of reconfiguration of neural connections on circuit topology and connection strength distribution. Using a model in which connections are randomly formed but maintained by Hebbian-like synaptic plasticity, the author shows that structural plasticity yields unimodal synaptic strength distributions and highly interconnected clusters of neurons that are responsive to the same stimulus, just as in cortical networks.

Using an established model of a self-organizing neural network with structural plasticity, Chapter 11, Structural Plasticity and the Generation of Bidirectional Connectivity, shows that any inhomogeneity in connection probability will lead to the observed overabundance of bidirectional connections in cortical microcircuits as compared with random networks. The authors discuss various factors influencing this overrepresentation, including spike timing-dependent synaptic plasticity of excitatory and inhibitory synapses.

Chapter 12, Spike Timing-Dependent Structural Plasticity of Multicontact Synaptic Connections, summarizes models of spike timing-dependent synaptic and structural plasticity, showing how cooperative synapse formation and the observed multisynapse connections in cortex can arise from the dependence of dendritic spine dynamics on pre- and postsynaptic spiking. In addition, the chapter demonstrates how cooperation between spines leads to stable synaptic connections with highly reliable synaptic transmission.

Chapter 13, Selection of Synaptic Connections by Wiring Plasticity for Robust Learning by Synaptic Weight Plasticity, shows that structural plasticity can generate functionally efficient connectivity structures. Using a simple model of an inference task, the authors found that a Hebbian form of structural plasticity selectively eliminates noninformative connections and so can enhance learning by synaptic plasticity. Furthermore, the generated connectivity was shown to be noise resilient, a property especially beneficial in rapidly changing environments.

5 STRUCTURAL PLASTICITY AND LEARNING AND MEMORY

Chapter 14, Within a Spine's Reach, reviews computational and experimental studies that address how dendritic spine dynamics can change synaptic connectivity and the dendritic distribution of synapses. The authors consider how spine dynamics might be harnessed to fine-tune information processing and storage, and discuss the importance of microscale connectivity for neurorealistic simulation and neuromorphic computing.

Chapter 15, Impact of Structural Plasticity on Memory Capacity, focuses on computational models that simulate learning using structural plasticity, showing that activity-dependent reallocation of synapses (rewiring) and dendritic remodeling enhance learning and storage capacity by exploiting nonlinear dendritic mechanisms. The authors also discuss efforts towards hardware implementation of models of structural plasticity.

Chapter 16, Long-Term Information Storage by the Interaction of Synaptic and Structural Plasticity, concentrates on multisynapse connections and the interactions between synaptic and structural plasticity. The chapter shows how synaptic connectivity can be controlled by external stimulation and how multisynapse connections enable long-term memory storage despite the continuous turnover of synapses observed in cortical networks.

Chapter 17, Impact of Structural Plasticity on Memory Formation and Decline, further discusses possible functions of structural plasticity for memory formation, showing that structural plasticity increases storage efficiency and stability of memories in Willshaw- and Hopfield-type network models. Moreover, the author demonstrates that models with structural plasticity can explain cognitive phenomena such as retrograde amnesia better than models based solely on synaptic plasticity.

6 NEUROGENESIS-RELATED STRUCTURAL PLASTICITY

Chapter 18, Adult Neurogenesis and Synaptic Rewiring in the Hippocampal Dentate Gyrus, describes a combined experimental and computational study on synaptic rewiring induced by the integration of adult-generated cells into the hippocampal circuitry. The chapter shows that higher rates of cell proliferation lead surprisingly to less synaptic rewiring. This experimental finding is explained by the model of homeostatic structural plasticity introduced in Chapter 7, A Detailed Model of Homeostatic Structural Plasticity Based on Dendritic Spine and Axonal Bouton Dynamics, which is here extended to incorporate neurogenesis and apoptosis.

Chapter 19, Modifications in Network Structure and Excitability May Drive Differential Activity-Dependent Integration of Granule Cells into Dentate Gyrus Circuits During Normal and Pathological Adult Neurogenesis, presents a computational model showing that both intrinsic cellular excitability and network connectivity, through their effect on network-wide activity patterns, critically influence the integration of new-born cells into established circuits. The model also reveals under what conditions aberrant structural integration of new cells may occur in chronic pathologies such as epilepsy.

Chapter 20, Computational Perspectives on Adult Neurogenesis, reviews a variety of computational models, ranging from simple abstract neural network models to complex biology-based models, that have been built to understand the functional role of adult neurogenesis. Proposed functions of neurogenesis include improving pattern separation, encoding temporal context, and learning novel information. The authors also introduce a neural modeling language called Neurons to Algorithms, suitable for modeling neurogenesis and structural plasticity in general.

Chapter 21, Restricted Boltzmann Machine Models of Hippocampal Coding and Neurogenesis, challenges the widely held assumption that neurogenesis increases pattern separation. The authors propose that the hippocampus forms a probabilistic, generative model of its input. Using Restricted Boltzmann Machine models, they show that networks with neurogenesis are more robust against interference but produce less pattern separation. Furthermore, they show that models with neurogenesis can learn representations of locations and sequential events.

7 STRUCTURAL PLASTICITY AND PATHOLOGY

In stroke, both the primary lesion site and adjacent and remote brain areas experience cell death and loss of connections. Chapter 22, Modeling the Impact of Lesions in the Brain, reviews computational models that explore the impact of such structural changes on brain topology and dynamics. In addition, the authors discuss how the brain may adapt to structural damage, and how their models can be applied to other brain disorders with aberrant connectivity, such as epilepsy, schizophrenia and Parkinson's disease.

Epilepsy can develop after a brain-damaging insult, such as stroke, through a number of molecular and cellular changes (epileptogenesis). Chapter 23, Network Models of Epilepsy-Related Pathological Structural and Functional Alterations in the Dentate Gyrus, reviews computational models that address the impact on brain dynamics of structural alterations in the adult hippocampus during epileptogenesis, such as hilar cell loss and mossy fiber sprouting. The models examine the relationship between pathological network reorganization and the emergence of hypersynchronous network activity.

Neuromotor recovery after stroke is the result of a complex interplay between activity-dependent reorganization of brain areas close to the lesion, the recruitment of new neural pathways, and the development of novel motor strategies. Chapter 24, Computational Models of Stroke Recovery, reviews various computational models of stroke recovery, relating them to mechanisms of synaptic and structural plasticity and discussing their implications for stroke treatment.

A treatment procedure that is increasingly used to alleviate the symptoms of degenerative brain disorders, such as Parkinson's disease, is deep brain stimulation (DBS). Chapter 25, Neural Plasticity in Human Brain Connectivity: The Effects of Deep Brain Stimulation, explores the changes in functional and structural connectivity induced by DBS and shows how these changes can be characterized and quantified by advanced graph theoretical measures and dynamic mean field models. The authors argue that DBS might have a restorative effect on damaged fiber tracts, opening up new avenues for the treatment of stroke.

8 OUTLOOK

The majority of plasticity studies in the past focused on synaptic plasticity rather than structural plasticity. With the advance of new imaging techniques, interest in structural plasticity has recently resurged. Many findings have revealed that the adult brain is capable of altering its wiring diagrams, but the driving forces underlying structural plasticity and its role in brain function remain unclear. This book reviews the current state of knowledge of structural plasticity and shows how unraveling the workings of structural plasticity may benefit from computational approaches. We hope that the book will inspire further research, both computational and experimental, and will help computational and experimental neuroscientists in their quest to understand the full adaptive potential of the adult brain.

Arjen van Ooyen and Markus Butz-Ostendorf

References

1. Butz M, Wörgötter F, van Ooyen A. Activity-dependent structural plasticity. *Brain Res Rev*. 2009;60(2):287–305.
2. Xu T, Yu X, Perlik AJ, et al. Rapid formation and selective stabilization of synapses for enduring motor memories. *Nature*. 2009;462(7275):915–919.
3. Toni N, Teng EM, Bushong EA, et al. Synapse formation on neurons born in the adult hippocampus. *Nat Neurosci*. 2007;10(6):727–734.
4. Keck T, Mrsic-Flogel TD, Vaz Afonso M, Eysel UT, Bonhoeffer T, Hübener M. Massive restructuring of neuronal circuits during functional reorganization of adult visual cortex. *Nat Neurosci*. 2008;11(10):1162–1167.
5. Yamahachi H, Marik SA, McManus JN, Denk W, Gilbert CD. Rapid axonal sprouting and pruning accompany functional reorganization in primary visual cortex. *Neuron*. 2009;64(5):719–729.
6. Brown CE, Boyd JD, Murphy TH. Longitudinal in vivo imaging reveals balanced and branch-specific remodeling of mature cortical pyramidal dendritic arbors after stroke. *J Cereb Blood Flow Metab*. 2010;30(4):783–791.
7. Deller T, Haas CA, Freiman TM, Phinney A, Jucker M, Frotscher M. Lesion-induced axonal sprouting in the central nervous system. *Adv Exp Med Biol*. 2006;557:101–121.
8. van Ooyen A. Using theoretical models to analyse neural development. *Nat Rev Neurosci*. 2011;12(6):311–326.
9. Marder E. Models identify hidden assumptions. *Nat Neurosci*. 2000;3:1198.
10. Abbott LF. Theoretical neuroscience rising. *Neuron*. 2008;60(3):489–495.

SECTION I

EXPERIMENTAL BACKGROUND

CHAPTER

1

Structural Plasticity and Cortical Connectivity

Vassilis Kehayas[1,2] *and Anthony Holtmaat*[1,3]

[1]University of Geneva, Geneva, Switzerland [2]Lemanic Neuroscience Doctoral School, Geneva, Switzerland [3]Chair Alain Rossier

OUTLINE

The Rewiring Brain
DOI: http://dx.doi.org/10.1016/B978-0-12-803784-3.00001-9

1 INTRODUCTION

Ever since early brain anatomists started to notice the intricacies of its microscopic morphology, they conjectured that its neurons and their synapses constitute the physical substrate of what we might today call cognition[1,2] (Fig. 1.1). This prompted the conclusion that neurons should possess the capacity to change their morphology and connections in order to adapt to the environment and enable learning. Indeed, during development there is an overproduction of synapses that wanes before animals enter adulthood.[3,4] However, it remained an open question for almost seven decades as to whether such changes can occur during an animal's adult life and to what extent they reflect cognitive processes. This delay was in part attributable to specious inferences regarding the morphological stability in the adult brain.[5] Additionally, technical limitations made the detection of small

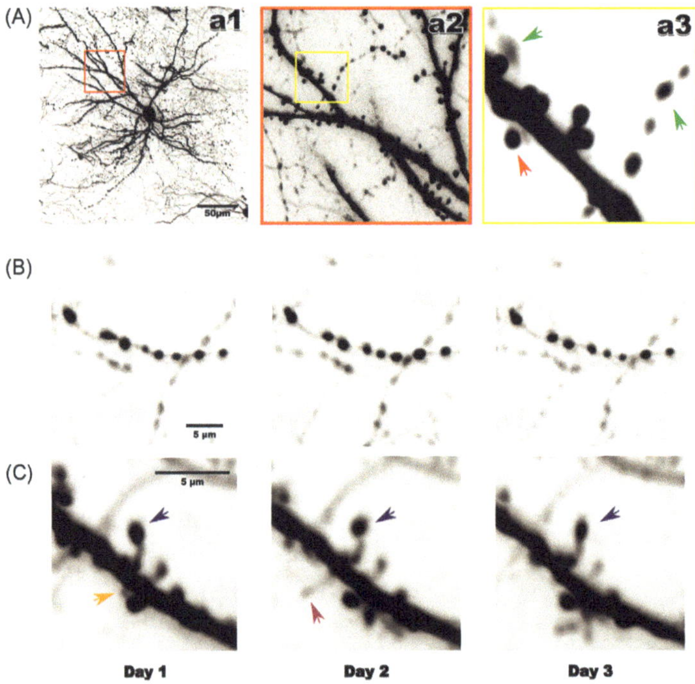

FIGURE 1.1 **Imaging of dendritic spines and axonal boutons.** (a1) An example of an enhanced green fluorescent protein (EGFP)-expressing cortical neuron in the mouse somatosensory cortex, imaged in vivo using two-photon laser scanning microscopy (max projection, top view). A dense cloud of axonal terminals and other neuropil structures can be seen in the background. (a2 and a3) Progressively higher magnification images in which dendritic spines (*red arrows*) and axonal boutons (*green arrows*) can be clearly seen. (B and C) An axon (B) and a dendrite (C) expressing EGFP were imaged for 3 consecutive days. Whereas no changes in the number of boutons have been observed in this example axon, some dendritic spines disappeared (*yellow arrow*), some appeared and disappeared (*magenta arrow*), some appeared and persisted (*cyan arrow*), and others remained stable (*blue arrow*) during this imaging period. Gray-scale colormaps are inverted and images in panels (a3, B, and C) have been processed for enhanced contrast.

and diffuse morphological changes nearly impossible. Methodological advances over the past few decades have enriched our knowledge of the potential for neurons to change their connectivity by allowing the dynamic observation of neuronal morphology and function in living animals.[6–9]

In this chapter, we will present the current state of knowledge of neuronal structural plasticity in the adult brain. We define neuronal *structural plasticity* as any change in neuronal morphology which, as such, has the potential to influence synaptic connectivity. We will mainly focus on cortical areas (cerebral cortex, hippocampus) of rodents, where most of the research on neuronal structural plasticity has been performed.

Structural plasticity can occur at various scales and in conjunction with several different physiological and pathological processes. A robust way in which the cortex could change the connectivity between neurons is by the addition or removal of neurons. Neurogenesis happens at a massive scale during development, but for long it was believed that this phenomenon does not occur in the adult brain.[10] We now know that adult neurogenesis does take place in particular brain areas, but only to a limited extent in cortical areas under physiological conditions.[10] Therefore, we will not further discuss this form of plasticity. Instead, we will focus on structural plasticity of preexisting neurons, which includes the large-scale expansion or shrinkage of axonal or dendritic arbors, as well as microscopic changes at the level of axonal boutons or dendritic spines. These microscopic structures form the majority of synapses in cortical areas.[4] As will be discussed later, alterations in their numbers or morphology could reflect drastic changes in connectivity and synaptic strength. Such forms of structural plasticity have been described under various experimental conditions in the cerebral cortex and hippocampus.

Processes that have been associated with neuronal structural plasticity include development,[11–13] aging,[14–16] exercise,[17,18] application of pharmacological agents,[19,20] stress,[15,21] and various pathological conditions.[22–25] Structural plasticity has also been induced using protocols of synaptic potentiation and depression, by alterations of experience, as well as in learning and memory tasks. Here, we will primarily focus on structural changes induced by these latter protocols, since they provide experimental conditions in which the link between the effect of the functional perturbation and changes of the underlying structural substrate should be in principle most direct and specific in the context of cortical connectivity.

In this context, we will discuss the extent to which different structural changes occur and whether some of them are preferentially related to particular experimental conditions. We will examine their dynamic properties and their causal determinants, especially those related to neuronal activity. Finally, we will attempt to derive general principles of neuronal structural plasticity that emerge from this body of research.

2 THE ROLE OF STRUCTURAL SYNAPTIC PLASTICITY IN HEBB'S THEORY OF CELL ASSEMBLIES

In his neuropsychological theory,[26] Donald O. Hebb proposed a mechanistic hypothesis for how structural synaptic changes could reflect altered functional states and accompany learning. At the time he published his theory, the prevalent view was that memory subsists by reverberating loops of activity in neuronal circuits,[26,27] which is in conflict with the fact that memory persists even when neuronal activity was interfered with. Hebb suggested that perception and learning are characterized by the formation of *cell assemblies*, which are created by the increase of synaptic connectivity within a subpopulation of neurons caused by their repeated or persistent coactivation. In his model, the prolonged activation of the cell assembly provides the time necessary for structural synaptic changes to occur. Thus a memory trace would transition from the ephemeral reverberating activity of the cell assembly to a long-lasting structural trace, which he termed the dual-trace mechanism. He believed that the most likely way in which this could happen is through the formation of new, or the enlargement of preexisting, synaptic contacts ("synaptic knobs"). However, he did not exclude a role for other processes (e.g., metabolic in nature) that affect the rhythmicity and/or the threshold of firing, and even acknowledged the possibility of a limited role for the motility of preexisting synaptic elements ("neurobiotaxis"). He also suggested that coactive axons that are in contact with the same postsynaptic neuron in nearby locations are more likely to fire the postsynaptic cell and could presumably induce structural changes by doing so. Thus both spatially and temporally coincident activities are implicated in Hebb's theory.

3 STRUCTURAL PLASTICITY FOLLOWING ENRICHED EXPERIENCE

Influenced by Hebb's theory and subsequent experimental work,[28] the next generation of neuroanatomists examined the effects of the animals' experience on synaptic structures. The first anatomical study to apply the enriched environment (EE) paradigm, in which animals are placed in cages that provide increased sensory stimuli (including social interactions), showed increased cortical weight as compared to individually housed littermates in an impoverished environment.[29] This increase was at least in part attributable to or accompanied by an increase in cortical thickness,[30–32] dendritic branching,[33–36] neuronal soma size,[32] synapse size,[37,38] the proportion of perforated synapses,[39] the number of synapses per neuron,[40] the frequency of multisynaptic boutons,[41] the number of dendritic spines,[42] and the number of glial cells.[30,31,43] Similar results have been found in the hippocampus.[44] It has also been reported that the distribution of interspine distances shifts to lower values after EE experience, suggesting that there might be some form of synaptic clustering associated with this form of plasticity.[45] Synaptic changes have been observed in cortical layers (L) 1 and 2/3,[37,38] 4,[38] and 5[42] in these post-mortem experiments. Effects on dendritic branching have been reported in L2/3, L4, and L5,[34,35] although terminal branches of basal dendrites appear to be primarily affected.[35]

Recent longitudinal imaging studies of cortical pyramidal cells in vivo have confirmed and extended some of these observations. They showed that EE experience, whether starting from birth or later, produces marked changes in spine dynamics.[46–48] An increase in the number of newly formed spines and lost preexisting spines is observed in both apical dendrites of L5 cells,[46–48] as well as in basal dendrites of L2/3 cells,[48] with a small fraction of the newly formed spines persisting for at least a month.[46,48] EE experience appears to leave axonal bouton turnover unaffected.[49]

Whether dendritic branching changes upon EE experience has so far not been specifically addressed in longitudinal imaging experiments in vivo.

Early studies utilizing the EE paradigm were the first to show that the adult brain retains the ability to adapt its structure in response to changes in the environment. The advantage of the experimental EE paradigm is that it resembles conditions that are relatively close to natural experience. However, at the same time this feature is a disadvantage. The multisensorial nature of the paradigm prohibits concrete conclusions concerning the patterns of neuronal activity that are associated with the observed structural plasticity. Experimental paradigms that involved more restricted manipulations have been used to probe more specific or subtle aspects of structural plasticity. These will be reviewed in the following sections.

4 STRUCTURAL PLASTICITY FOLLOWING SENSORY DEPRIVATION OR STIMULATION

The EE studies spurred an increase of interest in testing the effects of sensory deprivation on neuronal morphology. Early post-mortem studies found that many of the anatomical effects in the deprived areas of the cortex were the converse of those observed with EE experience: shrinkage of neuronal somata,[50] a decrease of cortical thickness,[50] and dendritic branching complexity,[51] as well as a decrease in the number of dendritic spines.[52] Sensory deprivation from birth was found to induce remodeling of dendritic trees in various sensory cortical areas.[51,53] In the adult cortex, this may still happen in concert with the functional topographic map reorganization that is typically observed upon sensory deprivation.[54–56]

Longitudinal in vivo imaging studies in adult animals have only partially reproduced the results from post-mortem experiments. They found no evidence for dendritic reorganization,[8,57,58] except for a reduction of terminal tip retractions in cortical L5 cells.[58] The discrepancy may be due to the fact that these studies were limited to apical dendrites, whereas the post-mortem studies have mostly detected such effects on basal dendrites of L2/3 pyramidal cells and L4 spiny stellate cells[54,56,59,60] and to a lesser extent, if at all, on apical dendrites of L2/3 pyramidals.[59,61]

Nonetheless, in vivo studies of spine dynamics have uncovered subtle but significant changes in spine dynamics. In particular, they showed that whisker or visual deprivation can increase the overall turnover of spines.[8,62] It can also change spine sizes and induce the formation of new spines,[63] prevent the naturally occurring spine elimination,[64,65] and affect the stability of newly formed[57,58] as well as of preexisting spines.[57] Local rules of dendritic plasticity may also be affected, as is evident from the disruption of clustered synapse potentiation that is normally observed in pyramidal cells.[66]

Most studies of structural plasticity in response to sensory deprivation have been focused on apical dendrites of L5 cortical pyramidal cells[8,57,62,64] (although see Ref.[58]). These effects are somewhat specific for certain subtypes of L5 cells, with complex-tufted pyramidal cells displaying a larger increase of new spine stabilization than simple-tufted pyramidal cells.[57] For spines on apical dendrites of L2/3 pyramidal neurons, the effects of sensory deprivation are less clear. One in vivo study in the primary visual cortex[63] showed that spine dynamics remain unaffected by monocular deprivation, whereas another study in the primary somatosensory cortex[58] presented evidence for increased persistence of newly formed spines upon whisker follicle ablation. However, it is important to note that due to the difference in the severity of the sensory deprivation (i.e., the mere reduction in visual input versus the permanent removal of the whisker sensory organs), these paradigms most likely engage distinct plasticity mechanisms. A recent in vivo study showed that a subset of L4 cells with an apical dendrite that extends toward L1 exhibits an increase in the loss of spines upon sensory deprivation.[67]

Post-mortem studies as well as studies in vivo are in agreement that some macroscopic reorganization of cortical axons, i.e., involving changes that span the size of a cortical column or more, can occur in superficial cortical layers upon long-term sensory deprivation in adult animals.[68–72] Specifically, higher axonal densities have been observed in deprived regions, likely due to axonal sprouting from neighboring, nondeprived, areas.[68–72] However, in another in vivo study, axons of L5 pyramidal cells were not affected by sensory deprivation.[62] Local axonal remodeling after sensory deprivation has also been reported, leading to an increased area of potential contact between connected pairs of L2/3 pyramidal cells in spared cortical areas as compared to unconnected pairs in the same area or connected pairs in control cortex.[60] In addition, it has been reported that, on average, the size of both pre- and postsynaptic elements, as well as their area of contact—albeit measured by diffraction-limited optical microscopy—were increased in the spared cortical area.[73]

A few studies have probed the effect of sensory stimulation on structural plasticity. Continuous visual stimulation with diffuse light has been found to induce an increase in dendritic spine densities on L4 and L5 cells of the visual cortex.[74] Prolonged passive whisker stimulation has also been shown to lead to a transient increase in synaptic density in the somatosensory cortex, which mostly concerned synapses on spines.[75] This effect on dendritic spines could be mimicked by direct stimulation of cortical cells by means of channelrhodopsin-2 (ChR2)-mediated optical activation.[76] On the other hand, L2/3 pyramidal cells appear to show no increase in spine numbers within the first few hours of whisker stimulation.[77] There are indications, however, that they may exhibit synaptic strengthening and increased branch-specific localized plasticity.[77] No evidence for dendritic or axonal rearrangements has been reported after sensory stimulation.

Collectively, sensory deprivation and stimulation paradigms have uncovered the potential of cortical neuronal networks to adapt their structure in response to changes in sensory input. One limitation of these experimental manipulations is that they represent extreme physiological states. In addition, in most of these studies the functional roles and the activity patterns of the synaptic partners that were undergoing the structural changes remained unknown. Therefore, it is difficult to appreciate how the structural changes were implicated in cortical network function and how they participated in optimizing the information processing of the circuits involved. Learning and memory tasks on the one hand, and LTP experiments on the other hand, have addressed some of these aspects.

5 STRUCTURAL PLASTICITY IN LEARNING AND MEMORY

So far, we have focused on structural plasticity that is associated with either generalized multisensory stimulation (i.e., enriched environment) or crude manipulations of the animals' experience (i.e., partial or complete sensory deprivation and overstimulation). However, in order to argue that structural changes are behaviorally relevant it needs to be determined whether they are causally linked to well-defined and specific cognitive functions, and whether they are restricted to cortical areas, neurons, or synapses that are known to be involved therein. Behavioral paradigms of learning and memory are increasingly used to assess the involvement of structural plasticity in these basic, yet important, aspects of cortical function.

A few early post-mortem studies have taken advantage of the hippocampus' involvement in spatial information processing and associative learning to study the relationship between structural plasticity and learning. Using the Morris' maze, a paradigm of spatial learning, a transient increase was found in the number of synapses[78] or spines[79] a few hours after training but not so 1 or more days later.[78–80] Synapses may tend to be more spatially clustered in this paradigm.[80] Using trace eyeblink conditioning, which is also dependent on hippocampus, spine numbers were shown to increase on basal dendrites,[81] but not apical dendrites[82] of CA1 pyramidal neurons. However, the postsynaptic density (PSD) area of spines with nonperforated synapses[82] and the number of multisynaptic boutons[83] did increase on apical dendrites. Lastly, a fear conditioning paradigm elicited an increase in spine density on both CA1 apical and basal dendrites within the first 24 hours.[84] This increase later subsided in CA1 and then started to be observed in the anterior cingulate cortex,[84] consistent with the view that the hippocampus sustains the early stages of memory formation, whereas the cortex becomes more involved in later stages.

Studies using similar protocols have been carried out in the cerebral cortex. A fear conditioning protocol based on the association of an auditory cue (conditioned stimulus, CS) and foot shocks (unconditioned stimulus, US), with freezing as a response (conditioned response, CR), induced a transient increase in the spine formation rates in the primary auditory cortex.[85] When CS and US were temporally unlinked, spine elimination was transiently increased.[85] In contrast, in the frontal association cortex fear conditioning led to an increase in the spine elimination rate, whereas fear extinction due to continued presentation of the CS without an US led to an increase in the spine formation rate.[86] Both phenomena were tightly correlated with the behavioral outcome (i.e., the level of CR).[86] Reconditioning of a CS led to the elimination of the spines that were formed upon extinction, whereas sequential conditioning to two different CS and subsequent extinction of one of them reinstalled new spines closer to where spines were eliminated during conditioning of the extinct CS as compared to the nonextinct CS.[86]

In primary somatosensory cortex, trace eyeblink conditioning caused a decrease in spine numbers in apical tufts of pyramidal cells[87] that correlated with the speed of learning. In the same area, an active whisker-mediated object localization task was accompanied by an increase in the number of new spines that persisted over the first few days of training.[88] The formation rate of this class of spines correlated with the behavioral performance of the animal.[88] Increased rates of spine elimination started to be observed a few days later, which led to a small net increase in spine density.[88]

Several studies have examined the effect of motor training on pyramidal cell dendritic spines in the motor cortex. Tasks involving skilled reaching, grasping, or running increased the formation of new spines.[46,47,65,89–93] This increase has been observed in apical dendrites of L5[46,47,89,91,93] and L2/3[65,92] pyramidal neurons. Increased densities of spines[94] or synapses[95] have also been reported for basal dendrites of L5 cells. Whereas this effect can start as early as 6 hours after training,[89] in some of these tasks spine elimination only started being observed at later stages of learning,[46,90,93] resulting in a modest initial increase in spine density.[93,96,97] The size[47] and stability of these new spines[46,47,89,93] were also positively affected by such motor learning tasks. Importantly, in these paradigms the rate of spine formation was also tightly correlated with behavioral performance.[46,93] Notably, a recent study has found that the selective shrinkage of synapses that were potentiated or newly formed after learning of a motor task led to a drop in behavioral performance.[92] This indicates that new spine formation and synapse potentiation are necessary for learning and memory. Spine elimination was also correlated with behavioral performance.[46]

The spatial organization of synaptic changes related to motor learning has also been studied in some detail. In a forelimb reaching task, new spines exclusively occurred in areas directly related to the task such as the contralateral motor cortex.[93] Moreover, structural synaptic changes within this area occurred preferentially in corticospinal neurons that are associated with the control of distal musculature and are thus engaged in the fine movements during grabbing, as opposed to corticospinal neurons that are associated with the control of proximal musculature.[94] Even further, it appears that individual dendritic branches of L5 pyramidal cells can be preferentially involved in different tasks, with one branch showing relatively increased levels of spine formation in one task and a sibling branch in another.[89,91] Such effects appear to be absent in L2/3 cells.[65] Both branch-specific spine formation and survival were sensitive to sleep deprivation,[89] and spine formation was facilitated by the occurrence of branch-specific calcium spikes.[91] Within a single dendritic branch, new spines were preferentially generated close to other new spines that were associated with the same task, and those that did grow near one another were more likely to survive for a longer period of time.[47]

Studies of learning-related axonal bouton dynamics are scarce. A recent study, however, has found evidence for increased bouton turnover in axons projecting from the orbitofrontal cortex to the dorsomedial frontal cortex in a paradigm that involves the detection of a nonpreferred odor linked to a reward by mice that had a pre-established preference to a different odor.[98]

6 STRUCTURAL PLASTICITY AND LONG-TERM FUNCTIONAL SYNAPTIC PLASTICITY

The studies that have been discussed so far can be interpreted as supporting the hypothesis that changes in neuronal activity are the underlying cause of structural plasticity. However, almost in their entirety, these studies lack any information on the features of neuronal activity that may act as the proximal cause for the observed structural changes. This gap is potentially bridged in studies of long-term potentiation (LTP) and depression

(LTD), in which the morphology of those synapses that were directly activated, potentiated, or depressed was tracked.

Post-mortem studies initially showed that spine head and spine neck enlargement were specifically induced in distal dendrites of dentate gyrus (DG) neurons by tetanic electrical stimulation of the lateral entorhinal cortex (EC) in vivo.[99–101] This paradigm takes advantage of the monosynaptic nature of this form of LTP,[102] and thus represents synapse-specific changes in network connectivity. Using a slightly different protocol where the medial part of the molecular layer of the DG was stimulated monosynaptically, the number of synapses,[103] the proportion of perforated synapses,[104] and the PSD surface area[105] were increased specifically in this area, with the majority of the effects occurring on large concave spines. A decrease in the number of synapses was found in nonstimulated adjacent areas, presumably due to heterosynaptic mechanisms.[91,92,103,104] An electron microscopy (EM) study of identified dendritic segments found the number of spines to be increased in DG after LTP induction in vivo, but no evidence for an overall increase of spine head volume, neck diameter, or spine length.[106] LTP induction in CA1 of the hippocampus in vitro showed an increase in shaft synapse densities[107,108] and perhaps an increase in the proportion of small spines,[108] but no change in spine or synapse size. Another study in the same area found no evidence of morphological changes whatsoever.[109]

Several methodological issues prevent drawing concrete conclusions from these studies. First, differences between the DG and CA1, or between the induction protocols (e.g., in vivo versus in vitro) may explain some of the conflicting evidence. Second, it has been shown that slicing of neural tissue produces abnormal spinogenesis,[110–112] which may mask potentially subtle LTP-related changes.[113] Along similar lines, studies applying multivariate population analyses in post-mortem material often lacked the statistical rigor that is required for unambiguous interpretation of the results. A more powerful approach would be if the potentiated or newly formed spines and synapses could be specifically identified and/or imaged longitudinally.

The identification of LTP-specific spine structural plasticity was improved using targeted stimulation of a few or single synapses, or by the detection of localized calcium fluctuations, in most cases combined with longitudinal optical imaging of the synaptic structures. These experiments show that spine enlargement takes place during the initial stages of LTP.[114–121] Depending on the protocol this can last throughout later stages as well.[114,115,117,119,120] The early stages of LTP are also characterized by a transient increase in the numbers of filopodia-like protrusions.[122] A more gradual increase in the number of spines that persist for several hours occurs in later stages.[123–125] Interestingly, an LTP stimulus that is applied to a spine immediately after it is formed can increase its size and also its probability of survival.[126] LTP was also found to increase the proportion of multi-synaptic boutons and perforated spines.[127] Other morphological changes associated with LTP are a decrease in spine neck length[128] and an increase in spine neck width,[121] both resulting in increased synaptic conductivity.

On the other end of the plasticity spectrum, low frequency stimulation, which causes LTD, induces spine shrinkage[117,129,130] and increases rates of spine retraction.[124] Axonal bouton densities have been shown to increase after LTD.[131] Boutons that appear or disappear were found to be smaller than the overall population of stable boutons, with disappearing boutons being in closer proximity to other boutons.[131]

Local dendritic plasticity rules have been studied in some detail using various LTP protocols. Such studies reveal that the potentiation of one spine also affects the potentiation levels of neighboring spines.[115,117,119,132] For example, if a stimulation protocol that does not by itself produce LTP is applied to a spine neighboring a recently potentiated spine, this one too can be potentiated to similar levels.[115,119,132] During the initial stages of the potentiation, these interspine interactions are local (~10 μm),[115] dependent on the temporal sequence of the two stimuli[115] as well as on the action of small diffusible molecules such as Ras,[132] but independent of protein synthesis[115,119] or calcium release from intracellular stores.[115] At later stages, the effects are more spread throughout the dendrite but still confined to single dendritic branches (~70 μm).[119] During these stages, they also have more loose temporal requirements and are dependent on protein synthesis.[119] Strong stimuli that activate many synapses within a hippocampal slice have been shown to promote biased spinogenesis close to spines that were activated by the stimulation.[118] Local plasticity rules are not restricted only to synaptic potentiation. Inactive spines that are located within a dendritic region containing multiple stimulated spines shrink and their synapses are weakened.[133]

7 ACTIVITY-DEPENDENT AND -INDEPENDENT STRUCTURAL SYNAPTIC PLASTICITY

To what extent are spine or synapse formation, elimination, stabilization, and destabilization dependent on activity? Information about the dynamics of axonal boutons is scarce but we know that despite their plasticity, they exhibit lower turnover rates as compared to dendritic spines.[134–136] Dendritic spines are formed continuously, often close to axonal boutons, and they can rapidly bear a functional synapse.[125,137–139] Longitudinal in vivo imaging of synapses and EM reconstructions of newly formed spines suggest that in most cases in the adult cortex spinogenesis precedes synaptogenesis.[137,140] A spine that acquires a synapse within a day is more likely to survive longer.[137,140] Conversely, a spine is more likely to be eliminated if it loses its PSD.[140,141] These results are corroborated by the finding that transient spines, i.e., those that appear and disappear within a period of minutes to a few days, are thinner than stable spines,[142,143] and most of them lack a synapse.[137,140]

New spines often grow toward large boutons that already bear one or more synapses, resulting in multisynaptic boutons.[73,125,137,138] Since stable synapses are not normally located on multisynaptic boutons,[125] this suggests that only one of the contacts survives (Fig. 1.3A). This process might be similar to what is seen for the synaptic integration of newborn neurons in the adult DG. On these neurons the fraction of spines contacting multisynaptic boutons runs down with the age of the neuron, which may be the result of spines competing with one another to be the sole recipient of synaptic input from a single bouton.[138] Additionally, the sizes of the two spines contacting a single multisynaptic bouton tend to be inversely correlated when one of them belongs to a newborn neuron.[138] However, when the two spines belong to the same dendrite, in the adult CA1, the sizes of the two spines are highly correlated.[144] Altogether, these studies suggest that dendrites grow spines in order to probe their environment for potential axonal partners. The fact that the preferred points of contact are often pre-established synaptic boutons suggests that presynaptic activity can play a role in directing their growth.

Support for exogenously driven spine formation comes from experiments in which excessive local exposure of a dendrite to glutamate can induce the formation of a new spine.[145,146] Such spines can rapidly recruit AMPA and NMDA receptors,[145] similarly to new spines that are formed spontaneously, suggesting that they can form a synapse. Glutamate-evoked de novo formation of spines is robust in very young brain tissue,[145,146] but appears not to be possible in acute slices from slightly older mice.[145] The finding that many new spines preferentially contact large multisynaptic boutons could be congruous with glutamate-evoked spine formation, since boutons that already bear a synapse may spill over glutamate.[146,147]

Both extrinsic and intrinsic factors have been shown to be involved in spine stabilization. Since the existence of a synapse on a spine is crucial for spine survival, and spine size positively correlates with synaptic currents,[148] it follows that spine (head) size is a major morphological correlate of spine age/spine survival.[149–151] Indeed, smaller spines are more susceptible to activity-dependent size changes as compared to larger spines.[149] At least for the largest of spines, the spine neck width is controlled independently of spine head size,[121,152] leaving them immune to further activity-dependent changes. Interestingly, larger spines tend to be spaced further away from each other,[153] and new spines are preferentially formed away from stable spines.[47] In addition, dendrites with low spine densities may experience relatively more spine additions as compared to spine-rich dendrites.[142] Together, these findings indicate that synaptic density is also homeostatically regulated by intrinsic mechanisms at the level of the dendrite. Total spine size of the whole spine population also appears to be homeostatically controlled in such a way that the total size of the population of spines remains largely unchanged.[149] However, extrinsic mechanisms dependent on the activity levels of neighboring spines can also positively or negatively affect the local spine density and the size of individual spines.[108,123]

8 STRUCTURAL PLASTICITY AND CORTICAL CONNECTIVITY

8.1 Large-Scale Structural Plasticity

In summary, the studies discussed earlier reveal several general features of neuronal structural plasticity in relation to sensory experience, learning, and functional synaptic plasticity. There is no clear evidence for substantial neurogenesis to underlie functional plasticity in the adult cortex. Similarly, large rearrangements of either dendritic or axonal trees are only observed upon drastic manipulations of the inputs to the network and, even then, the extent of those rearrangements is debatable. If any dendritic rearrangements occur, they appear to be mostly restricted to basal dendrites of L2/3 or to L4 cells upon long-term sensory deprivation. Changes in other parts of the dendritic tree appear to be restricted to small-scale protractions and retractions of terminal branch tips. More extensive in vivo studies of both apical and basal dendritic dynamics are required in order to determine the exact magnitude of such changes, as well as the conditions under which they might be evoked.

Most of the literature agrees that large-scale axonal changes in superficial cortical layers may occur after relatively long periods of sensory deprivation. However, these data should be cautiously interpreted, since tracer uptake, injection volumes, and/or fluorescent protein expression levels can be variable between experimental conditions. Therefore, these findings will need to be confirmed by longitudinal in vivo imaging studies of identified axons that constitutively express fluorescent proteins. Macroscopic structural rearrangements of axons have thus far not been reported under more physiological plasticity-inducing paradigms.

8.2 Microscopic Structural Plasticity and Cortical Connectivity

As seen above, the most abundantly reported forms of structural plasticity are changes at the level of synaptic structures such as spines and boutons. The possible consequences of this type of structural plasticity for neuronal networks become apparent when the statistical properties of cortical connectivity are taken into account. Axonal and dendritic trees in the cortex are profusely intermingled (Fig. 1.2A). At the mesoscopic scale, axons follow long and relatively straight trajectories,[4,154] and dendritic branches of individual cells are relatively sparsely distributed.[155] At the microscopic scale, dendritic spines and terminal axonal boutons increase the effective synaptic diameter of dendrites and axons, respectively, by allowing them to form synaptic connections, even when they are not in direct physical contact[156](Fig. 1.2A). These organizational features of the neuropil jointly enable an almost all-to-all potential connectivity within local circuits: it has been estimated that two neurons located within a few hundred micrometers of each other in the rat sensory cortex (incidentally, the size of a cortical column) most likely share at least one *potential synapse*, i.e., a location where an axon and a dendrite are close enough to be connected by the growth of a spine or a terminal bouton[157,158] (Fig. 1.2A, D). However, the fraction of all potential synapses that are being occupied by actual synapses, the *filling fraction*, is rather small in cortical areas of most species.[156] This leaves cortical neuronal circuits with a considerable potential to change their connectivity, as for every possible spine estimates suggest that there are at least three to four alternative axonal boutons with which they can connect, and vice versa.[156,159] The growth of a spine or a terminal bouton is an obviously more economical solution to the problem of reaching more synaptic partners in terms of wiring cost as compared to large-scale rearrangements of the neuropil. In fact, the optimization of wiring, in a way that synapse density is maximized while conductance time is minimized, may have been an important factor in the construction of cortical circuits throughout evolution.[160] Thus by utilizing the potential of changing the wiring diagram through microscopic structural plasticity, cortical neuronal circuits are able to achieve a relatively high degree of connectivity with minimal wiring requirements.

Based on the statistical distributions of axonal and dendritic trees, it can be predicted that most connected pairs of neurons would form just one synapse[4,157,161] (Fig. 1.2B). However, multisynaptic connections between pairs of neurons in cortical circuits have been empirically found to be overrepresented[162–165] (Fig. 1.2B). Indeed, the observed distributions of connections between cortical neurons are irreconcilable with a random formation and elimination of synapses between overlapping pre- and postsynaptic structures.[161,166] Instead, these empirical findings suggest that synapses are preferentially formed between synaptic partners that are already connected and eliminated from partners that are not connected with more than a critical number of synapses.[161,166] The fact

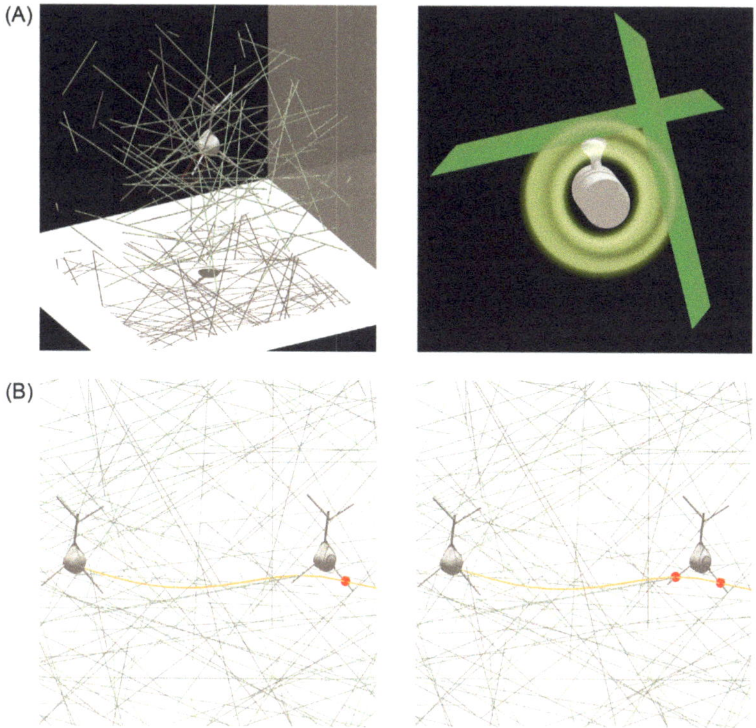

FIGURE 1.2 **Stationary synaptic connectivity.** (A) *Left*: Cortical pyramidal neurons are embedded in a dense network of local cortical axons. Due to their spatial arrangement, there is a high probability of points of overlap between the dendritic and axonal cloud (axons in *green*, soma and dendrites in *white*). *Right*: Spines extend a dendrite's effective diameter for potential connectivity (*green halo*). In places where an axon (*green*) passes close to a dendrite (*white*) within this diameter a potential synapse can be formed. (B) *Left*: Based on the statistical distributions of neuropil components, it can be expected for two local cortical pyramidal neurons that are synaptically connected to possess mostly one synapse with each other (*red sphere* on yellow axon). *Right*: It is found experimentally, however, that connected pairs form multiple synaptic connections with each other.

that neurons with similar functional properties are preferentially connected with each other[165] suggests that the biased connectivity observed is at least in part directed by activity-dependent, presumably Hebbian, mechanisms. However, the evidence reported in this paragraph needs to be interpreted with reference to the fact that local connections probably represent only a minority of the synaptic population,[167] and it cannot be excluded that long-range connections follow different connectivity rules.

8.3 Mechanisms of Microscopic Structural Plasticity Influencing Cortical Connectivity

How are these local connectivity patterns generated? Multisynaptic connections between two individual neurons could be achieved in different ways. Two alternative hypotheses can be considered. According to the *local plasticity hypothesis of synaptic*

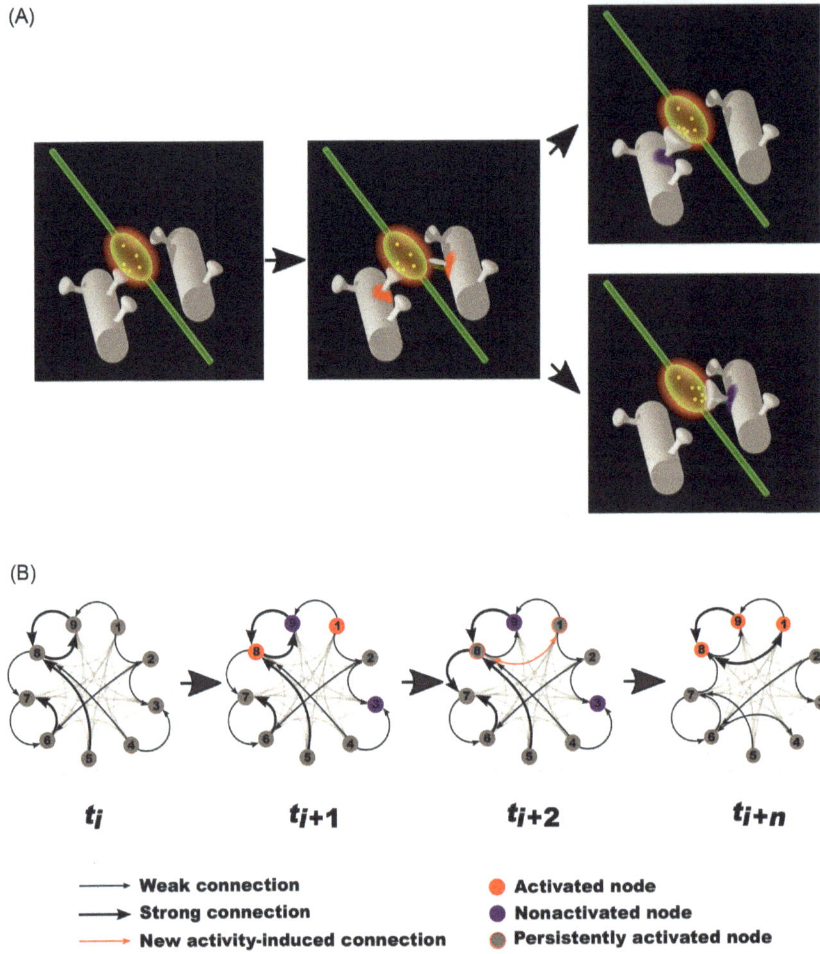

FIGURE 1.3 **Spatial and temporal aspects of structural plasticity.** (A) Spatial dynamics of structural plasticity. Two dendrites (*white*) and an axon (*green*) in close proximity are shown in a temporal sequence from left to right. One of the dendrites initially forms a synaptic connection through a spine (protrusion on dendrite) contacting a bouton (bulge in axon). When the bouton is activated, it can either induce the formation of new spines (middle panel, dendrite on the right) or stabilize preexisting spines (top right panel, dendrite on the left). A temporal and spatial window of opportunity is opened when these conditions are met for additional new spines, which may grow or stabilize in the vicinity of the formerly grown or stabilized spines (*red halos* on the dendrites) for a brief period of time (e.g., middle panel, transparent yellow spine on dendrite on the right). At the stationary state (rightmost panels), only some of those spines will be stabilized (enlarged "mushroom" spines). Those spines that have not established a strong synaptic connection and are in close proximity to spines that did get stabilized (*blue halos* on the dendrites) will face a higher chance to disappear (e.g., transparent blue spine, bottom right panel). (B) Temporal evolution of network connectivity under activity-dependent and independent control in a simplified network graph. Each panel represents the topology of the same network in a temporal sequence from left to right. Initially (t_i), a graph of 9 nodes representing randomly connected cells (numbered circles) in a network is shown. The thickness of the directed edges between nodes (*curved arrows*) indicates connection strength that could be seen as depicting synaptic weight of individual synapses and/or number of synaptic connections. For simplicity, only two "levels" of connection strength are shown ("weak" and "strong"). *Gray lines* in the

connectivity, one or more boutons of the same axonal segment could be contacting a single dendritic branch multiple times. Indeed, connected pairs show increased area of overlap within a single potential synapse site when examined with EM,[165] which would provide ample opportunity for multiple synapses to occur. The hypothesis is further supported by observations of increased spatial clustering of dendritic spines after EE experience, sensory stimulation, LTP, and learning, as we have already discussed. There are several ways in which this could happen.

First, MSB could arise, which, as we saw, are encountered more frequently after EE experience, LTP, or learning. Most MSB are statistically likely to form synapses with spines of different dendrites and of different cells.[137,138] However, it appears that after LTP MSB are connected preferentially with spines of the same dendrite.[127] Second, multiple synapses with different boutons of a single axon could form in close contact on the same dendritic segment, an observation which is not uncommon in EM preparations.[168] In one specific scenario of how these multiple contacts with the same dendrite could arise, spines and their synapses could split.[169] However, this has been found to be largely incompatible with most EM studies.[137,170,171] The independent formation of multiple synaptic connections between an axon and a dendrite is the most likely way.[137] Multiple synapses at a single axonal–dendritic intersection would thus bias estimates of actual connectivity based on measurements of potential synapses from light microscopy.

The *distributed plasticity hypothesis of synaptic connectivity* suggests that cell-wide, rather than local plasticity mechanisms are responsible for the observed preference of connected pairs to form multiple synapses. According to this hypothesis, multiple points of synaptic contact are formed in different branches of the postsynaptic cell. In stationary state circuits, where no plasticity is being induced, some evidence for both hypotheses can be found. For example, CA3 to CA1 connections in the hippocampus were found to be spatially biased to particular branches, rather lending support to the local plasticity hypothesis.[172] An example of distributed connectivity is perhaps provided by the dispersed position of synapses shared between functionally connected cortical L2/3 neurons.[164] Under both hypotheses, activity-dependent mechanisms such as STDP may account for the selective synaptogenesis that can result in the observed nonrandom cortical connectivity patterns[173–176] (Fig. 1.3B).

◀ background represent unrealized connections according to the all-to-all potential connectivity exhibited by cortical networks. At the second time point (t_{i+1}), some nodes are strongly or synchronously activated (red nodes, 1 and 8) while others are either inactivated or activated asynchronously (blue nodes, 3 and 9). In the simplest case, this could represent the similar tuning of both nodes to a common feature of the sensory parameter space. Activity-dependent mechanisms could induce the formation of new connections (red bidirectional edge between nodes 1 and 8, t_{i+2}) or strengthen preexisting connections. This could counterbalance activity-dependent or independent homeostatic mechanisms that continuously control the number and strength of a cell's synapses. After some more cycles of updating the connections, the network reaches a new stationary state (t_{i+n}), where the nodes that are repeatedly or persistently activated have created a subnetwork, a cell assembly, in which nodes are strongly interconnected with each other. Given the initial topology, the random generation of connections and/or several cycles of connection updates could prompt a node that was not initially activated (node 9) to develop strong interconnections with other activated nodes and thereby become part of the cell assembly. Nodes that do not participate in this cell assembly lack strong connections with the activated nodes. Note that the overall "sum" of the weights in the network has remained largely unchanged through the co-ordinated action of both activity-dependent and independent mechanisms of synaptic plasticity.

Based on the studies discussed in this chapter, we propose the following model for structural synaptic changes. New spines are continuously generated along dendrites. Spinogenesis appears to be favored in areas along the dendrite that experience low inter-spine competition and/or close to large boutons that already bear a synapse. Spinogenesis may continue to occur within the same area or in the vicinity of spines that are selectively potentiated. To what extent activity of the presynaptic bouton governs spinogenesis and whether synaptogenesis can also be an initiating factor for spine formation are questions that are still not fully investigated. In either case, synaptic competition between spines in close proximity or between those contacting the same axonal bouton ensues. During this process, the presence and the size of the synapse appear to be critical determinants of a spine's probability to stabilize (Fig. 1.3A).

Neurons that receive similar inputs will have a higher probability to display temporally congruent activity. This may favor the stabilization of synaptic inputs between them (and the destabilization of connections with other cells) through processes such as spike timing-dependent plasticity (STDP). Boutons that are highly or repeatedly activated might even promote de novo spinogenesis and have a higher probability to transiently form multiple synapses. As long as spines belonging to the activated cells and their synapses have not reached full maturity through activity-dependent mechanisms, they will remain susceptible to homeostatic as well as activity-dependent control. As a consequence, the repeated activation of a particular set of neurons by a relevant input to the network will bring the connectivity matrix closer to a stationary state, where neurons that were coactivated are preferentially connected with each other, forming a cell assembly (Fig. 1.3B). The resulting network containing the cell assembly will have incorporated a computational unit that is able to retrieve this activity state upon reactivation of even a subpopulation of the cell assembly with the same input. Presumably, during network activation such as in EE experience, sensory stimulation, and learning, some cell assemblies and their synaptic connections are selectively potentiated or new ones are formed, resulting in the imprint of the experience onto the synaptic connections of the network.

9 FUTURE PERSPECTIVES

There are still many aspects of neuronal structural plasticity that need further attention. For example, the relative potential for structural plasticity of pyramidal cells residing in different layers has not yet been fully explored. Additionally, it will be interesting to investigate differences in plasticity between different parts of the dendritic tree of pyramidal cells. It is not unreasonable to assume that given the, often distinct, origin of their inputs and the electrochemical differences of basal and apical dendrites, some degree of variability in their potential for structural plasticity exists. Since dendritic branches can operate as independent computational units[177] and thereby greatly enhance the information processing capacity of a cell,[178] it will also be interesting to investigate the conditions that give rise to the reported dendritic branch-specific structural plasticity. Furthermore, the temporal sequence of spinogenesis and synaptogenesis and the causal mechanisms that underlie such phenomena, whether activity-dependent or homeostatic, need to be further studied. At the circuit level, it will be important to discern the degree to which the observed synaptic dynamics lead to actual changes of the wiring diagram or if they rather represent the modification of preexisting connections between synaptic partners.

To answer these questions, it will be essential to separately monitor or control the activity of pre- and postsynaptic elements, while at the same time observing their morphology. Furthermore, it will be necessary to selectively label different neuronal assemblies and monitor their activity simultaneously. Technical tools that allow such experiments are emerging.[92,179–184] It will be crucial for these next steps that experiments are informed by and will update our theoretical understanding of structural plasticity in cortical networks.

Acknowledgments

We thank Karen Zito for her input and comments on the manuscript. A.H. is financially supported by grants from the Swiss National Science foundation (331003A_153448, CRSII3-154453), the National Centre for Competence in Research (NCCR) SYNAPSY financed by the Swiss National Science Foundation (51AU40_125759), the International Foundation for Research on Paraplegia and the Hans Wilsdorf Foundation (chair Alain Rossier). V.K. was supported by SyMBaD (EU FP7-PEOPLE-ITN Marie Curie, grant 238608).

References

1. Tanzi E. I fatti e le induzioni dell'odierna istologia del sistema nervoso. *Riv Sper Fren Med Leg.* 1893;19: 419–472.
2. Cajal SRY. The Croonian Lecture: La Fine Structure des Centres Nerveux. *Proc R Soc London.* 1894;55 (331–335):444–468.
3. Rakic P, Bourgeois J, Eckenhoff M, Zecevic N, Goldman-Rakic P. Concurrent overproduction of synapses in diverse regions of the primate cerebral cortex. *Science.* 1986;232(4747):232–235.
4. Braitenberg V, Schüz A. *Cortex: Statistics and Geometry of Neuronal Connectivity.* Berlin, Heidelberg: Springer; 1998.
5. Dunaevsky A, Woolley CS. In: Stuart G, Spruston N, Häusser M, eds. *Structural Plasticity of Dendrites.* Oxford University Press; 2007.
6. Svoboda K, Denk W, Kleinfeld D, Tank DW. In vivo dendritic calcium dynamics in neocortical pyramidal neurons. *Nature.* 1997;385(6612):161–165.
7. Lendvai B, Stern EA, Chen B, Svoboda K. Experience-dependent plasticity of dendritic spines in the developing rat barrel cortex in vivo. *Nature.* 2000;404(6780):876–881.
8. Trachtenberg JT, Chen BE, Knott GW, et al. Long-term in vivo imaging of experience-dependent synaptic plasticity in adult cortex. *Nature.* 2002;420(6917):788–794.
9. Grutzendler J, Kasthuri N, Gan W-B. Long-term dendritic spine stability in the adult cortex. *Nature.* 2002;420 (6917):812–816.
10. Ming G, Song H. Adult neurogenesis in the mammalian central nervous system. *Annu Rev Neurosci.* 2005; 28(1):223–250.
11. Wong ROL, Ghosh A. Activity-dependent regulation of dendritic growth and patterning. *Nat Rev Neurosci.* 2002;3(10):803–812.
12. Cohen-Cory S. The developing synapse: construction and modulation of synaptic structures and circuits. *Science.* 2002;298(5594):770–776.
13. Gogolla N, Galimberti I, Caroni P. Structural plasticity of axon terminals in the adult. *Curr Opin Neurobiol.* 2007;17(5):516–524.
14. Dumitriu D, Rapp PR, McEwen BS, Morrison JH. Estrogen and the aging brain: an elixir for the weary cortical network. *Ann N Y Acad Sci.* 2010;1204(1):104–112.
15. McEwen BS. Sex, stress and the hippocampus: allostasis, allostatic load and the aging process. *Neurobiol Aging.* 2002;23(5):921–939.
16. Petralia RS, Mattson MP, Yao PJ. Communication breakdown: the impact of ageing on synapse structure. *Ageing Res Rev.* 2014;14(1):31–42.

17. Draganski B, May A. Training-induced structural changes in the adult human brain. *Behav Brain Res.* 2008;192(1):137–142.
18. May A. Experience-dependent structural plasticity in the adult human brain. *Trends Cogn Sci.* 2011; 15(10):475–482.
19. Russo SJ, Dietz DM, Dumitriu D, Morrison JH, Malenka RC, Nestler EJ. The addicted synapse: mechanisms of synaptic and structural plasticity in nucleus accumbens. *Trends Neurosci.* 2010;33(6):267–276.
20. Robinson TE, Kolb B. Structural plasticity associated with exposure to drugs of abuse. *Neuropharmacology.* 2004;47:33–46.
21. Pittenger C, Duman RS. Stress, depression, and neuroplasticity: a convergence of mechanisms. *Neuropsychopharmacology.* 2008;33(1):88–109.
22. Fiala JC, Spacek J, Harris KM. Dendritic spine pathology: cause or consequence of neurological disorders? *Brain Res Rev.* 2002;39(1):29–54.
23. Galvez R, Greenough WT. Sequence of abnormal dendritic spine development in primary somatosensory cortex of a mouse model of the fragile X mental retardation syndrome. *Am J Med Genet Part A.* 2005;135A(2):155–160.
24. Papadopoulos CM, Tsai S-Y, Cheatwood JL, et al. Dendritic plasticity in the adult rat following middle cerebral artery occlusion and Nogo-a neutralization. *Cereb Cortex.* 2006;16(4):529–536.
25. Mostany R, Chowdhury TG, Johnston DG, Portonovo SA, Carmichael ST, Portera-Cailliau C. Local hemodynamics dictate long-term dendritic plasticity in peri-infarct cortex. *J Neurosci.* 2010;30(42):14116–14126.
26. Hebb DO. *The Organization of Behavior.* Routledge; 2002.
27. Bennett EL, Diamond MC, Krech D, Rosenzweig MR. Chemical and anatomical plasticity of brain. *Science.* 1964;146(3644):610–619.
28. Rosenzweig MR, Bennett EL. Psychobiology of plasticity: effects of training and experience on brain and behavior. *Behav Brain Res.* 1996;78(1):57–65.
29. Rosenzweig MR, Krech D, Bennett EL, Diamond MC. Effects of environmental complexity and training on brain chemistry and anatomy: a replication and extension. *J Comp Physiol Psychol.* 1962;55(6):429–437.
30. Diamond MC, Krech D, Rosenzweig MR. The effects of an enriched environment on the histology of the rat cerebral cortex. *J Comp Neurol.* 1964;123(1):111–119.
31. Diamond MC, Law F, Rhodes H, et al. Increases in cortical depth and glia numbers in rats subjected to enriched environment. *J Comp Neurol.* 1966;128(1):117–126.
32. Diamond MC, Lindner B, Raymond A. Extensive cortical depth measurements and neuron size increases in the cortex of environmentally enriched rats. *J Comp Neurol.* 1967;131(3):357–364.
33. Holloway RL. Dendritic branching: some preliminary results of training and complexity in rat visual cortex. *Brain Res.* 1966;2(4):393–396.
34. Volkmar FR, Greenough WT. Rearing complexity affects branching of dendrites in the visual cortex of the rat. *Science.* 1972;176(4042):1445–1447.
35. Greenough WT, Volkmar FR. Pattern of dendritic branching in occipital cortex of rats reared in complex environments. *Exp Neurol.* 1973;40(2):491–504.
36. Uylings HB, Kuypers K, Diamond MC, Veltman WA. Effects of differential environments on plasticity of dendrites of cortical pyramidal neurons in adult rats. *Exp Neurol.* 1978;62(3):658–677.
37. Mollgaard K, Diamond MC, Bennett EL, Rosenzweig MR, Lindner B. Quantitative synaptic changes with differential experience in rat brain. *Int J Neurosci.* 1971;2(3):113–127.
38. West RW, Greenough WT. Effect of environmental complexity on cortical synapses of rats: preliminary results. *Behav Biol.* 1972;7(2):279–284.
39. Greenough WT, West RW, DeVoogd TJ. Subsynaptic plate perforations: changes with age and experience in the rat. *Science.* 1978;202(4372):1096–1098.
40. Turner AM, Greenough WT. Differential rearing effects on rat visual cortex synapses. I. Synaptic and neuronal density and synapses per neuron. *Brain Res.* 1985;329(1–2):195–203.
41. Jones TA, Klintsova AY, Kilman VL, Sirevaag AM, Greenough WT. Induction of multiple synapses by experience in the visual cortex of adult rats. *Neurobiol Learn Mem.* 1997;68(1):13–20.
42. Globus A, Rosenzweig MR, Bennett EL, Diamond MC. Effects of differential experience on dendritic spine counts in rat cerebral cortex. *J Comp Physiol Psychol.* 1973;82(2):175–181.

43. Altman J, Das GD. Autoradiographic examination of the effects of enriched environment on the rate of glial multiplication in the adult rat brain. *Nature*. 1964;204(4964):1161–1163.
44. van Praag H, Kempermann G, Gage FH. Neural consequences of environmental enrichment. *Nat Rev Neurosci*. 2000;1(3):191–198.
45. Bian WJ, Miao WY, He SJ, Qiu Z, Yu X. Coordinated spine pruning and maturation mediated by inter-spine competition for cadherin/catenin complexes. *Cell*. 2015;162(4):808–822.
46. Yang G, Pan F, Gan W-B. Stably maintained dendritic spines are associated with lifelong memories. *Nature*. 2009;462(7275):920–924.
47. Fu M, Yu X, Lu J, Zuo Y. Repetitive motor learning induces coordinated formation of clustered dendritic spines in vivo. *Nature*. 2012;483(7387):92–95.
48. Jung CKE, Herms J. Structural dynamics of dendritic spines are influenced by an environmental enrichment: an in vivo imaging study. *Cereb Cortex*. 2014;24(2):377–384.
49. Qiao Q, Ma L, Li W, Tsai J-W, Yang G, Gan W-B. Long-term stability of axonal boutons in the mouse barrel cortex. *Dev Neurobiol*. 2016;76(3):252–261.
50. Gyllensten L, Malmfors T, Norrlin ML. Effect of visual deprivation on the optic centers of growing and adult mice. *J Comp Neurol*. 1965;124:149–160.
51. Coleman PD, Riesen AH. Evironmental effects on cortical dendritic fields. I. Rearing in the dark. *J Anat*. 1968;102(Pt 3):363–374.
52. Valverde P. Apical dendritic spines of the visual cortex and light deprivation in the mouse. *Exp Brain Res*. 1967;3(4):337–352.
53. Valverde F. Structural changes in the area striata of the mouse after enucleation. *Exp Brain Res*. 1968; 5(4):274–292.
54. Kossel A, Löwel S, Bolz J. Relationships between dendritic fields and functional architecture in striate cortex of normal and visually deprived cats. *J Neurosci*. 1995;15(5 Pt 2):3913–3926.
55. Hickmott PW, Steen PA. Large-scale changes in dendritic structure during reorganization of adult somatosensory cortex. *Nat Neurosci*. 2005;8(2):140–142.
56. Tailby C, Wright LL, Metha AB, Calford MB. Activity-dependent maintenance and growth of dendrites in adult cortex. *Proc Natl Acad Sci*. 2005;102(12):4631–4636.
57. Holtmaat A, Wilbrecht L, Knott GW, Welker E, Svoboda K. Experience-dependent and cell-type-specific spine growth in the neocortex. *Nature*. 2006;441(7096):979–983.
58. Schubert V, Lebrecht D, Holtmaat A. Peripheral deafferentation-driven functional somatosensory map shifts are associated with local, not large-scale dendritic structural plasticity. *J Neurosci*. 2013;33(22):9474–9487.
59. Churchill JD, Tharp JA, Wellman CL, Sengelaub DR, Garraghty PE. Morphological correlates of injury-induced reorganization in primate somatosensory cortex. *BMC Neurosci*. 2004;5(1):43.
60. Cheetham CEJ, Hammond MSL, McFarlane R, Finnerty GT. Altered sensory experience induces targeted rewiring of local excitatory connections in mature neocortex. *J Neurosci*. 2008;28(37):9249–9260.
61. Mataga N, Mizuguchi Y, Hensch TK. Experience-dependent pruning of dendritic spines in visual cortex by tissue plasminogen activator. *Neuron*. 2004;44(6):1031–1041.
62. Keck T, Mrsic-Flogel TD, Vaz Afonso M, Eysel UT, Bonhoeffer T, Hübener M. Massive restructuring of neuronal circuits during functional reorganization of adult visual cortex. *Nat Neurosci*. 2008;11(10):1162–1167.
63. Hofer SB, Mrsic-Flogel TD, Bonhoeffer T, Hübener M. Experience leaves a lasting structural trace in cortical circuits. *Nature*. 2009;457(7227):313–317.
64. Zuo Y, Yang G, Kwon E, Gan W-B. Long-term sensory deprivation prevents dendritic spine loss in primary somatosensory cortex. *Nature*. 2005;436(7048):261–265.
65. Ma L, Qiao Q, Tsai J-W, Yang G, Li W, Gan W-B. Experience-dependent plasticity of dendritic spines of layer 2/3 pyramidal neurons in the mouse cortex. *Dev Neurobiol*. 2016;76(3):277–286.
66. Makino H, Malinow R. Compartmentalized versus global synaptic plasticity on dendrites controlled by experience. *Neuron*. 2011;72(6):1001–1011.
67. Miquelajauregui A, Kribakaran S, Mostany R, Badaloni A, Consalez GG, Portera-Cailliau C. Layer 4 pyramidal neurons exhibit robust dendritic spine plasticity in vivo after input deprivation. *J Neurosci*. 2015; 35(18):7287–7294.
68. Darian-Smith C, Gilbert CD. Axonal sprouting accompanies functional reorganization in adult cat striate cortex. *Nature*. 1994;368(6473):737–740.

69. Florence SL, Taub HB, Kaas JH. Large-scale sprouting of cortical connections after peripheral injury in adult macaque monkeys. *Science*. 1998;282(5391):1117–1121.
70. Kossut M, Juliano SLL. Anatomical correlates of representational map reorganization induced by partial vibrissectomy in the barrel cortex of adult mice. *Neuroscience*. 1999;92(3):807–817.
71. Yamahachi H, Marik SA, McManus JNJ, Denk W, Gilbert CD. Rapid axonal sprouting and pruning accompany functional reorganization in primary visual cortex. *Neuron*. 2009;64(5):719–729.
72. Marik SA, Yamahachi H, McManus JNJ, Szabo G, Gilbert CD. Axonal dynamics of excitatory and inhibitory neurons in somatosensory cortex. *PLoS Biol*. 2010;8(6):e1000395.
73. Cheetham CEJ, Barnes SJ, Albieri G, Knott GW, Finnerty GT. Pansynaptic enlargement at adult cortical connections strengthened by experience. *Cereb Cortex*. 2014;24(2):521–531.
74. Parnavelas JG, Globus A, Kaups P. Continuous illumination from birth affects spine density of neurons in the visual cortex of the rat. *Exp Neurol*. 1973;40(3):742–747.
75. Knott GW, Quairiaux C, Genoud C, Welker E. Formation of dendritic spines with GABAergic synapses induced by whisker stimulation in adult mice. *Neuron*. 2002;34(2):265–273.
76. Wyatt RM, Tring E, Trachtenberg JT. Pattern and not magnitude of neural activity determines dendritic spine stability in awake mice. *Nat Neurosci*. 2012;15(7):949–951.
77. Zhang Y, Cudmore RH, Lin D-T, Linden DJ, Huganir RL. Visualization of NMDA receptor-dependent AMPA receptor synaptic plasticity in vivo. *Nat Neurosci*. 2015;18(3):402–407.
78. Eyre MD, Richter-Levin G, Avital A, Stewart MG. Morphological changes in hippocampal dentate gyrus synapses following spatial learning in rats are transient. *Eur J Neurosci*. 2003;17(9):1973–1980.
79. O'Malley A, O'Connell C, Murphy KJ, Regan CM. Transient spine density increases in the mid-molecular layer of hippocampal dentate gyrus accompany consolidation of a spatial learning task in the rodent. *Neuroscience*. 2000;99(2):229–232.
80. Rusakov DA, Davies HA, Harrison E, et al. Ultrastructural synaptic correlates of spatial learning in rat hippocampus. *Neuroscience*. 1997;80(1):69–77.
81. Leuner B, Falduto J, Shors TJ. Associative memory formation increases the observation of dendritic spines in the hippocampus. *J Neurosci*. 2003;23(2):659–665.
82. Geinisman Y, Disterhoft JF, Gundersen HJG, et al. Remodeling of hippocampal synapses after hippocampus-dependent associative learning. *J Comp Neurol*. 2000;417(1):49–59.
83. Geinisman Y, Berry RW, Disterhoft JF, Power JM, Van der Zee EA. Associative learning elicits the formation of multiple-synapse boutons. *J Neurosci*. 2001;21(15):5568–5573.
84. Restivo L, Vetere G, Bontempi B, Ammassari-Teule M. The formation of recent and remote memory is associated with time-dependent formation of dendritic spines in the hippocampus and anterior cingulate cortex. *J Neurosci*. 2009;29(25):8206–8214.
85. Moczulska KE, Tinter-Thiede J, Peter M, et al. Dynamics of dendritic spines in the mouse auditory cortex during memory formation and memory recall. *Proc Natl Acad Sci*. 2013;110(45):18315–18320.
86. Lai CSW, Franke TF, Gan W-B. Opposite effects of fear conditioning and extinction on dendritic spine remodelling. *Nature*. 2012;483(7387):87–91.
87. Joachimsthaler B, Brugger D, Skodras A, Schwarz C. Spine loss in primary somatosensory cortex during trace eyeblink conditioning. *J Neurosci*. 2015;35(9):3772–3781.
88. Kuhlman SJ, O'Connor DH, Fox K, Svoboda K. Structural plasticity within the barrel cortex during initial phases of whisker-dependent learning. *J Neurosci*. 2014;34(17):6078–6083.
89. Yang G, Lai CSW, Cichon J, Ma L, Li W, Gan W-B. Sleep promotes branch-specific formation of dendritic spines after learning. *Science*. 2014;344(6188):1173–1178.
90. Peters AJ, Chen SX, Komiyama T. Emergence of reproducible spatiotemporal activity during motor learning. *Nature*. 2014;510(7504):263–267.
91. Cichon J, Gan W-B. Branch-specific dendritic Ca^{2+} spikes cause persistent synaptic plasticity. *Nature*. 2015;520(7546):180–185.
92. Hayashi-Takagi A, Yagishita S, Nakamura M, et al. Labelling and optical erasure of synaptic memory traces in the motor cortex. *Nature*. 2015;525(7569):333–338.
93. Xu T, Yu X, Perlik AJ, et al. Rapid formation and selective stabilization of synapses for enduring motor memories. *Nature*. 2009;462(7275):915–919.

94. Wang L, Conner JM, Rickert J, Tuszynski MH. Structural plasticity within highly specific neuronal populations identifies a unique parcellation of motor learning in the adult brain. *Proc Natl Acad Sci*. 2011;108(6):2545–2550.
95. Kleim JA, Hogg TM, VandenBerg PM, Cooper NR, Bruneau R, Remple M. Cortical synaptogenesis and motor map reorganization occur during late, but not early, phase of motor skill learning. *J Neurosci*. 2004;24(3):628–633.
96. Kleim JA, Lussnig E, Schwarz ER, Comery TA, Greenough WT. Synaptogenesis and Fos expression in the motor cortex of the adult rat after motor skill learning. *J Neurosci*. 1996;16(14):4529–4535.
97. Kleim JA, Barbay S, Cooper NR, et al. Motor learning-dependent synaptogenesis is localized to functionally reorganized motor cortex. *Neurobiol Learn Mem*. 2002;77(1):63–77.
98. Johnson CM, Peckler H, Tai L-H, Wilbrecht L. Rule learning enhances structural plasticity of long-range axons in frontal cortex. *Nat Commun*. 2016;7(510):10785.
99. Van Harreveld A, Fifkova E. Swelling of dendritic spines in the fascia dentata after stimulation of the perforant fibers as a mechanism of post-tetanic potentiation. *Exp Neurol*. 1975;49(3):736–749.
100. Fifková E, Van Harreveld A. Long-lasting morphological changes in dendritic spines of dentate granular cells following stimulation of the entorhinal area. *J Neurocytol*. 1977;6(2):211–230.
101. Fifková E, Anderson CL. Stimulation-induced changes in dimensions of stalks of dendritic spines in the dentate molecular layer. *Exp Neurol*. 1981;74(2):621–627.
102. Bliss TV, Lomo T. Long-lasting potentiation of synaptic transmission in the dentate area of the anaesthetized rabbit following stimulation of the perforant path. *J Physiol*. 1973;232(2):331–356.
103. Desmond NL, Levy WB. Synaptic correlates of associative potentiation/depression: an ultrastructural study in the hippocampus. *Brain Res*. 1983;265(1):21–30.
104. Geinisman Y, DeToledo-Morrell L, Morrell F. Induction of long-term potentiation is associated with an increase in the number of axospinous synapses with segmented postsynaptic densities. *Brain Res*. 1991;566(1–2):77–88.
105. Desmond NL, Levy WB. Changes in the numerical density of synaptic contacts with long-term potentiation in the hippocampal dentate gyrus. *J Comp Neurol*. 1986;253(4):466–475.
106. Trommald M, Hulleberg G, Andersen P. Long-term potentiation is associated with new excitatory spine synapses on rat dentate granule cells. *Learn Mem*. 1996;3(2-3):218–228.
107. Lee K, Oliver M, Schottler F, Creager R, Lynch G. Ultrastructural effects of repetitive synaptic stimulation in the hippocampal slice preparation: a preliminary report. *Exp Neurol*. 1979;65(2):478–480.
108. Chang F-LF, Greenough WT. Transient and enduring morphological correlates of synaptic activity and efficacy change in the rat hippocampal slice. *Brain Res*. 1984;309(1):35–46.
109. Sorra KE, Harris KM. Stability in synapse number and size at 2 hr after long-term potentiation in hippocampal area CA1. *J Neurosci*. 1998;18(2):658–671.
110. Wenzel J, Otani S, Desmond NL, Levy WB. Rapid development of somatic spines in stratum granulosum of the adult hippocampus in vitro. *Brain Res*. 1994;656(1):127–134.
111. Kirov SA, Sorra KE, Harris KM. Slices have more synapses than perfusion-fixed hippocampus from both young and mature rats. *J Neurosci*. 1999;19(8):2876–2886.
112. Kirov SA, Harris KM. Dendrites are more spiny on mature hippocampal neurons when synapses are inactivated. *Nat Neurosci*. 1999;2(10):878–883.
113. Bourne JN, Harris KM. Coordination of size and number of excitatory and inhibitory synapses results in a balanced structural plasticity along mature hippocampal CA1 dendrites during LTP. *Hippocampus*. 2011;21(4):354–373.
114. Matsuzaki M, Honkura N, Ellis-Davies GCR, Kasai H. Structural basis of long-term potentiation in single dendritic spines. *Nature*. 2004;429(6993):761–766.
115. Harvey CD, Svoboda K. Locally dynamic synaptic learning rules in pyramidal neuron dendrites. *Nature*. 2007;450(7173):1195–1200.
116. Lang C, Barco A, Zablow L, Kandel ER, Siegelbaum S, Zakharenko SS. Transient expansion of synaptically connected dendritic spines upon induction of hippocampal long-term potentiation. *Proc Natl Acad Sci*. 2004;101(47):16665–16670.
117. Zhou Q, Homma KJ, Poo M. Shrinkage of dendritic spines associated with long-term depression of hippocampal synapses. *Neuron*. 2004;44(5):749–757.

118. De Roo M, Klauser P, Muller D. LTP promotes a selective long-term stabilization and clustering of dendritic spines. *PLoS Biol*. 2008;6(9):e219.
119. Govindarajan A, Israely I, Huang S-Y, Tonegawa S. The dendritic branch is the preferred integrative unit for protein synthesis-dependent LTP. *Neuron*. 2011;69(1):132–146.
120. Tanaka JI, Horiike Y, Matsuzaki M, Miyazaki T, Ellis-Davies GCR, Kasai H. Protein synthesis and neurotrophin-dependent structural plasticity of single dendritic spines. *Science*. 2008;319(5870):1683–1687.
121. Tønnesen J, Katona G, Rózsa B, Nägerl UV. Spine neck plasticity regulates compartmentalization of synapses. *Nat Neurosci*. 2014;17(5):678–685.
122. Maletic-Savatic M, Malinow R, Svoboda K. Rapid dendritic morphogenesis in CA1 hippocampal dendrites induced by synaptic activity. *Science*. 1999;283(5409):1923–1927.
123. Engert F, Bonhoeffer T. Dendritic spine changes associated with hippocampal long-term synaptic plasticity. *Nature*. 1999;399(6731):66–70.
124. Nägerl UV, Eberhorn N, Cambridge SB, Bonhoeffer T. Bidirectional activity-dependent morphological plasticity in hippocampal neurons. *Neuron*. 2004;44(5):759–767.
125. Nägerl UV, Köstinger G, Anderson JC, Martin KAC, Bonhoeffer T. Protracted synaptogenesis after activity-dependent spinogenesis in hippocampal neurons. *J Neurosci*. 2007;27(30):8149–8156.
126. Hill TC, Zito K. LTP-induced long-term stabilization of individual nascent dendritic spines. *J Neurosci*. 2013;33(2):678–686.
127. Toni N, Buchs PA, Nikonenko I, Bron CR, Muller D. LTP promotes formation of multiple spine synapses between a single axon terminal and a dendrite. *Nature*. 1999;402(6760):421–425.
128. Araya R, Vogels TP, Yuste R. Activity-dependent dendritic spine neck changes are correlated with synaptic strength. *Proc Natl Acad Sci*. 2014;111(28):E2895–E2904.
129. Okamoto K-I, Nagai T, Miyawaki A, Hayashi Y. Rapid and persistent modulation of actin dynamics regulates postsynaptic reorganization underlying bidirectional plasticity. *Nat Neurosci*. 2004;7(10):1104–1112.
130. Oh WC, Hill TC, Zito K. Synapse-specific and size-dependent mechanisms of spine structural plasticity accompanying synaptic weakening. *Proc Natl Acad Sci*. 2013;110(4):E305–E312.
131. Becker N, Wierenga CJ, Fonseca R, Bonhoeffer T, Nägerl UV. LTD induction causes morphological changes of presynaptic boutons and reduces their contacts with spines. *Neuron*. 2008;60(4):590–597.
132. Harvey CD, Yasuda R, Zhong H, Svoboda K. The spread of Ras activity triggered by activation of a single dendritic spine. *Science*. 2008;321(5885):136–140.
133. Oh WC, Parajuli LK, Zito K. Heterosynaptic structural plasticity on local dendritic segments of hippocampal CA1 neurons. *Cell Rep*. 2015;10(2):162–169.
134. De Paola V, Holtmaat A, Knott G, et al. Cell type-specific structural plasticity of axonal branches and boutons in the adult neocortex. *Neuron*. 2006;49(6):861–875.
135. Stettler DD, Yamahachi H, Li W, Denk W, Gilbert CD. Axons and synaptic boutons are highly dynamic in adult visual cortex. *Neuron*. 2006;49(6):877–887.
136. Majewska AK, Newton JR, Sur M. Remodeling of synaptic structure in sensory cortical areas in vivo. *J Neurosci*. 2006;26(11):3021–3029.
137. Knott GW, Holtmaat A, Wilbrecht L, Welker E, Svoboda K. Spine growth precedes synapse formation in the adult neocortex in vivo. *Nat Neurosci*. 2006;9(9):1117–1124.
138. Toni N, Teng EM, Bushong EA, et al. Synapse formation on neurons born in the adult hippocampus. *Nat Neurosci*. 2007;10(6):727–734.
139. Zito K, Scheuss V, Knott G, Hill T, Svoboda K. Rapid functional maturation of nascent dendritic spines. *Neuron*. 2009;61(2):247–258.
140. Cane M, Maco B, Knott G, Holtmaat A. The relationship between PSD-95 clustering and spine stability in vivo. *J Neurosci*. 2014;34(6):2075–2086.
141. Woods GF, Oh WC, Boudewyn LC, Mikula SK, Zito K. Loss of PSD-95 enrichment is not a prerequisite for spine retraction. *J Neurosci*. 2011;31(34):12129–12138.
142. Holtmaat AJGD, Trachtenberg JT, Wilbrecht L, et al. Transient and persistent dendritic spines in the neocortex in vivo. *Neuron*. 2005;45(2):279–291.
143. Zuo Y, Lin A, Chang P, Gan W-B. Development of long-term dendritic spine stability in diverse regions of cerebral cortex. *Neuron*. 2005;46(2):181–189.

144. Bartol TM, Bromer C, Kinney J, et al. Nanoconnectomic upper bound on the variability of synaptic plasticity. *Elife*. 2015;4:1–18.
145. Kwon H-B, Sabatini BL. Glutamate induces de novo growth of functional spines in developing cortex. *Nature*. 2011;474(7349):100–104.
146. Richards DA, Mateos JM, Hugel S, et al. Glutamate induces the rapid formation of spine head protrusions in hippocampal slice cultures. *Proc Natl Acad Sci*. 2005;102(17):6166–6171.
147. Arnth-Jensen N, Jabaudon D, Scanziani M. Cooperation between independent hippocampal synapses is controlled by glutamate uptake. *Nat Neurosci*. 2002;5(4):325–331.
148. Matsuzaki M, Ellis-Davies GC, Nemoto T, Miyashita Y, Iino M, Kasai H. Dendritic spine geometry is critical for AMPA receptor expression in hippocampal CA1 pyramidal neurons. *Nat Neurosci*. 2001;4(11):1086–1092.
149. Yasumatsu N, Matsuzaki M, Miyazaki T, Noguchi J, Kasai H. Principles of long-term dynamics of dendritic spines. *J Neurosci*. 2008;28(50):13592–13608.
150. Loewenstein Y, Kuras A, Rumpel S. Multiplicative dynamics underlie the emergence of the log-normal distribution of spine sizes in the neocortex in vivo. *J Neurosci*. 2011;31(26):9481–9488.
151. Loewenstein Y, Yanover U, Rumpel S. Predicting the dynamics of network connectivity in the neocortex. *J Neurosci*. 2015;35(36):12535–12544.
152. Arellano JI, Benavides-Piccione R, Defelipe J, Yuste R. Ultrastructure of dendritic spines: correlation between synaptic and spine morphologies. *Front Neurosci*. 2007;1(1):131–143.
153. Konur S, Rabinowitz D, Fenstermaker VL, Yuste R. Systematic regulation of spine sizes and densities in pyramidal neurons. *J Neurobiol*. 2003;56(2):95–112.
154. Stepanyants A, Tamás G, Chklovskii DB. Class-specific features of neuronal wiring. *Neuron*. 2004;43(2):251–259.
155. Wen Q, Stepanyants A, Elston GN, Grosberg AY, Chklovskii DB. Maximization of the connectivity repertoire as a statistical principle governing the shapes of dendritic arbors. *Proc Natl Acad Sci*. 2009;106(30):12536–12541.
156. Stepanyants A, Hof PR, Chklovskii DB. Geometry and structural plasticity of synaptic connectivity. *Neuron*. 2002;34(2):275–288.
157. Hellwig B. A quantitative analysis of the local connectivity between pyramidal neurons in layers 2/3 of the rat visual cortex. *Biol Cybern*. 2000;82(2):111–121.
158. Kalisman N, Silberberg G, Markram H. The neocortical microcircuit as a tabula rasa. *Proc Natl Acad Sci*. 2005;102(3):880–885.
159. Escobar G, Fares T, Stepanyants A. Structural plasticity of circuits in cortical neuropil. *J Neurosci*. 2008;28(34):8477–8488.
160. Chklovskii DB, Schikorski T, Stevens CF. Wiring optimization in cortical circuits. *Neuron*. 2002;34(3):341–347.
161. Fares T, Stepanyants A. Cooperative synapse formation in the neocortex. *Proc Natl Acad Sci*. 2009;106(38):16463–16468.
162. Markram H, Lübke J, Frotscher M, Roth A, Sakmann B. Physiology and anatomy of synaptic connections between thick tufted pyramidal neurones in the developing rat neocortex. *J Physiol*. 1997;500(2):409–440.
163. Feldmeyer D, Lübke J, Silver RA, Sakmann B. Synaptic connections between layer 4 spiny neurone layer 2/3 pyramidal cell pairs in juvenile rat barrel cortex: physiology and anatomy of interlaminar signalling within a cortical column. *J Physiol*. 2002;538(3):803–822.
164. Feldmeyer D, Lübke J, Sakmann B. Efficacy and connectivity of intracolumnar pairs of layer 2/3 pyramidal cells in the barrel cortex of juvenile rats. *J Physiol*. 2006;575(2):583–602.
165. Lee WA, Bonin V, Reed M, et al. Anatomy and function of an excitatory network in the visual cortex. *Nature*. 2016;532(7599):370–374.
166. Reimann MW, King JG, Muller EB, Ramaswamy S, Markram H. An algorithm to predict the connectome of neural microcircuits. *Front Comput Neurosci*. 2015;9(October):1–18.
167. Stepanyants A, Martinez LM, Ferecskó AS, Kisvárday ZF. The fractions of short- and long-range connections in the visual cortex. *Proc Natl Acad Sci*. 2009;106(9):3555–3560.
168. Kasthuri N, Hayworth KJ, Berger DR, et al. Saturated reconstruction of a volume of neocortex. *Cell*. 2015;162(3):648–661.
169. Lüscher C, Nicoll RA, Malenka RC, Muller D. Synaptic plasticity and dynamic modulation of the postsynaptic membrane. *Nat Neurosci*. 2000;3(6):545–550.

170. Sorra KE, Fiala JC, Harris KM. Critical assessment of the involvement of perforations, spinules, and spine branching in hippocampal synapse formation. *J Comp Neurol*. 1998;398(2):225–240.
171. Fiala JC, Allwardt B, Harris KM. Dendritic spines do not split during hippocampal LTP or maturation. *Nat Neurosci*. 2002;5(4):297–298.
172. Druckmann S, Feng L, Lee B, et al. Structured synaptic connectivity between hippocampal regions. *Neuron*. 2014;3(3):629–640.
173. Perin R, Berger TK, Markram H. A synaptic organizing principle for cortical neuronal groups. *Proc Natl Acad Sci*. 2011;108(13):5419–5424.
174. Legenstein R, Maass W. Branch-specific plasticity enables self-organization of nonlinear computation in single neurons. *J Neurosci*. 2011;31(30):10787–10802.
175. Zheng P, Dimitrakakis C, Triesch J. Network self-organization explains the statistics and dynamics of synaptic connection strengths in cortex. *PLoS Comput Biol*. 2013;9(1):e1002848.
176. Clopath C, Büsing L, Vasilaki E, Gerstner W. Connectivity reflects coding: a model of voltage-based STDP with homeostasis. *Nat Neurosci*. 2010;13(3):344–352.
177. Losonczy A, Makara JK, Magee JC. Compartmentalized dendritic plasticity and input feature storage in neurons. *Nature*. 2008;452(7186):436–441.
178. Poirazi P, Mel B. Impact of active dendrites and structural plasticity on the memory capacity of neural tissue. *Neuron*. 2001;29:779–796.
179. Kim J, Zhao T, Petralia RS, et al. mGRASP enables mapping mammalian synaptic connectivity with light microscopy. *Nat Methods*. 2012;9(1):96–102.
180. Wickersham IR, Finke S, Conzelmann K-K, Callaway EM. Retrograde neuronal tracing with a deletion-mutant rabies virus. *Nat Methods*. 2007;4(1):47–49.
181. Wickersham IR, Lyon DC, Barnard RJO, et al. Monosynaptic restriction of transsynaptic tracing from single, genetically targeted neurons. *Neuron*. 2007;53(5):639–647.
182. Häusser M. Optogenetics: the age of light. *Nat Methods*. 2014;11(10):1012–1014.
183. Chuong AS, Miri ML, Busskamp V, et al. Noninvasive optical inhibition with a red-shifted microbial rhodopsin. *Nat Neurosci*. 2014;17(8):1123–1129.
184. Akerboom J, Carreras Calderón N, Tian L, et al. Genetically encoded calcium indicators for multi-color neural activity imaging and combination with optogenetics. *Front Mol Neurosci*. 2013;6:2.

Further Reading

Dan Y, Poo MMM. Spike timing-dependent plasticity of neural circuits. *Neuron*. 2004;44(1):23–30.

CHAPTER

2

Structural Plasticity Induced by Adult Neurogenesis

Gregory W. Kirschen[1], Kurt A. Sailor[2] and Shaoyu Ge[1]

[1]Stony Brook University, Stony Brook, NY, United States [2]Pasteur Institute, Paris, France

OUTLINE

1 INTRODUCTION

A 5-day-old human embryo contains a small, seemingly uniform cluster of cells that, given adequate time, temperature, and biochemical building blocks, will give rise to all of the cell types, tissues, organs, and organ systems of a fully formed human in a remarkably precise and elegant progression. This cluster, termed the inner cell mass, contains the cells that make this entire process possible, namely the pluripotent embryonic stem cells (ESCs).[1–3] Through expression and dynamic DNA binding interactions of transcription factors including Oct3/4 and Nanog, these cells possess the capabilities of self-renewal and differentiation into the body's various cell types.[4–6] As pluripotent ESCs begin their

The Rewiring Brain
DOI: http://dx.doi.org/10.1016/B978-0-12-803784-3.00002-0

journey along particular developmental paths, their identities are pushed further along until they arrive at a fully differentiated state.

While most maturing cells of the fetus are destined for terminal differentiation by roughly the 10th week of gestation, a small subset delays their maturation and sets up residence in the various crevices of the growing body at an immature stage. These so-called adult stem cells (ASCs), which retain the capability to re-enter the cell cycle and differentiate into new cells of specific lineages over the lifespan, include the hematopoietic and endothelial stem cells of the bone marrow, the intestinal stem cells in the crypts of Lieberkühn of the gut epithelium, and the neural stem cells (NSCs) that line the lateral ventricles of the brain and the dentate gyrus (DG) of the hippocampus.[7,8] While non-neural ASCs are complex and dynamic in their own right, adult neural stem cells (ANSCs) are unique in that their environment and the properties of their offspring are incredibly plastic. Indeed, the electrophysiological, morphological, and signaling dynamics exhibited by adult-generated neurons are unmatched by any of the body's other ASC derivatives, and these properties serve to enhance the overall plasticity of the circuits in which adult-born neurons participate.

How newborn neurons are incorporated into the preexisting synaptic architecture of the adult brain, and how they in turn rewire the brain, are important topics to address. The surprising discovery of persistent postnatal neurogenesis in mammals, including humans, has served as the basis for an entirely new branch of neuroscience exploring the physiology of ANSCs and their progeny cells.[9–12] From a therapeutic standpoint, studying the potential regenerative and reparative properties of ANSCs may give us crucial insights in our endeavor to develop novel, targeted treatment strategies for a host of common debilitating neurological diseases such as Alzheimer disease, Parkinson disease, Huntington disease, and stroke, all of which exhibit cell damage and loss.

Throughout this chapter, we will trace adult neurogenesis by considering the interactions between newborn cells and their synaptic partners. Our goal is to review the experimental evidence for adult neurogenesis-induced rewiring, highlighting advances that have been made in understanding the unique advantages that these adult-generated cells confer to the plasticity and functioning of the mammalian brain. We will also come to appreciate the costs associated with maintaining a pool of ever-dividing progenitor cells (albeit at a decreasing rate over the lifespan), and constantly creating, pruning, and remodeling synapses between young and old neurons.

2 STRUCTURAL REWIRING INDUCED BY ADULT NEUROGENESIS: ANATOMICAL AND MORPHOLOGICAL EVIDENCE

The developing human fetal brain faces an incredible challenge: not only does it have to hastily generate roughly 100 billion neurons and an equal to slightly greater number of glial cells that the organism will carry forth for its life, but it must also establish the initial morass of local and remote connections.[13,14] Connectivity is substantially modified postnatally via neuronal apoptosis and synaptic pruning. Yet, in terms of basic layout, the infrastructure of the central nervous system (CNS) in place at birth bears a significant resemblance to that of the adult, with fundamental connections formed, gyri and fissures developed, neuronal migration nearly complete, and axonal track myelination underway.[15–18] The degree to which the CNS is developed in the neonate allows for many basic operations of the brain—emotion, sensory perception, and motor movement, for instance—that are present even at birth, though admittedly in a rudimentary form that will become refined with growth and experience. In addition, there are certain regions in the neonatal brain that will maintain enhanced plasticity and continuous structural/synaptic reorganization throughout the organism's life. These are the adult neurogenic regions.

The grossly established functional connectivity observed in many regions of the perinatal mammalian brain contrasts with that of the adult neurogenic regions, which include the subependymal layer of the subventricular zone (SVZ) of the lateral ventricles that gives rise to new interneurons of the olfactory bulb (OB), and the subgranular zone (SGZ) of the DG that gives rise to new dentate granule cells (DGCs) in the hippocampus. In these regions, rebirth, cell death, and reorganization constitute the norm, with accumulating evidence showing advantages of this turnover. The OB and hippocampus have similar functions; they are vital for certain types of learning and memory—faculties that require continuous formation, storage, removal, and retrieval of information.[19–23] Both regions receive lower order downstream input onto a population of adult-generated neurons that process and relay this input (in the DG), and shape the output (in the OB), contributing to the formation of complex representations of the environment.[24,25] It is plausible that sustained neurogenesis in adulthood exists as a solution to the problem of having to continuously absorb and bind new relevant information in an ever-changing, often unpredictable world. By allowing more *neurons* to integrate into the circuit, the potential storage capacity of the brain may be substantially increased, as opposed to simply adding more *synapses* or strengthening existing synapses among neurons already entrenched in the network. As opposed to individual synapses, neurons integrate a multitude of incoming inputs and change their output connections and properties in response. Thus the computational power embedded within a neuron is far greater than that of an individual synapse, perhaps explaining why the hippocampus and OB, which receive and integrate many proximal and distal inputs, utilize the strategy of neurogenesis.

In response to ongoing intrinsic processes and external input, new neurons in these brain regions reorganize their connected circuits, instilling in the system a greater capacity to encode and retrieve information from experiences in adulthood. With this larger picture in mind, we will explore the special properties of adult-generated neurons that allow them to enhance the plasticity of their connected circuits, beginning with a discussion of neuroblast migration, which creates architectural complexity in the rewiring brain.

From the standpoint of cell trafficking in the postnatal brain, there is probably no more impressive feat than that of adult-generated SVZ neuroblasts migrating tangentially to the relatively distant OB. This robust migration ability is also observed in so-called "resting" microglia that can roam the extracellular matrix, continuously surveying the microenvironment for injury and debris.[26] Still, unlike the overwhelming majority of neurons that are fixed at their location in the brain since the embryonic/perinatal period, adult-generated SVZ neuroblasts continuously migrate in chains in the rostral migratory stream (RMS), guided by "tubes" of glial cells, to their final destination in the OB.[9,27,28] Upon their arrival in the OB, neuroblasts primarily mature into one of two types of interneurons: GABAergic granule cells (GCs), which comprise the majority of newborn neurons, or GABAergic/dopaminergic periglomerular cells (PGCs).[27,29]

Migration of newborn cells in the DG, while shorter in distance as compared to that of the RMS, is no less complex. Original observations of newborn DGCs migrating from the hilus/subgranular zone of the DG into the granule cell layer revealed radial migration of these cells, perpendicular to the axis of the DG.[10,30–32] It was discovered that expression of the gap junction protein connexin43 (Cx43) by radial glial-like (RGL) cells (the neural progenitor cells, NPCs, of the DG[23,33,34]) is required not only for the proper radial migration of newborn DGCs, but also for newborn DGC survival, implying that direct electrical communication between newborn cells and putative "guidepost" RGL cells serves to direct newborn cells to their proper locations for integration and survival.[35,36] Recent work has uncovered a tangential dispersion phase that occurs prior to the radial migration phase.[37] Interestingly, neuronal precursors appeared to migrate tangentially along capillary vessels of the DG before ascending into the GCL. This finding complements earlier work demonstrating that the vasculature provides not only nutritional but also structural support to the neurogenic region, and in turn is remodeled by new neurons that stimulate angiogenic recruitment.[38] This is reminiscent to how exercised skeletal muscle secretes angiogenic factors to recruit an increased blood supply and sustain its high metabolic needs.[39]

The kinetics of postnatal neuroblast migration and integration dates back to the 1960s with "pulse-chase" labeling, which entails injecting radioactive or chemically altered nucleotides such as [^{3}H]-thymidine (^{3}H-dT) or 5-bromo-2′-deoxyuridine (BrdU) into animals to serially track dividing cells over time in fixed brain tissue.[9,10,31,40–42] An early study by Altman[9] found that neuroblasts migrate from their birthplace in the SVZ to a "checkpoint" within the subependymal layer of the OB over the course of about one week, and reach their final destination in the internal granule layer by the third week after birth. Cell type-specific tracking and manipulation approaches have since evolved, including fluorescent imaging of virally labeled dividing progenitor cells in fixed tissue sections, in cultured brain slices, and now even in the live, intact animal. Unlike the pulse-chase approach, in which only nuclei (and mitochondria) are labeled, the virally-mediated fluorescent reporter approach has the advantage of revealing the cellular architecture and connectivity over time, which has enhanced our understanding of the dynamic physiology of adult neurogenesis in both the DG and the SVZ/OB.

For fluorescent tagging of adult-born cells in the hippocampus, researchers have taken advantage of the retroviral lifecycle.[43] Because retroviruses are only able to insert their genomes into host cells when the nuclear envelope dissolves during the prophase of the cell cycle, only host cells undergoing mitosis will carry viral genes within their genome and pass these genes on to their progeny. Cell type-specific promoters can also be used to ensure that only cells of a certain lineage become labeled. Tashiro et al.[43] used a retroviral GFP-Cre recombinase approach to knock out the *N*-methyl D-aspartate receptor (NMDAR) of dividing cells in the DG of floxed-NR1KO mice, to study this receptor's importance in the cells' survival. By comparing the density of new neurons labeled with GFP (NR1KO) compared to red fluorescent protein control labeled cells (internal control), they found decreased numbers of GFP cells 1 week after viral injection, indicating this to be an early survival time point. Together with work from others, and us, this demonstrates the presence of a "critical period" where newborn neurons exhibit enhanced synaptic plasticity that relies on NMDAR glutamaterigic transmission.[44–46] The finding that knockout of NMDARs decreases newborn neuronal survival suggests that early synaptic transmission between newborn cells and the mature circuitry is essential for proper integration.

Along these lines and complementing these data, a systematic effort was undertaken to quantify the morphology of newborn DGCs across their development from 3 days to 14 months after birth,[47] the last time point being well past the time when they become electrophysiologically mature and functionally integrated, which occurs ~1–2 months postdivision.[48–52] By labeling newborn cells in the SGZ of the DG via stereotactic retrovirus injection, Zhao et al.[47] observed the initiation of spine formation as early as 12 days postinfection (dpi). Interestingly, dendrites that contained spines at 12 dpi had significantly more branch points and were longer than those lacking spines. This suggests that the process of synaptic integration begins at an early stage in DGC development, and that receiving synaptic input nurtures the growth of new dendrites. With regard to axonal elaboration, Zhao et al.[47] found that at 16 dpi, retrovirally tagged axons could be detected in the hilus, extending along the mossy fiber outflow tract to hippocampal area CA3. By day 56, these axons had migrated significantly farther and had successfully found their synaptic targets in CA3. Further work characterized the complete axonal output of adult-born cells to the CA3 from 10 to 77 days after birth and showed the steps of integration to be highly regular, suggesting tight control with intact guidance cues.[53] Thus the input and output connections of newborn DGCs appear to form on a timescale of roughly 2 months, as measured by gross morphological investigation.

Taking a more high-resolution look at these connections, we observe additional complexity. Adult-generated DGCs exhibit three types of dendritic spines: filopodia, whose striking features include their motility and long thin structure; thin spines referred to as "learning spines" that rapidly alter their shape, size, and number in response to activity; and mushroom spines, the relatively stable, thin-necked, bulbous-headed protrusions referred to as "memory spines."[54,55] Toni et al.[52] undertook a subsequent in-depth and high-resolution look at the synaptic development of newborn DGCs. In this study, the investigators scrutinized the process of synaptic integration of newborn DGCs with confocal microscopy, electron tomography, serial section electron microscopy, and live-cell imaging. After injecting a GFP-containing retrovirus into the DG of adult mice, they observed newborn cells receiving various types of synaptic contacts, including axosomatic, axodendritic, and axospinous synapses. First, to determine whether the afferents innervating newborn DGCs originated from the EC (the main synaptic input to the DG via the perforant pathway), they delivered an RFP-expressing lentiviral vector into the EC and identified points of contact between EC afferents and newborn DGCs (of note, lentiviruses will infect and tag both dividing and quiescent cells) and indeed, at 30 dpi, newborn DGCs had received EC input.

One intriguing finding that the investigators made related to the nature of the synaptic rewiring that occurred over the course of newborn DGC maturation and integration. Filopodial tips of newborn cells preferentially associated with *occupied* synaptic contacts between two mature cells, rather than free/unoccupied synapse sites. This affinity to associate with so-called "multiple synapse boutons" (MSBs) rather than "single synapse boutons" (SSBs) suggests that the filopodia of newborn DGCs possess some form of sensory mechanism that allows them not only to find synapses and set up connections but also the ability to sense which are occupied and which are unoccupied. In addition to becoming more stable (i.e., decreasing in motility) with age, it seems that the spines of newborn DGCs eventually outcompete preexisting DGCs' postsynaptic terminals rather than cohabitate, as over time, MSBs appear to mature into SSBs. Moreover, the functional significance of MSBs remains to be determined: they may be just an intermediate between old and new connections, or they may be important for hippocampal information processing. Still, these provoking findings give us hints about how the newborn cells begin to interact with the preexisting wiring as they find their inputs to successfully integrate. In fact, the newborn cells that do not outcompete their siblings in the competition for synaptic input undergo death by apoptosis, generally within the first few days after birth.[56] This emphasizes how critical new synapse formation is from the perspective of a newborn cell trying to survive in the competitive climate of the DG.

While the process of adult neurogenesis and integration has thus far been described largely via snapshots, with fixed tissue sections or electrophysiological recordings from newborn cells at different time points, we are moving increasingly toward the in vivo system to watch and record neurogenesis and integration in intact animals. For example, Mizrahi[57] infused a lentivirus-containing green fluorescent protein (GFP) vector into the SVZ and utilized two-photon confocal microscopy to image the morphological maturation of fluorescently labeled newborn neurons as they integrated into the OB. Interestingly, Mizrahi observed a remarkable degree of dendritic morphological change up to 3 months after adult-born GCs had entered the OB, under baseline conditions. By contrast, under conditions of olfactory enrichment, dendritic spine stability increases markedly[58] (described in detail, below), suggesting that lack of external environmental input may result in more evanescent contacts between adult-born interneurons and the mature circuitry.

In vivo imaging of newborn DGCs has proven to be more technically challenging than that of newborn OB interneurons due to the DG's deeper location within the brain. Nevertheless, with in vitro hippocampal slice culture as well as chronic endoscopic imaging of retrovirally labeled dividing cells, our group has observed structural cellular dynamics comparable to those described in the OB,[57] with the rapid formation and retraction of filopodia and tangential migration of adult-born cells in the DG (*unpublished observations*).

Studying the dynamics of newborn neurons has revealed key features about their synaptic integration, yet many questions remain unanswered. For example, we still do not know whether the synaptic input that newborn DGCs receive differs in some fundamental way from that of older DGCs. Do particular EC afferents and newborn neurons seek each other out to form stable connections, or are these synaptic partners somewhat promiscuous? Work by others and our own group has shown that under the right environmental contexts (e.g., an opportunity to form a new spatial memory), newly generated adult-born neurons that have properly established input and output connections are preferentially recruited over older DGCs during the weeks following their initial integration, implying a certain degree of engrained input specificity.[51,59,60] This may also be partially explained, however, by newborn DGCs' intrinsic properties.

3 NEWBORN NEURONS PROMOTE THEIR OWN INTEGRATION: ELECTROPHYSIOLOGICAL EVIDENCE

Apart from morphological studies of newborn DGCs and their synaptic contacts, electrophysiological characterization has demonstrated the properties of these cells and evidence of functional integration. Schmidt-Hieber et al.[61] compared the electrophysiological properties of newly born versus mature DGCs. They prepared acute hippocampal slices from adult rats, patched onto young and mature DGCs (determined by morphology and immunoreactivity for the immature marker polysialic acid neural cell adhesion molecule (PSA-NCAM)), and performed whole cell baseline recordings and measured potentiation after theta burst stimulation.[34,62] While resting membrane potentials were comparable between cells of the two age ranges, mature cells fired reliable trains of action potentials (APs) during current pulses, while young cells fired much fewer APs per pulse, implying that immature cells may have already started to develop connections. More surprising, however, was the finding that the threshold for LTP induction was significantly lower in newborn versus mature DGCs, suggesting that despite their immaturity, newly born neurons strengthen their newly formed connections as they begin to incorporate into the hippocampal circuit.

Following this study, Espósito et al.[50] recorded from retrovirally labeled newborn neurons starting as early as 1 dpi, and followed the developmental trajectory of these cells as they integrated into the circuit. Although cells between 1 and 7 days old were intrinsically silent and unable to respond to evoked postsynaptic currents, the investigators identified so-called "immature spikes" from spineless 2-week-old DGCs, resulting from slow GABAergic excitatory input to these cells, reminiscent of the role of GABA as an excitatory neurotransmitter during embryonic development.[63–65] Subsequent studies have confirmed this developmental switch from using GABA to using glutamate as the principal excitatory neurotransmitter that occurs as adult-born DGCs mature. Glutamatergic synaptic input and AMPA receptor expression emerge by the first month after neuronal birth.[44,45] At around the same developmental time, there is also downregulation of the sodium–potassium–chloride cotransporter 1 (NaKCC1), which immature DGCs use to keep intracellular chloride high, as well as upregulation of the potassium inward rectifier channel (Kir), which tends to decrease the neurons' input resistance.[66–68] Together, these developmental changes contribute to the role reversal of GABA from excitation in immature cells to inhibition in mature cells. As such, adult-generated mature DGCs become virtually indistinguishable from their prenatally and early postnatally derived counterparts, both in terms of electrophysiological properties as well as connectivity, although, as discussed earlier, the system is still able to identify which cells have been recently generated and hence favor their recruitment during memory formation and retrieval in adulthood.[51,69,70] In sum, the hyperplastic properties of newborn DGCs allow them to efficiently form synaptic connections with mature cells of the EC, and once connected to the network, their activity increasingly resembles that of the preexisting mature DGCs, allowing them to contribute to the overall integrity of the circuit.

Synaptic integration of newborn neurons has likewise been investigated in-depth in the OB, which receives a steady supply of new interneurons throughout life. Unlike in the case of the DG, however, the OB has two predominant types of adult-born neurons that are localized in separate OB layers and carry out different roles once integrated into the circuitry.[71,72] PGCs reside in the glomerular layer of the OB and primarily make dendrodendritic synapses with M/T cells (the principal output neurons of the OB, projecting to the olfactory cortex) and both PGCs and M/T cells receive input from olfactory sensory neuron (OSN) axons from the olfactory epithelium. When newborn GCs reach the end of the RMS and disperse radially into the GCL of the OB, they extend their dendrites to form dendrodendritic synapses onto the lateral dendrites of M/T cells. These contacts are localized to the external plexiform layer, allowing GCs (either perinatally derived or newly born) to sense and inhibit the output activity of the M/T cells.[73,74] GCs also receive strong excitatory input from olfactory cortex that synapses onto their proximal dendritic domains, allowing for top-down processing to occur.[75] This network connectivity enables PGCs to modulate sensory input, while GCs modulate or filter the output. Since GCs receive significant top-down input, their modulation of the OB output is highly regulated by cognitive states during encoding, e.g., attention, reward, and context. Importantly, gain-of-function occurs when adult-generated OB interneurons are stimulated, causing an animal to more rapidly learn a difficult odor discrimination task, whereas stimulating the preexisting, early postnatal-derived population had no effect, suggesting a specialized role in learning for adult-born neurons.[20,21,76]

The synaptic integration of adult-generated PGCs and GCs follows similar trajectories, beginning during the migratory period, although the physiology of adult-generated GCs has been more extensively studied than that of adult-generated PGCs, likely due to the fact that GCs outnumber PGCs by roughly 20:1.[77,78] Although ~1 week old, migrating neuroblasts do not contribute actively to network activity (lacking Na^+ currents and the ability to generate action potentials), these cells are still sensitive to their chemical environments and begin the process of integration even before their integration into the OB.[79] They express functional $GABA_AR$ and α-amino-3-hydroxy-5-methyl-4-isoxazolepropionic acid receptors (AMPARs) during the migratory period, and start to express NMDARs by ~2 weeks of age as they reach the OB, at which point they elaborate dendrites and develop fast, voltage-gated Na^+ currents that elicit action potentials. Morphological maturation of PGCs occurs on a slightly longer timescale; these cells fully elaborate their dendrites by 4 weeks of age, receiving input from OSNs and M/T cells, and synapse onto both mitral cells and OSNs in the glomeruli.[79,80,81] Similar to newborn neurons in the DG, newborn neurons of the OB receive GABAergic, then glutamatergic input.[79,82] GABA is excitatory for newborn cells in both the SVZ and DG, and GABAergic input serves to enhance their excitability, thus facilitating integration. However, for migrating neuroblasts destined for the OB, GABA may play more of a signaling role than an excitation role, since these cells are electrically silent during their migration and will not form functional connections for several weeks.[79] In contrast, their counterparts in the DG, despite being largely silent during the first week after birth, start to display Na^+ currents and can generate GABA-induced, "immature" action potentials shortly thereafter.[46,50,61,77] So-called "synapse unsilencing" occurs in adult-born DGCs through GABA-induced excitation triggering AMPA receptor expression, allowing these cells to switch over to mature glutamatergic signaling.[44] Apart from the intrinsic electrophysiology of newborn neurons, the local microenvironment also plays a key role in accommodating these young cells into the existing circuits.

4 LOCAL MICROENVIRONMENTS SUPPORT ONGOING NEURONAL INTEGRATION

Several "behind-the-scenes" players critically regulate newborn cells' maturation and synaptic integration into existing circuits: astrocytes, microglial cells, and local interneurons. Hippocampal astrocytes not only encourage ANSCs to re-enter the cell cycle and proliferate but also push the newborn progeny toward the neuronal lineage through local release of signaling molecules such as adenosine triphosphate (ATP).[83,84] Interestingly, in the absence of astrocytes, cultured ANSCs seem to favor the oligodendrocyte lineage.[84] Additionally, astrocytic vesicles contain compounds that influence synaptic transmission in the DG. For example, astrocyte-derived glutamate potentiates excitatory transmission between perforant pathway afferents and DGCs.[85] Likewise, Sultan et al.[86] recently discovered that blocking astrocyte vesicular release results in aberrant synaptic integration of newborn neurons adjacent to these astrocytes. These investigators genetically introduced either botulinum toxin (which directly cleaves one of the components of the vesicle-docking SNARE protein complex) or a dominant negative SNARE protein into hippocampal astrocytes. They found that newborn cells in the vicinity of vesicle-blocked astrocytes harbored shorter apical dendrites with fewer dendritic protrusions, exhibited diminished NMDAR- and AMPAR-mediated synaptic transmission, and were electrophysiologically more immature than age-matched control adult-generated DGCs, as evidenced by their higher NMDA/AMPA excitatory postsynaptic current (EPSC) ratio. Moreover, newborn DGCs in astrocyte vesicle-blocked mice were less likely to survive by 30 days as compared to those in a normal astrocyte environment. The effects on dendritic morphology were partially restored by the reintroduction of D-serine, which was reduced in astrocyte vesicle-blocked extracellular medium, suggesting that this molecule is a key astrocyte-released mediator of newborn DGC maturation and integration. One remaining question is whether these astrocytes are pre/perinatally derived or adult-derived, as multipotent neural precursors in the DG retain the capability to divide asymmetrically and give rise to new astrocytes in adulthood.[87,88] However, it seems more likely that mature, pre/perinatally born astrocytes are most important, as these astrocytes widely outnumber newborn astrocytes in the adult brain, and the astrocytic fate potential of multipotent ANSCs in the DG declines with age.[89,90] Nevertheless, astrocytes are clearly essential for both fate determination and integration of new neurons.

On the other hand, microglia also regulate adult neurogenesis and synaptic remodeling, but just as their role in CNS homeostasis is distinct from that of astrocytes, so too is their specific contribution to maintenance of neurogenic homeostasis. We have so far considered the birth, survival, and integration of newborn neurons in the process of adult neurogenesis, but in fact the majority of newly generated cells die by apoptosis if they do not successfully integrate into the circuit (as much as 50% within the first ~3 days after division[91]). This presents the brain with a significant custodial problem; namely, having to clean up the cellular debris from continuously dying neurons. Newborn cells are especially likely to die during the transition from transiently amplifying neuroprogenitor to neuroblast cell (days 1–4), and the swift removal of their apoptotic bodies is facilitated by the phagocytic action of microglia in the SGZ as well as in the SVZ and RMS.[56,92] Not only do microglia remove dead cells, but they also affect ANSC proliferation. Microglia are in close physical proximity to *Sox2*- and *GFAP*-expressing ANSCs, and microglial processes contact ANSCs.[93] Moreover, activation of their toll-like receptor 9 signaling pathway can curb the aberrant increase in hippocampal neurogenesis observed upon seizure induction. Conversely, microglia (and microglia-conditioned medium) have been shown to enhance the neurogenic capacity of ANSCs in a cultured neurosphere model of SVZ neurogenesis[94] via cytokine release.[95]

Microglia are also responsible for pruning at the level of the synapse. While autophagocytic (self-eating) remodeling appears to be the predominant mechanism for *axonal* rewiring,[96,97] pruning in dendrites appears to be under the control of microglia.[98,99] Parkhurst et al.[99] depleted microglia by targeting expression of the diphtheria toxin receptor (DTR) to CX_3CR1^{CreER}-expressing cells (those of the microglia/macrophage lineage), a strategy that caused selective apoptosis of DTR-expressing cells upon diphtheria toxin induction. They observed disruptions in hippocampal glutamatergic synaptic protein levels as well as decreased learning-dependent dendritic spine turnover; effects mediated by microglia-derived brain-derived neurotrophic factor (BDNF) signaling. Microglial signals may thus help to promote maintenance of firing rates at a certain level in the face of synaptic rewiring, or rewiring homeostasis, in the hippocampus, a concept we will explore further, below.

Finally, newborn DGC integration is regulated by local DG interneurons. Aside from the canonical glutamatergic perforant pathway inputs from the EC onto DGCs, significant innervation by local GABAergic interneurons also modulates circuit-level activity in the DG.[100] In fact, newborn DGCs establish synapses with local GABAergic parvalbumin + interneurons (PV-INs) before establishing those with the EC.[101] The relationship between mature and immature DGCs with these interneurons differs fundamentally, however. Whereas PV-INs form a feedback inhibition circuit onto mature DGCs, adult-generated young (4-week-old) DGCs are not as susceptible to this form of feedback inhibition.[102] This may help to explain why newborn DGCs are preferentially recruited during the formation of new spatial memories in adulthood.[51,59,60] Essentially, while older cells are relatively suppressed via rapid silencing by local PV-INs, younger cells have a lower threshold to fire with the additive effects of GABAergic depolarization and a relative lack of activity-dependent PV-IN suppression. Unlike older DGCs that have already set up their sparsely activated connections, newborn cells fire relatively indiscriminately, allowing them to participate in ongoing network activity during memory acquisition.[103] Thus the neurogenic microenvironment helps to guide new adult-generated neurons along during their maturation and while they establish functional synaptic connections (Fig. 2.1).

So far, we have considered the physiology of adult-born neurons as it relates to existing synaptic circuitry. Newborn cells display special properties that facilitate their integration, including the abilities to migrate, dynamically probe the microenvironment, and establish functional connections with preexisting presynaptic targets, in addition to exhibiting a low threshold for strengthening their synapses. Moreover, their local environment actively promotes their continuous integration. With these ideas in mind, we now move on to discussing the contribution of these new neurons to the overall plasticity of the networks to which they are connected.

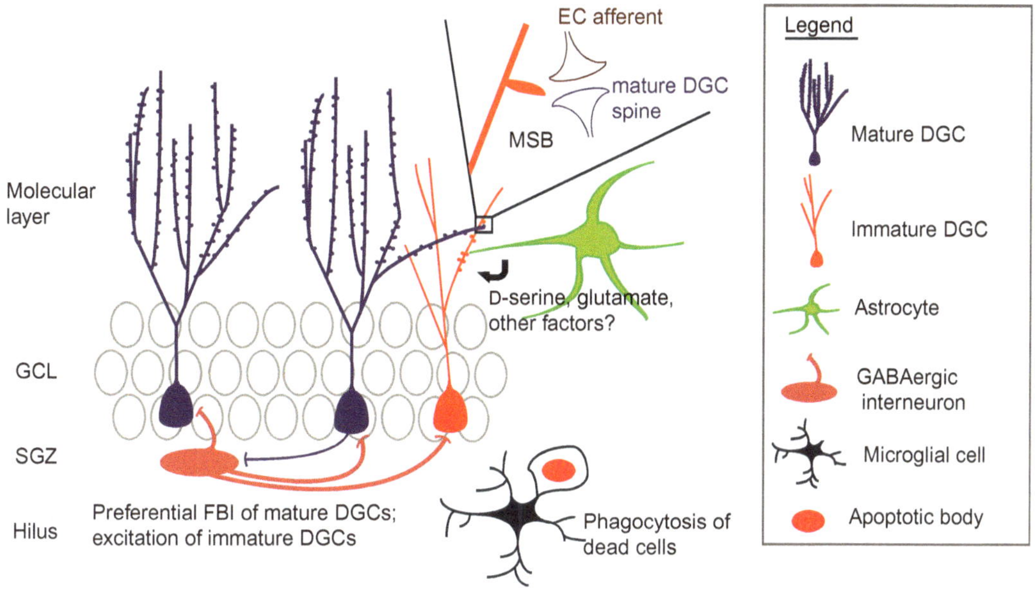

FIGURE 2.1 **Network-intrinsic control of newborn neuronal synaptic integration in the adult dentate gyrus.** Newly generated dentate granule cells (DGCs) migrate into the granule cell layer (GCL) of the dentate gyrus from the subgranular zone (SGZ) and extend their apical dendrites into the molecular layer to receive afferent innervation from entorhinal cortex (EC). Their dendritic spines, whose development depends on local astrocytic release of D-serine and glutamate, preferentially associate with preexisting synapses over "free" synaptic sites, forming multiple synapse boutons (MSBs). The successful integration of adult-born DGCs into the circuit is promoted by a combination of their low threshold for LTP and strong feedback inhibition (FBI) onto more mature DGCs by local GABAergic interneurons. In addition, homeostatic removal of the apoptotic bodies of newborn cells that fail to integrate is carried out by local microglia phagocytosis.

5 SYNAPTIC REWIRING BY NEW NEURONS: BALANCING THE FIRING BUDGET

Having shown that structural rewiring continues to include new neurons throughout adulthood in the mammalian brain, how can we examine the influence that adult-generated neurons have on network rewiring? A conceptually straightforward approach that has been taken to address this fundamental question is to simply alter the number of new neurons entering the circuit and observe the consequences. These experiments have been accomplished using environmental, pharmacological, and increasingly precise genetic approaches.

As far as experience-dependent rewiring goes, commonly employed experimental paradigms involve exposing rodents to so-called enriched environments (EEs) or impoverished environments (IEs), and/or increasing their daily doses of aerobic exercise. Both EE exposure and exercise lead to increases in hippocampal neurogenesis, and thus allow researchers to examine the synaptic rewiring that occurs as a result of generating more neurons in adulthood.[44,104–107] An intriguing observation made by Butz et al.[108] was the inverse relationship that exists between the rates of hippocampal neurogenesis and synaptic rewiring, as measured by axonal lysosome accumulation; comparing gerbils reared in EEs to those reared in IEs and/or those treated with methamphetamine (known to acutely decrease adult hippocampal neurogenesis[109]). With the development of a novel neural network model of connectivity among young, adult-generated DGCs and local glutamatergic and GABAergic innervation that corroborated their experimental findings, the investigators argue that the DG adapts to maintain a certain *firing homeostasis*[110,111] under different neurogenic states, while maintaining the average firing rate of its neurons.

While an inverse relationship between neurogenesis and synaptogenesis was noted when looking acutely at the proliferation of newborn cells, a recent study examining the chronic synaptic connectivity changes resulting from exposure to an EE revealed an increased number and a more diverse cadre of inputs onto newly divided cells. This increase in inputs makes sense for reasons we will lay out soon. Here, Bergami et al.[112] used a monosynaptic tracing approach to identify presynaptic partners of newborn neurons after EE exposure. Briefly, by introducing the EnvA receptor (TVA) and rabies glycoprotein into dividing hippocampal cells via a retroviral vector also carrying the fluorescent marker DsRed, the investigators were able to subsequently infect these cells specifically with rabies virus (RABV) carrying GFP. RABV was then able to travel from postsynaptic to presynaptic terminals, and retrogradely infect (and fluorescently tag) the presynaptic neuron.[113] Thus this method allowed for the detailed examination of synaptic rewiring of a specific neuronal population at different time intervals after EE exposure. Intriguingly, they found that when an EE was provided during the second to sixth week after cell division (but not 1 week before to 3 weeks after, or between 9 and 13 weeks after division), local interneuron afferents from the DG made twice as many synapses onto newborn cells as those of nonenriched animals. Moreover, after 4 weeks of enrichment, newborn cells also received interneuron inputs from hippocampal areas CA1 and CA3, previously unappreciated sources of innervation. On the other hand, physical aerobic exercise, which also increases the rate of hippocampal neurogenesis in adulthood, resulted in a much less pronounced change in interneuron connectivity to newborn cells, suggesting that merely increasing the number of newborn DGCs is not sufficient to significantly alter synaptic remodeling in the DG, at least as far as interneuron input is concerned. Instead, the investigators argue, exposure to an EE locally rewires new neurons during a critical period of 2–6 weeks after neuronal birth, accelerating their establishment of inputs (and presumably, their synaptic maturation), as well as establishing new connections with other hippocampal subfields. While this may at first seem counter to the inverse relationship between neurogenesis and rewiring discussed above, we must keep in mind that the increase was in *GABAergic* innervation. Given that EE exposure promotes GABA-dependent synaptic maturation[44] and that mature DGCs are inhibited by GABA,[46,50] the increased number of newborn cells born and firing was likely offset by an overall increase in inhibitory tone onto the DG (Fig. 2.2). Still, the sum total of inputs to the DG was not directly assessed in this study (i.e., including those of mature, established DGCs). Hence, it remains to be determined how monosynaptic connectivity between mature DGCs and their inputs changes under enrichment-induced hippocampal neurogenesis.

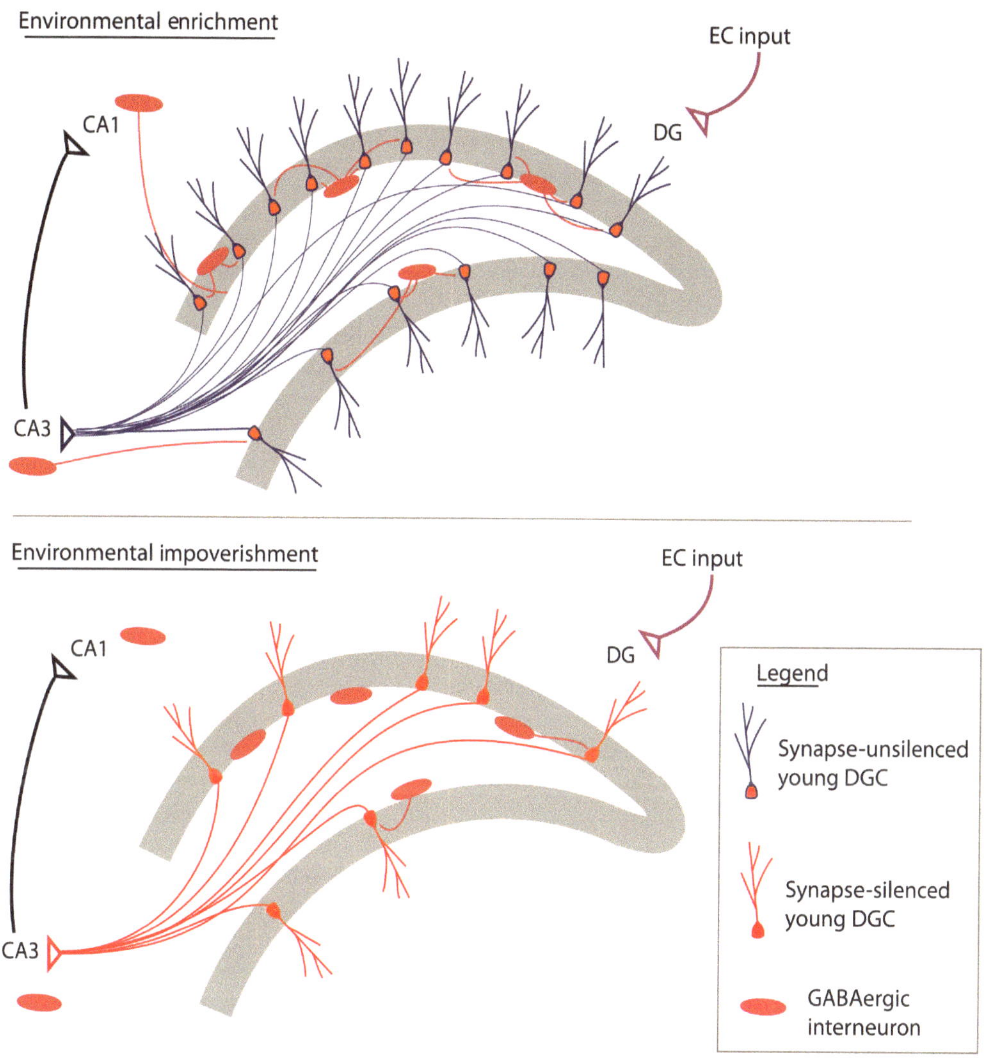

FIGURE 2.2 **Experience-induced adult neurogenesis shapes synaptic rewiring in the dentate gyrus.** Exposure to an enriched environment (top) increases the survival of young, adult-generated dentate granule cells (DGCs) and also facilitates maturational "synapse unsilencing" whereby GABA-induced excitation triggers AMPA receptor upregulation at the synapse, allowing DGCs to commence glutamatergic signaling. The increased number of newborn DGCs entering the circuit is paralleled by an increase in GABAergic interneuron innervation (both local and distal from areas CA3 and CA1). In this way, the dentate gyrus may buffer its overall firing activity in the face of an increased number of new neurons. By contrast, impoverished experience (bottom) leads to less survival of adult-born neurons, which exhibit delayed glutamatergic maturation and recruit weaker interneuron feedback inhibition. Triangles represent axon terminals.

However, findings from a prior investigation employing a more dramatic manipulation of hippocampal neurogenesis suggest that this alteration of inputs likely occurs at the global level of the DG circuit. Singer et al.[114] genetically ablated NPCs using a suicide gene-mediated approach. They introduced thymidine kinase (TK) into *Nestin+* NPCs of mice and subsequently treated mice with the antiviral drug gancyclovir (GCV), which becomes activated in TK-expressing cells and leads to apoptosis. After ablating NPCs, the investigators found elevated survival of cells born just prior to ablation, as well as reduced inhibitory input onto existing DGCs, as measured by decreased expression of the inhibitory synaptic marker vesicular GABA transporter (VGAT). Thus the DG likely exhibits less inhibition when neurogenesis rates are low and heightened compensatory inhibition when more new neurons enter the circuit. While more work is needed to elucidate the mechanisms underlying experience-dependent rewiring occurs in the DG and to map its functional consequences, a line of investigations has been undertaken in the OB to address these questions.

Whereas exercise and environmental enrichment strongly modulate adult hippocampal, but not OB neurogenesis,[105] olfactory input can profoundly influence OB neurogenesis, and thus has been used experimentally to understand the rewiring that takes place in the OB as a result of newborn interneurons. Olfactory deprivation in adulthood, commonly achieved by unilateral naris occlusion, leads to a pronounced decrease in newborn GC proliferation in the SVZ and survival in the ipsilateral, but not contralateral OB.[115–117] Conversely, olfactory enrichment (exposure to a diverse palate of odors) has been shown to increase the survival of newly generated neurons while also improving olfactory discrimination and odor memory.[19,118,119] Moreover, olfactory discrimination learning itself increases the survival of adult-generated OB neurons.[120] Like in the hippocampus with adult-born DGCs, adult-born OB interneurons also experience a critical period of activity-dependent synaptic integration. Kelsch et al.[121] performed unilateral naris occlusion in adult rats and found that GCs in the OB maturing during sensory deprivation exhibited a reduction in distal and basal dendritic glutamatergic synaptic inputs, but an increase in inputs at the proximal dendritic domains (likely reflecting increased local M/T innervation and/or top-down feedback innervation from olfactory cortex). Despite the relatively short and early 2-week period of sensory deprivation (between 14 and 28 days after cell birth), these changes remained stable over the ensuing month. By contrast, older adult-generated GCs that had already established their synaptic contacts prior to sensory deprivation exhibited no alterations in their distal or basal dendritic domains, but did experience increased inputs at their proximal domains. Interestingly, when the investigators genetically raised the intrinsic excitability of newborn GCs via introduction of a slowly depolarizing voltage-gated sodium channel, they found that normal synaptic development was not perturbed at any of the dendritic domains, but glutamatergic input was restored to normal levels in sensory-deprived animals.

On the other side of experience-dependent rewiring in the OB, enriched olfactory experiences have been employed to study adult-generated interneuron integration. One such investigation of adult-generated GCs and PGCs employed chronic two-photon imaging of lentivirally labeled cells before, during, and after olfactory enrichment. Here, Livneh and Mizrahi[58] found that adult-born, mature GCs and PGCs exhibited increased spine stability (with decreased turnover) upon odor enrichment that was maintained 1 week after the end of the exposure period. Similar findings were reported in the more naturalistic odor enrichment setting of the postpartum period, in which olfaction is known to be particularly heightened and influential on maternal behavior. Comparing lactating mother mice to naïve virgin controls, Kopel et al.[122] tagged newborn GCs with a lentivirus expressing GFP under the synaptophysin promoter and observed increased spine stability and lower spine density in the mothers versus the virgin controls. Despite the decrease in spine density, there was increased presynaptic synaptophysin in the mothers, suggesting enhanced integration and synapse maturation. Interestingly, spine dynamics of preexisting "resident" GCs was unaffected by parturition, pointing to a preferential effect on the immature cells. Thus in times when olfactory cues are abundant and evolutionarily salient, such as a mother bonding with and caring for her pups in the postpartum period, newborn GCs rapidly establish sparse, durable connections that may enhance information encoding and influence maternal behavior. That the preexisting OB circuitry does not rewire itself appreciably even in an enriched olfactory environment has been observed in another context.

To examine how experience-dependent changes in neurogenesis may percolate into the mature synaptic architecture of the OB, Mizrahi and Katz[123] fluorescently tagged M/T cells of adult mice and visualized their apical dendrites with two-photon microscopy after either olfactory enrichment or olfactory discrimination learning. After either 40 days of exposure to an odor-enriched environment or successful learning of an olfactory discrimination task, apical dendrites of M/T cells remained remarkably stable, without pronounced morphological changes. This synaptic stability was not impervious to manipulation, however, as application of the potent $GABA_A$ receptor blocker bicuculline resulted in synaptic turnover, as measured by pruning and sprouting events. The investigators concluded that overt structural changes in the apical dendrites of M/T cells may not be necessary for memory acquisition or storage, and argued that the inherent stability of the established glomerular circuitry may function to preserve its local organization. This stability may help to explain how animals are able to form new memories in adulthood without having to bear the cost of losing old ones. In addition, adult-born neurons, specifically granule cells, have increased survival with olfactory learning during a critical period between 18 and 30 dpi.[124] The experience-dependent neurogenic rewiring that occurs in the OB is depicted in Fig. 2.3.

We can draw several parallels between rewiring in the OB versus the DG. While the nature of the newborn neurons themselves is opposite (with DGCs forming excitatory, and OB interneurons forming inhibitory connections), the functional outcome of increased neurogenesis is actually quite similar in encouraging firing homeostasis in the face of increased circuit-derived activity. In the case of the OB, such activity arises from stimulated OSNs and is in part modulated directly by the adult-born inhibitory interneurons. In the DG, excitatory and hyperplastic adult-born DGCs alter connectivity by competing with mature DGCs at the perforant pathway, but firing homeostasis is maintained via the recruitment of local and distal inhibitory interneuron innervation. Yet both forms of adult neurogenesis promote information encoding and processing, and serve to improve pattern discrimination memory.[21]

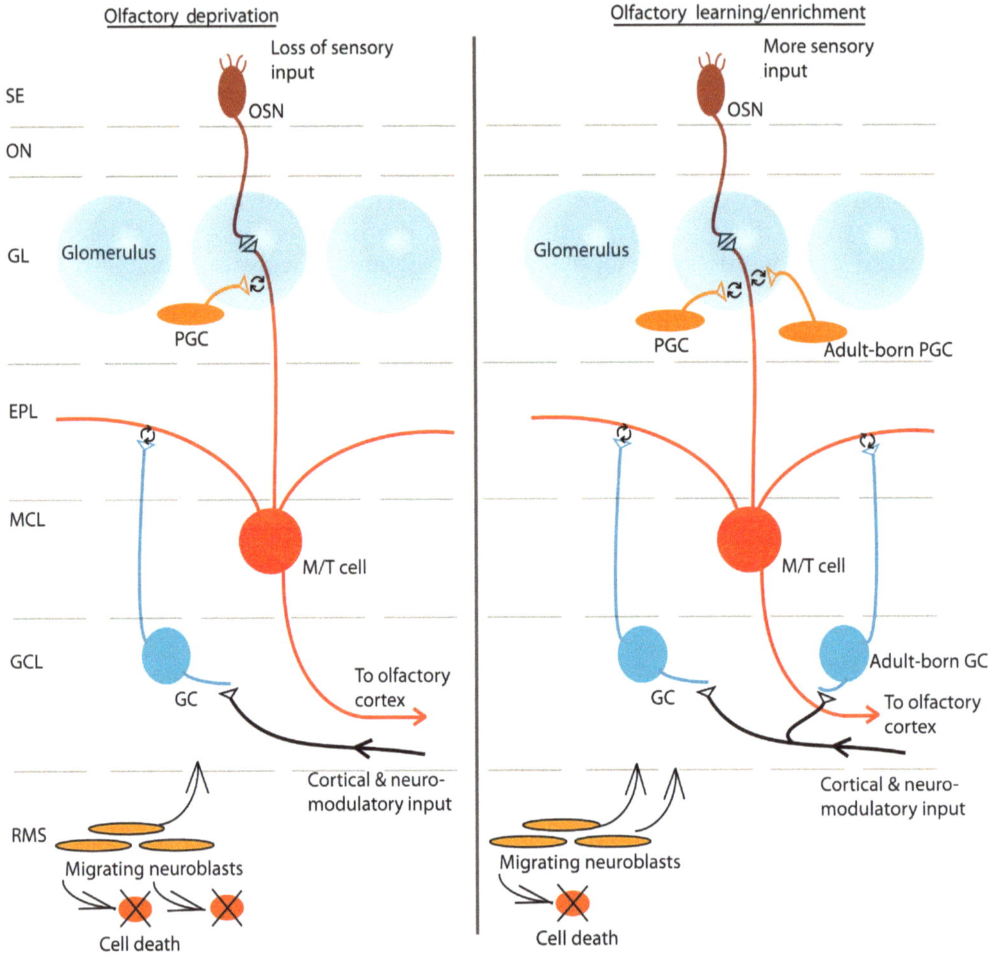

FIGURE 2.3 **Experience-induced adult neurogenesis enhances circuit modulation in the olfactory bulb.** Under conditions of olfactory deprivation, e.g., naris occlusion, fewer neuroblasts survive, dying by apoptosis in the subventricular zone (SVZ, not shown) or rostral migratory stream (RMS). Within the olfactory bulb (OB), existing granule cells (GCs) form reciprocal dendrodendritic synapses with mitral/tufted (M/T) cells (reciprocal arrows) and receive significant top-down input from olfactory cortex in addition to other modulatory inputs. Periglomerular cells (PGCs) also form reciprocal dendrodendritic synapses onto M/T cells. Olfactory discrimination learning and olfactory enrichment promote the survival and stable synapse formation of adult-born PGCs and GCs, the latter of which also recruits more distal inputs. The existing M/T cell glomerular dendritic structure remains relatively fixed despite the enriched experience. Adult-born PGCs and GCs serve to refine and tamp down activity of the OB at sensory input and network output, respectively. Other abbreviations, *SE*: sensory epithelium, *ON*: olfactory nerve, *GL*: glomerular layer, *EPL*: external plexiform layer, *MCL*: mitral cell layer, *GCL*: granule cell layer, *OSN*, olfactory sensory neuron.

6 CONCLUSION

In this chapter, we have taken an in-depth look at how a unique class of constantly dividing and integrating cells in the adult mammalian brain gives rise to an intricate rewiring dynamic. Though metabolically and organizationally costly, adult neurogenesis affords distinct computational advantages to systems involved in the processing and storing of various types of information. To make sense of an explosion of new experimental findings, the field of neurogenesis is increasingly employing computational neuroscience and mathematical modeling. Indeed, very recent publications that lend on these approaches have contributed importantly to our understanding of the workings and functional correlates of neurogenesis-induced plasticity. For instance, Danielson et al.[125] coupled two-photon calcium imaging of young and mature DGCs with spatial information and remapping analyses to determine the spatial tuning and contextual discrimination ability of individual units. Likewise, Temprana et al.[102] discovered a feedforward/feedback inhibitory interneuron network with adult-born DGCs and developed a computational model to describe how such inhibition could be useful in novel stimulus discrimination. As our knowledge of the complex circuitry and behavior of the neurogenic regions continues to grow, we look forward to the development of increasingly sophisticated and predictive computational models to enrich our understanding of the underlying biology.

Acknowledgment

We thank Roberto Vincis for his discussion regarding neurogenesis in the subventricular zone and synaptic rewiring in the olfactory bulb.

References

1. Tang F, Barbacioru C, Bao S, et al. Tracing the derivation of embryonic stem cells from the inner cell mass by single-cell RNA-Seq analysis. *Cell Stem Cell*. 2010;6(5):468–478.
2. Nichols J, Zevnik B, Anastassiadis K, et al. Formation of pluripotent stem cells in the mammalian embryo depends on the POU transcription factor Oct4. *Cell*. 1998;95(3):379–391.
3. Thomson JA, Itskovitz-Eldor J, Shapiro SS, et al. Embryonic stem cell lines derived from human blastocysts. *Science*. 1998;282(5391):1145–1147.
4. Niwa H, Miyazaki J, Smith AG. Quantitative expression of Oct-3/4 defines differentiation, dedifferentiation or self-renewal of ES cells. *Nat Genet*. 2000;24(4):372–376.
5. Loh YH, Wu Q, Chew JL, et al. The Oct4 and Nanog transcription network regulates pluripotency in mouse embryonic stem cells. *Nat Genet*. 2006;38(4):431–440.
6. Tsankov AM, Gu H, Akopian V, et al. Transcription factor binding dynamics during human ES cell differentiation. *Nature*. 2015;518(7539):344–349.
7. Young HE, Black Jr. AC. Adult stem cells. *Anat Rec A Discov Mol Cell Evol Biol*. 2004;276(1):75–102.
8. Johansson CB, Svensson M, Wallstedt L, Janson AM, Frisen J. Neural stem cells in the adult human brain. *Exp Cell Res*. 1999;253(2):733–736.
9. Altman J. Autoradiographic and histological studies of postnatal neurogenesis. IV. Cell proliferation and migration in the anterior forebrain, with special reference to persisting neurogenesis in the olfactory bulb. *J Comp Neurol*. 1969;137(4):433–457.
10. Altman J, Das GD. Autoradiographic and histological evidence of postnatal hippocampal neurogenesis in rats. *J Comp Neurol*. 1965;124(3):319–335.

11. Eriksson PS, Perfilieva E, Bjork-Eriksson T, et al. Neurogenesis in the adult human hippocampus. *Nat Med.* 1998;4(11):1313–1317.
12. Kuhn HG, Dickinson-Anson H, Gage FH. Neurogenesis in the dentate gyrus of the adult rat: age-related decrease of neuronal progenitor proliferation. *J Neurosci Off J Soc Neurosci.* 1996;16(6):2027–2033.
13. Azevedo FA, Carvalho LR, Grinberg LT, et al. Equal numbers of neuronal and nonneuronal cells make the human brain an isometrically scaled-up primate brain. *J Comp Neurol.* 2009;513(5):532–541.
14. Herculano-Houzel S, Collins CE, Wong P, Kaas JH. Cellular scaling rules for primate brains. *Proc Natl Acad Sci USA.* 2007;104(9):3562–3567.
15. Chi JG, Dooling EC, Gilles FH. Gyral development of the human brain. *Ann Neurol.* 1977;1(1):86–93.
16. Marin-Padilla M. Early prenatal ontogenesis of the cerebral cortex (neocortex) of the cat (*Felis domestica*). A Golgi study. I. The primordial neocortical organization. *Z Anat Entwick.* 1971;134(2):117–145.
17. Angevine Jr. JB, Sidman RL. Autoradiographic study of cell migration during histogenesis of cerebral cortex in the mouse. *Nature.* 1961;192:766–768.
18. Sternberger NH, Itoyama Y, Kies MW, Webster Hd. Immunocytochemical method to identify basic protein in myelin-forming oligodendrocytes of newborn rat C.N.S. *J Neurocytol.* 1978;7(2):251–263.
19. Rochefort C, Gheusi G, Vincent JD, Lledo PM. Enriched odor exposure increases the number of newborn neurons in the adult olfactory bulb and improves odor memory. *J Neurosci Off J Soc Neurosci.* 2002;22(7):2679–2689.
20. Sahay A, Scobie KN, Hill AS, et al. Increasing adult hippocampal neurogenesis is sufficient to improve pattern separation. *Nature.* 2011;472(7344):466–470.
21. Sahay A, Wilson DA, Hen R. Pattern separation: a common function for new neurons in hippocampus and olfactory bulb. *Neuron.* 2011;70(4):582–588.
22. Gheusi G, Cremer H, McLean H, Chazal G, Vincent JD, Lledo PM. Importance of newly generated neurons in the adult olfactory bulb for odor discrimination. *Proc Natl Acad Sci USA.* 2000;97(4):1823–1828.
23. Kempermann G, Jessberger S, Steiner B, Kronenberg G. Milestones of neuronal development in the adult hippocampus. *Trends Neurosci.* 2004;27(8):447–452.
24. Aimone JB, Wiles J, Gage FH. Potential role for adult neurogenesis in the encoding of time in new memories. *Nat Neurosci.* 2006;9(6):723–727.
25. Schoppa NE, Urban NN. Dendritic processing within olfactory bulb circuits. *Trends Neurosci.* 2003;26(9):501–506.
26. Nimmerjahn A, Kirchhoff F, Helmchen F. Resting microglial cells are highly dynamic surveillants of brain parenchyma in vivo. *Science.* 2005;308(5726):1314–1318.
27. Alvarez-Buylla A, Nottebohm F. Migration of young neurons in adult avian brain. *Nature.* 1988;335(6188):353–354.
28. Pencea V, Bingaman KD, Freedman LJ, Luskin MB. Neurogenesis in the subventricular zone and rostral migratory stream of the neonatal and adult primate forebrain. *Exp Neurol.* 2001;172(1):1–16.
29. Maher BJ, Westbrook GL. Co-transmission of dopamine and GABA in periglomerular cells. *J Neurophysiol.* 2008;99(3):1559–1564.
30. Altman J, Bayer SA. Horizontal compartmentation in the germinal matrices and intermediate zone of the embryonic rat cerebral cortex. *Exp Neurol.* 1990;107(1):36–47.
31. Altman J. Proliferation and migration of undifferentiated precursor cells in the rat during postnatal gliogenesis. *Exp Neurol.* 1966;16(3):263–278.
32. Duan X, Chang JH, Ge S, et al. Disrupted-in-schizophrenia 1 regulates integration of newly generated neurons in the adult brain. *Cell.* 2007;130(6):1146–1158.
33. Fukuda S, Kato F, Tozuka Y, Yamaguchi M, Miyamoto Y, Hisatsune T. Two distinct subpopulations of nestin-positive cells in adult mouse dentate gyrus. *J Neurosci Off J Soc Neurosci.* 2003;23(28):9357–9366.
34. Seki T, Arai Y. Temporal and spacial relationships between PSA-NCAM-expressing, newly generated granule cells, and radial glia-like cells in the adult dentate gyrus. *J Comp Neurol.* 1999;410(3):503–513.
35. Kunze A, Congreso MR, Hartmann C, et al. Connexin expression by radial glia-like cells is required for neurogenesis in the adult dentate gyrus. *Proc Natl Acad Sci USA.* 2009;106(27):11336–11341.
36. Liebmann M, Stahr A, Guenther M, Witte OW, Frahm C. Astrocytic Cx43 and Cx30 differentially modulate adult neurogenesis in mice. *Neurosci Lett.* 2013;545:40–45.

37. Sun GJ, Zhou Y, Stadel RP, et al. Tangential migration of neuronal precursors of glutamatergic neurons in the adult mammalian brain. *Proc Natl Acad Sci USA*. 2015;112(30):9484–9489.
38. Palmer TD, Willhoite AR, Gage FH. Vascular niche for adult hippocampal neurogenesis. *J Comp Neurol*. 2000;425(4):479–494.
39. Lloyd PG, Prior BM, Yang HT, Terjung RL. Angiogenic growth factor expression in rat skeletal muscle in response to exercise training. *Am J Physiol Heart Circ Physiol*. 2003;284(5):H1668–H1678.
40. Frazier-Cierpial L, Brunjes PC. Early postnatal cellular proliferation and survival in the olfactory bulb and rostral migratory stream of normal and unilaterally odor-deprived rats. *J Comp Neurol*. 1989;289(3):481–492.
41. Lois C, Garcia-Verdugo JM, Alvarez-Buylla A. Chain migration of neuronal precursors. *Science*. 1996;271 (5251):978–981.
42. Miller MW, Nowakowski RS. Use of bromodeoxyuridine-immunohistochemistry to examine the proliferation, migration and time of origin of cells in the central nervous system. *Brain Res*. 1988;457(1):44–52.
43. Tashiro A, Zhao C, Gage FH. Retrovirus-mediated single-cell gene knockout technique in adult newborn neurons in vivo. *Nat Protoc*. 2006;1(6):3049–3055.
44. Chancey JH, Adlaf EW, Sapp MC, Pugh PC, Wadiche JI, Overstreet-Wadiche LS. GABA depolarization is required for experience-dependent synapse unsilencing in adult-born neurons. *J Neurosci*. 2013;33 (15):6614–6622.
45. Ge S, Pradhan DA, Ming GL, Song H. GABA sets the tempo for activity-dependent adult neurogenesis. *Trends Neurosci*. 2007;30(1):1–8.
46. Ge S, Goh EL, Sailor KA, Kitabatake Y, Ming GL, Song H. GABA regulates synaptic integration of newly generated neurons in the adult brain. *Nature*. 2006;439(7076):589–593.
47. Zhao C, Teng EM, Summers Jr. RG, Ming GL, Gage FH. Distinct morphological stages of dentate granule neuron maturation in the adult mouse hippocampus. *J Neurosci Off J Soc Neurosci*. 2006;26(1):3–11.
48. Carlen M, Cassidy RM, Brismar H, Smith GA, Enquist LW, Frisen J. Functional integration of adult-born neurons. *Curr Biol CB*. 2002;12(7):606–608.
49. Jessberger S, Kempermann G. Adult-born hippocampal neurons mature into activity-dependent responsiveness. *Eur J Neurosci*. 2003;18(10):2707–2712.
50. Esposito MS, Piatti VC, Laplagne DA, et al. Neuronal differentiation in the adult hippocampus recapitulates embryonic development. *J Neurosci Off J Soc Neurosci*. 2005;25(44):10074–10086.
51. Kee N, Teixeira CM, Wang AH, Frankland PW. Preferential incorporation of adult-generated granule cells into spatial memory networks in the dentate gyrus. *Nat Neurosci*. 2007;10(3):355–362.
52. Toni N, Teng EM, Bushong EA, et al. Synapse formation on neurons born in the adult hippocampus. *Nat Neurosci*. 2007;10(6):727–734.
53. Sun GJ, Sailor KA, Mahmood QA, et al. Seamless reconstruction of intact adult-born neurons by serial end-block imaging reveals complex axonal guidance and development in the adult hippocampus. *J Neurosci*. 2013;33(28):11400–11411.
54. Petrak LJ, Harris KM, Kirov SA. Synaptogenesis on mature hippocampal dendrites occurs via filopodia and immature spines during blocked synaptic transmission. *J Comp Neurol*. 2005;484(2):183–190.
55. Bourne J, Harris KM. Do thin spines learn to be mushroom spines that remember? *Curr Opin Neurobiol*. 2007;17(3):381–386.
56. Sierra A, Encinas JM, Deudero JJ, et al. Microglia shape adult hippocampal neurogenesis through apoptosis-coupled phagocytosis. *Cell Stem Cell*. 2010;7(4):483–495.
57. Mizrahi A. Dendritic development and plasticity of adult-born neurons in the mouse olfactory bulb. *Nat Neurosci*. 2007;10(4):444–452.
58. Livneh Y, Mizrahi A. Experience-dependent plasticity of mature adult-born neurons. *Nat Neurosci*. 2012;15 (1):26–28.
59. Tashiro A, Makino H, Gage FH. Experience-specific functional modification of the dentate gyrus through adult neurogenesis: a critical period during an immature stage. *J Neurosci*. 2007;27(12):3252–3259.
60. Gu Y, Arruda-Carvalho M, Wang J, et al. Optical controlling reveals time-dependent roles for adult-born dentate granule cells. *Nat Neurosci*. 2012;15(12):1700–1706.
61. Schmidt-Hieber C, Jonas P, Bischofberger J. Enhanced synaptic plasticity in newly generated granule cells of the adult hippocampus. *Nature*. 2004;429(6988):184–187.

62. Seki T, Arai Y. Distribution and possible roles of the highly polysialylated neural cell adhesion molecule (NCAM-H) in the developing and adult central nervous system. *Neurosci Res*. 1993;17(4):265–290.
63. Ben-Ari Y. Excitatory actions of gaba during development: the nature of the nurture. *Nat Rev. Neurosci*. 2002;3 (9):728–739.
64. LoTurco JJ, Owens DF, Heath MJ, Davis MB, Kriegstein AR. GABA and glutamate depolarize cortical progenitor cells and inhibit DNA synthesis. *Neuron*. 1995;15(6):1287–1298.
65. Khazipov R, Esclapez M, Caillard O, et al. Early development of neuronal activity in the primate hippocampus in utero. *J Neurosci Off J Soc Neurosci*. 2001;21(24):9770–9781.
66. Mongiat LA, Esposito MS, Lombardi G, Schinder AF. Reliable activation of immature neurons in the adult hippocampus. *PLoS One*. 2009;4(4):e5320.
67. Khirug S, Yamada J, Afzalov R, Voipio J, Khiroug L, Kaila K. GABAergic depolarization of the axon initial segment in cortical principal neurons is caused by the Na–K–2Cl cotransporter NKCC1. *J Neurosci Off J Soc Neurosci*. 2008;28(18):4635–4639.
68. Tozuka Y, Fukuda S, Namba T, Seki T, Hisatsune T. GABAergic excitation promotes neuronal differentiation in adult hippocampal progenitor cells. *Neuron*. 2005;47(6):803–815.
69. Laplagne DA, Esposito MS, Piatti VC, et al. Functional convergence of neurons generated in the developing and adult hippocampus. *PLoS Biol*. 2006;4(12):e409.
70. Laplagne DA, Kamienkowski JE, Esposito MS, et al. Similar GABAergic inputs in dentate granule cells born during embryonic and adult neurogenesis. *Eur J Neurosci*. 2007;25(10):2973–2981.
71. Lledo PM, Merkle FT, Alvarez-Buylla A. Origin and function of olfactory bulb interneuron diversity. *Trends Neurosci*. 2008;31(8):392–400.
72. Doetsch F, Caille I, Lim DA, Garcia-Verdugo JM, Alvarez-Buylla A. Subventricular zone astrocytes are neural stem cells in the adult mammalian brain. *Cell*. 1999;97(6):703–716.
73. Nagayama S, Homma R, Imamura F. Neuronal organization of olfactory bulb circuits. *Front Neural Circuit*. 2014;8:98.
74. Lledo PM, Saghatelyan A. Integrating new neurons into the adult olfactory bulb: joining the network, life-death decisions, and the effects of sensory experience. *Trends Neurosci*. 2005;28(5):248–254.
75. Restrepo D, Doucette W, Whitesell JD, McTavish TS, Salcedo E. From the top down: flexible reading of a fragmented odor map. *Trends Neurosci*. 2009;32(10):525–531.
76. Alonso M, Lepousez G, Sebastien W, et al. Activation of adult-born neurons facilitates learning and memory. *Nat Neurosci*. 2012;15(6):897–904.
77. Lledo PM, Alonso M, Grubb MS. Adult neurogenesis and functional plasticity in neuronal circuits. *Nat Rev Neurosci*. 2006;7(3):179–193.
78. Lepousez G, Valley MT, Lledo PM. The impact of adult neurogenesis on olfactory bulb circuits and computations. *Annu Rev Physiol*. 2013;75:339–363.
79. Carleton A, Petreanu LT, Lansford R, Alvarez-Buylla A, Lledo PM. Becoming a new neuron in the adult olfactory bulb. *Nat Neurosci*. 2003;6(5):507–518.
80. Belluzzi O, Benedusi M, Ackman J, LoTurco JJ. Electrophysiological differentiation of new neurons in the olfactory bulb. *J Neurosci Off J Soc Neurosci*. 2003;23(32):10411–10418.
81. Bovetti S, Veyrac A, Peretto P, Fasolo A, De Marchis S. Olfactory enrichment influences adult neurogenesis modulating GAD67 and plasticity-related molecules expression in newborn cells of the olfactory bulb. *PLoS ONE*. 2009;4(7):e6359.
82. Imai T. Construction of functional neuronal circuitry in the olfactory bulb. *Semin Cell Develop Biol*. 2014;35:180–188.
83. Cao X, Li LP, Wang Q, et al. Astrocyte-derived ATP modulates depressive-like behaviors. *Nat Med*. 2013;19 (6):773–777.
84. Song H, Stevens CF, Gage FH. Astroglia induce neurogenesis from adult neural stem cells. *Nature*. 2002;417 (6884):39–44.
85. Jourdain P, Bergersen LH, Bhaukaurally K, et al. Glutamate exocytosis from astrocytes controls synaptic strength. *Nat Neurosci*. 2007;10(3):331–339.
86. Sultan S, Li L, Moss J, et al. Synaptic integration of adult-born hippocampal neurons is locally controlled by astrocytes. *Neuron*. 2015;88(5):957–972.

87. Rietze R, Poulin P, Weiss S. Mitotically active cells that generate neurons and astrocytes are present in multiple regions of the adult mouse hippocampus. *J Comp Neurol*. 2000;424(3):397–408.
88. Taupin P, Gage FH. Adult neurogenesis and neural stem cells of the central nervous system in mammals. *J Neurosci Res*. 2002;69(6):745–749.
89. Ge WP, Miyawaki A, Gage FH, Jan YN, Jan LY. Local generation of glia is a major astrocyte source in postnatal cortex. *Nature*. 2012;484(7394):376–380.
90. Krzisch M, Temprana SG, Mongiat LA, et al. Pre-existing astrocytes form functional perisynaptic processes on neurons generated in the adult hippocampus. *Brain Struct Funct*. 2015;220(4):2027–2042.
91. Mandyam CD, Harburg GC, Eisch AJ. Determination of key aspects of precursor cell proliferation, cell cycle length and kinetics in the adult mouse subgranular zone. *Neuroscience*. 2007;146(1):108–122.
92. Fourgeaud L, Traves PG, Tufail Y, et al. TAM receptors regulate multiple features of microglial physiology. *Nature*. 2016;532(7598):240–244.
93. Matsuda T, Murao N, Katano Y, et al. TLR9 signalling in microglia attenuates seizure-induced aberrant neurogenesis in the adult hippocampus. *Nat. Commun*. 2015;6:6514.
94. Walton NM, Sutter BM, Laywell ED, et al. Microglia instruct subventricular zone neurogenesis. *Glia*. 2006;54 (8):815–825.
95. Shigemoto-Mogami Y, Hoshikawa K, Goldman JE, Sekino Y, Sato K. Microglia enhance neurogenesis and oligodendrogenesis in the early postnatal subventricular zone. *J Neurosci Off J Soc Neurosci*. 2014;34 (6):2231–2243.
96. Dawirs RR, Teuchert-Noodt G, Hildebrandt K, Fei F. Granule cell proliferation and axon terminal degradation in the dentate gyrus of gerbils (*Meriones unguiculatus*) during maturation, adulthood and aging. *J Neural Transm*. 2000;107(6):639–647.
97. Dawirs RR, Teuchert-Noodt G, Kacza J. Naturally occurring degrading events in axon terminals of the dentate gyrus and stratum lucidum in the spiny mouse (*Acomys cahirinus*) during maturation, adulthood and aging. *Develop. Neurosci*. 1992;14(3):210–220.
98. Paolicelli RC, Bolasco G, Pagani F, et al. Synaptic pruning by microglia is necessary for normal brain development. *Science*. 2011;333(6048):1456–1458.
99. Parkhurst CN, Yang G, Ninan I, et al. Microglia promote learning-dependent synapse formation through brain-derived neurotrophic factor. *Cell*. 2013;155(7):1596–1609.
100. Sik A, Penttonen M, Buzsaki G. Interneurons in the hippocampal dentate gyrus: an in vivo intracellular study. *Eur J Neurosci*. 1997;9(3):573–588.
101. Deshpande A, Bergami M, Ghanem A, et al. Retrograde monosynaptic tracing reveals the temporal evolution of inputs onto new neurons in the adult dentate gyrus and olfactory bulb. *Proc Natl Acad Sci USA*. 2013;110 (12):E1152–E1161.
102. Temprana SG, Mongiat LA, Yang SM, et al. Delayed coupling to feedback inhibition during a critical period for the integration of adult-born granule cells. *Neuron*. 2015;85(1):116–130.
103. Aimone JB, Deng W, Gage FH. Resolving new memories: a critical look at the dentate gyrus, adult neurogenesis, and pattern separation. *Neuron*. 2011;70(4):589–596.
104. van Praag H, Kempermann G, Gage FH. Running increases cell proliferation and neurogenesis in the adult mouse dentate gyrus. *Nat Neurosci*. 1999;2(3):266–270.
105. Brown J, Cooper-Kuhn CM, Kempermann G, et al. Enriched environment and physical activity stimulate hippocampal but not olfactory bulb neurogenesis. *Eur J Neurosci*. 2003;17(10):2042–2046.
106. Kempermann G, Kuhn HG, Gage FH. More hippocampal neurons in adult mice living in an enriched environment. *Nature*. 1997;386(6624):493–495.
107. van Praag H, Kempermann G, Gage FH. Neural consequences of environmental enrichment. *Nat Rev Neurosci*. 2000;1(3):191–198.
108. Butz M, Teuchert-Noodt G, Grafen K, van Ooyen A. Inverse relationship between adult hippocampal cell proliferation and synaptic rewiring in the dentate gyrus. *Hippocampus*. 2008;18(9):879–898.
109. Teuchert-Noodt G, Dawirs RR, Hildebrandt K. Adult treatment with methamphetamine transiently decreases dentate granule cell proliferation in the gerbil hippocampus. *J Neural Transm*. 2000;107(2):133–143.
110. Van Ooyen A, Van Pelt J, Corner MA. Implications of activity dependent neurite outgrowth for neuronal morphology and network development. *J Theor Biol*. 1995;172(1):63–82.

111. Wolff JR, Missler M. Synaptic reorganization in developing and adult nervous systems. *Ann Anatomy Anatomischer Anz Off Organ Anatomisch Ges*. 1992;174(5):393–403.
112. Bergami M, Masserdotti G, Temprana SG, et al. A critical period for experience-dependent remodeling of adult-born neuron connectivity. *Neuron*. 2015;85(4):710–717.
113. Wickersham IR, Finke S, Conzelmann KK, Callaway EM. Retrograde neuronal tracing with a deletion-mutant rabies virus. *Nat Method*. 2007;4(1):47–49.
114. Singer BH, Gamelli AE, Fuller CL, Temme SJ, Parent JM, Murphy GG. Compensatory network changes in the dentate gyrus restore long-term potentiation following ablation of neurogenesis in young-adult mice. *Proc Natl Acad Sci USA*. 2011;108(13):5437–5442.
115. Corotto FS, Henegar JR, Maruniak JA. Odor deprivation leads to reduced neurogenesis and reduced neuronal survival in the olfactory bulb of the adult mouse. *Neuroscience*. 1994;61(4):739–744.
116. Mandairon N, Sacquet J, Jourdan F, Didier A. Long-term fate and distribution of newborn cells in the adult mouse olfactory bulb: influences of olfactory deprivation. *Neuroscience*. 2006;141(1):443–451.
117. Najbauer J, Leon M. Olfactory experience modulated apoptosis in the developing olfactory bulb. *Brain Res*. 1995;674(2):245–251.
118. Mandairon N, Stack C, Kiselycznyk C, Linster C. Enrichment to odors improves olfactory discrimination in adult rats. *Behav Neurosci*. 2006;120(1):173–179.
119. Mandairon N, Stack C, Linster C. Olfactory enrichment improves the recognition of individual components in mixtures. *Physiol Behav*. 2006;89(3):379–384.
120. Alonso M, Viollet C, Gabellec MM, Meas-Yedid V, Olivo-Marin JC, Lledo PM. Olfactory discrimination learning increases the survival of adult-born neurons in the olfactory bulb. *J Neurosci Off J Soc Neurosci*. 2006;26(41):10508–10513.
121. Kelsch W, Lin CW, Mosley CP, Lois C. A critical period for activity-dependent synaptic development during olfactory bulb adult neurogenesis. *J Neurosci Off J Soc Neurosci*. 2009;29(38):11852–11858.
122. Kopel H, Schechtman E, Groysman M, Mizrahi A. Enhanced synaptic integration of adult-born neurons in the olfactory bulb of lactating mothers. *J Neurosci*. 2012;32(22):7519–7527.
123. Mizrahi A, Katz LC. Dendritic stability in the adult olfactory bulb. *Nat Neurosci*. 2003;6(11):1201–1207.
124. Mouret A, Gheusi G, Gabellec MM, de Chaumont F, Olivo-Marin JC, Lledo PM. Learning and survival of newly generated neurons: when time matters. *J Neurosci*. 2008;28(45):11511–11516.
125. Danielson NB, Kaifosh P, Zaremba JD, et al. Distinct contribution of adult-born hippocampal granule cells to context encoding. *Neuron*. 2016;90(1):101–112.

CHAPTER

3

Structural Neural Plasticity During Stroke Recovery*

Kimberly Gerrow[1] *and Craig E. Brown*[1,2]

[1]University of Victoria, Victoria, BC, Canada [2]University of British Columbia, Vancouver, BC, Canada

OUTLINE

1 INTRODUCTION

Stroke is a focal neurologic deficit caused by a disruption in cerebral circulation. The majority of stroke cases are caused by an obstruction (e.g., blood clot) to the circulation (also known as ischemic stroke), rather than the rupture or hemorrhage of vessels. The worldwide prevalence of stroke is around 33 million people annually, which is the second-leading cause of death behind heart disease and the leading cause of disability in adults.[1] Effective treatment options are extremely limited for both types of stroke. In the case of ischemic stroke, treatment

*This work was supported by operating, salary, and equipment grants to C.E.B. from CIHR, Heart and Stroke Foundation of BC and Yukon, MSFHR, NSERC, and CFI.

DOI: http://dx.doi.org/10.1016/B978-0-12-803784-3.00003-2

with a thrombolytic agent such as tissue plasminogen activator (tPA) can limit brain damage due to stroke, but is only effective if given within a few hours from onset.[2] Currently, treatment initiated after the window for tPA has passed remains limited to physical and/or cognitive therapy, which provides some benefit to patients; though complete recovery is usually not expected.[3] The neurobiological mechanisms that underlie how people can regain function spontaneously or respond to physical rehabilitation are not well understood. However, the accumulated evidence from basic science studies over the past two to three decades has clearly demonstrated that the brain, even well after developmental processes have ceased, is capable of reorganizing its wiring diagram to adapt to extreme challenges such as stroke. For many basic and clinical scientists, understanding how the brain rewires after stroke and what mechanisms dictate this process represents our best hope for optimizing recovery from stroke.

Brain plasticity and repair after stroke can be broadly classified as functional or structural. Before we delve into the structural changes that occur after stroke, which is the focus of this chapter, we will first briefly discuss functional changes to provide the reader context (for a more extensive review, see Refs.[4,5]). Functional plasticity refers to the range of changes that are observed at the level of synapses, neural networks, and behavior of the organism. These can include changes in synaptic strength, neuronal excitability, regional brain activity patterns, as well as sensory, motor, and cognitive abilities. In the immediate aftermath of a stroke, failing energy production leads to dysfunction of ATP-dependent pumps.[6] Ionic gradients that maintain the resting membrane potential are lost along with extracellular potassium buffering, leading to neuronal depolarization, excessive glutamate release, aberrant activation of signal transduction pathways, and spreading depression like depolarizations that propagate throughout the stroke-affected hemisphere.[6,7] In the center of the ischemic stroke, a necrotic zone forms (referred to as the ischemic core or infarct) in tandem with immediate neurological impairments. In the areas neighboring the ischemic core, known as the peri-infarct zone, metabolically compromised but viable neurons become hyperexcitable to electrical- or sensory-evoked stimuli and show increased long-term potentiation.[8–10] For the next few days to weeks, the ischemic hemisphere shifts into a hypoexcitable state, mediated in part by a rise in $GABA_A$-mediated tonic inhibition.[11] On a macroscopic level, the loss of cortical excitability is evident in the fact that resting state or sensory-evoked cortical activity is suppressed in the stroke-affected hemisphere.[12–15] Meanwhile neuronal excitability in the undamaged hemisphere is enhanced temporarily during this time.[8,15] With respect to the restoration of neurological function, the greatest gains usually occur from 1 to 12 weeks following a stroke, although improvements can continue beyond this window. Functionally, this correlates best with the return of near normal cellular excitability and brain activation patterns in the stroke-affected hemisphere.[16] In summary, the peri-infarct region and connected regions are hot spots for functional neuronal plasticity.[5] Since structure influences function and vice versa, the remainder of this chapter will describe the plethora of structural changes that are presumed to mediate or at least reflect functional changes to neurons recovering from stroke.

2 ANIMAL MODELS OF STROKE

Much of what is currently known about structural plasticity has been derived from studying animal models of stroke. Therefore we will briefly discuss some of the different models used in experimental stroke research as well as the advantages and limitations intrinsic to each. Ischemic stroke can be reproduced in animals by several techniques: mechanical occlusion of arteries using surgical thread or clamps; photothrombotic occlusion where a photosensitive dye in the blood is activated with light[17]; or by microinjection of a potent vasoconstrictor such as endothelin-1.[18] Although the precise location of these ischemic stroke models can vary, in general, experimental studies usually target the middle cerebral artery (MCA) or branches of it. The reason for this is quite straightforward. First, the MCA or its branches is often the site of ischemic stroke in humans. Secondly, this vascular network supplies brain regions (e.g., primary motor and somatosensory cortex, striatum) that are crucial for sensory and motor function. Therefore, one can link changes in brain structure and function with how well the experimental animal recovers movement or sensory function in the limbs or face.

Each of these stroke models has its own advantages and disadvantages.[19] MCA occlusion models mimic a common type of stroke found in humans and have the advantage that reperfusion can be experimentally induced. However, MCA occlusion models include higher mortality rates and large unevenly distributed lesions that make functional deficits more variable and severe. Photothrombotic stroke generally leads to severe and irreversible ischemic damage, therefore studying reperfusion injury is not possible. Further, some studies suggest that photothrombotic stroke does not lead to changes in collateral blood flow and edema,[17,19] which normally occur in humans. However, photothrombosis is advantageous in that it produces focal lesions in specific cortical areas that lead to well-defined functional deficits. For instance, a recent study showed that photothrombotic occlusion of a single penetrating cortical arteriole could produce demonstrable functional deficits.[20] Furthermore, since photothrombosis can be induced through the skull or a cranial window, it is ideal for longitudinal light microscopic imaging of neuronal and vascular structure in vivo.[21] A detailed and precise analysis of circuit rearrangements in the peri-infarct cortex would be much more difficult, if not impossible in models with more variable stroke location and extent of injury. Microinjection of endothelin-1 also produces smaller lesions (although more variable in size than photothrombosis) that can be targeted to specific brain regions and offers the advantage of some reperfusion. However, in addition to its action on blood vessels, endothelin-1 also acts on receptors that are expressed on neurons and astrocytes in the brain.[22] Specifically, endothelin-1 itself is a potent activator of astrocyte growth and axonal sprouting.[23,24] In this sense, it is difficult to disentangle the effects of endothelin-1 on blood flow versus its actions on neurons and astrocytes. Despite the inherent flaws in each of these stroke models, they all capture elements of the human disease, thus providing great insight into the mechanisms of neural repair. The patterns of rewiring and recovery of function in these mouse models closely mimic those seen in human stroke survivors.[25,26]

3 AXONAL SPROUTING AND REWIRING CONNECTIONS

During development, axons grow and navigate over long distances to find their appropriate targets and are guided by the temporal and spatial patterns of several molecular cues.[27] A stereotyped wiring pattern is first established and is subsequently refined through neuronal activity. Axonal sprouting and regeneration after stroke partially resembles a recapitulation of this neurodevelopmental program: at both the cellular/structural and molecular levels.

Peri-infarct zones undergo striking changes in the expression of growth factors that are usually expressed during development.[28] Initial studies showed that growth-associated protein (GAP-43), which is expressed on axonal growth cones in development, was upregulated in peri-infarct tissue in the first 2 weeks after stroke.[29,30] The upregulation of GAP-43 was followed in time by a significant increase in the expression of the synaptic vesicle protein synaptophysin, suggesting that axonal growth was followed by synaptogenesis.[29] A few years later, Carmichael's group performed quantitative analysis of growth-associated gene expression in peri-infarct zones after distal occlusion of the MCA.[31–33] This study and others showed that growth promoting genes, particularly those associated with axon guidance,[34] were upregulated in different phases, some peaking relatively early (3–7 days poststroke: GDF-10), while others peaked later (14–28 days poststroke: SCG10, MARCKS, L1), or showed early and sustained activation (GAP-43, c-Jun). Inhibitory factors were also induced, albeit somewhat later after stroke, such as ephrinA5 or EphB1 receptor, semaphorin 3a, Nogo receptor complex signaling, and chondroitin sulfate proteoglycans (aggrecan, versican, neurocan). Of note, the expression profiles of these growth promoting and inhibitory factors were fundamentally different in the brains of young versus aged animals. In the aged brain, where injury induced axon growth is limited and delayed, growth genes were induced later, and growth inhibiting molecules turned on earlier.[31,32]

Estimates of gene and protein expression in peri-infarct and interconnected regions are crucial for a mechanistic understanding of axonal plasticity, but can only provide correlative and indirect evidence of axonal sprouting. A series of studies led by Carmichael provided a more direct, quantitative analysis of axonal sprouting in the poststroke brain. Using biotinylated dextran amines (BDAs) to trace projections to the peri-infarct cortex after distal MCA occlusion (that damaged portions of the whisker-barrel somatosensory cortex), they found a specific increase in retrogradely labeled neurons from nearby medial-anterior and posterior portions of the barrel field.[33] New local projections from within the barrel field were prominent 4 weeks after stroke and were not found in control hemispheres, nor described in published connections of the barrel cortex.[33] This shift in connectivity was also shown to apply to long-range corticocortical projection neurons following targeted damage to the primate motor cortex[35] or mouse somatosensory cortex.[13,32] In the former, Dancause et al. reported increased connections between the premotor area, a region they had previously shown undergoes functional remodeling, with Areas 1 and 2 of the primary somatosensory cortex.[35] In the latter study, Brown et al. analyzed stroke-induced changes to afferent projections to the peri-infarct forelimb motor cortex, a region where new forelimb sensory maps had emerged after stroke.[13] They discovered a significant increase in connections between motor forelimb cortex and more distant (~1–2 mm from injection site) posterior-medial regions such as the retrosplenial cortex and posterior parietal cortex. It should be noted that this study also revealed a profound loss of connections between the peri-infarct motor cortex and more lateral regions such as the whisker-barrel primary somatosensory cortex and secondary somatosensory cortex. While the consequences of this disconnection have not been well studied, it is likely that motor deficits after stroke in sensory motor cortex are based, in part, on a lack of communication between sensory and motor cortical regions.

From a mechanistic standpoint, the signaling pathways that mediate this sprouting response of intracortical connections are an active field of research, as well as the functional significance of this phenomenon as it relates to behavioral outcome. Since many studies have linked a particular signaling pathway with axonal sprouting after injury, we will restrict our focus to just a couple of elegant studies published in the last few years. Li et al. used the clever tactic of sequentially injecting fluorescent neuronal tracers to determine at a cellular level which neurons sprouted new connections after stroke.[32,36] These neurons were then isolated by laser capture microdissection and subjected to an unbiased whole genome expression analysis. As expected, several genes in sprouting neurons were identified as they related to axonal guidance, calcium signaling, extracellular matrix function, growth factors, transcription factors, and cytoskeletal modifying genes.[32] In particular, Atrx (alpha-thalassemia/mental retardation syndrome X-linked), Lingo1-NOGO receptor signaling, and growth/differentiation factor 10 (GDF-10) were upregulated in peri-infarct neurons. Since these signaling pathways could conceivably stimulate or inhibit the sprouting response, Li et al. used gain and loss of function experiments (siRNA; infusion of protein or blocking antibodies) to test their role after distal occlusion of the MCA.[32,36] As described previously, stroke normally leads to sprouting of ipsilesional motor cortex axons into premotor and somatosensory areas (Fig. 3.1A–G). Blocking Atrx or Lingo1/NOGO signaling in the peri-infarct cortex dramatically increased poststroke axonal sprouting (Fig. 3.1H,I), suggesting that these signaling pathways normally acted as a break on sprouting. On the other hand, blocking GDF-10 reduced the normal sprouting response after stroke while stimulating GDF-10 signaling significantly expanded the extent of sprouting. It is interesting to note that when axonal sprouting was stimulated with GDF-10 infusions, recovery of forelimb use and dexterity were improved relative to untreated stroke-affected mice, whereas when sprouting was blocked with GDF-10 siRNA, behavioral outcome was worse.[36] Collectively these studies show that the extent of axonal sprouting after stroke is dictated by a balance between growth promoting and inhibiting factors, which has direct consequences for recovery of function.

Stroke-related changes in connectivity are not limited solely to intracortical projections but also include long-range cortical output projections, interhemispheric callosal projections, and subcortical inputs. Rearrangement and sprouting of motor corticofugal projections to the brainstem or spinal cord have received considerable attention given that many research groups are seeking to improve recovery of motor function after stroke. While it may seem hard to imagine that projections to such distant sites could be affected by cortical stroke, a recent study has revealed that growth promoting factors such as GAP-43 and neurotrophin-3 are strongly upregulated in the cervical spinal cord following cortical photothrombotic stroke.[37] Corresponding with this, motor cortical projections from the undamaged hemisphere have been shown to sprout bilaterally into the cervical spinal cord[38] and red nucleus,[39] thereby providing a circuit where cortical neurons in the undamaged hemisphere could help control motor output in the stroke-affected limb. These sprouting axons appear to play a causal role in the recovery of forelimb motor function since reversible blockade of corticospinal fibers (using a pharmacogenetic/DREADD-based approach) transiently abolished forelimb grasping abilities in rats that had previously

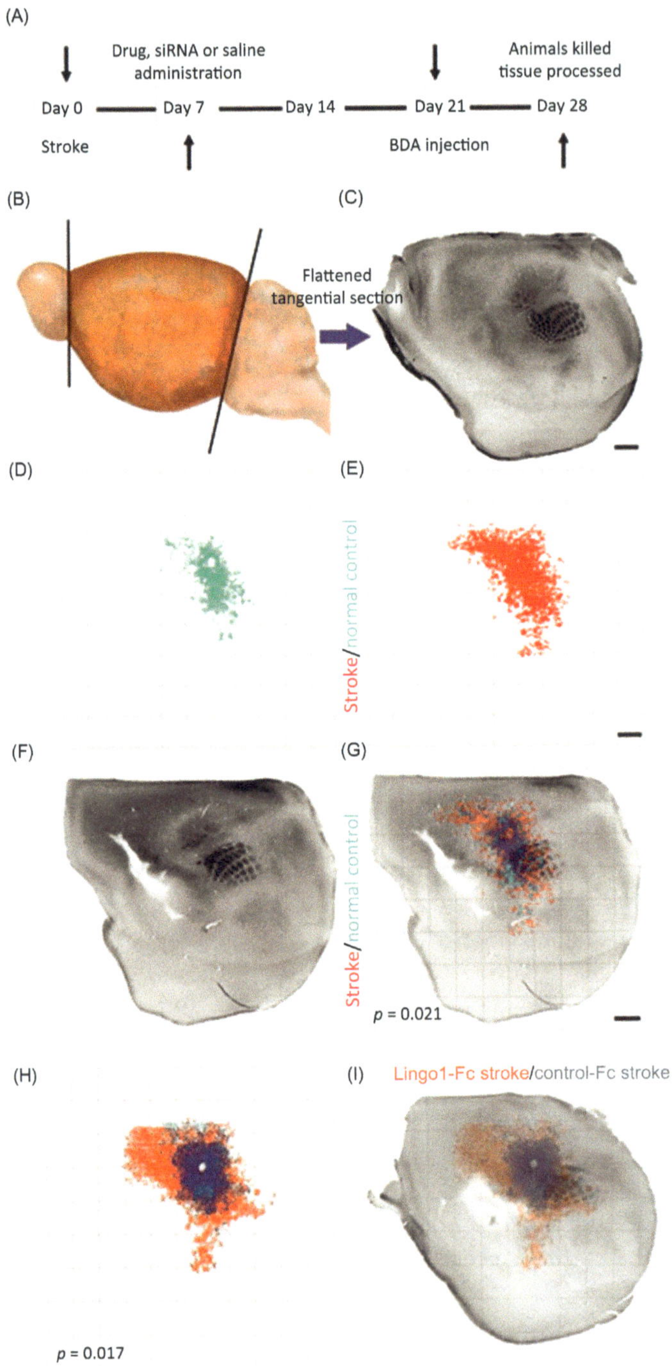
(A)
Drug, siRNA or saline administration
Animals killed tissue processed
Day 0
Day 7
Day 14
Day 21
Day 28
Stroke
BDA injection
(B)
(C)
Flattened tangential section
(D)
(E)
Stroke/normal control
(F)
(G)
Stroke/normal control
p = 0.021
(H)
(I)
Lingo1-Fc stroke/control-Fc stroke
p = 0.017

shown improved forelimb function after stroke.[40] In addition to these output projections, neurons in the undamaged cortex provide extensive interhemispheric connections with the peri-infarct region. Work from Chesselet and Carmichael's group showed that a cortical lesion could induce extensive transcallosal axonal sprouting into peri-infarct cortical and striatal regions.[41,42] Surprisingly, they discovered that the type of cortical lesion dictated the axonal sprouting response with thermocoagulation lesion but not aspiration lesion leading to sprouting. One interesting aspect of this sprouting response was that it was accompanied by synchronous low frequency activity (0.1–2 Hz) across both cerebral hemispheres, which was not apparent after the aspiration lesion. By blocking neuronal activity with tetrodotoxin, they found that synchronous cortical activity was indeed necessary for the spouting of transcallosal projections.[42] This finding agrees with many developmental studies showing that reducing neural activity disrupts proper thalamocortical or interhemispheric connectivity in the visual cortex.[43–45] Therefore, enhancing electrical activity in surviving circuits after stroke may play a beneficial role in promoting neural repair and rewiring.

When it comes to the effects of stroke on subcortical projections to the cerebral cortex, much less is known. What we do know from tracing studies is that thalamocortical projections are damaged following most types of cortical stroke.[33,46] Similarly, cholinergic or noradrenergic inputs to the cerebral cortex from the basal forebrain or locus cereleus, respectively, also appear damaged and significantly reduced in density.[47,48] However, to

FIGURE 3.1 **Quantitative connectional mapping of sprouting axons after stroke.** Stroke was produced in the whisker (barrel) field of the somatosensory cortex in young adult and aged animals by permanent occlusion of the distal middle cerebral artery (MCA), which induces a region of axonal sprouting in peri-infarct sensorimotor cortex. The whole genome expression profile of isolated neurons that sprout new connections after stroke was compared to that of adjacent, nonsprouting neurons. This "sprouting transcriptome" identified a neuronal growth program that consists of growth factor, cell adhesion, axonal guidance, and cytoskeletal modifying molecules that differed by age and time point. To determine the role of these three gene systems in axonal sprouting after stroke in vivo, a quantitative connectional mapping technique was used to statistically compare the motor cortical connections of peri-infarct cortex in control, stroke and stroke plus gain- or loss-of-function conditions. (A) Timeline for in vivo axonal tracer experiments. Animals received a sham or stroke surgery followed by siRNA or drug delivery 1 week later. Three weeks later, biotinylated dextran amine (BDA) was microinjected into forelimb motor cortex. (B and C) The cortex was then removed, flattened, tangentially cut (B) and stained for cytochrome oxidase (C) and BDA in the same sections. (D and E) The location of BDA-labeled axons was digitally mapped in *x,y* coordinates relative to the injection site. (D) The quantitative connectional map of forelimb sensorimotor connections in sham-operated mice. (E) The quantitative connectional map of forelimb sensorimotor cortex connections after stroke. (F) Cytochrome oxidase staining in layer 4 identifies the mouse somatosensory body map after stroke. (G) Each connectional map was registered to the somatosensory body map from the same brain to produce a group connectional and functional map of cortex. Axonal sprouting was identified when a pattern of cortical connections was precisely mapped and statistically different across treatment conditions. Dark blue indicates areas of dense overlap. (H) Lingo1 restricts cortical axonal sprouting after stroke. Quantitative connectional maps of forelimb motor connections show that Lingo1-Fc release into the peri-infarct cortex induced a significant different pattern of new connections (*red*) compared to the sprouting seen in controls (*blue*). (I) Registration of the motor cortical connections with the underlying cytochrome oxidase body map shows that Lingo1-Fc blocked sprouting lateral and rostral to the somatosensory cortex, in secondary somatosensory cortex, and motor cortex. Scale bars = 1 mm. Source: *Modified with permission from Li S, Overman JJ, Katsman D, et al. An age-related sprouting transcriptome provides molecular control of axonal sprouting after stroke. Nat Neurosci. 2010;13(12):1496–1504. Copyright by Nature Publishing Group.*

what extent these crucial sensory and neuromodulatory connections can sprout and recover lost contacts with cortical neurons is not well understood and warrants further investigation.

Up to this point, we have focused primarily on what neuronal tracing studies have revealed about short- and long-range changes in connectivity after stroke. However, there is substantial work describing changes to synaptic ultrastructure after ischemic cortical stroke. In 1974, Little et al. examined synaptic alterations up to 24 hours following cerebral infarction in squirrel monkeys.[49] For the first 90 minutes, they observed little change to synaptic ultrastructure except for an increase in extracellular space. By 3 hours, terminal boutons appeared shrunken with a specific reduction in axo-somatic but not axodendritic synapses. Surprisingly, the size and distribution of synaptic vesicles were unaltered at this time. From 6 to 12 hours poststroke, most terminal boutons were in poor condition being markedly shrunken, electron dense with synaptic vesicles distorted and depleted. Later electron microscopy studies extended the time frame of ultrastructural examination to surviving peri-infarct and contralateral cortical tissue in the days and weeks that follow stroke. In general, terminal boutons and synaptic vesicles declined precipitously in peri-infarct cortex for the first 4 days after stroke, which was followed by a progressive increase in number over the ensuing 12 weeks of recovery. In the undamaged contralateral hemisphere, rather than a loss of synapses, Jones et al. reported a progressive increase in synapse density and synapses per neuron by 30 days after stroke.[50] Furthermore, these structural changes were linked to increased use of the unaffected forepaw which typically manifests as a compensatory behavioral strategy after stroke.[51,52] While this compensatory strategy may aid in the ability to resume day-to-day activities, it may also have a cost to circuits in the peri-infarct cortex. For example, if rats are given a unilateral stroke and then encouraged to use the unaffected or "good" forepaw rather than the stroke impaired one, the functional representation of the forepaw in peri-infarct motor cortex shrinks (defined by intracortical microstimulation) along with motor performance.[53] Curiously, however, the number of axodendritic synapses within the peri-infarct region aberrantly increases. These findings suggest that underuse of the stroke-affected paw (and presumably reduced cortical activity in the stroke-affected hemisphere) leads to a compensatory addition of excitatory synapses, perhaps as a homeostatic mechanism to restore normal levels of excitability.

In summary, there is plenty of evidence from histological studies analyzing synaptic ultrastructure and neural connections to show that stroke alters the cortical wiring diagram. It is difficult to summarize all the reported changes in connectivity after stroke, in part because the manner with which connections change will depend on which region is damaged and the inherent anisotropy of connections that the damaged and surviving regions possess. However with that being said, some general principles emerge. For one, sprouting of new connections occurs primarily in regions that are functionally related to the one that is damaged.[13,35,54] For example, motor cortex projections sprout into somatosensory cortex (Fig. 3.1) or striatum suggesting that use or activity-dependent processes are important in guiding sprouting. Second, experimental manipulations that encourage axonal promoting correlate with improved recovery of sensory or motor function.[36] Third, cortical infarcts (even small ones) lead to a profound disconnection, not just for those projections that terminate in the damaged area but also those connections that run near the site of damage en route to other regions. This fact may explain why ministrokes in motor cortex can lead to widespread disruptions in ipsilesional corticospinal motor output (>1 mm from infarct border) despite these cells maintaining their intrinsic excitability.[55]

4 DENDRITIC ARBOR REMODELING

As one might expect, changes in axonal connectivity like those described earlier are accompanied by postsynaptic modifications of the dendrite. When describing changes to the structure of dendrites during and after a stroke, it is useful to think of it as a two-stage process. The first stage concerns the acute effects of ischemia on dendritic structure that happen within the first few minutes to hours after an ischemic event. The second phase pertains to the structural modifications that can occur over days to weeks after the survival of the dendrite is no longer in jeopardy.

When blood flow is acutely cut off to the brain, dendritic structure can change very rapidly depending on the degree of ischemia they experience. For neurons subjected to global ischemia or ones that are located in the ischemic core during a focal stroke (in both cases where blood flow is severely reduced), dendritic structure is disturbed within just a few minutes.[56–58] Imaging of fluorescently labeled dendrites in cell culture models or in vivo shows that dendrites appear as beads on a string when subjected to severe ischemia (Fig. 3.2).[57,59,60] The beaded appearance of dendrites is not simply an artifact of fluorescent protein aggregation since imaging studies using cytoplasmic fluorescent proteins (e.g., GFP, YFP) or lipophilic membrane-based tracers (e.g., DiI) reveal the exact same beaded structure.[57,59] Rather, this beading is likely a consequence of the derangement of microtubules[61] that serve as the backbone for dendrites.

When the loss of blood flow continues for an extended period of time, the beaded appearance of dendrites becomes more pronounced, and eventually the dendrite disintegrates, signifying the point when permanent damage has occurred. However, if a sufficient amount of blood flow is returned in time, the structure of dendrites can show an amazing degree of recovery. For example, a series of in vivo time-lapse imaging studies from the Murphy laboratory used both focal and global ischemia models to study the recovery of dendritic structure of layer 5 cortical neurons during and after ischemia.[56–58] In both models of ischemia, they convincingly demonstrated that severely beaded ischemic cortical dendrites could reestablish their normal structure within just a few minutes after restoring blood flow, albeit with subtle changes to dendritic spines (Fig. 3.2). Using similar experimental approaches, Sword et al.[62] showed that each wave of spreading depolarization that occurs after brain injury leads to immediate but reversible damage (i.e., beading) to dendritic structure in the pericontusional cortex. One interesting note about this phenomenon was that the likelihood for cortical dendrites to recover structure was reduced with each successive round of spreading depression. Collectively, these studies highlight the remarkable resiliency of dendritic structure after ischemic or traumatic brain injury, but also suggest there are limits to how much and for long dendrites can stave off permanent damage from ischemia.

After the initial ischemic insult when survival is no longer the primary issue, experimental evidence suggests that dendrites can undergo structural modifications during the weeks and months that follow a stroke. Remodeling of cortical dendritic length and branching

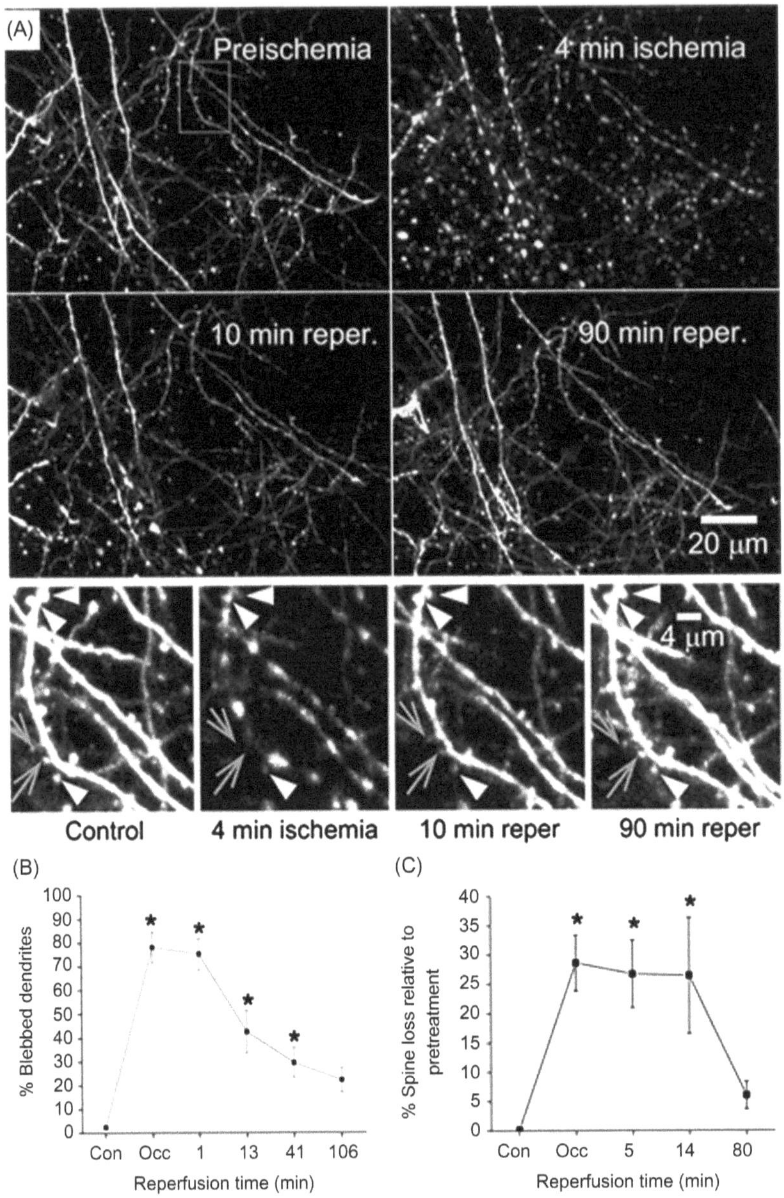

FIGURE 3.2 **A rapid and reversible loss of spiny dendritic structure during ischemia caused by bilateral carotid artery occlusion.** After establishing a model where reversible ischemia was produced during two-photon imaging, its impact on dendrite and spine structure in layer V neuron apical tufts was assessed. Dendritic structure was quantified within the first 100 μm of cortex in 16 XFP transgenic mice where both ischemia and reperfusion were confirmed. (A) Maximal intensity projection images of layer 5 neuron apical dendrites before and 4 minutes after ischemia and various time points during reperfusion (reper.). The *white boxed area* in the top left corresponds to a region in closeup images shown in the closeup panel below. Bottom, closeup views of top panels; *white arrowheads* show the location of dendritic spines present throughout the ischemia/reperfusion; *large*

could provide a framework for individual neurons to change their lines of communication and restore normal activity levels after stroke. This was first explored in studies using light microscopic examination of Golgi-Cox or DiI labeled neurons in peri-infarct cortex, and more remote but functionally connected locations such as the frontal, motor, and somatosensory cortex of the undamaged hemisphere. In the undamaged hemisphere, seminal work from Jones and Schallert reported that the number of dendritic branches in layer 5 neurons increased 2–4 weeks after forelimb sensorimotor cortex damage.[63] They went on to show that this increase in apical and basilar dendritic arborization in the undamaged hemisphere was related to overuse of the "good" or stroke spared forelimb, since immobilization of the good limb prevented changes in dendritic complexity.[51] These findings were generally supported by findings from the Corbett group who found significantly greater basilar (but not apical) dendritic branches in layer 5 neurons of the contralesional motor cortex.[64] Again, increased use of the unaffected or good forelimb modified dendritic plasticity, since animals given social housing or intensive motor rehabilitation displayed enhanced dendritic arbor growth in the undamaged hemisphere. By contrast, some other studies have been less supportive, reporting either no change[65–67] or a reduction[67,68] in the length and branching of pyramidal dendrites in the cingulate or sensorimotor cortex of the undamaged hemisphere after MCA occlusion or cortical lesion. Histological studies of dendritic structure in the stroke-affected hemisphere have been sparse in number although generally more in agreement. For example, two studies from the Kolb lab report a shortening of layer 5 and 3 cortical dendrites with fewer branches 2–4 months after cortical injury.[67,69] In conclusion, the evidence for dendritic growth or retraction after stroke using histological approaches has been mixed, particularly for the undamaged hemisphere.

One obvious limitation of classical studies that examined dendritic structure in fixed tissue specimens was that growth or retraction of dendrites was inferred based on sampling rather than directly visualized. In order to overcome this limitation, investigators have more recently employed longitudinal in vivo two-photon imaging to directly track changes in individual dendrites before and after an ischemic stroke.[70–72] When examining GFP-labeled layer 5 apical dendrites in the hemisphere opposite to MCA stroke, Johnston et al.[71] found no evidence of growth or even retraction of dendritic tips during the first 4 weeks of stroke recovery. Given this finding, it is hard to imagine widespread growth of pyramidal dendrites in the undamaged hemisphere, which agrees with previous imaging studies in undamaged somatosensory or visual cortex.[70,73–75] However, since the infarct zone in the Johnston et al. study was not expansive (infarct volume ranged from 2 to

gray arrows indicate transiently lost spines. (B) Average percentage of beaded dendrites during occlusion and the indicated average times of reperfusion. The change in blebbed dendrite percentages increased significantly after occlusion (~78% of dendrites) and gradually returned to baseline after 106 minutes of reperfusion (one-way ANOVA, $*p < 0.05$). *Con*, control; *Occ*, occlusion. (C) Changes in spine number relative to preischemia conditions in 10 different animals during occlusion and various time points after reperfusion. Although dendritic swelling was severe, careful inspection of 3D stacks revealed that most spines were still present on blebbed dendrites; however, on average ~30% of spines were transiently lost. A significant difference ($*p < 0.05$) was observed between control and occlusion, control and 5-minute reperfusion, and control and 14-minute reperfusion (one-way ANOVA). Source: *Figure modified with permission from Murphy TH, Li P, Betts K, Liu R. Two-photon imaging of stroke onset in vivo reveals that NMDA-receptor independent ischemic depolarization is the major cause of rapid reversible damage to dendrites and spines. J Neurosci. 2008;28(7):1756–1772. Copyright by the Society of Neuroscience.*

5 mm^3), one could argue that recruitment of structural plasticity mechanisms in the undamaged hemisphere may occur, but only after much larger insults.[76]

In the stroke-damaged hemisphere, two studies have tracked long-term structural changes to the apical dendrites of pyramidal neurons in vivo. The first study from Brown et al.[70] imaged layer 5 apical dendritic tips every 2 weeks for up to 6 weeks after a small photothrombotic stroke. This group found that stroke led to a significant increase in both extension and retraction of dendrites within the first 2 weeks of recovery (Fig. 3.3) but not at later time points. Of note, while an entire dendritic branch could retract in some cases, they did not find any evidence of new dendritic branch formation. Furthermore, the retraction of dendrites occurred preferentially from those dendrites projecting toward the site of infarction whereas those directed away, in some but not all instances extended (Fig. 3.3). Given that the activity of cortical neurons is dampened the closer one gets to the stroke border,[14,58] the retraction of dendrites proximal to the stroke and extension in the opposite direction may reflect a compensatory strategy to restore excitatory synaptic input.[77] Not long after this study, Mostany and Portera-Cailliau[72] employed similar imaging techniques to understand the nature of peri-infarct dendritic arbor remodeling for up

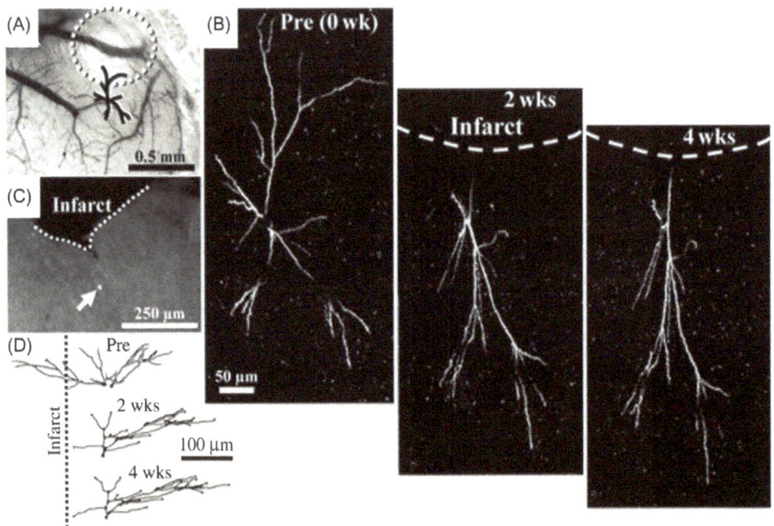

FIGURE 3.3 **Remodeling of dendritic arbors after stroke.** Pyramidal apical dendrites in the peri-infarct zone were imaged at 2-week intervals before and after photothrombotic stroke. Stroke notably altered the configuration of apical dendrites in the peri-infarct zone. The effect of stroke was especially prominent in the first 2 weeks of recovery and inversely correlated with the distance of each neuron to the infarct border. (A) Surface of the brain 2 weeks after stroke, e.g., as shown in B. (B) Time-lapse two-photon images of dendritic arbor (maximal intensity projection in the x–y plane) situated close to the infarct border (~145 μm). Note the immediate retraction of dendrites poststroke and gradual growth of the dendrites away from the site of infarction after stroke. (C) Postmortem identification of GFP-labeled neuron imaged in vivo. (D) Reconstructed drawings of the imaged dendrite shown from the side or the y–z axis. Source: *Figure modified with permission from Brown CE, Boyd JD, Murphy TH. Longitudinal in vivo imaging reveals balanced and branch-specific remodeling of mature cortical pyramidal dendritic arbors after stroke. J Cereb Blood Flow Metab. 2010;30(4):783–791. Copyright by the International Society of Cerebral Blood Flow and Metabolism.*

to 90 days after MCA stroke. In agreement with Brown et al., they did not find evidence of new branch formation after stroke and noted that much of the dendritic tip remodeling involved retraction. Contrary to the Brown et al. study, they found little evidence of tip growth. While the findings of the two in vivo imaging studies may seem at odds with respect to dendrite growth, they actually show mostly similar changes. The discrepancy lies in how each study measured and defined "growth." The Mostany and Portera-Cailliau[72] report measured changes in tip length from a landmark (e.g., a large spine) near the end of the dendrite whereas Brown et al.[70] focused on changes in length from the last branch point (much further away from the tip, sometimes 100s of microns). In both studies, there was evidence that dendrites can extend (i.e., become longer) after stroke when measuring from the last branch point. However, this extension appears to be more based on growth associated with stretching along the entire length of the terminal dendrite (possibly mechanical in nature) rather than classical "growth" from the tip of the dendrite as seen in development, although Brown et al.[70] did observe growth from the dendritic tip in a few instances. Subtle discrepancies such as these may also be explained by the fact that Brown et al.[70] studied dendrites within 0.5 mm of the stroke border, whereas Mostany and Portera-Cailliau[72] focused on dendrites 1–2.5 mm away.

In conclusion, the degree to which dendrites can remodel branches after stroke is still somewhat controversial after all these years. In the undamaged hemisphere, it would seem that the majority of evidence favors limited growth and remodeling of pyramidal dendrites, although there are dozens of reports suggesting that growth can be augmented by environmental or growth signaling factors. Meanwhile dendrites in the stroke-affected hemisphere do show some degree of plasticity although these modifications appear biased toward dendrite retraction, rather than growth.

5 DENDRITIC SPINE PLASTICITY

Spines are small protrusions from the dendritic shaft that receive the majority of excitatory synapses.[78] A strong positive correlation has been established between the presence of a dendritic spine and a functional synapse.[79] Furthermore, electron microscopy work has shown that the overwhelming majority of spines show classical anatomical features of a synapse (e.g., postsynaptic density (PSD) opposed by presynaptic bouton), with only 3%–4% of spines that clearly lack a synapse, usually associated with small "head-less" protrusions that resemble filopodia.[80] Therefore, imaging and analyzing dendritic spines have become a well-established proxy for estimating the number of excitatory synapses a neuron possesses, as well as dynamic aspects such as synapse formation, elimination, and stability.[73]

Spine changes in peri-infarct pyramidal neurons show both temporal and spatial relationship with the arrest of blood flow. At the onset of severe ischemia, the beaded appearance of dendrites in the infarct core makes most spines indiscernible. Similar to the fate described in the previous section, these pyramidal neuron dendrites and spines will quickly degenerate and reach the point of no return if blood flow is not restored. In the event where reperfusion occurs within minutes to an hour after stroke onset, spines reappear in generally the same location and density as before stroke (Fig. 3.2).[57,60] For neurons that experience a moderate reduction in cerebral blood flow (i.e., after small stroke or in hypoperfused regions surrounding the infarct core), changes to dendritic spine number and structure depend on time after stroke and distance to the ischemic border zone. For example, modeling moderate levels of ischemia (~50% reduction) with cortical endothelin-1 injection led to no discernible changes in spine number or stability over a few hours.[57] However, when examining dendritic spines in peri-infarct regions from 4 to 24 hours after photothrombotic stroke, spine numbers declined by 20%–40% relative to prestroke levels.[60,81] This loss of synaptic connections in the first day has also been demonstrated with ultrastructural[82] and longitudinal in vivo imaging approaches.[60,70,83] The loss of spines is accompanied by a concomitant decrease in synaptic markers, such as syntaxin, SNAP-25, vesicular GABA transporter (VGAT), and glutamate receptors,[84] suggesting a loss of functional synapses. Importantly, spine loss is restricted to the peri-infarct cortex although the precise size of this zone depends on the model of stroke employed. For instance after small photothrombotic stroke, most spine loss occurs within 500 μm from the infarct border and is not correlated with infarct volume but inversely with the distance from the infarct edge.[81,85] In addition to spine pruning, the average length of spines increased in peri-infarct cortex.[81] The reason behind this change in spine morphology remains unclear, but could be due to spine growth or may simply reflect the possibility that longer spines were more likely to survive ischemia. The latter explanation does not seem so far-fetched since an increase in distance from spine head to dendritic shaft may limit the propagation of damaging chemical (e.g., calcium) or electrical stimuli.[86]

Over the following days and weeks after a stroke, the density of dendritic spines in the peri-infarct cortex slowly recovers back to prestroke levels (Fig. 3.4A,B).[82,83,87,88] Longitudinal imaging of spines before and after ischemic stroke has revealed important clues as to how peri-infarct dendritic spines recover. First, spine turnover rates (comprised of spine gain and loss) are significantly elevated for several weeks after stroke.[13,83] During the first 1–2 weeks, turnover is skewed more toward spine loss, which is followed (from 2 to 5 weeks poststroke) by a period of increased spine formation (Fig. 3.4A–C).[13,83] Of note, the likelihood of newly formed spines persisting for up to 3 months after stroke was much greater for the later born spines (Fig. 3.4D).[83] Furthermore, spines that formed in peri-infarct regions with acutely lower blood flow levels (proximal to the infarct border where flow <70% of normal) tended to persist longer than spines formed in regions with more sufficient blood flow (>70%). Whether new persistent spines successfully integrate into functional circuits after stroke has not been formally tested. However, retrospective EM studies on newly formed spines in normal mice has shown that a large fraction of new spines do not bear a morphologically defined PSD (and thus are not likely functional) unless they persist for at least 4 days.[89] Collectively, these studies show that stroke initially leads to dendritic spine loss and instability on peri-infarct pyramidal neurons. After this period (e.g., from 2 weeks onward), spine formation rates increase and these spines tend to persist longer, depending on their location relative to the infarct and the acute level of ischemia they experienced. The mechanisms controlling dendritic spine dynamics in the peri-infarct zone remain ambiguous and beyond the scope of this review. However, since it is well known that manipulations of synaptic activity can influence dendritic filopodia/spine formation and stability,[90–92] stroke-related changes in cortical activity are certain to dictate some of these structural modifications.

In the unaffected hemisphere, whether dendritic spine remodeling is influenced by stroke remains controversial. For example, Takatsuru et al. imaged apical dendritic spine

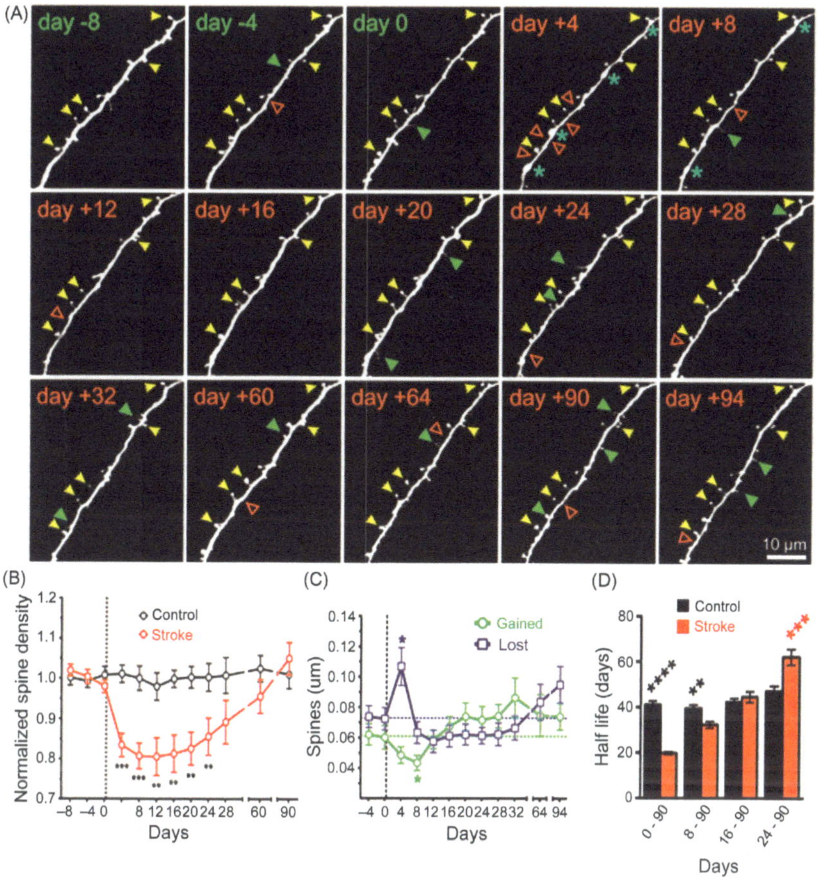

FIGURE 3.4 **Spine dynamics in peri-infarct cortex after stroke.** In vivo longitudinal imaging of apical dendrites from layer 5 pyramidal neurons in the peri-infarct cortex, 1–3 mm away from the edge of the infarct, up to 3 months after middle cerebral artery occlusion (MCAO) stroke. (A) High resolution two-photon images illustrate spine dynamics over time. Subsequent images show that spines may be categorized as persistent (*yellow arrow*), gained (*green arrow*), or lost (*open red arrow*). The day of imaging is marked in the upper left corner, and transient dendrite swelling post-MCAO noted by a *blue asterisk* at days +4 and +8. (B) Normalized spine density averaged across neurons in stroke mice (*red*) and control mice (*black*). There was a significant decrease in spine density immediately (+4 days) after stroke that peaked by +12 days and sustained for 3 weeks. Gradually, spine density returned to normal levels, such that by 1–2 months poststroke, spine density was indistinguishable from basal conditions or control animals. Data is shown as mean±SEM (two-way ANOVA followed by Bonferroni's multiple comparisons test: $^{**}p \leq 0.01$, $^{***}p < 0.001$). The vertical dashed line denotes time of MCAO. (C) Altered spine dynamics underlie the recovery of dendritic spines post-MCAO. The number of gained (*green*) versus lost (*blue*) spines per unit length of dendrite as a function of time before and after MCAO. The decrease in spine density after MCAO was due to an increase in the number of lost spines, and a coincidental decrease in the number of gained spines, compared to baseline. The period where spine density gradually recovered (+16 to 32 days) is associated with a reversal of this trend. Data is shown as mean±SEM ($^{*}p \leq 0.05$). (D) Half-life of spines in stroke (*red*) and control (*black*) animals at various time intervals after MCAO. The half-life of spines present on the day of MCAO was lower in the stroke group. In contrast, the half-life of spines present at +24 days was higher in the stroke group. This suggests that the recovery of spines is mediated not only by an increase in the number of gained spines but also by a stabilization of spines +24 day poststroke. Error bars represent SEM ($^{**}p \leq 0.01$,

dynamics over a 6-hour period in the undamaged hemisphere where their electrophysiological recordings had revealed an enhancement of sensory-evoked cortical responsiveness after photothrombotic stroke.[93] Here, they found a transient increase in dendritic spine formation and elimination at 1 week, but not 3 days, 2 or 4 weeks recovery. This study, however, contrasts with several histological studies that quantified dendritic spine density in contralesional pyramidal neurons and reported no effect of stroke.[66,67] Furthermore, a more recent two-photon imaging study by Johnston et al. found absolutely no changes in dendritic spine turnover in the undamaged hemisphere at any time point after MCA stroke, nor did they find any functional changes in cortical responsiveness with intrinsic signal optical imaging.[71] One important difference to note is that the Johnston et al. study imaged spines every 4 days whereas Takatsuru imaged spines over a 6-hour period, therefore, it is impossible to know in the latter study how long new spines persisted. As mentioned earlier, the formation of a spine with a functional synapse usually takes a few days to manifest, although spines can exhibit a PSD within 6 hours of formation.[94] Therefore, the functional relevance of spines that formed over a 6-hour period is uncertain and could represent a more transient population. However, even if the new spines in the Takatsuru paper were short lived, they cannot be dismissed as irrelevant since transient spines could temporarily destabilize existing synapses or strengthen ones through correlated activity.[95]

In conclusion, stroke clearly alters dendritic spine dynamics and stability in peri-infarct neurons, with much more limited effects in the undamaged hemisphere. Although remodeling of dendritic spines correlates spatially and temporally with the emergence of new receptive fields and cortical sensory/motor map representations in the peri-infarct region,[13,96] causal relationships between these phenomena have not been proven as yet. For instance, it is conceivable that increased dendritic spine formation and elimination in the peri-infarct cortex in the first few weeks after stroke may represent an intrinsic homeostatic response of neurons quieted due to being deprived of their afferent excitatory inputs or victims of enhanced intracortical inhibition.[11] Future studies that suppress or enhance dendritic spine stability, for example, through optogenetic means or cell-specific disruption of synaptic adhesion proteins, could help reveal whether this innate structural response to stroke is beneficial or maladaptive for the recovery of cortical network activity and ultimately sensory motor function of the affected limbs.

6 PERSPECTIVES AND FUTURE DIRECTIONS

6.1 Technological Developments

What we currently know about the effects of stroke on neuronal structure and connectivity is strongly influenced by what tools were available at the time. For instance, the development of transgenic mice that express GFP or YFP under the control of a Thy-1 promoter[97] catalyzed

***$p < 0.001$, ****$p < 10^{-6}$). Source: *Figure modified with permission from Mostany R, Chowdhury TG, Johnston DG, Portonovo SA, Carmichael ST, Portera-Cailliau C. Local hemodynamics dictate long-term dendritic plasticity in peri-infarct cortex. J Neurosci. 2010;30(42):14116–14126. Copyright by the Society of Neuroscience.*

many in vivo imaging studies over the past decade or so.[13,57,75,83,87,98] These time-lapse imaging studies have significantly expanded our understanding of the acute and long-term effects of stroke on dendritic structure. However, one obvious drawback has been that all of these imaging studies were focused almost exclusively on dendrites from excitatory pyramidal neurons in cortical layer 5. This technical limitation has been overcome with the explosion of new viral vectors or mouse lines that express Cre recombinase under the control of promoters that are expressed within specific cell types.[99] When crossed with a mouse line that expresses a Cre-dependent fluorescent reporter protein (tdtomato, YFP) or genetically encoded calcium sensor (e.g., GCamP), one can image the structure and functional activity of any cell type.[100] We expect future imaging studies to take full advantage of these new tools to uncover the impact of stroke on other important cell types in the brain, such as the diverse array of inhibitory interneurons that regulate cortical excitability, or thalamic neurons that play a central role in processing and relaying sensory information to the cortex. Lastly, the proliferation of mouse lines with Cre-dependent control of genes implicated in axonal/dendritic growth and synaptogenesis (e.g., "floxed" BDNF, cadherins, PTEN lines) will enable scientists to image the effect of gene knock down or overexpression on structural plasticity within specific cell types.[101] These types of studies would allow for more causal relationships to be established between structural and functional changes after stroke.

Not only can defined nucleotide changes be engineered into the genome of the mouse, but genetic switches can be designed to target expression or ablation of a gene (for which basic molecular information is available) to any tissue at any defined time. As mentioned in the preceding paragraph, the Cre-*LoxP* system is the most advanced technology for spatial and temporal control of genetic inactivation (for review, see Ref.[102]). The development of a new tool based on a bacterial CRISPR-associated protein-9 nuclease (Cas9) offers an efficient and reliable way to make precise, targeted changes to the genome, even in adulthood.[103,104] Several groups have now taken advantage of this method to edit a particular target gene, as a form of genetic therapy.[105]

6.2 Minimizing the Dark Side of Structural Plasticity Through Intelligent Intervention

Although there is general agreement among published studies that enhancing structural plasticity/remodeling correlates with improved stroke recovery, these approaches may also have a dark side. It is entirely possible that new stroke therapies that alter the molecular machinery governing dendritic stability or axonal growth may interfere with other crucial brain functions such as learning and memory, or lead to runaway excitation in the case of seizures. Such unintended consequences have occurred in preclinical spinal cord injury studies, where activation of axonal sprouting, which was thought to enhance plasticity and recovery, also induced neuropathic pain.[106] These results offer a word of caution to prospective treatments that seek to augment structural plasticity to improve stroke recovery.

6.3 Computational Modeling Studies

One approach to mitigate the likelihood of unintended consequences when tinkering with damaged brain circuits is to use modeling studies. A recent study by Butz and van

Ooyen[77] exploited the extensive literature on structural and functional cortical adaptations that occur following retinal lesions, to construct a rule for dendritic spine and axonal bouton formation. At a very basic level, their model posits that the propensity for neurons to sprout new connections is commensurate with its intrinsic activity level. Therefore, neurons in the lesion projection zone (LPZ) with lowest levels of activity would exhibit the greatest degree of remodeling, whereas those neurons with more normal activity levels would be much more stable. Indeed this model appears to predict much of the experimental data following retinal lesions. Whether this model would fully recapitulate experimental stroke data is uncertain (at least to our knowledge), but there are some interesting parallels one can draw between stroke and retinal lesions. We know from imaging and electrophysiological studies that peri-infarct neurons lose many of their afferent inputs and become hypoexcitable for several days to weeks following stroke.[11–15] In this sense, the peri-infarct zone would resemble the center LPZ or peri-LPZ. Experimentally we know that these neurons, like those in the LPZ, show an initial loss of dendritic spines that is followed by a progressive increase in spine formation and stability from 2 to 5 weeks recovery,[13,81,83] which a computational model might indeed predict. However, what structural and functional changes are occurring contemporaneously in inhibitory neurons is unknown and could disrupt predictions. Therefore, there is need to generate more experimental data to guide the parameters of a computational model so that we can accurately predict how network functions could be affected when we target certain aspects of a circuit with a therapy. To this end, we look forward to what future studies will reveal about the impact of stroke on the diverse array of cells and circuits that make up our ever changing brain.

References

1. Go AS, Mozaffarian D, Roger VL, et al. Executive summary: heart disease and stroke statistics—2013 update: a report from the American Heart Association. *Circulation.* 2013;127(1):143–152.
2. Adams HP, del Zoppo G, Alberts MJ, et al. Guidelines for the early management of adults with ischemic stroke. *Circulation.* 2007;115(20):e478–e534.
3. Lynch E, Hillier S, Cadilhac D. When should physical rehabilitation commence after stroke: a systematic review. *Int J Stroke.* 2014;9(4):468–478.
4. Murphy TH, Corbett D. Plasticity during stroke recovery: from synapse to behaviour. *Nat Rev Neurosci.* 2009;10(12):861–872.
5. Nudo RJ. Plasticity. *NeuroRx.* 2006;3(4):420–427.
6. Lipton P. Ischemic cell death in brain neurons. *Physiol Rev.* 1999;79(4):1431–1568.
7. Dirnagl U, Iadecola C, Moskowitz MA. Pathobiology of ischaemic stroke: an integrated view. *Trends Neurosci.* 1999;22(9):391–397.
8. Mohajerani MH, Aminoltejari K, Murphy TH. Targeted mini-strokes produce changes in interhemispheric sensory signal processing that are indicative of disinhibition within minutes. *Proc Natl Acad Sci USA.* 2011;108(22):E183–E191.
9. Fujioka H, Kaneko H, Suzuki SS, Mabuchi K. Hyperexcitability-associated rapid plasticity after a focal cerebral ischemia. *Stroke.* 2004;35(7):e346–e348.
10. Ding M-C, Wang Q, Lo EH, Stanley GB. Cortical excitation and inhibition following focal traumatic brain injury. *J Neurosci.* 2012;31(40):14085–14094.
11. Clarkson AN, Huang BS, MacIsaac SE, Mody I, Carmichael ST. Reducing excessive GABA-mediated tonic inhibition promotes functional recovery after stroke. *Nature.* 2010;468(7321):305–309.

12. Winship IR, Murphy TH. In vivo calcium imaging reveals functional rewiring of single somatosensory neurons after stroke. *J Neurosci*. 2008;28(26):6592–6606.
13. Brown CE, Aminoltejari K, Erb H, Winship IR, Murphy TH. In vivo voltage-sensitive dye imaging in adult mice reveals that somatosensory maps lost to stroke are replaced over weeks by new structural and functional circuits with prolonged modes of activation within both the peri-infarct zone and distant sites. *J Neurosci*. 2009;29(6):1719–1734.
14. Lim DH, LeDue JM, Mohajerani MH, Murphy TH. Optogenetic mapping after stroke reveals network-wide scaling of functional connections and heterogeneous recovery of the peri-infarct. *J Neurosci*. 2014;34(49):16455–16466.
15. Dijkhuizen RM, Ren J, Mandeville JB, et al. Functional magnetic resonance imaging of reorganization in rat brain after stroke. *Proc Natl Acad Sci USA*. 2001;98(22):12766–12771.
16. Cramer SC. Repairing the human brain after stroke. II. Restorative therapies. *Ann Neurol*. 2008;63(5):549–560.
17. Watson BD, Dietrich WD, Busto R, Wachtel MS, Ginsberg MD. Induction of reproducible brain infarction by photochemically initiated thrombosis. *Ann Neurol*. 1985;17(5):497–504.
18. Fuxe K, Bjelke B, Andbjer B, Grahn H, Rimondini R, Agnati LF. Endothelin-1 induced lesions of the frontoparietal cortex of the rat. A possible model of focal cortical ischemia. *Neuroreport*. 1997;8(11):2623–2629.
19. Carmichael ST. *Rodent Models of Focal Stroke: Size, Mechanism, and Purpose*. 2005;2:396–409.
20. Shih AY, Blinder P, Tsai PS, et al. The smallest stroke: occlusion of one penetrating vessel leads to infarction and a cognitive deficit. *Nat Neurosci*. 2013;16(1):55–63.
21. Jablonka JA, Burnat K, Witte OW, Kossut M. Remapping of the somatosensory cortex after a photothrombotic stroke: dynamics of the compensatory reorganization. *Neuroscience*. 2010;165(1):90–100.
22. Naidoo V, Naidoo S, Raidoo DM. Immunolocalisation of endothelin-1 in human brain. *J Chem Neuroanat*. 2004;27(3):193–200.
23. Uesugi M, Kasuya Y, Hama H, et al. Endogenous endothelin-1 initiates astrocytic growth after spinal cord injury. *Brain Res*. 1996;728(2):255–259.
24. Uesugi M, Kasuya Y, Hayashi K, Goto K. SB209670, a potent endothelin receptor antagonist, prevents or delays axonal degeneration after spinal cord injury. *Brain Res*. 1998;786(1–2):235–239.
25. Wu J, Quinlan EB, Dodakian L, et al. Connectivity measures are robust biomarkers of cortical function and plasticity after stroke. *Brain*. 2015;138(8):2359–2369.
26. Reddy H, Bendahan D, Lee MA, et al. An expanded cortical representation for hand movement after peripheral motor denervation. *J Neurol Neurosurg Psychiatry*. 2002;72:203–210.
27. Dudanova I, Klein R. Integration of guidance cues: parallel signaling and crosstalk. *Trends Neurosci*. 2013;36(5):295–304.
28. Carmichael ST. Cellular and molecular mechanisms of neural repair after stroke: making waves. *Ann Neurol*. 2006;59(5):735–742.
29. Stroemer RP, Kent TA, Hulsebosch CE. Enhanced neocortical neural sprouting, synaptogenesis, and behavioral recovery with D-amphetamine therapy after neocortical infarction in rats. *Stroke*. 1998;29(11):2381–2393.
30. Ng SC, de la Monte SM, Conboy GL, Karns LR, Fishman MC. Cloning of human GAP-43: growth association and ischemic resurgence. *Neuron*. 1988;1(2):133–139 [pii].
31. Li S, Carmichael ST. Growth-associated gene and protein expression in the region of axonal sprouting in the aged brain after stroke. *Neurobiol Dis*. 2006;23(2):362–373.
32. Li S, Overman JJ, Katsman D, et al. An age-related sprouting transcriptome provides molecular control of axonal sprouting after stroke. *Nat Neurosci*. 2010;13(12):1496–1504.
33. Thomas Carmichael S, Wei L, Rovainen CM, Woolsey TA. New patterns of intracortical projections after focal cortical stroke. *Neurobiol Dis*. 2001;8(5):910–922.
34. Urban ETR, Bury SD, Barbay HS, Guggenmos DJ, Dong Y, Nudo RJ. Gene expression changes of interconnected spared cortical neurons 7 days after ischemic infarct of the primary motor cortex in the rat. *Mol Cell Biochem*. 2012;369(1):267–286.
35. Dancause N, Barbay S, Frost SB, et al. Extensive cortical rewiring after brain injury. *J Neurosci*. 2005;25:10167–10179.
36. Kawaguchi R, Coppola G, St C. GDF10 is a signal for axonal sorouting and functional recovery after stroke. *Nat Neurosci*. 2015; (October).

37. Sist B, Fouad K, Winship IR. Plasticity beyond peri-infarct cortex: spinal up regulation of structural plasticity, neurotrophins, and inflammatory cytokines during recovery from cortical stroke. *Exp Neurol*. 2014;252:47–56.
38. Chen P, Goldberg DE, Kolb B, Lanser M, Benowitz LI. Inosine induces axonal rewiring and improves behavioral outcome after stroke. *Proc Natl Acad Sci USA*. 2002;99(13):9031–9036.
39. Papadopoulos CM, Tsai S-Y, Alsbiei T, O'Brien TE, Schwab ME, Kartje GL. Functional recovery and neuroanatomical plasticity following middle cerebral artery occlusion and IN-1 antibody treatment in the adult rat. *Ann Neurol*. 2002;51(4):433–441.
40. Wahl AS, Omlor W, Rubio JC, et al. Asynchronous therapy restores motor control by rewiring of the rat corticospinal tract after stroke. *Science*. 2014;344(6189):1250–1255.
41. Napieralski JA, Butler AK, Chesselet M-F. Anatomical and functional evidence for lesion-specific sprouting of corticostriatal input in the adult rat. *J Comp Neurol*. 1996;373(4):484–497.
42. Carmichael ST, Chesselet M-F. Synchronous neuronal activity is a signal for axonal sprouting after cortical lesions in the adult. *J Neurosci*. 2002;22(14):6062–6070.
43. Mizuno H, Hirano T, Tagawa Y. Evidence for activity-dependent cortical wiring: formation of interhemispheric connections in neonatal mouse visual cortex requires projection neuron activity. *J Neurosci*. 2007;27(25):6760–6770.
44. Herrmann K, Shatz CJ. Blockade of action potential activity alters initial arborization of thalamic axons within cortical layer 4. 1995;92(November):11244–11248.
45. Ruthazer ES, Stryker MP. The role of activity in the development of long-range horizontal connections in area 17 of the ferret. *J Neurosci*. 1996;16(22):7253–7269.
46. van der Zijden JP, van der Toorn A, van der Marel K, Dijkhuizen RM. Longitudinal in vivo MRI of alterations in perilesional tissue after transient ischemic stroke in rats. *Exp Neurol*. 2008;212(1):207–212.
47. Tajrine D, Garofalo L, Cuello AC, Ribeiro-da-Silva A. Responses of cortical noradrenergic and somatostinergic fibres and terminals to adjacent strokes and subsequent treatment with NGF and/or the ganglioside GM1. *J Neurosci Res*. 1997;50(4):627–642.
48. Garofalo L, Ribeiro-da-Silva A, Cuello AC. Potentiation of nerve growth factor-induced alterations in cholinergic fibre length and presynaptic terminal size in cortex of lesioned rats by the monosialoganglioside GM1. *Neuroscience*. 1993;57(1):21–40.
49. Little JR, Kerr FW, Sundt TJ. *Synoptic Alterations in Developing Cortical Infarction: An Experimental Investigation in Monkeys*. 1974;5:August.
50. Jones TA, Kleim JA, Greenough WT. Synaptogenesis and dendritic growth in the cortex opposite unilateral sensorimotor cortex damage in adult rats: a quantitative electron microscopic examination. *Brain Res*. 1996;733(1):142–148.
51. Jones TA, Schallert T. Use-dependent growth of pyramidal neurons after neocortical damage. *J Neurosci*. 1994;14(4):2140–2152.
52. Chu CJ, Jones TA. Experience-dependent structural plasticity in cortex heterotopic to focal sensorimotor cortical damage. *Exp Neurol*. 2000;166(2):403–414.
53. Kim SY, Allred RP, Adkins DL, et al. Experience with the "Good" limb induces aberrant synaptic plasticity in the perilesion cortex after stroke. *J Neurosci*. 2015;35(22):8604–8610.
54. Carmichael ST. Plasticity of cortical projections after stroke. *Neuroscience*. 2003;9(1):64–75.
55. Anenberg E, Arstikaitis P, Niitsu Y, et al. Ministrokes in channel rhodopsin-2 transgenic mice reveal widespread deficits in motor output despite maintenance of cortical neuronal excitability. *J Neurosci*. 2014;34(4):1094–1104.
56. Murphy TH, Li P, Betts K, Liu R. Two-photon imaging of stroke onset in vivo reveals that NMDA-receptor independent ischemic depolarization is the major cause of rapid reversible damage to dendrites and spines. *J Neurosci*. 2008;28(7):1756–1772.
57. Zhang S, Boyd J, Delaney K, Murphy TH. Rapid reversible changes in dendritic spine structure in vivo gated by the degree of ischemia. *J Neurosci*. 2005;25(22):5333–5338.
58. Li P, Murphy TH. Two-photon imaging during prolonged middle cerebral artery occlusion in mice reveals recovery of dendritic structure after reperfusion. *J Neurosci*. 2008;28(46):11970–11979.
59. Park JS, Bateman MC, Goldberg MP. Rapid alterations in dendrite morphology during sublethal hypoxia or glutamate receptor activation. *Neurobiol Dis*. 1996;3(3):215–227.

60. Zhang S, Murphy TH. Imaging the impact of cortical microcirculation on synaptic structure and sensory-evoked hemodynamic responses in vivo. *PLoS Biol*. 2007;5(5):e119.
61. Tomimoto H, Yanagihara T. Golgi electron microscopic study of the cerebral cortex after transient cerebral ischemia and reperfusion in the gerbil. *Neuroscience*. 1994;63(4):957–967.
62. Sword J, Masuda T, Croom D, Kirov SA. Evolution of neuronal and astroglial disruption in the peri-contusional cortex of mice revealed by in vivo two-photon imaging. *Brain*. 2013;136(5):1446–1461.
63. Jones TA, Schallert T. Overgrowth and pruning of dendrites in adult rats recovering from neocortical damage. *Brain Res*. 1992;581(1):156–160.
64. Biernaskie J, Corbett D. Enriched rehabilitative training promotes improved forelimb motor function and enhanced dendritic growth after focal ischemic injury. *J Neurosci*. 2001;21(14):5272–5280 [pii].
65. Prusky G, Whishaw IQ. Morphology of identified corticospinal cells in the rat following motor cortex injury: absence of use-dependent change. *Brain Res*. 1996;714(1–2):1–8 [pii].
66. Papadopoulos CM, Tsai S-Y, Cheatwood JL, et al. Dendritic plasticity in the adult rat following middle cerebral artery occlusion and Nogo-a neutralization. *Cereb Cortex*. 2006;16(4):529–536.
67. Gonzalez CLR, Kolb B. A comparison of different models of stroke on behaviour and brain morphology. *Eur J Neurosci*. 2003;18(7):1950–1962.
68. Jones TA, Liput DJ, Maresh EL, et al. Use-dependent dendritic regrowth is limited after unilateral controlled cortical impact to the forelimb sensorimotor cortex. *J Neurotrauma*. 2012;29(7):1455–1468.
69. Kolb B, Gorny G, Côte S, Ribeiro-da-Silva a, Cuello a C. Nerve growth factor stimulates growth of cortical pyramidal neurons in young adult rats. *Brain Res*. 1997;751(2):289–294.
70. Brown CE, Boyd JD, Murphy TH. Longitudinal in vivo imaging reveals balanced and branch-specific remodeling of mature cortical pyramidal dendritic arbors after stroke. *J Cereb Blood Flow Metab*. 2010;30(4):783–791.
71. Johnston DG, Denizet M, Mostany R, Portera-Cailliau C. Chronic in vivo imaging shows no evidence of dendritic plasticity or functional remapping in the contralesional cortex after stroke. *Cereb Cortex*. 2013;23(4):751–762.
72. Mostany R, Portera-Cailliau C. Absence of large-scale dendritic plasticity of layer 5 pyramidal neurons in peri-infarct cortex. *J Neurosci*. 2011;31(5):1734–1738.
73. Holtmaat A, Svoboda K. Experience-dependent structural synaptic plasticity in the mammalian brain. *Nat Rev Neurosci*. 2009;10(9):647–658.
74. Lee WC a, Huang H, Feng G, et al. Dynamic remodeling of dendritic arbors in GABAergic interneurons of adult visual cortex. *PLoS Biol*. 2006;4(2):271–280.
75. Trachtenberg JT, Chen BE, Knott GW, et al. Long-term in vivo imaging of experience-dependent synaptic plasticity in adult cortex. *Nature*. 2002;420(6917):788–794.
76. Touvykine B, Mansoori BK, Jean-Charles L, Deffeyes J, Quessy S, Dancause N. The effect of lesion size on the organization of the ipsilesional and contralesional motor cortex. *Neurorehabil Neural Repair*. 2015;29:1–13.
77. Butz M, van Ooyen A. A simple rule for dendritic spine and axonal bouton formation can account for cortical reorganization after focal retinal lesions. *PLoS Comput Biol*. 2013;9(10):39–43.
78. Arellano JI, Benavides-Piccione R, Defelipe J, Yuste R. Ultrastructure of dendritic spines: correlation between synaptic and spine morphologies. *Front Neurosci*. 2007;1(1):131–143.
79. Zito K, Scheuss V, Knott G, Hill T, Svoboda K. Rapid functional maturation of nascent dendritic spines. *Neuron*. 2009;61(2):247–258.
80. Arellano JI, Espinosa A, Fairén A, Yuste R, DeFelipe J. Non-synaptic dendritic spines in neocortex. *Neuroscience*. 2007;145(2):464–469.
81. Brown CE, Wong C, Murphy TH. Rapid morphologic plasticity of peri-infarct dendritic spines after focal ischemic stroke. *Stroke*. 2008;39(4):1286–1291.
82. Ito U, Kuroiwa T, Nagasao J, Kawakami E, Oyanagi K. Temporal profiles of axon terminals, synapses and spines in the ischemic penumbra of the cerebral cortex: ultrastructure of neuronal remodeling. *Stroke*. 2006;37(8):2134–2139.
83. Mostany R, Chowdhury TG, Johnston DG, Portonovo SA, Carmichael ST, Portera-Cailliau C. Local hemodynamics dictate long-term dendritic plasticity in peri-infarct cortex. *J Neurosci*. 2010;30(42):14116–14126.
84. Raghavendra Rao VL, Bowen KK, Dhodda VK, et al. Gene expression analysis of spontaneously hypertensive rat cerebral cortex following transient focal cerebral ischemia. *J Neurochem*. 2002;83(5):1072–1086.

85. Reeson P, Tennant KA, Gerrow K, et al. Delayed inhibition of VEGF signaling after stroke attenuates blood–brain barrier breakdown and improves functional recovery in a comorbidity-dependent manner. *J Neurosci*. 2015;35(13):5128–5143.
86. Adrian M, Kusters R, Wierenga CJ, Storm C, Hoogenraad CC, Kapitein LC. Barriers in the brain: resolving dendritic spine morphology and compartmentalization. *Front Neuroanat*. 2014;8:1–12.
87. Brown CE, Li P, Boyd JD, Delaney KR, Murphy TH. Extensive turnover of dendritic spines and vascular remodeling in cortical tissues recovering from stroke. *J Neurosci*. 2007;27(15):4101–4109.
88. Wu F, Catano M, Echeverry R, et al. Urokinase-type plasminogen activator promotes dendritic spine recovery and improves neurological outcome following ischemic stroke. *J Neurosci*. 2014;34(43):14219–14232.
89. Knott GW, Holtmaat A, Wilbrecht L, Welker E, Svoboda K. Spine growth precedes synapse formation in the adult neocortex in vivo. *Nat Neurosci*. 2006;9(9):1117–1124.
90. Nägerl UV, Eberhorn N, Cambridge SB, Bonhoeffer T. Bidirectional activity-dependent morphological plasticity in hippocampal neurons. *Neuron*. 2016;44(5):759–767.
91. Kirov SA, Harris KM. Dendrites are more spiny on mature hippocampal neurons when synapses are inactivated. *Nat Neurosci*. 1999;2(10):878–883.
92. Portera-Cailliau C, Pan DT, Yuste R. Activity-regulated dynamic behavior of early dendritic protrusions: evidence for different types of dendritic filopodia. *J Neurosci*. 2003;23(18):7129–7142.
93. Takatsuru Y, Fukumoto D, Yoshitomo M, Nemoto T, Tsukada H, Nabekura J. Neuronal circuit remodeling in the contralateral cortical hemisphere during functional recovery from cerebral infarction. *J Neurosci*. 2009;29(32):10081–10086.
94. Cane M, Maco B, Knott G, Holtmaat A. The relationship between PSD-95 clustering and spine stability in vivo. *J Neurosci*. 2014;34(6):2075–2086.
95. Barnes SJ, Cheetham CEJ. A role for short-lived synapses in adult cortex? *J Neurosci*. 2014;34(21):7044–7046.
96. Nudo RJ, Milliken GW. Reorganization of movement representations in primary motor cortex following focal ischemic infarcts in adult squirrel monkeys. *J Neurophysiol*. 1996;75(5):2144–2149.
97. Feng G, Mellor RH, Bernstein M, et al. Imaging neuronal subsets in transgenic mice expressing multiple spectral variants of GFP. *Neuron*. 2000;28(1):41–51.
98. Grutzendler J, Kasthuri N, Gan W-B. Long-term dendritic spine stability in the adult cortex. *Nature*. 2002;420 (6917):812–816.
99. Taniguchi H, He M, Wu P, et al. A resource of Cre driver lines for genetic targeting of GABAergic neurons in cerebral cortex. *Neuron*. 2011;71(6):995–1013.
100. Chen T-W, Wardill TJ, Sun Y, et al. Ultrasensitive fluorescent proteins for imaging neuronal activity. *Nature*. 2013;499(7458):295–300.
101. Chow DK, Groszer M, Pribadi M, et al. Laminar and compartmental regulation of dendritic growth in mature cortex. *Nat Neurosci*. 2009;12(2):116–118.
102. Gavériaux-Ruff C, Kieffer BL. Conditional gene targeting in the mouse nervous system: Insights into brain function and diseases. *Pharmacol Ther*. 2007;113(3):619–634.
103. Jinek M, Chylinski K, Fonfara I, Hauer M, Doudna JA, Charpentier E. A programmable dual-RNA-guided. *Science (80-)*. 2012;337:816–822.
104. Gasiunas G, Barrangou R, Horvath P, Siksnys V. Cas9-crRNA ribonucleoprotein complex mediates specific DNA cleavage for adaptive immunity in bacteria. *Proc Natl Acad Sci USA*. 2012;109(39):E2579–E2586.
105. Mali P, Yang L, Esvelt KM, et al. RNA-guided human genome engineering via Cas9. *Science*. 2013;339 (6121):823–826.
106. Dobkin BH, Carmichael ST. The specific requirements of neural repair trials for stroke. *Neurorehabil Neural Repair*. 2015;30(5):470–478.

CHAPTER

4

Is Lesion-Induced Synaptic Rewiring Driven by Activity Homeostasis?

Markus Butz-Ostendorf[1] *and Arjen van Ooyen*[2]

[1]Biomax Informatics AG, Planegg, Germany

[2]VU University Amsterdam, Amsterdam, The Netherlands

OUTLINE

The Rewiring Brain
DOI: http://dx.doi.org/10.1016/B978-0-12-803784-3.00004-4

1 INTRODUCTION

As shown by early neuroanatomical tracing studies[1–3] and recent high-resolution microscopy studies,[4–9] a lasting reduction in afferent activity to the brain, brought about by peripheral or central lesions (e.g., retinal lesions or stroke), triggers massive rewiring of synaptic connections. Synaptic rewiring is mediated by structural adaptations (structural plasticity), such as axon remodeling,[10] changes in spine numbers,[11] and synapse formation and deletion,[12] and contributes to the functional repair of brain networks. However, despite four decades of lesion studies, the principles governing lesion-induced structural plasticity remain poorly understood.

By modifying the input a neuron receives, lesions alter the level of neuronal electrical activity. An early in vivo rat study showed that reducing the level of electrical activity in cervical ganglion cells led to an increase in dendritic spine numbers,[13] enhancing the cells' potential to make synaptic connections with active neurons and, therefore, to restore activity. On the basis of this study, a role for activity homeostasis in synaptic rewiring was proposed,[14,15] but experimental limitations prevented further study into the guiding principles of synaptic rewiring.

Current experimental studies use in vivo time lapse imaging and advanced techniques for measuring electrical activity that enable detailed monitoring of dendritic spines,[4,16–18] axonal branches and boutons,[5,19] and electrical activity over extended periods of time.[20–23] Drawing on the results of such studies, here we hypothesize that the attempt of neurons to maintain their average electrical activity at a particular level (homeostatic regulation) may drive structural plasticity after loss of input caused by lesions. We define structural plasticity as encompassing all the structural changes, such as spine dynamics and axon remodeling, that lead to the formation or deletion of synapses. Structural plasticity can connect previously unconnected neurons, disconnect neurons, or change the number of synapses by which neurons are connected. By contrast, synaptic plasticity is defined as a change in the strength of existing synapses. Hebbian synaptic plasticity changes synapse strength depending on the correlation between pre- and postsynaptic activity,[24,25] whereas synaptic scaling (homeostatic synaptic plasticity) modifies the strengths of all the cell's incoming synapses so as to stabilize neuronal activity around some set-point value.[26]

In the next sections, we first present the current view on events following a lesion-induced loss of afferent activity. We then describe some limitations of the current view and present our hypothesis of homeostatic structural plasticity. We discuss in vitro and in vivo studies that may support our hypothesis and indicate how homeostatic structural plasticity could be tested experimentally. Lastly, we discuss the potential implications of homeostatic structural plasticity for brain function and neurological therapy.

2 CURRENT VIEW AND LIMITATIONS

Below, we present the prevailing view on cortical events following a loss of afferent activity, using retinal lesions and monocular deprivation as examples. We then describe some limitations of the current view.

2.1 Current View

Retinal lesions or monocular deprivation deprives a corresponding group of neurons (in the case of lesions, called lesion projection zone or LPZ) in the primary visual cortex of visual input, causing a drop in the neurons' level of electrical activity.[21,22,27–29] Activity levels then recover relatively quickly, over a couple of days,[21,22,29] by the strengthening and weakening of existing intracortical synapses through synaptic scaling[21] and Hebbian synaptic plasticity,[30] possibly in combination with a reduction in inhibition.[21,29] Not all neurons recover their activity, at least over the 2- to 3-day period during which the cortex is monitored.[29] In addition to changes in existing synapses, retinal lesions trigger massive structural changes such as axon sprouting and axonal bouton formation[5,31,32] and formation and deletion of dendritic spines.[4] By binding to boutons, newly formed spines may form transient synapses, which could become consolidated depending on the correlation between pre- and postsynaptic activity, i.e., according to Hebbian rules.[33–35] Although formation of new synapses may affect activity levels, the restoration of activity is believed to occur mainly through the relatively fast changes (hours to a few days) in the strength of existing synapses.

2.2 Limitations

The quick, partly recovery of activity in the visual cortex may not be stable and activity dynamics may continue beyond few days after deprivation. Some experimental studies indeed suggest that restoration of electrical activity is in fact a slow process taking place on a much longer timescale than homeostatic or Hebbian synaptic plasticity. One retinal lesion study showed that the level of electrical activity in the LPZ restored from the outside to the inside of the LPZ over a period of up to a year.[27] Such a slow restoration of activity was also evident from the neuronal activity markers c-fos and zif268.[28,36]

In addition, slow structural changes, including spine dynamics and ingrowth of axonal arbors from intact peri-LPZ to the LPZ,[5,31] may directly contribute to the restoration of activity. Although direct evidence is still lacking, the ingrowing axons[5,31] may form synapses with the newly formed spines in the LPZ[4] and so help restore activity in the LPZ. This view is supported by the observation that spreading of electrical activity from intact peri-LPZ to LPZ occurs only after more than 6 days postlesion,[20] a result that was interpreted to indicate that axon outgrowth and synapse formation with LPZ neurons are necessary for activity to propagate from intact to deprived areas.[20]

The current view on activity restoration following a loss of input relies on synaptic scaling and Hebbian rules. However, synaptic scaling can modify postsynaptic activity only when cells are connected to active presynaptic cells. Likewise, Hebbian rules can strengthen or consolidate synapses only after they are formed. When cells are unconnected, the formation of spines and boutons itself cannot be governed by Hebbian rules, as this would require information about both pre- and postsynaptic activity. Synapse formation and synaptic rewiring depend crucially on the availability of spines and boutons, but the factors regulating spine and bouton formation are little known.

3 HOMEOSTATIC STRUCTURAL PLASTICITY

Below, we present a hypothesis for the regulation of spine and bouton formation, and describe what we expect on the basis of this hypothesis for the events following a lesion-induced decrease in activity.

3.1 Hypothesis

We postulate that a persistent decrease in a neuron's electrical activity below a desired value (set-point) triggers the formation of spines and boutons (Fig. 4.1). Likewise, we

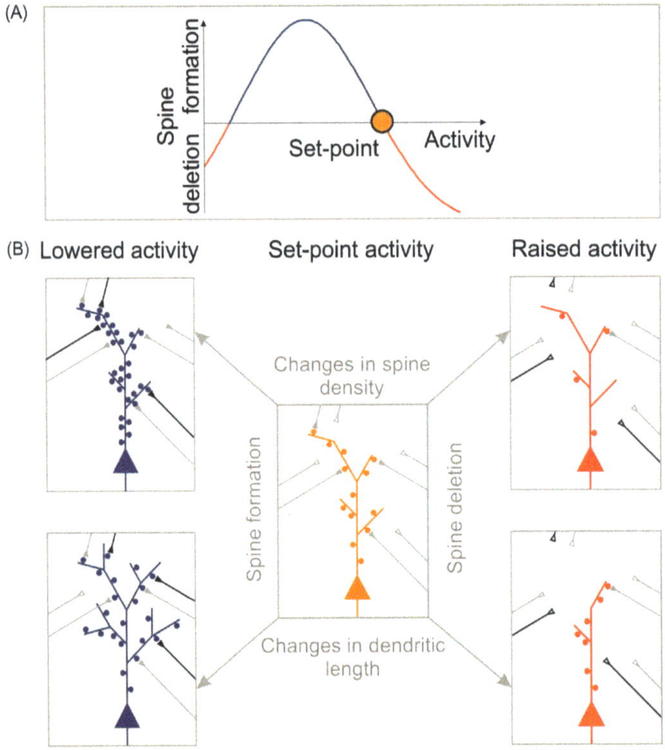

FIGURE 4.1 **Homeostatic structural plasticity.** Our hypothesis is that the attempt of neurons to maintain their average electrical activity at a desired level (homeostatic set-point) drives spine formation and deletion. (A) We postulate that neurons form new spines when their activity level is below the set-point and delete spines when their activity level is above the set-point or below a certain minimum level. (B) We further argue that a neuron can form or delete spines either through a change in spine density or through a change in dendritic length. The middle image shows a cell at set-point activity, showing spines (*filled circles*), axons (*straight lines with triangles*) from neighboring cells, and synaptic connections (*spines with filled axon triangles*). When the activity of the neuron is lowered below the set-point (*left*), new spines are formed through an increase in spine density (*top left*) or an increase in dendritic length (*bottom left*). The new spines enable axons from neighboring cells to make new synaptic connections (*bold filled triangles*), which may restore the neuron's activity level. When the activity of the neuron is raised above the set-point (*right*), spines are deleted through a decrease in spine density (*top right*) or a decrease in dendritic length (*bottom right*). Deletion of spines that are bound in synapses leads to the break-up of synaptic connections (*bold open triangles*), which may reduce the neuron's activity level.

hypothesize that a persistent increase in a neuron's electrical activity above the set-point results in deletion of spines and boutons. In addition, we assume that activity below a certain minimum level will also result in deletion of spines and boutons.

3.2 Expectations

We expect that after a lesion and the resulting drop in electrical activity, a neuron will increase its spine formation. A neuron may achieve this by directly upregulating de novo formation of spines,[37] attenuating the ongoing elimination of spines,[38] or retaining more of the spontaneously formed new spines.[16] The extra spines enable the neuron to make new synaptic input connections with active neurons, which may raise its level of electrical activity. Note that increased spine formation does not necessarily imply a higher total number of spines, because removal of spines (e.g., those that are not receiving input[39]) could counterbalance the increased formation. In addition to spine formation, we expect that a reduction in electrical activity triggers bouton formation.

By contrast, we expect that for persistently high levels of electrical activity, a neuron will delete spines and boutons. Deletion of spines that are bound in synapses breaks up existing connections, which may lower the neuron's level of electrical activity. Thus when a neuron deviates in either direction from a desired set-point activity, formation and deletion of spines have the potential to bring activity back to the set-point value (homeostasis). Such a cell-autonomous set-point value has been proposed before in the context of synaptic scaling[40] and has recently convincingly been demonstrated experimentally.[41]

A neuron may form or delete spines and boutons while keeping its dendritic and axonal length constant. Alternatively, a neuron could change spine or bouton number by changing its dendritic or axonal length, respectively (Fig. 4.1).

Spine and bouton formation triggered by low activity may be capable of restoring activity in the LPZ. The spines that would be formed in the LPZ as a result of the lesion-induced drop in activity could make new intracortical connections with boutons from active, peri-LPZ neurons and so raise the activity of LPZ neurons. However, restoration of activity could fail if activity drops below the minimum level for spine and bouton formation or if there are not enough boutons within reach with which the newly formed spines can make synapses. The latter could happen, e.g., if the LPZ is too big.

4 IN VITRO INDICATIONS FOR HOMEOSTATIC STRUCTURAL PLASTICITY

Because the impact of electrical activity on spine formation and neurite outgrowth may be seen clearer in vitro than in vivo owing to the fewer confounding network interactions in vitro, we first review in vitro studies. The prevalent view seems to be that in vitro spine formation and neurite outgrowth are triggered by increased electrical activity.[42–44] However, a large body of studies instead shows that low activity increases and high (but also very low) activity decreases spine number and neurite length.

4.1 Dendritic Spines

Blocking electrical activity in mature hippocampal[45–47] or cerebellar cultures[48] by tetrodotoxin (TTX) resulted in increased spine density (Fig. 4.2A). By contrast, long-term elevations in activity, as in epileptic hippocampal tissue cultures, led to a reversible reduction in spine numbers[49–51] (Fig. 4.2B). Similarly, spines reduced in length[52] or even completely disappeared[53] after excessive stimulation of NMDA receptors. Furthermore, high activity not only decreases spine number but can also reduce spine motility,[54] hindering the spines' ability to search for axons and to form new synapses.

4.2 Dendrite and Axon Outgrowth

Application of the excitatory neurotransmitter glutamate caused dendrites to retract, an effect that was mediated by calcium influx through voltage-dependent calcium channels.[55] Likewise, increased neuronal electrical activity, as in epileptic seizures, resulted in dendritic retraction.[51,56] During development, low concentrations of glutamate allowed dendritic outgrowth, whereas elevated concentrations prevented outgrowth or caused retraction.[57]

Axons may respond to changes in electrical activity in the same way as dendrites. Reducing electrical activity through application of the NMDA receptor blocker MK-801 resulted in considerable axonal outgrowth and sprouting.[58] By contrast, raising activity by stimulation of kainate receptors caused axons to retract[59] (Fig. 4.2D) or resulted in the immobility of axonal filopodia,[60,61] hampering the axon's ability to find dendrites and to form synapses.[60]

4.3 Minimum Activity for Spine Formation and Neurite Outgrowth

As described earlier, lowering activity may induce spine formation and neurite outgrowth. However, when activity drops below a minimum level, spines may be deleted and neurites retracted.[62] A lasting reduction in activity promoted neurite outgrowth,[58] but a complete loss of activity by blocking both NMDA and AMPA receptors resulted in a loss of dendritic spines[63] and prevention of axonal outgrowth.[58] Thus the effect of electrical activity may be described by a bell-shaped curve, with an optimal activity level for growth, a homeostatic set-point above which, and a minimum level below which, neurites retract and spines are deleted (Fig. 4.1A). A bell-shaped response curve has indeed been observed for the formation of spine head protrusions[47] (Fig. 4.2C). Such a bell-shaped curve may account for the seemingly conflicting results on the effect of electrical activity.[42,43,46,48] When electrical activity is around the set-point, an increase in activity may elicit spine deletion and neurite retraction, whereas when activity is low, the same increase may trigger spine formation and neurite outgrowth. In addition, the duration of the activity change may be important. A short-lasting change in activity may have another effect on spine formation and neurite outgrowth than a long-lasting, lesion-induced change.

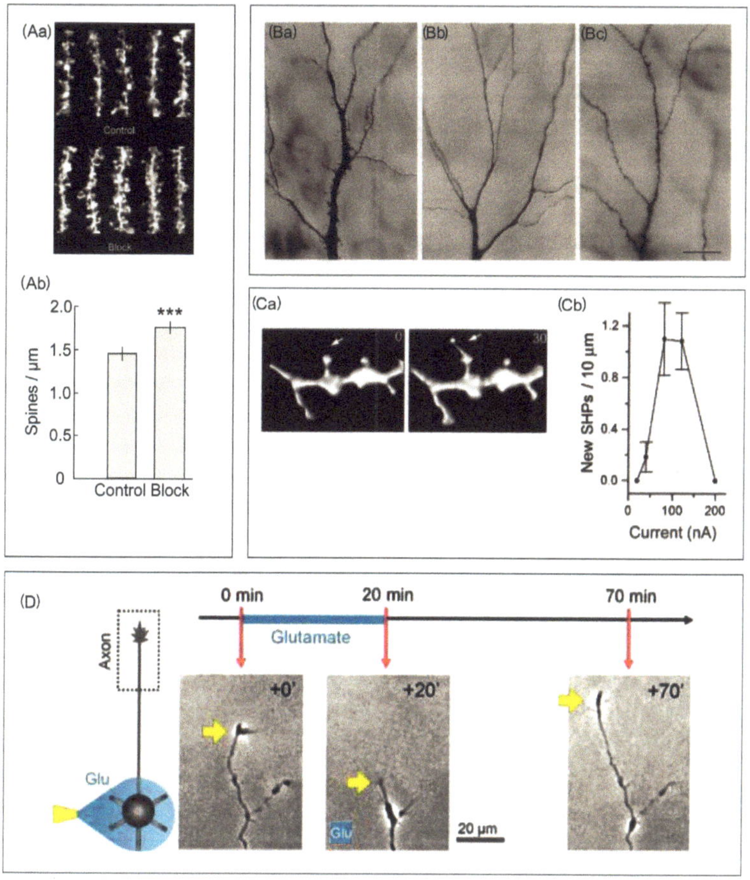

FIGURE 4.2 **In vitro indications for homeostatic structural plasticity.** Various findings indicate that low electrical activity increases and high activity decreases spine number and axon length. (A) Pyramidal neurons in hippocampal slice cultures became more spiny when electrical activity was blocked by TTX.[45,46] Cultures were taken from adult rats (62–80 days). (Aa) *Top row*: Five dendritic segments with spines from control situation. *Bottom row*: Dendritic segments and spines under TTX condition. (Ab) Dendritic spine densities were significantly higher under TTX condition. (B) High electrical activity due to epileptic seizures evoked by application of the $GABA_A$ receptor antagonist picrotoxin reversibly decreased spine densities[49] in hippocampal slice cultures. Cultures were taken from neonatal rat pups and incubated for 12–20 days before picrotoxin application. Picrotoxin caused cells to spontaneously exhibit paroxysmal depolarization shifts over several hours. (Ba) Pyramidal cell from control culture with normal spine density. (Bb) Reduced spine density after 3 days of picrotoxin treatment. (Bc) Spine density restored to normal levels when, after 3 days of picrotoxin treatment, cultures were put back in control medium for 1 week. (C) Spine head protrusions (SHPs) on pyramidal cells from hippocampal slice cultures formed in a window of reduced electrical activity.[47] Cultures were taken from 6-day-old mice and kept 3–6 weeks in vitro before TTX application. (Ca) New SHPs were seen within 30 minutes after TTX application. SHPs may have formed in response to a TTX-induced reduction in glutamate release from nearby presynaptic terminals. (Cb) SHP formation has a bell-shaped dependence on the amount of iontophoretically applied exogenous glutamate (amount of glutamate is proportional to the iontophoresis current). (D) Twenty minutes of perisomatic application of glutamate caused axons of isolated dentate granual cells to retract.[59] Cells were taken from 3-day-old rats. Axon retraction was mediated by intracellular calcium waves propagating from the soma to the axon terminals. Axons started to grow again when glutamate application was stopped. Source: *Reproduced with permission from (A) Kirov SA, Harris KM. Dendrites are more spiny on mature hippocampal neurons when synapses are inactivated. Nat Neurosci. 1999;2(10):878–883, Nature Publishing Group; (B) Müller M, Gähwiler BH, Rietschin L, Thompson SM. Reversible loss of dendritic spines and altered excitability after chronic epilepsy in hippocampal slice cultures. Proc Natl Acad Sci USA. 1993;90(1):257–261, copyright (1993) National Academy of Sciences, USA; (C) Richards DA, Mateos JM, Hugel S et al. Glutamate induces the rapid formation of spine head protrusions in hippocampal slice cultures. Proc Natl Acad Sci USA. 2005;102(17):6166–6171, copyright (2005) National Academy of Sciences, USA; (D) Yamada RX, Sasaki T, Ichikawa J, Koyama R, Matsuki N, Ikegaya Y. Long-range axonal calcium sweep induces axon retraction. J Neurosci. 2008;28(18):4613–4618, Society of Neuroscience.*

5 IN VIVO INDICATIONS FOR HOMEOSTATIC STRUCTURAL PLASTICITY

Below, we review in vivo studies in visual cortex, barrel cortex, and stroke that suggest that a long-term reduction in electrical activity triggers spine formation and promotes neurite outgrowth, leading to connectivity reorganization and activity restoration. In visual and barrel cortex studies, cortical cells have reduced peripheral input from the eyes or whiskers, respectively; whereas in stroke studies, cells have reduced central input from the stroke site.

5.1 Visual Cortex

Loss of visual input caused by monocular deprivation resulted in increased spine numbers in the primary visual cortex[37] (Fig. 4.3A). The newly formed spines likely established synapses with neurons in ocular dominance columns of the nondeprived eye,[37] thereby increasing the activity of the deprived columns.[21]

In focal retinal lesions experiments, a compact group of neurons in the primary visual cortex (the lesion projection zone or LPZ) is deprived of visual input.[4,64–66] In one such experiment, the loss of input triggered extensive formation and deletion of spines[4] (Fig. 4.4Aa), although the total number of spines did not increase. As indicated by the enhanced occurrence of persistent spines (Fig. 4.4Ab), many of the newly formed spines became stabilized,[4] probably by becoming bound into synapses. Consolidation of new synapses may have involved LTP,[34,67] and deletion of existing synapses may have entailed LTD.[35,39] We conjecture that LPZ neurons established new afferent synapses with active, peri-LPZ neurons while deleting afferent synapses with inactive, LPZ neurons, thereby keeping the total number of spines more or less constant. Other studies indeed showed that axon collaterals sprouted from the peri-LPZ into the LPZ[5,31] (Fig. 4.4B), enabling synapse formation of LPZ neurons with peri-LPZ neurons. The peri-LPZ lost some horizontal input from the LPZ and therefore may have had a mildly reduced level of electrical activity, which could have induced the axonal outgrowth.[68]

In another study, it was shown that the level of electrical activity in the LPZ, which drops after the lesion, slowly restored from the outside to the inside over a period of up to a year, with the border between normal and reduced activity slowly migrating toward the center of the LPZ.[27] This slow restoration of activity, also evident from the neuronal activity markers c-fos and zif268,[28,36] is more likely driven by structural plasticity than by homeostatic synaptic plasticity[21,22] or intrinsic neuronal plasticity,[40] as the latter two act on faster timescales of hours to a few days. Several studies suggest that the restoration of activity may be brought about by changes in intracortical connectivity.[20,32,69,70] The ingrowing axons from the peri-LPZ[5,31] may form synapses with the newly formed spines in the LPZ[4] and so slowly restore activity in the LPZ. Furthermore, the observation that electrical stimulation in the peri-LPZ can evoke suprathreshold activation in the LPZ only after more than 6 days postlesion[20] (Fig. 4.4C) also suggests that axon outgrowth and synapse formation are involved.

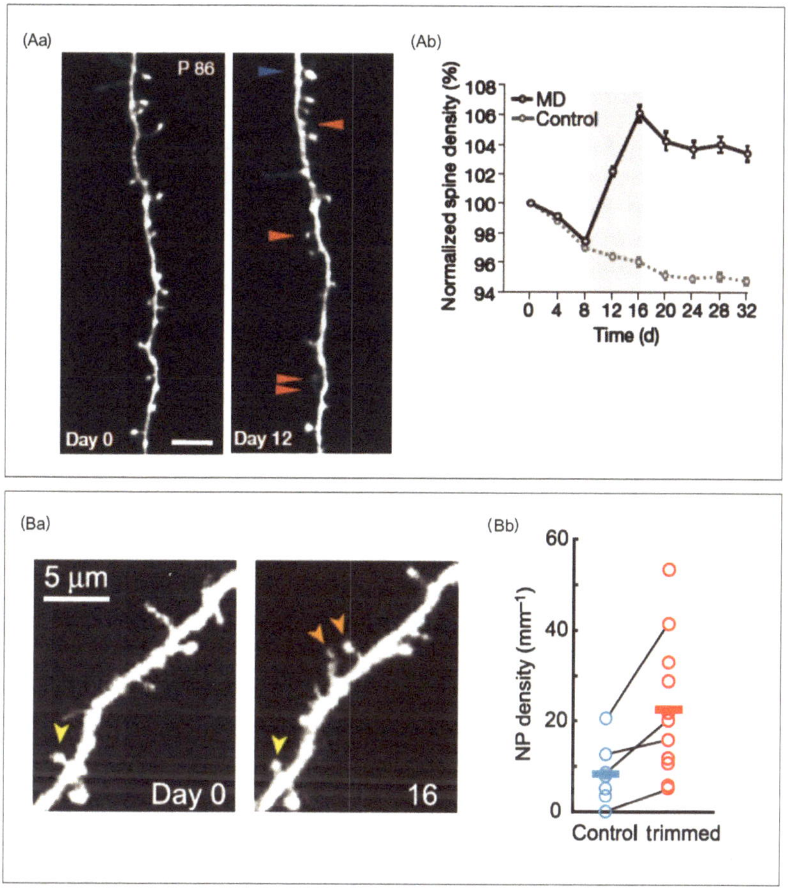

FIGURE 4.3 **Indications for homeostatic structural plasticity from sensory deprivation experiments.** Reduced electrical activity caused by sensory deprivation enhances spine formation in visual and barrel cortex. (A) Spine density in visual cortex increased after monocular deprivation in mature mice (postnatal day 86).[37] Deprivation began at day 8 of the experiment until day 16 (*gray area* in Ab). (Aa) Dendrite of a pyramidal cell in visual cortex. *Red arrows* point to newly formed spines 4 days after deprivation; the *blue arrow* points to a lost spine. (Ab) Normalized spine density after monocular deprivation (MD) and in controls. (B) Number of persistent spines in barrel cortex increased after whisker trimming in adult mice owing to an increased consolidation of newly formed spines.[16] Whiskers were trimmed at day 8 of the experiment. (Ba) Dendrite of a pyramidal cell in barrel cortex. *Yellow arrow* indicates a spine that was present before (day 0) until at least 8 days after whisker trimming (day 16); *orange arrows* indicate new persistent spines. (Bb) Density of new persistent spines (NP). Every circle (*blue*: control; *red*: whisker trimmed) represents a single pyramidal cell. Horizontal bars indicate group averages. *Black lines* connect paired experiments, in which a complete control period of 8 days was followed by 20 days of deprivation. Source: *Reproduced with permission from (A) Hofer SB, Mrsic-Flogel TD, Bonhoeffer T, Hübener M. Experience leaves a lasting structural trace in cortical circuits. Nature. 2009;457(7227):313–317, Nature Publishing Group; (B) Holtmaat A, Wilbrecht L, Knott GW, Welker E, Svoboda K. Experience-dependent and cell-type-specific spine growth in the neocortex. Nature. 2006;441(7096):979–983, Nature Publishing Group.*

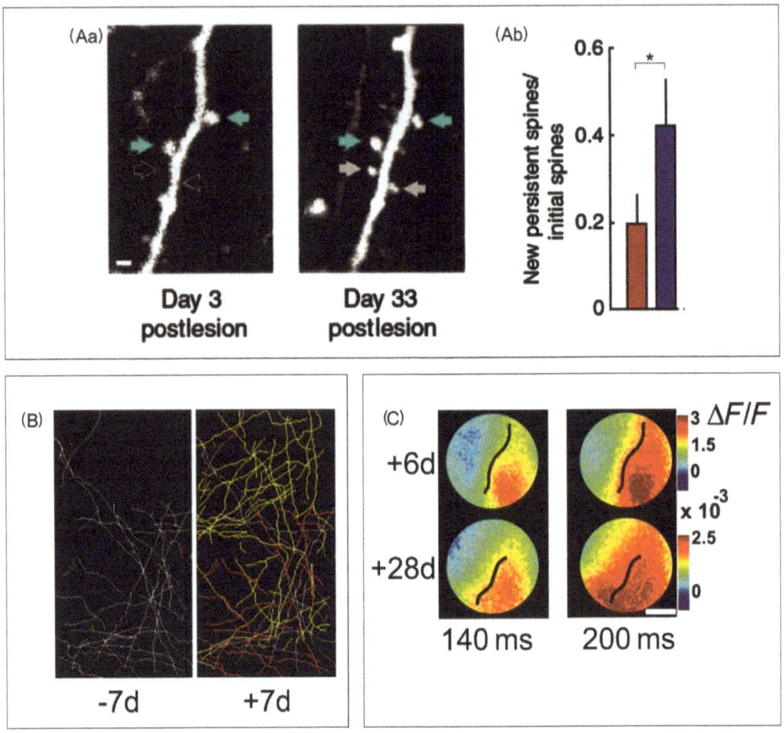

FIGURE 4.4 **Indications for homeostatic structural plasticity from retinal lesion experiments.** Focal retinal lesions, which deprive an area in visual cortex (the lesion project zone or LPZ) of visual input, trigger cortical spine formation and growth of new axonal branches. (A) Spine formation in the LPZ was enhanced after focal retinal lesions in adult mice.[4] (Aa) Dendrite of pyramidal cell in visual cortex. *Blue arrows* point to spines that were present before the lesion and at 3 and 33 days after the lesion. *Gray arrows* indicate spines that were formed after the lesion and were present at day 33 but not yet at day 3. All marked spines were persistent, being maintained over the entire 2-month duration of the experiment. (Ab) The ratio of numbers of new persistent spines to initial spines significantly increased after the lesion. (B) In adult monkeys, focal retinal lesions led to a substantial sprouting of axonal branches of cortical neurons in the peri-LPZ and ingrowth of the new fibers into the LPZ.[5] Persistent axonal branches 7 days before and 7 days after the lesion are shown in *gray*; newly formed axonal branches are shown in *yellow* and pruned branches in *red*. (C) Activity dynamics in visual cortex after a focal retinal lesion in adult rats.[20] At 28 days after the lesion (*bottom row*), but not at 6 days after the lesion (*top row*), electrical stimulation in the peri-LPZ (area at the right of the *black line*) caused a spread of activity into the LPZ (area at the left of the *black line*), indicating that connectivity from peri-LPZ to LPZ changed during the time period after the lesion. Activity was measured by voltage-sensitive dye imaging and is shown 140 ms and 200 ms after stimulation. Source: *Reproduced with permission from (A) Keck T, Mrsic-Flogel TD, Vaz Afonso M, Eysel UT, Bonhoeffer T, Hübener M. Massive restructuring of neuronal circuits during functional reorganization of adult visual cortex. Nat Neurosci. 2008;11(10):1162–1167, Nature Publishing Group; (B) Yamahachi H, Marik SA, McManus JN, Denk W, Gilbert CD. Rapid axonal sprouting and pruning accompany functional reorganization in primary visual cortex. Neuron. 2009;64 (5):719–729, Elsevier; (C) Palagina G, Eysel UT, Jancke D. Strengthening of lateral activation in adult rat visual cortex after retinal lesions captured with voltage-sensitive dye imaging in vivo. Proc Natl Acad Sci USA. 2009;106 (21):8743–8747, National Academy of Sciences, USA.*

5.2 Barrel Cortex

Reducing the activity of barrel cortex neurons by whisker clipping led to an increased number of new persistent spines[16,18] (Fig. 4.3B). The increase was due not to a higher rate of de novo spine formation but to a higher rate of consolidation of newly formed spines.[16] Similarly, in another study, whisker trimming prevented spine loss by reducing the rate of ongoing spine elimination, not by increasing the rate of de novo spine formation.[38] We postulate that by creating extra spines, either by keeping more of the newly formed spines[16] or by attenuating ongoing spine elimination,[38] neurons deprived of input can connect to neurons from intact barrels and thereby compensate for their loss of input. By contrast, raising the activity of barrel cortex neurons by whisker stimulation caused pruning of excitatory dendritic spines and an increase in inhibitory synaptic contacts.[71]

Dendritic morphology is generally stable in the mature brain.[11,72] In mouse barrel cortex, whisker ablation-induced loss of activity triggered formation of new dendritic spines but no changes in dendritic structure.[73] In rat barrel cortex, however, whisker ablation initiated dendritic outgrowth.[74] Since increasing the dendrite length may lead to formation of new spines and synapses, dendritic outgrowth could help restore neuronal electrical activity. Besides dendritic outgrowth, whisker clipping can trigger axonal outgrowth.[75]

5.3 Stroke

Neurons with afferent connections from the stroke site are deprived of input and can experience a significant reduction in electrical activity.[6,7,76–79] Input-deprived neurons create new spines,[6,80,81] enabling the formation of synapses with sprouting axons from intact surrounding tissue.[6,82–85] During stroke recovery, functional connectivity reorganizes and activity partly restores.[86] Restoration of activity may be brought about by synaptic scaling[86] and loss of inhibition,[87] but also by the formation of new excitatory synapses with intact tissue.[6]

By contrast, temporarily increased activity after stroke, so-called spreading depression, causes a reversible loss of dendritic spines in the peri-infarct area.[88] As seen in resected human brain tissue, the even higher electrical activity in temporal lobe epilepsy results in massive cell death as well as loss of dendritic structure[89] and synaptic connectivity[90] of surviving neurons.

Changes in dendritic morphology are sometimes observed after stroke. In adult mice, focal cortical stroke led to strongly enhanced rewiring of dendritic arbors.[91] Dendrites in the intact cortex withdrew their branches from the infarct and extended branches toward intact areas.[91] Interestingly, dendritic remodeling was inversely correlated with distance from the infarct,[80,81] suggesting that the loss of input from the infarct was compensated for by new synaptic contacts with the intact cortex. The farther a cell is from the infarct, the less it will be affected by a loss of input, and the less remodeling would be needed. In another study, however, no evidence was found for growth of dendrites in the peri-infarct area.[92]

Neurons distant from the stroke site may, via long-range connections, experience altered input (diaschisis),[93,94] which could trigger changes in dendritic morphology. After a stroke in one hemisphere, neurons in the contralesional hemisphere were found to have increased dendritic length,[95,96] possibly induced by the loss of colossal input from the infarct site. However, such contralesional growth of dendrites is not always observed.[97]

6 EXPERIMENTAL TESTING OF HOMEOSTATIC STRUCTURAL PLASTICITY

Structural plasticity is a slow process,[27] so spine, bouton, and activity dynamics should be monitored beyond the first few days postlesion. Furthermore, the different processes affecting spine (and bouton) numbers should be well distinguished. The de novo formation of spines should be discriminated from the consolidation of spines and the ongoing deletion of spines. According to our hypothesis, an increase in spine formation events or a decrease in spine deletion events in response to reduced neuronal electrical activity will be an indication of homeostatic structural plasticity. Electrical activity should be lowered persistently and well away from its normal or set-point value. Around the normal value, the number of spines may predominantly be governed by Hebbian consolidation of spines[98] rather than by homeostatic structural plasticity.

6.1 Growth Curves

To test the postulated growth curves for bouton and spine formation (Fig. 4.1A), one could use two-photon glutamate uncaging[34] or optogenetics[99] to alter electrical activity in vivo. Gradually decreasing the cells' average electrical activity should promote spine formation, whereas increasing it should promote spine deletion.

Intracellular calcium mediates the effect of electrical activity on neuronal morphology.[100–103] To determine the growth curves more quantitatively, one could challenge in vitro neurons with rises in intracellular calcium of varying amplitudes by calcium uncaging[101] and measure the impact on neurite outgrowth and spine formation. This may also reveal differences between axons and dendrites or between different cell types. For example, neurons characterized by a high firing rate may have their homeostatic set-point at a higher activity level than neurons that fire less frequently.[41] In addition, excitatory and inhibitory neurons may have the same or different growth curves.

6.2 Activity Restoration

The direct involvement of homeostatic structural plasticity in activity restoration after lesions can be tested in several ways. We predict that in retinal lesion experiments, blocking spine formation,[104] axon outgrowth, or synapse formation[105] will prevent the long-term restoration of firing rates in the LPZ and the remapping of cortical representations.

To test whether a long-lasting drop in electrical activity from a desired set-point is indeed the trigger for morphological changes and activity restoration, one could artificially increase the average electrical activity in the LPZ by optogenetics,[99] two-photon glumamate uncaging,[34] or chronically implanted electrode arrays.[23] The increase in electrical activity should prevent LPZ neurons from creating new dendritic spines and, therefore, thwart synapse formation and activity restoration in the LPZ.

Since we expect that synaptic scaling is not the only mechanism responsible for activity restoration, we predict that activity keeps changing well beyond the first few days postlesion as a result of compensatory structural adaptations. Moreover, we predict that restoration of activity will still take place in knockouts that block synaptic scaling.[106] Lastly, to show that activity does not merely increase owing to disinhibition, one could perform in vitro or in vivo lesion experiments under blockade of GABA receptors.

7 DISCUSSION

Below, we discuss the relation of homeostatic structural plasticity to other forms of plasticity, as well as the potential implications of homeostatic structural plasticity for cortical remapping, maladaptive compensatory responses, neurodegeneration, and the treatment of stroke.

7.1 Relation to Other Forms of Plasticity

We envisage that homeostatic structural plasticity operates in conjunction with synaptic plasticity and other forms of structural plasticity. In general, structural plasticity creates or deletes spines and synapses, whereas synaptic plasticity can modify the strength of newly formed synapses. These two forms of plasticity interact.[37] Once synapses become strengthened by Hebbian synaptic plasticity (LTP), spines grow in size and transform in shape.[33,34,107] Such potentiated synapses, and their spines, are less likely to be deleted than synapses weakened by LTD.[35,39]

In addition, Hebbian synaptic plasticity can influence synapse formation, as spine formation is enhanced in the vicinity of synapses potentiated by LTP.[108] Such spine formation can result in duplication of the potentiated synapse if its presynaptic neuron binds to the newly formed spine, effectively increasing the overall connection strength in a structural manner. This LTP-induced synapse formation might be referred to as Hebbian structural plasticity and may have an important role in motor learning.[109] Other studies, however, reported that synapse strengthening hampers spine formation.[110,111]

Homeostatic structural plasticity could help restore activity when homeostatic synaptic plasticity (synaptic scaling) alone may not be sufficient, as with large drops in activity caused by lesions. Synapse strength cannot increase indefinitely and when cells are not connected to active presynaptic neurons, synaptic scaling will not be able to restore postsynaptic activity. Homeostatic structural plasticity, which can lead to new connections, may help input-deprived neurons find new sources of activity. We propose a hierarchical organization of homeostatic mechanisms, with fast synaptic mechanisms constantly in place and slow structural mechanisms setting in when activity levels are persistently outside their desired range.

7.2 Cortical Remapping

The reshaping of cortical maps following deafferentation is a well-known phenomenon[3,65,112] that is generally interpreted as resulting from Hebbian synaptic plasticity.[113] By modifying anatomical connectivity, structural plasticity could complement this Hebbian synaptic plasticity.

Like visual cortical remapping after retinal lesions,[4] somatotopic remapping following digit or limb amputations may partly be governed by homeostatic structural plasticity. Amputations dramatically disturb electrical activity in the somatosensory cortex and give rise to cortical rewiring resulting in enlarged representations of neighboring body parts into the silenced representation of the amputated digit or limb.[3,112,114] Such cortical remapping may also have an important role in functional reorganization after stroke.[115] If stroke-induced cortical lesions leave somatosensory representations partially intact, remapping may extend neighboring representations into impaired representations, and may so contribute to functional recovery. Interestingly, spine turnover, dendritic rerouting, and axonal sprouting co-occur with cortical remapping after stroke.[6,80,81,91]

7.3 Maladaptive Responses

In our view, homeostatic structural plasticity is not a goal-directed process that necessarily results in recovery of activity or regaining of impaired function. A neuron changes spine and bouton formation depending solely on its own activity level, and establishes connections depending only on the availability of spines and boutons on other cells, which could lead to changes in network connectivity that are in fact maladaptive. Homeostatic structural plasticity might therefore also have a role in the etiology of such neurological symptoms as phantom pain[116–119] writers' or musicians' cramp,[120,121] tinnitus after hearing loss,[122] and spasticity following spinal cord injury.[123] In spinal cord injury, alpha motor neurons that suffer from a loss of input from the (partially) damaged pyramidal tract increase their spine numbers[123] and may restore activity by whatever axonal branches are available for synapse formation. Synapse formation with peripheral proprio- or nociceptive afferents may then give rise to the typical spasticity syndrome. Late-onset epilepsy after stroke could likewise result from homeostatically driven network rewiring.[124–126]

7.4 Neurodegeneration

In neurodegenerative disorders, such as Alzheimer's disease and multiple sclerosis (MS), the damage to cells, axons, and dendrites may move neurons away from their homeostatic set-point of electrical activity, triggering compensatory structural changes. In Alzheimer's disease, cell death[127] decreases the input to surviving neurons, possibly initiating dendritic remodeling,[128] including dendritic prouting.[129,130]

In MS, axons (white matter) are lost as a result of demyelination. MS typically shows a biphasic progression, with a primary phase in which the brain is still able to recover from relapses, and a secondary, irreversible phase characterized by accelerated decline.[131,132] Interestingly, not only white matter but also gray matter (dendrites and neurons) is affected in MS,[133,134] and the course of gray matter degeneration, but not that of white matter degeneration, reflects the secondary phase of accelerated decline.[132] However, the cause of this sudden increase in gray matter degeneration is unknown. We hypothesize that a steady loss of axons disturbs the homeostasis of neuronal activity, triggering compensatory changes that can restore the network initially but when axon loss becomes too severe initiate a sudden degeneration of gray matter. Neuronal activity may indeed be involved in the entire course of MS. In the early stages of MS, cortical networks show increased synchronization of network activity,[135] while glutamate release is permanently enhanced.[136] Furthermore, neurons loose synaptic input,[137] and reorganization of brain networks takes place predominantly in the earliest phases of MS, when the brain may try to counterbalance neuronal and axonal damage.[138]

7.5 Neurological Therapy

Current therapies of stroke rehabilitation principally aim at promoting Hebbian synaptic plasticity by stimulating neural and motor activity. Homeostatic structural plasticity, however, implies that pauses in treatment may have a cardinal role, too. During pauses, when activity is low, homeostatic structural plasticity can trigger the formation of new synaptic circuits that may later be reinforced by LTP during subsequent stimulation periods. A modeling study[139] showed that after lesions, pauses in stimulation led to more stable network repair than continuous stimulation. Interestingly, clinical studies found that the effect of continued neurological treatment ceases[140] and that intensive physical therapy shortly after stroke can be detrimental.[141,142] The notion of homeostatic structural plasticity may also help guide recent methods for promoting stroke recovery, such as pharmacological interventions[143] and noninvasive brain stimulation (tDC and TMS).[144,145] Indeed, therapeutic approaches based on homeostatic plasticity have recently been proposed by clinicians.[146]

7.6 Computational Modeling

To explore the potential consequences of homeostatic structural plasticity, we recently built a computational model of structural plasticity,[125] partly based on earlier models,[14,147–151] that describes each neuron as having separate axonal elements (boutons) and dendritic elements (spines) (see also Chapters 7, 8, and 18 of this book). In the model, synapses are formed by random combination of axonal and dendritic elements. Neurons generate new elements when neuronal electrical activity is below a desired set-point value, and delete elements, including those bound in synapses, when activity is above the set-point or below a certain minimum level. Axonal or dendritic elements that remain after their synaptic counterparts have been deleted become again available for synapse formation. The model was used to explore the response of a network to a focal loss of external input.[125] The simulations revealed that as a result of the lesion, activity dropped in the network but that the ensuing changes in the number of axonal and dendritic elements led to a rewiring of the network and a restoration of activity, with spine dynamics and connectivity changes resembling those observed experimentally.[125,126]

8 CONCLUSION

The majority of plasticity studies over the past two decades have focused on synaptic plasticity[25,152] rather than on structural plasticity. Theoretical work has underlined the potential of Hebbian synaptic plasticity for learning[153] and cortical remapping.[113] With the progress in synaptic plasticity research,[154] the field moved largely away from structural plasticity. Only recently have new imaging techniques made it possible to monitor spine and synapse dynamics in the living brain and has interest in structural plasticity resurged, but the principles governing structural plasticity remain unclear. We hypothesized here that spine and bouton formation may be driven by activity homeostasis, and that this homeostatic structural plasticity may account, at least partly, for the connectivity and activity changes that occur in response to lesions.[1,4,5,31]

In homeostatic structural plasticity, neurons, by adapting their number of synaptic contact sites (spines, boutons), endeavor to change their connectivity so as to reach and maintain a desired level of neuronal activity. Homeostatic structural plasticity does not require information about pre- and postsynaptic activity, as does Hebbian synaptic plasticity (synapse-centric plasticity), but only needs the local activity state of the neuron (neuron-centric plasticity). In general, homeostatic structural plasticity may act as organizing principle driving both the formation of networks[149,150,155–157] and the compensatory structural changes following loss of input or neurons.[125,126] Further investigating this form of plasticity, both experimentally and computationally, may deepen our understanding of the brain's response to injury and might ultimately lead to novel treatments for stimulating brain repair after stroke or neurodegeneration.

References

1. Raisman G. Neuronal plasticity in the septal nuclei of the adult rat. *Brain Res*. 1969;14(1):25–48.
2. Cotman CW, Nieto-Sampedro M. Cell biology of synaptic plasticity. *Science*. 1984;225(4668):1287–1294.
3. Merzenich MM, Nelson RJ, Stryker MP, Cynader MS, Schoppmann A, Zook JM. Somatosensory cortical map changes following digit amputation in adult monkeys. *J Comp Neurol*. 1984;224(4):591–605.
4. Keck T, Mrsic-Flogel TD, Vaz Afonso M, Eysel UT, Bonhoeffer T, Hübener M. Massive restructuring of neuronal circuits during functional reorganization of adult visual cortex. *Nat Neurosci*. 2008;11(10):1162–1167.
5. Yamahachi H, Marik SA, McManus JN, Denk W, Gilbert CD. Rapid axonal sprouting and pruning accompany functional reorganization in primary visual cortex. *Neuron*. 2009;64(5):719–729.
6. Brown CE, Aminoltejari K, Erb H, Winship IR, Murphy TH. In vivo voltage-sensitive dye imaging in adult mice reveals that somatosensory maps lost to stroke are replaced over weeks by new structural and functional circuits with prolonged modes of activation within both the peri-infarct zone and distant sites. *J Neurosci*. 2009;29(6):1719–1734.
7. Winship IR, Murphy TH. Remapping the somatosensory cortex after stroke: insight from imaging the synapse to network. *Neuroscientist*. 2009;15(5):507–524.
8. Hinman JD, Rasband MN, Carmichael ST. Remodeling of the axon initial segment after focal cortical and white matter stroke. *Stroke*. 2013;44(1):182–189.
9. Benowitz LI, Carmichael ST. Promoting axonal rewiring to improve outcome after stroke. *Neurobiol Dis*. 2010;37(2):259–266.
10. De Paola V, Holtmaat A, Knott G, et al. Cell type-specific structural plasticity of axonal branches and boutons in the adult neocortex. *Neuron*. 2006;49(6):861–875.
11. Trachtenberg JT, Chen BE, Knott GW, et al. Long-term in vivo imaging of experience-dependent synaptic plasticity in adult cortex. *Nature*. 2002;420(6917):788–794.

12. Chklovskii DB, Mel BW, Svoboda K. Cortical rewiring and information storage. *Nature*. 2004;431 (7010):782–788.
13. Wolff JR, Joó F, Dames W. Plasticity in dendrites shown by continuous GABA administration in superior cervical ganglion of adult rat. *Nature*. 1978;274(5666):72–74.
14. Wolff JR, Wagner GP. Selforganization in synaptogenesis: interaction between the formation of excitatory and inhibitory synapses. In: Basar E, Flohr H, Hermann Haken HC, Mandell AJ, eds. *Synergetics of the Brain*. Berlin, Heidelberg, New York, Tokyo: Springer; 1983:50–59.
15. Teuchert-Noodt G. Neuronal degeneration and reorganization: a mutual principle in pathological and in healthy interactions of limbic and prefrontal circuits. *J Neural Transm Suppl*. 2000;60:315–333.
16. Holtmaat A, Wilbrecht L, Knott GW, Welker E, Svoboda K. Experience-dependent and cell-type-specific spine growth in the neocortex. *Nature*. 2006;441(7096):979–983.
17. Keck T, Scheuss V, Jacobsen RI, et al. Loss of sensory input causes rapid structural changes of inhibitory neurons in adult mouse visual cortex. *Neuron*. 2011;71(5):869–882.
18. Wilbrecht L, Holtmaat A, Wright N, Fox K, Svoboda K. Structural plasticity underlies experience-dependent functional plasticity of cortical circuits. *J Neurosci*. 2010;30(14):4927–4932.
19. Marik SA, Yamahachi H, Meyer zum Alten Borgloh S, Gilbert CD. Large-scale axonal reorganization of inhibitory neurons following retinal lesions. *J Neurosci*. 2014;34(5):1625–1632.
20. Palagina G, Eysel UT, Jancke D. Strengthening of lateral activation in adult rat visual cortex after retinal lesions captured with voltage-sensitive dye imaging in vivo. *Proc Natl Acad Sci USA*. 2009;106 (21):8743–8747.
21. Hengen KB, Lambo ME, van Hooser SD, Katz DB, Turrigiano GG. Firing rate homeostasis in visual cortex of freely behaving rodents. *Neuron*. 2013;80(2):335–342.
22. Keck T, Keller GB, Jacobsen RI, Eysel UT, Bonhoeffer T, Hübener M. Synaptic scaling and homeostatic plasticity in the mouse visual cortex in vivo. *Neuron*. 2013;80(2):327–334.
23. Abe H, McManus JN, Ramalingam N, et al. Adult cortical plasticity studied with chronically implanted electrode arrays. *J Neurosci*. 2015;35(6):2778–2790.
24. Hebb DO. *The Organization of Behavior*. New York: Wiley & Sons; 1949.
25. Bliss TV, Lomo T. Plasticity in a monosynaptic cortical pathway. *J Physiol*. 1970;207(2):61P.
26. Turrigiano GG, Leslie KR, Desai NS, Rutherford LC, Nelson SB. Activity-dependent scaling of quantal amplitude in neocortical neurons. *Nature*. 1998;391(6670):892–896.
27. Giannikopoulos DV, Eysel UT. Dynamics and specificity of cortical map reorganization after retinal lesions. *Proc Natl Acad Sci USA*. 2006;103(28):10805–10810.
28. Hu TT, Laeremans A, Eysel UT, Cnops L, Arckens L. Analysis of c-fos and zif268 expression reveals time-dependent changes in activity inside and outside the lesion projection zone in adult cat area 17 after retinal lesions. *Cereb Cortex*. 2009;19(12):2982–2992.
29. Barnes SJ, Sammons RP, Jacobsen RI, Mackie J, Keller GB, Keck T. Subnetwork-specific homeostatic plasticity in mouse visual cortex in vivo. *Neuron*. 2015;86(5):1290–1303.
30. Young JM, Waleszczyk WJ, Wang C, Calford MB, Dreher B, Obermayer K. Cortical reorganization consistent with spike timing-but not correlation-dependent plasticity. *Nat Neurosci*. 2007;10(7):887–895.
31. Darian-Smith C, Gilbert CD. Axonal sprouting accompanies functional reorganization in adult cat striate cortex. *Nature*. 1994;368(6473):737–740.
32. Darian-Smith C, Gilbert CD. Topographic reorganization in the striate cortex of the adult cat and monkey is cortically mediated. *J Neurosci*. 1995;15(3 Pt 1):1631–1647.
33. Engert F, Bonhoeffer T. Dendritic spine changes associated with hippocampal long-term synaptic plasticity. *Nature*. 1999;399(6731):66–70.
34. Matsuzaki M, Honkura N, Ellis-Davies GC, Kasai H. Structural basis of long-term potentiation in single dendritic spines. *Nature*. 2004;429(6993):761–766.
35. Becker N, Wierenga CJ, Fonseca R, Bonhoeffer T, Nägerl UV. LTD induction causes morphological changes of presynaptic boutons and reduces their contacts with spines. *Neuron*. 2008;60(4):590–597.
36. Hu TT, Van der Gucht E, Eysel UT, Arckens L. Retinal lesions induce layer-specific Fos expression changes in cat area 17. *Exp Brain Res*. 2010;205(1):139–144.
37. Hofer SB, Mrsic-Flogel TD, Bonhoeffer T, Hübener M. Experience leaves a lasting structural trace in cortical circuits. *Nature*. 2009;457(7227):313–317.

38. Zuo Y, Yang G, Kwon E, Gan WB. Long-term sensory deprivation prevents dendritic spine loss in primary somatosensory cortex. *Nature*. 2005;436(7048):261–265.
39. Wiegert JS, Oertner TG. Long-term depression triggers the selective elimination of weakly integrated synapses. *Proc Natl Acad Sci USA*. 2013;110(47):E4510–E4519.
40. Maffei A, Turrigiano GG. Multiple modes of network homeostasis in visual cortical layer 2/3. *J Neurosci*. 2008;28(17):4377–4384.
41. Hengen KB, Torrado Pacheco A, McGregor JN, Van Hooser SD, Turrigiano GG. Neuronal firing rate homeostasis is inhibited by sleep and promoted by wake. *Cell*. 2016;165(1):180–191.
42. Papa M, Segal M. Morphological plasticity in dendritic spines of cultured hippocampal neurons. *Neuroscience*. 1996;71(4):1005–1011.
43. Annis CM, O'Dowd DK, Robertson RT. Activity-dependent regulation of dendritic spine density on cortical pyramidal neurons in organotypic slice cultures. *J Neurobiol*. 1994;25(12):1483–1493.
44. Segal M, Greenberger V, Korkotian E. Formation of dendritic spines in cultured striatal neurons depends on excitatory afferent activity. *Eur J Neurosci*. 2003;17(12):2573–2585.
45. Kirov SA, Harris KM. Dendrites are more spiny on mature hippocampal neurons when synapses are inactivated. *Nat Neurosci*. 1999;2(10):878–883.
46. Kirov SA, Goddard CA, Harris KM. Age-dependence in the homeostatic upregulation of hippocampal dendritic spine number during blocked synaptic transmission. *Neuropharmacology*. 2004;47(5):640–648.
47. Richards DA, Mateos JM, Hugel S, et al. Glutamate induces the rapid formation of spine head protrusions in hippocampal slice cultures. *Proc Natl Acad Sci USA*. 2005;102(17):6166–6171.
48. Bravin M, Morando L, Vercelli A, Rossi F, Strata P. Control of spine formation by electrical activity in the adult rat cerebellum. *Proc Natl Acad Sci USA*. 1999;96(4):1704–1709.
49. Müller M, Gähwiler BH, Rietschin L, Thompson SM. Reversible loss of dendritic spines and altered excitability after chronic epilepsy in hippocampal slice cultures. *Proc Natl Acad Sci USA*. 1993;90(1):257–261.
50. Drakew A, Müller M, Gähwiler BH, Thompson SM, Frotscher M. Spine loss in experimental epilepsy: quantitative light and electron microscopic analysis of intracellularly stained CA3 pyramidal cells in hippocampal slice cultures. *Neuroscience*. 1996;70(1):31–45.
51. Jiang M, Lee CL, Smith KL, Swann JW. Spine loss and other persistent alterations of hippocampal pyramidal cell dendrites in a model of early-onset epilepsy. *J Neurosci*. 1998;18(20):8356–8368.
52. Segal M. Morphological alterations in dendritic spines of rat hippocampal neurons exposed to *N*-methyl-D-aspartate. *Neurosci Lett*. 1995;193(2):73–76.
53. Halpain S, Hipolito A, Saffer L. Regulation of F-actin stability in dendritic spines by glutamate receptors and calcineurin. *J Neurosci*. 1998;18(23):9835–9844.
54. Fischer M, Kaech S, Wagner U, Brinkhaus H, Matus A. Glutamate receptors regulate actin-based plasticity in dendritic spines. *Nat Neurosci*. 2000;3(9):887–894.
55. Mattson MP, Dou P, Kater SB. Outgrowth-regulating actions of glutamate in isolated hippocampal pyramidal neurons. *J Neurosci*. 1988;8(6):2087–2100.
56. Multani P, Myers RH, Blume HW, Schomer DL, Sotrel A. Neocortical dendritic pathology in human partial epilepsy: a quantitative Golgi study. *Epilepsia*. 1994;35(4):728–736.
57. Mattson MP. Neurotransmitters in the regulation of neuronal cytoarchitecture. *Brain Res*. 1988;472(2):179–212.
58. McKinney RA, Lüthi A, Bandtlow CE, Gähwiler BH, Thompson SM. Selective glutamate receptor antagonists can induce or prevent axonal sprouting in rat hippocampal slice cultures. *Proc Natl Acad Sci USA*. 1999;96(20):11631–11636.
59. Yamada RX, Sasaki T, Ichikawa J, Koyama R, Matsuki N, Ikegaya Y. Long-range axonal calcium sweep induces axon retraction. *J Neurosci*. 2008;28(18):4613–4618.
60. Chang S, De Camilli P. Glutamate regulates actin-based motility in axonal filopodia. *Nat Neurosci*. 2001;4(8):787–793.
61. Tashiro A, Dunaevsky A, Blazeski R, Mason CA, Yuste R. Bidirectional regulation of hippocampal mossy fiber filopodial motility by kainate receptors: a two-step model of synaptogenesis. *Neuron*. 2003;38(5):773–784.
62. al-Mohanna FA, Cave J, Bolsover SR. A narrow window of intracellular calcium concentration is optimal for neurite outgrowth in rat sensory neurones. *Brain Res Dev Brain Res*. 1992;70(2):287–290.
63. McKinney RA, Capogna M, Dürr R, Gähwiler BH, Thompson SM. Miniature synaptic events maintain dendritic spines via AMPA receptor activation. *Nat Neurosci*. 1999;2(1):44–49.

64. Eysel UT, Gonzalez-Aguilar F, Mayer U. A functional sign of reorganization in the visual system of adult cats: lateral geniculate neurons with displaced receptive fields after lesions of the nasal retina. *Brain Res.* 1980;181(2):285–300.
65. Kaas JH, Krubitzer LA, Chino YM, Langston AL, Polley EH, Blair N. Reorganization of retinotopic cortical maps in adult mammals after lesions of the retina. *Science.* 1990;248(4952):229–231.
66. Heinen SJ, Skavenski AA. Recovery of visual responses in foveal V1 neurons following bilateral foveal lesions in adult monkey. *Exp Brain Res.* 1991;83(3):670–674.
67. Yang G, Pan F, Gan WB. Stably maintained dendritic spines are associated with lifelong memories. *Nature.* 2009;462(7275):920–924.
68. Enes J, Langwieser N, Ruschel J, et al. Electrical activity suppresses axon growth through Ca(v)1.2 channels in adult primary sensory neurons. *Curr Biol.* 2010;20(13):1154–1164.
69. Calford MB, Wright LL, Metha AB, Taglianetti V. Topographic plasticity in primary visual cortex is mediated by local corticocortical connections. *J Neurosci.* 2003;23(16):6434–6442.
70. Eysel UT. Functional reconnections without new axonal growth in a partially denervated visual relay nucleus. *Nature.* 1982;299(5882):442–444.
71. Knott GW, Quairiaux C, Genoud C, Welker E. Formation of dendritic spines with GABAergic synapses induced by whisker stimulation in adult mice. *Neuron.* 2002;34(2):265–273.
72. Lee WC, Huang H, Feng G, et al. Dynamic remodeling of dendritic arbors in GABAergic interneurons of adult visual cortex. *PLoS Biol.* 2006;4(2):e29.
73. Schubert V, Lebrecht D, Holtmaat A. Peripheral deafferentation-driven functional somatosensory map shifts are associated with local, not large-scale dendritic structural plasticity. *J Neurosci.* 2013;33(22):9474–9487.
74. Tailby C, Wright LL, Metha AB, Calford MB. Activity-dependent maintenance and growth of dendrites in adult cortex. *Proc Natl Acad Sci USA.* 2005;102(12):4631–4636.
75. Marik SA, Yamahachi H, McManus JN, Szabo G, Gilbert CD. Axonal dynamics of excitatory and inhibitory neurons in somatosensory cortex. *PLoS Biol.* 2010;8(6):e1000395.
76. Gao TM, Pulsinelli WA, Xu ZC. Changes in membrane properties of CA1 pyramidal neurons after transient forebrain ischemia in vivo. *Neuroscience.* 1999;90(3):771–780.
77. Bolay H, Gürsoy-Ozdemir Y, Sara Y, Onur R, Can A, Dalkara T. Persistent defect in transmitter release and synapsin phosphorylation in cerebral cortex after transient moderate ischemic injury. *Stroke.* 2002;33(5):1369–1375.
78. Carmichael ST, Tatsukawa K, Katsman D, Tsuyuguchi N, Kornblum HI. Evolution of diaschisis in a focal stroke model. *Stroke.* 2004;35(3):758–763.
79. Alstott J, Breakspear M, Hagmann P, Cammoun L, Sporns O. Modeling the impact of lesions in the human brain. *PLoS Comput Biol.* 2009;5(6):e1000408.
80. Brown CE, Li P, Boyd JD, Delaney KR, Murphy TH. Extensive turnover of dendritic spines and vascular remodeling in cortical tissues recovering from stroke. *J Neurosci.* 2007;27(15):4101–4109.
81. Brown CE, Wong C, Murphy TH. Rapid morphologic plasticity of peri-infarct dendritic spines after focal ischemic stroke. *Stroke.* 2008;39(4):1286–1291.
82. Carmichael ST. Plasticity of cortical projections after stroke. *Neuroscientist.* 2003;9(1):64–75.
83. Stroemer RP, Kent TA, Hulsebosch CE. Neocortical neural sprouting, synaptogenesis, and behavioral recovery after neocortical infarction in rats. *Stroke.* 1995;26(11):2135–2144.
84. Lee JK, Kim JE, Sivula M, Strittmatter SM. Nogo receptor antagonism promotes stroke recovery by enhancing axonal plasticity. *J Neurosci.* 2004;24(27):6209–6217.
85. Carmichael ST, Wei L, Rovainen CM, Woolsey TA. New patterns of intracortical projections after focal cortical stroke. *Neurobiol Dis.* 2001;8(5):910–922.
86. Lim DH, LeDue JM, Mohajerani MH, Murphy TH. Optogenetic mapping after stroke reveals network-wide scaling of functional connections and heterogeneous recovery of the peri-infarct. *J Neurosci.* 2014;34(49):16455–16466.
87. Schiene K, Bruehl C, Zilles K, et al. Neuronal hyperexcitability and reduction of GABAA-receptor expression in the surround of cerebral photothrombosis. *J Cereb Blood Flow Metab.* 1996;16(5):906–914.
88. Risher WC, Ard D, Yuan J, Kirov SA. Recurrent spontaneous spreading depolarizations facilitate acute dendritic injury in the ischemic penumbra. *J Neurosci.* 2010;30(29):9859–9868.

89. Tóth K, Eross L, Vajda J, Halász P, Freund TF, Maglóczky Z. Loss and reorganization of calretinin-containing interneurons in the epileptic human hippocampus. *Brain*. 2010;133(9):2763–2777.
90. Alonso-Nanclares L, Kastanauskaite A, Rodriguez JR, Gonzalez-Soriano J, Defelipe J. A stereological study of synapse number in the epileptic human hippocampus. *Front Neuroanat*. 2011;5:8.
91. Brown CE, Boyd JD, Murphy TH. Longitudinal in vivo imaging reveals balanced and branch-specific remodeling of mature cortical pyramidal dendritic arbors after stroke. *J Cereb Blood Flow Metab*. 2010;30(4):783–791.
92. Mostany R, Portera-Cailliau C. Absence of large-scale dendritic plasticity of layer 5 pyramidal neurons in peri-infarct cortex. *J Neurosci*. 2011;31(5):1734–1738.
93. Bütefisch CM, Netz J, Wessling M, Seitz RJ, Hömberg V. Remote changes in cortical excitability after stroke. *Brain*. 2003;126(Pt 2):470–481.
94. Carrera E, Tononi G. Diaschisis: past, present, future. *Brain*. 2014;137(Pt 9):2408–2422.
95. Biernaskie J, Corbett D. Enriched rehabilitative training promotes improved forelimb motor function and enhanced dendritic growth after focal ischemic injury. *J Neurosci*. 2001;21(14):5272–5280.
96. Jones TA, Schallert T. Overgrowth and pruning of dendrites in adult rats recovering from neocortical damage. *Brain Res*. 1992;581(1):156–160.
97. Johnston DG, Denizet M, Mostany R, Portera-Cailliau C. Chronic in vivo imaging shows no evidence of dendritic plasticity or functional remapping in the contralesional cortex after stroke. *Cereb Cortex*. 2013;23(4):751–762.
98. Wyatt RM, Tring E, Trachtenberg JT. Pattern and not magnitude of neural activity determines dendritic spine stability in awake mice. *Nat Neurosci*. 2012;15(7):949–951.
99. Fenno L, Yizhar O, Deisseroth K. The development and application of optogenetics. *Annu Rev Neurosci*. 2011;34:389–412.
100. Kater SB, Mattson MP, Cohan C, Connor J. Calcium regulation of the neuronal growth cone. *Trends Neurosci*. 1988;11(7):315–321.
101. Lohmann C, Wong RO. Regulation of dendritic growth and plasticity by local and global calcium dynamics. *Cell Calcium*. 2005;37(5):403–409.
102. Ghiretti AE, Paradis S. Molecular mechanisms of activity-dependent changes in dendritic morphology: role of RGK proteins. *Trends Neurosci*. 2014;37(7):399–407.
103. Konur S, Ghosh A. Calcium signaling and the control of dendritic development. *Neuron*. 2005;46(3):401–405.
104. Cole CJ, Mercaldo V, Restivo L, et al. MEF2 negatively regulates learning-induced structural plasticity and memory formation. *Nat Neurosci*. 2012;15(9):1255–1264.
105. Eroglu C, Allen NJ, Susman MW, et al. Gabapentin receptor alpha2delta-1 is a neuronal thrombospondin receptor responsible for excitatory CNS synaptogenesis. *Cell*. 2009;139(2):380–392.
106. Kaneko M, Stellwagen D, Malenka RC, Stryker MP. Tumor necrosis factor-alpha mediates one component of competitive, experience-dependent plasticity in developing visual cortex. *Neuron*. 2008;58(5):673–680.
107. Knott GW, Holtmaat A, Wilbrecht L, Welker E, Svoboda K. Spine growth precedes synapse formation in the adult neocortex in vivo. *Nat Neurosci*. 2006;9(9):1117–1124.
108. Toni N, Buchs PA, Nikonenko I, Bron CR, Muller D. LTP promotes formation of multiple spine synapses between a single axon terminal and a dendrite. *Nature*. 1999;402(6760):421–425.
109. Xu T, Yu X, Perlik AJ, et al. Rapid formation and selective stabilization of synapses for enduring motor memories. *Nature*. 2009;462(7275):915–919.
110. Vlachos A, Helias M, Becker D, Diesmann M, Deller T. NMDA-receptor inhibition increases spine stability of denervated mouse dentate granule cells and accelerates spine density recovery following entorhinal denervation in vitro. *Neurobiol Dis*. 2013;59:267–276.
111. Bourne JN, Harris KM. Coordination of size and number of excitatory and inhibitory synapses results in a balanced structural plasticity along mature hippocampal CA1 dendrites during LTP. *Hippocampus*. 2011;21(4):354–373.
112. Merzenich MM, Kaas JH, Wall J, Nelson RJ, Sur M, Felleman D. Topographic reorganization of somatosensory cortical areas 3b and 1 in adult monkeys following restricted deafferentation. *Neuroscience*. 1983;8(1):33–55.
113. Kohonen T. Cortical maps. *Nature*. 1990;346(6279):24.
114. Merzenich MM, Kaas JH, Wall JT, Sur M, Nelson RJ, Felleman DJ. Progression of change following median nerve section in the cortical representation of the hand in areas 3b and 1 in adult owl and squirrel monkeys. *Neuroscience*. 1983;10(3):639–665.

115. Zepeda A, Vaca L, Arias C, Sengpiel F. Reorganization of visual cortical maps after focal ischemic lesions. *J Cereb Blood Flow Metab*. 2003;23(7):811–820.
116. Flor H, Elbert T, Knecht S, et al. Phantom-limb pain as a perceptual correlate of cortical reorganization following arm amputation. *Nature*. 1995;375(6531):482–484.
117. Ramachandran VS, Altschuler EL. The use of visual feedback, in particular mirror visual feedback, in restoring brain function. *Brain*. 2009;132(Pt 7):1693–1710.
118. Gustin SM, Peck CC, Cheney LB, Macey PM, Murray GM, Henderson LA. Pain and plasticity: is chronic pain always associated with somatosensory cortex activity and reorganization? *J Neurosci*. 2012;32 (43):14874–14884.
119. Makin TR, Scholz J, Filippini N, Henderson Slater D, Tracey I, Johansen-Berg H. Phantom pain is associated with preserved structure and function in the former hand area. *Nat Commun*. 2013;4:1570.
120. Quartarone A, Bagnato S, Rizzo V, et al. Abnormal associative plasticity of the human motor cortex in writer's cramp. *Brain*. 2003;126(Pt 12):2586–2596.
121. Tamura Y, Ueki Y, Lin P, et al. Disordered plasticity in the primary somatosensory cortex in focal hand dystonia. *Brain*. 2009;132(Pt 3):749–755.
122. Eggermont JJ, Roberts LE. The neuroscience of tinnitus. *Trends Neurosci*. 2004;27(11):676–682.
123. Bandaru SP, Liu S, Waxman SG, Tan AM. Dendritic spine dysgenesis contributes to hyperreflexia after spinal cord injury. *J Neurophysiol*. 2015;113(5):1598–1615.
124. van Oss C, van Ooyen A. Effects of inhibition on neural network development through activity-dependent neurite outgrowth. *J Theor Biol*. 1997;185(2):263–280.
125. Butz M, van Ooyen A. A simple rule for dendritic spine and axonal bouton formation can account for cortical reorganization after focal retinal lesions. *PLoS Comput Biol*. 2013;9(10):e1003259.
126. Butz M, Steenbuck ID, van Ooyen A. Homeostatic structural plasticity can account for topology changes following deafferentation and focal stroke. *Front Neuroanat*. 2014;8:115.
127. Behl C. Apoptosis and Alzheimer's disease. *J Neural Transm (Vienna)*. 2000;107(11):1325–1344.
128. Moolman DL, Vitolo OV, Vonsattel JP, Shelanski ML. Dendrite and dendritic spine alterations in Alzheimer models. *J Neurocytol*. 2004;33(3):377–387.
129. Scheibel AB, Tomiyasu U. Dendritic sprouting in Alzheimer's presenile dementia. *Exp Neurol*. 1978;60(1):1–8.
130. Masliah E, Mallory M, Hansen L, et al. Patterns of aberrant sprouting in Alzheimer's disease. *Neuron*. 1991;6 (5):729–739.
131. Barkhof F. The clinico-radiological paradox in multiple sclerosis revisited. *Curr Opin Neurol*. 2002;15 (3):239–245.
132. Fisher E, Lee JC, Nakamura K, Rudick RA. Gray matter atrophy in multiple sclerosis: a longitudinal study. *Ann Neurol*. 2008;64(3):255–265.
133. Geurts JJ, Barkhof F. Grey matter pathology in multiple sclerosis. *Lancet Neurol*. 2008;7(9):841–851.
134. Geurts JJ, Stys PK, Minagar A, Amor S, Zivadinov R. Gray matter pathology in (chronic) MS: modern views on an early observation. *J Neurol Sci*. 2009;282(1-2):12–20.
135. Roosendaal SD, Schoonheim MM, Hulst HE, et al. Resting state networks change in clinically isolated syndrome. *Brain*. 2010;133(Pt 6):1612–1621.
136. Geurts JJ, Wolswijk G, Bö L, et al. Altered expression patterns of group I and II metabotropic glutamate receptors in multiple sclerosis. *Brain*. 2003;126(Pt 8):1755–1766.
137. Trapp BD, Wujek JR, Criste GA, et al. Evidence for synaptic stripping by cortical microglia. *Glia*. 2007;55 (4):360–368.
138. Mainero C, Caramia F, Pozzilli C, et al. fMRI evidence of brain reorganization during attention and memory tasks in multiple sclerosis. *Neuroimage*. 2004;21(3):858–867.
139. Butz M, van Ooyen A, Wörgötter F. A model for cortical rewiring following deafferentation and focal stroke. *Front Comput Neurosci*. 2009;3:10.
140. Liepert J, Uhde I, Gräf S, Leidner O, Weiller C. Motor cortex plasticity during forced-use therapy in stroke patients: a preliminary study. *J Neurol*. 2001;248(4):315–321.
141. Dromerick AW, Lang CE, Birkenmeier RL, et al. Very early constraint-induced movement during stroke rehabilitation (VECTORS): a single-center RCT. *Neurology*. 2009;73(3):195–201.
142. Murphy TH, Corbett D. Plasticity during stroke recovery: from synapse to behaviour. *Nat Rev Neurosci*. 2009;10(12):861–872.

143. Rösser N, Flöel A. Pharmacological enhancement of motor recovery in subacute and chronic stroke. *NeuroRehabilitation*. 2008;23(1):95–103.
144. Bolognini N, Pascual-Leone A, Fregni F. Using non-invasive brain stimulation to augment motor training-induced plasticity. *J Neuroeng Rehabil*. 2009;6:8.
145. Hummel F, Celnik P, Giraux P, et al. Effects of non-invasive cortical stimulation on skilled motor function in chronic stroke. *Brain*. 2005;128(Pt 3):490–499.
146. Takeuchi N, Izumi S. Combinations of stroke neurorehabilitation to facilitate motor recovery: perspectives on Hebbian plasticity and homeostatic metaplasticity. *Front Hum Neurosci*. 2015;9:349.
147. Dammasch IE, Wagner GP, Wolff JR. Self-stabilization of neuronal networks. I. The compensation algorithm for synaptogenesis. *Biol Cybern*. 1986;54(4-5):211–222.
148. Dammasch IE, Wagner GP, Wolff JR. Self-stabilization of neuronal networks. II. Stability conditions for synaptogenesis. *Biol Cybern*. 1988;58(3):149–158.
149. van Ooyen A, van Pelt J. Activity-dependent outgrowth of neurons and overshoot phenomena in developing neural networks. *J Theor Biol*. 1994;167:27–43.
150. van Ooyen A, van Pelt J, Corner MA. Implications of activity dependent neurite outgrowth for neuronal morphology and network development. *J Theor Biol*. 1995;172(1):63–82.
151. van Ooyen A. Using theoretical models to analyse neural development. *Nat Rev Neurosci*. 2011;12(6):311–326.
152. Bi GQ, Poo MM. Synaptic modifications in cultured hippocampal neurons: dependence on spike timing, synaptic strength, and postsynaptic cell type. *J Neurosci*. 1998;18(24):10464–10472.
153. Hopfield JJ. Neural networks and physical systems with emergent collective computational abilities. *Proc Natl Acad Sci USA*. 1982;79(8):2554–2558.
154. Ho VM, Lee JA, Martin KC. The cell biology of synaptic plasticity. *Science*. 2011;334(6056):623–628.
155. Tetzlaff C, Okujeni S, Egert U, Wörgötter F, Butz M. Self-organized criticality in developing neuronal networks. *PLoS Comput Biol*. 2010;6(12):e1001013.
156. Butz M, Steenbuck ID, van Ooyen A. Homeostatic structural plasticity increases the efficiency of small-world networks. *Front Synaptic Neurosci*. 2014;6:7.
157. van Ooyen A. Activity-dependent neural network development. *Network Comput Neural Systems*. 1994;5:401–423.

SECTION II

HOMEOSTATIC STRUCTURAL PLASTICITY

CHAPTER

5

Network Formation Through Activity-Dependent Neurite Outgrowth: A Review of a Simple Model of Homeostatic Structural Plasticity

Arjen van Ooyen

VU University Amsterdam, Amsterdam, The Netherlands

OUTLINE

The Rewiring Brain
DOI: http://dx.doi.org/10.1016/B978-0-12-803784-3.00005-6

1 INTRODUCTION

Neuronal electrical activity plays an important role in the development of neurons into neural networks. Many processes that determine network connectivity and neuronal function are, on a variety of time scales and levels of organization, modulated by electrical activity. Cell death, cell differentiation, neurite outgrowth and branching, synapse formation and elimination, and ion channel and neurotransmitter expression are all influenced by electrical activity.[1–3] This activity-dependent maturation begins even before the onset of sensory responses, driven by intrinsically generated patterns of electrical activity.[4]

Electrical activity is a prominent factor not only during development but also in adulthood. Alterations in afferent activity in the mature brain, such as those caused by peripheral or central lesions (retinal lesions or stroke, for example), trigger extensive axon and dendrite remodeling and changes in synapse numbers, leading to massive adaptations in network connectivity.[5–9]

This chapter focuses on activity-dependent neurite outgrowth, a form of activity-dependent structural plasticity.[10] Structural plasticity is defined as encompassing all the structural changes that lead to the formation or deletion of synapses, such as neurite elongation and retraction and changes in dendritic spine numbers.[10] Both during development and in adulthood, the elongation, branching, and retraction of neurites (axons and dendrites) are under control of electrical activity.[5,7,11,12] Electrical activity exerts its effect on neurite outgrowth by modifying the level of intracellular calcium, particularly in the growth cone, a specialized structure at the tip of outgrowing neurites. Calcium, which enters the cell through voltage- and ligand-gated calcium channels, is the principal regulator of growth cone motility and neurite outgrowth.[11,13–15] Thus any factor that can change calcium influx, such as action potential firing, may also modulate neurite outgrowth.[14,16]

During development, a high intracellular calcium concentration, caused by membrane depolarization, a high neuronal firing rate, or stimulation by excitatory neurotransmitters, arrests neurite outgrowth or even causes retraction. Conversely, a low calcium concentration, due to a low firing rate, hyperpolarization, or inhibitory neurotransmitters, promotes neurite elongation.[17–21] In the adult nervous system, neurite outgrowth may be similarly affected by neuronal electrical activity. Raising activity causes axons to retract[22] or results in immobility of axonal filopodia,[23] whereas lowering electrical activity triggers axonal outgrowth.[24] Likewise, axons in the visual cortex start sprouting when cortical activity is reduced as a result of a focal retinal lesion.[7] Dendrites may respond to changes in electrical activity in the same way as axons. Increased neuronal activity caused by epileptic seizures results in dendritic retraction,[25] whereas decreased cortical activity induced by whisker ablation[5] or stroke[8] elicits dendritic outgrowth and remodeling.

Thus, both during development and in adulthood, the way in which electrical activity modulates neurite outgrowth seems to contribute to keeping neuronal electrical activity at a particular level (homeostasis). When the electrical activity of a neuron is above a desired value (homeostatic set-point), its neurites retract, breaking-up synaptic connections and so reducing neuronal activity. Conversely, when activity is below the desired value, neurites grow out, making new synaptic connections and so raising the neuron's activity. When the level of electrical activity or intracellular calcium is below a minimum level, however, neurites may also retract. In the hippocampus, e.g., a decrease in electrical activity promotes axonal sprouting, whereas a complete block of activity has no growth-inducing effect and even prevents outgrowth.[26] Thus an optimal window of calcium or electrical activity may be necessary for neurite outgrowth, below and above which neurites retract.[11,21,27–31]

Activity-dependent neurite outgrowth leads to reciprocal interactions between network activity and connectivity. Changes in activity affect neurite outgrowth and thus synaptic connectivity, which in turn changes neuron and network activity. To help unravel the potential consequences of these interactions, we built the computational model that is described in this chapter,[32–35] which is one of the first models of homeostatic structural plasticity.[36] In the model, the neurite extensions of each neuron are represented by a circular neuritic field, which expands when the neuron's electrical activity is below a homeostatic set-point and retracts when the neuron's activity is above the set-point. Neurons connect synaptically when their neuritic fields overlap. In this chapter, I review studies that have employed the model and compare model results with empirical data on network formation and reorganization. I show that the model, simple as it is, has many emergent properties that are relevant for both neural development and adult plasticity.

2 MODEL

2.1 Overview

Our reason for building the neuritic field model was to explore what phenomena could emerge from activity-dependent neurite outgrowth. The model was inspired in part by developing cultures of dissociated cortex cells, in which initially disconnected cells assemble themselves, without external input, into synaptically connected network by neurite outgrowth and synaptogenesis.[37–40] In the model, growing neurons are described as expanding neuritic fields, representing both axons and dendrites, and neurons become synaptically connected when their neuritic fields overlap, with a connection strength proportional to the area of overlap. The outgrowth of each neuron depends on its own level of electrical activity. The neuritic field expands when the neuron's electrical activity is below a set-point, thereby increasing the field overlap and connectivity with other neurons. Conversely, the neuritic field retracts when activity is above the set-point, thereby reducing the connectivity with other neurons. Thus a reciprocal influence exists between electrical activity (fast dynamics) and outgrowth (slow dynamics): electrical activity determines outgrowth, while in turn outgrowth alters connectivity and thus activity. Through these interactions between outgrowth and activity, the initially disconnected neurons organize themselves into a synaptically connected network, guided only by the activity that is generated by the network itself; there is no external input.

2.2 Neuron Model

The electrical activity of the neurons is described by the shunting model.[41] In this model, excitatory inputs drive the membrane potential toward a maximum (the excitatory saturation potential), whereas inhibitory inputs drive the membrane potential toward a minimum (the inhibitory saturation potential). Transformed into dimensionless equations, the model becomes[33]:

$$\frac{dX_i}{dT} = -X_i + (1 - X_i)\sum_{k=1}^{N} W_{ik}F(X_k) - (H + X_i)\sum_{l=1}^{M} W_{il}F(Y_l) \tag{5.1}$$

$$\frac{dY_j}{dT} = -Y_j + (1 - Y_j)\sum_{k=1}^{N} W_{jk}F(X_k) - (H + Y_j)\sum_{l=1}^{M} W_{jl}F(Y_l) \tag{5.2}$$

where X_i and Y_j are the membrane potentials of, respectively, the excitatory neuron i and the inhibitory neuron j, expressed in units of excitatory saturation potential; T is time, expressed in units of membrane time constant; N and M are the number of excitatory and inhibitory neurons, respectively; and H is the ratio of inhibitory to excitatory saturation potential. The terms $(1 - X_i)$ and $(1 - Y_j)$ imply that excitatory inputs drive the membrane potential toward 1. Similarly, the terms $(H + X_i)$ and $(H + Y_j)$ imply that inhibitory inputs drive the membrane potential toward $-H$. The Ws denote the connection strengths (all $W \geq 0$), with k and l the indices of the excitatory and inhibitory driver neurons, respectively; and i and j the indices of the excitatory and inhibitory target neurons, respectively. The term $F(X_i)$ is the firing rate, which is a sigmoidal function of the membrane potential:

$$F(X_i) = \frac{1}{1 + e^{(\theta - X_i)/\alpha}} \tag{5.3}$$

where α determines the steepness of the function and θ represents the firing threshold. The low firing rate when the membrane potential is subthreshold reflects spontaneous neuronal activity.

2.3 Outgrowth and Connectivity

Neurons are located at random positions on a two-dimensional surface. Each neuron has a circular neuritic field, the radius of which is variable. When the fields of neurons i and j overlap, both neurons become connected with a strength

$$W_{ij} = A_{ij}S \tag{5.4}$$

where $A_{ij} = A_{ji}$ is the area of overlap, representing the total number of synapses formed reciprocally between neurons i and j; and S is a constant of proportionality, representing the strength of a single synapse. Synaptic strength may be taken to depend on the type of connection: S^{ee}, S^{ei}, S^{ie}, and S^{ii}, where, e.g., S^{ei} is the inhibitory-to-excitatory synaptic strength.

For both excitatory and inhibitory neurons, the change in neuritic field size depends on the neuron's own firing rate:

$$\frac{dR_i}{dT} = \rho G[F(X_i)] \tag{5.5}$$

where R_i is the radius of the neuritic field of neuron i and ρ determines the rate of outgrowth. The outgrowth function G (Fig. 5.1) is defined as

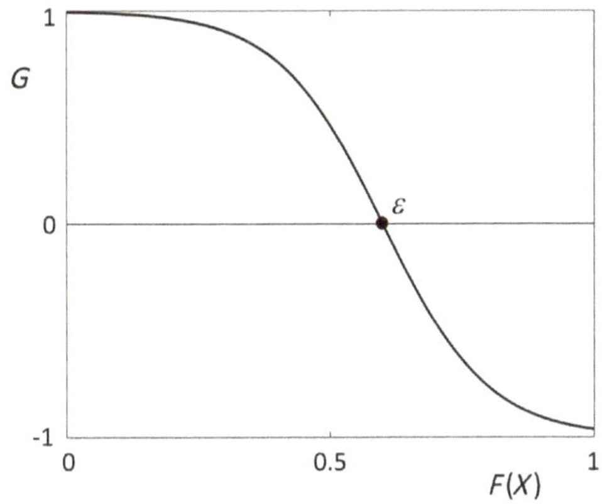

FIGURE 5.1 **Outgrowth function.** The function G (see Eq. 5.6) describes how the outgrowth of a neuritic field depends on the neuron's firing rate $F(X)$. A neuritic field expands when the firing rate is lower than ε and shrinks when the firing rate is higher than ε. At the homeostatic set-point ε, the neuritic field remains constant.

$$G[F(X_i)] = 1 - \frac{2}{1 + e^{[\varepsilon - F(X_i)]/\beta}} \tag{5.6}$$

where ε is the homeostatic set-point, i.e., the value of $F(X_i)$ for which $G = 0$; and β determines the steepness of the function. Depending on the value of $F(X_i)$, a neuritic field grows out $[G > 0 \text{ if } F(X_i) < \varepsilon]$, retracts $[G < 0 \text{ if } F(X_i) > \varepsilon]$, or remains constant $[G = 0 \text{ if } F(X_i) = \varepsilon]$. In biological neurons, the effect of electrical activity on neurite outgrowth is mediated by calcium,[11,13–15] with the concentration of intracellular calcium acting as indicator of firing rate.[42–44]

2.4 Parameters

For the studies conducted by Van Ooyen and co-workers,[32–35,45] the fraction $M/(N + M)$ of inhibitory cells, if present, is in the range of 0.1–0.2.[46] The nominal values of the other parameters are $\rho = 0.0001$, $H = 0.1$, $\theta = 0.5$, $\alpha = 0.1$, $\beta = 0.1$, and $\varepsilon = 0.6$, for both excitatory and inhibitory neurons. Neurite outgrowth is on a time scale of days or weeks,[37,47] so connectivity is quasi-stationary on the time scale of membrane potential dynamics. The value of the outgrowth rate ρ is small enough for the quasi-stationarity approximation to be valid but not so small that it unnecessarily slows down the simulations. The neurons are initialized with no or small neuritic fields and with zero membrane potentials.

3 RESULTS

3.1 Network Assembly, Overshoot, and Homeostasis

Since the neurons are initialized with no or small neuritic fields, most neurons are initially disconnected or organized in small, isolated clusters (Fig. 5.2A). Consequently,

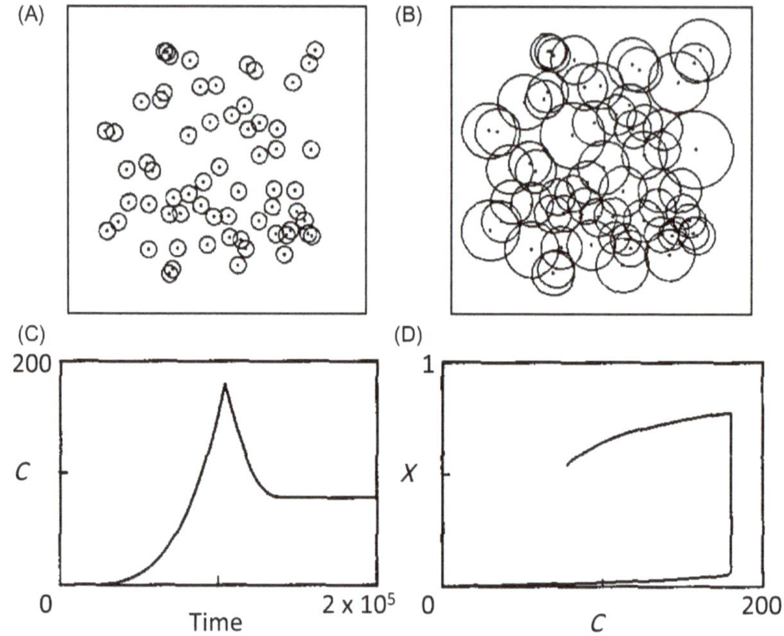

FIGURE 5.2 **Network assembly.** In this example, all cells are excitatory. (A) Early stage of network development. Neuritic fields are small, connectivity is low, and cells have a low level of electrical activity. (B) Network at equilibrium. The electrical activity of all cells is at the homeostatic set-point, and the neuritic field sizes remain constant. (C) Development of network connectivity $C = (1/2)\sum_{i=1,k=1}^{N} A_{ik}$ = total area of overlap (see Eq. 5.4) over time. (D) Network-averaged membrane potential X against network connectivity C. Electrical activity is initially low, so connectivity increases. When connectivity is strong enough, activity abruptly jumps to a much higher level. This level exceeds the homeostatic set-point, so connectivity and activity then decrease until activity is at the homeostatic set-point. Source: *Reproduced with permission from van Ooyen A, van Pelt J. Activity-dependent outgrowth of neurons and overshoot phenomena in developing neural networks. J Theor Biol. 1994;167:27–43. Elsevier.*

neuronal firing rates $F(X_i)$ are below the homeostatic set-point ε, so neuritic fields start expanding. As the neurons grow, they begin to form more and stronger connections, linking neurons together and slowly raising the level of activity in the network. At some degree of connectivity, network activity abruptly jumps to a much higher level (Fig. 5.2D). The activity of the neurons is then so high that $F(X_i) > \varepsilon$. As a consequence, neuritic field size and connectivity start decreasing and activity drops. As neurons adjust the size of their neuritic fields and react to the adjustments of their neighbors, the network eventually reaches a stable equilibrium in which the connectivity between cells is such that for all cells $F(X_i) = \varepsilon$ and neuritic fields and connectivity no longer change (Fig. 5.2B). The neurons thus self-organize, via a transient phase of high connectivity (overshoot) (Fig. 5.2C), into a stable network and achieve network-wide activity homeostasis, even though they can monitor only their own activity level. The neurons adapt to the local cell density, with neurons acquiring small neuritic fields in areas with a high cell density and large fields in areas with a low cell density (Fig. 5.2B). Because of differences in cell density, the developmental course of neuritic field size and electrical activity also varies among cells.[32]

The assembly of initially unconnected model neurons into a connected network strongly resembles development in cultures of dissociated cortex cells, with respect to both activity and connectivity.[37,47–50] The first 3 weeks in vitro show a phase of steady neurite outgrowth and synapse formation,[37,48] with neuron firing and network activity abruptly appearing within a window of a few days[49] and network structure exhibiting a transition from local to global connectivity.[50] In the next week, this is followed by a substantial elimination of synapses until a stable connectivity level is reached.[37,48] Thus, as in the model, there is a transient overproduction (overshoot) of synapses during network development. Overproduction of structural elements followed by pruning is a general feature of nervous system development,[51] in vitro and in vivo, and occurs, e.g., with respect to dendritic length,[52] number of dendrites and axons,[53] and synapse numbers.[54] Variation in dendritic growth among individual cells is also seen in the cerebral cortex, with cells displaying clear, minor, or no overshoot in dendritic length.[52] Neurite outgrowth and synapse elimination in culture are controlled by electrical activity. Similarly to what is observed in the model,[32] chronic blockade of electrical activity enhances neurite outgrowth[55] and prevents subsequent synapse pruning.[37]

3.2 Relationship Between Activity and Connectivity

For a purely excitatory network ($M = 0$), the relationship between activity and connectivity and the changes in activity and connectivity during development can be predicted directly from Eq. (5.1).[32] For a given connectivity matrix **W**, the equilibrium points of X_i are solutions of

$$0 = -X_i + (1 - X_i)\sum_{k=1}^{N} W_{ik}F(X_k) \qquad (5.7)$$

If all cells have the same ε and the variations in X_i are small relative to the average membrane potential $\overline{X}$ of the network, then

$$0 = -\overline{X} + (1 - \overline{X})\overline{W}F(\overline{X}) \qquad (5.8)$$

where $\overline{W}$ is the average connection strength. Rewriting this equation gives

$$\overline{W} = \frac{\overline{X}}{(1 - \overline{X})F(\overline{X})} \qquad 0 \leq \overline{X} < 1 \qquad (5.9)$$

Eq. (5.9), which defines a manifold (Fig. 5.3), provides the equilibrium value(s) of $\overline{X}$ for a given, fixed value of $\overline{W}$. Equilibrium states on branch CD of the manifold are unstable with respect to $\overline{X}$; equilibrium states on branches ABC and DF are stable. Because changes in $\overline{W}$ are slow, being caused by neuritic field outgrowth and retraction, $\overline{W}$ can be considered quasi-stationary on the time scale of membrane potential dynamics. That is, in the time that $\overline{X}$ relaxes to its equilibrium value, $\overline{W}$ hardly changes. In other words, at any given value for $\overline{W}$, $\overline{X}$ is at its equilibrium value. The slow evolution of $\overline{X}$, i.e., the changes in $\overline{X}$ that are brought about by changes in $\overline{W}$, therefore take place along the manifold.

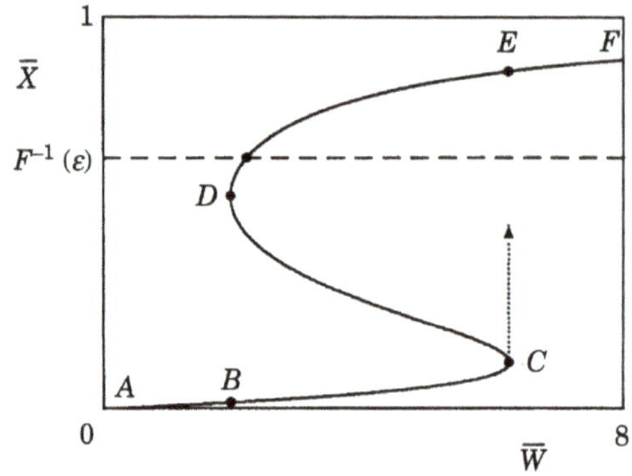

FIGURE 5.3 **Relationship between activity and connectivity.** The manifold (see Eq. 5.9) defines the equilibrium value(s) of the network-averaged membrane potential $\overline{X}$ for a given, fixed value of the network-averaged connectivity $\overline{W}$ in a purely excitatory network. Equilibrium values on branch CD are unstable with respect to $\overline{X}$; equilibrium values on branches ABC and DF are stable. The intersection point with the line $\overline{X} = F^{-1}(\varepsilon)$ (F^{-1} is the inverse of the firing rate function, see Eq. 5.3) is the equilibrium state of the whole system, at which $\overline{W}$ remains constant. See further text. Source: *Reproduced with permission from van Ooyen A, van Pelt J, Corner MA. Implications of activity dependent neurite outgrowth for neuronal morphology and network development. J Theor Biol. 1995;172 (1):63–82. Elsevier.*

If for all cells $F(X_i) = \varepsilon$, the neuritic fields, and therefore $\overline{W}$, remain constant (see Eqs. 5.5 and 5.6). Thus, at the intersection point with the line $\overline{X} = F^{-1}(\varepsilon)$ (F^{-1} is the inverse of F), $\overline{W}$ remains constant; above and below that line, it decreases and increases, respectively. Suppose that the intersection point is on branch DE (Fig. 5.3). During development, connectivity and activity are initially low, so $\overline{W}$ increases, and $\overline{X}$ follows the branch ABC until it reaches C, at which point it jumps to branch DE. However, $\overline{X}$ is then so high that the neuritic fields begin to retract and $\overline{W}$ to decrease until $\overline{X}$, moving along branch DE, reaches the intersection point. Thus, in order to arrive at an intersection point on branch DE, a developing network has to go through a phase in which connectivity is higher than in the final situation (overshoot).

A network that increases its connectivity starting from a quiescent state follows another trajectory (ABC) than a network that decreases its connectivity starting from an active state (FED), a phenomenon called hysteresis. In other words, a higher connectivity is needed to trigger activity in a quiescent network than to sustain it once the network has been activated. The existence of the hysteresis loop (i.e., the S-shaped curve of Fig. 5.3) hinges upon the firing rate function F having a (soft or hard) firing threshold and low spontaneous activity for subthreshold membrane potentials.[32] Interestingly, a very similar hysteresis loop between activity and connectivity was found in a detailed spiking and conductance-based large-scale network model of the human cortex.[56]

3.3 Slow Fluctuations in Activity

The level of electrical activity above which neurites retract may be different for different types of neurons,[11,57] which, translated to the model, means that there may be variation among cells in ε. When such variation is present, the network can exhibit complex periodic behavior, with individual cells displaying slow oscillations in electrical activity that differ in frequency and amplitude.[34] Note that these oscillations are caused by the outgrowth and retraction of neuritic fields and thus occur on a time scale of days. Slow fluctuations in the electrical activity of individual cells, with periods of increased activity sometimes lasting as long as several days, have also been observed in developing cultures of dissociated cortex cells.[58]

3.4 Effect of Inhibition on Overshoot

In networks that contain both excitatory and inhibitory cells, overshoot still takes place and can even be enhanced.[33] To counterbalance inhibition, the cells need a higher excitatory connectivity, and thus a longer developmental time, to reach the activity level at which connectivity starts declining. In developing cultures of dissociated cortex cells, blockade of inhibitory GABAergic transmission indeed advances synapse elimination.[48]

Furthermore, in mixed networks, as opposed to purely excitatory networks, the decline in overall connectivity can be considerably delayed relative to the onset of network activity.[33] In parts of the network with many inhibitory cells, excitatory cells can still be growing out, while in parts with fewer inhibitory cells they are already retracting. Cells that receive inhibition become activated later and will retract later than cells without inhibition. These differences in growth mean that the average connectivity can still increase markedly after the onset of network activity. A delayed decline in connectivity relative to the onset of network activity is also observed in developing cultures.[37]

3.5 Multiple Equilibrium States

In contrast with purely excitatory networks, mixed networks do not under all initial conditions go to the same global end state with respect to electrical activity and connectivity. In a network with moderate inhibition and an initial connectivity that is higher than a certain level, connectivity will not be pruned but, instead, will continue to increase[33] (Fig. 5.4). As the excitatory connectivity increases, so does the inhibitory connectivity, which keeps activity low and stimulates further outgrowth. Moreover, if inhibitory cell density is high, inhibitory cells form connections among themselves, inhibiting each other's activity but stimulating each other's outgrowth, leading to more inhibition, lower activity, and further outgrowth of cells.

In a simplified model, it was shown that mixed networks have two equilibrium states: an attractor A at a relatively low level of connectivity and an attractor B at high connectivity[35] (Fig. 5.5A and B). In attractor A, the cells have a constant level of electrical activity, whereas in attractor B they exhibit fast oscillations in activity (Fig. 5.5C) (resulting from interactions between excitation and inhibition,[59,60] not from neuritic field outgrowth and retraction). Although activity oscillates, the time-averaged activity in attractor B is at the set-point level, so there are no net changes in neuritic field sizes or connectivity

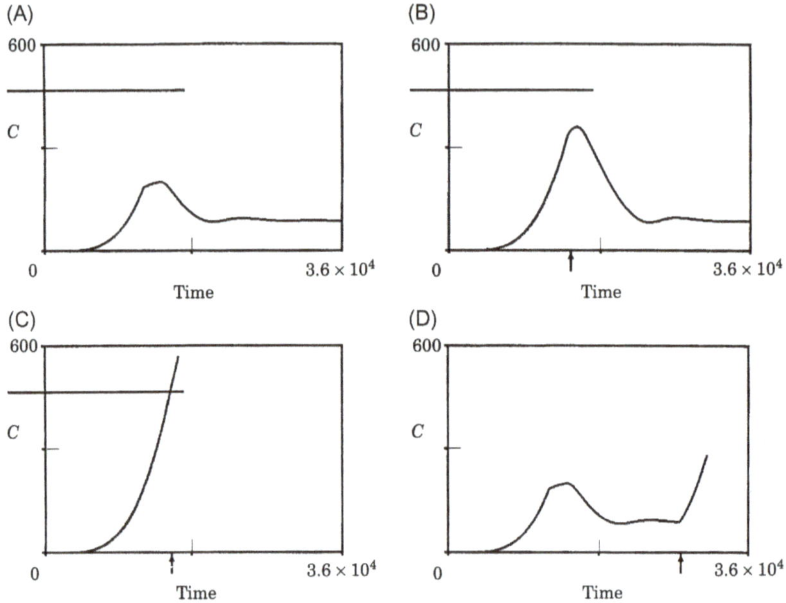

FIGURE 5.4 **Failure of connectivity pruning.** (A) Normal development. Total connectivity $C = \sum_{p=1,q=1}^{N+M} A_{pq}$ (see Eq. 5.4). (B) Electrical activity is blocked until the time indicated by the *arrow*. During the activity block, connectivity increases. Once activity is allowed to return, connectivity is pruned back to its normal equilibrium value. (C) If electrical activity is blocked for a longer period so that connectivity becomes higher than the critical value indicated by the *horizontal line*, connectivity is not pruned but continues to increase once activity returns. (D) Activity is blocked in a normally developed network at the time indicated by the *arrow*. Source: *Reproduced with permission from van Ooyen A, van Pelt J, Corner MA. Implications of activity dependent neurite outgrowth for neuronal morphology and network development. J Theor Biol. 1995;172(1):63–82. Elsevier.*

(Fig. 5.5D). Attractor B may be interpreted as a pathological state, with epileptic-like activity. During normal development, with a low initial level of connectivity, the network ends up, via an overshoot in connectivity, in attractor A (Fig. 5.5B). However, with high initial connectivity, which could be brought about by temporarily blocking electrical activity, the network goes to attractor B (Fig. 5.5B). Furthermore, the higher the level of inhibition during development (number of inhibitory cells, strength of inhibitory synapses), the more likely the network is to end up in attractor B.

The latter finding may provide an explanation for the experimental observation that hypoxic-ischemic encephalopathy in rat pups, which causes a preferential loss of excitatory cells and synapses (and thus a relatively high level of inhibition), can lead to epileptiform activity later on in adulthood.[61,62] Similarly, enhanced rather than reduced tonic $GABA_A$ inhibition is found in typical absence epilepsy.[63] The presence of two stable attractors may also explain that following chronic blockade of electrical activity in developing cultures of dissociated cortex cells (resulting in a high density of synapses), synapse elimination fails to occur once the block is removed and activity returns to control levels.[55,64]

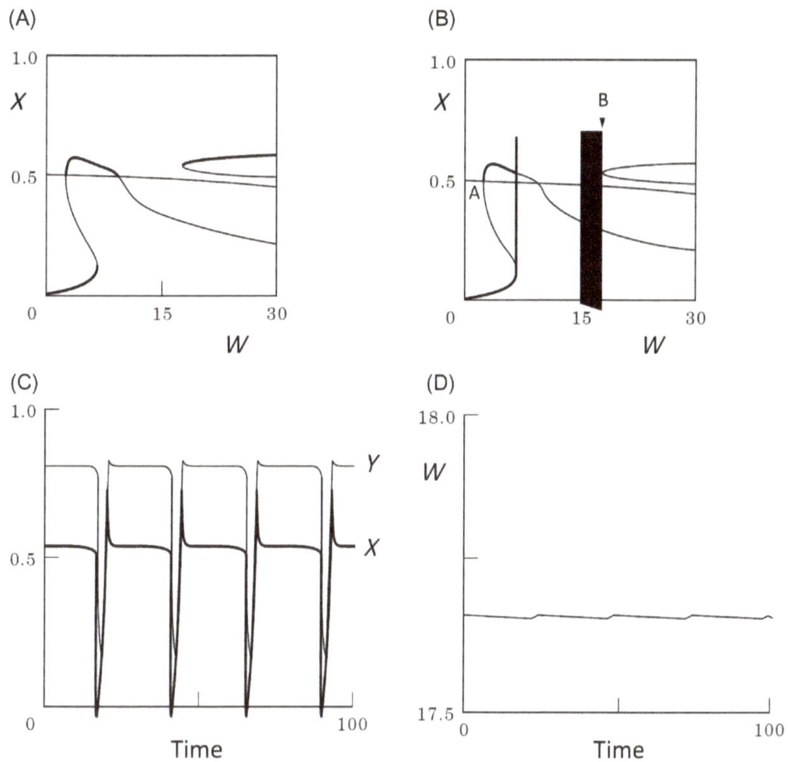

FIGURE 5.5 **Two attractors.** (A) The manifold shows the equilibrium value(s) of the excitatory activity X for a given, fixed value of the connectivity W in a simplified network with excitation and inhibition[35] (compare with Fig. 5.3). The *bold lines* indicate stable and the *thin lines* unstable points with respect to X. The *thin, nearly horizontal line* denotes points where W is at equilibrium. (B) The *bold lines* now depict development over time, not stability as in (A). The whole system has two attractors. With a low initial level of connectivity ($W = 0$), the network ends up, via an overshoot in connectivity, in a point attractor (marked with an A), in which W and X are constant. With a high initial level of connectivity ($W = 15$), the network ends up, via oscillations in X and a slow increase in W (since the changes in W are much slower than those in X, no separate oscillations are visible), in a limit cycle attractor (marked with a B and an *arrow*), in which W is nearly constant but X oscillates. (C) Excitatory activity X and inhibitory activity Y in the limit cycle attractor. (D) Connectivity W in the limit cycle attractor. Source: *Reproduced with permission from van Oss C, van Ooyen A. Effects of inhibition on neural network development through activity-dependent neurite outgrowth. J Theor Biol. 1997;185(2):263–280. Elsevier.*

3.6 Differentiation Between Excitatory and Inhibitory Cells

Although in the model there are no intrinsic differences between excitatory and inhibitory cells with respect to outgrowth rules, they nevertheless differentiate, with the neuritic fields of inhibitory cells becoming smaller than those of excitatory cells[33] (Fig. 5.6A–C). The mechanism is as follows. Each cell will attain a neuritic field size for which the input from overlapping cells is such that $F(X_i) = \varepsilon$. An excitatory cell that receives inhibition therefore needs more excitatory input than a cell that is not inhibited, and thus grows a larger neuritic field. As a result, each inhibitory cell will become

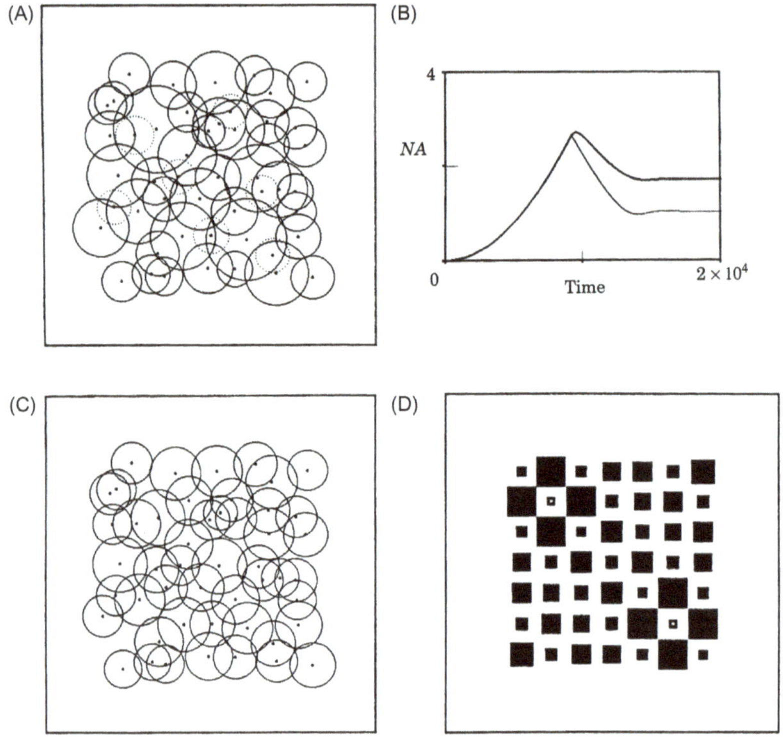

FIGURE 5.6 **Influence of inhibitory cells.** (A) Mature network showing the neuritic fields of excitatory cells (*continuous lines*) and inhibitory cells (*dotted lines*). (B) The average neuritic field area of excitatory cells (*thick line*) and that of inhibitory cells (*thin line*) over time. Cell size starts differentiating once the network has become active. (C) Same placing of cells as in (A), but all former inhibitory cells are now excitatory. (D) Cells on grid positions. The diameter of a square is proportional to the area of the neuritic field (scaled to the maximum area found in the network). Cells with a *white dot* in the middle are inhibitory. All networks have torus boundary conditions. Source: *Reproduced with permission from van Ooyen A, van Pelt J, Corner MA. Implications of activity dependent neurite outgrowth for neuronal morphology and network development. J Theor Biol. 1995;172(1):63–82. Elsevier.*

surrounded by large excitatory cells, whereas the inhibitory cell itself can remain small because a small neuritic field already yields sufficient overlap with its large surrounding cells. In other words, an inhibitory cell becomes small by increasing the size of its direct neighbors. Differences in cell size emerge irrespective of initial size: inhibitory cells may start at the same size as excitatory cells or may be introduced in a well-advanced or mature excitatory network.

In the cerebral cortex, the dendrites and axons of inhibitory neurons are, on the whole, indeed smaller than those of excitatory neurons.[65] Pyramidal cells, which are excitatory, have large apical dendrites and long axons, and often form long-range connections.[66] Nonpyramidal cells, which are generally smaller, are mostly inhibitory and mainly make local connections.[46]

3.7 Patchy Connectivity Structure

By inducing outgrowth, inhibitory cells enhance the degree of connectivity among their neighboring cells.[33] To obtain sufficient input, excitatory cells that receive inhibition need to grow larger neuritic fields and hence become more tightly connected among each other than uninhibited cells. In other words, inhibitory cells impose a kind of clustered connectivity structure onto the network (Fig. 5.6D), somewhat resembling the patchy organization of lateral connections observed in the neocortex.[67,68]

Guided by intrinsically generated electrical activity,[69] the cortex self-organizes from a diffuse network into a network with local neuronal clusters of corticocortical connections.[70] Inhibitory neurons may help shape these patchy connectivity patterns.[71] A conceptual model has been proposed in which basket interneurons (inhibitory cells) veto pyramidal neurons (excitatory cells) to form connections within circular domains.[71] The model requires pyramidal axons to step over a zone of inhibition, which indeed is the effect of inhibition in the neuritic field model, as inhibition favors outgrowth. In the same vein, double bouquet cells, another type of inhibitory cells, have been suggested to impose a microcolumnar organization upon the cerebral cortex.[72]

In the neuritic field model, when inhibitory cells are in the position that they can connect among themselves, they inhibit each other's activity but stimulate each other's outgrowth,[33] leading to the formation of long-range inhibitory connections, as found in the cortex.[73,74] Thus the distribution of inhibitory cells has a large impact on the connectivity structure (neuronal clusters, long-range inhibition) that arises in the network. Excitatory and inhibitory synaptic strength (S; see Eq. 5.4) also influence network structure.[32] When synaptic strength is low, cells develop into a single interconnected network. When synaptic strength is high, separate, mutually disconnected subnetworks emerge, because contact with fewer cells is then sufficient to reach set-point activity.

3.8 Self-Repair of Connectivity After Lesions

After cell loss (which, in the brain, could be caused by stroke or neurodegeneration), the remaining cells, especially those in the neighborhood of the deleted cells, lose connections and consequently undergo a drop in activity [$F(X_i) < \varepsilon$]. This triggers neuritic field outgrowth and formation of new connections until activity is restored [$F(X_i) = \varepsilon$]. The more cells are deleted, the larger the compensatory increase in neuritic field size, and the bigger the average neuritic field size in the network.[33]

Some brain regions lose neurons through old age, and in these regions there is indeed a steady increase of dendritic extent per neuron.[75] Accordingly, no increase in dendritic extent occurs in brain regions that do not lose neurons with age.[76] Stroke also leads to cell loss and deprives neurons surrounding the stroke site of horizontal input.[8,9] Similarly to what is observed in the model, the deprived neurons remodel their dendritic arbors, withdrawing branches from the infarct and extending branches toward intact areas.[8,77] In this way, cells may form new connections and restore their level of activity. In agreement with such a process of compensatory sprouting, dendritic remodeling is inversely correlated with distance from the infarct.[8] The farther a cell is from the infarct, the less it will be affected by a loss of input, and the less remodeling is needed.

3.9 Neurogenesis-Induced Network Reorganization

The model network can also cope with insertion of new cells (neurogenesis) in a mature network, in which electrical activity is at the homeostatic set-point. The newly added cells grow out and become electrically active as a result of the input they receive from the network. At the same time, the activity of the preexisting cells, especially those surrounding the newly inserted cells, increases above the set-point because of the extra input they receive from the new cells. These changes in activity trigger adjustments in neuritic field size and connectivity of both new and preexisting cells, until the activity of all cells has returned to the homeostatic set-point. In the adult brain, insertion of newborn cells takes place in the dentate gyrus of the hippocampus,[78] and, as in the model, induces structural adaptations in preexisting circuits as cells become synaptically integrated.[79,80]

3.10 Neuronal Death During Development

Experimental findings have shown that neurite retraction occurs not only when activity is above a set-point value but also when it is below a certain minimum value.[11,26] If a growth function is used in which neurite retraction also takes place when neuronal activity is very low, the initial activity (connectivity) should be high enough to stimulate outgrowth.[45] However, even when initial activity is high enough, some cells, after transiently growing out, may still lose all of their connections and become silent (such cells are effectively "dead"), resembling the neuronal death seen during normal neural development.[81] Interestingly, experimental studies showed that neurons are especially vulnerable to cell death when their intracellular calcium concentration is substantially below resting level,[82] possibly as a result of diminished calcium influx due to low electrical activity.

3.11 Differentiation of Intrinsic Properties

In addition to neurite outgrowth, the maximal conductances of ion channels are regulated by activity.[83,84] Activity modifies the maximal conductances of ion channels in such a way that excitability is decreased when neuronal activity is high and increased when neuronal activity is low. In a model that extends the neuritic field model with activity-regulated maximal conductances, and that uses spiking neurons rather than firing rates (see Eq. 5.3), cells self-assemble into a coupled network exhibiting a rich repertoire of activity patterns.[85] Although all the cells are governed by the same activity-dependent rules for growth and conductance change, they nevertheless differentiate, with cells acquiring different firing patterns and distinct mixtures of conductances. Thus the interplay between outgrowth and changes in intrinsic properties leads to the formation of a highly differentiated network.

3.12 Self-Organized Criticality

Experiments have uncovered an intriguing dynamical state characterized by so-called neuronal avalanches in a variety of neural systems, including acute and cultured cortical slices,[86,87] developing cultures of dissociated cortex cells,[40] the developing retina,[88] and the neocortex in vivo.[89] Neuronal avalanches are spontaneous bursts of activity that have power-law size and duration distributions.[86,87] The number of events of a given size (e.g., in terms of number of electrodes on which activity is recorded) falls as the size to the power $-3/2$, and the number of events of a given duration falls as the duration to the power -2. Power laws typically emerge in systems when they are close to a transition in behavior (i.e., when they are critical). Simple models have shown that the experimentally observed power laws can arise if connectivity is such that every neuron that fires an action potential causes, on average, one other neuron to fire.[90] With this connectivity, the network is critical in the sense that activity neither dies out nor increases over time.

How do networks develop and maintain patterns of connectivity that satisfy this criticality condition? In a version of the neuritic field model in which neuronal activity is generated by a Poisson spiking model,[44] the activity bursts in the network at equilibrium, when neuritic field sizes and connectivity are stable, exhibit the same power-law dependencies for size (as measured in number of action potentials) and duration as observed experimentally.[86,87] The property of the model that neurons grow out when activity is low and withdraw when activity is high forces the network to find a middle ground between all-to-all connectivity (producing excessive activity) and local connectivity (producing insufficient activity), and this is what provides the potential for critical, power-law behavior.[44] The self-organization of the model into a critical state with neuronal avalanches has also been demonstrated analytically.[91] Moreover, the neuritic field model can account for the different activity stages through which developing cultures of dissociated cortex cells traverse until they reach criticality.[92] In the mature brain, criticality may enhance information coding and transmission.[93–95]

3.13 Retinal Mosaics

The vertebrate retina contains several cell types distributed in multiple layers, with each cell type typically occupying only a single layer.[96] Furthermore, cells of a given type are positioned semiregularly within a layer, forming what is known as a retinal mosaic, so named because of the way the cell bodies and their dendrites tile the surface. Retinal mosaics ensure that the visual field is uniformly sampled and that there are no holes in visual space.

One of the mechanisms underlying mosaic formation is lateral migration.[96,97] During development, retinal cells arrive randomly spaced in their destination layer, but then start to move laterally within the layer, possibly as a result of neurite interactions as cells grow out,[98–100] until they form a regular mosaic. This mechanism was investigated in a version of the neuritic field model that was extended with an equation for lateral movement.[101] In the extended model, cell bodies repel each other in proportion to their neuritic overlap:

$$\frac{d\mathbf{C}_i}{dT} = \eta \sum_{k=1}^{N} u(\mathbf{C}_i - \mathbf{C}_k) W_{ik} \tag{5.10}$$

where $\mathbf{C}_i$ is a vector denoting the two-dimensional position of cell i; η determines the rate of movement; N is the total number of cells; W_{ik} is the connection strength from cell k to

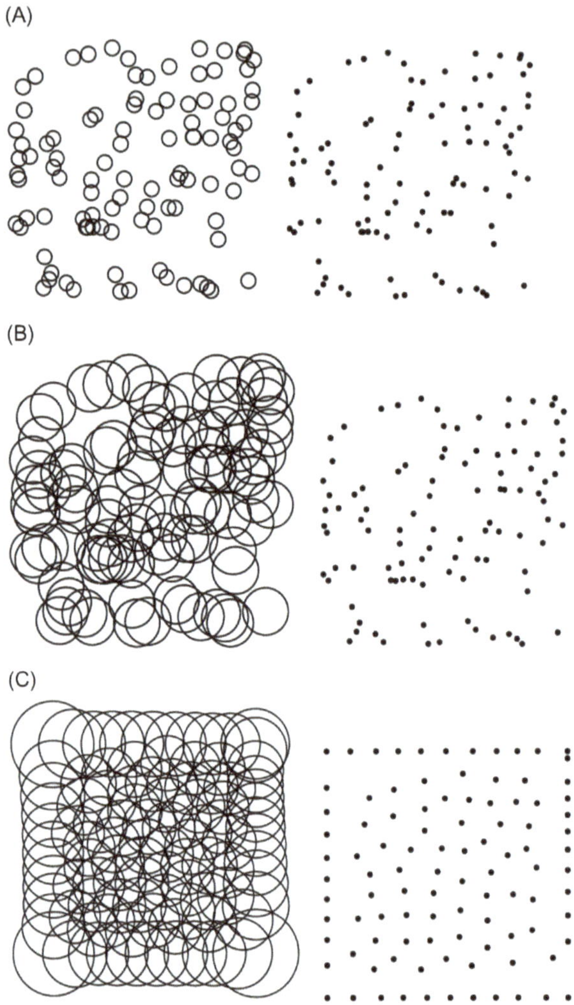

FIGURE 5.7 **Retinal mosaic development.** On the left, the neuritic fields; on the right, the positions of the cell bodies. The same network is shown at three stages of development. All cells are excitatory. (A) Neuritic fields have grown out, but cells have not yet moved, so the mosaic is still random. (B) Neuritic fields have grown enough to cause some small cell movement and a slight increase in mosaic regularity. (C) Neuritic sizes are uniform and the mosaic is highly regular. Cells at the border become larger because the network was not simulated with torus boundary conditions. Source: *Reproduced with permission from Eglen SJ, van Ooyen A, Willshaw DJ. Lateral cell movement driven by dendritic interactions is sufficient to form retinal mosaics. Network. 2000;11(1):103–118. Taylor & Francis Ltd., www.tandfonline.com.*

cell i; and $u(\mathbf{V})$ is the vector $\mathbf{V}$ normalized to unit length, except that $u(\mathbf{0}) = \mathbf{0}$. Elements of $\mathbf{C}_i$ are bounded to keep each cell body within the surface. All cells are excitatory.

Initially, cells are randomly distributed across the surface (Fig. 5.7A). As neuritic fields grow out, cells gradually begin to repel each other, slowly settling into a regular

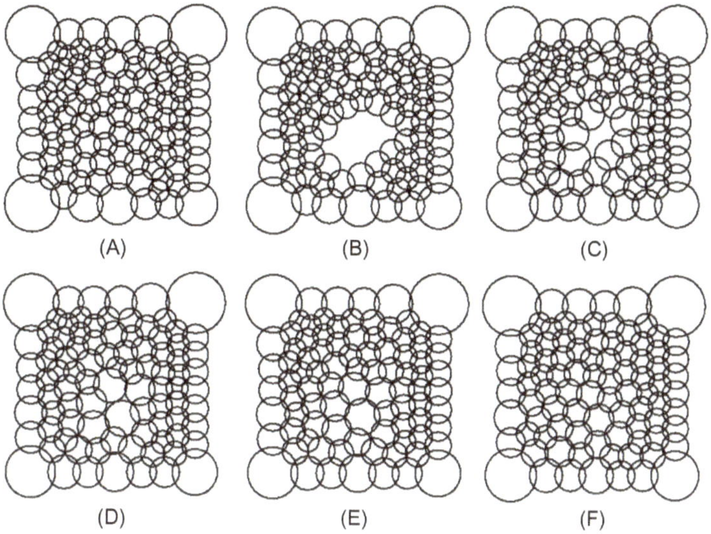

FIGURE 5.8 **Lesioning retinal cells.** After the network has developed (A), 10 central cells are removed (B). The neighboring cells move into this vacated area and expand to cover up the hole (C and D). All the cells in the network increase their neuritic fields a little in order to adapt to the new situation (E and F). All cells are excitatory. Source: *Still frames reproduced with permission from www.anc.ed.ac.uk/~stephen/mosaics/index.html, Stephen Eglen.*

hexagonal-like mosaic layout (Fig. 5.7B and C), in which cell position and neuritic field size remain stationary.[101] The amount of cell movement is small and in line with what is observed experimentally.[98] The regularity of the mosaic, however, is higher than observed experimentally, but reducing the precision with which cells detect neuritic overlap can remedy this.[101] The network can cope with changes in the number of cells.[96] If cells are added, the network automatically reorganizes, with cells getting closer together but maintaining an equally regular pattern. Similarly, if cells are deleted (i.e., a local lesion), the neighboring cells move in and expand to occupy the empty area, restoring mosaic regularity (Fig. 5.8), in agreement with experimental findings.[102]

Cell mosaics are not restricted to the retina and have, e.g., also been discovered in rat cerebellum and avian tectum.[103] Lateral cell movement due to neurite interactions may also in these cases be a possible mechanism for mosaic formation.

3.14 Developmental Changes in Burst Patterns

A computational study into the development of firing patterns applied the neuritic field model to simulate the self-assembly of 10,000 neurons, both excitatory and inhibitory, into a connected network.[104] Instead of the firing rate model (see Eq. 5.3), an integrate-and-fire model[105] was used, as well as a model of synaptic facilitation and depression.[106,107] Synaptic facilitation is the short-term enhancement of synaptic strength due to the prolonged elevation of presynaptic calcium levels following synaptic activity. Synaptic depression is the short-term decrease in synaptic strength as a result of the depletion of neurotransmitter vesicles after synaptic activity. The focus of the study was on bursts of activity, where bursts are defined as periods of activity in which the network-average per neuron firing rate is higher than a given threshold. As connectivity developed in the model, bursts initially had a low firing rate and long duration, but then gradually evolved into more intense and shorter ones later on,[104] precisely as had been observed in developing cultures of dissociated cortex cells.[58] Thus the experimentally observed change in burst shape during development can be explained by activity-dependent neurite outgrowth and dynamical synapses, without the need for such features as LTP/LTD, detailed neuronal morphology, or specific patterns of connectivity.

3.15 Developmental Transitions in Cognition

An important question in psychology is how developmental transitions in cognition can arise even though brain growth may be gradual. Children show stages in development during which their cognitive abilities remain more or less stable,[108] interspersed with rapid transitions when cognition changes discontinuously and qualitatively new knowledge is acquired. In the network governed by the neuritic field model, electrical activity shows such a discontinuous change as connectivity gradually increases. Beyond a critical connectivity value, the network suddenly jumps from a state of low activity to a state of high activity (Figs. 5.2D and 5.3). The implications of this growth-related bifurcation for cognition have been studied[109,110] in Exact ART,[111] a neural network model based on the Adaptive Resonance Theory (ART).[112] ART specifies network structures and learning rules for unsupervised classification of input patterns. In an ART network, which uses the shunting model as neuron model (Eqs. 5.1 and 5.2), input patterns that are alike according to some criterion are classified into one category. The network constructs categories during the presentation of input patterns and remains adaptable to unknown patterns. Increasing the range and strength of excitatory and inhibitory connections in Exact ART revealed that the network exhibits qualitatively different dynamical regimes affecting the way learned categories are represented.[109,110] At low connectivity, representations of learned categories are mainly local, whereas at high connectivity they are mostly distributed. A change in representation is an important condition for learning qualitatively new knowledge. Thus a gradual, continuous developmental change in connectivity could lead to a relatively rapid, discontinuous change in cognitive abilities.

4 DISCUSSION

By adapting their neuritic field size, neurons in the model endeavor to reach and maintain a desired level (homeostatic set-point) of electrical activity. Neuritic field size expands when the neuron's electrical activity is below the set-point and shrinks when the neuron's electrical activity exceeds the set-point. By increasing its neuritic field size, the neuron enhances its afferent connectivity and potentially also its level of electrical activity; by decreasing its neuritic field size, the neuron achieves the opposite. In this way, each neuron is pursuing its own goal of attaining a set-point level of activity, with the resulting network connectivity arising from many neurons simultaneously seeking to support their activity need.[1] This local striving for a desired level of electrical activity leads to many, seemingly unrelated, emergent properties: global homeostasis of network activity, even though neurons can monitor only their own activity level; a transient phase of high network connectivity (overshoot); developmental neuronal death; self-organized criticality; patchy network connectivity; retinal mosaics; formation of a normal or a pathological network (epileptic-like oscillatory electrical activity) depending on early level of inhibition, with too much inhibition leading to an abnormal mature network; differentiation in neuritic size between excitatory and inhibitory cells; differentiation in intrinsic firing patterns; developmental changes in network bursts; developmental transitions in cognition; self-repair of connectivity and restoration of activity following cell loss (stroke, neurodegeneration); and reorganization of connectivity during integration of newborn cells (adult neurogenesis).

The model shows the power of bottom-up modeling. In a bottom-up approach, instead of starting with a particular goal or predescribed behavior that the model should reproduce, one explores the emergent properties and phenomena resulting from the interactions among the constituents of the model.[3,113,114] Without this type of modeling, it would have been difficult to surmise that all of the above listed findings could have been different manifestations of the same underlying process, namely homeostatic activity-dependent neurite outgrowth.

The results of the model do not depend critically on the neuritic fields being described as circles. Other approaches for characterizing the neuritic extent of a neuron give essentially the same results.[32] Also, modeling axonal and dendritic fields separately, in which case connectivity among excitatory cells and among inhibitory cells is no longer necessarily symmetrical, does not alter the main findings.[33] Even assuming that the length of neurites is constant and that only the numbers of dendritic spines and axonal boutons, the pre- and postsynaptic parts of synapses, change in a homeostatic manner, would produce similar results, with spine and bouton number instead of neurite extent characterizing the size of a neuron (see further below). The precise form of the outgrowth function is also not essential, as long as high activity induces neuritic field shrinkage and low activity induces neuritic field expansion. The results are also robust when, instead of the shunting model, another model is used to describe neuronal electrical activity. For example, in a detailed spiking and conductance-based model of the human cortex,[56] the same type of hysteresis loop between activity and connectivity was found as in the neuritic field model,[32] implying that the results that rely on hysteresis, such as overshoot, can also be observed in more detailed models.

4.1 Future Experimental Studies

To determine the relationship between neuronal electrical activity and rate of neurite outgrowth (growth function, see Eq. 5.6) more quantitatively, one could continuously stimulate neurons at different frequencies and measure their morphological response. Alternatively, since intracellular calcium mediates the effect of electrical activity on neurite outgrowth,[11,13–15] one could challenge neurons with different rises in intracellular calcium by means of calcium uncaging.[13] This may also reveal differences in growth response between axons and dendrites or between different cell types (excitatory, inhibitory). For example, different types of neurons may have their homeostatic set-point at different levels of electrical activity or intracellular calcium. Further open questions concern the way the set-point is encoded biologically and how cells measure the deviation between the actual level of activity or calcium and the set-point level.[115] Lastly, in interpreting the results from experimental studies, one should realize that with an optimal calcium window for neurite outgrowth, an increase in calcium or electrical activity could lead to either stimulation or suppression of outgrowth, depending on the resting value of calcium.

To correlate activity history with neuronal morphology, one could simultaneously monitor electrical activity and morphology in developing networks in vitro or in vivo, using advanced techniques such as optogenetics, two-photon microscopy, calcium imaging, and multielectrode arrays, which enable measuring activity and neurite morphology over extended periods of time.[7,116–119] With these techniques, many specific questions can be investigated to test the model results. Do the slow fluctuations in electrical activity as observed in developing cultures of dissociated cortex cells[58] correlate with periodic changes in neurite outgrowth, as in the model[34]? Does synaptic input from inhibitory cells increase the dendritic extent of excitatory cells, as the model predicts[33]? Similarly, do excitatory cells in networks in which inhibition is blocked show less variation in dendritic extent than excitatory cells in networks with intact inhibition? Does cell loss cause a drop in activity and trigger a compensatory increase in the axonal and dendritic extents of the remaining cells, as seen in the model[33]?

In the model, external input, from sources outside the network, changes the distribution of neuritic field sizes.[32,33] If a cell receives external excitatory input, it needs less input from other cells in the network to attain the set-point level of activity and hence can do with a smaller neuritic field. Does external excitatory input indeed give rise to smaller dendrites in developing or mature networks? Does continual external input prevent networks from reaching critical connectivity? A network that has grown in the presence of external input is not expected to be in a critical state when the external stimulation is removed. Likewise, does external stimulation of the developing retina influence mosaic regularity, since retinal cells that receive external excitatory input would grow out less and repel other cells less than retinal cells without external input?

4.2 Future Modeling Studies

In the model, no distinction is made between axons and dendrites. One way to do this is to provide each neuron with two neuritic fields, one for the axonal extensions and one for the dendritic extensions.[92] If wanted, the axonal and dendritic fields can be given different growth parameters or functions. Simulations with separate axonal and dendritic fields show that the results presented here do not critically depend on neurons having a single neuritic field.[33]

A further step toward a more detailed description of growing neurons is to employ, not circular neuritic fields, but spatial functions describing the density of the neuron's axonal and dendritic branches. The connectivity between two neurons can then be calculated by estimating the number of synapses formed in the area of spatial overlap between the axonal and dendritic density fields.[120]

Instead of using fields, one can model the actual morphology of elongating and branching neurites, as in a compartmental modeling approach[121–123] or as in the program NETMORPH.[124,125] In NETMORPH, a simulation tool for building synaptically connected networks with realistic neuron morphologies, axonal and dendritic morphologies are created by stochastic rules for the behavior of individual growth cones, the structures at the terminal tips of neurites that mediate elongation and branching. The growth of terminal branches can be made dependent on the time-averaged local membrane potential at the growth cone. This membrane potential will be influenced by both the firing rate at the soma and the local synaptic potentials at the neurite.

Using neuritic fields is one way to abstract away from detailed neuronal morphology. Another way is to describe each neuron as having separate axonal synaptic elements (representing boutons) and dendritic synaptic elements (representing spines), which can combine to form synapses.[126,127] In the synaptic element model, with growth rules partly inspired by the neuritic field model, neurons generate new elements when neuronal electrical activity is below the set-point, and delete elements, including those bound in synapses, when activity is above the set-point or below a certain minimum level. A growth function with a minimum and a set-point level was also used in the neuritic field model.[45]

A promising further line of modeling work is to combine activity-dependent neurite outgrowth with other activity-dependent properties, such as intrinsic ion channel conductances, which are also regulated in a homeostatic manner.[84,85,128] Even early developmental processes, such as differentiation of cell identity, are partly governed by electrical activity.[129] For example, during development, GABA-induced depolarization (GABA is excitatory early on) is necessary for proper excitatory synapse formation and dendritic development,[130] a mechanism that may contribute to balancing excitation and inhibition. Electrical activity influences neuronal identity not only during development but also in adulthood: the relative proportions of different types of inhibitory interneurons are continuously adjusted in response to neuronal activity.[131] In the neuritic field model, the identity of cells, i.e., whether they are excitatory or inhibitory, is fixed and has to be set before-hand. It would be interesting to make this aspect of network development also dependent on electrical activity and part of the self-organization process. One can envisage a model of self-organization in which neurite outgrowth, intrinsic cellular conductances, and ratio of inhibitory to excitatory cells are all controlled by electrical activity in a homeostatic manner. Such a model may also include homeostatic and Hebbian synaptic plasticity.[83,132] The interactions among these different types of activity-dependent processes, operating at different time and spatial scales, may generate extremely rich dynamics.[133,134]

The possible roles of activity-dependent neurite outgrowth, or structural plasticity in general, in cognitive processes such as learning and memory have not been studied extensively.[109,110] A form of memory could occur when, after the application of external input, the networks were pushed into a different stationary state with respect to neuritic field sizes and connectivity. For this to be feasible, however, different solutions of neuritic field size distributions must exist in which the electrical activity of all cells is at the homeostatic set-point. Whether these multiple equilibrium points exist for the circular neuritic field model is unknown.

Homeostatic activity-dependent neurite outgrowth provides a mechanism by which unconnected neurons can self-organize into a connected network and offers a way of automatically creating, without parameter tuning, balanced connectivity in large-scale networks in which neuronal activity is neither too high nor too low. In general, activity-dependent neurite outgrowth may act as a central organizing principle driving both the development of networks and the reactive and compensatory structural changes induced by neurogenesis, deafferentiation, and neurodegeneration in adulthood.

References

1. van Ooyen A. Activity-dependent neural network development. *Network Comput Neural Syst*. 1994;5:401–423.
2. van Ooyen A. *Modeling Neural Development*. Cambridge, MA; London: MIT Press; 2003.
3. van Ooyen A. Using theoretical models to analyse neural development. *Nat Rev Neurosci*. 2011;12(6):311–326.
4. Corner MA, van Pelt J, Wolters PS, Baker RE, Nuytinck RH. Physiological effects of sustained blockade of excitatory synaptic transmission on spontaneously active developing neuronal networks—an inquiry into the reciprocal linkage between intrinsic biorhythms and neuroplasticity in early ontogeny. *Neurosci Biobehav Rev*. 2002;26(2):127–185.
5. Tailby C, Wright LL, Metha AB, Calford MB. Activity-dependent maintenance and growth of dendrites in adult cortex. *Proc Natl Acad Sci USA*. 2005;102(12):4631–4636.
6. Keck T, Mrsic-Flogel TD, Vaz Afonso M, Eysel UT, Bonhoeffer T, Hübener M. Massive restructuring of neuronal circuits during functional reorganization of adult visual cortex. *Nat Neurosci*. 2008;11(10):1162–1167.
7. Yamahachi H, Marik SA, McManus JN, Denk W, Gilbert CD. Rapid axonal sprouting and pruning accompany functional reorganization in primary visual cortex. *Neuron*. 2009;64(5):719–729.
8. Brown CE, Aminoltejari K, Erb H, Winship IR, Murphy TH. In vivo voltage-sensitive dye imaging in adult mice reveals that somatosensory maps lost to stroke are replaced over weeks by new structural and functional circuits with prolonged modes of activation within both the peri-infarct zone and distant sites. *J Neurosci*. 2009;29(6):1719–1734.
9. Winship IR, Murphy TH. Remapping the somatosensory cortex after stroke: insight from imaging the synapse to network. *Neuroscientist*. 2009;15(5):507–524.
10. Butz M, Wörgötter F, van Ooyen A. Activity-dependent structural plasticity. *Brain Res Rev*. 2009;60 (2):287–305.
11. Kater SB, Mattson MP, Cohan C, Connor J. Calcium regulation of the neuronal growth cone. *Trends Neurosci*. 1988;11(7):315–321.
12. Mattson MP. Neurotransmitters in the regulation of neuronal cytoarchitecture. *Brain Res*. 1988;472(2):179–212.
13. Lohmann C, Wong RO. Regulation of dendritic growth and plasticity by local and global calcium dynamics. *Cell Calcium*. 2005;37(5):403–409.
14. Konur S, Ghosh A. Calcium signaling and the control of dendritic development. *Neuron*. 2005;46(3):401–405.
15. Ghiretti AE, Moore AR, Brenner RG, et al. Rem2 is an activity-dependent negative regulator of dendritic complexity in vivo. *J Neurosci*. 2014;34(2):392–407.
16. McAllister AK. Cellular and molecular mechanisms of dendrite growth. *Cereb Cortex*. 2000;10(10):963–973.
17. Cohan CS, Kater SB. Suppression of neurite elongation and growth cone motility by electrical activity. *Science*. 1986;232(4758):1638–1640.
18. Mattson MP, Dou P, Kater SB. Outgrowth-regulating actions of glutamate in isolated hippocampal pyramidal neurons. *J Neurosci*. 1988;8(6):2087–2100.
19. Mattson MP, Kater SB. Excitatory and inhibitory neurotransmitters in the generation and degeneration of hippocampal neuroarchitecture. *Brain Res*. 1989;478(2):337–348.
20. Fields RD, Neale EA, Nelson PG. Effects of patterned electrical activity on neurite outgrowth from mouse sensory neurons. *J Neurosci*. 1990;10(9):2950–2964.

21. Hui K, Fei GH, Saab BJ, Su J, Roder JC, Feng ZP. Neuronal calcium sensor-1 modulation of optimal calcium level for neurite outgrowth. *Development*. 2007;134(24):4479–4489.
22. Yamada RX, Sasaki T, Ichikawa J, Koyama R, Matsuki N, Ikegaya Y. Long-range axonal calcium sweep induces axon retraction. *J Neurosci*. 2008;28(18):4613–4618.
23. Tashiro A, Dunaevsky A, Blazeski R, Mason CA, Yuste R. Bidirectional regulation of hippocampal mossy fiber filopodial motility by kainate receptors: a two-step model of synaptogenesis. *Neuron*. 2003;38 (5):773–784.
24. Enes J, Langwieser N, Ruschel J, et al. Electrical activity suppresses axon growth through Ca(v)1.2 channels in adult primary sensory neurons. *Curr Biol*. 2010;20(13):1154–1164.
25. Jiang M, Lee CL, Smith KL, Swann JW. Spine loss and other persistent alterations of hippocampal pyramidal cell dendrites in a model of early-onset epilepsy. *J Neurosci*. 1998;18(20):8356–8368.
26. McKinney RA, Lüthi A, Bandtlow CE, Gähwiler BH, Thompson SM. Selective glutamate receptor antagonists can induce or prevent axonal sprouting in rat hippocampal slice cultures. *Proc Natl Acad Sci USA*. 1999;96 (20):11631–11636.
27. Lipton SA, Kater SB. Neurotransmitter regulation of neuronal outgrowth, plasticity and survival. *Trends Neurosci*. 1989;12(7):265–270.
28. Kater SB, Mills LR. Regulation of growth cone behavior by calcium. *J Neurosci*. 1991;11(4):891–899.
29. al-Mohanna FA, Cave J, Bolsover SR. A narrow window of intracellular calcium concentration is optimal for neurite outgrowth in rat sensory neurones. *Brain Res Dev Brain Res*. 1992;70(2):287–290.
30. Ramakers GJ, Avci B, van Hulten P, et al. The role of calcium signaling in early axonal and dendritic morphogenesis of rat cerebral cortex neurons under non-stimulated growth conditions. *Brain Res Dev Brain Res*. 2001;126(2):163–172.
31. Henley J, Poo MM. Guiding neuronal growth cones using Ca^{2+} signals. *Trends Cell Biol*. 2004;14(6):320–330.
32. van Ooyen A, van Pelt J. Activity-dependent outgrowth of neurons and overshoot phenomena in developing neural networks. *J Theor Biol*. 1994;167:27–43.
33. van Ooyen A, van Pelt J, Corner MA. Implications of activity dependent neurite outgrowth for neuronal morphology and network development. *J Theor Biol*. 1995;172(1):63–82.
34. van Ooyen A, van Pelt J. Complex periodic behaviour in a neural network model with activity-dependent neurite outgrowth. *J Theor Biol*. 1996;179(3):229–242.
35. van Oss C, van Ooyen A. Effects of inhibition on neural network development through activity-dependent neurite outgrowth. *J Theor Biol*. 1997;185(2):263–280.
36. Butz M, Steenbuck ID, van Ooyen A. Homeostatic structural plasticity can account for topology changes following deafferentation and focal stroke. *Front Neuroanat*. 2014;8:115.
37. van Huizen F, Romijn HJ, Habets AM. Synaptogenesis in rat cerebral cortex cultures is affected during chronic blockade of spontaneous bioelectric activity by tetrodotoxin. *Brain Res*. 1985;351(1):67–80.
38. Marom S, Shahaf G. Development, learning and memory in large random networks of cortical neurons: lessons beyond anatomy. *Q Rev Biophys*. 2002;35(1):63–87.
39. Wagenaar DA, Pine J, Potter SM. An extremely rich repertoire of bursting patterns during the development of cortical cultures. *BMC Neurosci*. 2006;7:11.
40. Pasquale V, Massobrio P, Bologna LL, Chiappalone M, Martinoia S. Self-organization and neuronal avalanches in networks of dissociated cortical neurons. *Neuroscience*. 2008;153(4):1354–1369.
41. Grossberg S. Nonlinear neural networks: principles, mechanisms and architectures. *Neural Networks*. 1988;1:17–61.
42. Aizenman CD, Manis PB, Linden DJ. Polarity of long-term synaptic gain change is related to postsynaptic spike firing at a cerebellar inhibitory synapse. *Neuron*. 1998;21(4):827–835.
43. Soto-Treviño C, Thoroughman KA, Marder E, Abbott LF. Activity-dependent modification of inhibitory synapses in models of rhythmic neural networks. *Nat Neurosci*. 2001;4(3):297–303.
44. Abbott LF, Rohrkemper R. A simple growth model constructs critical avalanche networks. *Prog Brain Res*. 2007;165:13–19.
45. van Ooyen A, Pakdaman K, Houweling AR, van Pelt J, Vibert JF. Network connectivity changes through activity-dependent neurite outgrowth. *Neural Process Lett*. 1996;3:123–130.
46. Markram H, Toledo-Rodriguez M, Wang Y, Gupta A, Silberberg G, Wu C. Interneurons of the neocortical inhibitory system. *Nat Rev Neurosci*. 2004;5(10):793–807.

47. Schilling K, Dickinson MH, Connor JA, Morgan JI. Electrical activity in cerebellar cultures determines Purkinje cell dendritic growth patterns. *Neuron*. 1991;7(6):891–902.
48. van Huizen F, Romijn HJ, Habets AM, van den Hooff P. Accelerated neural network formation in rat cerebral cortex cultures chronically disinhibited with picrotoxin. *Exp Neurol*. 1987;97(2):280–288.
49. Habets AM, van Dongen AM, van Huizen F, Corner MA. Spontaneous neuronal firing patterns in fetal rat cortical networks during development in vitro: a quantitative analysis. *Exp Brain Res*. 1987;69(1):43–52.
50. Soriano J, Rodríguez Martínez M, Tlusty T, Moses E. Development of input connections in neural cultures. *Proc Natl Acad Sci USA*. 2008;105(37):13758–13763.
51. Purves D, Lichtman JW. Elimination of synapses in the developing nervous system. *Science*. 1980;210 (4466):153–157.
52. Uylings HBM, van Eden CG, Parnavelas JG, Kalsbeek A. The prenatal and postnatal development of rat cerebral cortex. In: Kolb B, Trees RC, eds. *The Cerebral Cortex of the Rat*. Cambridge, MA: MIT Press; 1990.
53. Gorgels TG, De Kort EJ, Van Aanholt HT, Nieuwenhuys R. A quantitative analysis of the development of the pyramidal tract in the cervical spinal cord in the rat. *Anat Embryol (Berl)*. 1989;179(4):377–385.
54. O'Kusky JR. Synapse elimination in the developing visual cortex: a morphometric analysis in normal and dark-reared cats. *Brain Res*. 1985;354(1):81–91.
55. van Huizen F, Romijn HJ. Tetrodotoxin enhances initial neurite outgrowth from fetal rat cerebral cortex cells in vitro. *Brain Res*. 1987;408(1–2):271–274.
56. Deco G, Ponce-Alvarez A, Mantini D, Romani GL, Hagmann P, Corbetta M. Resting-state functional connectivity emerges from structurally and dynamically shaped slow linear fluctuations. *J Neurosci*. 2013;33 (27):11239–11252.
57. Guthrie PB, Mattson MP, Mills L, Kater SB. Calcium homeostasis in molluscan and mammalian neurons: neuron-selective set-point of calcium rest concentrations. Paper presented at: *Society for Neuroscience* 1988.
58. van Pelt J, Wolters PS, Corner MA, Rutten WL, Ramakers GJ. Long-term characterization of firing dynamics of spontaneous bursts in cultured neural networks. *IEEE Trans Biomed Eng*. 2004;51(11):2051–2062.
59. Börgers C, Kopell N. Effects of noisy drive on rhythms in networks of excitatory and inhibitory neurons. *Neural Comput*. 2005;17(3):557–608.
60. Whittington MA, Traub RD, Kopell N, Ermentrout B, Buhl EH. Inhibition-based rhythms: experimental and mathematical observations on network dynamics. *Int J Psychophysiol*. 2000;38(3):315–336.
61. Romijn HJ, van Marle J, Janszen AW. Permanent increase of the GAD67/synaptophysin ratio in rat cerebral cortex nerve endings as a result of hypoxic ischemic encephalopathy sustained in early postnatal life: a confocal laser scanning microscopic study. *Brain Res*. 1993;630(1–2):315–329.
62. Romijn HJ, Voskuyl RA, Coenen AM. Hypoxic-ischemic encephalopathy sustained in early postnatal life may result in permanent epileptic activity and an altered cortical convulsive threshold in rat. *Epilepsy Res*. 1994;17(1):31–42.
63. Cope DW, Di Giovanni G, Fyson SJ, et al. Enhanced tonic GABAA inhibition in typical absence epilepsy. *Nat Med*. 2009;15(12):1392–1398.
64. van Huizen F, Romijn HJ, Corner MA. Indications for a critical period for synapse elimination in developing rat cerebral cortex cultures. *Brain Res*. 1987;428(1):1–6.
65. Abeles M. *Corticonics: Neural Circuits of the Cerebral Cortex*. Cambridge: CUP; 1991.
66. Spruston N. Pyramidal neurons: dendritic structure and synaptic integration. *Nat Rev Neurosci*. 2008;9 (3):206–221.
67. Lund JS, Angelucci A, Bressloff PC. Anatomical substrates for functional columns in macaque monkey primary visual cortex. *Cereb Cortex*. 2003;13(1):15–24.
68. Muir DR, Douglas RJ. From neural arbors to daisies. *Cereb Cortex*. 2011;21(5):1118–1133.
69. Dupont E, Hanganu IL, Kilb W, Hirsch S, Luhmann HJ. Rapid developmental switch in the mechanisms driving early cortical columnar networks. *Nature*. 2006;439(7072):79–83.
70. Sonty RV, Juliano SL. Development of intrinsic connections in cat somatosensory cortex. *J Comp Neurol*. 1997;384(4):501–516.
71. Lund JS, Yoshioka T, Levitt JB. Comparison of intrinsic connectivity in different areas of macaque monkey cerebral cortex. *Cereb Cortex*. 1993;3(2):148–162.
72. DeFelipe J, Hendry SH, Hashikawa T, Molinari M, Jones EG. A microcolumnar structure of monkey cerebral cortex revealed by immunocytochemical studies of double bouquet cell axons. *Neuroscience*. 1990;37 (3):655–673.

73. McDonald CT, Burkhalter A. Organization of long-range inhibitory connections with rat visual cortex. *J Neurosci*. 1993;13(2):768–781.
74. Tamamaki N, Tomioka R. Long-range GABAergic connections distributed throughout the neocortex and their possible function. *Front Neurosci*. 2010;4:202.
75. Coleman PD, Flood DG. Dendritic proliferation in the aging brain as a compensatory repair mechanism. *Prog Brain Res*. 1986;70:227–237.
76. Coleman PD, Buell SJ, Magagna L, Flood DG, Curcio CA. Stability of dendrites in cortical barrels of C57BL/6N mice between 4 and 45 months. *Neurobiol Aging*. 1986;7(2):101–105.
77. Brown CE, Wong C, Murphy TH. Rapid morphologic plasticity of peri-infarct dendritic spines after focal ischemic stroke. *Stroke*. 2008;39(4):1286–1291.
78. Ming GL, Song H. Adult neurogenesis in the mammalian central nervous system. *Annu Rev Neurosci*. 2005;28:223–250.
79. Mongiat LA, Schinder AF. Adult neurogenesis and the plasticity of the dentate gyrus network. *Eur J Neurosci*. 2011;33(6):1055–1061.
80. Ge S, Yang CH, Hsu KS, Ming GL, Song H. A critical period for enhanced synaptic plasticity in newly generated neurons of the adult brain. *Neuron*. 2007;54(4):559–566.
81. Oppenheim RW. Cell death during development of the nervous system. *Annu Rev Neurosci*. 1991;14:453–501.
82. Franklin JL, Johnson EM. Suppression of programmed neuronal death by sustained elevation of cytoplasmic calcium. *Trends Neurosci*. 1992;15(12):501–508.
83. Turrigiano GG, Nelson SB. Hebb and homeostasis in neuronal plasticity. *Curr Opin Neurobiol*. 2000;10(3):358–364.
84. Liu Z, Golowasch J, Marder E, Abbott LF. A model neuron with activity-dependent conductances regulated by multiple calcium sensors. *J Neurosci*. 1998;18(7):2309–2320.
85. Abbott LF, Jensen O. Self-organizing circuits of model neurons. In: Bower J, ed. *Computational Neuroscience, Trends in Research*. New York: Plenum; 1997:227–230.
86. Beggs JM, Plenz D. Neuronal avalanches in neocortical circuits. *J Neurosci*. 2003;23(35):11167–11177.
87. Beggs JM, Plenz D. Neuronal avalanches are diverse and precise activity patterns that are stable for many hours in cortical slice cultures. *J Neurosci*. 2004;24(22):5216–5229.
88. Hennig MH, Adams C, Willshaw D, Sernagor E. Early-stage waves in the retinal network emerge close to a critical state transition between local and global functional connectivity. *J Neurosci*. 2009;29(4):1077–1086.
89. Petermann T, Thiagarajan TC, Lebedev MA, Nicolelis MA, Chialvo DR, Plenz D. Spontaneous cortical activity in awake monkeys composed of neuronal avalanches. *Proc Natl Acad Sci USA*. 2009;106(37):15921–15926.
90. Zapperi S, Bækgaard Lauritsen K, Stanley HE. Self-organized branching processes: mean-field theory for avalanches. *Phys Rev Lett*. 1995;75(22):4071–4074.
91. van den Akker B, Ibarz B, Memmesheimer RM. Self-organized criticality in a model for developing neural networks. *BMC Neurosci*. 2011;12(Suppl. 1):221.
92. Tetzlaff C, Okujeni S, Egert U, Wörgötter F, Butz M. Self-organized criticality in developing neuronal networks. *PLoS Comput Biol*. 2010;6(12):e1001013.
93. Plenz D, Thiagarajan TC. The organizing principles of neuronal avalanches: cell assemblies in the cortex? *Trends Neurosci*. 2007;30(3):101–110.
94. Kinouchi O, Copelli M. Optimal dynamical range of excitable networks at criticality. *Nature Phys*. 2006;2:348–352.
95. Haldeman C, Beggs JM. Critical branching captures activity in living neural networks and maximizes the number of metastable States. *Phys Rev Lett*. 2005;94(5):058101.
96. Eglen SJ. Development of regular cellular spacing in the retina: theoretical models. *Math Med Biol*. 2006;23(2):79–99.
97. Eglen SJ. Cellular spacing: analysis and modelling of retinal mosaics. In: Le Novere N, ed. *Computational Systems Neurobiology*. Springer; 2012:365–385.
98. Reese BE, Necessary BD, Tam PP, Faulkner-Jones B, Tan SS. Clonal expansion and cell dispersion in the developing mouse retina. *Eur J Neurosci*. 1999;11(8):2965–2978.
99. Galli-Resta L, Novelli E, Viegi A. Dynamic microtubule-dependent interactions position homotypic neurones in regular monolayered arrays during retinal development. *Development*. 2002;129(16):3803–3814.

100. Fuerst PG, Koizumi A, Masland RH, Burgess RW. Neurite arborization and mosaic spacing in the mouse retina require DSCAM. *Nature*. 2008;451(7177):470–474.
101. Eglen SJ, van Ooyen A, Willshaw DJ. Lateral cell movement driven by dendritic interactions is sufficient to form retinal mosaics. *Network*. 2000;11(1):103–118.
102. Perry VH, Linden R. Evidence for dendritic competition in the developing retina. *Nature*. 1982;297(5868):683–685.
103. Cook JE, Chalupa LM. Retinal mosaics: new insights into an old concept. *Trends Neurosci*. 2000;23(1):26–34.
104. Kawasaki F, Stiber M. A simple model of cortical culture growth: burst property dependence on network composition and activity. *Biol Cybern*. 2014;108(4):423–443.
105. Abbott LF. Lapicque's introduction of the integrate-and-fire model neuron. *Brain Res Bull*. 1907;50(5–6):303–304:1999.
106. Markram H, Wang Y, Tsodyks M. Differential signaling via the same axon of neocortical pyramidal neurons. *Proc Natl Acad Sci USA*. 1998;95(9):5323–5328.
107. Tsodyks M, Pawelzik K, Markram H. Neural networks with dynamic synapses. *Neural Comput*. 1998;10(4):821–835.
108. Piaget J, Inhelder B. *The Psychology of the Child*. 8 ed. New York: Basic Books; 1969.
109. Raijmakers ME, Molenaar PC. Modeling developmental transitions in adaptive resonance theory. *Dev Sci*. 2004;7(2):149–157.
110. Raijmakers MEJ, Molenaar PCM. Modelling developmental transitions in neural networks: bifurcations in an adaptive resonance theory model. In: Mareschal D, Sirois S, Westermann G, Johnson MH, eds. *Neuroconstructivism: Perspectives and Prospects*. Vol. 2. Oxford: Oxford University Press; 2007:99–128.
111. Molenaar PC, Raijmakers ME. Exact ART: a complete implementation of an ART network. *Neural Networks*. 1997;10(4):649–669.
112. Grossberg S. Adaptive pattern classification and universal recoding: II. feedback, expectation, olfaction, illusions. *Biol Cybern*. 1976;23(4):187–202.
113. Hogeweg P, Hesper B. An adaptive, selfmodifying, non goal directed modelling approach. In: Elzas MS, Oren TI, Zeigler BP, eds. *Modelling and Simulation Methodology: Knowledge Systems Paradigms*. Amsterdam North Holland; 1989:77–92.
114. Boissel JP, Ribba B, Grenier E, Chapuisat G, Dronne MA. Modelling methodology in physiopathology. *Prog Biophys Mol Biol*. 2008;97(1):28–39.
115. Houweling AR, van Ooyen A. Homeostasis at multiple spatial and temporal scales. In: Squire L, ed. *New Encyclopedia of Neuroscience*. Amsterdam: Elsevier Press; 2008.
116. Marik SA, Yamahachi H, Meyer zum Alten Borgloh S, Gilbert CD. Large-scale axonal reorganization of inhibitory neurons following retinal lesions. *J Neurosci*. 2014;34(5):1625–1632.
117. Abe H, McManus JN, Ramalingam N, et al. Adult cortical plasticity studied with chronically implanted electrode arrays. *J Neurosci*. 2015;35(6):2778–2790.
118. Keck T, Keller GB, Jacobsen RI, Eysel UT, Bonhoeffer T, Hübener M. Synaptic scaling and homeostatic plasticity in the mouse visual cortex in vivo. *Neuron*. 2013;80(2):327–334.
119. Hengen KB, Lambo ME, van Hooser SD, Katz DB, Turrigiano GG. Firing rate homeostasis in visual cortex of freely behaving rodents. *Neuron*. 2013;80(2):335–342.
120. van Pelt J, van Ooyen A. Estimating neuronal connectivity from axonal and dendritic density fields. *Front Comput Neurosci*. 2013;7:160.
121. Hely TA, Graham B, van Ooyen A. A computational model of dendrite elongation and branching based on MAP2 phosphorylation. *J Theor Biol*. 2001;210(3):375–384.
122. Graham BP, van Ooyen A. Compartmental models of growing neurites. *Neurocomputing*. 2001;38-40:31–36.
123. Hjorth JJ, van Pelt J, Mansvelder HD, van Ooyen A. Competitive dynamics during resource-driven neurite outgrowth. *PLoS ONE*. 2014;9(2):e86741.
124. Koene RA, Tijms B, van Hees P, et al. NETMORPH: a framework for the stochastic generation of large scale neuronal networks with realistic neuron morphologies. *Neuroinformatics*. 2009;7(3):195–210.
125. van Ooyen A, Carnell A, de Ridder S, et al. Independently outgrowing neurons and geometry-based synapse formation produce networks with realistic synaptic connectivity. *PLoS One*. 2014;9(1):e85858.
126. Butz M, van Ooyen A. A simple rule for dendritic spine and axonal bouton formation can account for cortical reorganization after focal retinal lesions. *PLoS Comput Biol*. 2013;9(10):e1003259.

127. Dammasch IE, Wagner GP, Wolff JR. Self-stabilization of neuronal networks. I. The compensation algorithm for synaptogenesis. *Biol Cybern*. 1986;54(4-5):211–222.
128. LeMasson G, Marder E, Abbott LF. Activity-dependent regulation of conductances in model neurons. *Science*. 1993;259(5103):1915–1917.
129. Spitzer NC. Electrical activity in early neuronal development. *Nature*. 2006;444(7120):707–712.
130. Wang DD, Kriegstein AR. GABA regulates excitatory synapse formation in the neocortex via NMDA receptor activation. *J Neurosci*. 2008;28(21):5547–5558.
131. Dehorter N, Ciceri G, Bartolini G, Lim L, del Pino I, Marín O. Tuning of fast-spiking interneuron properties by an activity-dependent transcriptional switch. *Science*. 2015;349(6253):1216–1220.
132. Turrigiano GG. The self-tuning neuron: synaptic scaling of excitatory synapses. *Cell*. 2008;135(3):422–435.
133. Tetzlaff C, Kolodziejski C, Markelic I, Wörgötter F. Time scales of memory, learning, and plasticity. *Biol Cybern*. 2012;106(11–12):715–726.
134. Fauth M, Wörgötter F, Tetzlaff C. The formation of multi-synaptic connections by the interaction of synaptic and structural plasticity and their functional consequences. *PLoS Comput Biol*. 2015;11(1):e1004031.

CHAPTER

6

Clustered Arrangement of Inhibitory Neurons Can Lead to Oscillatory Dynamics in a Model of Activity-Dependent Structural Plasticity

Rosanna C. Barnard[1], *Istvan Z. Kiss*[1], *Simon F. Farmer*[2] *and Luc Berthouze*[1]

[1]University of Sussex, Falmer, United Kingdom [2]Institute of Neurology, London, United Kingdom

OUTLINE

The Rewiring Brain
DOI: http://dx.doi.org/10.1016/B978-0-12-803784-3.00006-8

1 INTRODUCTION

Neuronal networks continuously undergo changes in their connectivity. This structural plasticity starts with development and continues throughout adulthood. It is typically supported by experience-dependent mechanisms. In computational neuroscience, a number of mathematical models of activity-dependent structural plasticity have been developed to further our understanding of the formation, function, and underlying structure of neuronal networks.[1] These models operate on the assumption that network structure results from a number of activity-dependent processes such as neurite outgrowth and growth cone behavior,[2–4] and naturally occurring cell death.[5,6] It has been shown that combining the Shunting model,[7] describing the neuronal activity of a group of neurons, with terms describing activity-dependent outgrowth, in which initially disconnected neurons organize themselves into a network under the influence of endogenous activity, results in a model that can display complex periodic behavior of electrical activity and connectivity, including small-amplitude stable oscillations and intermittent unstable burst-like oscillations.[8,9] A unifying assumption of this so-called Extended Shunting model (ES model hereafter) is the homeostatic regulation of neuronal excitability toward a target level, a phenomenon which is considered ubiquitous in both development and adulthood.[10,11] Such a mechanism allows neurons to maintain stability of function whilst undergoing significant anatomical and functional changes and it plays a key role in network growth and stability.

Early neural network formation depends on a number of factors: (1) neurotrophic substances exerting chemoattraction and chemorepulsion, (2) neurotransmitters, most notably glutamate (an excitatory transmitter), GABA (classically inhibitory transmitter but may be temporarily excitatory during early development), glycine, and taurine, and (3) neural network activity. It is important to note that neurotransmitters can act directly on the network growth through modulation of intracellular calcium[12] and also indirectly through synaptic mechanisms causing depolarization (excitation) and hyperpolarization (inhibition) of the target neuron and thus modulation of neuronal activity.[13–15] It is the third mechanism of network activity-dependent growth that the ES model considers and in particular the balance between excitation and inhibition in a neuronal network, which plays a crucial role in determining the stability and computational effectiveness of the network.[16] Animal and human studies have shown that the disruption of excitatory and inhibitory neurotransmitter levels results in neural migration disorders and epilepsy[15]; furthermore it is noted that some human genetic disorders are associated with excessive inhibition in the brain.[17] Previous work involving the ES model emphasized the importance of the balance between excitation and inhibition, suggesting that a network developing under conditions of relatively high inhibition is more likely to end up in a "pathological" state with oscillatory behavior, whereas a network developing under a low level of inhibition during its development enables the system to move to the "normal" state, where the network structure reaches a stable end state with stable levels of activity.[9] The oscillatory behavior induced by moderate proportions of inhibition was thought to be a result of interactions between excitatory and inhibitory activities, occurring on the timescale of the electrical activity.[18]

Central to this chapter is another factor that has been identified as affecting system behavior beyond the proportion of inhibitory neurons, namely, the spatial arrangement of inhibitory neurons.[18] The developmental migration of inhibitory neurons within the central nervous system was thought to be spatially random.[19,20] Recent studies, however, demonstrate that during neural development clusters of inhibitory neurons form. The result of this inhibitory clustering is the production of higher amplitude inhibitory postsynaptic potentials (IPSPs) within excitatory neurons.[21,22]

Under neural activity homeostasis, neural inhibition will induce outgrowth and allow disconnected subnetworks to become connected. We therefore hypothesize that the spatial distribution of inhibitory neurons will play an important role in the interaction between structure and function. When highly clustered inhibitory cells (with self-inhibition) become electrically inhibited, outgrowth of inhibitory neurons is stimulated (as their activity is below the homeostatic target value), inducing long-range inhibition within the whole network. This in turn means that these networks develop stronger excitatory–excitatory, excitatory–inhibitory (and inhibitory–excitatory) connectivity, and as a result, possibly increased network instability.

A comprehensive account of the extent to which the spatial organization of excitatory and inhibitory neurons influences global system behavior and the structural characteristics of the resulting neuronal networks is lacking. Such an account is made necessary by a growing body of evidence suggesting the presence of nonrandom network attributes within neuronal network structures, including modular community structure, short characteristic path length (associated with high global efficiency of information transfer), high levels of clustering (i.e., the propensity for neighbors of a random neuron to share a connection, associated with robustness to random error), and degree distributions compatible with the existence of highly connected hub neurons.[23–25] In this chapter, we investigate how much influence the spatial arrangement of excitatory and inhibitory neurons, and more specifically, the spatial clustering of inhibitory neurons, has on the global system dynamics, network structure, and electrical activity in the ES model.

In what follows, we begin by outlining the differential equations governing the activity-dependent neuronal network growth model, giving a detailed description of each model element, and explaining how these elements combine to create a system in which each and every neuron strives to maintain a desired level of electrical activity (homeostasis). We explain the choice of model parameters, describe the implementation of the numerical scheme used to solve the resulting system of ordinary differential equations (ODEs), and outline the terminologies and notation used throughout this chapter. We define novel methods for measuring the extent of inhibitory clustering within one- and two-dimensional (1D and 2D) lattice networks composed of various neuron arrangements. The lattice networks have periodic boundary conditions, i.e., they are mapped onto a torus, thus approximating large systems. In Section 5, we show that the proportion of inhibitory neurons alone is a necessary but not sufficient condition for inducing network instability characterized by oscillatory behavior of the connectivity and electrical activity. By considering 1D and 2D networks we are able to demonstrate that in general, higher levels of spatial clustering of inhibition induce oscillatory behaviors, whilst lower levels of inhibitory clustering lead to a final, stable end state. We look at the qualitatively distinct global system behaviors produced by the ES model and attempt to understand whether different system behaviors translate into different underlying network structures. This is done by using network theoretical measures and by comparing these across networks corresponding to different system behaviors. We conclude with a discussion of our findings including neurodevelopmental parallels and suggestions for future experimental and computational work in the field.

2 MODEL

This work is based on the model of structural plasticity proposed by Van Ooyen et al.[18] It extends the Shunting model proposed by Grossberg[7] by incorporating activity-dependent outgrowth terms. This allows a network containing excitatory and inhibitory neurons to be grown from initial conditions. Each neuron is assigned a unique spatial location (see Section 4.2) and modeled with a variable radius, denoting the area of its neuritic (axonal and dendritic) extensions. When the circular fields of two neurons overlap, those neurons become connected with a strength proportional to the area of overlap, multiplied by the appropriate synaptic efficacy term, determined by the type (excitatory or inhibitory) of the two neurons.

Precisely, the system is governed by the following dimensionless coupled differential equations describing the rate of change of electrical activity and neuritic radii of N excitatory and M inhibitory neurons denoted $(X_i)_{i=1,2,\ldots,N}$ and $\left(Y_j\right)_{j=N+1,N+2,\ldots,N+M}$ respectively:

$$\frac{\mathrm{d}X_i}{\mathrm{d}T} = -X_i + (1 - X_i)\sum_{k=1}^{N} W_{ik}F(X_k) - (H + X_i)\sum_{l=N+1}^{N+M} W_{il}F(Y_l) \tag{6.1}$$

$$\frac{\mathrm{d}Y_j}{\mathrm{d}T} = -Y_j + \left(1 - Y_j\right)\sum_{k=1}^{N} W_{jk}F(Y_k) - \left(H + Y_j\right)\sum_{l=N+1}^{N+M} W_{jl}F(Y_l) \tag{6.2}$$

$$\frac{\mathrm{d}R_i}{\mathrm{d}T} = \rho G(F(X_i)) \tag{6.3}$$

$$\frac{\mathrm{d}R_j}{\mathrm{d}T} = \rho G\left(F\left(Y_j\right)\right). \tag{6.4}$$

Eqs. (6.1) and (6.2) correspond to the original Shunting model, where X_i and Y_j refer to the electrical activity of excitatory neuron i and inhibitory neuron j, respectively, expressed in units of a saturation value (finite maximum). Electrical activities have a finite minimum, i.e., $-H$ times the saturation value. A change in electrical activity of a neuron is influenced by its own electrical activity and that of any other neuron it shares a connection with. The strength of the connection between neurons i and j at any point in time is given by element W_{ij} of the (symmetric) weight matrix $\mathbf{W}$. In this weight matrix, self-overlap terms are set to zero ($W_{ii} = 0, \forall i \in \{1, \ldots, N+M\}$) and nondiagonal elements are calculated as

$$W_{ij} = A_{ij}S_{[i][j]} \tag{6.5}$$

where A_{ij} describes the area of overlap between the circular neuritic fields of neurons i and j, and $S_{[i][j]}$ describes a synaptic strength parameter from a neuron of type (excitatory or inhibitory) $[j]$ to a neuron of type $[i]$.

TABLE 6.1 Default Values for All Model Parameters Used in Simulations

Parameter name	Parameter description	Default parameter value
ρ	Rate of growth of neuritic field	0.0001
θ	Firing threshold for firing rate function F	0.5
α	Slope of firing rate function F	0.1
β	Slope of growth function G	0.1
ε	Threshold for growth function G	0.6
H	Ratio between minimum and maximum electrical activity values	0.1
S_{ee}	Excitatory to excitatory synaptic efficacy	1
S_{ei}	Inhibitory to excitatory synaptic efficacy	1
S_{ie}	Excitatory to inhibitory synaptic efficacy	1
S_{ii}	Inhibitory to inhibitory synaptic efficacy	1

Eqs. (6.3) and (6.4) result from extending the Shunting model to incorporate activity dependency of neurite outgrowth. R_i and R_j denote the radii of the neuritic field of excitatory neuron $i \in \{1, \ldots, N\}$ and inhibitory neuron $j \in \{N+1, \ldots, L=N+M\}$, respectively. Parameter ρ describes the rate of outgrowth of the neuritic radii (ρ is constant in our analyses) and the functions F and G are continuous saturating functions:

$$F(u) = \frac{1}{1 + e^{(\theta-u)/\alpha}} \tag{6.6}$$

$$G(u) = 1 - \frac{2}{1 + e^{(\varepsilon-u)/\beta}}. \tag{6.7}$$

The function F can be thought of as governing the mean firing rate of each neuron, with values in the bounded range $(0, 1)$. The function G describes the growth rate of a neuron as a function of the neuron's firing rate and has values in the range $(-1, 1)$. The role and default values of parameters θ, α, ε, and β are given in Table 6.1. The value of ε has great influence on system behavior, since each neuron attempts to maintain electrical activity equal to ε. When a neuron's electrical activity stabilizes at ε, the growth function G tends to zero and the neuritic radius will cease to change. If all neurons in the system are able to maintain an electrical activity equal to ε simultaneously, a system steady state is reached.

3 MODEL IMPLEMENTATION

All simulations were implemented in the MATLAB environment, using a variable-order stiff ODE solver (*ode15s*). Initial conditions were specified, consisting of the spatial location of all neurons within the system and the value of each neuron's radius and electrical activity at time zero. In the 1D model, neurons were placed onto lattice positions on a line with periodic boundary conditions, leaving a nominal distance of 1 between each neuron pair.

In the 2D model, a square number $L = N + M$ of neurons were assigned positions on a square lattice with periodic boundary conditions, with every adjacent pair of neurons being separated by nominal distance 1. In both 1D and 2D models, initial electrical activity values were set to zero for all neurons. Different scenarios of initial radii values were used as specified in each of the relevant sections. These included:

- Randomly (uniformly) generated values in the range $(0, 0.6)$ creating an initially largely disconnected network
- Identical values (radii = 0.4) creating a totally disconnected network
- Identical values (radii = 0.6) for a 1D ring network in which each neuron is only connected to its two immediate neighbors.

We imposed that all neuritic radii be greater than or equal to zero at all times. The ODE system was solved until system behavior could be clearly identified (typically up to a time $> 10^5$).

4 METHODS

4.1 Parameter Choice

Our analyses used similar parameters to those used by Van Ooyen et al.[18] and are summarized in Table 6.1. They differed in two respects. First, whereas Van Ooyen et al. set the proportion of inhibitory neurons in the range of 10%–20%, in accordance with in vivo observations in various brain regions,[18] we systematically varied this proportion in order to provide a fuller picture of system behavior in relation to the balance between excitation and inhibition (see also the next subsection). Second, for simplicity, we kept the synaptic efficacy parameters ($S_{ee}, S_{ie}, S_{ei}, \text{and} S_{ii}$) identical throughout.

4.2 Spatial Arrangement of Excitatory and Inhibitory Neurons

Since connectivity is determined by the amount of overlap between neuritic fields centered over fixed neuron locations, it follows that the joint degree[1] distributions for X (excitatory) and Y (inhibitory) neurons will strongly depend on the spatial arrangement of these neurons, and particularly, whether they are well mixed or clustered—an operational definition of this clustering will be given in the following subsection. We used a systematic approach to building 1D arrangements of excitatory and inhibitory neurons assigned to lattice positions. We define a *tile* as a single row of $l \in \mathbb{Z}^+$ neurons, with nominal distance 1 between consecutive neurons. Then, complex 1D networks with periodic boundary conditions could be formed by duplicating tiles a specified number of times. For example, a network of size $L = 12$ with a $1/3$ proportion of inhibitory neurons could consist of a single tile of length $l = 12$ with 4 inhibitory neurons or 4 identical tiles of length $l = 3$ with 1 inhibitory neuron each (see example neuron arrangement in

[1]The degree of a node in a network is the number of connections it has to other nodes in the network. The joint degree distribution describes the probability distribution of connections between nodes of degree k_1 and nodes of degree k_2 over the whole network.

subsequent paragraph). The use of periodic boundary conditions means that any pair of neurons in the network will be at most $L/2 = (1/2 \times \text{number of neurons in the network})$ distance away from one another. 2D arrangements of N excitatory and M inhibitory neurons were generated in MATLAB using random permutations of a vector containing N zeros and M ones, and reshaping the resulting column vector of size $L \times 1$ into a matrix of size $\sqrt{L} \times \sqrt{L}$.

When representing spatial arrangements throughout the text in this chapter, we use filled circles to denote inhibitory neurons and empty circles to denote excitatory neurons, e.g., ●○○●○○●○○●○○. In figure legends, neuron arrangements are represented using red circles in various hues to denote inhibitory neurons and blue circles in various hues to denote excitatory neurons. The use of unique colors enables the reader to make a direct comparison between each neuron's spatial location and the associated dynamics of their neuritic radius and electrical activity.

4.3 Measuring Spatial Clustering of Inhibition in 1D and 2D Neuron Arrangements

Measuring the clustering of inhibitory neurons in spatial arrangements of both excitatory and inhibitory neurons is not straightforward, and we are not aware of any established method for quantifying it. In what follows, we define measures for clustering in 1D and 2D lattice networks as defined previously.

4.3.1 *1D Inhibitory Clustering Measure*

We define a measure of inhibitory clustering in a 1D network of neurons assigned to lattice positions with periodic boundary conditions as

$$C_{1D} = \frac{1}{D}\left(\sum_{i=1}^{D} \frac{1}{i^2} N_{i-1}\right) \tag{6.8}$$

where D is the shortest distance between the first and last inhibitory neurons in the network, taking periodic boundary conditions into account, and N_i denotes the number of pairs of inhibitory neurons with i excitatory neurons separating them. N_i is consistently counted without multiplicity, meaning that the total N_i count will always be equal to $(M-1)$ with M inhibitory neurons, and any N_i pairs of inhibitory neurons should not have another inhibitory neuron in between them. For example, network ●●○○●○○○○○ has $D=4$, $N_0=1$ (the only adjacent pair of inhibitory neurons), $N_2=1$ (the single pair of inhibitory neurons separated by 2 excitatory neurons), and $N_1=N_3=N_4=0$ leading to a clustering value of $C_{1D} = \frac{1}{4}\left(\frac{1}{1}1 + \frac{1}{3^2}1\right) = \frac{10}{36}$. The inhibitory clustering measure takes value 1 when the greatest possible level of inhibitory clustering is achieved in a network, i.e., when all inhibitory neurons are adjacent to one another and hence $N_0 = D$, and its value tends toward zero as the extent of inhibitory clustering decreases. The contribution of inhibitory neuron pairs decays with the number of excitatory neurons separating them. The weighting factor was selected as a trade-off between an insufficient penalty for the spreading out of inhibitory neurons (e.g., $1/i$) and insensitivity to subtle variations in the structure of the network (e.g., $1/i^3$), hence our choice of $1/i^2$.

4.3.2 2D Inhibitory Clustering Measure

We define a measure of inhibitory clustering for a 2D lattice network containing L neurons with M inhibitory neurons and periodic boundary conditions as follows:

$$C_{2D} = 1 - \frac{WH}{L}\left[\sum_{R=r_1}^{r_n} N_{WH}\left(\frac{R_W R_H}{(WH-M)}\right)^2\right] \tag{6.9}$$

where one firstly computes the smallest possible "covering rectangle" of width W and height H which covers all M lattice positions of inhibitory neurons, taking continuous boundary conditions into account, followed by decomposing all WH neurons contained *within* the covering rectangle into $n \in Z^+$ inhibitory-empty rectangles of various dimensions, from the largest possible rectangle of size $(W-1)\times H$ or $W\times(H-1)$, to the smallest possible rectangle of size 1×1, containing a single neuron. Each inhibitory-empty rectangle is counted without replacement. R_W and R_H denote the width and height of rectangle R and the term N_{WH} denotes the *number* of inhibitory-empty rectangles of size $R_W \times R_H$ contained within the covering rectangle. If a covering rectangle contains no excitatory neurons, the summation term will be equal to zero and hence $C_{2D} = 1$.

5 RESULTS

5.1 1D Results

5.1.1 The Proportion of Inhibitory Neurons—A Necessary Condition for Global System Behavior

In our analyses, it was found that the proportion of inhibition alone cannot accurately predict the system's global behavior. Simulations with tile size 13 were implemented where the proportion of inhibition in the system was set to $4/13 \approx 30.77\%$ with all other parameters held constant, bar the placement of neurons. For the initial conditions, all electrical activity values were set to zero, and all neuritic radii values were set to 0.4, creating a wholly disconnected network. A number of spatial arrangements were considered, randomly picked from a set of unique tile arrangements generated by a MATLAB script that selected at random M out of L positions for inhibitory neurons, and placed N excitatory neurons onto the remaining lattice positions (note that a more systematic approach is used in Section 5.1.2).

Three qualitatively distinct global behaviors are found (Fig. 6.1), namely (1) stable periodic oscillations of small amplitude of both network structure and electrical activity (Fig. 6.1A and B), (2) large amplitude unstable oscillations in the network structure and unstable burst-like oscillations in electrical activity (Fig. 6.1C and D), and (3) complete stabilization of both network structure and electrical activity (Fig. 6.1E and F), all occurring after an initial period of overshoot and pruning of connections.

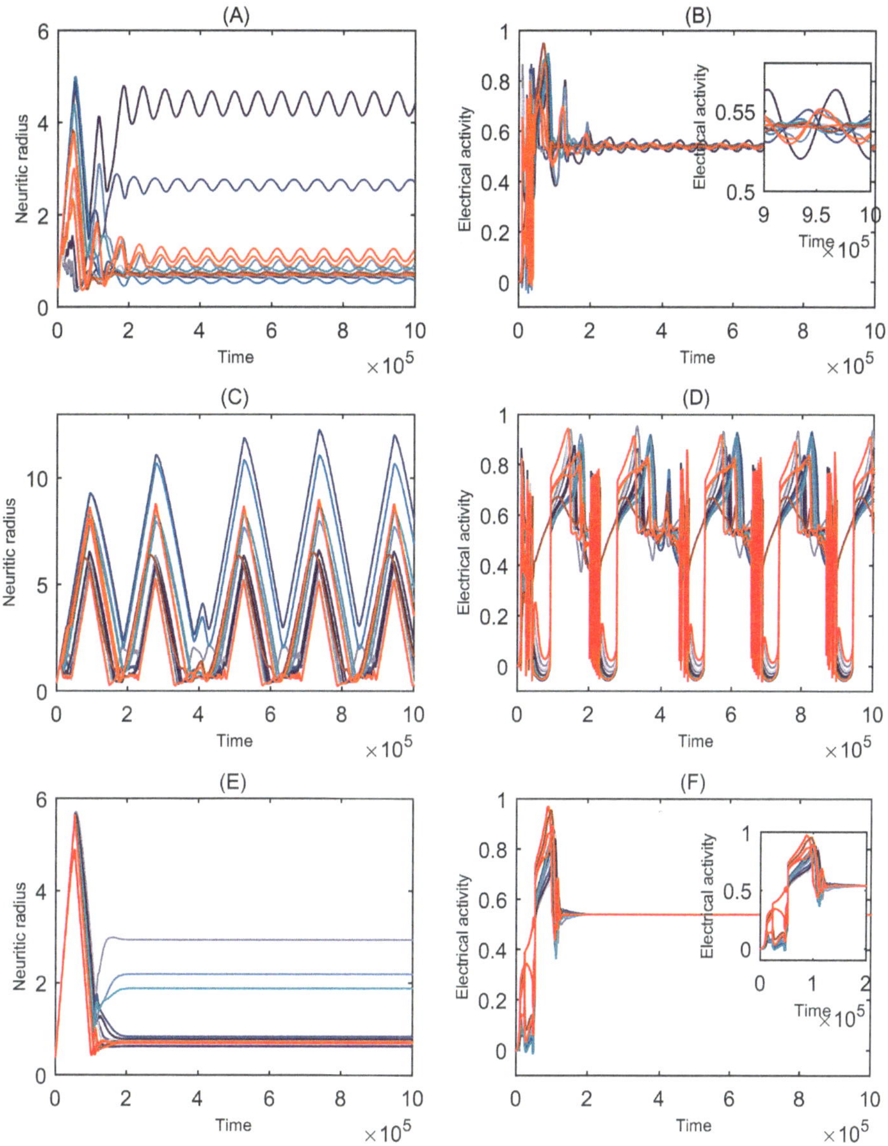

FIGURE 6.1 **Global system behavior of mixed 1D lattice networks with various spatial arrangements of neurons.** In all cases, $L=13$, $N=9$, $M=4$ and default parameters listed in Table 6.1 are used. The networks are initially wholly disconnected and neurons are at rest. Figures on the *left-hand side* depict the dynamics of the individual neuritic radii values; figures on the *right-hand side* depict the dynamics of the individual electrical activity values. Neuron arrangements are as follows: ●●○○●●●●●●●●● (A and B), ●●●○○●●●●●●●● (C and D), ●●●●○○○●●●●●● (E and F).

Since all networks were initialized with identical neuritic field sizes and electrical activities, we conclude that varying the spatial arrangement of neurons within the network can induce distinct global behaviors. The proportion of inhibition alone may be a necessary but not *sufficient* condition to induce oscillatory system behavior. The relationship between E/I ratio, spatial arrangement, and the occurrence of oscillations in the ES model is nontrivial and cannot be attributed to any single characteristic. As pointed out by Van Ooyen and Van Pelt,[8] oscillations can arise as a result of the hysteresis loop formed by the coupling between network structure and electrical activity in the ES model, or due to fluctuations in the *effective* E/I ratio, i.e., the dynamically emerging ratio of total excitatory to inhibitory connectivity, irrespective of the fixed number of excitatory and inhibitory neurons in the arrangement.

5.1.2 The Effect of Inhibitory Clustering on Global System Behavior in 1D

The occurrence of different global behaviors when the proportion of inhibitory neurons in a tile is fixed and simulations of various spatial arrangements are run (Fig. 6.1) suggests that some feature of spatial organization of neurons (and specifically inhibitory neurons) could impact the presence of oscillatory behavior. As most models typically assume a well-mixed population, here, we focus on the possibility that interinhibitory connectivity induced by spatial clustering of inhibitory neurons could be an important determinant.

Simulations with fixed tile size and number of inhibitory neurons, but various levels of inhibitory clustering, revealed distinct global behaviors and a trend suggesting that networks with a higher clustering measure are more likely to experience oscillatory behaviors (Table 6.2). In networks containing three inhibitory neurons and seven excitatory neurons (Fig. 6.2), the most extreme oscillatory behavior is observed when inhibitory clustering is at its greatest (Fig. 6.2A and B). In this case, the network structure and electrical activities experienced a bout of periodic oscillations, before developing unstable burst-like oscillations of much greater amplitude. As the extent of inhibitory clustering decreases slightly, the system experiences stable oscillations with a smaller amplitude, in both network structure and electrical activity (Fig. 6.2C and D). Here, an idiosyncratic behavior of the ES model is observed whereby the largest neuritic field in the network continues to grow whilst experiencing stable oscillations (Fig. 6.2C). This behavior results from the fact that once the neuron's field completely overlaps those of every other neuron in the network, further growth has no effect on the connection strengths to itself and others and outgrowth continues unbounded. Eventually, the inhibitory neurons become more evenly distributed across the lattice positions and the neuritic radii and their associated electrical activity experience complete stabilization (Fig. 6.2E–H). An exhaustive list (taking symmetries and reflections into account) of all possible spatial configurations for tile sizes 8, 10, and 14, containing 2, 3, and 4 inhibitory neurons, respectively, also suggests that higher inhibitory clustering increases the likelihood of oscillatory behavior being observed (Table 6.2).

A systematic mapping of the relationship between inhibitory clustering and the proportion of inhibitory neurons reveal the emergence of distinct regimes (Fig. 6.3). There are no significant differences for the regimes observed between simulations with initially disconnected networks (radii = 0.4, Fig. 6.3A) and those starting with a regular ring network (radii = 0.6, Fig. 6.3B). Networks with lower proportions of inhibitory neurons and lower clustering values are more likely to experience complete stabilization of electrical activity

TABLE 6.2 Global System Behavior and Clustering Measure for Mixed 1D Simulations with Various Spatial Arrangements

ID	Spatial arrangement	Tile size	%I	Behavior	1D clustering measure
1	●●○○○○○○	8	25	SO	1
2	●○●○○○○○	8	25	S	0.125
3	●○○●○○○○	8	25	S	0.03704
4	●○○○●○○○	8	25	S	0.015625
5	●●●○○○○○○○	10	30	UO	1
6	●●○●○○○○○○	10	30	SO	0.4167
7	●●○○●○○○○○	10	30	SO	0.2778
8	●○●○●○○○○○	10	30	S	0.125
9	●●○○○●○○○○	10	30	SO	0.2125
10	●○●○○●○○○○	10	30	S	0.07222
11	●○●○○○●○○○	10	30	S	0.05208
12	●○○●○○●○○○	10	30	S	0.03704
13	●●●●○○○○○○○○○○	14	28.57	SO	1
14	●●●○●○○○○○○○○○	14	28.57	UO	0.5625
15	●●○●●○○○○○○○○○	14	28.57	UO	0.5625
16	●●●○○●○○○○○○○○	14	28.57	UO	0.4222
17	●●○●○●○○○○○○○○	14	28.57	S	0.3
18	●●○○●●○○○○○○○○	14	28.57	SO	0.4222
19	●○●●○●○○○○○○○○	14	28.57	S, TO	0.3
20	●●●○○○●○○○○○○○	14	28.57	UO	0.3438
21	●●○●○○●○○○○○○○	14	28.57	S	0.2269
22	●●○○●○●○○○○○○○	14	28.57	SO	0.2269
23	●●○○○●●○○○○○○○	14	28.57	S	0.3438
24	●○●●○○●○○○○○○○	14	28.57	S	0.2269
25	●○●○●○●○○○○○○○	14	28.57	S	0.125
26	●●●○○○○●○○○○○○	14	28.57	UO	0.2914
27	●●○●○○○●○○○○○○	14	28.57	S	0.1875
28	●●○○●○○●○○○○○○	14	28.57	S, TO	0.1746
29	●●○○○●○●○○○○○○	14	28.57	S	0.1875
30	●●○○○○●●○○○○○○	14	28.57	S	0.2914

(Continued)

TABLE 6.2 (Continued)

ID	Spatial arrangement	Tile size	%*I*	Behavior	1D clustering measure
31	●○●●○○○●○○○○○○	14	28.57	S	0.1875
32	●○●○●○○●○○○○○○	14	28.57	S	0.08730
33	●○●○○●○●○○○○○○	14	28.57	S	0.08730
34	●○○●●○○●○○○○○○	14	28.57	S	0.1746
35	●●●○○○○○●○○○○○	14	28.57	SO (burst-like)	0.2535
36	●●○●○○○○●○○○○○	14	28.57	S	0.1613
37	●●○○●○○○●○○○○○	14	28.57	S	0.1467
38	●●○○○●○○●○○○○○	14	28.57	S	0.1467
39	●●○○○○●○●○○○○○	14	28.57	S	0.1613
40	●●○○○○○●●○○○○○	14	28.57	UO	0.2535
41	●○●●○○○○●○○○○○	14	28.57	S	0.1613
42	●○●○●○○○●○○○○○	14	28.57	S	0.07031
43	●○●○○●○○●○○○○○	14	28.57	S	0.05903
44	●○●○○○●○●○○○○○	14	28.57	S	0.07031
45	●○○●●○○○●○○○○○	14	28.57	S	0.1467
46	●○○●○●○○●○○○○○	14	28.57	S	0.05903

Default values were used on initially disconnected networks. S denotes stabilization of the network and activity, SO denotes stable oscillatory behavior, UO denotes unstable oscillatory behavior, and TO denotes transient oscillatory behavior. All 1D clustering measures are given as exact values or to four significant figures.

and network structure. However, a handful of configurations with *high* inhibitory clustering and low proportions of inhibitory neurons also experienced complete stabilization. In general, as inhibitory clustering increases, systems developed oscillatory behaviors. We observe that the oscillatory regime also extends into slightly higher proportions of inhibitory neurons than the stabilization regime. Past the oscillatory regime, with the highest proportions of inhibitory neurons, and a range of inhibitory clustering values, we see systems experience unbounded growth.

5.1.3 Is There an Optimal Proportion of Inhibition for Inducing Oscillatory Behavior in 1D Networks Containing Inhibitory Clusters?

When high levels of inhibitory clustering are considered and networks are made up of repeated tile arrangements, thus creating network structures with one or more inhibitory clusters, a proportion of inhibitory neurons in the range [20%, 50%] induced oscillatory behavior in both network structure and electrical activity (Fig. 6.4). Within this range, some configurations experienced stable oscillations of network structure and electrical activity with small amplitudes, whilst the majority of other simulations experienced

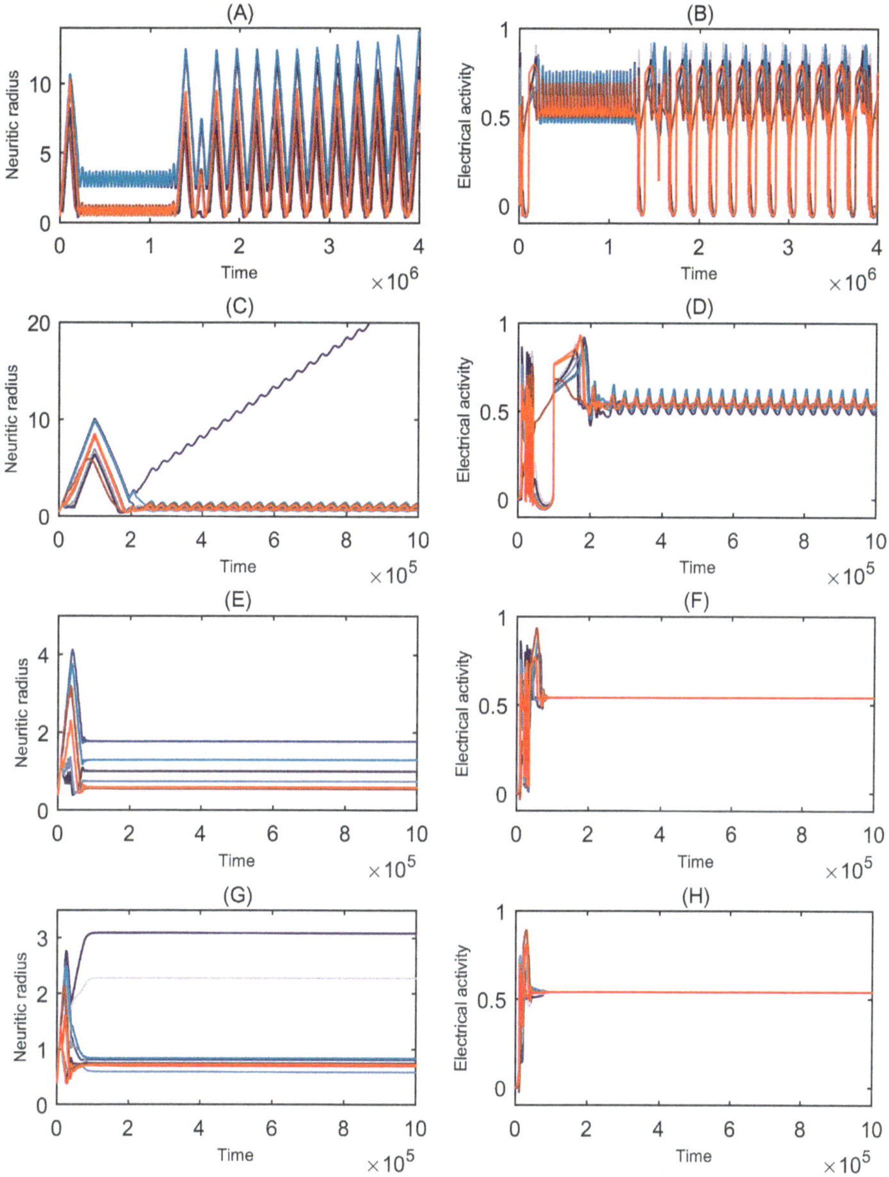

FIGURE 6.2 **Global system behavior of mixed 1D lattice networks with various levels of inhibitory clustering.** In all cases, $L = 10$, $N = 7$, $M = 3$ and default parameters listed in Table 6.1 are used. The networks are initially wholly disconnected and neurons are at rest. Figures on the *left-hand side* depict the dynamics of the individual neuritic radii values; figures on the *right-hand side* depict the dynamics of the individual electrical activity values. Neuron arrangements are as follows: (A and B), (C and D), (E and F), (G and H).

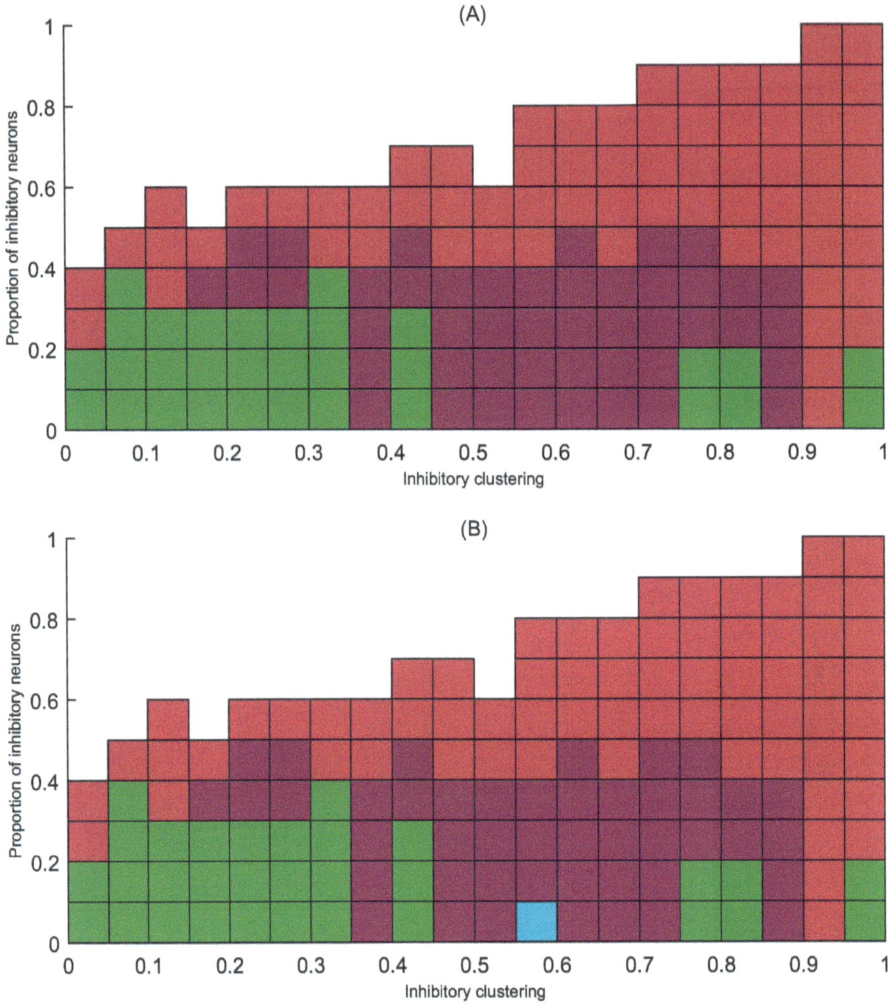

FIGURE 6.3 **1D system behaviors for growth simulation set-ups with various proportions of inhibitory neurons and levels of inhibitory clustering.** *Green boxes* denote stabilization of network structure and electrical activity. *Purple boxes* denote large unstable oscillations in network structure and unstable burst-like oscillations in electrical activity. *Blue boxes* denote stable oscillatory behavior of neuritic radii and electrical activities. *Pink boxes* denote unbounded growth examples, where the system was unable to reach the desired level of electrical activity for all neurons, with all neuritic fields growing indefinitely. *White areas* did not have any suitable configurations to run. In panel (A), the networks were initially disconnected. In panel (B), the networks started as regular ring networks.

sustained unstable burst-like oscillatory behavior of electrical activity and sustained unstable oscillations of neuritic radii values with larger amplitudes. For proportions of inhibition below 20%, all simulations experienced complete stabilization of network structure and electrical activity. With a proportion of inhibition above 50%, the systems were unable to achieve the desired level of electrical activity for every neuron ($\varepsilon = 0.6$). Neurons which are unable to achieve an electrical activity of ε experience unbounded growth of

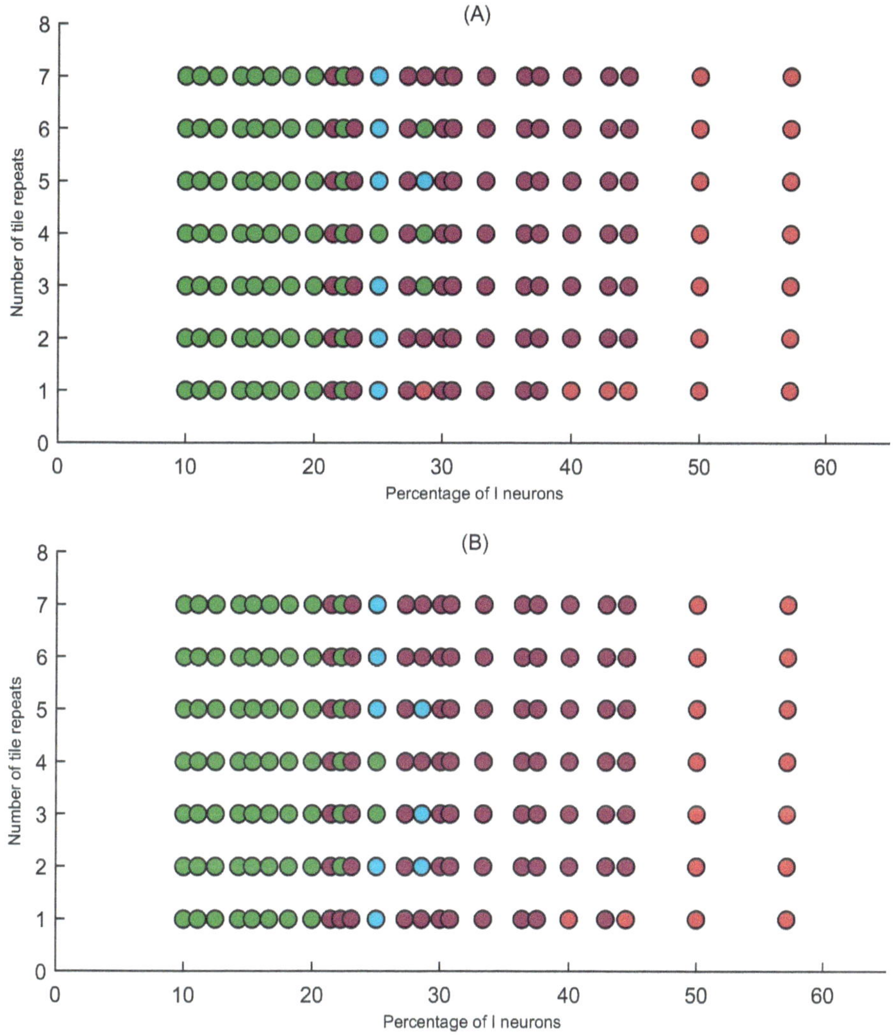

FIGURE 6.4 **1D system behaviors as a function of the percentage of inhibitory cells (*horizontal axis*) and number of tile repeats (*vertical axis*).** All tile arrangements are contiguous. Tiles are repeated in an adjacent manner to maintain a 1D system. *Green circles* denote networks experiencing complete stabilization. *Blue circles* denote periodic (stable) oscillatory behavior. *Purple circles* denote unstable oscillatory behavior. *Pink circles* denote simulations where all neuritic fields grow unbounded because the desired level of electrical activity cannot be reached. In panel (A), the networks were initially disconnected. In panel (B), the networks started with sparse connectivity, with neuritic radii values randomly generated in the interval $(0, 0.6)$ (identical seed of value 10 for all networks). All other parameters were as listed in Table 6.1.

their neuritic field, with their electrical activity stabilizing at a level below the homeostatic set-point. In such a situation, a stable network end state is unachievable. We see very similar regimes emerge when the starting network is either totally disconnected (Fig. 6.4A) or sparsely connected (Fig. 6.4B). In both cases, we observe a transition between stabilization and oscillatory behaviors as the proportion of inhibition exceeds 20%, and between oscillatory behaviors and unbounded growth as the proportion of inhibitory neurons exceeds that of excitatory neurons.

5.1.4 Network Analysis of 1D Global Behavior Types

To gain a deeper understanding of how distinct global network behaviors are characterized in terms of network theoretical measures, we plotted neuritic radii and electrical activity dynamics, alongside corresponding network structure visualizations and degree distributions, for three simulation configurations with distinct global behaviors (Figs. 6.5–6.7). Using the subgraph counting algorithm described by Ritchie et al.,[26] we performed motif analysis on networks at specific time points from each simulation to determine if global behavior types are also characterized by specific network motifs (Table 6.3). Here, we list the behavior types observed and describe each in terms of structural and electrical dynamics, and we discuss other noteworthy characteristics.

TOTAL STABILIZATION

Networks which experience total stabilization of their structure and electrical activity (Fig. 6.5) undergo a period of rapid overshoot in the number of connections, until a critical level of network connectivity is reached (Fig. 6.5A). At this stage, a rapid pruning of connections occurs, with all neuritic radii shrinking at a similar rate. Simultaneously, we observe an overshoot followed by a steep reduction in the electrical activity for all neurons (Fig. 6.5B). Following periods of overshoot and pruning, the neuritic radius and electrical activity for all neurons tend quickly toward their equilibrium values. Once all values have stabilized, we look to a visual representation of the final network structure (Fig. 6.5C) and the corresponding degree distribution (Fig. 6.5D) to understand the structural elements in more detail. In areas containing inhibitory neurons, neighboring excitatory neurons have large field sizes, and areas of high inhibitory clustering seem to coincide with even larger excitatory neighbors. The corresponding degree distribution suggests that the introduction of inhibitory neurons has resulted in a long-tailed degree distribution, with the presence of excitatory hub nodes with high degree that are able to contact the majority of the remaining neurons in the system. We note that the inhibitory clustering measure for the network structure (○○●○●○○●○○○●○), which experienced complete stabilization of both network structure and electrical activity, is 0.04707. Looking at the resulting network in terms of network motifs present within the structure (Table 6.3), we observe that the majority of motifs at stabilization contain open loops as opposed to closed loops.

STABLE OSCILLATIONS

Networks developing stable oscillatory behavior appear to experience relatively minor changes in their structure over time (Fig. 6.6). The associated electrical activity dynamics for this scenario (Fig. 6.1B) show very similar behavior to that of the neuritic radii values. After the initial period of overshoot and pruning of network connections, the stable oscillatory regime is reached, and both network structure and electrical activity experience stable oscillations of similar, small amplitudes. Once the system has reached the oscillatory regime, the network structure has a long-tailed degree distribution,

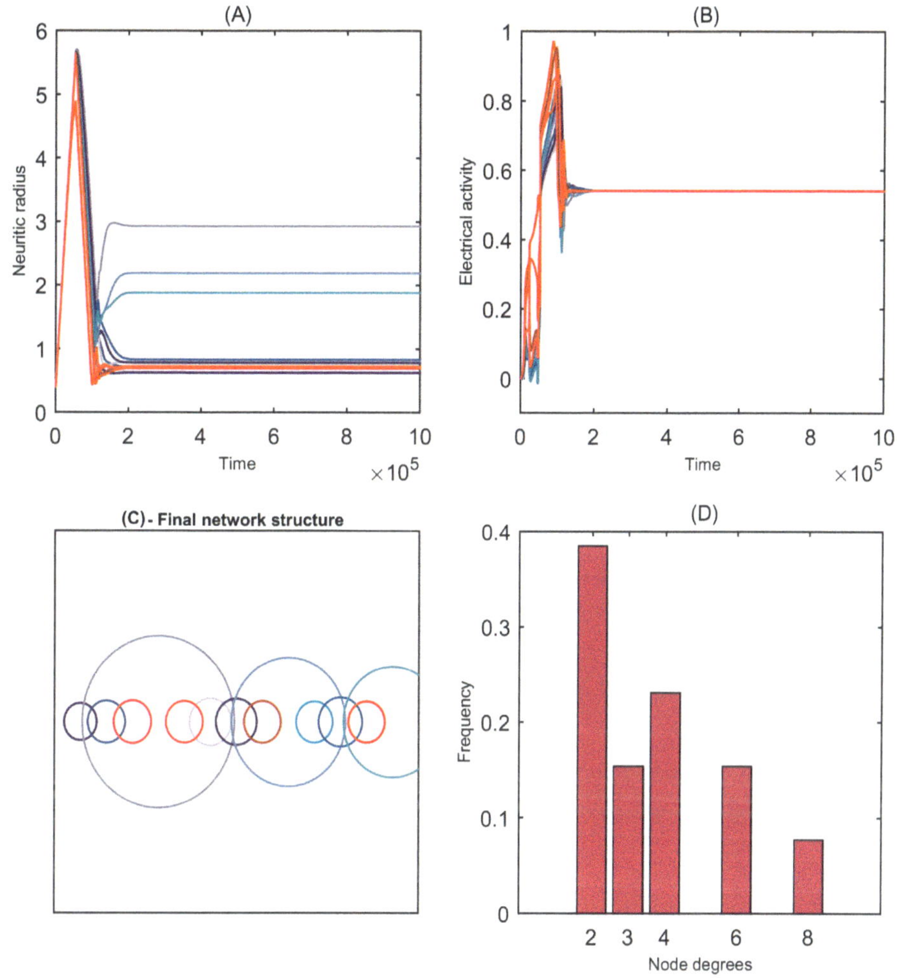

FIGURE 6.5 **Global system behavior and analysis of the associated network structure at equilibrium.** This is the same simulation as shown in Fig. 6.1E and F, with the spatial arrangement ●●●●● ●●●●●●●. Panel (A) depicts the dynamics of the neuritic radii, (B) depicts the dynamics of the electrical activity values, (C) is a visualization of the network's structure at the end of the simulation, where *circles with shades of blue* denote the field of an excitatory neuron and *circles with shades of red* denote the field of an inhibitory neuron. Panel (D) shows the degree distribution of all neurons at equilibrium.

indicating the existence of hub nodes. Looking at the dynamics of the two largest neuritic radii (both excitatory neurons neighboring one or more inhibitory neurons), we observe that their oscillations are almost out of phase with one another; it appears that it is the interaction between these two large neuritic fields which holds the system in a stable oscillatory regime. The inhibitory clustering measure for the network structure (○○○○○●●○○●○●○) is 0.2269. Furthermore, we note that the network structure during the low point in the oscillation of one neuritic radius (Fig. 6.6) contains very similar motifs to the

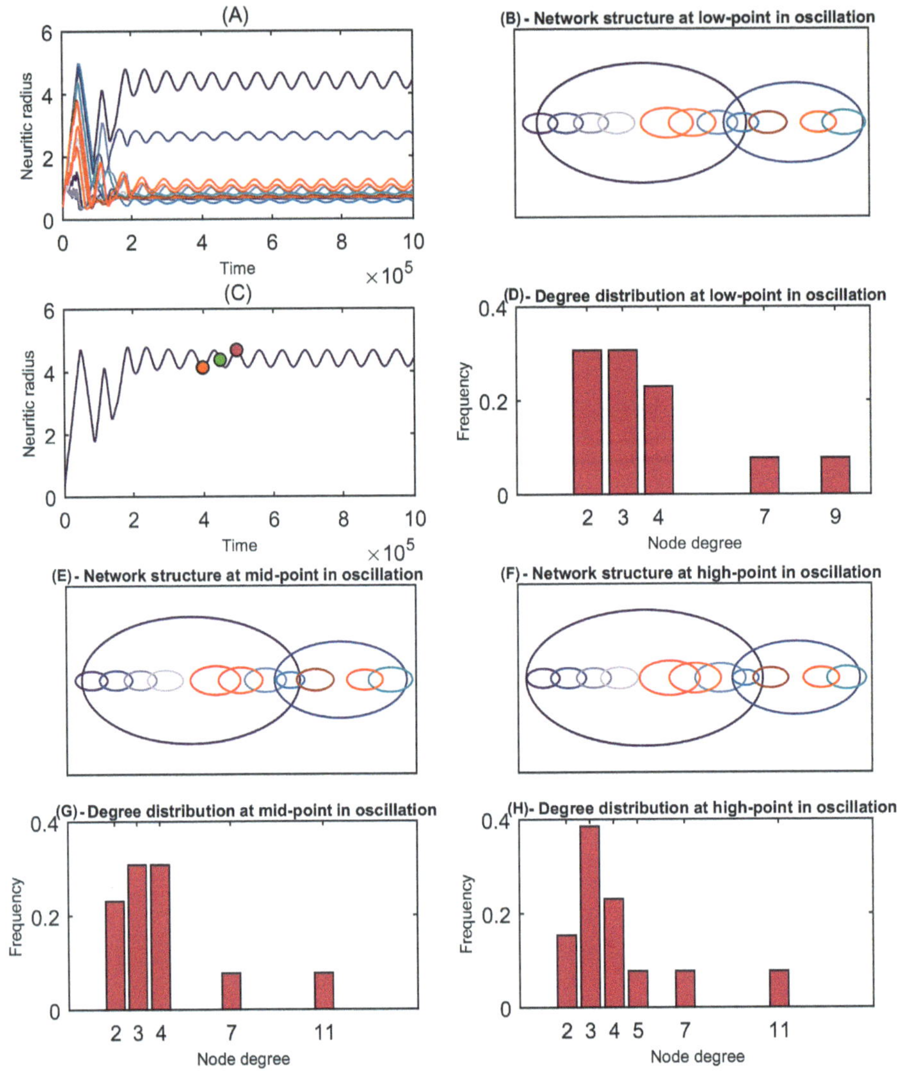

FIGURE 6.6 **Network structure and degree distribution of the simulation shown in Fig. 6.1A and B, at three time points during the periodic oscillations.** See colored markers in panel (C), where a *red circle* denotes a low point in an oscillation, a *green circle* denotes a mid-point in an oscillation, and a *pink circle* denotes a high-point in an oscillation. The spatial arrangement of neurons is ●●●●●●●●●●●●● with the network being initially disconnected. Panels (B) and (D) show the network structure and associated degree distribution at a low point in an oscillation, panels (E) and (G) show the network structure and associated degree distribution at a mid-point in an oscillation, and panels (F) and (H) show the network structure and associated degree distribution at a high-point of an oscillation shown in panel (C).

final network structure of the simulation which experienced complete stabilization (Fig. 6.5), but by the mid- and high points in the oscillations, the majority of motifs increase in number consistent with increased network connectivity (Table 6.3).

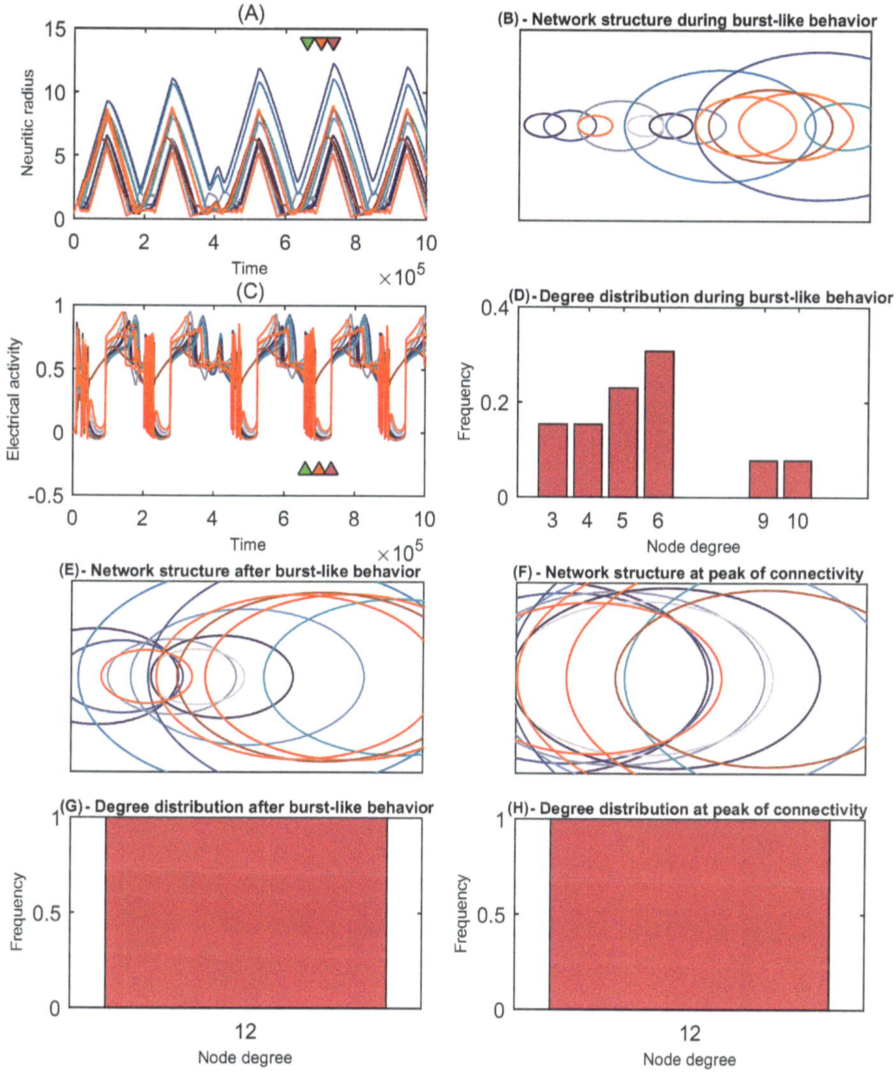

FIGURE 6.7 **Network structure and degree distribution of the simulation shown in Fig. 6.1C and D, at three time points during the unstable oscillations.** See colored markers in panel (C), where a *green triangle* denotes a time during burst-like behavior of electrical activity, a *red triangle* denotes a time immediately after the burst-like behavior, and a *pink triangle* denotes a time where the network is experiencing a peak of connectivity (corresponding triangles are also shown in panel A). The spatial arrangement of neurons is ●●●● ●●●●●●●● with the network initially disconnected. Panels (B) and (D) show the network structure and associated degree distribution during the burst-like behavior. Panels (E) and (G) depict the network structure and associated degree distribution after the burst-like behavior and panels (F) and (H) show the network structure and associated degree distribution at the peak of connectivity.

TABLE 6.3 Network Motif Counts at Various Points of the Simulations in Figs. 6.5–6.7

Reference	Fig. 6.5	Fig. 6.6	Fig. 6.6	Fig. 6.6	Fig. 6.7	Fig. 6.7	Fig. 6.7
Time	**1,000,000**	**397,061**	**448,310**	**496,220**	**659,940**	**700,040**	**733,720**
Motif				**Motif count**			
	1	1	3	4	17	715	715
	11	12	18	19	45	0	0
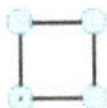	0	0	0	0	2	0	0
	0	0	0	0	0	0	0
	23	39	75	69	21	0	0
	13	13	17	19	38	286	286
Triangles involved in complete squares	4	4	11	15	36	286	286
Triangles involved in diagonal squares	12	13	17	17	36	0	0
Triangles not involved in complete/diagonal squares	0	0	0	0	0	0	0
	48	52	64	64	75	0	0
	51	43	28	31	77	0	0
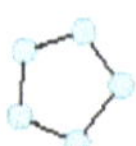	0	0	0	0	1	0	0
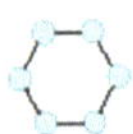	0	0	0	0	1	0	0

Network motifs are displayed here in two dimensions for ease of visualization, but all structures considered are 1D lattice structures with periodic boundaries.

UNSTABLE OSCILLATIONS

Networks experiencing sustained unstable oscillatory behavior undergo far more extreme changes in their structure, level of connectivity, and electrical activity over time (Fig. 6.7). From the start of the simulation, all neuritic radii endure unstable oscillations with large amplitudes, mostly in synchronization with one another (i.e., all neuritic fields in the system are expanding or retracting simultaneously). During each increase in field sizes, we see bursting behavior of the electrical activity values, followed immediately afterwards by a return to the resting electrical activity (equal to zero in the ES model). Coinciding with the peak of each oscillation in field size, we see a rapid increase in electrical activity for the majority of neurons. Following this, we observe unstable movements of the electrical activity values and the neuritic fields reduce in size, before the bursting behavior commences once again. In terms of network attributes, during unstable burst-like oscillations it is clear that the network structure alternates between full and partial connectivity at various points during the oscillations. During the bursting behavior of the electrical activity, the network structure is not fully connected, with a long-tailed degree distribution that implies the existence of excitatory hub nodes. Sometime after the bursting behavior, the network structure becomes fully connected once again and the cycle of behavior continues. The inhibitory clustering measure for the network structure (○○●○○○○○●●●○○) is 0.2914. Looking at motif analysis of the network structure at various time points (Table 6.3), it is immediately obvious that the changes in number and distribution of network motifs during the time course of this simulation are much more extreme when compared with other behavior types we analyzed in this section. At time 659, 940, when the electrical activity values are exhibiting burst-like behavior (Fig. 6.7), the network contains a wide array of motifs, from triangles and squares to pentagons and even hexagons. However, once the network becomes fully connected (after the bursting behavior and at the peak of network connectivity), we see that the network is entirely made up of fully connected squares (i.e., high clustering).

UNBOUNDED GROWTH

In some network configurations, it is possible for a single neuron not to be able to reach the desired level of activity ε. In such case, its neuritic field keeps expanding seeking activity homeostasis. However, once the neuritic field overlaps the neuritic fields of all remaining neurons, further increases in field size no longer yield any extra input. Hence, whilst the electrical activity stabilizes at a level below the homeostatic set-point, the growth behavior continues. As mentioned previously, this behavior is an idiosyncrasy of the ES model; it is not biologically relevant and therefore does not warrant further analysis.

5.2 2D Results

5.2.1 The Effect of Inhibitory Clustering on Global System Behavior in 2D

Growth simulations of various neuron arrangements on 2D lattice structures were implemented, and in all cases we observed either complete stabilization of network structure and electrical activity, or stable oscillations in network structure and electrical activity. Complete stabilization and stable, periodic oscillations both occurred after the typical initial overshoot and pruning periods (Fig. 6.8). No 2D growth simulations were observed to exhibit burst-like behavior of electrical activity and large unstable oscillations in network structure, as seen previously in 1D growth simulations, e.g., Fig. 6.2A and B. Although oscillatory 2D arrangements tended to have larger 2D clustering values, it was still possible to detect deviations from this general rule, e.g., see arrangements 51 and 52 in Table 6.4. Examples such as this suggest that the 2D inhibitory clustering measure (9) is overly sensitive to how inhibitory neurons are organized *within* the covering rectangle. Instead, when considering only the component of the 2D inhibitory clustering measure which characterizes the area of the smallest area within which all inhibitory neurons are found, i.e., WH/M, and testing various arrangements of $L=36$ neurons, with $M=6,7,8,9,10,11$ inhibitory neurons (and therefore varying levels of inhibition between 16.67% and 30.56%), we find a robust relationship whereby neuron arrangements with smaller WH/M values are more likely to experience oscillatory behavior (Fig. 6.9). Our results also show that, for a given WH/M value, networks with larger M, and hence a higher proportion of inhibitory neurons, are more likely to experience oscillatory behavior. Intuitively the observed relationship between WH/M and the presence of oscillatory behavior is logical, since the smallest possible value $WH/M=1$ signifies that the covering rectangle is of equal size to the total number of M inhibitory neurons and contains no excitatory neurons, meaning that the inhibitory neurons are maximally clustered, e.g., arrangement 47 in Table 6.4. However, even with a fixed number of inhibitory neurons, maximal clustering can be achieved in multiple ways and can exhibit distinct outcomes. For example, for $M=6$, maximal clustering can be achieved using a single row of six neurons or two rows of three neurons. When these arrangements were compared in growth simulations of a 2D network of 36 neurons with all else held constant, the single line of six inhibitory neurons induced complete stabilization, whereas the network containing a single inhibitory cluster of two lines of three neurons experienced stable oscillations in both network structure and electrical activity.

6 DISCUSSION

The main findings of our simulations relate to system stability and the effect of spatial clustering of inhibitory neurons. We demonstrate that subtle alterations in 1D and 2D arrangements of excitatory and inhibitory neurons can have marked effects on global behavior of electrical activity and network connectivity in the spatially embedded ES model. In particular, we find that lattice networks containing highly clustered inhibitory neurons are more likely to display oscillatory dynamics of both electrical activity and network structure than those where excitatory and inhibitory neurons are more evenly distributed across all lattice positions.

Instability is mostly evident in 1D networks, where burst-like behavior of electrical activity and large unstable oscillations in network structure can be observed. Results of 2D network simulations tend to be more stable; no burst-like behaviors are observed in 2D, suggesting that networks in higher spatial dimensions either mask or protect against the emergence of the most pathological network behaviors. It is important to note that although we describe periodic oscillations as being stable, this type of behavior does not necessarily imply a healthy network outcome.

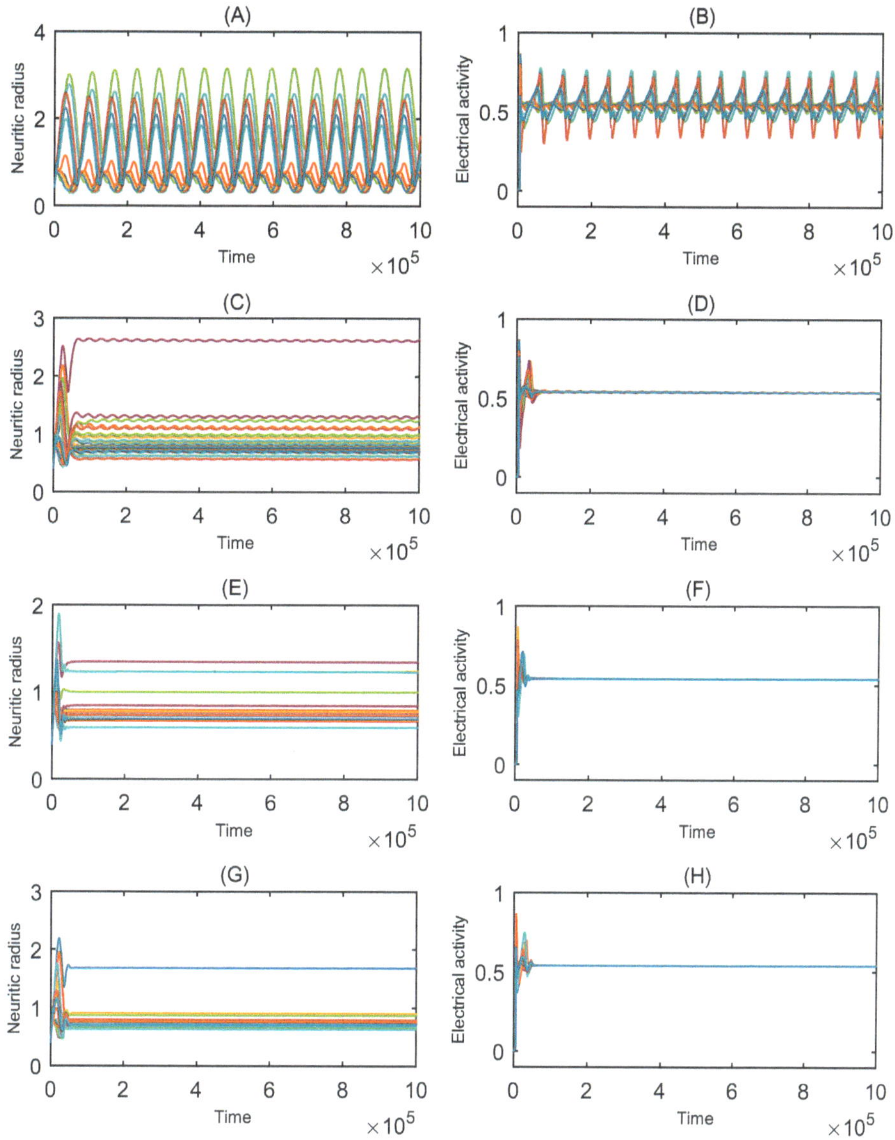

FIGURE 6.8 **Global system behavior of mixed 2D lattice networks with various levels of inhibitory clustering.** In all cases, $L = 36$, $N = 30$, $M = 6$ and default parameters listed in Table 6.1 are used. The networks are initially wholly disconnected and neurons are at rest. Figures on the *left-hand side* depict the dynamics of the individual neuritic radii values; figures on the *right-hand side* depict the dynamics of the individual electrical activity values. Panels (A) and (B) depict the dynamics of arrangement 47 in Table 6.4, panels (C) and (D) depict the dynamics of arrangement 51 in Table 6.4, panels (E) and (F) depict the dynamics of arrangement 49 in Table 6.4, and panels (G) and (H) depict the dynamics of arrangement 50 in Table 6.4.

TABLE 6.4 Global System Behavior and Clustering Measure for Mixed 2D Simulations with Various Spatial Arrangements

ID	Spatial arrangement	L	M	Behavior	2D clustering
47	● ● ● ○ ○ ○ ● ● ● ○	36	6	Stable oscillatory	1
48	● ● ○ ○ ○ ○ ○ ○ ○ ○ ○ ○ ● ○ ○ ○ ○ ○ ○ ○ ○ ● ● ○ ○ ○ ○ ○ ○ ○ ○ ○ ○ ● ○ ○	36	6	Stabilization	0.8965
49	● ● ○ ○ ● ○ ● ○ ○ ● ● ○	36	6	Stabilization, transient oscillations	0.7222
50	○ ○ ○ ● ○ ○ ● ● ● ● ○ ○ ● ○	36	6	Stabilization	0.6667
51	● ● ● ○ ○ ○ ● ○ ● ○ ○ ○ ○ ○ ● ○	36	6	Stable oscillatory	0.8056
52	○ ○ ● ○ ○ ○ ○ ○ ○ ● ○ ○ ○ ○ ● ● ○ ○ ○ ○ ○ ○ ● ○ ○ ○ ● ○ ○ ○ ○ ○ ○ ○ ○ ○	36	6	Stabilization	0.9794

Default values were used on initially disconnected networks.

No attempt has been made here to simulate the extraordinarily complex 3D structure of the developing and mature mammalian cerebral cortex using the spatially embedded ES model. Such results may lead to further dilution of the spatial effects given that higher dimensions lead to greater overall connectivity. This may in turn move the system closer to a well-mixed population where mean field models with fewer equations can provide accurate approximations. Nevertheless, there are interesting parallels between the ES

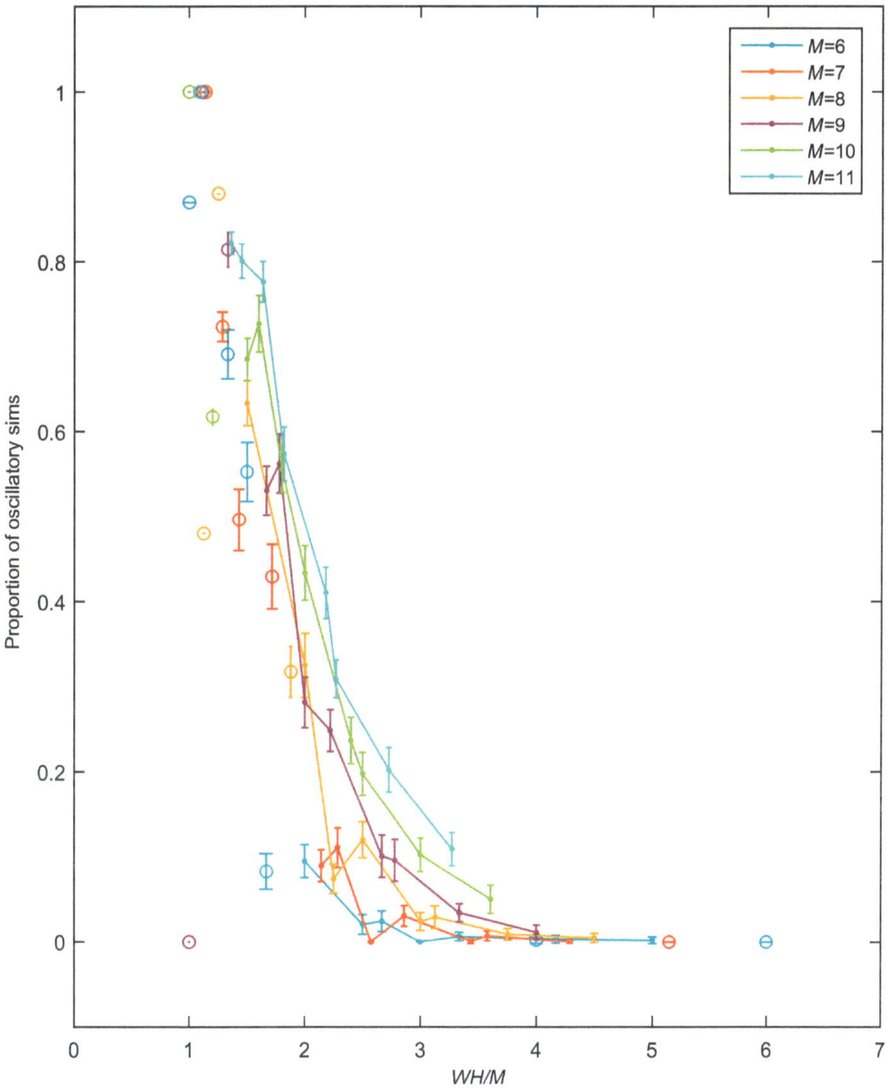

FIGURE 6.9 **The proportion of 2D neural network growth simulations depicting oscillatory behavior of electrical activity and network connectivity for a range of *WH*/*M* values, where the terms W, H, and M are defined in Eq. (6.9).** Growth simulations consisted of networks of $L = 36$ neurons with $M = 6, 7, 8, 9, 10, 11$ inhibitory neurons, and default parameters listed in Table 6.1 were used. Networks were initially wholly disconnected and neurons were at rest. For each M, 2 million neuron arrangements were generated at random, from which the unique set of *WH* values was extracted. For each (M, *WH*) pair, 200 neuron arrangements were selected at random. Growth simulations of all 200 arrangements were implemented and the result tested for behavior through thresholding of the norm of the variance of the last 10 data points, where the norm was taken over all L neurons. Thresholds were empirically determined and confirmed visually. The procedure of picking 200 arrangements for each (M, *WH*) pair was repeated 50 times to obtain the mean and standard deviation, displayed in the plot using error bars. Empty circles denote (M, *WH*) pairs for which less than 200 arrangements were available for testing.

model and the neurodevelopmental literature which will be discussed later. If it is a general rule that increased spatial dimension leads to greater tolerance of inhibitory clustering effects on network stability, then it is possible that 3D network activity-dependent network growth will be even more robust than 2D network growth.

Oscillatory behavior in network structure, especially an unstable one, has been considered, for the most part, to be an undesired outcome as it renders the network *fragile* and therefore more vulnerable to the development of a pathological dynamical outcome. Modeling work by Ehrens et al., e.g., suggested that seizures can manifest through small perturbations in the synaptic connections that render the entire cortical network unstable.[27] An alternative view is that instability of network structure that increases networks' sensitivity to random perturbations might allow a network to depart from an undesirable state, thus providing a means for resilience and adaptation. It is difficult to explore such a possibility with the present model due to the lack of separation between network structure and electrical activity. Whilst the authors of the original model were able to observe distinct oscillations in electrical activity and network structure, the simulations reported in this chapter, in order to pursue a particular research question, used a narrower set of parameters in which there was no separation of timescales.

6.1 Comparison to Previous Modeling Results

Previous studies of the ES model have postulated that outgrowth and interactions between excitation and inhibition can lead to complex patterns of development in individual cells.[18] Excitation increases activity but inhibits outgrowth, whereas inhibition does the opposite.[18] When dealing with a purely excitatory model, with individual neurons assigned various outgrowth rates (the parameter ρ in the ES model), complex periodic behavior of connectivity and electrical activity can be observed, with Van Ooyen and Van Pelt concluding that the precise network behavior depends on the spatial distribution of the cells and the distribution of the outgrowth properties over the cells.[8] A later study with a simple two-neuron $E-I$ model showed that the introduction of inhibition could account for network bistability.[9] It concluded that purely excitatory networks result in the same final attractor regardless of initial conditions, whereas mixed (excitatory and inhibitory) networks do not necessarily do so, i.e., they can be bistable. Further this study shed light on the excitation–inhibition balance, demonstrating that networks developing under conditions of relatively high inhibition result in a "pathological" state with oscillatory activity whereas a low level of inhibition during the initial stage of development enables the system to develop toward the "normal" stable end state.[9]

Our work reinforces the idea that networks with higher levels of inhibition are more likely to experience oscillatory dynamics of and on the network. However, through revealing distinct global behaviors for different spatial arrangements with a fixed number of excitatory and inhibitory neurons, we have established that the global behavioral dynamics and structural characteristics of the network rely not only on the balance between excitation and inhibition but also on the spatial arrangement of neurons within the network (and, to a limited extent, the initial conditions). In particular, we have offered that spatial clustering of inhibitory neurons, which we have characterized through measures of inhibitory clustering in 1D and 2D, is an important determinant of the network dynamics. Networks with higher levels of inhibitory clustering are more likely to undergo oscillatory dynamics in electrical activity and network structure than networks with low levels of inhibitory clustering.

There is little published literature available on the effects of increased inhibitory clustering on neuronal network structure and dynamics. In studies of the ES model, inhibitory neurons are found to (1) impose a structure on neighboring excitatory neurons, helping them to connect to different parts of a structure by inducing outgrowth and (2) increase the overall degree of connectivity in a network.[18] Van Ooyen et al. proposed that the number and distribution of inhibitory neurons is important: when inhibitory neurons are able to contact one another (i.e., our clustering measures Eqs. (6.8) and (6.9) become larger), they are electrically inhibited (self-inhibition), but their outgrowth will become stimulated.[18] This causes the ultimate level of inhibition in such a system to be higher than in a network without self-inhibition. It has also been noted that long-range inhibition is obtained when inhibitory cells occur in a clustered fashion and are able to stimulate one another's outgrowth.[18] More recently, Litwin-Kumar and Doiron investigated the implications of *excitatory* clustering on cortical activity, finding that even modest clustering substantially changed the behavior of the network activity, introducing slow dynamics during which clusters of neurons experience temporary changes in their firing rate.[28] In a network structure consisting of solely excitatory and inhibitory neurons, excitatory clustering and inhibitory clustering will be somewhat congruous concepts, and, in this respect, our results support and complement those of Litwin-Kumar and Doiron.[28]

To test the hypothesis that network structures with high levels of inhibitory clustering are more likely to undergo oscillatory behavior, we proposed measures of inhibitory clustering within 1D tile networks and 2D lattice networks in which neurons are assigned to lattice positions and networks are taken to have continuous boundary conditions. The fundamental idea that spatial arrangements affect dynamical behavior was highlighted in the original work but not systematically investigated.[18]

6.2 1D Network Behaviors

Performing growth simulations of 1D network arrangements, and in particular using the notion of tiles, means that an exhaustive investigation of the 1D network was possible. As it stands, our 1D measure (8) captures the extent of inhibitory clustering within a tile network and provides ease of comparison across tile network structures of different sizes. Even within 1D networks, there is scope for improvement, however. Whilst the measure takes into account the closeness of $(M-1)$ inhibitory pairs with M inhibitory neurons, it does not measure the *order* of these pairs. Using the current measure, it is therefore possible to find distinct spatial arrangements which produce an identical clustering measure but experience different global behaviors. For example, arrangements 21 and 22 in Table 6.2 have the same clustering measure, but one network experiences complete stabilization whilst the other undergoes stable oscillations.

The 1D inhibitory clustering measure allowed us to map out the relationship between inhibitory clustering and proportion of inhibition in a 1D system, along with the associated dynamics of and on the circular tile networks. Irrespective of whether the networks were initially connected or disconnected, we observed a clear structure (see Fig. 6.3A and B, respectively). However, it is important to note that the different regimes are extremely coarse grained: for each segment plotted, we ran one individual simulation, where the proportion of neurons was contained within an interval of width 10% and the clustering measure was contained within an interval of length 0.05. In order to obtain sufficient coverage of all segments, we considered various network sizes, a range of values for M and N, and various spatial arrangements. We therefore expected to observe increased variability on the boundaries of areas delimiting different regimes, and we treated the regimes produced with caution, focusing more on the qualitative results than on specific behaviors.

A similar structure is observed again when plotting the proportion of inhibitory neurons in a contiguous tile versus the number of adjacent tile repeats combined to form the entire 1D network (Fig. 6.4). The idea of having adjacent tile repeats is a similar concept to that explored in a modeling study by Malagarriga et al. in which mixed cortical macrocolumns, operating in a partially synchronized irregular regime, are seen to affect the dynamics of the network itself.[29] The authors found that in this type of mesoscopic network, excitation and inhibition spontaneously segregate, with some columns acting mainly in an excitatory manner while some others had predominantly an inhibitory effect on their neighbors. They also reported that hub nodes (i.e., nodes with high degree) are preferentially inhibitory, contrasting with our finding that increased inhibitory clustering induces *excitatory* neurons of high degree. Whilst we do not suggest that excitatory hubs are physiologically likely, it is worth mentioning that in our networks, hubs are an emergent property; whereas, in the above study, hubs are the result of imposing a scale-free network structure on the model. Furthermore, in agreement with Malagarriga et al.,[29] when we analyzed the characteristics of neuronal networks exhibiting distinct global behaviors, we observed some emerging segregation of excitation and inhibition, suggesting that clustered inhibitory neurons induce a secondary network "layer" of high degree excitatory neurons which could be responsible for the transport of information across the whole network. We acknowledge that the 1D networks studied here are biologically unrealistic due to their size and therefore there are limited conclusions to be drawn regarding the presence of network motifs and degree distributions. Nevertheless, our results speak to recent results from mathematical work by Ritchie et al. showing that in networks with identical degree distribution and clustering coefficient, differences in higher order structure (as assessed by varied subgraph counts) can have marked effects on the dynamics running *on* the networks.[26] Network analysis on simulations with larger networks consisting of 2D and 3D lattices with periodic boundary conditions would provide more insight into the emergence of specific network motifs and their function.

6.3 2D Network Behaviors

Increasing dimensionality leads to increasing complexity, such that an exhaustive list of 2D neural network growth simulations has not been attempted here. Further, designing the 2D inhibitory clustering measure (9) using a similar approach to the 1D measure (8) led to a loss of sensitivity and a not-always-monotonic relationship between the measure (9) and the probability of oscillatory dynamics being observed; see, e.g., arrangements 48 and 51 in Table 6.4. In fact, using simpler components of our 2D clustering measure to compute WH/M revealed a far clearer relationship between the arrangement of neurons and the global behavior of the network and its electrical activity (Fig. 6.9), where in general, configurations with smaller values WH/M were more likely to be oscillatory. In contrast to this relationship, we found some 2D arrangements contained such large clusters of inhibitory neurons, that the effects of this clustered inhibition appeared to be eliminated, and the system experienced complete stabilization (see, e.g., Fig. 6.9, $M=9$, $WH/M=1$, where 9 inhibitory neurons were arranged into a 3×3 grid of neurons and hence were maximally clustered, but the system experienced complete stabilization of network structure and electrical activity—note that the inhibitory neuron in the center of such an arrangement is completely surrounded by inhibitory neurons). This finding suggests that self-inhibition can have nontrivial effects and more work is needed to refine methods for measuring spatial clustering of neurons and to understand how local structure of neurons within the covering rectangle will determine global outcome.

Various approaches can be considered either to improve the sensitivity of the 2D clustering measure (Eq. 6.9) or to propose novel methods for quantifying clustering of neuron locations in mixed neuronal arrangements. If one considers inhibitory neurons as clusters in a "sea" of excitatory neurons, the average cluster size of the cluster size distribution coupled with intercluster distances may provide an improved picture of how the spatial arrangement of neurons influences the outcome at system and network level. Furthermore, topological or geometric measures such as the perimeter of an inhibitory clusters may add insights.

6.4 Future Modeling Studies

The ES model uses circular neuritic fields to model the area of influence of a single neuron. One consequence of this is that all neurons connect to their direct neighbors before connecting to neurons located further away topologically. As a result, the networks cannot develop the small-world properties observed in neuronal networks. Furthermore, restricting the state space of neuron locations to equally spaced lattice positions within a given area is not biologically realistic and greatly reduces the possible arrangements of neurons.

Analytical work on the spatially embedded ES model (especially in more than 1D) is not possible without engaging in coarse-grain modeling that would inevitably involve removing the very characteristic that we seek to study, namely, spatial embedding. Indeed, this has been the case for all analytical works carried out on this type of model to date, e.g., Van Ooyen and Van Pelt.[8] However, as pointed out by Wolf et al.,[30] "whenever the essential ingredients of a phenomenon need to be identified or when a qualitatively novel type of behavior demands conceptual advancement, idealization—even counterfactual idealization, that is the neglect of known features—is required."

Future modeling studies should investigate the effects of inhibitory clustering using an activity-dependent growth model where neurons are assigned to random locations within some state space, have a variable and discrete number of synaptic elements, and various synapse formation rules can be implemented (e.g., random, distance-dependent, or otherwise), in order to create specific network attributes. Such a modeling approach was shown to conserve small-world network properties in a study investigating cortical reorganization after focal retinal lesions.[31]

6.5 Future Experimental Studies

Until recently it was not appreciated that within the developing mesencephalon, there exist inhibitory interneurons whose lineage results in their being spatially clustered rather than randomly distributed.[21,22] It is not known how such an arrangement could impact on in vivo activity-dependent network development, but models such as the one explored in this chapter point to clustering as having a crucial role in network structure and function. Our results especially in 2D suggest that some degree of inhibitory clustering can be tolerated by the network. Thus in vivo it may be that network stability is maintained whilst allowing for the benefits of inhibitory clustering.

Networks in which there is inhibition of inhibition exist in vivo. The physiological effect of such connectivity is to increase network excitation. The fact that such networks can be detected in the mature nervous system again indicates that during development a certain level of inhibitory to inhibitory connectivity is tolerated. In vivo such arrangements are being actively studied but their functional implications remain poorly understood.[32] Interestingly, inhibitory to inhibitory interactions are seen only in certain inhibitory neuron subtypes of the extremely diverse inhibitory interneuron network.[33,34] Functionally such arrangements within the sensory system have been shown to increase salience of signal detection.[32,35] How such networks emerge during in vivo development and how inhibition of inhibition affects network activity-dependent growth remains to be explored.

Inhibitory neurons in the mature nervous system show greatest synaptic plasticity.[36] In relation to the ES model, it is interesting to note that in vivo the on-going shaping of network structure is crucially dependent on neural inhibition. In the mature nervous system, inhibitory synapses are more dynamic in response to the sensory environment than excitatory synapses.[36] Furthermore, suppression of inhibition leads to more exuberant yet less focused connectivity, and therefore in sensory systems a less functional connectivity.[13,36,37] Future models may explore how through inhibitory mechanisms, dynamic network stability is achieved, i.e., a network that can rapidly remodel to reflect changes in environment experience yet one in which the essential structure and dynamics are preserved.

Excessive inhibition and disruption of inhibition leads to network pathology. As discussed, the results of our modeling study indicate that a certain level of inhibitory clustering can be tolerated and result in a stable network. However, even small changes in the arrangement or proximity between inhibitory neurons can lead to instability. In vivo we suggest that early clustering of inhibitory neurons that inhibit one another during the activity-dependent stage of neural development will lead to excessive connectivity and network instability. Whilst it is unknown how inhibitory clustering might play out in vivo studies of neural migration disorders there are strong hints that such a mechanism or a similar mechanism may be of importance.[38] We suggest that there should be particular emphasis placed on the detection in vivo of inhibitory neuron clusters and an exploration of their potential role in network pathology. Finally, in an era where clinical neuroscience is exploring neural plasticity therapy as a way of overcoming the effects of brain injury, a focus on neural inhibition as a potential target for inducing favorable network remodeling [17] will be informed by dynamic neural network growth models in which both the quantity and the spatial arrangement of inhibition are important variables.

6.6 Concluding Remarks

Simulations involving the ES model suggest that the complex dynamical behavior neuronal networks experience is influenced by a combination of factors, including the balance between excitation and inhibition, the spatial topology of those excitatory and inhibitory neurons, and the connections between them. There are too many intrinsic limitations to the ES model to be allowed to make strong statements about cortical development. Nevertheless, this work suggests that modeling studies should be careful to attribute particular dynamic properties to the E/I ratio as such properties may be meaningful in well-mixed populations but insufficiently descriptive in the kind of structured networks found in physiological systems. This certainly highlights the need for further work in developing tractable mathematical models on structured networks, where balancing the need for analytical tractability versus model realism is key in order to identify clear dependencies between model ingredients and outcomes.

Acknowledgment

Rosanna Barnard acknowledges funding for her PhD studies from EPSRC (Engineering and Physical Sciences Research Council) EP/M506667/1. Simon Farmer acknowledges funding support from the UCLH Biomedical Research Centre.

References

1. Pellegrini GL, de Arcangelis L, Herrmann HJ, Perrone-Capano C. Activity-dependent neural network model on scale-free networks. *Phys Rev E*. 2007;76(1):016107.
2. Cohan CS, Kater SB. Suppression of neurite elongation and growth cone motility by electrical activity. *Science*. 1986;232(4758):1638–1640.
3. Fields DR, Neale EA, Nelson PG. Effects of patterned electrical activity on neurite outgrowth from mouse sensory neurons. *J Neurosci*. 1990;10(9):2950–2964.
4. Schilling K, Dickinson MH, Connor JA, Morgan JI. Electrical activity in cerebellar cultures determines Purkinje cell dendritic growth patterns. *Neuron*. 1991;7(6):891–902.
5. Oppenheim RW. Cell death during development of the nervous system. *Annu Rev Neurosci*. 1991;14 (1):453–501.
6. Ferrer I, Soriano E, Del Rio JA, Alcantara S, Auladell C. Cell death and removal in the cerebral cortex during development. *Progr Neurobiol*. 1992;39(1):1–43.
7. Grossberg S. Nonlinear neural networks: Principles, mechanisms, and architectures. *Neural Networks*. 1988;1 (1):17–61.
8. Van Ooyen A, Van Pelt J. Complex periodic behaviour in a neural network model with activity-dependent neurite outgrowth. *J Theoret Biol*. 1996;179(3):229–242.
9. Van Oss C, Van Ooyen A. Effects of inhibition on neural network development through activity-dependent neurite outgrowth. *J Theoret Biol*. 1997;185(2):263–280.
10. Turrigiano GG, Nelson SB. Homeostatic plasticity in the developing nervous system. *Nat Rev Neurosci*. 2004;5 (2):97–107.
11. Hengen KB, Lambo ME, Van Hooser SD, Katz DB, Turrigiano GG. Firing rate homeostasis in visual cortex of freely behaving rodents. *Neuron*. 2013;80(2):335–342.
12. Henley J, Poo Mm. Guiding neuronal growth cones using Ca^{2+} signals. *Trends Cell Biol*. 2004;14(6):320–330.
13. Dong W, Aizenman CD. A competition-based mechanism mediates developmental refinement of tectal neuron receptive fields. *J Neurosci*. 2012;32(47):16872–16879.
14. Yamamoto N, López-Bendito G. Shaping brain connections through spontaneous neural activity. *Eur J Neurosci*. 2012;35(10):1595–1604.
15. Luhmann HJ, Fukuda A, Kilb W. Control of cortical neuronal migration by glutamate and GABA. *Front Cell Neurosci*. 2015;9:4.
16. Haider B, Duque A, Hasenstaub AR, McCormick DA. Neocortical network activity in vivo is generated through a dynamic balance of excitation and inhibition. *J Neurosci*. 2006;26(17):4535–4545.

17. Ganguly K, Poo Mm. Activity-dependent neural plasticity from bench to bedside. *Neuron*. 2013;80(3):729–741.
18. Van Ooyen A, Van Pelt J, Corner MA. Implications of activity dependent neurite outgrowth for neuronal morphology and network development. *J Theoret Biol*. 1995;172(1):63–82.
19. Ang ES, Haydar TF, Gluncic V, Rakic P. Four-dimensional migratory coordinates of GABAergic interneurons in the developing mouse cortex. *J Neurosci*. 2003;23(13):5805–5815.
20. Tanaka DH, Yanagida M, Zhu Y, et al. Random walk behavior of migrating cortical interneurons in the marginal zone: time-lapse analysis in flat-mount cortex. *J Neurosci*. 2009;29(5):1300–1311.
21. Brown KN, Chen S, Han Z, et al. Clonal production and organization of inhibitory interneurons in the neocortex. *Science*. 2011;334(6055):480–486.
22. Ciceri G, Dehorter N, Sols I, Huang ZJ, Maravall M, Marín O. Lineage-specific laminar organization of cortical GABAergic interneurons. *Nat Neurosci*. 2013;16(9):1199–1210.
23. Bullmore E, Sporns O. Complex brain networks: graph theoretical analysis of structural and functional systems. *Nat Rev Neurosci*. 2009;10(3):186–198.
24. Sporns O. The non-random brain: efficiency, economy, and complex dynamics. *Front Computat Neurosci*. 2011;5:5.
25. Van den Heuvel MP, Sporns O. Network hubs in the human brain. *Trends Cogn Sci*. 2013;17(12):683–696.
26. Ritchie M, Berthouze L, House T, Kiss IZ. Higher-order structure and epidemic dynamics in clustered networks. *J Theoret Biol*. 2014;348:21–32.
27. Ehrens D, Sritharan D, Sarma SV. Closed-loop control of a fragile network: application to seizure-like dynamics of an epilepsy model. *Front Neurosci*. 2015;9:58.
28. Litwin-Kumar A, Doiron B. Slow dynamics and high variability in balanced cortical networks with clustered connections. *Nat Neurosci*. 2012;15(11):1498–1505.
29. Malagarriga D, Villa AE, Garcia-Ojalvo J, Pons AJ. Mesoscopic segregation of excitation and inhibition in a brain network model. *PLoS Comput Biol*. 2015;11(2):e1004007.
30. Wolf F, Engelken R, Puelma-Touzel M, Weidinger JDF, Neef A. Dynamical models of cortical circuits. *Curr Opin Neurobiol*. 2014;25:228–236.
31. Butz M, van Ooyen A. A simple rule for dendritic spine and axonal bouton formation can account for cortical reorganization after focal retinal lesions. *PLoS Comput Biol*. 2013;9(10):e1003259.
32. Pi HJ, Hangya B, Kvitsiani D, Sanders JI, Huang ZJ, Kepecs A. Cortical interneurons that specialize in disinhibitory control. *Nature*. 2013;503(7477):521–524.
33. Ma Y, Hu H, Agmon A. Short-term plasticity of unitary inhibitory-to-inhibitory synapses depends on the presynaptic interneuron subtype. *J Neurosci*. 2012;32(3):983–988.
34. Pfeffer CK, Xue M, He M, Huang ZJ, Scanziani M. Inhibition of inhibition in visual cortex: the logic of connections between molecularly distinct interneurons. *Nat Neurosci*. 2013;16(8):1068–1076.
35. Jiang X, Wang G, Lee AJ, Stornetta RL, Zhu JJ. The organization of two new cortical interneuronal circuits. *Nat Neurosci*. 2013;16(2):210–218.
36. Villa KL, Berry KP, Subramanian J, et al. Inhibitory synapses are repeatedly assembled and removed at persistent sites in vivo. *Neuron*. 2016;89(4):756–769.
37. Fredj NB, Hammond S, Otsuna H, Chien CB, Burrone J, Meyer MP. Synaptic activity and activity-dependent competition regulates axon arbor maturation, growth arrest, and territory in the retinotectal projection. *J Neurosci*. 2010;30(32):10939–10951.
38. Luhmann HJ, Sinning A, Yang JW, et al. Spontaneous neuronal activity in developing neocortical networks: from single cells to large-scale interactions. *Front Neural Circuits*. 2016;10.

CHAPTER

7

A Detailed Model of Homeostatic Structural Plasticity Based on Dendritic Spine and Axonal Bouton Dynamics

Markus Butz-Ostendorf[1] *and Arjen van Ooyen*[2]

[1]Biomax Informatics AG, Planegg, Germany [2]VU University Amsterdam, Amsterdam, The Netherlands

OUTLINE

1 INTRODUCTION

Raisman[1] was the first to show that the brain rewires its connectivity in response to lesions. He demonstrated that cutting one of the two fiber bundles innervating the septal nuclei in the forebrain caused the remaining fiber tract to sprout new axon collaterals and

The Rewiring Brain
DOI: http://dx.doi.org/10.1016/B978-0-12-803784-3.00007-X

to replace the disconnected tract. With this early study, Raisman paved the way for over five decades of lesion studies. Lesion studies serve two important goals. First, they reveal the brain's potential for self-repair after lesions (e.g., stroke) and for compensatory reorganization during neurodegeneration (e.g., Alzheimer's disease). Second, lesion studies help to understand how experience creates memory traces in the brain's connectivity. Peripheral lesions, such as monocular deprivation, cause a change in experience, triggering alterations in connectivity.[2,3] This second goal of lesion studies is nowadays better addressed by time-lapse in vivo imaging in freely behaving animals.[4]

In the early 1980s, Eysel et al.[5] developed a well-reproducible technique for accurately destroying a small circumscribed area of the retina. Owing to the retinotopic organization of the visual cortex, such a peripheral lesion removes the input of a precisely locatable area in the primary visual cortex. Although the peripheral lesion initially leaves the retinocortical and corticocortical connections intact, the brain responds with massive structural reorganization, e.g., axonal sprouting[6,7] and dendritic spine turnover,[8] in the affected cortical area. This structural reorganization goes along with functional reorganization of the visual cortex.[9,10] In a similar manner, digit amputations cause a cortical remapping of the digit representations in the primary somatosensory cortex such that the remaining fingers enlarge their cortical representations into the representation of the lost digit.[11–13]

Wolff et al.[14] were probably the first to realize the impact of electrical activity on brain reorganization after lesions. They showed that locally applying GABA to the cervical ganglion of rats, thereby lowering the level of electrical activity, induced compensatory dendritic sprouting and spinogenesis, enhancing the cells' potential to make new synaptic connections and, therefore, to restore activity. Wolff et al. used light and electron microscopy to study reactive changes in the dendritic and axonal structures that make up the synapse, so-called postsynaptic and presynaptic elements.[15] Inspired by these early observations, Wolff postulated the Compensation Theory of Synaptogenesis. This theory states that neurons respond locally with an increase in postsynaptic elements if their electrical activity is below a certain set-point and with a decrease in postsynaptic elements if the electrical activity becomes too high. According to the Compensation Theory, activity may also exert morphogenetic effects on axons.

On the basis of the Compensation Theory, Dammasch and coworkers[16–18] formulated the Compensation Algorithm, which, in contrast to existing models of synaptic plasticity used in Hopfield[19] and Willshaw networks,[20] for the first time modeled the synapse as consisting of pre- and postsynaptic elements that develop independently and merge randomly to form synapses. Dammasch showed that the Compensation Algorithm can lead to highly specific formation of synapses, even emulating some forms of Hebbian learning,[21] although pre- and postsynaptic elements depend solely on the cell's level of electrical activity and pre- and postsynaptic elements merge purely randomly. The Compensation Algorithm was a main inspiration for the Model of Structural Plasticity (MSP)[22] introduced in this chapter.

Independently, Van Ooyen and Van Pelt developed the Neurite Outgrowth Model in the mid-1990s[23–26] in order to simulate the growth of cell cultures (see Chapter 5 of this book). In their model, the neuritic fields (i.e., axonal and dendritic extensions) of dissociated cultured neurons are approximated by circles, with a soma at the center of each circle. Similarly to the Compensation Algorithm, the circles grow out and retract in an activity-dependent, homeostatic manner. If activity is lower than a certain set-point, the radius of the cell's circle increases; if activity becomes higher than the set-point, the radius decreases. If the circles overlap, cells become bi-directionally connected with a connectivity proportional to the area of overlap. The radius of the circle remains constant if the cell's activity is at the set-point. Owing to its simplicity, the Neurite Outgrowth Model became very popular and has recently been used to study self-organized criticality in developing cell cultures.[27,28]

The MSP combines features of the Compensation Algorithm[16] and the Neurite Outgrowth Model.[23] It adopts the idea of synaptic elements from the Compensation Algorithm, while the spatial organization, with distance-dependent synapse formation, is taken from the Neurite Outgrowth Model. The attractiveness of MSP is that it is formulated at the right degree of detail in order to simulate recent in vivo imaging data on dendritic spine and axonal bouton dynamics[7,8,29–31] and distance-dependent synapse formation,[32] avoiding the costly simulation of detailed dendritic morphologies with multicompartment models.[33]

Many experimental findings, both in vitro[34–41] and in vivo,[2,8,29,30,42–45] suggest that a persistent decrease in a neuron's electrical activity below a set-point value triggers the formation of spines and boutons, whereas a persistent increase in electrical activity above the set-point or below a certain minimum level results in deletion of spines and boutons. Thus when a neuron deviates in either direction from a desired set-point activity, formation and deletion of spines and boutons, accompanied by synapse formation and deletion, have the potential to bring activity back to the set-point value (homeostasis). MSP can implement such activity-dependent spine and bouton dynamics and can in principle be used with any spatial network topology, from a one-dimensional[46] or two-dimensional topology[22] up to a layered cortical organization[47] or a full 3D topology with intra- and interarea connectivities (in prep). Moreover, the implementation of MSP in the freely available and highly parallelized modeling framework NEST (www.nest-initiative.org) makes it feasible to create large-scale networks.[47]

In the next sections, we describe MSP and its main results and compare them with experimental findings. In the Discussion, we provide some perspectives on future experimental and modeling studies. The text of this chapter is partly based on Butz and Van Ooyen.[22]

2 MODEL

First, a general overview of MSP is given, followed by a detailed description of the different components of the model.

2.1 The Model at a Glance

From a theoretical point of view, the probability that two adjacent neurons form new synapses or break existing ones depends on the number of synaptic contact possibilities that each neuron has, i.e., the number of axonal boutons and dendritic spines. In principle, changes in spine and bouton number can be caused directly by changes in the numerical density of spines and boutons or indirectly by neurite outgrowth and retraction. Since we are interested in the effective change in synaptic contact possibilities, we abstract away from detailed neuronal morphology by using point neurons that carry sets of synaptic elements representing axonal boutons (axonal elements), excitatory dendritic spines (excitatory dendritic elements), and inhibitory postsynaptic densities (inhibitory dendritic elements) (Fig. 7.1B). Excitatory and inhibitory neurons form exclusively excitatory or inhibitory axonal elements, respectively, but can express both excitatory and inhibitory dendritic elements. Synaptic elements develop independently of potential contact partners[15,48] and connect only to matching synaptic counterparts,[49] which motivated our idea of rendering axonal and dendritic elements as "plugs" and "sockets," respectively (Fig. 7.1B). Synapses are formed by merging corresponding synaptic elements or are deleted when synaptic elements are lost. Thus activity-dependent changes in the number of synaptic elements can alter synaptic connectivity.

The execution of the model proceeds as follows. We continuously determine for each neuron its electrical activity on a millisecond time scale t. Electrical activity causes influx of calcium. Calcium concentration decays exponentially to zero, so the neuron's intracellular calcium concentration effectively comes to represent a time-averaged level of its electrical activity. The intracellular calcium concentration then drives continuous changes in the neuron's number of synaptic elements.[50] In general, we use a growth rule by which each neuron creates new synaptic elements when its electrical activity (calcium concentration) is below a given homeostatic set-point and deletes synaptic elements when its activity is above this set-point or below a certain minimum value (Fig. 7.2A). Structural changes, i.e., changes in the number of synaptic elements, are much slower than changes in electrical activity, so reorganization of network connectivity and the dynamics of electrical activity take place on different time scales and do not directly interfere with each other. Because formation and deletion of synaptic elements occur continuously over time (reflecting an implicit growth process) but the formation and breaking of synapses are singular events, we update connectivity at discrete time steps. Since changes in synaptic elements are slow and the algorithm for updating connectivity is computationally costly, we update connectivity only every 100 ms. The structural dynamics are slow enough to be comparable to changes in neuronal morphology and cortical connectivity over periods of days to several months. However, we do not unnecessarily slow down structural plasticity to the point that the entire simulation time would sum up to weeks or months in milliseconds.

In every connectivity update, we determine for each neuron how many vacant synaptic elements (i.e., not bound in a synapse) are available for synapse formation, and in case the number of synaptic elements decreased, which of its synaptic elements, either vacant or bound in a synapse, should be removed. In every connectivity update, deletion of bound synaptic elements immediately causes the breaking of synapses. Vacant synaptic elements spontaneously decay with a certain time constant. For synapse formation, we randomly distribute vacant synaptic elements on complementary synaptic elements, with excitatory axonal elements binding to excitatory dendritic elements, and inhibitory axonal elements connecting to inhibitory dendritic elements (Fig. 7.1B). Synapse formation is additionally dependent on the (Euclidean) distance between neurons, making it more likely that neurons connect to neighboring neurons than to remote neurons. This scheme is repeated for every update in connectivity.

An important feature of MSP is that after the breaking of a synapse, the complementary synaptic element remains, i.e., the axonal element if the dendritic element was deleted or the dendritic element if the axonal element was deleted. The vacant synaptic element is available for synapse formation again, so synaptic connections can rewire.

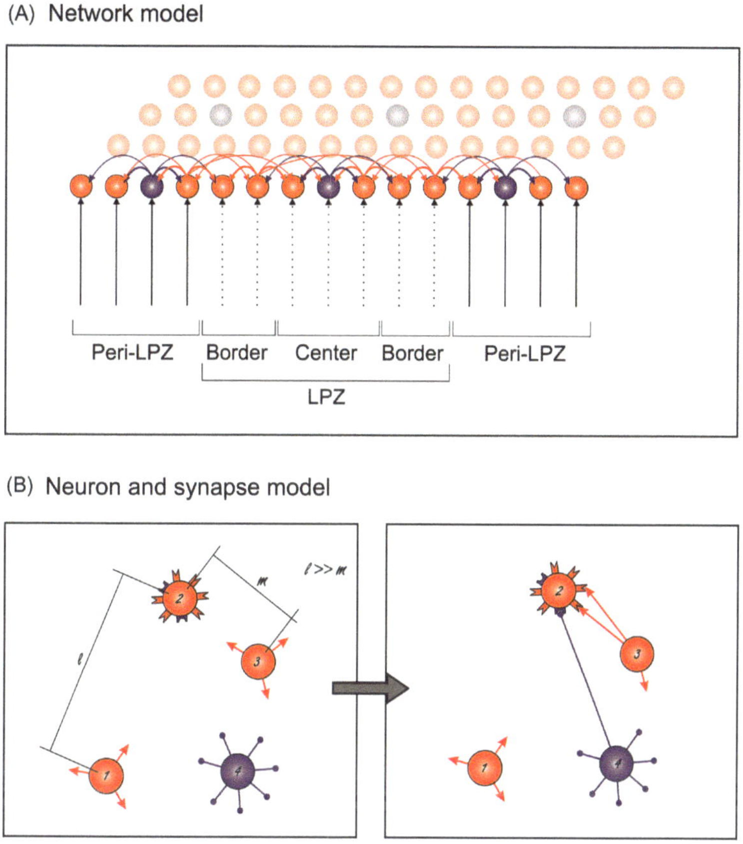

FIGURE 7.1 **Schematic representation of the Model of Structural Plasticity (MSP).** (A) The two-dimensional network consists of excitatory (*red*) and inhibitory neurons (*blue*), which receive both external input (*vertical connections*) and input from other neurons in the network (*horizontal connections*). Neurons are interconnected by directed synapses that have a fixed strength. The linewidth indicates the number of synapses between neurons. A circumscribed loss of external input defines the lesion projection zone (LPZ), which can be subdivided into the center and the border. The peri-LPZ consists of neurons that surround the LPZ. (B) Synapses consist of recombinable axonal (plugs) and dendritic elements (sockets). Axonal and dendritic elements are either excitatory (*red*) or inhibitory (*blue*). Complementary axonal and dendritic elements merge to form synapses. The probability that a synapse is formed between two neurons depends on the number of elements that each neuron has and on the Euclidean distance between the neurons. Thus synapse formation from neuron 3 to 2 with distance m is more likely than from neuron 1 to 2 with distance l, given that $l \gg m$ and neurons 1 and 3 have the same number of vacant axonal elements. Source: *Figure reproduced from Butz M, van Ooyen A. A simple rule for dendritic spine and axonal bouton formation can account for cortical reorganization after focal retinal lesions. PLoS Comput Biol. 2013;9(10):e1003259.*

2.2 Electrical Activity

In principle, the outcomes and predictions of MSP do not depend on the choice of neuron model. That is, the same structural dynamics can, with some adjustments of the model parameters, be obtained with any neuron model for electrical activity. Hence, several different neuron models have been used in different versions of MSP. The early

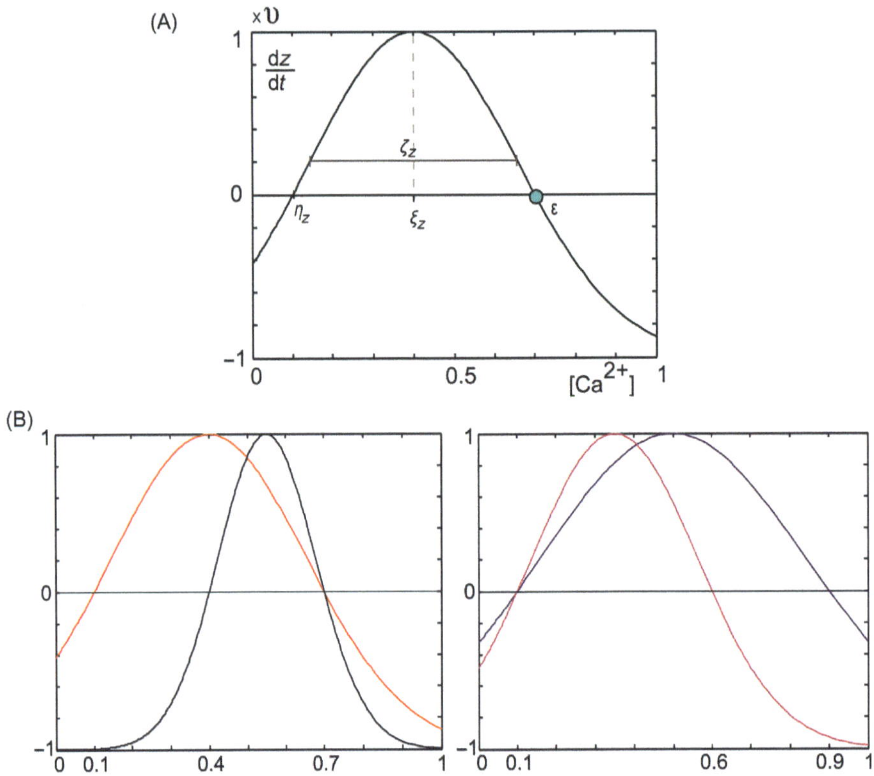

FIGURE 7.2 **Growth curves.** (A) The graph depicts the principal shape of the growth curves, giving the change dz/dt (with ν denoting the growth rate) in each type of synaptic element z as a function of the intracellular calcium concentration $[Ca^{2+}]$. Parameter ξ_z is the center and ζ_z is the width of the Gaussian curve. Parameters η_z (unstable fixed point) and ε (stable fixed point or homeostatic set-point) are the intersections with the x-axis. (B) (*Left panel*) Two growth curves with different values of η_z but the same value of ε. In our study on cortical reorganization after focal retinal lesions[22] (see Figs. 7.3 and 7.4), the *red curve* was used for the dendritic elements of each neuron and the *black curve* for the axonal elements of each neuron. (*Right panel*) Two growth curves with different values of ε but the same value of η_z. These growth curves were used for different cell types in a large-scale model of the development of a cortical column.[47] Note that the growth rules for the axonal and dendritic elements of each individual neuron must have the same value of ε (see text). Source: *Panel A reproduced from Butz M, van Ooyen A. A simple rule for dendritic spine and axonal bouton formation can account for cortical reorganization after focal retinal lesions. PLoS Comput Biol. 2013;9(10):e1003259.*

Compensation Algorithm used McCulloch-Pitts neurons, a simple spiking neuron model with a discrete time scale. MSP in the original version[22] used Izhikevich neurons,[51] a spiking neuron model with a wide range of biologically realistic firing patterns yet more cost efficient than, e.g., Hodgkin–Huxley neurons.[52] In the Izhikevich model, the membrane potential v of a neuron is computed by two differential equations:

$$\frac{dv}{dt} = k_1 v^2 + k_2 v + k_3 - u + I \tag{7.1}$$

$$\frac{du}{dt} = a(bv - u) \tag{7.2}$$

where v is in mV, time t is in ms, $k_1 = 0.04\ \text{mV}^{-1}\ \text{ms}^{-1}$, $k_2 = 5\ \text{ms}^{-1}$, $k_3 = 140\ \text{mV ms}^{-1}$, u is a membrane recovery variable, and I represents the input to the neuron. Each time the membrane potential exceeds 30 mV (i.e., fires a spike), v and u are reset to their resting values:

$$\text{if } v \geq 30\ \text{mV, then } \begin{cases} v \leftarrow c \\ u \leftarrow u + d \end{cases} \tag{7.3}$$

Parameters were set to $a = 0.1\ \text{ms}^{-1}$, $b = 0.2\ \text{ms}^{-1}$, $c = -65\ \text{mV}$, and $d = 2\ \text{mV ms}^{-1}$. Diaz-Pier et al.[47] used simple integrate-and-fire neurons for large-scale simulations, which produced on the whole the same structural dynamics as simulations with Izhikevich neurons. Importantly, routines in NEST for structural plasticity can be combined with any spiking neuron model available in NEST.

Each neuron receives input $I = I^{ext} + I^{syn}$ that comprises synaptic input I^{syn} from other neurons in the network (horizontal input from adjacent and distant cortical neurons) and external input I^{ext} (Fig. 7.1A). Input I^{ext} subsumes all input (e.g., sensory input) that is not modeled explicitly. As in the original version of the neuron model,[51] we define the external input I^{ext} as a noisy current (white noise). Explicitly modeled intra-area or interarea network connectivity is represented by $C_{i,j}$, denoting the number of synapses from neuron j to neuron i. If a synapse exists, it has a fixed strength of $1\ \text{mV ms}^{-1}$. Neurons can be either excitatory or inhibitory, with excitatory and inhibitory neurons differing only in the sign of synaptic transmission.

2.3 Growth Rules

The growth rules determine how the number of synaptic elements of a neuron changes depending on electrical activity.

2.3.1 Definitions and Time Scales

We define A_j as the number of axonal elements per neuron, and D_i^{ex} and D_i^{in} as the number of excitatory and inhibitory dendritic elements per neuron, respectively. Axonal elements are either excitatory or inhibitory, depending on the type of hosting neuron, whereas each type of neuron can have both excitatory and inhibitory dendritic elements. In the numerical integration of the growth rules, A_j, D_i^{ex}, and D_i^{in} are treated as real-valued variables, but when synaptic elements are deleted or used for synapse formation, the values of A_j, D_i^{ex}, and D_i^{in} are rounded off to their smallest integer values.

In general, the change in number of synaptic elements is proportional to $\nu \times G(f(x(t)))$, where ν is the growth rate, G is the growth function, $x(t)$ is a measure of the electrical activity of the neuron, and f is some converting function. Changes in number of synaptic elements are slow compared with fluctuations in electrical activity. A growth rate of $\nu = 10^{-4}\ \text{ms}^{-1}$ is sufficiently low in order to separate the two time scales of electrical activity and structural changes. That means that on a fast time scale, activity changes while network connectivity can be considered quasi-stable. On a slow time scale, connectivity changes while activity is at the equilibrium belonging to the current connectivity.

2.3.2 *Measures of Electrical Activity*

The function f smoothes the fluctuations of electrical activity $x(t)$ over time. The Compensation Algorithm takes the postsynaptic membrane potential for $x(t)$ and uses a sigmoidal function for f to compute the firing probability of the neuron. For a given (static) network connectivity, $f(x(t))$ converges toward an equilibrium value, which, if it is smaller or bigger than a given homeostatic set-point, will instigate structural changes.

MSP uses the firing rate as a measure of electrical activity $x(t)$ of each neuron and the intracellular calcium concentration $[Ca^{2+}]_i$ as a smoothed and time-averaged signal $f(x(t))$. In the implementation of MSP in Butz and van Ooyen,[22] the calcium concentration $[Ca^{2+}]_i$ increases by a fixed amount $\beta = 0.001\ \text{ms}^{-1}$ every time neuron i fires an action potential, and otherwise decreases exponentially to zero with decay time $\tau_{Ca} = 10{,}000$ ms.[27] The concentration decay is of the same magnitude as measured experimentally.[53] With the Izhikevich model as neuron model, the change in intracellular calcium concentration is thus given by

$$\frac{d[Ca^{2+}]_i}{dt} = \begin{cases} -\dfrac{[Ca^{2+}]_i}{\tau_{Ca}} + \beta & \text{if } v \geq 30\ \text{mV} \\ -\dfrac{[Ca^{2+}]_i}{\tau_{Ca}} & \text{otherwise} \end{cases} \tag{7.4}$$

Biologically, electrical activity, through changes in the level of intracellular calcium, modulates actin and tubulin polymerization and therefore neurite outgrowth[54] and spine formation.[34–37] If the average electrical activity of a neuron exceeds some maximum level, it will withdraw dendritic spines[38,39] and retract neurite branches,[55] thereby reducing connectivity and hence activity. If activity becomes lower, the neuron will generate vacant synaptic elements,[35] thus potentially increasing connectivity and activity. However, if activity falls below some minimum level, outgrowth and synapse formation will be halted.[8,56–58] These findings are further supported by Richards et al.,[36] showing that reducing the level of electrical activity in hippocampal slice cultures increases the number of spine head protrusions, whereas this increase does not occur with a complete block of activity. Earlier studies[41,59] are in line with these results. Thus homeostatic regulation in combination with a minimal level of electrical activity required for outgrowth may serve as a guiding principle for the formation of synaptic elements, i.e., for neurite outgrowth[54,60] and spine and bouton formation.[61,62]

2.3.3 *General Requirements for Growth Rules*

The growth rules G, which are defined as differential equations, can be linear, sigmoidal, or Gaussian functions of $f(x(t))$. The Compensation Algorithm[16] used linear growth rules, the Neurite Outgrowth Model[23] sigmoidal growth rules, and the MSP algorithm[22] Gaussian growth rules. There are some requirements for G in order for the network to reach a homeostatic equilibrium. The growth rule needs to have at least one stable fixed point, i.e., the growth curve must intersect the x-axis with a negative slope in the vicinity of the intersection (Fig. 7.2A). This is the homeostatic set-point of electrical activity (or calcium concentration), and the growth rules for the different types of synaptic elements of an individual neuron must all have the same set-point. Otherwise the growth of synaptic elements of an individual neuron will not converge. By contrast, different types of neurons can have different set-points (Fig. 7.2B, right panel). For example, a variety of set-points at high and very low levels of electrical activity are used in a model of the self--organizing growth of the connectivity of a cortical column.[47]

Networks with identical growth rules (linear or sigmoidal) with one stable fixed point for all synaptic elements generally develop toward an equilibrium in which the activity of all neurons is at the homeostatic set-point of electrical activity. This situation is in some cases biologically unrealistic. Especially after lesions, networks do not always restore electrical activity completely. For example, after large peripheral lesions, with large cortical areas deprived of input, not all neurons will return to set-point activity.[8,63,64] Gaussian growth rules were introduced[22] to overcome this shortcoming. With Gaussian growth curves, there is a second, unstable fixed point at the left side of the homeostatic set-point (Fig. 7.2A). This unstable fixed point serves as a minimum level of electrical activity below which neurons fail to restore activity to the set-point. To let networks grow from scratch with Gaussian growth rules, we need a higher external input I^{ext} in the beginning of the simulation, in order to raise activities above the unstable fixed point. Once neurons are close to the set-point, I^{ext} can be reduced again, and neuronal activity could then drop below the unstable fixed point.

2.3.4 *Finding Appropriate Growth Rules*

Finding appropriate growth rules, especially for axonal elements, has always been a challenge. Wolff and Wagner phrased their postulate on axonal growth rather vaguely in their Compensation Theory,[65] stating that increasing electrical activity during development promotes axonal growth. It remained undefined for what initial activity levels this statement holds. Dammasch et al.[16] modeled activity-dependent changes in axonal elements with a linear function such that activity levels higher than the homeostatic set-point promote axonal growth:

$$G_{A_j}(f(x(t))) = \frac{\mathrm{d}A_j}{\mathrm{d}t} = f(x(t)) - \varepsilon \tag{7.5}$$

where $\mathrm{d}A_j/dt$ is the change in number of axonal elements and ε is the homeostatic set-point. However, with this axonal growth rule, networks cannot grow from scratch but require an initial level of connectivity, which may subsequently be balanced homeostatically. Using growth rules, either linear or sigmoidal, with a negative slope in the vicinity of the set-point would overcome this shortcoming but would not reflect the observation that increasing activity promotes outgrowth.

In principle, Gaussian growth rules are biologically more realistic than linear or sigmoidal rules because they define maximal growth not for the activity extremes but for some optimal range between the unstable and stable fixed points. By varying the location of the unstable fixed point (the stable fixed point must be the same for all types of synaptic elements of an individual neuron), axonal and dendritic elements can have different activity levels of optimal growth (Fig. 7.2B, left panel). The Gaussian growth rules are given by

$$G(f(x(t))) = \frac{dz_i}{dt} = \nu\left(2e^{-(([Ca^{2+}]_i - \xi_z)/\zeta_z)^2} - 1\right) \tag{7.6}$$

with

$$\xi_z = \frac{\eta_z + \varepsilon}{2}$$

$$\zeta_z = \frac{\eta_z - \varepsilon}{2\sqrt{-\ln(1/2)}}$$

where $z \in \{A, D^{ex}, D^{in}\}$ is the real-valued number of synaptic elements, ν is the growth rate, ξ_z is the center of the Gaussian growth curve (i.e., the level of electrical activity with optimal growth), ζ_z is the width of the Gaussian curve, η_z is the unstable fixed point, and ε is the homeostatic set-point (Fig. 7.2A). By choosing z, we obtain the growth curve of a particular type of synaptic element for each neuron i. Eq. (7.6) applies to a neuron's total number of synaptic elements of the respective type regardless of whether the elements are bound in a synapse or unbound.

When growth rules with higher activity levels for optimal axonal growth than for optimal dendritic growth are used (Fig. 7.2B, left panel), increasing activity during development (when activity is still low) indeed promotes axonal growth. At the same time, networks can grow out from scratch if initial activity levels are at least temporarily higher than the unstable fixed points. Simulations revealed that growth rules with higher levels for optimal axonal than for optimal dendritic growth produced biologically more realistic results after lesions than identical growth rules for axonal and dendritic elements.[22]

Modeling the change in synaptic elements of inhibitory neurons poses the same kind of problem as for axonal elements. Experimental data showed that reducing the level of activity causes a rapid loss of dendritic spines on inhibitory neurons.[66] Like axonal growth rules, dendritic growth rules with a local positive slope in the vicinity of the homeostatic set-point, which would be required for modeling the decrease in dendritic elements on inhibitory neurons at low activity levels, do not enable the formation of inhibitory synapses during development. A possible solution here again is the use of Gaussian growth rules with different optimal activity levels for excitatory and inhibitory neurons. In some simulations, we gave synaptic elements of inhibitory neurons higher unstable and stable fixed points than the synaptic elements of excitatory neurons, which produced the desired behavior (see Chapter 8). Even with set-points matching the realistic firing rates of different types of excitatory and inhibitory neurons in the different layers of a cortical column, MSP could reliably grow networks toward a homeostatic equilibrium.[47]

2.4 Synapse Formation and Network Topology

In each connectivity update (i.e., every 100 ms), we compute the number of vacant axonal and dendritic elements. Recall that excitatory neurons exclusively form excitatory axonal elements and inhibitory neurons exclusively inhibitory axonal elements (Dale's law). Excitatory and inhibitory dendritic elements are formed by both types of neurons. As a result of the growth rules, the real-valued number of synaptic elements can increase.

In order to combine vacant axonal and dendrite elements with each other, we round off the real-valued number of synaptic elements to the next smaller integer value. Combining vacant synaptic elements can be done in several ways. We can randomly choose a synaptic element from the set of vacant excitatory axonal elements and one from the set of vacant excitatory dendritic elements (Fig. 7.1B), and so on, until one set becomes empty (once a vacant synaptic element has been selected for synapse formation, it is no longer vacant and is removed from the set of vacant elements). The same procedure is then applied to vacant inhibitory synaptic elements.

Synapse formation can be made dependent on the (Euclidean) distance between two neurons (Fig. 7.1B). For this, we place neurons with some jitter on a two-dimensional grid. We use two-dimensional Gaussian curves $K_{i,j}$ centered at each neuron j and define the likelihood of synapse formation for every Euclidean distance toward any other neuron i in the network:

$$K_{i,j} = e^{-\frac{(\mathrm{pos}_{xj}-\mathrm{pos}_{xi})^2+(\mathrm{pos}_{yj}-\mathrm{pos}_{yi})^2}{\sigma^2}} \tag{7.7}$$

where pos_{xi} is the x-coordinate and pos_{yi} is the y-coordinate of the location of postsynaptic neuron i, and pos_{xj} and pos_{yj} the coordinates of presynaptic neuron j.

The value of $K_{i,j}$ is then taken into account when synaptic elements are combined. For example, we compared a random number from a uniform distribution with $K_{i,j}$. If this number is smaller than or equal to $K_{i,j}$, the synapse is formed; otherwise not and the chosen synaptic elements remain vacant. Once new synapses are formed, the connectivity values $C_{i,j}$ are updated. The kernel $K_{i,j}$ is, of course, not limited to Gaussian curves. Any probability function can be used, including measured connection probabilities from anatomical intra- or interarea connectivity.

2.5 Synapse Deletion

As a result of the growth rules, the real-valued number of synaptic elements can decrease. If the decrease is big enough to reduce the integer number of elements (after rounding off to the next smaller integer), then synaptic elements will be deleted. The MSP algorithm determines randomly which of the synaptic elements of a neuron will be deleted, with vacant and bound synaptic elements having an equal probability of being selected. If a bound synaptic element is deleted, the synapse breaks apart immediately, i.e., $C_{i,j} - 1$, but the complementary synaptic element remains and becomes vacant again. It can rebind to a different synaptic element in the next connectivity update. In addition, we introduce a spontaneous, activity-independent decay of vacant synaptic elements in order to limit the maximal number of vacant elements in the model.

3 MODEL RESULTS

Introducing Gaussian growth rules in MSP leads to increased biological plausibility compared with the original Compensation Algorithm with linear growth rules. Gaussian growth rules further enable network formation from scratch, making it possible to compare MSP with developing dissociated cell cultures.[24,28] MSP is robust and even allows for network formation with sets of growth rules that have different high and low fixed points for different cell types,[47] a scenario that is more realistic than all cell types having identical fixed points. This robustness means that MSP can be used for biologically realistic simulations. For instance, the structural connectivity that is measured experimentally[67] does generally not resemble a homeostatically balanced network configuration. By allowing MSP to make minor changes in the connectivity so that cell activities move toward their homeostatic set-points, we will obtain a slightly modified circuit connectivity that could, e.g., be used in functional simulations.

MSP has been widely used to account for cortical events following lesion-induced reductions in afferent activity, as in retinal lesions or cortical lesions such as stroke. In retinal lesions, a corresponding area in the primary visual cortex (the lesion projection zone or LPZ) is deprived of visual input, causing a drop in the neurons' electrical activity.[10,63,68–70] Likewise, in stroke, cortical neurons with afferent connections originating from the stroke site lose input and experience a drop in electrical activity.[71–76] In model networks, MSP can, even with simple linear growth rules, restore electrical activities to a predefined set-point by rewiring cortical network connectivity after a loss of input[22,77,78] (Fig. 7.3). In one model study, we tested how external stimulation of activity could promote network repair after deafferentation.[77] Interestingly, we found that the positive effect of external stimulation, which may mimic direct cortical stimulation or physical training, saturated after about 30 days. Pauses in stimulation of about 10%–20% of the total duration of the stimulation optimally prevented this saturation and led to even stronger synapse formation than continuous stimulation. Even more remarkably, network reorganization after paused stimulation remained stable even when the stimulation was permanently turned off, whereas networks after turning off continuous stimulation returned to a state similar to the one right after the lesion.

The way in which networks rewire their connectivity after a loss of input depends crucially on the availability of vacant synaptic elements, which in turn is regulated by the activity-dependent growth rules. With linear, sigmoidal, or Gaussian growth rules that are identical for axons and dendrites and that enable the formation of synaptic elements at low levels of activity, neurons at low activity levels form both dendritic and axonal elements at the same time. Hence, activity restoration comes about by strongly increased connectivity between the input-deprived neurons, amplifying the remaining activity after the loss of input. Literally, the deprived neurons pull themselves by their own boot straps, which lead to strong recurrent connections within the input-deprived area and pronounced oscillatory neuronal activity.[22]

Using different axonal and dendritic Gaussian growth rules with a low unstable fixed point for dendritic elements and a higher unstable fixed point for axonal elements (Fig. 7.2B, left panel), we observe interesting cooperative effects between the intact and input-deprived areas in the network[22] (Fig. 7.3). Neurons in the input-deprived area form dendritic elements but no axonal elements when activity is lower than the minimal level of activity (i.e., the unstable fixed point) for axonal element formation. Neurons in the intact area surrounding the deprived area have slightly reduced levels of activity, because they lose input from the deprived area. With this slightly reduced activity, these neurons are in an optimal activity range for axonal element formation (to some extent they form

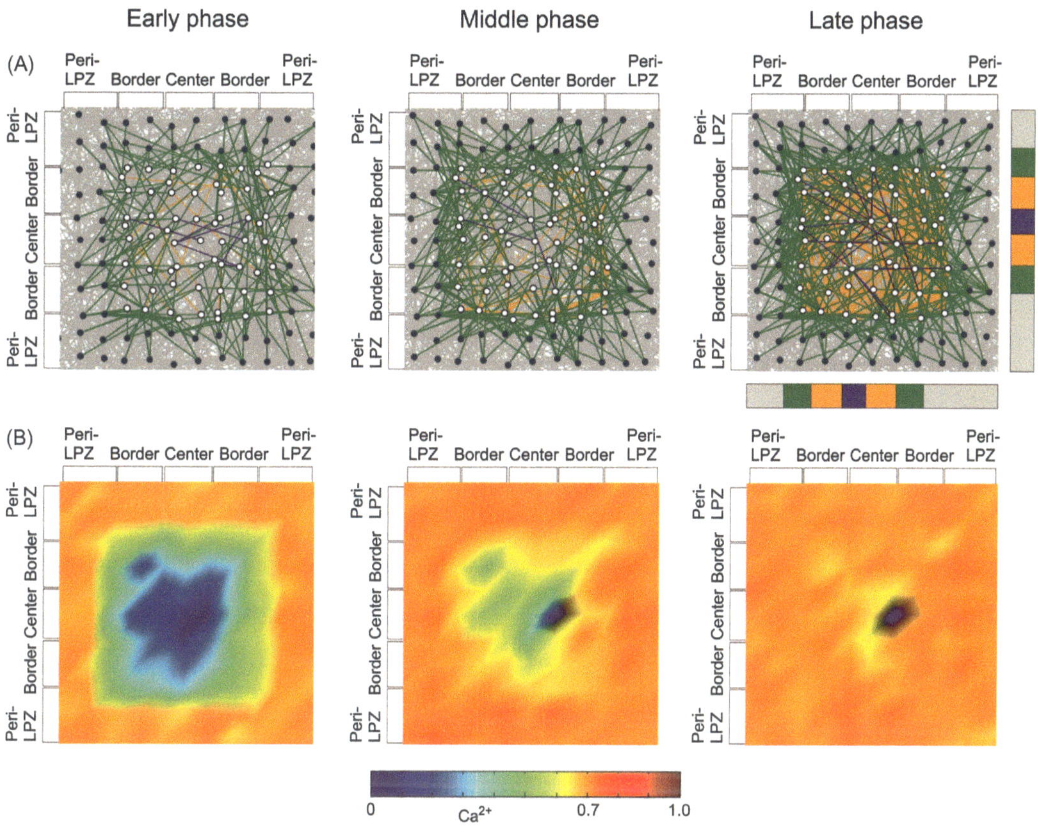

FIGURE 7.3 **Network rewiring in the model leads to a sequential increase in electrical activity from the border to the center of the lesion projection zone.** (A) Color-labeled are those synaptic connections that are newly formed after the lesion and impinge on neurons anywhere in the lesion projection zone (LPZ). Preexisting and other connections are in *gray*, covering almost the entire background. Connections originating from neurons in the peri-LPZ are labeled *green*, from the border of the LPZ *yellow*, and from the center *blue*. In the early phase, new connections are formed mainly from the peri-LPZ to the border of the LPZ. Owing to fluctuations in activity, transient connections may also occur in the border or the center but usually disappear again later. In the middle phase, connections are formed from the border to the center, and finally in the late phase connections are added that originate from the center. (B) In the early phase, new connections raise the electrical activity in the border of the LPZ and subsequently in the center. Depending on the size of the LPZ, neurons in the center may fail to recover. Intracellular calcium concentration (reflecting a neuron's electrical activity) of individual neurons is interpolated to show the spatial distribution of activity in the LPZ and peri-LPZ. Panels (A) and (B) are close-ups of the combined area of the LPZ and peri-LPZ. The relative size of the LPZ (border: *orange*; center: *blue*) and the peri-LPZ (*green*) to the entire network (*gray*) is indicated by the *vertical and horizontal bars* next to the right panel in (A). Total number of neurons (excitatory and inhibitory): LPZ: 62 neurons; peri-LPZ: 71 neurons. The whole network has 400 neurons, of which 80% are excitatory and 20% are inhibitory. Source: *Figure reproduced from Butz M, van Ooyen A. A simple rule for dendritic spine and axonal bouton formation can account for cortical reorganization after focal retinal lesions. PLoS Comput Biol. 2013;9(10):e1003259.*

dendritic elements, too). Vacant axonal elements of these neurons connect to vacant dendritic elements of neurons that lost their input. The newly formed synapses convey electrical activity from the intact area to the deprived area (Fig. 7.3). In this way, the front between high and low activity moves slowly (on the time scale of synapse formation) from the outside to inside of the deprived area and, if the deprived area is not too big, restores activity in the whole input-deprived area.[22]

3.1 Comparing MSP Results With Experimental Data

Interestingly, after focal retinal lesions (e.g., in mice), one can exactly see such a "closure" of the LPZ, with restoration of electrical activity from the outside to inside of the LPZ.[10] The LPZ heels like a wound by axons growing from the intact area into the LPZ,[6,7] forming synapses with newly formed dendritic spines on the deafferented neurons.[8]

There are also remarkable similarities between the time course of events in cortical networks after focal retinal lesions and that in the model network after loss of external input. As the model predicts, the strongest replacement of dendritic spines occurs in the center of the LPZ.[8] Spines that are connected to axons that do not provide activity anymore after the lesion, i.e., those that belong to deprived neurons, are being lost and replaced by vacant spines, which can bind to the newly formed axons. The model further predicts a sprouting of axon collaterals from the intact areas into the LPZ, with a transient overshoot in axon density and subsequent pruning—phenomena that were indeed reported in monkeys.[7] Interestingly, rewiring of connectivity and restoration of activity in the model do not require any form of Hebbian plasticity, suggesting that activity restoration in cortical networks may also be brought about by homeostatic structural plasticity, with Hebbian synaptic plasticity consolidating the anatomical changes in connectivity.

Activity restoration in the model goes along with a remapping of input representations[22] (Fig. 7.4). We conjecture that changes in structural connectivity may also underlie the cortical remappings observed after, e.g., focal retinal lesions[8] or digit amputation.[13] The traditional view is that Hebbian plasticity adapts cortical mappings.[79] By contrast, the insights gleaned from the way MSP rewires networks after loss of input suggest that connections from the intact areas—first to the border of the LPZ and subsequently from the border to the center—transport activity into the LPZ (Fig. 7.4). Thereby, adjacent representations in the intact areas enlarge into the deprived area, very much like what is observed in cortical areas after deafferentation. In fact, it has been shown that newly formed horizontal connections convey suprathreshold activity from intact areas into the LPZ.[80]

Note that the time course of events, such as the restoration of activity from the outside to inside of the LPZ, is crucial for the remapping to occur in the model. It is merely one parameter in the model, namely the minimal level of activity for growing axonal elements (i.e., the value of the unstable fixed point), that distinguishes between functional network reorganization and network reorganization producing strongly recurrent connectivity and pronounced oscillatory activity.[22,78] It is well possible that the latter scenario may underlie the emergence of late onset epilepsy observed after lesions.[81–83]

MSP can also be applied to central lesions. Instead of a circumscribed area losing its external input after, e.g., focal retinal lesions, central lesions such as stroke deprive neighboring areas (penumbra) and remote areas of input (the phenomenon that remote brain regions are affected by a lesion is called diaschisis[84,85]). Whereas the primary lesion site is permanently lost, secondarily affected areas are subject to structural and functional reorganization.[71,72] A phenomenon that is commonly observed after focal stroke is the

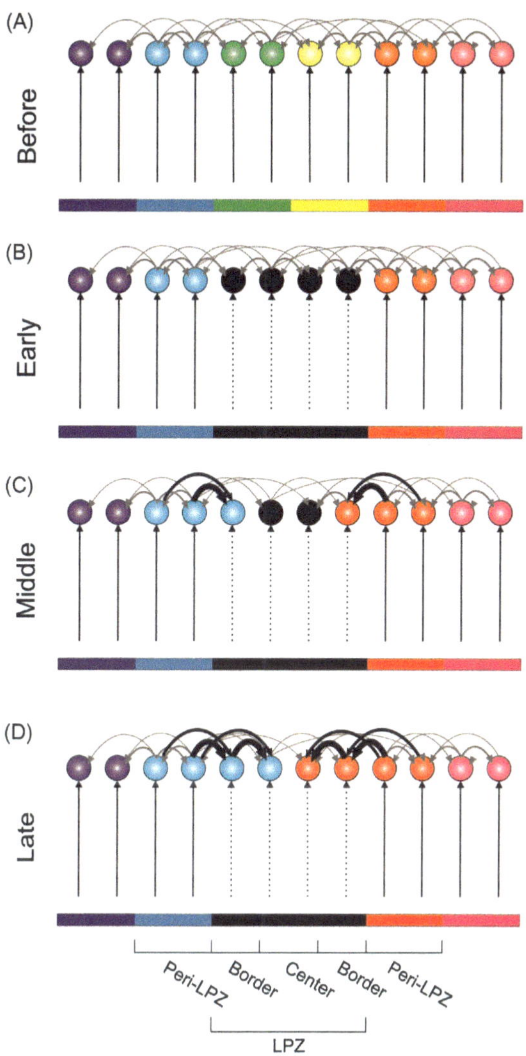

FIGURE 7.4 **Schematic figure showing how structural plasticity in the model gives rise to remapping of input representations.** New horizontal connections are formed between neurons in the peri-LPZ and the border of the LPZ, as well as between the border and the center of the LPZ. These new connections are responsible for the functional network reorganization. (A) Network before lesion. Colors indicate spatial locations in an input layer (e.g., the retina) and the strongest neuronal response in the network (e.g., primary visual cortex). Vertical input is in *black*, horizontal recurrent connections are in *gray*. Linewidth indicates the number of synapses. (B) *Black bar* indicates a lesion in the input layer. Vertical input to the corresponding area in the network is permanently removed (*dashed arrow lines*). Silent neurons in the LPZ early after the lesion are labeled *black*. (C) Additional synapses are formed at the rim of the LPZ (*bold black arrows*) in the middle phase after the lesion. Owing to these additional horizontal synapses, neurons in the border become active in response to adjacent input representations (*blue and red*). (D) Late after the lesion, additional synapses from the border to the center contribute to the activity of the center neurons, which now also become sensitive to adjacent input representations (*blue and red*). Source: *Figure reproduced from Butz M, van Ooyen A. A simple rule for dendritic spine and axonal bouton formation can account for cortical reorganization after focal retinal lesions. PLoS Comput Biol. 2013;9(10):e1003259.*

reorientation of dendritic branches away from the lesion site toward intact areas.[86] MSP predicts that dendritic reorientation may be the consequence of local differences in electrical activity and homeostatic growth rules for synaptic elements. Thus dendritic reorientation does not necessarily need goal-directed processes or spatial guidance cues but could emerge from purely local processes alone.

So far, MSP has been applied to mesoscopic networks, i.e., networks with a diameter of up to 1 mm.[22,78] Interestingly, the topological changes in MSP networks after a focal loss of input resemble changes in macroscopic interarea connectivity after focal stroke. Both mesoscopic model networks[78] and macroscopic brain networks[87,88] became more random after a lesion, while areas that were deprived of input and restored their activity had increased betweenness centrality.[78] This is a counterintuitive result, as betweenness centrality is a measure for the importance of a node, and random networks have nodes with low betweenness centrality.[89] Thus the occurrence of increased betweenness centrality in more random networks points to a specific reorganization process. Remarkably, patients recovering from a focal stroke indeed showed increasingly random brain networks, while particular areas (diaschisis) increased their betweenness centrality.[88]

4 DISCUSSION

With MSP, we have shown that after a lesion-induced reduction in activity, networks can restore their activity by means of local, homeostatic growth rules for structural plasticity.[22,78] As a result of these growth rules and the network rewiring they bring about, spatial input presentations become remapped after a lesion. Importantly, the remapping and the spatial and temporal changes in network activity and axonal and dendritic elements are similar to those observed experimentally after focal lesions.[6–10,80,86] We therefore postulate that activity-dependent, homeostatic structural plasticity may serve as a fundamental principle by which brain networks build and reshape their circuitry.[90–92]

4.1 Future Experimental Studies

Electrical activity influences the intracellular calcium concentration,[53,54] which in turn influences neuronal morphogenesis.[50,93–95] Simplified rules that describe these highly nonlinear processes are very much needed. Although the growth rules implemented in MSP are inspired by experimental observations, their precise shape and applicability to in vivo situations have still to be validated. To test the postulated growth curves, one could use two-photon glutamate uncaging[96] or optogenetics[97] to alter electrical activity in vivo and monitor the impact on bouton and spine formation.

To determine the growth curves more quantitatively, one could challenge in vitro neurons with rises in intracellular calcium of varying amplitudes by calcium uncaging[50] and measure the impact on neurite outgrowth and spine formation. This may also reveal differences between axons and dendrites or between different cell types. For example, neurons characterized by a high firing rate may have their homeostatic set-point at a higher activity level than neurons that fire less frequently.[98] In addition, excitatory and inhibitory neurons may have the same or different growth curves.

The comparison between MSP and experimental data on spine and bouton dynamics and cortical reorganization after focal retinal lesions revealed that the best matching results were obtained for Gaussian growth rules with a higher unstable fixed point for axonal elements than for dendritic elements[22] (Fig. 7.2B, left panel). From these postulated growth rules, we can derive predictions that can be tested experimentally. The most important prediction from MSP is that after a loss of input, homeostatic structural plasticity is the predominant mechanism for activity restoration. Therefore, we predict that in retinal lesion experiments, blocking spine formation,[99] axon outgrowth, or synapse formation[100] will prevent the long-term restoration of firing rates in the LPZ and the remapping of cortical representations. Furthermore, we expect that restoration of activity will still take place in knockouts that block synaptic scaling[101] but leave structural plasticity intact.

According to our postulated growth rules, spines and boutons should optimally form at activity levels slightly below the homeostatic set-point and retract if electrical activity exceeds the set-point (Fig. 7.2). In addition, spines and boutons need a minimum level of electrical activity to form. If activity drops below this level, structural reorganization will fail to occur and activity will not be restored. This need for a minimum activity level has so far not been discussed in the experimental literature on lesion-induced plasticity and could explain why in some cases activity restoration fails.[8] Optogenetics in combination with time-lapse imaging might reveal the precise dependency of a neuron's morphogenesis on its local firing rate.

The model simulations further predict the potential effects of external stimulation on network repair after loss of input. A direct testable consequence of the growth rules of MSP is that artificially bringing the activity of the input-deprived neurons back to or beyond the set-point by optogenetics,[97] two-photon glutamate uncaging[96] or chronically implanted electrode arrays[102] would prevent neuronal morphogenesis and structural network reorganization. Conversely, external stimulation may also promote network repair, especially with big lesions, if it drives low activity back to the optimal regime. Lastly, model simulations may indicate risk factors for the emergence of strongly recurrent connectivity and associated epileptic-like network activity following recovery from lesions such as stroke.[22] Ideally, model studies may predict the optimal temporal and spatial stimulation patterns that will best support recovery.[77]

4.2 Future Modeling Studies

With MSP we proposed an algorithm that can both grow and self-organize developing neuronal networks[103] and rebalance electrical activity in a mature network after a central or peripheral lesion.[22,78] The introduction of Gaussian growth rules was the key to obtaining biologically more plausible structural dynamics after lesions, such as outside-in activity restoration for small peripheral lesions, remapping of spatially organized inputs, massive dendritic element turnover and axonal sprouting, overshoot and pruning.

An open question is how inhibitory circuits reshape during development and after lesions. A first modeling study was done with higher homeostatic set-points for inhibitory than for excitatory neurons,[47] with promising results that need to be assessed further. Higher set-points for inhibitory neurons could contribute to critical periods during development.[104] The interplay between inhibitory neurons with high set-points and excitatory neurons with lower set-points could be as follows. As soon as inhibitory neurons approach their higher set-point activity, they inhibit and hence lower the activity of excitatory neurons, which in turn would respond with enhanced growth of synaptic elements. In that way, inhibition could be the trigger for critical periods in time[26] and may also influence the spatial layout of connections.[24] Indeed, influencing the level of inhibition in developing cortical networks can shift the onset of the critical period and even change the structural layout of ocular dominance columns.[104]

With the implementation of MSP in the simulation framework NEST for spiking neurons (www.nest.initiative.org), large-scale networks can be set up with up to 10^6 neurons that can form and delete synapses. Such large networks are necessary if the response to lesions in multiple cortical columns with different cell types is to be studied. Large-scale networks are also required for studying the structural changes in intra- and interarea connectivity, e.g., in order to better predict the consequences of diaschisis after stroke. Moreover, NEST enables investigating the interplay between structural plasticity and the large variety of synaptic plasticity mechanisms described in the literature. For example, synapse deletion in MSP could be made dependent on synaptic strength so as to model the interaction between synaptic and structural plasticity.[105,106] Importantly, network models with homeostatic structural plasticity will better reflect the full plastic potential of the brain than network models containing only synaptic plasticity.

References

1. Raisman G. Neuronal plasticity in the septal nuclei of the adult rat. *Brain Res*. 1969;14(1):25–48.
2. Hofer SB, Mrsic-Flogel TD, Bonhoeffer T, Hübener M. Experience leaves a lasting structural trace in cortical circuits. *Nature*. 2009;457(7227):313–317.
3. Hofer SB, Mrsic-Flogel TD, Bonhoeffer T, Hübener M. Prior experience enhances plasticity in adult visual cortex. *Nat Neurosci*. 2006;9(1):127–132.
4. Fu M, Zuo Y. Experience-dependent structural plasticity in the cortex. *Trends Neurosci*. 2011;34(4):177–187.
5. Eysel UT, Gonzalez-Aguilar F, Mayer U. A functional sign of reorganization in the visual system of adult cats: lateral geniculate neurons with displaced receptive fields after lesions of the nasal retina. *Brain Res*. 1980;181(2):285–300.
6. Darian-Smith C, Gilbert CD. Axonal sprouting accompanies functional reorganization in adult cat striate cortex. *Nature*. 1994;368(6473):737–740.
7. Yamahachi H, Marik SA, McManus JN, Denk W, Gilbert CD. Rapid axonal sprouting and pruning accompany functional reorganization in primary visual cortex. *Neuron*. 2009;64(5):719–729.
8. Keck T, Mrsic-Flogel TD, Vaz Afonso M, Eysel UT, Bonhoeffer T, Hübener M. Massive restructuring of neuronal circuits during functional reorganization of adult visual cortex. *Nat Neurosci*. 2008;11(10):1162–1167.
9. Kaas JH, Krubitzer LA, Chino YM, Langston AL, Polley EH, Blair N. Reorganization of retinotopic cortical maps in adult mammals after lesions of the retina. *Science*. 1990;248(4952):229–231.
10. Giannikopoulos DV, Eysel UT. Dynamics and specificity of cortical map reorganization after retinal lesions. *Proc Natl Acad Sci USA*. 2006;103(28):10805–10810.
11. Merzenich MM, Kaas JH, Wall JT, Sur M, Nelson RJ, Felleman DJ. Progression of change following median nerve section in the cortical representation of the hand in areas 3b and 1 in adult owl and squirrel monkeys. *Neuroscience*. 1983;10(3):639–665.
12. Merzenich MM, Kaas JH, Wall J, Nelson RJ, Sur M, Felleman D. Topographic reorganization of somatosensory cortical areas 3b and 1 in adult monkeys following restricted deafferentation. *Neuroscience*. 1983;8(1):33–55.
13. Merzenich MM, Nelson RJ, Stryker MP, Cynader MS, Schoppmann A, Zook JM. Somatosensory cortical map changes following digit amputation in adult monkeys. *J Comp Neurol*. 1984;224(4):591–605.

14. Wolff JR, Joó F, Dames W. Plasticity in dendrites shown by continuous GABA administration in superior cervical ganglion of adult rat. *Nature*. 1978;274(5666):72–74.
15. Wolff JR, Laskawi R, Spatz WB, Missler M. Structural dynamics of synapses and synaptic components. *Behav Brain Res*. 1995;66(1–2):13–20.
16. Dammasch IE, Wagner GP, Wolff JR. Self-stabilization of neuronal networks. I. The compensation algorithm for synaptogenesis. *Biol Cybern*. 1986;54(4–5):211–222.
17. Dammasch IE, Wagner GP, Wolff JR. Self-stabilization of neuronal networks. II. Stability conditions for synaptogenesis. *Biol Cybern*. 1988;58(3):149–158.
18. Cromme LJ, Dammasch IE. Compensation type algorithms for neural nets: stability and convergence. *J Math Biol*. 1989;27(3):327–340.
19. Hopfield JJ. Neural networks and physical systems with emergent collective computational abilities. *Proc Natl Acad Sci USA*. 1982;79(8):2554–2558.
20. Willshaw DJ, Buneman OP, Longuet-Higgins HC. Non-holographic associative memory. *Nature*. 1969;222(5197):960–962.
21. Dammasch IE. Structural realization of a Hebb-type learning rule. In: Cotterill RMJ, ed. *Models of Brain Functions*. Cambridge University Press; 1989.
22. Butz M, van Ooyen A. A simple rule for dendritic spine and axonal bouton formation can account for cortical reorganization after focal retinal lesions. *PLoS Comput Biol*. 2013;9(10):e1003259.
23. van Ooyen A, van Pelt J. Activity-dependent outgrowth of neurons and overshoot phenomena in developing neural networks. *J Theor Biol*. 1994;167:27–43.
24. van Ooyen A, van Pelt J, Corner MA. Implications of activity dependent neurite outgrowth for neuronal morphology and network development. *J Theor Biol*. 1995;172(1):63–82.
25. van Ooyen A, van Pelt J. Complex periodic behaviour in a neural network model with activity-dependent neurite outgrowth. *J Theor Biol*. 1996;179(3):229–242.
26. van Oss C, van Ooyen A. Effects of inhibition on neural network development through activity-dependent neurite outgrowth. *J Theor Biol*. 1997;185(2):263–280.
27. Abbott LF, Rohrkemper R. A simple growth model constructs critical avalanche networks. *Prog Brain Res*. 2007;165:13–19.
28. Tetzlaff C, Okujeni S, Egert U, Wörgötter F, Butz M. Self-organized criticality in developing neuronal networks. *PLoS Comput Biol*. 2010;6(12):e1001013.
29. Holtmaat A, Wilbrecht L, Knott GW, Welker E, Svoboda K. Experience-dependent and cell-type-specific spine growth in the neocortex. *Nature*. 2006;441(7096):979–983.
30. Wilbrecht L, Holtmaat A, Wright N, Fox K, Svoboda K. Structural plasticity underlies experience-dependent functional plasticity of cortical circuits. *J Neurosci*. 2010;30(14):4927–4932.
31. Marik SA, Yamahachi H, Meyer zum Alten Borgloh S, Gilbert CD. Large-scale axonal reorganization of inhibitory neurons following retinal lesions. *J Neurosci*. 2014;34(5):1625–1632.
32. Stepanyants A, Hof PR, Chklovskii DB. Geometry and structural plasticity of synaptic connectivity. *Neuron*. 2002;34(2):275–288.
33. Hines ML, Carnevale NT. The NEURON simulation environment. *Neural Comput*. 1997;9(6):1179–1209.
34. Kirov SA, Harris KM. Dendrites are more spiny on mature hippocampal neurons when synapses are inactivated. *Nat Neurosci*. 1999;2(10):878–883.
35. Kirov SA, Goddard CA, Harris KM. Age-dependence in the homeostatic upregulation of hippocampal dendritic spine number during blocked synaptic transmission. *Neuropharmacology*. 2004;47(5):640–648.
36. Richards DA, Mateos JM, Hugel S, et al. Glutamate induces the rapid formation of spine head protrusions in hippocampal slice cultures. *Proc Natl Acad Sci USA*. 2005;102(17):6166–6171.
37. Bravin M, Morando L, Vercelli A, Rossi F, Strata P. Control of spine formation by electrical activity in the adult rat cerebellum. *Proc Natl Acad Sci USA*. 1999;96(4):1704–1709.
38. Müller M, Gähwiler BH, Rietschin L, Thompson SM. Reversible loss of dendritic spines and altered excitability after chronic epilepsy in hippocampal slice cultures. *Proc Natl Acad Sci USA*. 1993;90(1):257–261.
39. Drakew A, Müller M, Gähwiler BH, Thompson SM, Frotscher M. Spine loss in experimental epilepsy: quantitative light and electron microscopic analysis of intracellularly stained CA3 pyramidal cells in hippocampal slice cultures. *Neuroscience*. 1996;70(1):31–45.

40. Jiang M, Lee CL, Smith KL, Swann JW. Spine loss and other persistent alterations of hippocampal pyramidal cell dendrites in a model of early-onset epilepsy. *J Neurosci*. 1998;18(20):8356–8368.
41. McKinney RA, Capogna M, Dürr R, Gähwiler BH, Thompson SM. Miniature synaptic events maintain dendritic spines via AMPA receptor activation. *Nat Neurosci*. 1999;2(1):44–49.
42. Zuo Y, Yang G, Kwon E, Gan WB. Long-term sensory deprivation prevents dendritic spine loss in primary somatosensory cortex. *Nature*. 2005;436(7048):261–265.
43. Schubert V, Lebrecht D, Holtmaat A. Peripheral deafferentation-driven functional somatosensory map shifts are associated with local, not large-scale dendritic structural plasticity. *J Neurosci*. 2013;33(22):9474–9487.
44. Brown CE, Li P, Boyd JD, Delaney KR, Murphy TH. Extensive turnover of dendritic spines and vascular remodeling in cortical tissues recovering from stroke. *J Neurosci*. 2007;27(15):4101–4109.
45. Brown CE, Wong C, Murphy TH. Rapid morphologic plasticity of peri-infarct dendritic spines after focal ischemic stroke. *Stroke*. 2008;39(4):1286–1291.
46. Butz M, Lehmann K, Dammasch IE, Teuchert-Noodt G. A theoretical network model to analyse neurogenesis and synaptogenesis in the dentate gyrus. *Neural Networks*. 2006;19(10):1490–1505.
47. Diaz-Pier S, Naveau M, Butz-Ostendorf M, Morrison A. Automatic generation of connectivity for large-scale neuronal network models through structural plasticity. *Front Neuroanat*. 2016;10:57.
48. Knott GW, Holtmaat A, Wilbrecht L, Welker E, Svoboda K. Spine growth precedes synapse formation in the adult neocortex in vivo. *Nat Neurosci*. 2006;9(9):1117–1124.
49. Lohmann C, Finski A, Bonhoeffer T. Local calcium transients regulate the spontaneous motility of dendritic filopodia. *Nat Neurosci*. 2005;8(3):305–312.
50. Lohmann C, Wong RO. Regulation of dendritic growth and plasticity by local and global calcium dynamics. *Cell Calcium*. 2005;37(5):403–409.
51. Izhikevich EM. Simple model of spiking neurons. *IEEE Trans Neural Networks*. 2003;14(6):1569–1572.
52. Izhikevich EM. Which model to use for cortical spiking neurons? *IEEE Trans Neural Networks*. 2004;15(5):1063–1070.
53. Hofer SB, Ko H, Pichler B, et al. Differential connectivity and response dynamics of excitatory and inhibitory neurons in visual cortex. *Nat Neurosci*. 2011;14(8):1045–1052.
54. Lipton SA, Kater SB. Neurotransmitter regulation of neuronal outgrowth, plasticity and survival. *Trends Neurosci*. 1989;12(7):265–270.
55. Wu GY, Cline HT. Stabilization of dendritic arbor structure in vivo by CaMKII. *Science*. 1998;279(5348):222–226.
56. al-Mohanna FA, Cave J, Bolsover SR. A narrow window of intracellular calcium concentration is optimal for neurite outgrowth in rat sensory neurones. *Brain Res Dev Brain Res*. 1992;70(2):287–290.
57. Ramakers GJ, Avci B, van Hulten P, et al. The role of calcium signaling in early axonal and dendritic morphogenesis of rat cerebral cortex neurons under non-stimulated growth conditions. *Brain Res Dev Brain Res*. 2001;126(2):163–172.
58. Groc L, Petanjek Z, Gustafsson B, Ben-Ari Y, Hanse E, Khazipov R. In vivo blockade of neural activity alters dendritic development of neonatal CA1 pyramidal cells. *Eur J Neurosci*. 2002;16(10):1931–1938.
59. Dalva MB, Ghosh A, Shatz CJ. Independent control of dendritic and axonal form in the developing lateral geniculate nucleus. *J Neurosci*. 1994;14(6):3588–3602.
60. McKinney RA, Lüthi A, Bandtlow CE, Gähwiler BH, Thompson SM. Selective glutamate receptor antagonists can induce or prevent axonal sprouting in rat hippocampal slice cultures. *Proc Natl Acad Sci USA*. 1999;96(20):11631–11636.
61. Rao A, Craig AM. Signaling between the actin cytoskeleton and the postsynaptic density of dendritic spines. *Hippocampus*. 2000;10(5):527–541.
62. Halpain S. Actin and the agile spine: how and why do dendritic spines dance? *Trends Neurosci*. 2000;23(4):141–146.
63. Barnes SJ, Sammons RP, Jacobsen RI, Mackie J, Keller GB, Keck T. Subnetwork-specific homeostatic plasticity in mouse visual cortex in vivo. *Neuron*. 2015;86(5):1290–1303.
64. Smirnakis SM, Brewer AA, Schmid MC, et al. Lack of long-term cortical reorganization after macaque retinal lesions. *Nature*. 2005;435(7040):300–307.
65. Wolff JR, Wagner GP. Selforganization in synaptogenesis: interaction between the formation of excitatory and inhibitory synapses. In: Basar E, Flohr H, Hermann Haken HC, Mandell AJ, eds. *Synergetics of the Brain*. Berlin, Heidelberg, New York, Tokyo: Springer; 1983:50–59.

66. Keck T, Scheuss V, Jacobsen RI, et al. Loss of sensory input causes rapid structural changes of inhibitory neurons in adult mouse visual cortex. *Neuron*. 2011;71(5):869–882.
67. Binzegger T, Douglas RJ, Martin KA. A quantitative map of the circuit of cat primary visual cortex. *J Neurosci*. 2004;24(39):8441–8453.
68. Hu TT, Laeremans A, Eysel UT, Cnops L, Arckens L. Analysis of c-fos and zif268 expression reveals time-dependent changes in activity inside and outside the lesion projection zone in adult cat area 17 after retinal lesions. *Cereb Cortex*. 2009;19(12):2982–2992.
69. Keck T, Keller GB, Jacobsen RI, Eysel UT, Bonhoeffer T, Hübener M. Synaptic scaling and homeostatic plasticity in the mouse visual cortex in vivo. *Neuron*. 2013;80(2):327–334.
70. Hengen KB, Lambo ME, van Hooser SD, Katz DB, Turrigiano GG. Firing rate homeostasis in visual cortex of freely behaving rodents. *Neuron*. 2013;80(2):335–342.
71. Brown CE, Aminoltejari K, Erb H, Winship IR, Murphy TH. In vivo voltage-sensitive dye imaging in adult mice reveals that somatosensory maps lost to stroke are replaced over weeks by new structural and functional circuits with prolonged modes of activation within both the peri-infarct zone and distant sites. *J Neurosci*. 2009;29(6):1719–1734.
72. Winship IR, Murphy TH. Remapping the somatosensory cortex after stroke: insight from imaging the synapse to network. *Neuroscientist*. 2009;15(5):507–524.
73. Gao TM, Pulsinelli WA, Xu ZC. Changes in membrane properties of CA1 pyramidal neurons after transient forebrain ischemia in vivo. *Neuroscience*. 1999;90(3):771–780.
74. Bolay H, Gürsoy-Ozdemir Y, Sara Y, Onur R, Can A, Dalkara T. Persistent defect in transmitter release and synapsin phosphorylation in cerebral cortex after transient moderate ischemic injury. *Stroke*. 2002;33(5):1369–1375.
75. Carmichael ST, Tatsukawa K, Katsman D, Tsuyuguchi N, Kornblum HI. Evolution of diaschisis in a focal stroke model. *Stroke*. 2004;35(3):758–763.
76. Alstott J, Breakspear M, Hagmann P, Cammoun L, Sporns O. Modeling the impact of lesions in the human brain. *PLoS Comput Biol*. 2009;5(6):e1000408.
77. Butz M, van Ooyen A, Wörgötter F. A model for cortical rewiring following deafferentation and focal stroke. *Front Comput Neurosci*. 2009;3:10.
78. Butz M, Steenbuck ID, van Ooyen A. Homeostatic structural plasticity can account for topology changes following deafferentation and focal stroke. *Front Neuroanat*. 2014;8:115.
79. Young JM, Waleszczyk WJ, Wang C, Calford MB, Dreher B, Obermayer K. Cortical reorganization consistent with spike timing-but not correlation-dependent plasticity. *Nat Neurosci*. 2007;10(7):887–895.
80. Palagina G, Eysel UT, Jancke D. Strengthening of lateral activation in adult rat visual cortex after retinal lesions captured with voltage-sensitive dye imaging in vivo. *Proc Natl Acad Sci USA*. 2009;106(21):8743–8747.
81. Topolnik L, Steriade M, Timofeev I. Partial cortical deafferentation promotes development of paroxysmal activity. *Cereb Cortex*. 2003;13(8):883–893.
82. Avramescu S, Timofeev I. Synaptic strength modulation after cortical trauma: a role in epileptogenesis. *J Neurosci*. 2008;28(27):6760–6772.
83. Volman V, Bazhenov M, Sejnowski TJ. Pattern of trauma determines the threshold for epileptic activity in a model of cortical deafferentation. *Proc Natl Acad Sci USA*. 2011;108(37):15402–15407.
84. Carrera E, Tononi G. Diaschisis: past, present, future. *Brain*. 2014;137(Pt 9):2408–2422.
85. Bütefisch CM, Netz J, Wessling M, Seitz RJ, Hömberg V. Remote changes in cortical excitability after stroke. *Brain*. 2003;126(Pt 2):470–481.
86. Brown CE, Boyd JD, Murphy TH. Longitudinal in vivo imaging reveals balanced and branch-specific remodeling of mature cortical pyramidal dendritic arbors after stroke. *J Cereb Blood Flow Metab*. 2010;30(4):783–791.
87. Rehme AK, Grefkes C. Cerebral network disorders after stroke: evidence from imaging-based connectivity analyses of active and resting brain states in humans. *J Physiol*. 2013;591(1):17–31.
88. Wang L, Yu C, Chen H, et al. Dynamic functional reorganization of the motor execution network after stroke. *Brain*. 2010;133(Pt 4):1224–1238.
89. Rubinov M, Sporns O. Complex network measures of brain connectivity: uses and interpretations. *Neuroimage*. 2010;52(3):1059–1069.
90. Butz M, Wörgötter F, van Ooyen A. Activity-dependent structural plasticity. *Brain Res Rev*. 2009;60(2):287–305.
91. van Ooyen A. Activity-dependent neural network development. *Network Computat Neural Syst*. 1994;5:401–423.

92. van Ooyen A. Using theoretical models to analyse neural development. *Nat Rev Neurosci.* 2011;12(6):311–326.
93. Kater SB, Mattson MP, Cohan C, Connor J. Calcium regulation of the neuronal growth cone. *Trends Neurosci.* 1988;11(7):315–321.
94. Ghiretti AE, Paradis S. Molecular mechanisms of activity-dependent changes in dendritic morphology: role of RGK proteins. *Trends Neurosci.* 2014;37(7):399–407.
95. Konur S, Ghosh A. Calcium signaling and the control of dendritic development. *Neuron.* 2005;46(3):401–405.
96. Matsuzaki M, Honkura N, Ellis-Davies GC, Kasai H. Structural basis of long-term potentiation in single dendritic spines. *Nature.* 2004;429(6993):761–766.
97. Fenno L, Yizhar O, Deisseroth K. The development and application of optogenetics. *Annu Rev Neurosci.* 2011;34:389–412.
98. Hengen KB, Torrado Pacheco A, McGregor JN, Van Hooser SD, Turrigiano GG. Neuronal firing rate homeostasis is inhibited by sleep and promoted by wake. *Cell.* 2016;165(1):180–191.
99. Cole CJ, Mercaldo V, Restivo L, et al. MEF2 negatively regulates learning-induced structural plasticity and memory formation. *Nat Neurosci.* 2012;15(9):1255–1264.
100. Eroglu C, Allen NJ, Susman MW, et al. Gabapentin receptor alpha2delta-1 is a neuronal thrombospondin receptor responsible for excitatory CNS synaptogenesis. *Cell.* 2009;139(2):380–392.
101. Kaneko M, Stellwagen D, Malenka RC, Stryker MP. Tumor necrosis factor-alpha mediates one component of competitive, experience-dependent plasticity in developing visual cortex. *Neuron.* 2008;58(5):673–680.
102. Abe H, McManus JN, Ramalingam N, et al. Adult cortical plasticity studied with chronically implanted electrode arrays. *J Neurosci.* 2015;35(6):2778–2790.
103. Butz M, Steenbuck ID, van Ooyen A. Homeostatic structural plasticity increases the efficiency of small-world networks. *Front Synaptic Neurosci.* 2014;6:7.
104. Hensch TK. Critical period plasticity in local cortical circuits. *Nat Rev Neurosci.* 2005;6(11):877–888.
105. Becker N, Wierenga CJ, Fonseca R, Bonhoeffer T, Nägerl UV. LTD induction causes morphological changes of presynaptic boutons and reduces their contacts with spines. *Neuron.* 2008;60(4):590–597.
106. Wiegert JS, Oertner TG. Long-term depression triggers the selective elimination of weakly integrated synapses. *Proc Natl Acad Sci USA.* 2013;110(47):E4510–E4519.

CHAPTER

8

Critical Periods Emerge from Homeostatic Structural Plasticity in a Full-Scale Model of the Developing Cortical Column

Sebastian Rinke[1], Mikaël Naveau[2,3], Felix Wolf[1] and Markus Butz-Ostendorf[4]

[1]Technische Universität Darmstadt, Darmstadt, Germany [2]Jülich Aachen Research Alliance, Forschungszentrum Jülich, Jülich, Germany [3]Caen Normandy University, Caen, France [4]Biomax Informatics AG, Planegg, Germany

OUTLINE

The Rewiring Brain
DOI: http://dx.doi.org/10.1016/B978-0-12-803784-3.00008-1

1 INTRODUCTION

Models do not become automatically better when one merely increases their size, and in general simple models are more powerful than very detailed models because they illustrate core principles and require only a minimum number of model assumptions. However, there can be situations in which models with a large number of neurons are absolutely needed. This is the case if one wants to study dynamics that would not arise in small networks such as physiological activity. Other examples are models that aim to bridge different spatial scales, e.g., models that couple networks of multiple cortical layers or link a microscopic and a meso- or macroscopic level of description. The latter situation occurs if one wants to model rewiring of local synaptic connectivity in one brain area and at the same time global connectivity (between brain areas) after a focal stroke. In both cases, one needs a reasonable number of neurons per subnetwork multiplied by the number of coupled networks. Networks with a few million neurons are the minimum to address the latter case.

So far, large-scale neuronal network models used fixed synaptic connectivity with plasticity modeled as changes in the weight of synapses (rather than forming, deleting, or rewiring synaptic connections).[1] Especially as synaptic connectivity needs to be sparse (since number of connections grows quadratically with neuron numbers) in large-scale networks, it matters where synaptic connections are placed for cognitive performance (see Chapter 17) and having no possibility to rewire connections is a real disadvantage. Moreover, initializing the layout of connections in the beginning of a simulation is a tremendous combinatorial problem and to obtain a functional network connectivity requires many assumptions and fine tuning.

Deploying the Model of Structural Plasticity (MSP,[2] see Chapter 7) for large-scale neuronal networks provides at least three important benefits compared to traditional approaches for setting up large-scale neuronal networks. First, by MSP networks can grow from scratch. In MSP, each neuron creates new synaptic elements when its level of electrical activity is below a homeostatic set-point and deletes synaptic elements when its activity is above the set-point or below a certain minimum level. Synapses are formed by merging pre- and postsynaptic elements. In this way, a large-scale network can develop network connectivity itself with a minimum of and biologically easily defensible assumptions. Second, the network grows into a balanced state as MSP has homeostatic properties making fine tuning of arbitrarily set connectivity obsolete. Finally, by MSP the network has self-repairing capabilities making it an interesting modeling approach for simulating all sorts of lesioning experiments.

Here we describe two approaches for integrating MSP into massive-parallel large-scale neuronal network environments. The first is the well-known modeling environment NEST for large-scale spiking neuronal networks (www.nest-initiative.org). In two examples, a two-population network executable on a normal workstation and a cortical column with 100,000 neurons running on a supercomputer with 25 nodes and 24 cores per node, we demonstrate the power of simulating MSP with NEST. Integrating MSP into NEST enables us to grow networks of up to 10^5 neurons from scratch on a supercomputer,[3] which is highly sufficient for the aforementioned cases. Although NEST is actually able to run much larger networks, it is by design optimized for static connectivity and the attempt to parallelize MSP in NEST is quickly beaten by the limits of global communication deployed in NEST. We therefore describe a second algorithmic scenario which lifts this limitation and further increases the scalability of MSP through approximation. In particular, we adapt the Barnes–Hut *n*-body method, which is known from particle physics, to implement MSP.[4]

The biological impact of simulating structural plasticity in large-scale networks is to better understand the growth of large neuronal assembles such as the cortical column. Can simple building rules generate such highly complex structures? So far, this question could not be addressed in networks with only a few hundred neurons as the population of inhibitory neurons (usually only 10%–20% of the total neuron number) was too small to see significant results by numeric simulations. By choosing different homeostatic set-points for each group of neurons, and simulating 100,000 neurons in total in the virtual cortical column, we were able to show that MSP can self-organize balanced cortical connectivity in and between multiple cortical layers with excitatory and inhibitory neurons with very different activity levels.[3] Simulations further revealed that the interplay between high activity levels of inhibitory neurons and lower activity levels of excitatory neurons leads to the emergence of critical periods during development.

2 MSP IN A NUTSHELL

In MSP, each neuron carries separate axonal synaptic elements (representing boutons) and dendritic synaptic elements (representing spines) which can combine to form synapses. A neuron generates new elements when its electrical activity is below a homeostatic set-point and deletes elements, including those bound in synapses, when activity is above the set-point or below a certain minimum level. MSP proceeds in three reoccurring steps: (1) update in electrical activity and intracellular calcium concentration as a measure for the current average in electrical activity of the individual neuron, (2) update of synaptic elements according to defined local activity-dependent growth rules in MSP, and (3) update of network connectivity. Activity-dependent decrease in synaptic elements causes the breaking of existing synapses if the synaptic element to be deleted is bound in a synapse. Synapse formation results from connecting vacant matching synaptic elements (excitatory pre- with excitatory postsynaptic elements and inhibitory pre- with inhibitory postsynaptic elements). After the breakup of a synapse caused by the loss of a synaptic element, the remaining synaptic counterpart becomes vacant and available for synapse formation again.

MSP can be formalized by the following set of equations and parameters. Intracellular calcium is calculated as follows:

$$\frac{dCa^{2+}}{dt} = \begin{cases} -\frac{Ca^{2+}(t)}{\tau} + \beta & \text{if the neuron fires} \\ -\frac{Ca^{2+}(t)}{\tau} & \text{otherwise} \end{cases} \tag{8.1}$$

with β the influx of calcium per spike and τ the decay time of calcium. The described examples were all simulated with the following Gaussian growth rules:

$$\frac{dz}{dt} = \nu\left(2\exp\left(-\frac{Ca(t) - \xi}{\zeta}\right) - 1\right) \tag{8.2}$$

where z is the number of synaptic elements of a particular type (pre- or postsynaptic, excitatory or inhibitory), $\xi = (\eta + \epsilon)/2$, $\zeta = (\epsilon - \eta)/2\sqrt{\ln 2}$, and ν is the growth rate. η is the left intersection of the curve with the x-axis, representing the minimum calcium concentration required for synaptic element formation. ϵ, the right intersection, is the homeostatic set-point neurons strive toward. NEST also provides linear growth rules to be chosen (not discussed here). If further growth rules are needed, their implementation can be requested from the NEST consortium. In the described examples, formation and deletion of synapses is independent of the spatial distance of neurons. Although NEST offers an environment to set up spatial networks, the integration with the MSP framework is still work in progress and was not available at the time this study was completed.

3 MSP IMPLEMENTATION IN NEST

We used the simulation environment for spiking neuronal networks NEST, which provides routines to simulate structural plasticity since version 2.8. These routines implement the MSP algorithm[2] as described in Chapter 7 of this book. The structural plasticity simulation framework in NEST is closely linked with the core of the NEST environment (Fig. 8.1A) enabling the combination of MSP with any neuron model for electrical activity provided by NEST. Further, all functionality of NEST for generating and recording electrical activity are fully available to structural plasticity simulations. The challenge of integrating structural plasticity into NEST was to endow NEST with a variable set of synapses during runtime. Additionally, the model of a synapse in NEST was extended by the concept of synaptic elements. (As in biological reality, the synapse is not mere a weight factor but consists of a pair of matching pre- and postsynaptic elements.) Every synaptic element has its own individual activity-dependent growth rule. Thereby, the user can choose individual target levels of electrical activity for the synaptic elements of every type of neuron (e.g., excitatory or inhibitory neurons in the different layers of the virtual cortical column)

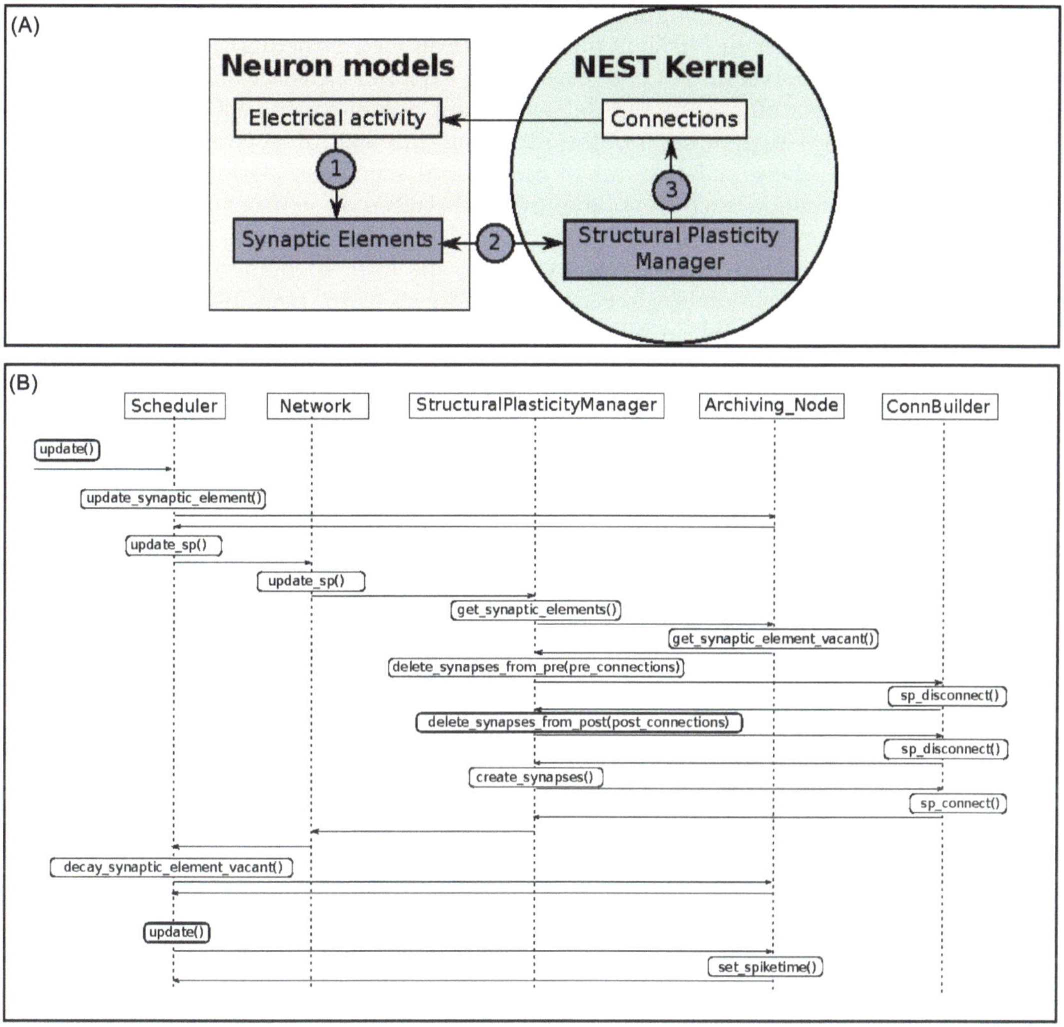

FIGURE 8.1 **Implementation of MSP in NEST.** (A) All methods for simulating spiking neuron models in NEST can be used in combination with MSP. The update in synaptic elements is implemented locally at the level of the neuron model (*square*) (which allows efficient parallelization of the update in synaptic elements). The update in connectivity requires global knowledge of the current connectivity and the currently available amount of vacant synaptic elements for synapse formation and is therefore implemented at the level of the nest kernel (*circle*) which uses global communication to update the number of synaptic connections during runtime of the simulation. (B) Chart of new classes and routines to implement MSP in NEST. Classes are indicated by *squares*; routines by *squares with rounded corners*. *Arrows* indicate the interaction between routines. Refer main text for the description of classes and routines. Source: *Reproduced from Diaz-Pier S, Naveau M, Butz-Ostendorf M, Morrison A. Automatic generation of connectivity for large-scale neuronal network models through structural plasticity. Front Neuroan. 2016;10:57.*

in the simulation. The key to updating the number of synaptic elements and synaptic connections between neurons in NEST is the newly developed Structural Plasticity Manager developed by Mikaël Naveau.[3] A C++ class using MPI and OpenMP for parallel computing collects the number of created and deleted synaptic elements of each neuron and adds and deletes synapses depending on the changing number of synaptic elements. The Structural Plasticity Manager hooks up to the general scheduler algorithm in NEST that communicates information such as spiking events between all neurons using global communication. Since growth of synaptic elements with $\nu = 10^{-4}$ elements per millisecond is much slower than the electrical activity dynamics, the implementation of MSP does not significantly slow down simulation speed in networks up to 10^5 neurons.

The implementation of MSP in NEST calls a series of functions that closely resembles the original MSP algorithm and neatly integrates into the existing modeling framework NEST for networks of spiking neurons (Fig. 8.1B). The scheduler calls the update() function which orchestrates the update in electrical activity. Further preexisting NEST functions are executed that generate the electrical activity of all neurons distributed and parallelized over the available cores and record the intracellular calcium concentration as a measure for the current average electrical activity of the neuron. The scheduler further calls the local function update_synaptic_elements() that computes the current number of synaptic elements for each neuron and the resulting number of vacant synaptic elements and those that need to be deleted. The presented implementation of NEST requires global communication as neurons must inform each other of the availability of vacant synaptic elements at a time as well as the removal (as synaptic counterparts become vacant again) and the formation of synapses (as synapse formation decreases the number of vacant synaptic elements). Therefore, the function update_sp() (sp = structural plasticity) is called on network level and subsequently triggers the respective functions like get_synaptic_elements_vacant(), delete_synapses_from_pre() and -post(), and create_synapses(). Delete and create functions directly communicate with the modified connection builder in NEST that now can be executed during runtime and makes variable synapse sets first possible.

4 CRITICAL PERIODS IN A SELF-ORGANIZING TWO-POPULATION NETWORK

We simulated a network of 1000 leaky-integrate-and-fire neurons with 80% excitatory and 20% inhibitory neurons. We used individual growth rules with different set-points of target activity for the excitatory and inhibitory population of neurons. Growth rules were defined by $\eta_e = 0.0$, $\epsilon_e = 0.05$, and $\nu = 10^{-4}$ elements per millisecond for the synaptic elements of all excitatory neurons (with subscripts indicating the type of neuron e: excitatory and i: inhibitory) and by $\eta_i = 0.0$, $\epsilon_i = 0.2$, and $\nu = 10^{-4}$ elements per millisecond for the synaptic elements of all inhibitory neurons. Initial connectivity between all neurons was set to zero. We ran the simulation for 3000 seconds with an integration step of 0.1 milliseconds and updates in connectivity taking place every 100 milliseconds. During the entire simulation, we grew network connectivity according to the Gaussian growth rules defined by the MSP algorithm. After 3000 seconds, the simulation had reached a stable equilibrium in electrical activity in terms of firing rate and in number of synaptic connections.

5 INHIBITION TRIGGERS THE ONSET OF CRITICAL PERIODS

The course of network development for excitatory and inhibitory neurons can be segregated in characteristic phases (Fig. 8.2). Due to the shape of the growth curve, excitatory neurons initially grow synaptic elements a little faster than inhibitory neurons (A) and quickly reach their set-point of activity after a brief overshoot (B) and rebound (C). The growth dynamic of the excitatory neurons is superimposed by the influence of the forming inhibitory network. As soon as the inhibitory connections exert a certain impact on the excitatory neurons lowering their activity significantly below their set-point (D), a secondary, activity-dependent phase of amplified growth in excitatory connections is triggered. The amplified growth resembles a critical period as known from juvenile brain development. A critical period is a phase during brain development in which plastic changes are possible that are not possible outside this period.[5] Plastic changes during the critical period are brought about by pronounced synapse formation and deletion[6] while during adulthood plastic changes predominantly manifest itself as adaptations in the strength of existing synapses.[7] The phase of pronounced growth (E) is followed by pruning of connections. The whole developmental network rewiring continues until inhibitory neurons have reached their equilibrium (F). Hence, the inhibitory neurons in the network open and close the "critical period" in this simple model network.

6 GROWING A VIRTUAL CORTICAL COLUMN FROM SCRATCH

The simulation of a virtual column used 80,000 neurons according to Potjans and Diesmann.[8] We initialized connectivity among all neurons with zero and used MSP with Gaussian growth rules to grow network connectivity in an activity-dependent manner with synapse formation being equally likely between all neurons. We compared the resulting connectivity with the previous cortical column model from the literature.[8] The parameter set for this simulation is specified in Table 8.1. Note that not all parameter configurations allow for a stable network formation. A thorough investigation of the parameter regimes leading to stable, unstable, or oscillatory growth is subject to ongoing research. Simulations were performed on JUROPATEST (70 nodes with two Intel Xeon E5-2695 v3 (14 cores each) at 2.3 GHz and 128 GiB RAM, running Scientific Linux 6.5) and JURECA (260 nodes with two Intel Xeon E5-2680 v3 (12 cores each) and 128 GiB RAM, running CentOS 7).[3] We ran each simulation until a stable homeostatic equilibrium in electrical activity and network connectivity was reached. This was the case at 700 simulated seconds which took about 24 hours on JURECA with 25 nodes and 24 cores per node.

7 LOW TARGET ACTIVITY LEVELS IMPOSE PRONOUNCED SYNAPTIC REWIRING

For the chosen parameter set, all eight neuron populations reached a balanced network state and their individual target activity level (Fig. 8.3). Excitatory neurons in superficial

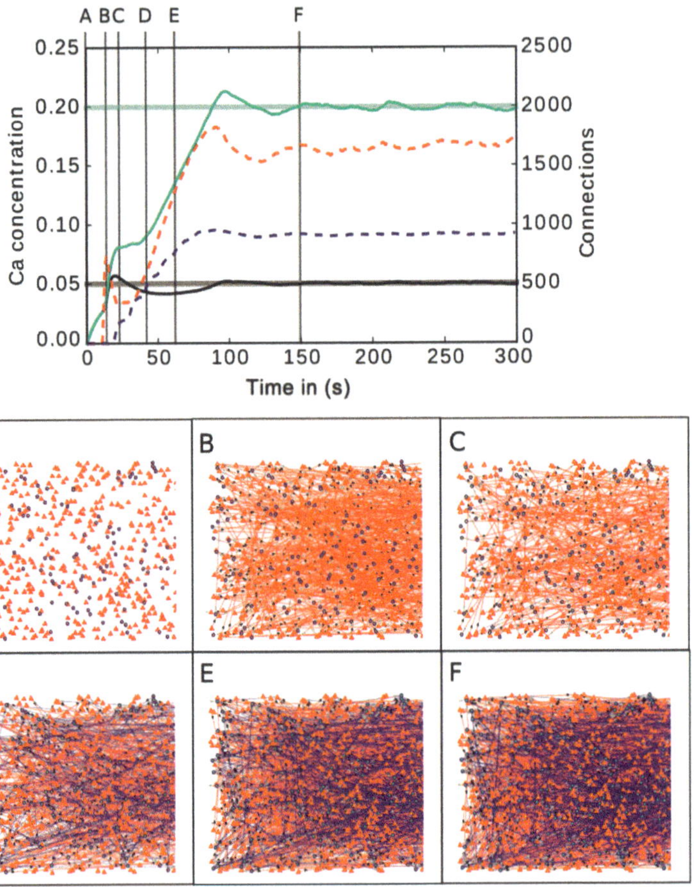

FIGURE 8.2 **Growth of a two-population network.** Top panel: Changes in calcium concentration and total number of connections in the network over time. Time is given in seconds. *Gray horizontal bar* indicates the set-point of the excitatory neuron population while the *turquoise bar* indicates the set-point of the inhibitory neurons. Correspondingly, the *black and turquois lines* show the course of the averaged calcium concentration for excitatory and inhibitory neuron populations. The *dashed curves* show the changing total number of excitatory (*red*) and inhibitory connections (*blue*) in the entire network over time. A–F indicate a series of characteristic time points during network formation as explained in the main text. Bottom panel: Diagrams A–F display a snapshot of the network connectivity at the time points (A–F) marked in the top panel. Point neurons are placed randomly on a two-dimensional grid. *Red triangles* indicate the position of the excitatory neurons and *blue circles* the position of inhibitory neurons, respectively. Correspondingly, *red lines* indicate excitatory connections and *blue lines* indicate inhibitory connections. Source: *Reproduced from Diaz-Pier S, Naveau M, Butz-Ostendorf M, Morrison A. Automatic generation of connectivity for large-scale neuronal network models through structural plasticity. Front Neuroan. 2016;10:57.*

cortical layers with very low activity levels balanced their activities first and onward maintained a stable equilibrium while average activities in the other layers were still dramatically fluctuating. Maintaining a stable equilibrium in the superficial layers goes along with pronounced synaptic rewiring of their afferent excitatory and inhibitory connections. The

TABLE 8.1 Parameter Specification for the Virtual Column Model

Parameters	Values
Growth curve type	Gaussian
Growth rate excitatory dendritic	0.0001
Growth rate excitatory axonal	0.00018
Growth rate inhibitory dendritic	0.0001
Growth rate inhibitory axonal	0.00025
η excitatory	0.0
η inhibitory	0.0
ϵ excitatory L23 dendritic	0.05
ϵ inhibitory L23 dendritic	0.25
ϵ excitatory L23 axonal	0.05
ϵ inhibitory L23 axonal	0.2
ϵ excitatory L4 dendritic	0.26
ϵ inhibitory L4 dendritic	0.45
ϵ excitatory L4 axonal	0.26
ϵ inhibitory L4 axonal	0.45
ϵ excitatory L5 dendritic	0.55
ϵ inhibitory L5 dendritic	0.5
ϵ excitatory L5 axonal	0.55
ϵ inhibitory L5 axonal	0.5
ϵ excitatory L6 dendritic	0.35
ϵ inhibitory L6 dendritic	0.59
ϵ excitatory L6 axonal	0.35
ϵ inhibitory L6 axonal	0.59

From Diaz-Pier S, Naveau M, Butz-Ostendorf M, Morrison A. Automatic generation of connectivity for large-scale neuronal network models through structural plasticity. Front Neuroan. 2016;10:57.

homeostatic growth rules permanently compensate for the dramatic changes in input activity neurons in superficial layers are bombarded with from the other, not yet balanced layers. By contrast outgoing connections from deep and middle layers are much less reorganizing. The changing input to the superficial layers from deep and middle layers is therefore brought about by the strong activity dynamics in the other layers rather than by structural dynamics. The different extent of structural dynamics in the different layers is additionally a direct consequence of the shape of the growth curves. The closer low and

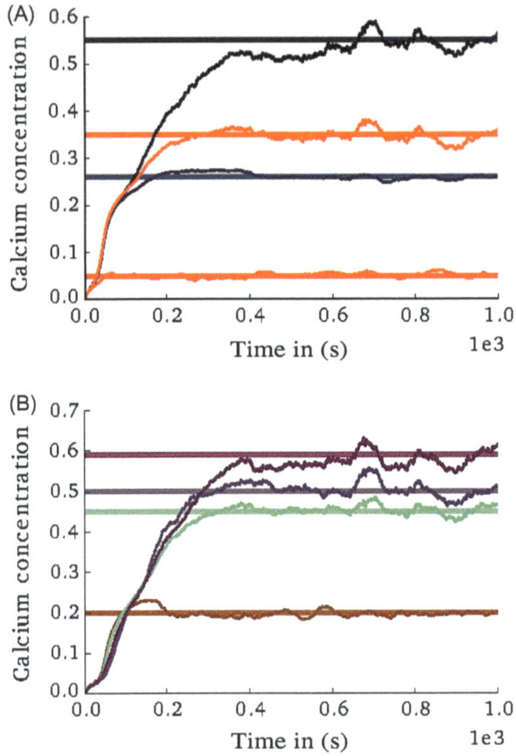

FIGURE 8.3 **Course of calcium concentration in the detailed cortical column model.** (A) Average calcium concentration of every excitatory neuron population. The colors indicate the ensemble of excitatory neurons in the different cortical layers: *red*—layers II/III; *dark blue*—layer IV; *black*—layer V; *orange*—layer VI. (B) Average calcium concentration of every inhibitory neuron population. The colors indicate the ensemble of inhibitory neurons in the different cortical layers: *brown*—layers II/III; *cyan*—layer IV; *gray*—layer V; *purple*—layer VI. Source: *Reproduced from Diaz-Pier S, Naveau M, Butz-Ostendorf M, Morrison A. Automatic generation of connectivity for large-scale neuronal network models through structural plasticity. Front Neuroan. 2016;10:57.*

high fix points are (which is the case for a low set-point), the steeper is the slope of the growth curve in vicinity of the set-point, hence the stronger is the change in synaptic elements when activities fluctuate around the set-point. In superficial layers, the response to perturbations of the homeostasis in electrical activity consequently exerts stronger rewiring than in middle and deep layers with higher target levels in electrical activity.

The interplay between excitation and inhibition in the cortical column simulation resembles the critical period phenomenon that was already visible in the simplified two-layer scenario (Fig. 8.4). The effect of the critical period is most pronounced for recurrent excitatory and inhibitory connections of the superficial layers. The network of superficial cortical layers II and III has thus the highest and longest plastic potential by synaptic rewiring. This is remarkable as the effect is exclusively brought about by the target level of electrical activity of each neuron population while all other parameters especially the growth rate are identical.

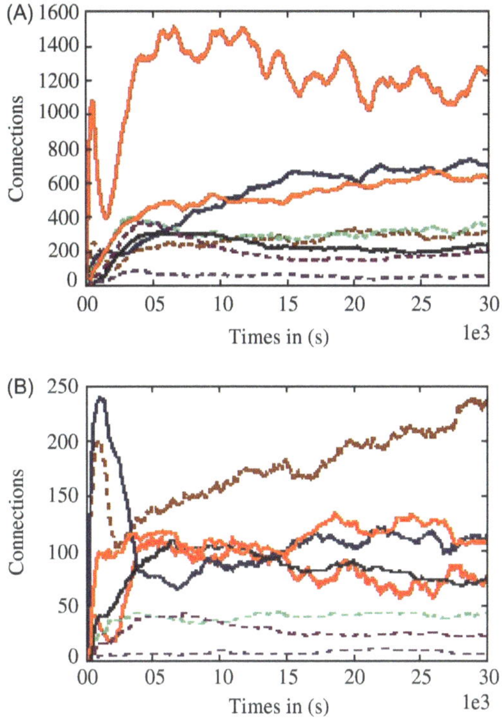

FIGURE 8.4 **Connections over time.** As an example, we show developing outgoing connectivity from (A) excitatory and (B) inhibitory neuron populations of layers II/III to all populations in the other layers. *Colors and line styles* indicate the target population of each connection. *Solid lines* represent excitatory target populations while *dashed lines* indicate inhibitory target populations. Color code as in Fig. 8.3. Source: *Reproduced from Diaz-Pier S, Naveau M, Butz-Ostendorf M, Morrison A. Automatic generation of connectivity for large-scale neuronal network models through structural plasticity. Front Neuroan. 2016;10:57.*

8 COMPARISON BETWEEN SELF-ORGANIZING AND RECONSTRUCTED CONNECTIVITY

A previous cortical column model[8] used experimental data[9] to set up synaptic connectivity between all neurons of the different cortical layers. Here, we compared the reconstructed connectivity of this model with the self-organized connectivity from the MSP simulation in NEST. To compare the total number of synapses between all layers and cell types of the two models, we linearly normalized total numbers and plotted them as concentric circles (Fig. 8.5). Many self-organized connections show an almost perfect match with the reconstructed model which is most surprising as the only assumption that was made was the target activity. Hence, it seems that the anatomical strength of the connections between some layers immediately follows from the target activity level of their principle neurons. This behavior is especially seen for the deep and superficial layers IV and VI. The connectivity of the superficial layers in the

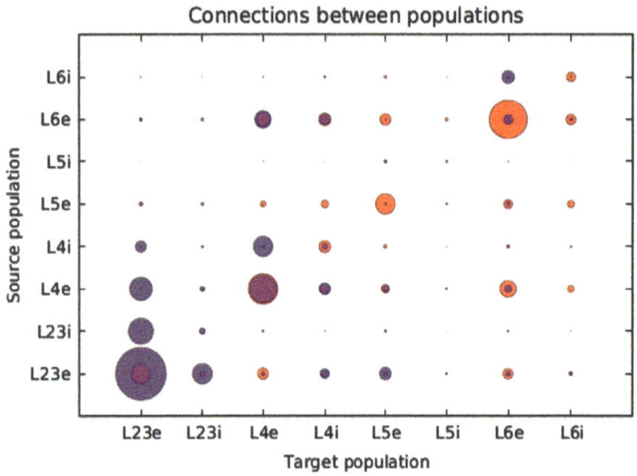

FIGURE 8.5 **Comparison between modeled and biological connectivity.** Every *circle* represents the normalized connectivity between all combinations of cortical layers (layers II/III, IV, V, and VI). "e" and "i" indicate excitatory and inhibitory connections, respectively. *Blue circles* represent the resulting connectivity in the model at equilibrium while *red circles* represent biological connectivity according to Potjans and Diesmann.[8] Source: *Reproduced from Diaz-Pier S, Naveau M, Butz-Ostendorf M, Morrison A. Automatic generation of connectivity for large-scale neuronal network models through structural plasticity. Front Neuroan. 2016;10:57.*

reconstructed model is surprisingly high regarding the low target activity. A much weaker connectivity would be sufficient to reach that level of activity. The self-organizing connectivity especially within the superficial layer was already quite strong, too, when the snapshot was taken and was still growing. As the strong connectivity in the self-organized network is a consequence of the interplay of developing inhibition and excitation (see explanations on emerging "critical periods" in the model before), we may conclude that also in the biological cortical column, excitatory connectivity is shaped by inhibition during critical periods.

9 SCALABILITY LIMITATION OF MSP

In the previous sections, we discussed a parallel implementation of MSP in the NEST neural network simulator[3] and showed results of structural plasticity simulations with up to 10^5 neurons. However, MSP's computational complexity in terms of the number of neurons prevents it from scaling to larger neuron counts. In the following, we give an overview of the three iteratively executed steps of MSP together with their computational complexity: (1) update of electrical activity, (2) update of synaptic elements, and (3) update of connectivity.

9.1 Update of Electrical Activity

The electrical activity of every neuron is continuously calculated on a millisecond timescale. This step considers every neuron and thus runs in time $O(n)$ linear with the number of neurons.

9.2 Update of Synaptic Elements

Based on a set-point for the calcium concentration, a homeostatic growth rule determines for each neuron when axonal and dendritic synaptic elements are created or deleted. Every neuron is considered once, which results in linear cost $O(n)$ of this step.

9.3 Update of Connectivity

Depending on the number of synaptic elements, existing synapses are deleted and new synapses are formed. Synapse deletion runs in linear time $O(n)$ as every neuron is examined to decide which of its synapses have to be deleted.

To decide which pairs of axonal and dendritic elements form a synapse, the model considers all pairs of neurons with a vacant axonal element on one side of the pair and a vacant dendritic element on the other. We call a neuron with a vacant axonal element *source* neuron. The *target* neuron which the source neuron's axonal element will try to connect to is determined by considering every neuron as a potential target and calculating the probability of establishing a connection. The probability depends on the distance between source and target, and the number of unbound dendritic elements available at the target. In this way, every vacant axonal element selects a target neuron. Given that every neuron creates a certain amount of both axonal and dendritic elements, finally all pairs of neurons have to be considered. Thus the execution time of this step grows quadratically $O(n^2)$ with the number of neurons, which becomes prohibitive when the number of neurons grows beyond a hundred thousand. Later in this chapter, we present execution times of MSP which show its limited scalability to neuron counts in the order of 10^5. Note that the human brain has about 10^{11} neurons.

10 A SCALABLE ALGORITHM FOR MSP

The scalability of MSP is limited by the quadratic computational complexity of calculating the probabilities in the synapse creation phase. A similar challenge is known from particle physics where in n-body problems pairs of bodies have to be considered for force calculations. To improve their scalability, powerful approximation methods have been developed.[10,11] Their principle is based on the observation that particles distant enough from a target particle do not need to be considered individually for force calculations. The most influential algorithms are Barnes–Hut[10] and the Fast Multipole Method[11] (FMM). Unfortunately, they cannot be applied to simulating structural plasticity directly. While they calculate the force exerted on (Barnes–Hut) or the potential of (FMM) each body, we need to select pairs of neurons (bodies). Additionally, n-body simulations subject each particle to force calculation in every iteration. In the brain, however, the set of neurons with vacant axonal elements varies and only for this mutable set vacant dendritic elements have to be found. On the other hand, the approximation of advanced n-body methods seems to better mimic biological behavior in the brain. While trying to find a vacant dendrite, a neuron's vacant axon has only partial knowledge of other available neurons. In this section, we present a scalable approximation method for simulating structural plasticity based on MSP.[4] The algorithm, an adaptation of Barnes–Hut, reduces the complexity of MSP from $O(n^2)$ to $O(n \log^2 n)$.

Similarly to Barnes–Hut, we combine distant neurons into groups whenever possible instead of considering them individually. Neurons in the same group have a similar distance to the source neuron. We represent a group of neurons through a virtual neuron whose

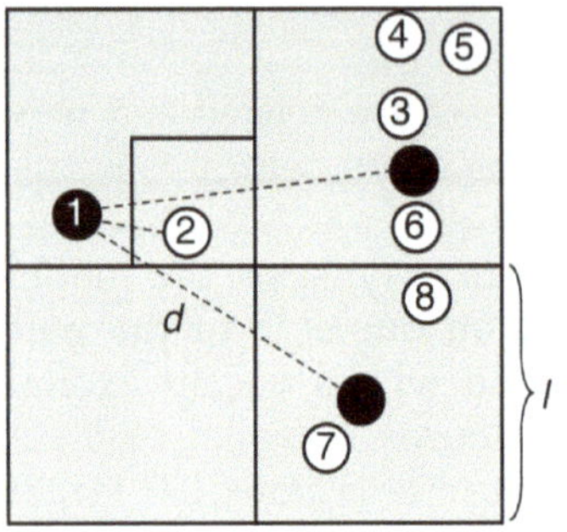

FIGURE 8.6 **A two-dimensional example of grouping neurons.** Neurons 3–6 and 7–8 are in the same respective squares. They form two groups. The length of their squares is denoted by l, which can be seen as the spatial extent of the group. The two groups are represented by a virtual neuron (*black solid circle*). The source neuron is neuron 1. Instead of considering all individual neurons as target neuron candidates, grouping reduces the work at this stage to considering only two virtual neurons and two normal neurons (neuron 2 and the source neuron itself). To avoid autapses, the source neuron's probability is set to zero. A *dashed line* depicts the distance d from the source neuron to the other neurons under consideration, which are neuron 2 and the two virtual neurons.

position is a linear combination of the positions of the group members. Weight factors position the virtual neuron closer to neurons with many vacant dendritic elements. The number of vacant dendritic elements of the virtual neuron is the sum of vacant dendritic elements present in the entire group. This approach resembles the concept of the center of mass in gravitational versions of the Barnes–Hut algorithm. Only neurons close to the source neuron are considered individually because they differ more in their relative distance to the source neuron. Otherwise, the probability error, caused by using the averaged position of the virtual neuron, could become too large. Fig. 8.6 shows an example. Neuron 2 is too close to the source neuron and hence not considered as part of a group. Below, we explain the three steps of the algorithm: (1) tree construction, (2) tree update, and (3) target neuron selection.

10.1 Tree Construction

Similarly to the Barnes–Hut algorithm, we start by forming a tree of neuron groups. However, compared to particles in an n-body simulation, our neurons do not move. Thus the tree is created only once at the start of the simulation. The tree construction proceeds as follows: given a cube that contains all neurons in our simulation domain, we create a spatial tree representation of the domain step by step. If the domain contains more than one neuron, we subdivide it into eight cubical subdomains of the same size. Each of the eight subdomains is then recursively subdivided if it contains more than one neuron. The recursion ends when every subdomain contains at most one neuron. As a result, we obtain an octree (i.e., a tree with at most eight children per node) with the root representing the cube containing all neurons. Its children are the eight subdomains that it was divided into and so on. Every leaf in the tree represents a single neuron, every inner node represents a virtual neuron in its subdomain. This hierarchy of subdomains defines the groups of neurons that we need. For ease of illustration, we have used two-dimensional examples in our figures. In comparison to three dimensions, we have subdivided a domain into four squares and the resulting tree is a quadtree (i.e., a tree with at most four children per node). Fig. 8.7 shows the final subdivision of a domain and the resulting tree. The depth of the tree depends on the distribution of the neurons. Although organized in layers, they are still rather homogeneously distributed. For this reason, we assume that the depth of the tree is $\Theta(\log n)$. As in Barnes–Hut, we create the tree by successively inserting all neurons into the tree. Since our tree is of depth $\Theta(\log n)$, tree creation takes $O(n \log n)$.

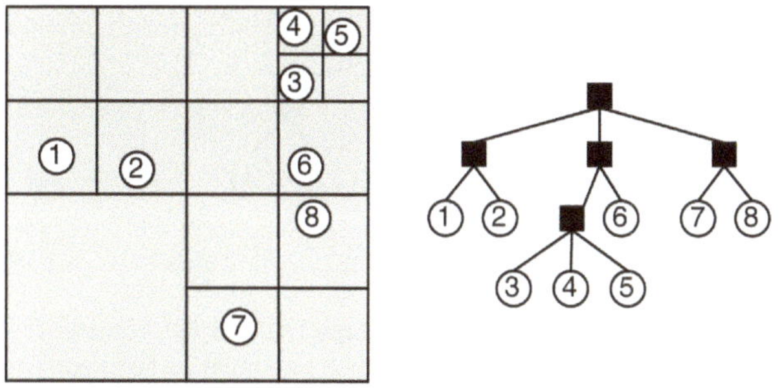

FIGURE 8.7 **A two-dimensional tree construction example.** On the left, we see the subdivision of the simulation domain. On the right, we see the resulting tree. Neurons are numbered from 1 to 8. The inner nodes in the tree are virtual neurons, while the leaves are real neurons belonging to their subdomains.

10.2 Tree Update

Neurons do not move, but the number of their vacant dendritic elements is subject to change. For this reason, the tree has to be updated before creating new synapses. For every leaf (i.e., real neuron), we store the current number of vacant dendritic elements. For every inner node (virtual neuron), we not only update the number of vacant elements but also the position of its virtual neuron. The number of vacant dendritic elements is simply the sum of those available on its (direct) children. Let v be a virtual neuron. Then the number of vacant dendritic elements is $D_v = \Sigma_{i \in \text{Children}} D_i$. The position is a linear combination of the positions of its children and their vacant dendritic elements. After updating its vacant element count, the x-coordinate of the virtual neuron v is calculated as $x_v = 1/D_v \, \Sigma_{i \in \text{Children}} x_i D_i$. The y and z-coordinates are obtained in a similar way. We update the information in the tree bottom-up from the leaves to the root via postorder traversal, which takes time $O(n)$.

10.3 Target Neuron Selection

After a tree update, we form new synapses by finding a target neuron for every vacant axonal element. To minimize the number of probability calculations, we already decide at the coarser level of neuron groups which neurons the source neuron will not connect to and which we therefore do not need to consider any further. If the source neuron decides to connect to a virtual neuron, we unfold the group it represents. This makes all its (virtual) constituent neurons visible, from which we again select one. Every group selection decreases the number of target neuron candidates. The recursion ends once a single real target neuron has been selected. Selecting a target neuron for a given source neuron means choosing a path from the root to a leaf. To decide which subdomains we consider as a whole on the path down the tree, we use the acceptance criterion (AC) of the Barnes–Hut method. Let d be the distance from the source neuron to the virtual neuron. Let l denote the length of the virtual neuron's subdomain. If $l/d < \theta$, we calculate a single connection probability for the entire subdomain. Otherwise, we unfold it and recursively apply the AC to its constituent subdomains. Here, θ is a configurable precision parameter that ensures that subdomains for which we calculate probabilities are distant enough from the source neuron in relation to their size. Note that a subdomain can be unfolded for two reasons, either because it has been selected to form a connection or because it does not satisfy the AC. The use of the Barnes–Hut AC is motivated by Rinke et al.[4]

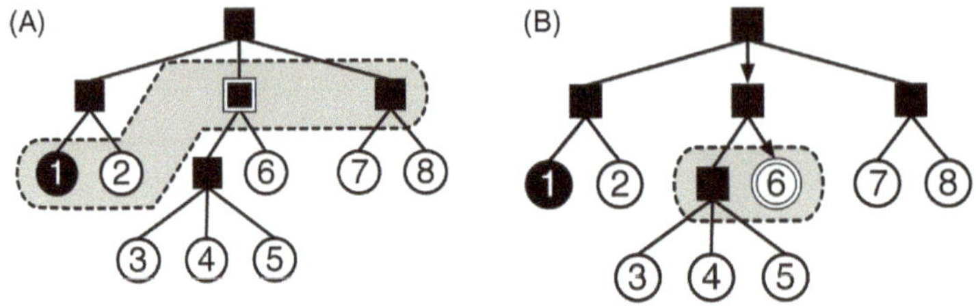

FIGURE 8.8 **A two-dimensional example of target neuron selection in two steps (A) and (B).** Neuron 1 is the source neuron. *Areas shaded in gray* identify the set of (virtual) neurons from which one must be selected. The selection is framed. In (B), *arrows* indicate the path from the root to the target neuron 6.

Fig. 8.8 continues our previous example. Starting from the root, the AC is applied. Because the root does not satisfy $l/d < \theta$ for its domain, we need to unfold it. The same is true for its first child. At this point, the recursion stops because both its children are leaves. The other two children of the root satisfy the AC and remain closed for now. The gray-shaded area marks the first set of nodes for which we now calculate probabilities. The difference to the original MSP is that we only consider a subset of neurons, with some of them being virtual. Based on their probabilities, we select one neuron from this subset. It is the second child of the root (framed square), a virtual neuron. In the next step (Fig. 8.8B), we unfold the subdomain of the selected neuron and apply the AC to one virtual neuron and neuron 6. Both are accepted (gray area). Note that a real neuron is trivially accepted since unfolding is not possible. We calculate the connection probabilities and select one neuron, which is neuron 6. It is a real neuron and thus the target neuron for source neuron 1. Now, the selection terminates at this point.

The complexity of the target neuron selection depends on the number of nodes to consider. With $\theta = 0$, all subdomains are unfolded and we consider every neuron for a given source neuron. That is, the algorithm is exact and behaves as the original MSP with complexity $O(n^2)$. For $\theta > 0$, we start considering groups of neurons. Here, the complexity depends on the depth of the tree, which we assume to be $\Theta(\log n)$, as previously stated. When randomly selecting nodes on the path down from the root to the target neuron, after every selection the depth of the remaining subtree is reduced in the worst case by one only. That is, we have to perform $\Theta(\log n)$ steps (depth of the tree) until we find a target neuron. To determine the complexity of each step, we follow the argument of Barnes and Hut for homogeneously distributed particles.[10] In particular, when increasing the number of neurons, the new additional subdomains not containing the source neuron incur a certain amount of extra probability calculations. This amount depends on θ but not on the number of neurons. Consequently, increasing the number of neurons by a constant factor only increases the number of probability calculations by an additive constant (for θ). That is, the complexity of each step is $O(\log n)$. Therefore, it takes time proportional to $O(\log^2 n)$ to find a target neuron for one source neuron. Under the biologically motivated assumption that the tree is of depth $\Theta(\log n)$, the complexity of finding a target neuron for every neuron in one connectivity update is therefore $O(n \log^2 n)$.

```
Create empty network without synapses                          ▷O(n)
Initialize number of synaptic elements per neuron              ▷O(n)
Construct tree from domain                                     ▷O(n log n)
for i ← 1, simulation steps do
    UPDATEELECTRICALACTIVITY                                   ▷O(n)
    UPDATESYNAPTICELEMENTS                                     ▷O(n)
    if i mod 100 = 0 then
        UPDATECONNECTIVITY {                                   ▷O(n log² n)
            Delete synapses & update network                   ▷O(n)
            Create synapses {
                Update tree                                    ▷O(n)
                Find target neuron for every vacant axonal element   ▷O(n log² n)
                Update network                                 ▷O(n)
            }
        }
    end if
end for
```

FIGURE 8.9 **The simulation flow of MSP when using the hierarchical algorithm.** The complexity of each step is shown on the right. The number of simulation steps is a parameter provided by the user.

10.4 Error Analysis

By choosing $\theta > 0$, we reduce the complexity and derive from the exact probabilities of the original MSP, which introduces an error. Unfortunately, it seems difficult to determine analytically which effect the error has on the actual structure of the resulting neural network. Hence, we run simulations with different precision parameters and analyze their effect on the resulting networks in Section 11. To ensure that the approximated networks are still sufficiently precise, a practical usage guideline for the scalable algorithm is to set $\theta \leq 0.5$. More details on the approximation error of the scalable algorithm are discussed by Rinke et al.[4]

10.5 Summary

The execution flow of the scalable MSP algorithm is depicted in Fig. 8.9. The steps within the for loop are executed in every iteration. As can be seen, the connectivity update, which depends on the number of synaptic elements, has the greatest complexity. As synaptic elements grow much slower compared to the frequency at which neurons fire, the connectivity update is run only every 100 simulation steps. However, as the experimental results in Section 11 show, even a single connectivity update using the exact MSP takes as long as about 50 minutes for 10^6 neurons, which makes it the clear bottleneck. This is why we must update the connectivity more quickly. This algorithm brings the complexity from $O(n^2)$ down to $O(n \log^2 n)$, which now becomes the overall computational complexity of the approximated MSP.

11 RESULTS OF THE SCALABLE ALGORITHM

Let us now evaluate the algorithm in terms of performance and impact of the approximation on the result quality. Our simulation parameters correspond to layer 5A of the rat cortex,[12] which is about 500 μm thick with a density of 54,500 neurons per mm^3. Of these, 20% are inhibitory, the remaining 80% are excitatory. We randomly distribute the neurons in a volume of height 500 μm and let the other two dimensions grow with the number of neurons. Initially, no synapses exist. Every neuron is initialized with vacant synaptic elements: one excitatory and one inhibitory dendritic element plus one axonal element. The type of the axonal element depends on the type of the neuron. Excitatory neurons form excitatory axonal elements while inhibitory neurons form inhibitory axonal elements. However, both neuron populations grow excitatory and inhibitory dendritic elements. Synapses are only possible between synaptic elements of the same type. That is, excitatory axonal elements only connect to excitatory dendritic elements. A similar rule applies to inhibitory synaptic elements. In our experiments,[4] we compare parallel versions of the original MSP algorithm and the scalable MSP approximation algorithm. In both cases, neurons are evenly distributed across processes. Our compute nodes are equipped with two Intel Xeon E5-2670 (8 cores each) and 32 GiB RAM. The cluster network is InfiniBand FDR.

11.1 Performance

With the exception of synapse formation, which is accelerated via approximation, the time-step loop of the scalable algorithm performs the same steps as the original MSP algorithm. Another difference is the construction of the octree, which the scalable algorithm performs once before the simulation starts. Nevertheless, even for 10^7 neurons tree construction does not exceed 2 minutes. We do not include this time in our measurements as it is a one-off expense. Fig. 8.10A and B shows weak scaling results for one connectivity update for the original MSP and the scalable algorithm. We simulate the very first connectivity update where no synapses yet exist and every neuron has vacant elements as described earlier. That is, every neuron is trying to find a target neuron for its vacant axonal element. The original MSP needs about 50 minutes for 10^6 neurons, while the scalable algorithm terminates in 2.5 minutes. On the other hand, even for high precision with $\theta = 0.1$, the scalable algorithm is still in the range of 5 minutes for 10^7 neurons. This large number of neurons is practically out of reach for the original MSP. The results also show that reducing the precision of the approximation method from 0.1 to 0.2 can help to further reduce the execution time by at least a factor of 2.5. Fig. 8.10C depicts strong scaling results for the scalable method. The smaller decrease in execution time for 32 cores is consistent across measurements (10 for 10^5 neurons, 5 for 10^6). We believe the reason for this dent can be found in the hardware topology of the target system. Although only responsible for a subset of the neurons, in both implementations each process stores the data of all neurons. Otherwise, the design of the octree would have been more complicated. That is, the scalable algorithm replicates the entire octree on each process. This is why the total number of neurons in our experiments is limited by the amount of memory available per compute node. To improve on this limit, a memory-optimized implementation of the scalable algorithm could use a distributed octree that truly partitions the neurons and only stores locally those neurons a process is responsible for. This will enable the scalable algorithm to perform even larger simulations of structural plasticity than in our experiments.

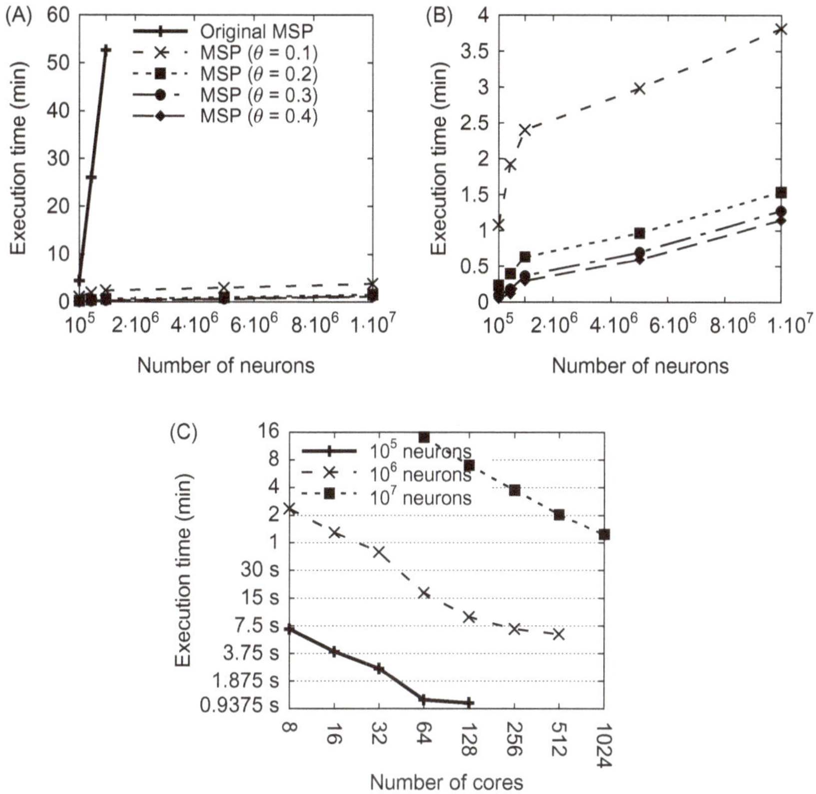

FIGURE 8.10 **Scaling results.** (A and B) Weak scaling execution times with different precision parameters. The legend in (A) also applies to (B). The number of neurons per core (MPI process) is 10^4. The numbers of neurons (cores) are: 10^5 (10), 5×10^5 (50), 10^6 (100), 5×10^6 (500), 10^7 (1000). (C) Strong scaling execution time with $\theta = 0.3$.

11.2 Accuracy

To validate the accuracy of the scalable algorithm, we compare the neural networks it generates to those of the original MSP. We consider a neural network as a weighted directed graph where neurons are the vertices and synapses are the directed edges pointing from a source neuron to a target neuron. The number of synapses reaching from the source to the target is the edge weight. Our comparison is based on the following graph topology metrics that Butz et al.[13] use to describe the structure of neural networks: (1) number of edges, (2) average Euclidean distance, (3) average shortest path length, (4) global efficiency, (5) average betweenness centrality, and (6) average clustering coefficient. We compare networks of the original MSP and the scalable algorithm with 10^3, 10^4, and 10^5 neurons. The quadratic computational complexity of some of the graph metrics prevents us from evaluating them for larger neuron counts. We start with an empty network and initially vacant synaptic elements. The biological simulation time is 6×10^6 ms, which

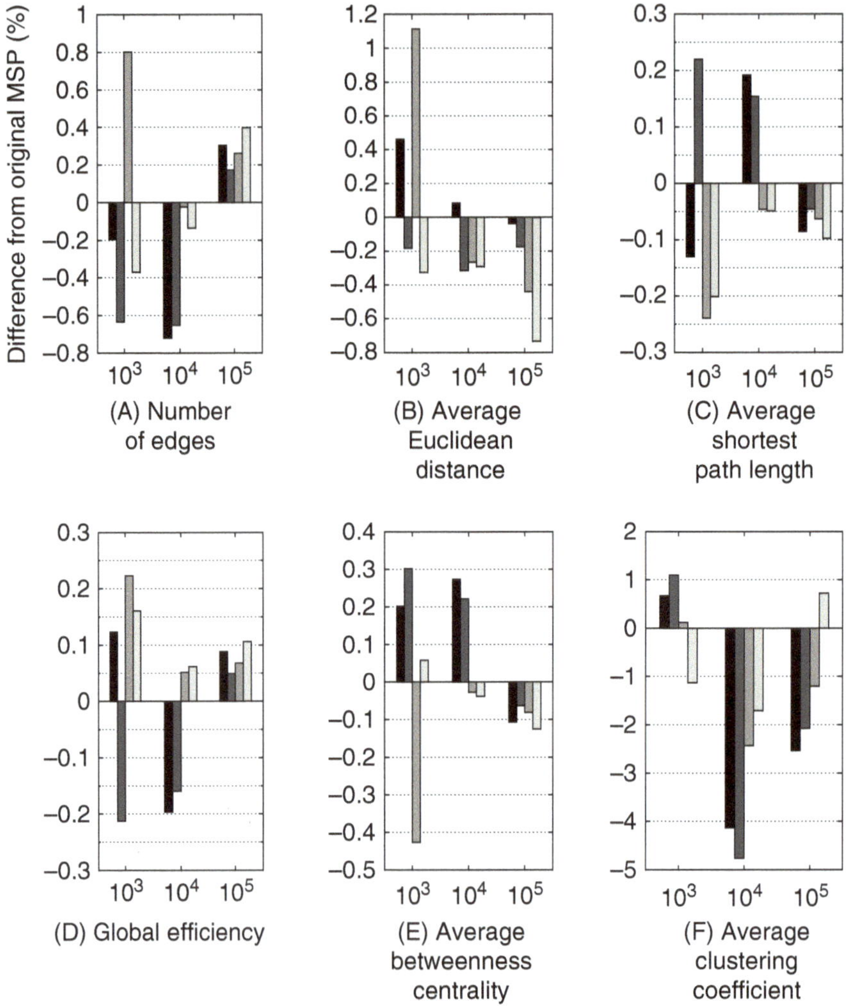

FIGURE 8.11 **Network comparison between the original MSP and the scalable algorithm for 10^3, 10^4, 10^5 neurons.** The bars at each cluster correspond from left to right to $\theta = 0.1$, 0.2, 0.3, 0.4.

allows 60,000 connectivity updates. At this time, the neuronal electrical activity and the network have reached their equilibrium and change only insignificantly. Fig. 8.11 shows the metrics of the networks generated by the scalable algorithm relative to those of the original MSP. Except for the average clustering coefficient, the networks produced by the scalable algorithm differ only by about 1% from the original MSP. This is even true for low precisions. However, for 10^4 neurons the average clustering coefficient differs by about 5% with even a small $\theta = 0.2$. To investigate this case further, we calculate the average and standard deviation of the metrics over 11 simulations for both algorithms (Fig. 8.12). We use a different random number seed for every run. Interestingly, even for the exact MSP, individual average clustering coefficients vary by about 5% across measurements. That is,

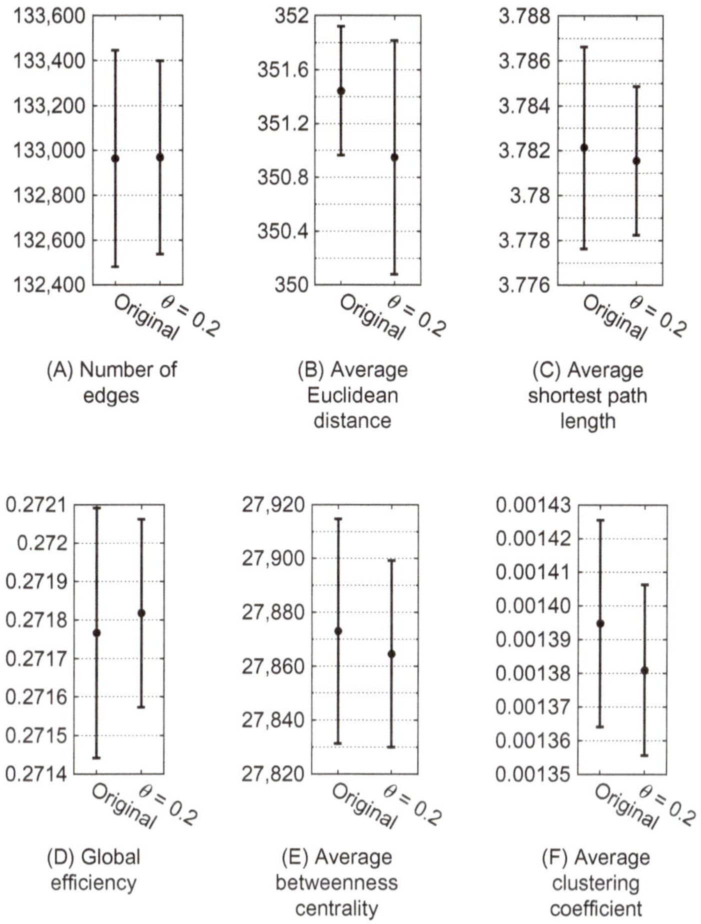

FIGURE 8.12 **Network comparison between the original MSP and the scalable algorithm with $\theta = 0.2$ for 10^4 neurons.** The results per method are based on 11 simulations with a different random number seed for each run.

the 5% difference can also be found between different runs of the exact MSP. Fig. 8.12 gives an overview of how the results vary. It can be seen that the difference between the average of the exact MSP and the scalable algorithm is below 1% for all metrics. Also the standard deviations are similar except for the average Euclidean distance. However, although the "spread" around the average is different for this metric, the actual average of the scalable method only differs by 0.14%. Hence, we do not consider the difference in the standard deviation as significant. These experiments support the claim that our approximated networks are still precise enough to represent neural networks of the exact MSP. Nevertheless, to account for the variation of the results due to MSP's probabilistic approach, several simulation runs are necessary to reliably capture the essential structure of a neural network at the end of the simulation. This is true for both, the exact MSP and the approximation of it. Thus the scalability of the MSP algorithm is even more critical.

12 DISCUSSION

Implementing structural plasticity adds a new dimension of plastic reorganization to large-scale neuronal network simulations, with a wide range of new applications. With the MSP algorithm in NEST, it has become possible to grow any type of spiking neuronal network from scratch. The presented approach overcomes the tedious work of fitting connectivity parameters to set up balanced large-scale networks for detailed models of tissue cultures or mesoscopic networks of a few million neurons. The approach of Rinke et al. reviewed here goes a significant step beyond this. With 10^7 simulated neurons, the networks are already two orders of magnitude bigger than what is currently possible in NEST, and the approach paves the way for much bigger and even full-size brain simulations that can for the first time self-organize their connectivity. The main advantage over the NEST implementation is that the adaptation of the Barnes–Hut algorithm deployed here for the update in connectivity eliminates the quadratic complexity of MSP through approximation. Thus simulations with approximations of MSP of 10^7 neurons and beyond are now possible.

However, already with the current implementation of MSP in NEST, it has become possible to study the reciprocal dynamics of growing excitatory and inhibitory cells in neuronal networks that was before impossible to study in smaller networks.[13] It was known already from experiments that the timing of critical periods is regulated by inhibition.[5] However, the complete explanation of how inhibition exerts this timing effect is still lacking. The large-scale simulation of self-organizing excitatory and inhibitory networks now revealed the neurons' need for homeostasis in electrical activity as a potential mechanism for the occurrence of critical periods during development. In the model, inhibitory neurons strive toward higher activity target levels than excitatory neurons. While inhibitory neurons increase their firing rates toward the set-point, they stronger inhibit the population of excitatory neurons, which in turn have to counterbalance the increasing inhibition by forming extra excitatory input synapses. A phase of pronounced formation of excitatory synapses (and subsequent deletion due to overshoot behavior), the "critical period" in the model, first closes when the increase in inhibition stops (when inhibitory neurons have reached their target activity). Only after inhibition stops changing can excitatory neurons finally adapt their excitatory input synapse number to reach the target activity under the present, stable inhibition. Hence, temporarily blocking inhibition in the model would postpone the onset and the extent of the critical period, as it suspends the driving force for creating extra excitatory synapses, whereas increasing inhibition would have the opposite effect. Experimentally it has indeed been shown that reducing inhibition, e.g., by dark rearing or by deletion of the GABA-synthetic enzyme GAD65, postpones the critical period of ocular dominance plasticity, whereas increasing inhibition by GABA agonists such as benzodiazepines[5] opens the critical period earlier. Moreover, the overexpressed growth of excitatory connectivity under increased inhibition is visible in vivo by the resulting broader ocular dominance columns following diazepam (GABA agonist) application.[14] The different set-points of excitatory and inhibitory cells, together with Gaussian growth rules, may be the last missing piece in understanding ocular dominance plasticity. The model also illustrates that critical periods in ocular dominance plasticity can reopen later in adulthood when the homeostatic equilibrium in electrical activity is significantly disturbed, e.g., after brain lesions.[15] Interestingly, diazepam is used in stroke therapy to trigger critical periods in brain repair by increasing inhibition.[16]

Simulating structural plasticity in large-scale networks of 8×10^4 neurons made it possible to apply the growth principles responsible for the occurrence of critical periods to a full-scale cortical column network. We endowed every population of neurons with an individual target level of electrical activity and studied emerging connectivity. First of all, it is remarkable that the individual neuron populations indeed reached a stable equilibrium of their target activity level. Whether neurons reach the activity set-point depends on the availability of axonal elements provided by other neurons. If these neurons have a low target activity and already reached their set-point, other neurons possibly fail to balance their activities. Even more remarkable is the observation that quite a few connections between some neuron populations match almost perfectly the biological example.[8] We conclude from this result that the determining factor for intracolumn connectivity may be the target level of electrical activity of the connected (principle) neuron populations. We even observed a phase of pronounced synapse formation especially in the superficial layers, which resembles the "critical period" we saw in the simple two-population model. The superficial layers are an association site for intra- and intercortical connections and show a distinct plastic potential for synaptic rewiring.[17] Synaptogenesis in superficial layers is further assumed to play an important role in skill learning[18] even in the mature brain. Having a model that generates critical periods during network formation especially in structures known to play a predominant role in learning could create interesting opportunities for creating novel biologically inspired self-learning artificial systems.

12.1 Future Experimental Studies

The simulations described here have deepened the understanding of how homeostatic structural plasticity shapes cortical networks during development. In the model, critical periods emerge from the interplay of excitatory and inhibitory neurons with different target activity levels. This finding implicates several experimentally testable predictions. Every effect that moderately lowers the electrical activity of excitatory neurons during development should be suited to open a critical period during development. Conversely, increasing electrical activity should prevent the promoting effect of benzodiazepines during development. Further experimental studies are needed to backup the postulated growth rules and, e.g., by manipulating the activity of individual groups of neurons, to study the contribution of different cell types to the emergence of critical periods in different functional systems (such as sensory or motor cortex).

12.2 Future Modeling Studies

Being able to simulate structural plasticity in networks with even higher neuron numbers far beyond 10^7 neurons is an important step toward full-scale brain simulations. We see the highest potential of such detailed models in the simulation of brain lesions and the prediction of brain repair for optimized treatment. The ultimate goal will be to create an individualized treatment plan for each stroke patient from simulations based on real patient's neuroimaging data. Today functional imaging of electrical activity, e.g., after stroke, is readily available. High resolution data on the patient's connectivity is much more difficult to obtain. Here, full-scale simulations with self-organizing connectivity could fill an important gap in the available data by providing individual connectivity based on high resolution activity brain scans. Having a solid working model of the patient's brain activity and connectivity state, one may test different treatment options against the model before applying them to the patient.

13 CONCLUSION

We focused on the parallelization and scalability of a well-understood algorithm of structural plasticity, MSP,[2,6,13,19–21] in order to enable the rewiring of large-scale neuronal networks. We used a simple two-population network model of excitatory and inhibitory neurons that is large enough to study the interplay between developing excitatory and inhibitory subnetworks. We demonstrated that the homeostatic growth principle, which we already showed to be responsible for network repair after lesions,[2] can account for the occurrence of critical periods during development. Having understood the growth dynamics in the simple model, we extended the network model to the size of a cortical column containing multiple neuron populations with individual homeostatic set-points. We could identify critical periods also in the development of this much more complex network architecture. The resulting architecture showed impressive similarities with the biological example. The obtained results provide strong support for the postulated growth rules for axonal and dendritic synaptic structures. With this preparatory work at hand, we might even create larger network simulations that aim toward the complexity of the full brain. Using a highly innovative approach inspired by particle physics, we were able to run simulations that go far beyond what is possible in today's available plasticity models. The described approach paves the way for eventually growing an entire human in silico with all implications and opportunities this creates for future artificial systems.

Acknowledgments

This project has received funding from the European Union's Horizon 2020 Research and Innovation Programme under Grant Agreement No. 720270 (HBP SGA1). This work was also partially supported by the Helmholtz Association through the Helmholtz Portfolio Theme "Supercomputing and Modeling for the Human Brain" and its Initiative and Networking Fund, and by the Jülich Aachen Research Alliance (JARA).

References

[1] Kunkel S, Schmidt M, Eppler J, et al. Spiking network simulation code for petascale computers. *Front. Neuroinf.* 2014;8(78).

[2] Butz M, van Ooyen A. A simple rule for dendritic spine and axonal bouton formation can account for cortical reorganization after focal retinal lesions. *PLoS Comput. Biol.* October 2013;9(10):1–21.

[3] Diaz-Pier S, Naveau M, Butz-Ostendorf M, Morrison A. Automatic generation of connectivity for large-scale neuronal network models through structural plasticity. *Front. Neuroan.* 2016;10(57).

[4] Rinke S, Butz-Ostendorf M, Hermanns M-A, Naveau M, Wolf F. A scalable algorithm for simulating the structural plasticity of the brain. In: *Proceedings of the 28th International Symposium on Computer Architecture and High Performance Computing (SBAC-PAD)*, Los Angeles, CA, USA, October 2016, pp. 1–8.

[5] Hensch TK. Critical period plasticity in local cortical circuits. *Nat. Rev. Neurosci.* November 2005;6 (11):877–888.

[6] Butz M, Woergoetter F, van Ooyen A. Activity-dependent structural plasticity. *Brain Res Rev.* May 2009;60 (2):287–305.

[7] Fauth M, Tetzlaff C. Opposing effects of neuronal activity on structural plasticity. *Front Neuroanat.* 2016;10:75.

[8] Potjans TC, Diesmann M. The cell-type specific cortical microcircuit: relating structure and activity in a full-scale spiking network model. *Cereb Cortex.* March 2014;24(3):785–806.

[9] Binzegger T, Douglas RJ, Martin KA. A quantitative map of the circuit of cat primary visual cortex. *J Neurosci.* September 2004;24(39):8441–8453.

[10] Barnes J, Hut P. A hierarchical O(N log N) force-calculation algorithm. *Nature.* 1986;324(6096).

[11] Greengard L, Rokhlin V. A fast algorithm for particle simulations. *J. Comput. Phys.* 1987;73(2).

[12] Meyer H, Schwarz D, Wimmer V, et al. Inhibitory interneurons in a cortical column form hot zones of inhibition in layers 2 and 5a. *Proc. Nat. Acad. Sci.* 2011;108(40).

[13] Butz M, Steenbuck ID, van Ooyen A. Homeostatic structural plasticity increases the efficiency of small-world networks. *Front Synaptic Neurosci.* 2014;6:7.

[14] Hensch TK, Stryker MP. Columnar architecture sculpted by GABA circuits in developing cat visual cortex. *Science.* March 2004;303(5664):1678–1681.

[15] Arckens L, Schweigart G, Qu Y, et al. Cooperative changes in GABA, glutamate and activity levels: the missing link in cortical plasticity. *Eur J Neurosci.* December 2000;12(12):4222–4232.

[16] Lodder J, Luijckx G, van Raak L, Kessels F. Diazepam treatment to increase the cerebral GABAergic activity in acute stroke: a feasibility study in 104 patients. *Cerebrovasc Dis.* 2000;10(6):437–440.

[17] Neufeld J, Teuchert-Noodt G, Grafen K, Winter Y, Witte AV. Synapse plasticity in motor, sensory, and limbo-prefrontal cortex areas as measured by degrading axon terminals in an environment model of gerbils (*Meriones unguiculatus*). *Neural Plast.* 2009;:281561:2009.

[18] Xu T, Yu X, Perlik AJ, et al. Rapid formation and selective stabilization of synapses for enduring motor memories. *Nature.* December 2009;462(7275):915–919.

[19] Butz M, Teuchert-Noodt G, Grafen K, van Ooyen A. Inverse relationship between adult hippocampal cell proliferation and synaptic rewiring in the dentate gyrus. *Hippocampus.* 2008;18(9):879–898.

[20] Butz M, van Ooyen A, Woergoetter F. A model for cortical rewiring following deafferentation and focal stroke. *Front Comput Neurosci.* 2009;3:10.

[21] Butz M, Steenbuck ID, van Ooyen A. Homeostatic structural plasticity can account for topology changes following deafferentation and focal stroke. *Front Neuroanat.* 2014;8:115.

CHAPTER

9

Lesion-Induced Dendritic Remodeling as a New Mechanism of Homeostatic Structural Plasticity in the Adult Brain

Steffen Platschek[1], *Hermann Cuntz*[1,2,3,†], *Thomas Deller*[1,†] *and Peter Jedlicka*[1,†]

[1]Goethe University Frankfurt, Frankfurt, Germany [2]Ernst Strüngmann Institute (ESI) for Neuroscience in Cooperation with Max Planck Society, Frankfurt, Germany [3]Frankfurt Institute for Advanced Studies, Frankfurt, Germany

OUTLINE

[†]Joint senior authors.

The Rewiring Brain
DOI: http://dx.doi.org/10.1016/B978-0-12-803784-3.00009-3

1 INTRODUCTION

Homeostatic adaptation mechanisms of the brain have received considerable attention in the last two decades.[1–6] Most studies have focused on homeostatic synaptic and intrinsic plasticity, which are powerful compensatory tools for maintaining the firing of neurons in the face of perturbations. Homeostatic synaptic plasticity includes synaptic scaling[7,8] and homeostatic forms of metaplasticity[9–12] and heterosynaptic plasticity.[11,13] Both homeostatic synaptic and intrinsic plasticity are predominantly mediated by compensatory adjustments in the distribution and/or expression of synaptic receptors or channels[14–16] and usually do not involve structural changes of axons, spines, or dendrites. However, such homeostatic structural plasticity of axons and spines has been described and implicated in maintaining the stability of neural circuits that undergo rewiring and reorganization.[17–20] In contrast to structural plasticity of spines and axons, structural plasticity of dendrites[21,22] and its potential homeostatic features have not yet been sufficiently explored. In this chapter, first, we are going to review recent studies of structural dendritic changes focusing on functional homeostasis of the nervous system. In the second part, we will be describing our own computational approach revealing a new form of homeostatic dendritic plasticity.

Changes in the length or complexity of the dendritic tree have been observed under a variety of conditions. Most studies focused on dendritic plasticity during development (for reviews, see Refs. [24–26]). However, the adult nervous system also displays considerable dendritic plasticity (for review see [22]) induced by learning,[24–26] deafferentation, or stroke,[27–35] exposure to enriched environment,[36,37] or chronic application of various drugs.[38,39] One of the first observations of dendrite remodeling in the adult healthy brain has been made in inhibitory interneurons from the visual cortex.[40,41] The question of homeostatic effects implemented by compensatory alterations in dendritic morphology has only been raised recently. Tripodi and colleagues (2008)[42] observed that developing neurons (anterior comer cell (aCC) motor neurons) in *Drosophila* embryos reacted to the change in the number of synaptic inputs by adjusting the size of dendrites (see also Ref. [43]). Whereas a genetically induced increase in the density of presynaptic sites led to a reduction in the size of dendrites, a block of presynaptic release or a block of presynaptic neurotransmitter synthesis induced an increase in the size of dendrites. The observed dendritic plasticity in aCC motor neurons was mediated by protein kinase A signaling. The authors interpreted their findings in a homeostatic manner. They proposed that developing neurons fine-tune the growth of their dendrites to optimize the number of their synaptic connections.[42] Neurons in *Drosophila* embryos seem to be compensating the reduction in presynaptic activity by expanding their dendritic trees. The expansion of dendritic trees leads potentially to more synaptic input, thereby compensating for reduced presynaptic activity and therefore restoring neuronal activity. On the other hand, these neurons try to compensate excessive supply of presynaptic connections by reducing the extent of their dendritic arbors. Similar observations have been made by Yuan and colleagues (2011)[44] in nerve cells of the developing visual system in *Drosophila* larvae. They studied activity-dependent dendritic plasticity of larval ventral lateral neurons in the visual system, and their experiments revealed that increased exposure to light led to shortening of neuronal dendritic arbor. In contrast, decreased exposure to light induced expansion of dendrites. Similarly, decreasing presynaptic activity increased dendrite length. These effects were mediated by cyclic adenosine monophosphate signaling. Changes in the number of synaptic inputs followed changes in dendritic size, with larger dendrites having increased number of inputs and shrunk dendrites having smaller number of inputs. Interestingly, increasing postsynaptic activity of studied neurons also expanded their dendritic trees, while reducing postsynaptic activity had the opposite effect. The authors concluded that activity-dependent dendritic plasticity in the developing larval visual system reflects homeostatic structural plasticity possibly underlying compensatory adjustments of light sensitivity in response to alterations in visual input.[18,44] In summary, the work in *Drosophila* neurons represents one of the first studies of homeostatic dendritic plasticity during development supporting previous computational predictions.[45] It shows that growth/shrinkage of dendrites following reduced/increased presynaptic activity may enable addition/removal of synapses and thus achieve compensatory adjustment of the total synaptic drive (but see also Refs. [46,47] for opposite effects in developing tectal neurons of *Xenopus*).

Whereas the studies by Tripodi et al. (2008)[42] and Yuan et al. (2011)[44] focused on the developing central nervous system (CNS) (see also Ref. [48]), homeostatic structural plasticity of dendrites has not yet been specifically studied in the adult brain. Although there are some hints for homeostatic effects of dendritic remodeling in the existing literature (reviewed by Refs. [19,49]), so far this issue has not received much attention and has not been explicitly addressed. Therefore, by using a computational approach, we have recently started analyzing potential homeostatic roles of dendritic plasticity in dentate granule cells and other cell types of the adult nervous system.[50] In our simulations we have focused on denervation-induced dendritic retraction and have addressed the key question whether denervation-induced remodeling is functionally detrimental or homeostatic for denervated neurons (Fig. 9.1). The presented data are of clinical relevance

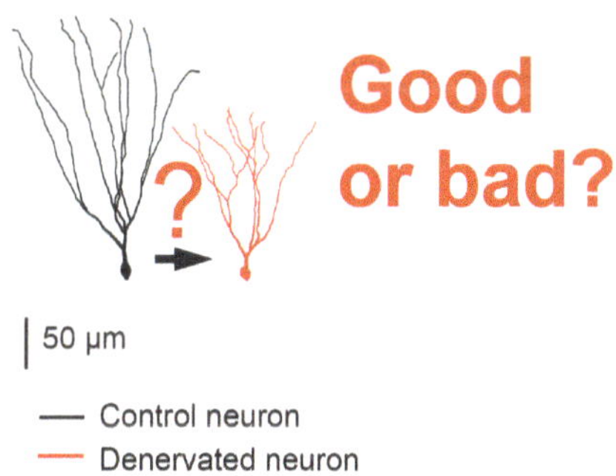

FIGURE 9.1 **Functional role of dendritic retraction is not well understood.** Denervation of neurons in the adult brain leads to remodeling and retraction of their dendrites. We started exploring functional consequences of this kind of dendritic plasticity. The schematic depicts two examples of reconstructed dentate granule cells. Note that denervation of granule cell dendrites caused a significant reduction in dendritic length and a loss of distal dendritic segments (red neuron, 90 days post lesion). Source: *Adapted from Platschek S, Cuntz H, Vuksic M, Deller T, Jedlicka P. A general homeostatic principle following lesion induced dendritic remodeling. Acta Neuropathol Commun. 2016;4:19. doi:10.1186/s40478-016-0285-8.*

because they offer a novel view on the effect of structural changes of denervated neurons and thus on potential interventions in patients with brain injury. The results we are presenting in this chapter have been previously published by *Acta Neuropathologica Communications* in an open-access article[50] under the Creative Commons Attribution License 4.0 (http://creativecommons.org/licenses/by/4.0/).

2 MODEL

In this section, first, we are going to describe the experimental model system and then we will be introducing computer models providing insights on the functional relevance of structural plasticity following denervation of the adult dentate gyrus.

Entorhinal cortex is the origin of the perforant path, i.e., the entorhinodentate and entorhinoammonic projections (see Ref. [51] for a general review on the rodent hippocampal anatomy). The majority of entorhinodentate perforant path fibers terminate in a laminar fashion on the dendrites of neurons in the dentate gyrus.[52] The dentate gyrus consists of three layers[51] (Fig. 9.2): granule cell layer, molecular layer (subdivided into inner and outer molecular layer, in this review), and hilus (polymorph layer). Principal neurons in the dentate gyrus are called granule cells and their somata form the granule cell layer. Granule cells are highly polarized cells: their dendrites extend into the molecular layer, whereas their axons terminate in the hilus and in stratum lucidum of area CA3 of Ammon's horn. Entorhinodentate afferents terminate in the outer two-thirds of the molecular layer where

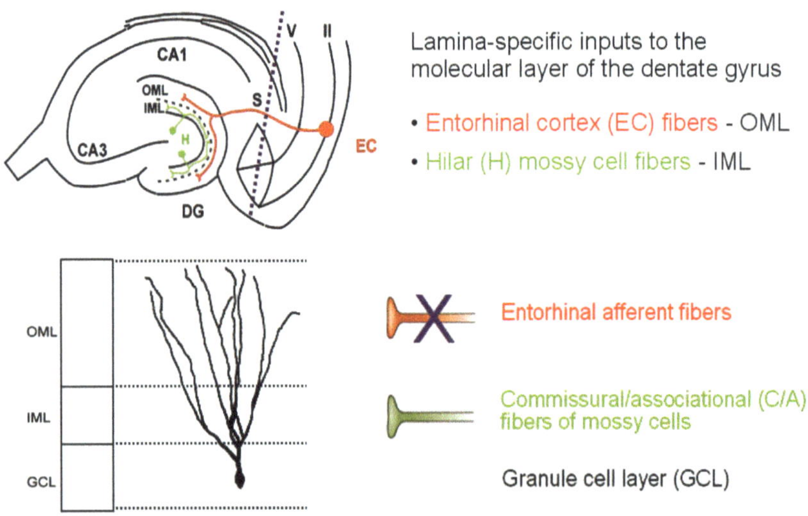

FIGURE 9.2 **Entorhinal cortex lesion.** Upper panel: The *dashed line* indicates the lesion, which transects the entorhinodentate fibers (*red*) to the OML of the DG (see Ref. [53]). Hilar mossy cells (*green*) terminate in the IML. Lower panel: Denervation of the OML by cutting entorhinal inputs leads to strong retraction of distal dendritic segments in denervated granule cells (not shown) as compared to healthy controls. *OML*, outer molecular layer; *IML*, inner molecular layer; *DG*, dentate gyrus.

they primarily innervate the distal dendritic segments of granule cells but also a much smaller number of GABAergic interneurons.[51] Most entorhinodentate fibers terminating on granule cells form excitatory synaptic inputs with dendritic spines, whereas entorhinodentate fibers terminating on GABAergic neurons form shaft synapses with these neurons.

Denervation of neurons in the adult dentate gyrus leads to a profound reorganization of their denervated dendritic tree.[53–55] Denervation-induced dendritic remodeling has been studied extensively using a classical experimental model of neuronal denervation and reorganization, the "entorhinal cortex lesion (ECL) model."[53–57] Upon ECL, entorhinal afferents to dentate granule cells anterogradely degenerate, and denervated granule cells remodel their dendritic tree.[32–35,58] Dendritic denervation shrinks the granule cell arbor and results in a loss and/or retraction of distal granule cell segments. Even after long survival times the granule cell tree remains shrunken, and the complexity of the dendritic tree is reduced[32,59] (Fig. 9.2). Such denervation-induced dendritic retraction has been considered pathological or "atrophic."[59–61] Here we describe our computational modeling results that indicate that, surprisingly, the retraction of dendrites may contribute to homeostatic restoration of physiological properties of denervated dentate granule cells.

To investigate the function of dendritic retraction, we used established anatomically and biophysically realistic compartmental models of 3D-reconstructed healthy and denervated granule cells.[62] To create compartmental models for control and denervated dentate granule cells, we employed full 3D reconstructions of neurons with long-standing (90 days), i.e., chronic, lesions displaying typical signs of neuronal atrophy and dendritic remodeling[32] and combined them with published values of passive parameters.[63] We used R_a (specific axial resistance) of 194 Ω cm, R_m (specific membrane resistance) of 38 kΩ cm^2, and C_m (specific membrane capacitance) of 1.01 μF cm^{-2}. R_m and C_m were scaled to model implicitly the spine membrane surface area. Our passive models reproduced experimentally determined values for input resistance and membrane time constant of typical granule cells in mice.[63] To extract the unique effects of morphological changes in the form of dendritic retraction, we maintained the same passive parameters in both control and denervated granule cells. The only difference between the two groups of neurons consisted in their different 3D morphology. Membrane time constant was not significantly different between the groups. However, as expected, input resistance was higher in denervated neurons with retracted dendrites. To insert active channels, we used a published active model of dentate granule cells.[64] Alternatively, we also used a different set of active biophysical properties[65] but the results were similar. To assess synaptic input–output curves in control and denervated granule cells, we simulated input synaptic activity. Excitatory synaptic responses were modeled using a bi-exponential synaptic conductance with reversal potential of 0 mV, maximum conductance of 0.5 nS, and rise time and decay time of 0.2 and 2.5 Ms, respectively.[63] In the first days after denervation, synapse and spine densities decreased[32–34,66] but returned back to control levels within ~2 weeks[32] because of terminal proliferation and sprouting of surviving afferent fibers.[66,67] Hence, in the case we analyzed, i.e., granule cells with long-standing (90 days) lesions, we distributed synaptic inputs with the same density (3±0.2 synapses/μm) on the dendritic tree as on the dendritic trees of nonlesioned control cells (see Ref. [32]). Synaptic inputs were activated by stochastic (Poisson) presynaptic spike trains.

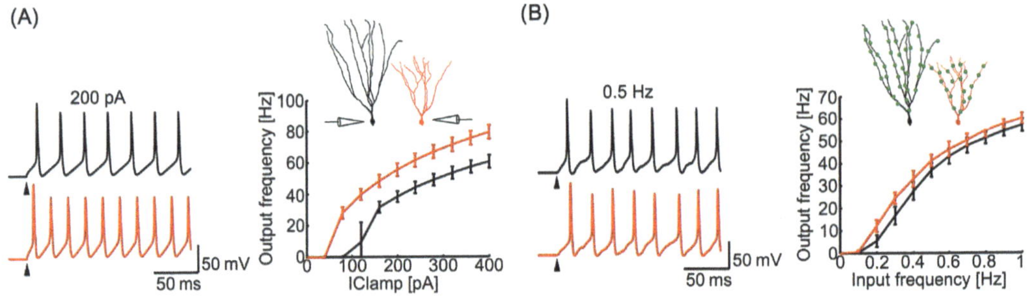

FIGURE 9.3 **Dendritic retraction leads to a precise homeostatic regulation of excitability in simulated granule cells.** (A) *Left*: Somatic membrane voltage traces in active compartmental models of a control (*black*) and a denervated (*red*) granule cell in response to +200 pA somatic current injections. Note an increase in the number of evoked action potentials in the denervated granule cell. *Right*: Due to higher input resistance, compartmental models of denervated cells displayed higher firing rates following current injections in the soma. (B) *Left*: Somatic membrane voltage traces in response to stochastic activation of dendritic synapses at 0.5 Hz in the active compartmental model. In line with spine data from Vuksic et al. (2011), excitatory synapses were distributed on dendrites of granule cells using the same synaptic density in both groups. *Right*: Corresponding firing rates in response to increasing frequency of stochastic synaptic activation. The *arrows* in (A) and (B) indicate the start of the stimulation. Note that the greater excitability (observed in (A)) effectively compensated for the smaller number of synapses on the shorter dendrites of denervated neurons. Source: *Adapted from Platschek S, Cuntz H, Vuksic M, Deller T, Jedlicka P. A general homeostatic principle following lesion induced dendritic remodeling. Acta Neuropathol Commun. 2016;4:19. doi:10.1186/s40478-016-0285-8.*

3 RESULTS

Simulations of input–output curves following somatic current injections revealed an increased excitability of denervated cells with remodeled and shortened dendrites as compared to controls (Fig. 9.3A). This was not surprising because increased excitability can be explained as a consequence of shorter total dendritic length and higher input resistance in denervated cells.[68,69] As pointed out in the previous section, denervated neurons with retracted dendrites have similar spine densities as control neurons,[32] which implies that the dendrites of denervated granule cells have a smaller total number of synapses than nondenervated cells. Remarkably, in simulations of input–output curves following stochastic activation of synaptic inputs, higher input resistance was able to compensate for the smaller total number of synapses leading to homeostatic maintenance of firing rates in denervated granule cells (Fig. 9.3B). Thus, the shortening of dendrites may be able to restore the normal firing pattern of granule cells and thereby normalize information throughput to the dentate gyrus.

Importantly, the predictions of compartmental models in reconstructed morphologies have been corroborated by simulations in models using synthetic morphologies. Artificial morphologies were generated by a new method of morphological modeling developed by Dr. Hermann Cuntz.[70] Morphological modeling based on optimal wiring principles[70–73] allowed us to generalize the key finding of excitability homeostasis to other cell types and dendritic trees (Fig. 9.4).

In conclusion, computer models predicted that structural plasticity of dendrites as reflected in their denervation-induced retraction might serve as a homeostatic mechanism that precisely maintains input–output firing relations.

4 DISCUSSION

In our simulations we addressed the fundamental question how dendritic retraction induced by denervation affects input–output function of dentate granule cells. We have shown that dendritic retraction might help maintain synaptically driven activity of a denervated neuron.

The traditional view of dendritic retraction regards it as an aspect of pathology associated with lesion-induced denervation and/or cell damage.[59–61] Based on our simulations, we here suggest that dendritic retraction and remodeling might not only be detrimental but could also foster functional regeneration following a partial injury or partial denervation. The extent of dendritic remodeling may play an important role, however, whether dendritic retraction can act in a homeostatic fashion. Whereas massive dendritic atrophy is

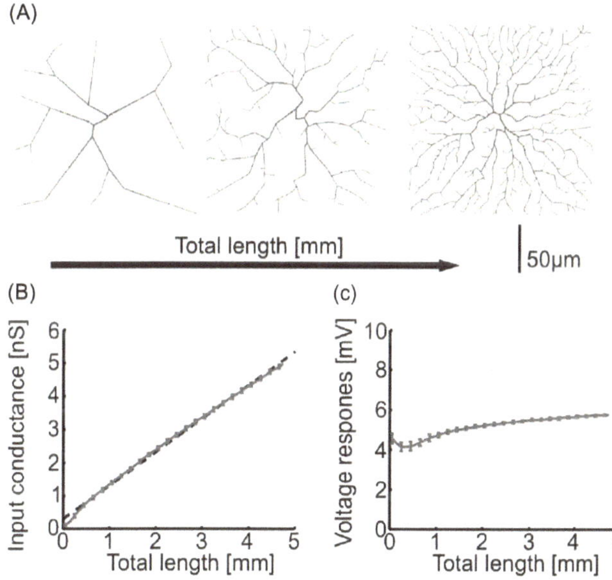

FIGURE 9.4 **General principles for the homeostatic adjustment of neuronal input–output relationship by dendritic plasticity.** To generalize the computational results from dentate granule cells (see Fig. 9.3) to other cell types, we have generated synthetic morphologies with a wide range of complexity and total dendritic length using morphological modeling based on optimal wiring principles.[70] (A) Examples of three morphological models with increasing dendritic complexity and total length (from the left to right) grown in a square dendritic field. (B) Input conductance increases with total dendritic length (i.e., with membrane surface) in the morphological models (*grey line*). *Dashed line* represents a linear fit. (C) Somatic voltage response following dendritic current injection remains similar across different dendrite lengths. This is due to a linear relation between total dendrite length and input conductance (B) and a similar linear relation between total dendrite length and total number of synapses (or current sources). Smaller dendritic trees have lower input conductance and higher input resistance associated with higher intrinsic excitability that compensates their lower number of synaptic inputs. This leads to a somatic response (evoked by synaptic activity or dendritic current injection) that is independent of total length. Source: *Adapted from Platschek S, Cuntz H, Vuksic M, Deller T, Jedlicka P. A general homeostatic principle following lesion induced dendritic remodeling. Acta Neuropathol Commun. 2016;4:19. doi:10.1186/s40478-016-0285-8.*

very likely a sign of severe neuropathology, limited dendritic retraction could help a denervated nerve cell maintain its responsiveness to its remaining synapses.

The novel mechanism we have recently described[50] and are reviewing in this chapter might be complementary to other forms of homeostatic structural plasticity.[17–19,74] Mild loss of presynaptic inputs or decrease in presynaptic activity leads to low neuronal activity that can be restored by a compensatory growth of new branches and spines[19] (Fig. 9.5). This would be an example of a complete compensation by homeostatic mechanisms resulting in overall maintenance of a mildly remodeled cell. A more extensive loss of presynaptic inputs, e.g., following denervation as a result of stroke, neurodegeneration, or lesion, cannot be fully compensated by sprouting of terminals and a reinnervation of dendrites. Under such more extreme conditions, an efficient strategy for restoring the control level of activity could be the retraction of denervated portions of dendrites associated with a compensatory increase in neuronal excitability (Fig. 9.5). Of note, denervated neurons possess also other mechanisms of homeostatic plasticity such as strengthening of remaining synapses (i.e., local synaptic scaling[62,75]) or partial recovery of spine loss by enhanced stability

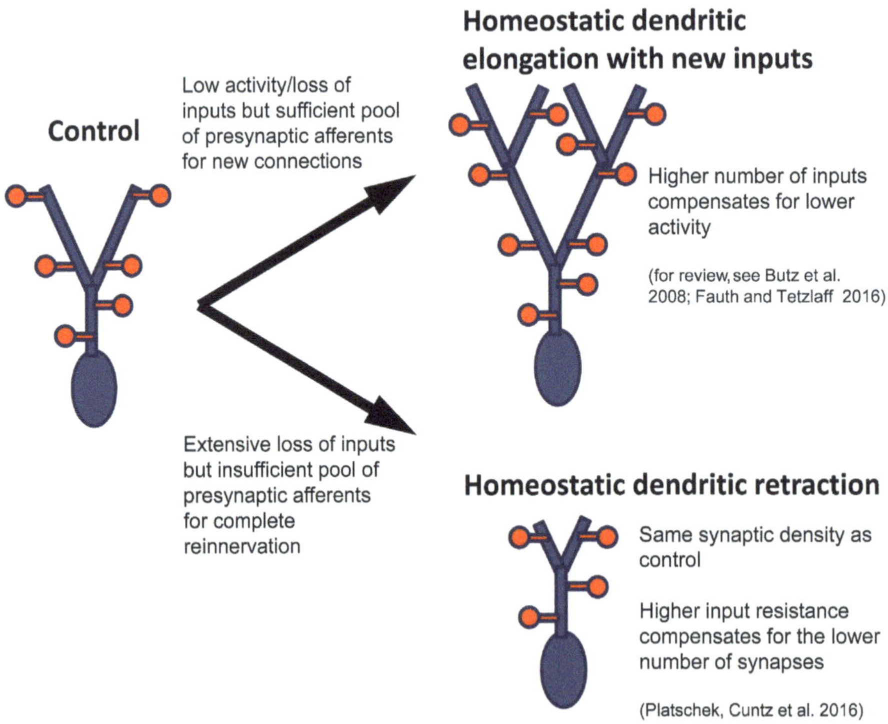

FIGURE 9.5 **Two different mechanisms for the homeostatic adjustment of low neuronal activity by dendritic plasticity.** Low activity induces an elongation of the dendritic tree and a compensatory growth of new spines (synaptic connections). However, under conditions of an incomplete renewal of lost inputs, dendritic retraction is employed by neurons to decrease their size, increase their excitability, and thereby restore their original activity. Neurons might use both of these two strategies: if the first strategy fails, then the second strategy is initiated.

of newly formed dendritic spines.[76,77] How these synaptic homeostatic mechanisms interact with dendritic homeostatic plasticity or with nonhomeostatic (Hebbian) plasticity mechanisms[19] is not yet clear and remains to be elucidated.

5 FUTURE EXPERIMENTAL STUDIES

Is dendritic retraction beneficial only for the activity homeostasis of a denervated neuron or could it contribute also to restoring the activity of an entire network? This question has not yet been clarified and needs to be addressed in new experiments and network simulations. Studies using ECL have shown that the loss of entorhinal afferents induces dendritic retraction not only in granule cells[32,35] but also affects dendrites of inhibitory parvalbumin-positive basket cells.[78] This indicates that the retraction of dendrites might affect excitability of excitatory and inhibitory neurons following their denervation. What is the net effect of such changes on excitation/inhibition balance and dentate network activity? Spontaneous firing of granule cells decreases after ECL but returns to its control level 1 week after ECL.[79] However, similar data on firing properties of inhibitory interneurons are missing, and therefore it is difficult to assess whether network dynamics of the dentate gyrus fully recovers. Field potential recordings in slices from lesioned rats suggested a long-lasting shift of excitation/inhibition balance towards inhibition upon ECL.[80]

Further in vitro and in vivo measurements combined with detailed simulations are needed to fully understand cellular and synaptic mechanisms of circuit adaptations in the dentate gyrus following ECL. Patch-clamp recordings of passive properties (input resistance and membrane time constant) and excitatory postsynaptic currents combined with 3D-confocal reconstructions can be used to acquire electrotonic and synaptic parameters along with dendritic morphologies from the same dentate granule cell. This would allow for establishing correlations between dendritic length, input resistance, and input–output function of granule cells. Measured correlations can be compared to computationally predicted correlations.[50] In vivo field potential recordings and in vitro patch-clamp recordings should be used in combination to explore the somatic and synaptic input–output relationship and firing of normal and denervated granule cells displaying dendritic retraction.

6 FUTURE MODELING STUDIES

Morphological modeling[70] combined with compartmental modeling will be a useful tool for addressing open issues of lesion-induced plasticity because it has a great potential of generating thousands of realistic dendritic morphologies that can be used for large-scale network models. A 1:1 full-scale version of the rat dentate gyrus containing 1 million different dentate granule cell morphologies has been recently created[81] (Fig. 9.6). This morphological model consists of granule cells and would require adding basket cells and other cell types. By using a similar approach, one could build a morphological model of the de-entorhinated dentate gyrus with dendritic retraction. After converting morphologies into active compartmental models,[82–85] it would become feasible to predict the effects of dendritic retraction on excitation–inhibition balance and network activity. Network

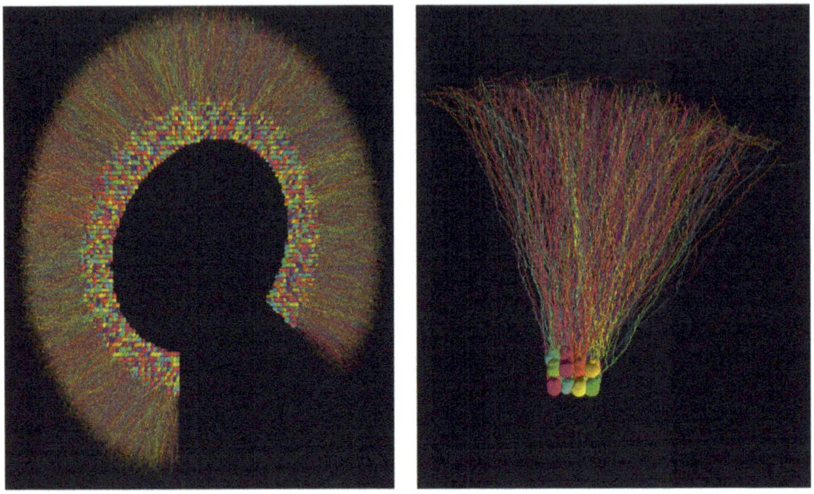

FIGURE 9.6 **A unique full-scale morphological model of the rat dentate gyrus containing the entire population of granule cells with realistic (diverse) dendritic morphologies.** One million dentate granule cells were synthesized using morphological modeling[70] by growing their dendritic trees within their characteristic dendritic fields. *Left*: Rendering of complete morphologies for all granule cells in a 20 μm transverse slice from the center of the model dentate gyrus. *Right*: Rendering of selected granule cells from the slice depicted on the left. Implementing dendritic retraction in morphological and compartmental models of the dentate gyrus will be a useful tool for studying its functional consequences at the network level. Source: *Adapted from Schneider CJ, Cuntz H, Soltesz I. Linking macroscopic with microscopic neuroanatomy using synthetic neuronal populations. PLoS Comput Biol. 2014;10(10):e1003921. doi:10.1371/journal.pcbi.1003921.*

modeling will be crucial for addressing the question what amount of dendritic retraction is most effective to restore network activity to its physiological pattern.

Another important extension of the dentate gyrus model would be adding dendritic morphologies of newly born granule cells.[86] Transection of entorhinal afferents (ECL) induces dendritic retraction not only in mature perinatally born granule cells but also in immature adult-born granule cells.[57,87] Adult-born granule cells are known to have distinct electrophysiological properties and network connectivity[86,88] and display increased proliferation as well as layer-specific spine formation and synaptic potentiation following ECL.[87] Thus, developing realistic morphological and compartmental models of adult-born granule cells[89] will be important to understand network changes following ECL.

7 A GENERAL PRINCIPLE FOR HOMEOSTATIC DENDRITIC PLASTICITY

Homeostatic adjustment of neuronal excitability by dendritic morphology is not limited to dendritic arbors in the dentate gyrus or hippocampus. By using a general morphological model (Fig. 9.4), we have shown that the enhancement of excitability is not only present in the specific case of denervated granule cells but also in any dendritic morphology.

The retraction of dendritic regions will lead to an increased input resistance. If synapse density does not change, the number of synaptic inputs decreases linearly with the length of dendrite. A change in the number of synapses gets canceled out by a change in input resistance leading to the firing rate being independent of the length of dendrites when all synapses are stochastically activated. This mechanism will enable any dendritic tree that undergoes structural remodeling including extension or shortening of its dendrites to adjust its responsiveness to synaptic bombardment. Importantly, such a mechanism represents a default feature of all dendritic trees that undergo shortening or lengthening of their dendrites (e.g., also during development), as long as synaptic density remains constant and dendrites are largely passive. Of course, many neurons have highly active dendrites and perform complex local computations going beyond simple input–output functions. However, we propose that the above-described simple dendritic compensation is available for all cells providing a robust, passive homeostatic backbone that can nevertheless be flexibly modified according to specific computational needs in a given circuit. Local adjustment of active properties, synaptic density, or inhibition in single dendritic branches can be used by neurons to enhance or suppress the impact of local synaptic activation. Specific experiments in various neurons in different areas have to be performed to assess whether a dendritic tree of a given cell type behaves as a simple linear input–output device or whether it uses additional nonlinear mechanisms to keep its complex computations independent of its size. For instance, recent data from *Drosophila* motoneurons suggest that by using simple dendritic compensation, dendrites become dispensable for synaptically driven basic activity but remain important for more complex functions.[90] In line with our simulations, these experiments have provided evidence that synaptic input to *Drosophila* motoneurons with altered dendritic architecture is able to evoke normal firing activity during behavior. This suggests the presence of a simple homeostatic mechanism coupling dendritic plasticity to robust maintenance of firing rate.

8 POTENTIAL SYNERGY WITH HOMEOSTATIC STRUCTURAL PLASTICITY OF THE AXON INITIAL SEGMENT

A recently discovered form of structural plasticity of the axon initial segment (AIS) might operate in line with the above-described structural dendritic plasticity supporting the compensation for the loss of afferent synaptic inputs.[91,92] In the avian auditory system, it has been shown that a removal of afferent input activity by cochlea ablation leads to an elongation of the AIS. The increased length of the AIS is accompanied by an altered distribution of Na^{+}[91]and K^{+} channels[92] associated with a homeostatic increase in neuronal excitability compensating for the decreased afferent activity (for review, see Refs. [74,93]). An opposite form of structural plasticity of the AIS can be induced by stimulation of hippocampal neurons causing a shift and shortening of AIS that leads to a decrease in neuronal excitability.[94–96] Interestingly, a recent computational study has shown that the location and the length of the AIS interact with dendritic morphology in determining neuronal excitability[97] (see also Ref. [98]). Whereas a long AIS at a distal location enhances excitability of neurons with larger dendritic arbors, a proximal AIS of intermediate length enhances excitability of neurons with smaller dendritic arbors. In summary, in addition to dendritic retraction, dentate granule cells may increase their excitability upon denervation also by changes in the location and length of their AIS.[96] Future studies should explore this possibility and look for evidence of lesion-induced AIS plasticity in granule cells.

9 CLINICAL RELEVANCE

In summary, homeostatic dendritic changes in the adult brain represent an important but largely unexplored form of neuronal plasticity. This form of plasticity may be of considerable relevance for neurons in the CNS to respond to perturbations and lesions of all kinds. Dendritic homeostatic plasticity provides nerve cells with a robust input–output function because neuronal output depends on synaptic input activity rather than on the absolute size of their dendritic trees. In this way, neurons can maintain their input–output relationship despite variations in their size. If the mechanisms underlying this form of structural plasticity can be unraveled, new targets for the treatment of neurological or psychiatric disorders associated with structural changes of dendrites are conceivable, including stroke,[28–30,99–101] Alzheimer's disease,[102,103] Huntington's disease,[104,105] addiction,[38] as well as depression, stress, and fear disorders.[106–109] Altogether, recent research suggests that in case of denervation, supportive interventions should be focusing not only on promoting fast recovery of synaptic loss but also on facilitating homeostatic forms of synaptic, dendritic plasticity, and AIS plasticity.

Acknowledgments

The work was supported by Deutsche Forschungsgemeinschaft (CRC 1080 to T.D.), by the Alzheimer Forschung Initiative eV (AFI No. 15038 to P.J.), by a BMBF grant (Germany–United States Collaboration in Computational Neuroscience to P.J., No. 01GQ1203A) and by a BMBF grant (No. 01GQ1406—Bernstein Award 2013 to H.C.).

References

1. Davis GW. Homeostatic control of neural activity: from phenomenology to molecular design. *Annu Rev Neurosci*. 2006;29:307–323.
2. Maffei A, Fontanini A. Network homeostasis: a matter of coordination. *Curr Opin Neurobiol*. 2009;19(2):168–173.
3. Williams A, O'Leary T, Marder E. Homeostatic regulation of neuronal excitability. *Scholarpedia*. 2013;8(1):1656.
4. Marder E, Goaillard J-M. Variability, compensation and homeostasis in neuron and network function. *Nat Rev Neurosci*. 2006;7(7):563–574.
5. Davis GW, Bezprozvanny I. Maintaining the stability of neural function: a homeostatic hypothesis. *Annu Rev Physiol*. 2001;63:847–869.
6. Davis G. Homeostatic signaling and the stabilization of neural function. *Neuron*. 2013;80(3):718–728.
7. Turrigiano GG. The self-tuning neuron: synaptic scaling of excitatory synapses. *Cell*. 2008;135(3):422–435.
8. Turrigiano G. Homeostatic synaptic plasticity: local and global mechanisms for stabilizing neuronal function. *Cold Spring Harb Perspect Biol*. 2012;4(1):a005736.
9. Abraham WC. Metaplasticity: tuning synapses and networks for plasticity. *Nat Rev Neurosci*. 2008;9(5):387.
10. Hulme SR, Jones OD, Abraham WC. Emerging roles of metaplasticity in behaviour and disease. *Trends Neurosci*. 2013;36(6):353–362.
11. Jedlicka P, Benuskova L, Abraham WC. A voltage-based STDP rule combined with fast BCM-like metaplasticity accounts for LTP and concurrent "heterosynaptic" LTD in the dentate gyrus in vivo. *PLoS Comput Biol*. 2015;11(11):e1004588.

12. Yger P, Gilson M. Models of metaplasticity: a review of concepts. *Front Comput Neurosci.* 2015;9:138.
13. Chistiakova M, Bannon NM, Chen J-Y, Bazhenov M, Volgushev M. Homeostatic role of heterosynaptic plasticity: models and experiments. *Front Comput Neurosci.* 2015;9:89.
14. Desai NS. Homeostatic plasticity in the CNS: synaptic and intrinsic forms. *J Physiol.* 2003;97(4):391–402.
15. Watt AJ, Desai NS. Homeostatic plasticity and STDP: keeping a neuron's cool in a fluctuating world. *Front Synaptic Neurosci.* 2010;2(June):5.
16. Turrigiano G. Too many cooks? Intrinsic and synaptic homeostatic mechanisms in cortical circuit refinement. *Annu Rev Neurosci.* 2011;34:89–103.
17. Butz M, Wörgötter F, van Ooyen A. Activity-dependent structural plasticity. *Brain Res Rev.* 2009;60(2):287–305.
18. Yin J, Yuan Q. Structural homeostasis in the nervous system: a balancing act for wiring plasticity and stability. *Front Cell Neurosci.* 2014;8(January):439.
19. Fauth M, Tetzlaff C. Opposing effects of neuronal activity on structural plasticity. *Front Neuroanat.* 2016;10(June):1–18.
20. Grubb MS, Shu Y, Kuba H, Rasband MN, Wimmer VC, Bender KJ. Short- and long-term plasticity at the axon initial segment. *J Neurosci.* 2011;31(45):16049–16055.
21. Tavosanis G. Dendritic structural plasticity. *Dev Neurobiol.* 2012;72(1):73–86.
22. Hickmott PW, Ethell IM. Dendritic plasticity in the adult neocortex. *Neurosci.* 2006;12(1):16–28.
23. Reese BE, Keeley PW, Lee SCS, Whitney IE. Developmental plasticity of dendritic morphology and the establishment of coverage and connectivity in the outer retina. *Dev Neurobiol.* 2011;71(12):1273–1285.
24. Gloor C, Luft AR, Hosp JA. Biphasic plasticity of dendritic fields in layer V motor neurons in response to motor learning. *Neurobiol Learn Mem.* 2015;125:189–194.
25. Yau SY, Lau BWM, Tong JB, et al. Hippocampal neurogenesis and dendritic plasticity support running-improved spatial learning and depression-like behaviour in stressed rats. *PLoS One.* 2011;6(9).
26. Lemaire V, Tronel S, Montaron M-F, Fabre A, Dugast E, Abrous DN. Long-lasting plasticity of hippocampal adult-born neurons. *J Neurosci.* 2012;32(9):3101–3108.
27. Andres RH, Horie N, Slikker W, et al. Human neural stem cells enhance structural plasticity and axonal transport in the ischaemic brain. *Brain.* 2011;134(6):1777–1789.
28. Frauenknecht K, Diederich K, Leukel P, et al. Functional improvement after photothrombotic stroke in rats is associated with different patterns of dendritic plasticity after G-CSF Treatment and G-CSF treatment combined with concomitant or sequential constraint-induced movement therapy. *PLoS One.* 2016;11(1):1–12.
29. Mostany R, Portera-Cailliau C. Absence of large-scale dendritic plasticity of layer 5 pyramidal neurons in peri-infarct cortex. *J Neurosci.* 2011;31(5):1734–1738.
30. Ueno Y, Chopp M, Zhang L, et al. Axonal outgrowth and dendritic plasticity in the cortical peri-infarct area after experimental stroke. *Stroke.* 2012;43(8):2221–2228.
31. Hickmott PW, Steen PA. Large-scale changes in dendritic structure during reorganization of adult somatosensory cortex. *Nat Neurosci.* 2005;8(2):140–142.
32. Vuksic M, Del Turco D, Vlachos A, et al. Unilateral entorhinal denervation leads to long-lasting dendritic alterations of mouse hippocampal granule cells. *Exp Neurol.* 2011;230(2):176–185.
33. Parnavelas JG, Lynch G, Brecha N, Cotman CW, Globus A. Spine loss and regrowth in hippocampus following deafferentation. *Nature.* 1974;248:71–73.
34. Caceres A, Steward O. Dendritic reorganization in the denervated dentate gyrus of the rat following entorhinal cortical lesions: A Golgi and electron microscopic analysis. *J Comp Neurol.* 1983;214(4):387–403.
35. Diekmann S, Ohm TG, Nitsch R. Long-lasting transneuronal changes in rat dentate granule cell dendrites after entorhinal cortex lesion. A combined intracellular injection and electron microscopy study. *Brain Pathol.* 1996;6(3):205–215.
36. Lauterborn JC, Palmer LC, Jia Y, et al. Chronic ampakine treatments stimulate dendritic growth and promote learning in middle-aged rats. *J Neurosci.* 2016;36(5):1636–1646.
37. Gelfo F, De Bartolo P, Giovine A, Petrosini L, Leggio MG. Layer and regional effects of environmental enrichment on the pyramidal neuron morphology of the rat. *Neurobiol Learn Mem.* 2009;91(4):353–365.
38. Li J, Zhang L, Chen Z, et al. Cocaine activates Rac1 to control structural and behavioral plasticity in caudate putamen. *Neurobiol Dis.* 2015;75:159–176.
39. Gonzalez CLR, Gharbawie OA, Whishaw IQ, Kolb B. Nicotine stimulates dendritic arborization in motor cortex and improves concurrent motor skill but impairs subsequent motor learning. *Synapse.* 2005;55(3):183–191.

40. Lee W-CA, Huang H, Feng G, et al. Dynamic remodeling of dendritic arbors in GABAergic interneurons of adult visual cortex. *PLoS Biol*. 2006;4(2):e29.
41. Chen JL, Flanders GH, Lee W-CA, Lin WC, Nedivi E. Inhibitory dendrite dynamics as a general feature of the adult cortical microcircuit. *J Neurosci*. 2011;31(35):12437–12443.
42. Tripodi M, Evers JF, Mauss A, Bate M, Landgraf M. Structural homeostasis: compensatory adjustments of dendritic arbor geometry in response to variations of synaptic input. *PLoS Biol*. 2008;6(10):e260.
43. Zwart MF, Randlett O, Evers JF, Landgraf M. Dendritic growth gated by a steroid hormone receptor underlies increases in activity in the developing Drosophila locomotor system. *Proc Natl Acad Sci*. 2013;110(40): E3878–E3887.
44. Yuan Q, Xiang Y, Yan Z, Han C, Jan LY, Jan YN. Light-induced structural and functional plasticity in Drosophila larval visual system. *Science*. 2011;333(6048):1458–1462.
45. Van Ooyen A, Van Pelt J, Corner MA. Implications of activity dependent neurite outgrowth for neuronal morphology and network development. *J Theor Biol*. 1995;172(1):63–82.
46. Sin WC, Haas K, Ruthazer ES, Cline HT. Dendrite growth increased by visual activity requires NMDA receptor and Rho GTPases. *Nature*. 2002;419(6906):475–480.
47. He H, Shen W, Hiramoto M, Cline HT. Experience-dependent bimodal plasticity of inhibitory neurons in early development. *Neuron*. 2016;90(6):1203–1214.
48. Kremer MC, Christiansen F, Leiss F, et al. Structural long-term changes at mushroom body input synapses. *Curr Biol*. 2010;20:1938–1944.
49. Butz M, van Ooyen A, Wörgötter F. A model for cortical rewiring following deafferentation and focal stroke. *Front Comput Neurosci*. 2009;3(August):10.
50. Platschek S, Cuntz H, Vuksic M, Deller T, Jedlicka P. A general homeostatic principle following lesion induced dendritic remodeling. *Acta Neuropathol Commun*. 2016;4:19.
51. Amaral DG, Scharfman HE, Lavenex P. The dentate gyrus: fundamental neuroanatomical organization (dentate gyrus for dummies). *Prog Brain Res*. 2007;163:3–22.
52. Deller T. The anatomical organization of the rat fascia dentata: new aspects of laminar organization as revealed by anterograde tracing with Phaseolus vulgaris-Luecoagglutinin (PHAL). *Anat Embryol (Berl)*. 1998;197(2):89–103.
53. Deller T, Del Turco D, Rappert A, Bechmann I. Structural reorganization of the dentate gyrus following entorhinal denervation: species differences between rat and mouse. *Prog Brain Res*. 2007;163:501–528.
54. Steward O. Reorganization of neuronal circuitry following central nervous system trauma: naturally occurring processes and opportunities for therapeutic intervention. In: Salzman SK, Faden AI, eds. *The Neurobiology of Central Nervous System Trauma*. New York: NY: Oxford University Press; 1994:266–287.
55. Deller T, Frotscher M. Lesion-induced plasticity of central neurons: sprouting of single fibres in the rat hippocampus after unilateral entorhinal cortex lesion. *Prog Neurobiol*. 1997;53(6):687–727.
56. Cotman CW, Nadler JV. Reactive synaptogenesis in the hippocampus. In: Cotman CW, ed. *Neuronal Plasticity*. New York: Raven Press; 1978:227–271.
57. Perederiy JV, Westbrook GL. Structural plasticity in the dentate gyrus- revisiting a classic injury model. *Front Neural Circuits*. 2013;7(February):17.
58. Schauwecker PE, McNeill TH. Dendritic remodeling of dentate granule cells following a combined entorhinal cortex/fimbria fornix lesion. *Exp Neurol*. 1996;141:145–153.
59. Willems LM, Zahn N, Ferreirós N, et al. Sphingosine-1-phosphate receptor inhibition prevents denervation-induced dendritic atrophy. *Acta Neuropathol Commun*. 2016;4:28.
60. Deitch JS, Rubel EW. Rapid changes in ultrastructure during deafferentation-induced dendritic atrophy. *J Comp Neurol*. 1989;281(2):234–258.
61. Soha JM, Herrup K. Abnormal Purkinje cell dendrites in lurcher chimeric mice result from a deafferentation-induced atrophy. *J Neurobiol*. 1996;29(3):330–340.
62. Vlachos A, Becker D, Jedlicka P, Winkels R, Roeper J, Deller T. Entorhinal denervation induces homeostatic synaptic scaling of excitatory postsynapses of dentate granule cells in mouse organotypic slice cultures. *PLoS One*. 2012;7(3):e32883.
63. Schmidt-Hieber C, Jonas P, Bischofberger J. Subthreshold dendritic signal processing and coincidence detection in dentate gyrus granule cells. *J Neurosci*. 2007;27(31):8430–8441.
64. Schmidt-Hieber C, Bischofberger J. Fast sodium channel gating supports localized and efficient axonal action potential initiation. *J Neurosci*. 2010;30(30):10233–10242.

65. Aradi I, Holmes WR. Role of multiple calcium and calcium-dependent conductances in regulation of hippocampal dentate granule cell excitability. *J Comput Neurosci*. 1999;6(3):215–235.
66. Steward O, Vinsant SL, Davis L. The process of reinnervation in the dentate gyrus of adult rats: an ultrastructural study of changes in presynaptic terminals as a result of sprouting. *J Comp Neurol*. 1988;267(2):203–210.
67. Steward O, Vinsant SL. Collateral projections of cells in the surviving entorhinal area which reinnervate the dentate gyrus of the rat following unilateral entorhinal lesions. *Brain Res*. 1978;149(1):216–222.
68. van Elburg R a J, van Ooyen A. Impact of dendritic size and dendritic topology on burst firing in pyramidal cells. *PLoS Comput Biol*. 2010;6(5):e1000781.
69. van Ooyen A, Duijnhouwer J, Remme MWH, van Pelt J. The effect of dendritic topology on firing patterns in model neurons. *Network*. 2002;13(3):311–325.
70. Cuntz H, Forstner F, Borst A, Häusser M. One rule to grow them all: a general theory of neuronal branching and its practical application. *PLoS Comput Biol*. 2010;6(8).
71. Cuntz H, Borst A, Segev I. Optimization principles of dendritic structure. *Theor Biol Med Model*. 2007;4:21.
72. Cuntz H, Forstner F, Haag J, Borst A. The morphological identity of insect dendrites. *PLoS Comput Biol*. 2008;4 (12):e1000251.
73. Cuntz H, Mathy A, Häusser M. A scaling law derived from optimal dendritic wiring. *Proc Natl Acad Sci USA*. 2012;109(27):11014–11018.
74. Adachi R, Yamada R, Kuba H. Plasticity of the axonal trigger zone. *Neuroscientist*. 2015;21(3):255–265.
75. Vlachos A, Ikenberg B, Lenz M, et al. Synaptopodin regulates denervation-induced homeostatic synaptic plasticity. *Proc Natl Acad Sci*. 2013;110(20):8242–8247.
76. Vlachos A, Bas Orth C, Schneider G, Deller T. Time-lapse imaging of granule cells in mouse entorhino-hippocampal slice cultures reveals changes in spine stability after entorhinal denervation. *J Comp Neurol*. 2012;520(9):1891–1902.
77. Vlachos A, Helias M, Becker D, Diesmann M, Deller T. Neurobiology of Disease NMDA-receptor inhibition increases spine stability of denervated mouse dentate granule cells and accelerates spine density recovery following entorhinal denervation in vitro. *Neurobiol Dis*. 2013;59:267–276.
78. Nitsch R, Frotscher M. Transneuronal changes in dendrites of GABAergic parvalbumin-containing neurons of the rat fascia dentata following entorhinal lesion. *Hippocampus*. 1993;3(4):481–490.
79. Reeves TM, Steward O. Changes in the firing properties of neurons in the dentate gyrus with denervation and reinnervation: implications for behavioral recovery. *Exp Neurol*. 1988;102(1):37–49.
80. Clusmann H, Nitsch R, Heinemann U. Long lasting functional alterations in the rat dentate gyrus following entorhinal cortex lesion: a current source density analysis. *Neuroscience*. 1994;61(4):805–815.
81. Schneider CJ, Cuntz H, Soltesz I. Linking macroscopic with microscopic neuroanatomy using synthetic neuronal populations. *PLoS Comput Biol*. 2014;10(10):e1003921.
82. Morgan RJ, Soltesz I. Nonrandom connectivity of the epileptic dentate gyrus predicts a major role for neuronal hubs in seizures. *Proc Natl Acad Sci USA*. 2008;105(16):6179–6184.
83. Schneider CJ, Bezaire M, Soltesz I. Toward a full-scale computational model of the rat dentate gyrus. *Front Neural Circuits*. 2012;6(November):83.
84. Jedlicka P, Deller T, Schwarzacher SW. Computational modeling of GABAA receptor-mediated paired-pulse inhibition in the dentate gyrus. *J Comput Neurosci*. 2010;29(3):509–519.
85. Jedlicka P, Hoon M, Papadopoulos T, et al. Increased dentate gyrus excitability in neuroligin-2-deficient mice in vivo. *Cereb Cortex*. 2010.
86. Piatti VC, Ewell LA, Leutgeb JK. Neurogenesis in the dentate gyrus: carrying the message or dictating the tone. *Front Neurosci*. 2013;7(April):50.
87. Perederiy JV, Luikart BW, Washburn EK, Schnell E, Westbrook GL. Neural injury alters proliferation and integration of adult-generated neurons in the dentate gyrus. *J Neurosci*. 2013;33(11):4754–4767.
88. Toni N, Schinder AF. Maturation and functional integration of new granule cells into the adult hippocampus. *Cold Spring Harb Perspect Biol*. 2016;8(1):a018903.
89. Beining M, Mongiat LA, Schwarzacher SW, Cuntz H, Jedlicka P. A consistent electrophysiological model of dentate granule cells based on pharmacology to study adult-born neurons. *Bernstein Conf*. 2015.
90. Ryglewski S, Kadas D, Hutchinson K, Schuetzler N, Vonhoff F, Duch C. Dendrites are dispensable for basic motoneuron function but essential for fine tuning of behavior. *Proc Natl Acad Sci USA*. 2014;111(50): 18049–18054.

91. Kuba H, Oichi Y, Ohmori H. Presynaptic activity regulates Na(+) channel distribution at the axon initial segment. *Nature*. 2010;465(7301):1075–1078.
92. Kuba H, Yamada R, Ishiguro G, Adachi R. Redistribution of Kv1 and Kv7 enhances neuronal excitability during structural axon initial segment plasticity. *Nat Commun*. 2015;6:8815.
93. Kuba H. Structural tuning and plasticity of the axon initial segment in auditory neurons. *J Physiol*. 2012;590 (22):5571–5579.
94. Grubb MS, Burrone J. Activity-dependent relocation of the axon initial segment fine-tunes neuronal excitability. *Nature*. 2010;465(7301):1070–1074.
95. Wefelmeyer W, Cattaert D, Burrone J. Activity-dependent mismatch between axo-axonic synapses and the axon initial segment controls neuronal output. *Proc Natl Acad Sci USA*. 2015;112(31):9757–9762.
96. Evans MD, Dumitrescu AS, Kruijssen DLH, Taylor SE, Grubb MS. rapid modulation of axon initial segment length influences repetitive spike firing. *Cell Rep*. 2015;13(6):1233–1245.
97. Gulledge AT, Bravo JJ. Neuron morphology influences axon initial segment plasticity. *eNeuro*. 2016;3(1).
98. Harty RC, Kim TH, Thomas EA, et al. Axon initial segment structural plasticity in animal models of genetic and acquired epilepsy. *Epilepsy Res*. 2013;105(3):272–279.
99. Adkins-Muir DL, Jones TA. Cortical electrical stimulation combined with rehabilitative training: enhanced functional recovery and dendritic plasticity following focal cortical ischemia in rats. *Neurol Res*. 2003;25 (8):780–788.
100. Papadopoulos CM, Tsai S-Y, Cheatwood JL, et al. Dendritic plasticity in the adult rat following middle cerebral artery occlusion and Nogo—a neutralization. *Cereb Cortex*. 2006;16(4):529–536.
101. Mostany R, Chowdhury TG, Johnston DG, Portonovo SA, Carmichael ST, Portera-Cailliau C. Local hemodynamics dictate long-term dendritic plasticity in peri-infarct cortex. *J Neurosci*. 2010;30(42):14116–14126.
102. Jiang X, Chai GS, Wang ZH, et al. CaMKII-dependent dendrite ramification and spine generation promote spatial training-induced memory improvement in a rat model of sporadic Alzheimer's disease. *Neurobiol Aging*. 2015;36(2):867–876.
103. Alpár A, Naumann N, Ueberham U, Arendt T, Gärtner U. Deprivation-induced dendritic shrinkage might be oppositely affected by the expression of wild-type and mutated human amyloid precursor protein. *J Neurosci Res*. 2009;87(8):1813–1822.
104. Sotrel A, Williams RS, Kaufmann WE, Myers RH. Evidence for neuronal degeneration and dendritic plasticity in cortical pyramidal neurons of Huntington's disease: a quantitative Golgi study. *Neurology*. 1993;43 (10):2088–2096.
105. Spires TL, Grote HE, Garry S, et al. Dendritic spine pathology and deficits in experience-dependent dendritic plasticity in R6/1 Huntington's disease transgenic mice. *Eur J Neurosci*. 2004;19(10):2799–2807.
106. Mateus-Pinheiro A, Patrício P, Bessa JM, Sousa N, Pinto L. Cell genesis and dendritic plasticity: a neuroplastic pas de deux in the onset and remission from depression. *Mol Psychiatry*. 2013;18(7):748–750.
107. Morais M, Santos PA, Mateus-Pinheiro A, et al. The effects of chronic stress on hippocampal adult neurogenesis and dendritic plasticity are reversed by selective MAO-A inhibition. *J Psychopharmacol*. 2014;28 (12):1178–1183.
108. Heinrichs SC, Leite-Morris KA, Guy MD, Goldberg LR, Young AJ, Kaplan GB. Dendritic structural plasticity in the basolateral amygdala after fear conditioning and its extinction in mice. *Behav Brain Res*. 2013;248:80–84.
109. Shams S, Pawluski JL, Chatterjee-Chakraborty M, Oatley H, Mastroianni A, Fleming AS. Dendritic morphology in the striatum and hypothalamus differentially exhibits experience-dependent changes in response to maternal care and early social isolation. *Behav Brain Res*. 2012;233(1):79–89.

SECTION III

STRUCTURAL PLASTICITY AND CONNECTIVITY

CHAPTER

10

The Role of Structural Plasticity in Producing Nonrandom Neural Connectivity

Paul Miller

Brandeis University, Waltham, MA, United States

OUTLINE

The Rewiring Brain
DOI: http://dx.doi.org/10.1016/B978-0-12-803784-3.00010-X

1 INTRODUCTION

Plasticity in neural circuits is a fascinating area of science, because each neuron responds only to its own inputs and other local signals; yet, as an ensemble, the circuits generated by such responses to low-level signals acquire the structure necessary to perform the high-level functions of an animal's brain. While evolution has ensured that essential structures develop in the appropriate manner, unlike human-made devices for computation and information processing, the mammalian brain has no master plan or predesigned circuitry. This absence of a fixed wiring plan provides our brains the freedom to adapt to the stimuli around us. For example, we learn to recognize our parents' faces or to understand a language that we grow up hearing or to write with a pen.

One type of plasticity that has proven to be surprisingly useful in generating networks capable of representing patterns of stimuli is Hebbian plasticity. In 1948 Donald Hebb made his well-known prescient suggestion[1] that:

> When an axon of cell A is near enough to excite a cell B and repeatedly or persistently takes part in firing it, some growth process or metabolic change takes place in one or both cells such that A's efficiency, as one of the cells firing B, is increased.

In particular, in the framework of Hopfield networks,[2,3] pattern completion and pattern separation are possible via Hebbian-like plasticity mechanisms. When a particular set of inputs excites a pattern of activity within a circuit, then Hebbian-like plasticity mechanisms cause the connections between coactive neurons within the circuit to strengthen. Such within-circuit strengthening of connections can imprint on the circuit patterns of activity that correspond to prior stimuli. If a similar, but perhaps corrupted, stimulus appears, then that prior pattern of activity can be recalled in a process known as pattern completion. If a pattern is presented that is between those of two earlier stimuli, then the network can respond by reproducing the activity of just one of the earlier stimuli (typically the closer one) in a process known as pattern separation.

One important distinction should be made between the type of plasticity used powerfully within Hopfield networks and the original formulation of Hebb's role. That distinction is based on whether causality matters. In Hebb's rule, for cell A to impact the firing of cell B, then any spike by cell A should occur before the spike of cell B. However, in Hopfield networks, the important factor is the coactivity during a stimulus, with no temporal asymmetry in the firing pattern. In particular, Hopfield networks have symmetric connections such that whenever the connection from cell A to cell B strengthens, so too by an identical amount, the connection from cell B to cell A strengthens.

Experimentally, changes in connection strength wrought by coactivity (or postsynaptic depolarization as a proxy for postsynaptic activity) and by precise timing have been characterized. The earliest studies of synaptic plasticity in mammalian brains in the 1970s[4] demonstrated that postsynaptic depolarization coupled with presynaptic spikes could produce a strengthening of the synaptic connection. Later, at the end of the 1990s, studies demonstrated that the precise order of spikes determined the type of plasticity.[5,6] In particular, in what is known as spike-timing dependent plasticity (STDP), as a direct validation of Hebb's rule, Bi and Poo demonstrated that if the presynaptic neuron repeatedly fired before the postsynaptic neuron, then the synaptic efficacy increased, whereas if the presynaptic neuron repeatedly fired after the postsynaptic neuron, then the synaptic efficacy decreased.[6] The timescale for STDP, being on the order of 20 ms for the separation of spike times, nicely matches the causal window suggested by Hebb, as the impact of the spike of one cell via its excitatory postsynaptic current on the membrane potential of another cell lasts on the order of 10 ms.

The two somewhat distinct rules for synaptic plasticity can be combined in what are known as triplet rules.[7] In one example of a triplet rule, the potentiation expected by standard STDP only occurs if the postsynaptic cell not only fires after the presynaptic cell but also within a broader time window of a prior postsynaptic spike. This process leads to a rate-dependence, as if the postsynaptic cell is firing rapidly, the potentiating events occur more often. Thus if the neurons are coactive with random spike times, at low rates the connections more often depress than potentiate, whereas at high rates they more often potentiate than depress. Such a rule enables the circuit to extract the types of static pattern typical of Hopfield networks and also to represent sequences if the neurons are excited in a regular order.[8,9]

1.1 Structural Plasticity

The review of plasticity in the previous section is based on functional plasticity—changes in the strength of an existing synapse. Yet, since the 2000s, more and more evidence has accrued to demonstrate the disappearance of synapses and appearance of new ones in a process known as structural plasticity.[10–14]

From the perspective of circuit modeling alone, it might appear that structural plasticity makes little difference to circuit function. That is, because in a model one can assume full connectivity—every neuron can be connected to every other neuron—and let functional plasticity drive the connection strength to zero for all those connections that get removed in structural plasticity. In fact, it may be the structural plasticity that is related to functional plasticity in this manner—those synapses whose strengths get diminished are the ones that get removed.

However, when one considers the biology—or even the physics—of the situation, it is clear that space is at a high premium within mammalian cortices, and that the bundles of processes—axonal and dendritic arbors—needed to produce the essential connectivity are almost space-filling. Therefore, when connectivity—even locally between cells located within the densest parts of each other's dendritic and axonal arbors—is rarely above 10%, then it is clear that the circuit is constrained in the number of connections that can be made. Moreover, given the metabolic cost to a cell of maintaining each synapse, with the continual need to replace the proteins lost by turnover, recycle vesicles, and resupply neurotransmitter, the costs of keeping a structural connection with no functional benefit mean that such connections should be removed for efficiency.

The converse although, is that in terms of adaptability, a connection that is of no use today might become essential tomorrow. Therefore, rather than a simple pruning-down of connections over development, it is beneficial for new connections to be sought and tried out over the course of an animal's lifetime in case they turn out to be useful. There is some evidence for this process, not only in the observations of new connections appearing while others disappear but also in that the lifetime of strong synapses appears to be much greater than that of weak synapses.[15] If we assume that the Hebbian-like plasticity mechanisms of the previous section are designed to strengthen "useful" connections and weaken those that are unbeneficial, then a removal of weak connections corresponds to removing those that have no benefit. In our implementation of structural plasticity, we follow a rule with this principle in mind.

1.2 Changing Connections in a Stable Manner

When a connection between a pair of neurons is lost (Fig. 10.1A), it is important for circuit stability that, at least on average, a new connection arises within the circuit. At first glance, it might appear that both neurons in the pair, once they separate, should seek a new partner, so they each maintain their number of either incoming or outgoing connections. However, if that were the case, then a single lost connection would be replaced by two new connections, and over time the network would become completely densely connected. Therefore, some sort of structural homeostasis—a balance between the rate of loss and addition of connections—is needed.

A number of possibilities are available for selection of new connections in a manner that maintains the total number of connections in the network at a fixed level. First (Model 1), it is possible that the postsynaptic cell controls the number of inputs that it receives, so that when it loses an input, a new spine emerges from the dendrite to seek a new input from a passing axon (Fig. 10.1B). Alternatively (Model 2), the presynaptic cell's axon could move in search of a new partner (Fig. 10.1C). Third (Model 3), it could be that both axon

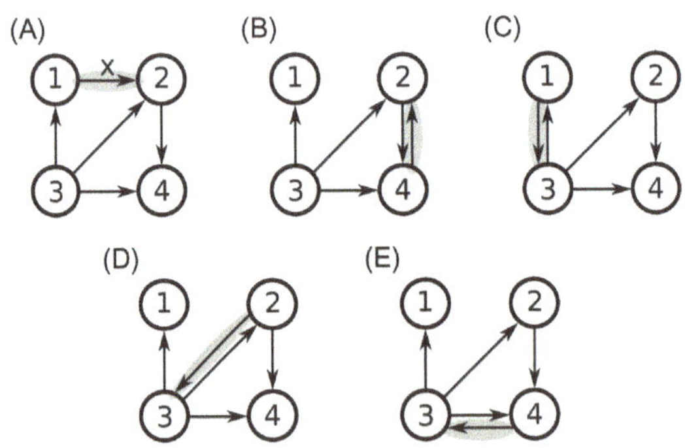

FIGURE 10.1 **Methods for generating new connections via structural plasticity.** If a connection from neuron 1 to neuron 2 is lost as in (A), then several methods for maintenance of total number of connections can be studied. (B) The postsynaptic neuron (2) receives input from a randomly chosen new presynaptic neuron (4). (C) The presynaptic neuron (1) provides input to a new randomly chosen postsynaptic neuron (3). (D) A new connection arises from a randomly chosen presynaptic cell (2) to a randomly chosen postsynaptic cell (3), maintaining a constant total number of connections in the circuit but without balance at the single neuron level. (E) A new connection is chosen with a likelihood that is maximum if it is from the presynaptic cell with the fewest outgoing synapses (4) to the postsynaptic cell with the fewest incoming synapses (3).

and dendrite seek new partners but at a reduced rate such that half of the time one replaces the partner and half of the time the other does (i.e., either Fig. 10.1B or Fig. 10.1C). A fourth alternative (Model 4) is that a completely random connection in the circuit arises whenever (on average) a connection is lost (Fig. 10.1D)—although this would require a global measurement to ensure the rate of addition of connections matched the rate of loss. Finally (Model 5), the random insertion of connections could be at a rate that is specific to the pair of neurons involved in the new connection, so that the probability of finding a new presynaptic (or postsynaptic) partner per unit time scales down with the number of such partners the neuron already possesses (Fig. 10.1E).

The different rules for maintaining balance in the number of connections amidst the ongoing randomness that is inevitable in the selection of new connections have consequences on the steady-state structure of the neural circuit. We explore these different consequences in forthcoming sections of this chapter.

1.3 Accounting for Nonrandom Features of Neural Circuits

It has long been known that neural circuits are not randomly connected. Many invertebrates that have been studied have neural circuits whose topography is fixed within a species such that individual cells can be identified and the wiring diagram specified.[16] Even in mammals with the most complex brains, the connection patterns at the macroscopic scale—between brain regions or subregions such as cortical areas—are fixed across members of a species. Moreover, even at the level of a small circuit, stereotypical connection patterns exist such as the laminar structure of cerebral cortex, with preferential connections with other cells within each of layers 2–4, while layer 4 neurons receive most thalamic feedforward input, while layer 5 neurons provide most output to other brain areas.[17] Finally, specific classes of neurons project and receive inputs from particular subsets of other cell types, rather than promiscuously. This is particularly true for the many types of inhibitory neurons that have been classified,[18–20] but also between classes of pyramidal cells.[21] Such connectivity patterns are likely regulated genetically, in particular during the early stages of development.[22]

However, if one focuses on a particular type of connection—such as between excitatory pyramidal cells—within a particular layer and within a small enough area, then one might expect the connectivity to be random. The topography of cellular processes at the fine level of detail—where and in which direction they bend or branch, from which side of the dendrite shaft a spine protrudes—certainly looks highly variable and random from cell to cell. However, when the connectivity structure at this finest of levels is evaluated, it turns out that the structure is nonrandom. In particular, when specific connection patterns among small groups of cells called motifs are counted, one finds an overabundance of certain motifs indicative of assemblies or "cliques" of cells more likely to connect with each other than with cells outside of their clique.[23–26] Similar nonrandom features typical of small-world networks, such as "hub" neurons, are found when connections are inferred via information transfer between cultured cells.[27]

Moreover, if the neural responses of neurons are measured, those neurons within a clique or assembly are more likely to respond to the same stimuli than are neurons from different cliques.[28] Since the stimulus-learning properties of Hopfield networks described in the Introduction section of this chapter rely upon the increased connection strength between units with similar functional responses, this observation motivated us[29] and others[30] to test whether Hebbian-like structural plasticity in a sparse recurrent network could account for all the observations concerning cliques in cortical circuits.

2 DETAILS OF OUR MODEL

In this section we summarize the model and the training protocol used. A more complete description can be found in our prior works.[29,31]

We consider a local association circuit in which excitatory neurons are connected randomly with an independent 10% probability for each potential connection. Inhibitory neurons are also included, with 25% connection probability to and from each other and excitatory cells.

2.1 Heterogeneous Inputs to the Circuit

An important feature that we consider in our models is the distribution and pattern of inputs to neurons in the association circuit. Since we are interested in general adaptive properties of cortical circuits—how they can come to represent specific novel patterns of inputs—we assume that the novel input combinations to be learned reach the circuit through random and overlapping projections. In particular, we assess the ability of the circuit to generate stimulus-pair specificity, by simulating a task in which the stimuli arrive as pairs of inputs (A + B, C + D, A + D, and B + C). Each individual input (i.e., "A", "B", "C", or "D") is represented by a spike train arising from a set of independent cells, with each independent cell projecting randomly to a set of neurons in the associative circuit that is uncorrelated with the projections of other input cells. We parametrically vary two properties of these inputs. First, the connection probability for any individual input cell can vary between 1/20 and 1/2. Second, the number of independent input cells per stimulus can vary from 2 up to 20 (Fig. 10.2). The lower the number of input cells, the greater the correlations within the inputs. That is, choice of two input cells with a large connection probability is equivalent to a large number of input cells with highly correlated projections.

The initial circuit contains little paired stimulus selectivity, because the randomness and large overlap of individual inputs is exacerbated when the stimuli arrive as pairs of inputs. That is, before training most neurons do not respond with high firing rates to one paired stimulus yet with low firing rates to all other paired stimuli. However, Hebbian-like plasticity rules—both from input cells to the local circuit and within the local circuit—accentuate the initially small differences in responses so that for many input configurations the network produces strong stimulus-pair selectivity after training. Indeed, the level of stimulus-pair selectivity can be sufficient to account for behavioral performance in tasks that require paired stimulus selectivity.[29,31]

2.2 Types of Functional Plasticity

The strengths of existing synapses within our network altered as a function of the spike trains of the neurons during each trial. We tested in our networks the effects of different combinations of plasticity mechanisms, including homeostatic scaling,[32–34] Hebbian-like plasticity between excitatory cells,[7] and long-term potentiation of inhibition (LTPi) onto excitatory cells.[35]

Homeostatic synaptic scaling of excitatory inputs onto neurons has been established both in vitro[33] and in vivo.[36] We implemented such scaling, following the methods of others,[34] by assuming each neuron has a goal or target firing rate. If the neuron is firing at a rate below its goal, then the strengths of all excitatory synapses onto the neuron increase multiplicatively. Whereas if the neuron is firing at a rate below its goal, then the strengths of all excitatory synapses decrease by the same factor.

While experimental measurements of homeostasis often require strong manipulations over a long period of time—such as preventing all activity in a cortical slice for a day or more—it is essential in models with ongoing Hebbian-like plasticity to incorporate stronger or faster homeostasis to combat circuit instability. While a reconciliation of these differences is a matter of ongoing research, it is possible that Hebbian-like plasticity is bounded on the timescale of a few hours, so that homeostasis has a chance to catch up between bouts of rapid Hebbian-like plasticity that more quickly reaches its limit. For simulations that would mean we are condensing many days of training into a single session in an unrealistic manner, but perhaps with a similar net effect on the circuit.

We consider two types of Hebbian-like functional plasticity between excitatory cells. STDP[6] produces a potentiation when a presynaptic neuron spikes before the postsynaptic neuron but a depression on any reciprocal synapse that exists since for the same pair of spikes the role of presynaptic and postsynaptic partners are reversed for a reciprocal connection. Triplet-STDP,[7] as mentioned in the Introduction section, includes an additional term that produces a rate dependence, so that coactive neurons strengthen their connections reciprocally. Full details of the rules are provided in Bourjaily and Miller (2011).[31]

LTPi[35] causes the connection from an inhibitory neuron to an excitatory neuron to increase in strength whenever the presynaptic inhibitory neuron produces a spike and the postsynaptic excitatory neuron is depolarized but does not spike within a short time window (of 25 ms). The spike of the excitatory neuron acts as a veto of the inhibitory plasticity. We assume in our model that neurons are depolarized enough being within the active circuit, so that LTPi generates cross-inhibition—when one subset of neurons is strongly activated by the stimulus, the active inhibitory cells provide enhanced inhibition to the excitatory neurons outside of the active subset, because those within the subset veto the potentiation more often.

2.3 Simulating Structural Plasticity

In our previous work,[31] we studied three different mechanisms. In the first mechanism, Model 1, whenever a synapse was lost (Fig. 10.1A), the postsynaptic neuron obtained a new, randomly chosen presynaptic partner (Fig. 10.1B). This mechanism is akin to a conservation of the number of dendritic spines, as when one disappears a new one seeks a

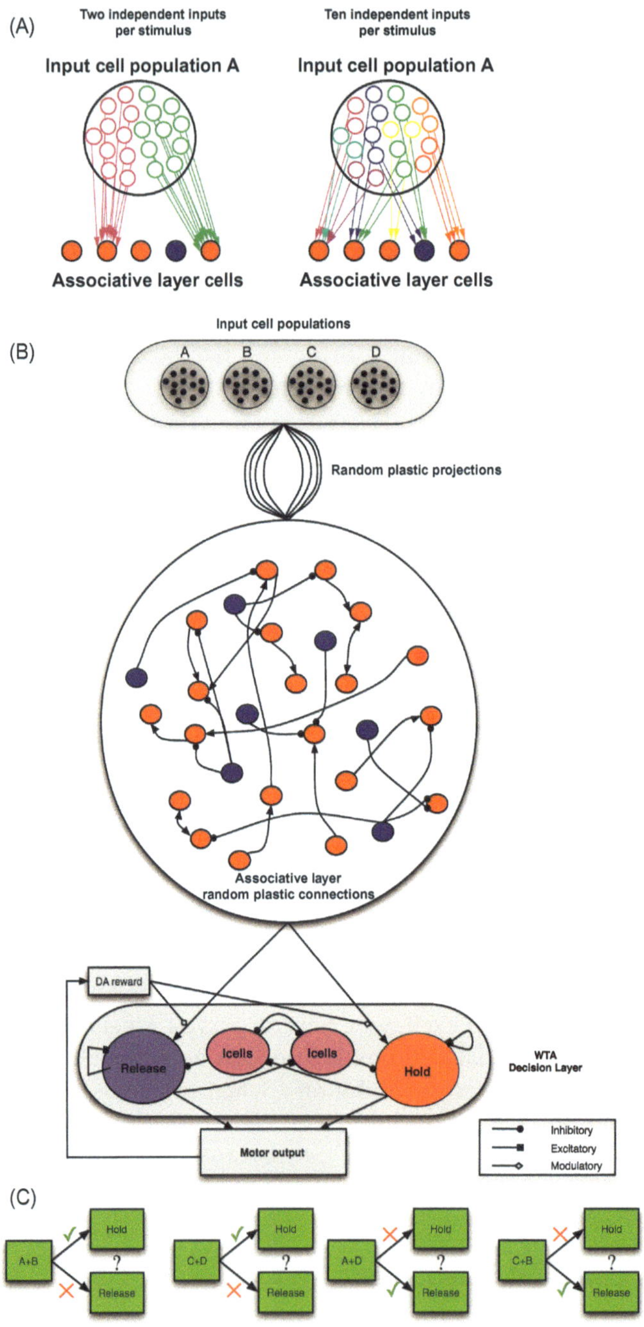

FIGURE 10.2 **Network architecture and inputs.** (A) Example of input configurations from a single stimulus-responsive input population. Each input population is represented by 20 cells, which output-independent Poisson spike trains via correlated connections. The total number of independent sets of connections ranges across our simulations from 2 to 20. Poisson input groups randomly project to a sparse-random recurrent network of

passing axon with which it connects. In the second mechanism, Model 2, whenever a synapse was lost, the presynaptic cell obtained a new, randomly chosen postsynaptic partner (Fig. 10.1C). Thus, Model 2 corresponds to a conservation of the number of axonal terminals. In the third mechanism, Model 3, either a new presynaptic partner was found for the postsynaptic cell or a new postsynaptic partner was found for the presynaptic cell. That is, with 50% probability the replacement followed Model 1 and with 50% probability the replacement followed Model 2.

We consider here two further models for replacement of synapses. First, in Model 4, we simply assume random replacement so that whenever a synapse disappears we choose both presynaptic cell and postsynaptic cell randomly from those without a preexisting connection and add the synaptic connection between them.

Finally, in Model 5, we included a homeostatic term in the rule for structural plasticity. When choosing the new presynaptic and postsynaptic cell, we weighted the choice according to the number of existing postsynaptic partners and presynaptic partners respectively. In particular, the probability, $P_i^{(pre)}$, of the cell, i, being chosen as a presynaptic cell scaled exponentially with its current number, $N_i^{(post)}$, of postsynaptic partners as $P_i^{(pre)} \propto \exp\left(-N_i^{(post)}/2\right)$. For example, this means that if one cell has six more postsynaptic connections than another, it is 20 times less likely to be chosen as the presynaptic partner.

excitatory (*red*) and inhibitory cell. Input projection probability ranges from 1/20 (sparse) to 1/2 (dense), with input connections selected independently between each set of independent input cells per input population. *Left:* With two independent sets of inputs per stimulus population and low input connection probability (1/5 is shown), associative layer cells receive highly correlated inputs that can be sparse and selective to a single stimulus. *Right:* With 10 independent inputs per stimulus population, associative layer cells receive more uniform (therefore, less selective) and less correlated inputs. (B) Complete network structure. Excitatory-to-excitatory connections (*arrows*) and inhibitory-to-excitatory connections (*balls*) are probabilistic and plastic. All-to-all inhibitory-to-inhibitory synapses are also present but not plastic. In the relevant simulations, STDP or triplet STDP occurs at excitatory-to-excitatory and input-to-excitatory synapses, whereas LTPi occurs at inhibitory-to-excitatory synapses. Inhibition is feedforward only (i.e., the network does not include recurrent excitatory-to-inhibitory synapses). Excitatory cells from the associative layer project all-to-all, with initially equal synaptic strength to excitatory cells in both the Hold and Release pools of the decision-making network. The decision-making network consists of two excitatory pools with strong intrapool recurrent connections, which compete via cross-inhibition between pools.[47] Strong intrapool recurrent excitation ensures bistability for each pool, while the cross-inhibition generates winner-take-all (*WTA*) dynamics such that only one population can be active following the stimulus, resulting in one decision. Whether the motor output (based on the decision of hold versus release) is correct for the corresponding cue, determines the presence of dopamine (*DA*) at the input synapses, according to the rules of the task in Fig. 10.2. (C) Logic of the task is an example of biconditional discrimination,[48] in which two of the four possible stimuli (A, B, C, and D) are presented simultaneously to a subject. If either both A and B are present or neither is present, the subject should make one response (such as *release* a lever). If either A or B but not both are present, the subject should make an alternative response (such as *hold* the lever until the end of the trial). Neurons must generate responses to specific stimulus-pairs (e.g., A + B) to perform this task successfully. A response to a single stimulus (e.g., A) is not sufficient to drive the correct response in one pairing without activating the incorrect response for the opposite pairing of that stimulus. Each one of the two simultaneously presented stimuli generates inputs that impinge on the associative network in the manner depicted in panel A. *STDP*, spike-timing dependent plasticity; *LTPi*, long-term potentiation of inhibition. Source: *Panel A was originally published in figure 1 of Bourjaily MA, Miller P. Excitatory, inhibitory, and structural plasticity produce correlated connectivity in random networks trained to solve paired-stimulus tasks. Front Comput Neurosci. 2011;5:37, whereas Panels B and C were originally published as figure 1 of Bourjaily MA, Miller P. Synaptic plasticity and connectivity requirements to produce stimulus-pair specific responses in recurrent networks of spiking neurons. PLoS Computational Biology. 2011;7(2):e1001091.*

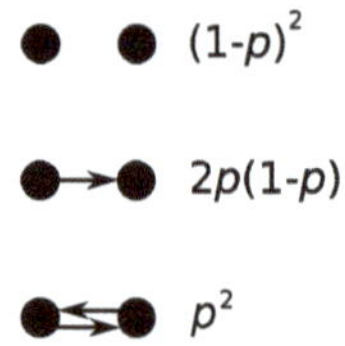

FIGURE 10.3 **Expected probability of finding two cells with each of the three possible 2-cell motifs if all connections have equal likelihood (no correlations present).** In the example networks without self-connections, the connection probability is taken as $p = \frac{N_{\text{Conns}}}{N(N-1)}$, where N_{Conns} is the number of connections and N is the number of cells. In cortical slices taken from rats, more bidirectional connections are found than expected by chance, so the abundance of the disconnected motif (*top*) and bidirectionally connected motif (*bottom*) are greater than chance, whereas the abundance of the singly connected motif (*center*) is below chance.[23,24]

2.4 Measuring the Excess Abundance of Motifs

When assessing whether particular motifs are present in the circuit in excess of that expected by chance, it is important to calculate the expected abundance appropriately. For example, when considering motifs of two cells, three motifs are possible, corresponding to zero, one, or two connections between the cells (Fig. 10.3). Given the total number of connections and the total number of potential connections, the baseline connection probability is calculated. If this value is denoted as p, then the expected numbers for the motifs with zero, one, and two connections are respectively $N_{Tot}(1-p)^2$, $2N_{Tot}p(1-p)$, and $N_{Tot}p^2$, where $N_{Tot} = N(N-1)$ is the total number of ordered pairs and N is the number of cells (and we omit self-connections).

Deviation from the expected number of motifs can be converted to a z-score by using the result for the binomial distribution, with number of samples N_{Tot} and the remaining factor the sampling probability for each motif, p_{Bin} (e.g., $p_{Bin} = (1-p)^2$ for the motif with zero connections) and evaluating the standard deviation as $\sigma_{Bin} = \sqrt{N_{Tot}p_{Bin}(1-p_{Bin})}$.

When considering 3-cell motifs (Fig. 10.4), it is important to discount the excess abundance already present in 2-cell motifs (Fig. 10.3). For example, if the 2-cell motifs already demonstrate an excess abundance of bidirectional connections, then certain 3-cell motifs (e.g., those numbered 3, 7–9, or 12–16) are already expected to be in excess. For the observed abundances of 3-cell motifs to provide additional information, we take the expected abundances to be that calculated given the observed (not expected) abundances of 2-cell motifs. The general procedure is to progress through higher order motifs of more and more cells in this manner, after accounting for all the abundances of lower order motifs. Using such methods, it has been estimated that cliques of on the order of 50 neurons are present in rat somatosensory cortex.[24,25]

2.5 Training Protocol

In the results presented here, the four inputs (A, B, C, and D) each with random projections as described above, are presented in four pairs (A + B, C + D, A + D, and B + C).

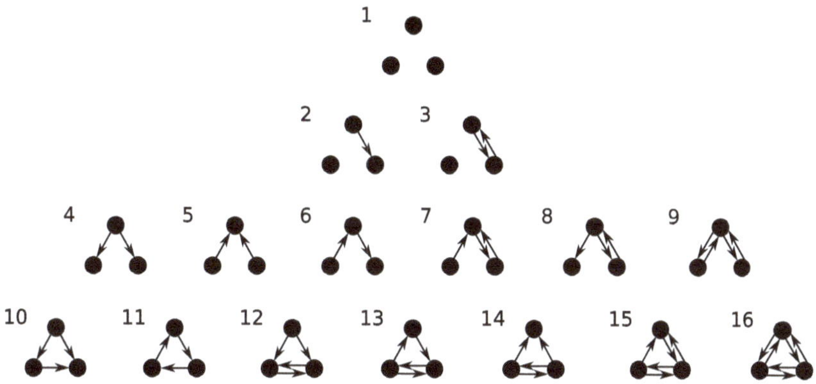

FIGURE 10.4 **The set of motifs possible for the connection patterns between three cells.** In each row, from top to bottom, the number of connected cell pairs increases from 0 to all 3. The motifs in the bottom row with three connected cells contribute to the measure of cliquishness. Excess of motif 4, but not of motif 5, observed in cortical slices[23,24] provides evidence for the underlying structural plasticity mechanisms (Fig. 10.11).

Five hundred sets of the four pairs (2000 total stimuli) are provided as inputs to the network, each for 0.5 s, with random ordering of the stimulus-pairs within each set.[29,31] Plasticity is ongoing throughout such training.

A separate decision-making circuit is connected to the association circuit whose properties are discussed in this chapter. The role of the decision-making circuit, whose internal connections are fixed, but whose inputs from the association circuit are modified by a reward-based plasticity rule,[29,31,37,38] is to assess whether the association circuit contains sufficiently reliable and pair-specific patterns of neural activity to solve a linearly nonseparable task. That is, the pairs "A + B" and "C + D" should produce one response, whereas the pairs "A + D" and "B + C" should produce the opposite response, even though the individual components of the paired stimuli are identical.

3 BEHAVIOR OF THE TRAINED NETWORK

After training the network for 2000 trials, we assessed how different plasticity mechanisms impact the network's performance in the paired association task. We found that including and varying the type of structural plasticity (left column) produced little difference from the network with no structural plasticity (Fig. 10.5B) when functional plasticity is left unchanged. However, removal of any one of homeostatic scaling of synaptic weights (Fig. 10.5D) or inhibitory plasticity (Fig. 10.5F), or the rate-dependence of STDP via the triplet term (Fig. 10.5H) had a strongly detrimental impact on performance.

Homeostasis of synaptic weights is essential for successful network function. Without homeostasis, the network becomes less selective rather than more selective over time under all parameters tested. This occurs because the set of less active neurons become silent as their inputs synapses weaken, while the set of more active neurons become highly

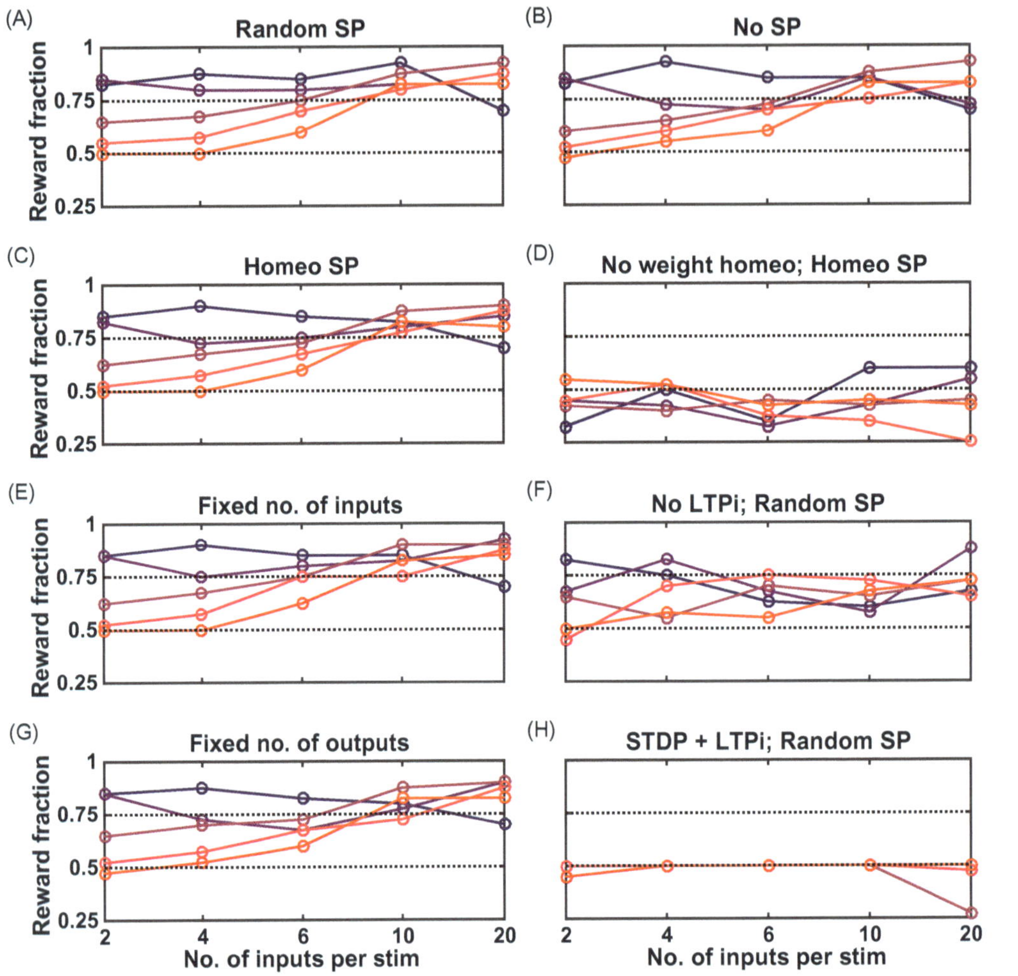

FIGURE 10.5 **Functional plasticity rules, but not structural plasticity, determine the network's ability to solve a linearly nonseparable paired association task.** Each panel depicts the network's performance in solving the task across the 25 different input configurations. The x-coordinate represents the number of inputs per stimulus, representing a decreasing correlation between input cells (Fig. 10.2). The color of curves from blue through red indicates greater connection probability per input cell to the associative layer cell (1/2, 1/3, 1/5, 1/10, 1/20) and thus increasing density of connections. Except where noted, all networks are trained with triplet (rate-dependent) STDP plus LTPi plus homeostatic synaptic scaling. (A) Structural plasticity (*SP*) with random replacement of lost synapses (model 4). (B) No structural plasticity. (C) Homeostatic structural plasticity (model 5). (D) No homeostatic scaling of synaptic weights, and homeostatic structural plasticity. (E) Structural plasticity with fixed number of inputs to each cell (Model 1, postsynaptic cells unchanged when a synapse is loss). (F) No inhibitory plasticity (LTPi removed), and random structural plasticity. (G) Structural plasticity with fixed number of outputs from each cell (Model 2, presynaptic cells unchanged when a synapse is loss). (H) Standard STDP used to update synaptic strengths with no triplet term providing rate dependence, together with inhibitory plasticity and random structural plasticity. *STDP*, spike-timing dependent plasticity; *LTPi*, long-term potentiation of inhibition.

responsive to all stimuli as their inputs strengthen, particularly with each other. While the fraction of cells in each subset depends on the input statistics (which determines initial firing rates) and the parameters of the plasticity mechanisms (which determine a soft threshold for synaptic potentiation versus depression), neither ubiquitously responsive nor ubiquitously unresponsive cells are useful in solving discrimination tasks.

Therefore, in the subsequent sections, we focus on how structural plasticity impacts the connectivity of the network, while the synaptic weights are modified by the successful combination of homeostatic scaling of synaptic weights, LTPi, and triplet (rate-dependent) STDP.

The input parameters leading to the best overall performance were a connection probability of 1/5 per input group, with 20 independent input groups per stimulus. Thus, the mean number of inputs per stimulus was 4 (8 per stimulus pair) with a standard deviation (from the binomial distribution) of 1.8 (2.5 per stimulus pair). Thus before training a typical neuron in the association circuit would be weakly selective to stimulus pairs, perhaps receiving 10–11 active inputs from its optimal stimulus pair, 7–9 active inputs when only one of its strongest stimuli is present, and 5–6 active inputs from its weakest stimulus pair.

Such weak initial selectivity was inadequate for solving the task. However, following training, the enhancement of connection strengths between neurons sharing the same optimal stimulus pair, combined with increased inhibition from neurons with other optimal stimulus pairs, led to enhanced selectivity from a large number of neurons (Fig. 10.6). The increased selectivity led to the fraction of correct responses in the task increasing from a chance level of 50%–90%. We focus on networks with these input parameters in the subsequent sections.

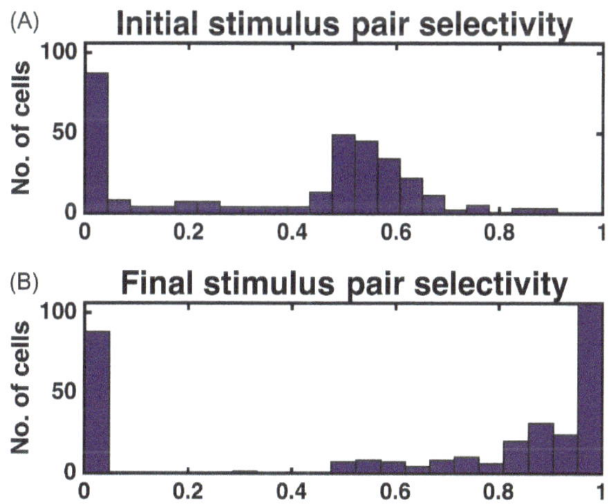

FIGURE 10.6 **Selectivity to paired stimuli increases following training and unsupervised learning.** Selectivity is defined here by recording the maximum response of a neuron to a stimulus pair and the mean response to the other three stimulus pairs and is calculated as (maximum − mean-of-others) / (maximum + mean-of-others). (A) Initial network before plasticity. (B) Network after 500 sets of the four stimuli while undergoing triplet STDP, LTPi, and homeostatic synaptic scaling combined with random structural plasticity (Model 4 as in Fig. 10.5A). *STDP*, spike-timing dependent plasticity; *LTPi*, long-term potentiation of inhibition.

4 STRUCTURAL PLASTICITY PRODUCES HIGHLY INTERCONNECTED ASSEMBLIES OF FUNCTIONALLY SIMILAR CELLS

The loss of connections between neurons that do not respond to the same stimuli, combined with their random replacement, leads over time to an excess of connections between coactive neurons. This is revealed first as an increase in the number of bidirectionally connected neurons (Fig. 10.7, blue curve). If two neurons are coactive, then any connection between them is likely to remain. If a second connection is produced between them, then that second connection is also likely to remain. Conversely, if two neurons are rarely coactive, then any connection between them is likely to disappear and any new connection produced between them is also likely to disappear quickly. Thus unidirectional connections between neurons are more rare than expected, with an excess of disconnected and bidirectionally connected neurons.

A similar argument leads to an excess of triplets of connections that are connected across all three pairs (an excess of motifs 10–16, Fig. 10.4). Such excess is an example of clustering or cliquishness in small-world network theory. If one neuron, A, is connected to two other neurons, B and C, then the likelihood is greater than chance that neurons B and C are also connected. Thus motifs 4–9 occur overall at below-chance levels, while motifs 10–16 overall occur at above-chance levels. Fig. 10.7 (red line) demonstrates the increasing abundance of the connected triplets that arises during training with structural plasticity.

Another analysis that can be used to examine the degree of clustering—and indeed the size of clusters—is to assess the connection probability between two cells as a function of the number of shared neighbors. To this end, in our fully trained network after 2000 trials, for each pair of neuron, we assessed whether they were connected (in either direction) and counted the number of other neurons to which they both had at least one connection.

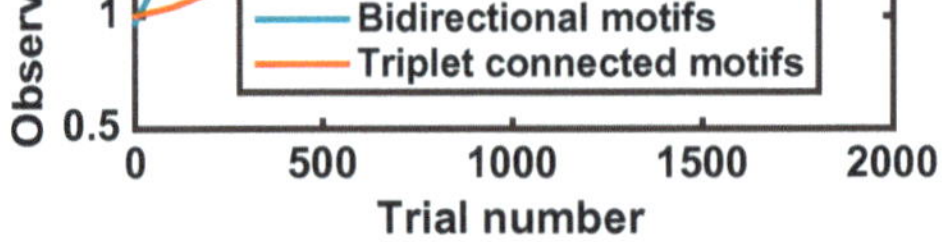

FIGURE 10.7 **Formation of clusters over training with structural plasticity.** The ratio of the observed number of bidirectionally coupled neurons (*blue*) and the number of triplets with at least one connection between each pair (*red*) both increase relative to the numbers expected by chance over many hundreds of trials of training with structural plasticity. Result is for a network with the homeostatic form of structural plasticity (Model 5, as in Fig. 10.5C) combined with triplet STDP, LTPi, and synaptic weight homeostasis (see also Bourjaily and Miller, 2011[31]). *STDP*, spike-timing dependent plasticity; *LTPi*, long-term potentiation of inhibition.

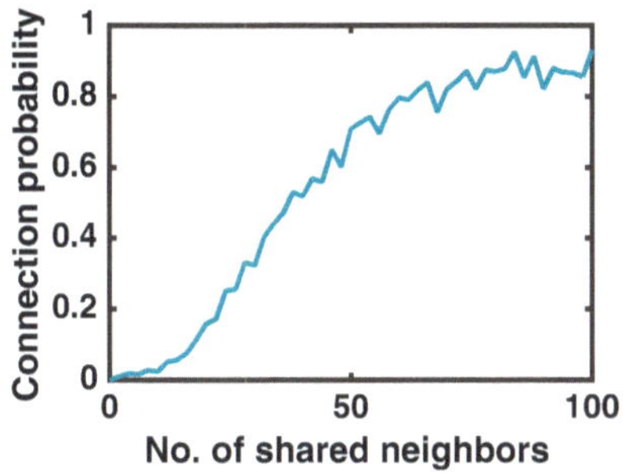

FIGURE 10.8 **Connection probability between two neurons increases with the number of shared connected neurons.** The number of shared neighbors is indicative of whether two neurons are in the same assembly, which also indicates their connection probability. However, even with just four assemblies, the relationship is continuous, whereas if connection probability were entirely defined by the assembly membership of each neuron in a pair, then one might expect stepwise changes, in particular, a big jump in connection probability between pairs in separate assemblies to pairs in the same assembly. Results shown are for the network after 500 sets of the four stimuli while undergoing triplet STDP, LTPi, and homeostatic synaptic scaling combined with random structural plasticity (Model 4, as in Fig. 10.5A). *STDP*, spike-timing dependent plasticity; *LTPi*, long-term potentiation of inhibition.

We then group all pairs with the same number of common connections (actually binning to combine 0–1; 2–3; 4–5; ... to improve power) and plotted the fraction of the group for which the pairs were connected as a function of the number of common neighbors. As can be seen from Fig. 10.8, the connection probability increases steadily with number of common neighbors, saturating between 50 and 100, which is a typical cluster size in a network with 320 excitatory cells.

Our hypothesis was that clustering of cells into assemblies in our network arose directly from their functional responses, with similarly responsive neurons clustering together and separating from clusters of neurons with other responses. Fig. 10.9 depicts the topology expected to arise in this manner, where neurons are colored according to the stimulus to which they respond most strongly.

To test that clustering did correlate with function in this manner, we considered all groups of three neurons that contained a connection between each of the three pairs (i.e., all trios of cells contributing to motifs 10–16 of Fig. 10.4). For such connected triplets, we labeled the neurons as i, j, and k, then accumulated the number with each of the $4 \times 4 \times 4$ combinations of preferred stimuli. That is, if one such connected triplet had cell i most responsive to stimulus pair A + B, cell j most responsive to stimulus pair C + D, and cell k most responsive to stimulus pair A + D, it would contribute a count of 1 to that entry in the $4 \times 4 \times 4$ grid of stimulus combinations. We found (Fig. 10.10) that within the vast majority of such connected triplets, the three neurons responded most to the same stimulus pair, so the example mentioned above would be a rare case (in fact, it appears dark, as the square in the third column of the second row of the top-left grid in Fig. 10.10). By contrast, squares on the diagonal of each 4×4 grid, each corresponding to the three neurons sharing the same preferred stimulus pair, are ubiquitously the lightest, indicating that they represent the majority of triplets.

Our simulation results are also compatible with cortical data demonstrating preferential connections between highly active neurons.[39] In the final trained network, we have a high

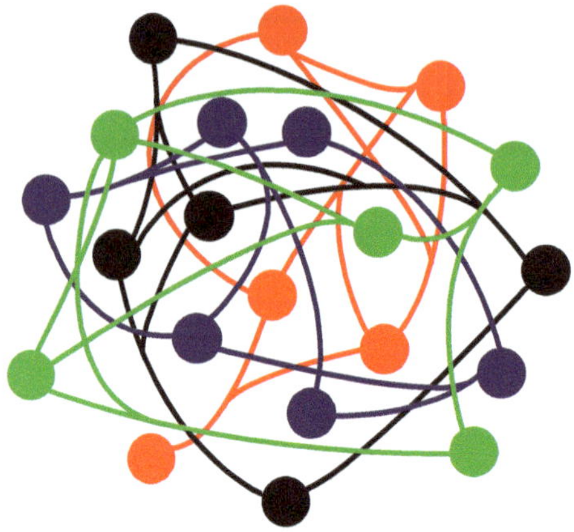

FIGURE 10.9 **Cartoon of the assemblies that can arise from Hebbian-like structural plasticity.** Four cliques are depicted in distinct colors, each corresponding to neurons selective to one of the four stimulus pairs. Neurons with similar stimulus responses (same color) connect predominantly with each other. Such positive feedback heightens selectivity to those trained stimuli within the network.

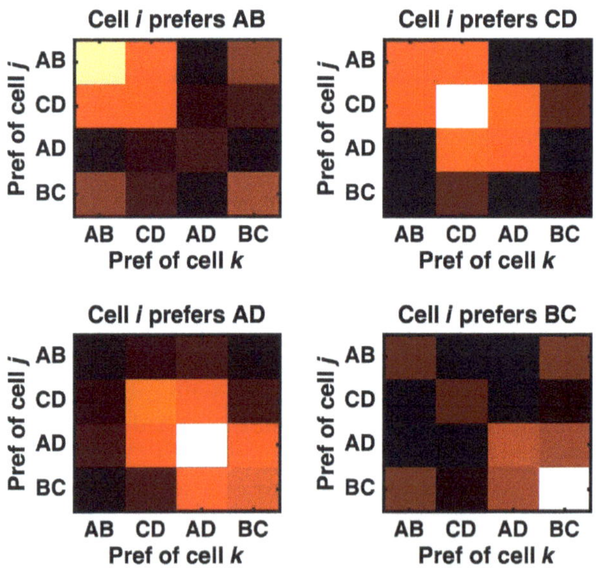

FIGURE 10.10 **Tuning preferences of neurons in connected triplets are mostly identical.** The number of connected triplets are indicated in color (*white* = most, *dark* = least) according to the combination of preferred stimulus responses of the three neurons in the triplet. The greatest numbers of connected triplets (*light squares*) fall on the diagonal of the $4 \times 4 \times 4$ matrix where all three neurons respond most strongly to the same stimulus pair. This result is indicative of the connectivity structure sketched in Fig. 10.9 (see also Bourjaily and Miller, 2011[29]).

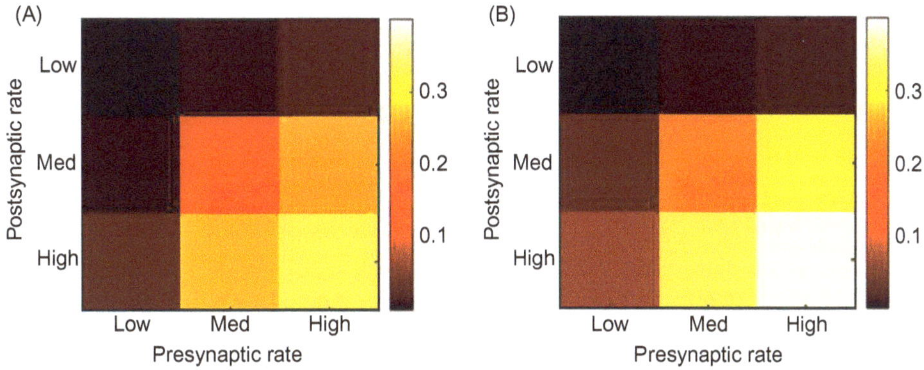

FIGURE 10.11 **Connection probability between neurons depends on their firing rate.** (A) Connection probability between neurons after 2000 trials with our standard model. Cells are grouped into low (<10 Hz), medium (10–30 Hz), and high (>30 Hz) mean evoked responses. (B) Connection probability between neurons after 400 trials with Model 3 of structural plasticity in the absence of stimulus input. Cells are grouped into low (<4 Hz), medium (4–12 Hz), and high (>12 Hz) mean spontaneous activity.

correlation (>0.9 in some cases) between the number of incoming or outgoing connections and the mean firing rate in response to our stimuli. Moreover, if we group neurons into three levels of mean firing rate (<10 Hz = Low; 10–30 Hz = Medium; >30 Hz = High), then the connection probability between neurons of high rate is double than that between neurons of medium rate (0.3 vs 0.15, Fig. 10.11A), whereas the connection probability between neurons of low rate (the majority of which are silent) is almost zero. While we are measuring stimulus-evoked responses, the cortical data are based on spontaneous activity[39]; because the structural changes of our network arise only from simulations of evoked activity, the models of structural plasticity that we simulate can account for the in vivo data.

Indeed, when we compare the structure of a network trained without stimulus input undergoing identical functional plasticity and Model 3 of structural plasticity,[29] we find similar results to those found via stimulus-evoked stimulations. If the correlation is 0.95 between firing rate of a neuron and each number of incoming and outgoing connections and with altered groups of firing rate (<4 Hz = Low; 4–12 Hz = Medium; >12 Hz = High), then the connection probability between neurons of high rate is almost double than that between neurons of medium rate (0.39 vs 0.21, Fig. 10.11B), whereas the connection probability between neurons of low rate (the majority of which are silent) is almost zero.

5 DIFFERENCES IN NETWORK TOPOGRAPHY ACROSS STRUCTURAL PLASTICITY MECHANISMS

In all of our mechanisms for structural plasticity we keep the same rule for loss of a synapse but investigate how different rules for gain of a new synapse (see Fig. 10.1 and Ref. [31]) impacts network structure.

Significant differences in the abundance of certain network motifs arose across Models 1–3. We focus on the 3-neuron motifs 4 and 5, corresponding to a triplet of cells with one neuron projecting to both of the other two (motif 4) or receiving input from both of the other two (motif 5) with no other connections within the triplet (Fig. 10.3). These motifs are worth considering carefully, because not only we see significant differences across our models but also in at least two separate sets of experimental data by different groups an excess of motif 4 was observed without any excess of motif 5.[23,24]

We find that Model 1 produces a strong excess of motif 4 and a small deficiency of motif 5 compared to that expected by chance (Fig. 10.12E). The reason is that for those neurons most responsive to stimuli, when they form a connection with a postsynaptic cell, the connection is likely to remain. For other, less active neurons, they begin to lose axonal projections as the postsynaptic cells seek new partners. The net effect is that independent postsynaptic cells adjust their dendritic spines until they find active presynaptic neurons that are highly responsive to the stimuli they respond to. If those postsynaptic cells respond to similar stimuli, they are more likely to find the same presynaptic partner. However, those highly responsive neurons do not receive more inputs than other cells, so there is no mechanism to produce an abundance of motif 5. Because the total number of motifs is fixed, whenever there is a mechanism to overexpress one motif, the number of others must decrease from the chance level.

Conversely Model 2 produces a strong excess of motif 5 and a small deficiency of motif 4 (Fig. 10.12G). The reasoning follows that of the above paragraph, but now presynaptic neurons seek new partners and settle stably more often than chance upon the most responsive postsynaptic partners. Therefore, unconnected presynaptic cells are more likely than chance to provide input to the same highly responsive postsynaptic cell.

Interestingly, with Model 3 one might expect symmetry between motifs 4 and 5, as Model 1 and Model 2 are used for replacing synapses just as often as each other. However, the equal combination of both mechanisms produces a configuration much more like Model 1, with an excess of motif 4 and a deficiency of motif 5. The explanation lies in the asymmetry of the mechanism for loss of synapses. A synapse is more likely to decrease in strength and be removed, if a postsynaptic neuron is active while the presynaptic neuron is inactive than if the postsynaptic neuron is inactive while the presynaptic neuron is active. Therefore, the tendency for highly responsive cells to have more postsynaptic partners remains in Model 3 and an excess of motif 4 appears.[31]

The results of random replacement (Model 4) are almost identical—across all motif types—to that of Model 3 mentioned earlier. In particular, an abundance of motif 4 arises when Model 4 is used (Fig. 10.11A).

One unsurprising result of Model 5, which includes a homeostatic term in the structural plasticity mechanism (see Fig. 10.1 and the Ref. [31]), is the maintenance of a smaller variation across cells in the number of connections they receive or send. However, the net effect of an increase in the number of triply connected motifs—an example of high clustering or cliquishness that arises in all models—is little impacted by homeostasis in the number of connections per cell. Moreover, Fig. 10.11C indicates that the pattern of 3-neuron motifs, particularly the overabundance of motif 4 and dearth of motifs 2 and 3 matches with that of Model 4 (Fig. 10.12A).

We did find that inclusion of the weight-dependent normalizing factor in the synaptic potentiation term was necessary to produce the excess of motif 4. Without such a term

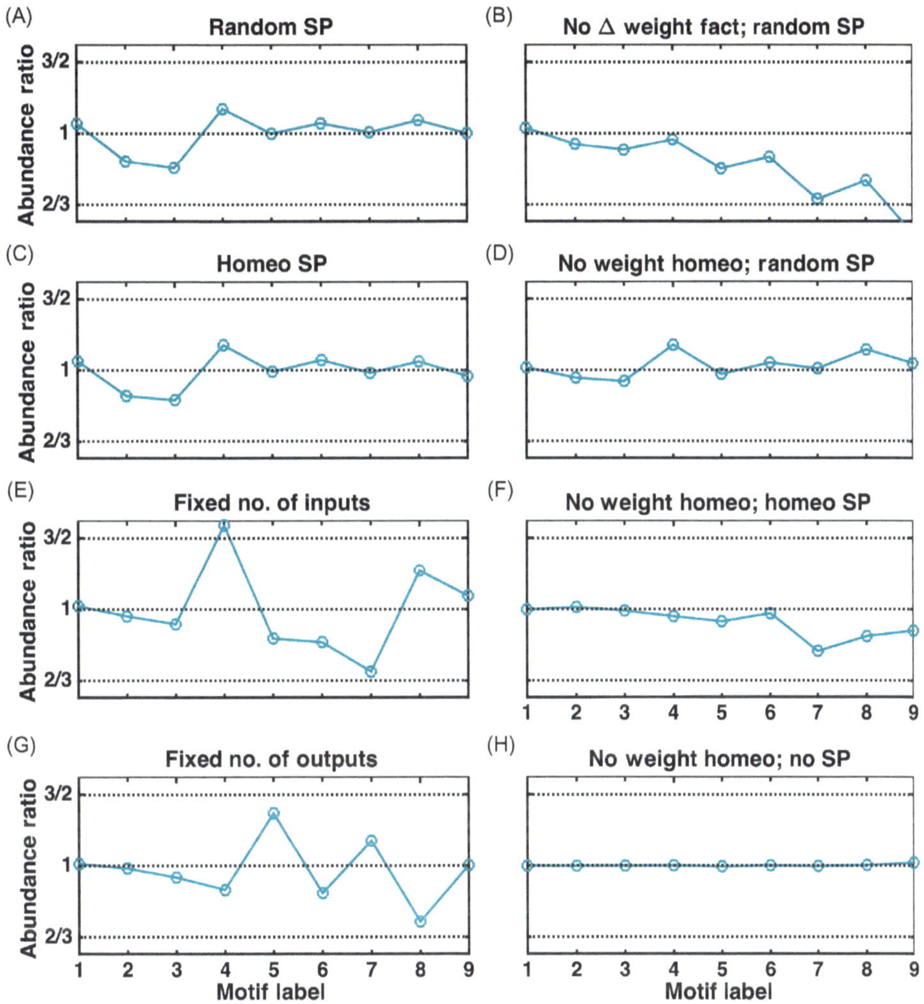

FIGURE 10.12 **Variations in the abundance of motifs can indicate the structural plasticity mechanism present.** Abundances of motifs 1–9 (Fig. 10.4) are plotted for trained networks following 2000 trials with different mechanisms for structural and functional plasticity. In experimental data, motif 4 is found in excess abundance but not motif 5. Ratio of the actual number to the expected number of each motif is plotted on a log scale, after 2000 trials with the given plasticity mechanisms. (A) The standard network with weight-normalized triplet STDP, LTPi, and homeostatic scaling of synaptic strengths combined with structural plasticity using random replacement of connections (Model 4). (B) Similar to (A) but with removal of the weight-dependent factor in the synaptic potentiation term. (C) Similar to (A) but the structural plasticity rule is homeostatic in the number of inputs and outputs per cell (Model 5). (D) Similar to (A) but without the firing rate–dependent homeostatic term to scale all synaptic strengths of a neuron. (E) Similar to (A) but using the structural plasticity rule that maintains a fixed number of inputs to each cell (Model 1). (F) Similar to (C) but without the firing rate–dependent homeostatic term to scale all synaptic strengths of a neuron. (G) Similar to (A) but using the structural plasticity rule that maintains a fixed number of outputs from each cell (Model 2). (H) A network with no structural plasticity (so motifs maintain their initial random distribution). *STDP*, spike-timing dependent plasticity; *LTPi*, long-term potentiation of inhibition.

(Fig. 10.12B), all triplets of neurons in which just one or two of the pairs were connected became underrepresented. Such a result is compatible with a reduction in the final selectivity of neurons when such a weight-normalizing term is included. Without the weight-normalizing term, he distinct groups of cells become more clustered than observed in cortical data when using our standard input configuration (20 groups with input probability of 1/5). However, if we use a sparser network (20 groups with input probability of 1/10), then random structural plasticity with no weight-dependent synaptic potentiation factor does also generate a final network with excess of motif 4 (data not shown).

6 CHANGES IN THE CONNECTIVITY PATTERN IMPACT THE DISTRIBUTION OF CONNECTION STRENGTHS

A long-standing question in the field of synaptic plasticity is whether the empirical rules for changes in connection strength as a function of neural activity are compatible with the observed distribution of synaptic strengths. Any discrepancy would be problematic, because it would suggest the rules for changes in synaptic strength measured in vitro are not the same as or do not fully describe those that sculpt cortical structure in vivo. That is, in principle, observed changes in neural connections produced by correlated activity as measured in vitro, combined with the observed correlations in neural activity measured in vivo, should, when simulated, lead to the observed distribution of neural connection strengths produced in vivo (and later measured in cortical slices). If they do not, then questions are raised about the ability of in vitro protocols to account for in vivo brain development. Or perhaps, a more subtle effect is missing.

A particular challenge is that the empirical rules for changes in connection strength are Hebbian-like. Hebbian-like rules produce positive feedback as they produce a strengthening of synapses that are already effective—where effective means a presynaptic spike being likely to help cause a postsynaptic spike—and such strengthening makes them even more effective. Thus strong synapses get stronger, while weak synapses get weaker. The resulting distribution of synaptic strengths in a circuit inevitably becomes bimodal, with synapses clustered around their maximum possible strength or around zero.[40] Yet, observed distributions of synaptic strengths are unimodal.

A proposed solution to the conundrum is a multiplicative rule for synaptic plasticity.[41,42] In this rule, any increase in synaptic strength wrought by correlated activity gains a multiplicative factor of $\left(W^{(\mathrm{max})} - W_{ij}\right)/W^{(\mathrm{max})}$, where $W^{(\mathrm{max})}$ is the maximum possible synaptic strength, so that weak synapses when potentiated gain a bigger boost, whereas strong synapses get less boost when potentiated. Depression gains no such factor. Addition of the multiplicative term to a synaptic plasticity rule is not at odds with empirical measurements, because the data used to generate the rules are obtained from many synapses after normalizing changes by the baseline synaptic strength. Such an adjustment to synaptic plasticity rules does indeed convert a bimodal into a unimodal distribution of synaptic strengths, at least in feedforward circuits.[42]

However, we found that in our recurrent circuit in the absence of structural plasticity, addition of such a multiplicative term was insufficient to generate a unimodal distribution of synaptic strengths (Fig. 10.13B,H). Surprisingly, when we added structural plasticity to

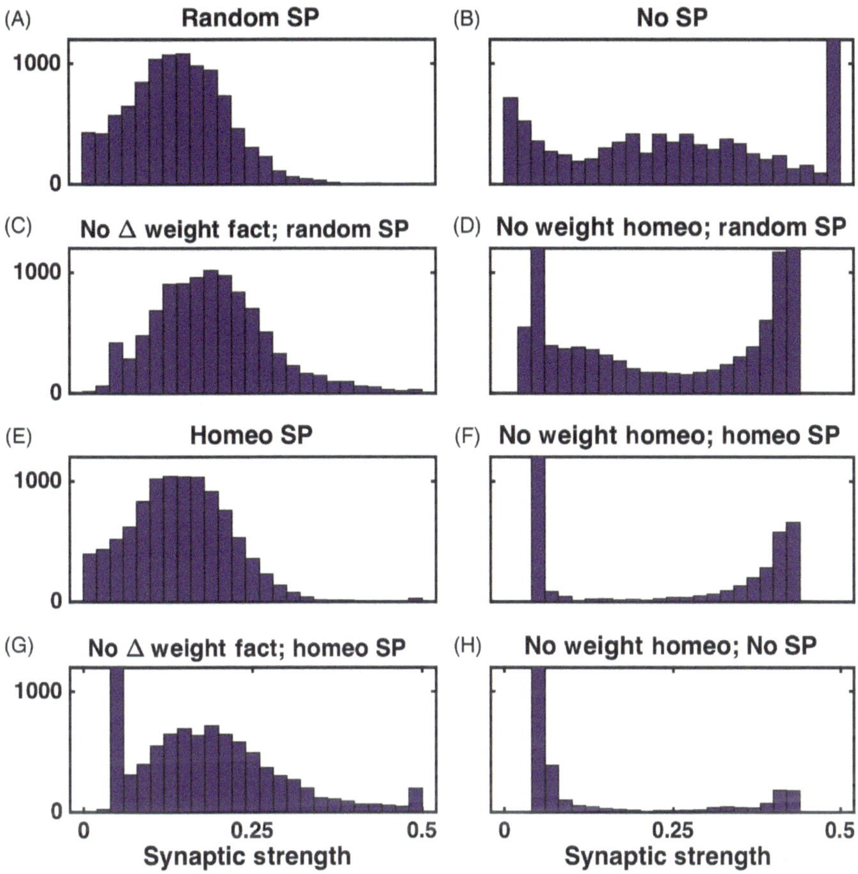

FIGURE 10.13 **Impact of structural plasticity on the distribution of synaptic strengths.** The final distribution of synaptic strengths across all connections in the network is depicted following a variety of combinations of structural plasticity and plasticity of synaptic strength. All networks represented in the left column (A, C, E, and G) include structural plasticity and produce (mostly) a unimodal distribution of synaptic strengths. All networks represented in the right column (B, D, F, and H) either lack structural plasticity or cell-wide homeostatic scaling of synaptic strengths. Although they all include a weight-dependent normalization term in the synaptic potentiation rule, they produce strongly bimodal distributions of synaptic strength. (A) The standard network with weight-normalized triplet STDP, LTPi, and homeostatic scaling of synaptic strengths combined with structural plasticity using random replacement of connections (Model 4). (B) Similar to (A) but without the structural plasticity term. (C) Similar to (A) but with removal of the weight-dependent factor in the synaptic potentiation term. (D) Similar to (A) but without the firing rate-dependent homeostatic term to scale all synaptic strengths of a neuron. (E) Similar to (A) but the structural plasticity rule is homeostatic in the number of inputs and outputs per cell (Model 5). (F) Similar to (E) but without the firing rate-dependent homeostatic term to scale all synaptic strengths of a neuron. (G) Similar to (E) but with removal of the weight-dependent factor in the synaptic potentiation term. (H) Similar to (A) but with neither structural plasticity nor the firing rate-dependent homeostatic term to scale all synaptic strengths of a neuron. *STDP*, spike-timing dependent plasticity; *LTPi*, long-term potentiation of inhibition.

the network, a unimodal distribution did arise (Fig. 10.13A,E). This was true for Models 3–5 of structural plasticity (Fig. 10.1D,E) but not for Models 1–2, in which either a neuron's set of inputs or a neuron's set of outputs remained unchanged (Fig. 10.1B,C). Moreover, even if the plasticity rule for synaptic potentiation did not include a normalizing multiplicative term, addition of structural plasticity to the network was sufficient to generate a unimodal distribution of synaptic strengths (Fig. 10.13C,E).

Upon further investigation, we found that if we did not include the firing rate-dependent homeostatic plasticity, which operated equally on all the synapses of any neuron, the distribution of synaptic strengths became bimodal, even when structural plasticity was present (Fig. 10.13D,F). Thus homeostatic synaptic scaling—a term with no direct impact on the distribution of strengths of synaptic inputs to an individual neuron, as it just scales the distribution—when combined with structural plasticity would convert a bimodal network distribution of synaptic strengths to a unimodal network distribution.

7 DISCUSSION

7.1 Future Experimental Studies

When clusters or assemblies of neurons are formed by experience, the size of the clusters and the number of them in any circuit depend on several factors. In particular, the fraction of neurons receiving strong input when a stimulus is presented determines whether the coding will be dense, producing a few clusters, or sparse, allowing for formation of many clusters. Similarly, the number of distinct and uncorrelated sets of input received by the circuit determines what fraction of the total number of possible clusters form. That is, if an animal is trained with a small number of stimuli for many weeks or months, it is likely that the receptive circuitry produces a small number of clusters, whereas a naturally living animal who experiences a much broader range of stimuli (in any modality) would gain a larger number of such clusters. Therefore, it would be revealing in future studies to assess how this fine-scale structure of cortical connectivity varies across cortical regions and depends on the richness of an animal's experience.

The final network structure, in which more active neurons have more connections and stronger ones, both incoming and outgoing, compared to less active neurons, produces the plaid-like connectivity structure[29] found by others[43] seeking to explain the lognormal distribution of synaptic weights. Such network structure is compatible with the observations in cortex of a strongly interconnected subset of highly active neurons.[39] Whether a correlation exists between mean strength of inputs/outputs and the total number of inputs/outputs of neurons is an open question worthy of further investigation.

Finally, whether the unimodal, lognormal-looking distribution of synaptic strengths is produced only when synapses are combined across multiple cells, or is present at the single-cell level, is a matter for future experimental work. The result would strongly constrain models that attempt to explain this feature of the cortical data.

7.2 Future Modeling Studies

An unexpected result from our simulations was the ineffectiveness of adding a multiplicative normalization factor in producing a unimodal distribution of synaptic weights. Perhaps, even more surprisingly, structural plasticity—even without such a normalization factor—did produce the observed unimodal distribution. In a previous work, a normalization term such as $(W_{max} - W_{ij})/W_{max}$, included when synaptic strengths, W_{ij}, were increased by STDP, led to a stable unimodal distribution of synaptic strengths onto a neuron receiving uncorrelated Poisson input.[42] However, in the recurrent circuit, the correlations in the network activity overwhelmed this term, making it ineffective. On the contrary, inclusion of structural plasticity, even without such a weight-normalization term, did give rise to a unimodal distribution of synaptic weights (Fig. 10.13). The reason for such an unexpected result is in part that weak synapses were removed and replaced by new ones that became stronger, and in part that transmission via the strongest synapses was partially correlated with transmission via other synapses, which also became potentiated as often as they were depressed. However, further modeling studies will be valuable in elucidating the exact requirements and necessary contributions of structural plasticity in producing a unimodal—or even more accurately, a lognormal—distribution of synaptic strengths. Furthermore, the reasons for differences in the synaptic weight distribution produced in recurrent networks compared to feedforward networks can be revealed by further modeling studies.

SUMMARY

The idea that Hebbian plasticity can generate assemblies or clusters of strongly connected neurons is an old one[1] (for a review, see Palm et al., 2014[44]). Here we investigate an extension of Hebb's idea, in which the consequence of a lack of Hebbian strengthening of a connection is the increase in likelihood that the connection is lost. According to simulations,[29,31,45] such structural plasticity produces the assemblies or clusters of preferentially connected neurons that are observed in mammalian cortex. These results imply that observations of a higher degree of interconnections between subgroups of neurons with similar responses[28] can be attributed to experience-driven changes and need not be evidence for the innate programming as suggested by some.[24,46]

References

1. Hebb DO. *Organization of behavior*. New York: Wiley; 1949.
2. Hopfield JJ. Neural networks and physical systems with emergent collective computational abilities. *Proc Natl Acad Sci USA*. 1982;79:2554–2558.
3. Hopfield JJ. Neurons with graded response have collective computational properties like those of two-state neurons. *Proc Natl Acad Sci USA*. 1984;81:3088–3092.
4. Bliss TVP, Lømo T. Long-lasting potentiation of synaptic transmission in the dentate area of the anaesthetized rabbit following stimulation of the perforant path. *J Physiol London*. 1973;232:331–356.
5. Markram H, Lubke J, Frotscher M, Sakmann B. Regulation of synaptic efficacy by coincidence of postsynaptic APs and EPSPs. *Science*. 1997;275:213–215.

6. Bi GQ, Poo MM. Synaptic modifications in cultured hippocampal neurons: dependence on spike timing, synaptic strength, and postsynaptic cell type. *J Neurosci*. 1998;18(24):10464–10472.
7. Pfister JP, Gerstner W. Triplets of spikes in a model of spike timing-dependent plasticity. *J Neurosci*. 2006;26 (38):9673–9682.
8. Clopath C, Busing L, Vasilaki E, Gerstner W. Connectivity reflects coding: a model of voltage-based STDP with homeostasis. *Nat Neurosci*. 2010;13(3):344–352.
9. Clopath C, Gerstner W. Voltage and spike timing interact in STDP—a unified model. *Front Synaptic Neurosci*. 2010;2:25.
10. Becker N, Wierenga CJ, Fonseca R, Bonhoeffer T, Nagerl UV. LTD induction causes morphological changes of presynaptic boutons and reduces their contacts with spines. *Neuron*. 2008;60(4):590–597.
11. Butz M, Wörgötter F, van Ooyen A. Activity-dependent structural plasticity. *Brain Res Rev*. 2009;60 (2):287–305.
12. Holtmaat A, Svoboda K. Experience-dependent structural synaptic plasticity in the mammalian brain. *Nat Rev Neurosci*. 2009;10(9):647–658.
13. De Paola V, Holtmaat A, Knott G, et al. Cell type-specific structural plasticity of axonal branches and boutons in the adult neocortex. *Neuron*. 2006;49(6):861–875.
14. Lendvai B, Stern EA, Chen B, Svoboda K. Experience-dependent plasticity of dendritic spines in the developing rat barrel cortex in vivo. *Nature*. 2000;404(6780):876–881.
15. Holtmaat A, Wilbrecht L, Knott GW, Welker E, Svoboda K. Experience-dependent and cell-type-specific spine growth in the neocortex. *Nature*. 2006;441(7096):979–983.
16. White JG, Southgate E, Thomson JN, Brenner S. The structure of the nervous system of the nematode Caenorhabditis elegans. *Philos Trans R Soc Lond B Biol Sci*. 1986;314(1165):1–340.
17. Lefort S, Tomm C, Floyd Sarria JC, Petersen CC. The excitatory neuronal network of the C2 barrel column in mouse primary somatosensory cortex. *Neuron*. 2009;61(2):301–316.
18. Ferrante M, Tahvildari B, Duque A, et al. Distinct functional groups emerge from the intrinsic properties of molecularly identified entorhinal interneurons and principal cells. *Cereb Cortex*. 2016.
19. Xu X, Callaway EM. Laminar specificity of functional input to distinct types of inhibitory cortical neurons. *J Neurosci*. 2009;29(1):70–85.
20. Dantzker JL, Callaway EM. Laminar sources of synaptic input to cortical inhibitory interneurons and pyramidal neurons. *Nat Neurosci*. 2000;3(7):701–707.
21. Wang Y, Markram H, Goodman PH, Berger TK, Ma J, Goldman-Rakic PS. Heterogeneity in the pyramidal network of the medial prefrontal cortex. *Nat Neurosci*. 2006;9(4):534–542.
22. Dantzker JL, Callaway EM. The development of local, layer-specific visual cortical axons in the absence of extrinsic influences and intrinsic activity. *J Neurosci*. 1998;18(11):4145–4154.
23. Song S, Sjöström PJ, Reigl M, Nelson S, Chklovskii DB. Highly nonrandom features of synaptic connectivity in local cortical circuits. *PLoS Biol*. 2005;3(3):e68.
24. Perin R, Berger TK, Markram H. A synaptic organizing principle for cortical neuronal groups. *Proc Natl Acad Sci USA*. 2011;108(13):5419–5424.
25. Perin R, Telefont M, Markram H. Computing the size and number of neuronal clusters in local circuits. *Front Neuroanat*. 2013;7:1.
26. Yoshimura Y, Dantzker JL, Callaway EM. Excitatory cortical neurons form fine-scale functional networks. *Nature*. 2005;433(7028):868–873.
27. Timme NM, Ito S, Myroshnychenko M, et al. High-degree neurons feed cortical computations. *PLoS Comput Biol*. 2016;12(5):e1004858.
28. Bharmauria V, Bachatene L, Cattan S, et al. Network-selectivity and stimulus-discrimination in the primary visual cortex: cell-assembly dynamics. *Eur J Neurosci*. 2016;43(2):204–219.
29. Bourjaily MA, Miller P. Synaptic plasticity and connectivity requirements to produce stimulus-pair specific responses in recurrent networks of spiking neurons. *PLoS Computational Biology*. 2011;7(2):e1001091.
30. Effenberger F, Jost J, Levina A. Self-organization in balanced state networks by STDP and homeostatic plasticity. *PLoS Comput Biol*. 2015;11(9):e1004420.
31. Bourjaily MA, Miller P. Excitatory, inhibitory, and structural plasticity produce correlated connectivity in random networks trained to solve paired-stimulus tasks. *Front Comput Neurosci*. 2011;5:37.
32. Turrigiano GG. The self-tuning neuron: synaptic scaling of excitatory synapses. *Cell*. 2008;135(3):422–435.

33. Turrigiano GG, Leslie KR, Desai NS, Rutherford LC, Nelson SB. Activity-dependent scaling of quantal amplitude in neocortical neurons. *Nature*. 1998;391(6670):892–896.
34. Renart A, Song P, Wang XJ. Robust spatial working memory through homeostatic synaptic scaling in heterogeneous cortical networks. *Neuron*. 2003;38:473–485.
35. Maffei A, Nataraj K, Nelson SB, Turrigiano GG. Potentiation of cortical inhibition by visual deprivation. *Nature*. 2006;443(7107):81–84.
36. Desai NS, Cudmore RH, Nelson SB, Turrigiano GG. Critical periods for experience-dependent synaptic scaling in visual cortex. *Nat Neurosci*. 2002;5(8):783–789.
37. Soltani A, Wang XJ. From biophysics to cognition: reward-dependent adaptive choice behavior. *Curr Opin Neurobiol*. 2008;18(2):209–216.
38. Soltani A, Wang XJ. Synaptic computation underlying probabilistic inference. *Nat Neurosci*. 2010;13 (1):112–119.
39. Yassin L, Benedetti BL, Jouhanneau JS, Wen JA, Poulet JF, Barth AL. An embedded subnetwork of highly active neurons in the neocortex. *Neuron*. 2010;68(6):1043–1050.
40. Song S, Miller KD, Abbott LF. Competitive Hebbian learning through spike-time-dependent synaptic plasticity. *Nat Neurosci*. 2000;3:919–926.
41. van Rossum MC, Shippi M, Barrett AB. Soft-bound synaptic plasticity increases storage capacity. *PLoS Comput Biol*. 2012;8(12):e1002836.
42. van Rossum MC, Bi GQ, Turrigiano GG. Stable Hebbian learning from spike timing-dependent plasticity. *J Neurosci*. 2000;20(23):8812–8821.
43. Koulakov AA, Hromadka T, Zador AM. Correlated connectivity and the distribution of firing rates in the neocortex. *J Neurosci*. 2009;29(12):3685–3694.
44. Palm G, Knoblauch A, Hauser F, Schuz A. Cell assemblies in the cerebral cortex. *Biol Cybern*. 2014;108 (5):559–572.
45. Litwin-Kumar A, Doiron B. Formation and maintenance of neuronal assemblies through synaptic plasticity. *Nat Commun*. 2014;5:5319.
46. Markram H, Perin R. Innate neural assemblies for lego memory. *Front Neural Circuits*. 2011;5:6.
47. Wang XJ. Probabilistic decision making by slow reverberation in cortical circuits. *Neuron*. 2002;36:955–968.
48. Sanderson DJ, Pearce JM, Kyd RJ, Aggleton JP. The importance of the rat hippocampus for learning the structure of visual arrays. *Eur J Neurosci*. 2006;24(6):1781–1788.

CHAPTER

11

Structural Plasticity and the Generation of Bidirectional Connectivity

Daniel Miner[1], Felix Z. Hoffmann[1,2], Florence Kleberg[1] and Jochen Triesch[1]

[1]Frankfurt Institute for Advanced Studies, Frankfurt am Main, Germany [2]Max Planck Institute for Brain Research, Frankfurt am Main, Germany

OUTLINE

1 INTRODUCTION

The recurrent microcircuitry of the cortex has, historically, often been approximated as a random graph, but recent experimental evidence[1–3] suggests that the graphs of individual cortical layers are strongly nonrandom. Similarly, recent experimental observations have demonstrated subtle order in the dynamics of dendritic spines,[4–7] and therefore,

The Rewiring Brain
DOI: http://dx.doi.org/10.1016/B978-0-12-803784-3.00011-1

effectively, synaptic connections. At the same time, recent advances in computational models featuring recurrent cortical circuitry, multiple forms of plasticity, and synaptogenesis[8–10] have allowed novel predictions of new and comparisons with old data acquired in the aforementioned experimental studies.

One such recent family of models is the so-called SORN, or Self-Organizing Recurrent Neural Network, set of models.[9–11] Though individual members of the family vary in detail, the common components include an excitatory–inhibitory population of neurons in which the recurrent connections of the excitatory pool are plastic and are imbued with both additive Hebbian spike-timing-dependent plasticity (STDP) and multiplicative synaptic normalization. Other potential features include intrinsic homeostatic plasticity, inhibitory STDP, short-term plasticity, and structural plasticity, the lattermost of which is of primary relevance here.

A 2013 study using a member of this model family[9] incorporated a simple form of structural plasticity into the model, which led to several important observations. These observations included a lognormal-like distribution of synaptic efficacies, a phenomenon in which stronger synapses remained more stable over time, and a power-law distribution of synaptic lifetimes, all of which have either previously or subsequently been experimentally observed.[2,5,7]

A further extension to this model by Miner and Triesch in 2016[10] replaced the binary neurons with leaky integrate-and-fire (LIF) neurons and added a modification to the structural plasticity in which the neurons were embedded in a two-dimensional surface and part of the structural plasticity was made distance dependent. Within this framework, several important observations were made. The modifications led to an overrepresentation of bidirectional synapse pairs, as has been observed in the cortex.[1,2] Additionally, they seeded an overrepresentation of highly connected triangular subgraphs or motifs in the network, which was then tuned by the other synaptic plasticity mechanisms, resulting in a pattern of subgraph overrepresentation similar to what has been seen in experiments.[2] Finally (and the modifications to the structural plasticity were not strictly necessary for this part), it was noted that the power-law slope of the synaptic lifetime distribution could be manipulated by changing the effective pruning rate (which could hypothetically be achieved in several ways such as biasing STDP toward depression or potentiation).

Focusing on the matter of the above-random bidirectional connectivity, the level of overrepresentation generated by the model, which was about a factor of two[10] did not account fully for the degree of overrepresentation observed in most experiments, noted to be between three[1] and four[2,12] (though one study measured the overrepresentation to be at chance level for the juvenile mouse barrel cortex[13]). In this chapter, we will explain why this simple topologically influenced structural plasticity necessarily leads to an overabundance of bidirectional connections between neurons, and discuss what additional factors may come into play in the creation of bidirectional connections in the cortex that may account for the full degree of overrepresentation.

2 SELF-ORGANIZATION OF RECURRENT CORTICAL WIRING

Traditionally, recurrent cortical microcircuitry has often been modeled as a random directed graph. As stated, however, the recurrent local wiring of the cortex differs significantly from that of a random directed graph, possessing complex and nonrandom structures.[1–3] One notable feature (among many) of this nonrandomness is an overabundance of bidirectional connections.

The microcircuitry of the cortex, as a biologically grown system, must, in some sense of the term, necessarily self-organize. The process of self-organization can be thought of as the development of complex structure via the interaction of a small set of relatively simple rules. Self-organizing neural networks are a natural platform for modeling the development of cortical structure. Work on the original SORN model was first published by Lazar et al. in 2009,[11] in the context of reservoir computing.[14] The original network model consisted of coupled populations of excitatory and inhibitory binary neurons, with sparse recurrent connections between the excitatory units. The excitatory units were imbued with a form of intrinsic plasticity, in which the firing threshold of an individual neuron was adjusted as a function of its recent activity history so that that neuron approached a target activity rate. The recurrent connections between them were imbued with a binarized form of STDP, as well as a form of synaptic scaling or normalization in which all incoming weights to each unit were multiplicatively scaled so that their sum maintained a constant value. This plasticity within the recurrent network was entirely unsupervised (though a supervised readout was used to interpret the network's state).

In the presence of this plasticity, the SORN heartily outperformed static reservoirs with supervised readouts in sequence storage and recall, prediction, and inference tasks.[11,15] Subsequently, it was adopted with some minor modifications—namely, intrinsic noise and a simple form of structural plasticity which deleted zero-strength connections and added weak new connections at a constant rate—to explain certain properties of synaptic statistics and dynamics.[9] This model successfully reproduced both the biologically observed heavy-tailed distribution of synaptic weights[2] and the phenomenon by which stronger synapses tend to remain more stable over time.[5] It also predicted a power-law distribution of synaptic lifetimes, a phenomenon which was subsequently experimentally observed.[7] However, it also tended to produce feedforward network structures,[16] arguing against the self-organization of an overrepresentation of bidirectional connections in such a model. This is, in many ways, an intuitive result—classical STDP depresses and potentiates based on the ordering of pre- and postsynaptic spike pairings, so it is only natural that in the absence of mitigating factors, it will potentiate connections in one direction while depressing those in the other.

An evolution of this model, first published by Miner and Triesch in 2016,[10] was able to increase the degree of biological realism in the model and provide a powerful competing factor to drive the formation of bidirectional connections. The biological realism was firstly increased by transitioning the model to an LIF-based framework (resulting in the name "LIF-SORN"), and adapting the various plasticity mechanisms accordingly. The other important change was the embedding of the neural population in a 1 mm × 1 mm two-dimensional sheet (approximating a small section of a cortical layer) and the corresponding addition of a distance-dependent bias in the structural plasticity mechanism by which closer neurons were more likely to connect[10] (as has been experimentally observed[3]).

The structural plasticity operates as follows: every second of simulation time, two operations take place. The first is a scan of the sparse connection matrix in the simulation to look for connections with an efficacy of less than 0.001 μV. Any such connections are removed from the network. The second is the insertion of weak new synapses (0.001 mV).

This is done at an approximately constant rate, which has been empirically determined to achieve an overall local connection fraction of 0.1. At each iteration, the number of random insertions is chosen by drawing (and rounding to the nearest integer value) from a Gaussian with a mean equal to the determined optimal rate. Once the number of insertions has been drawn, the new synapses are randomly selected from the list of all potential synapses. This is not done uniformly, however, but with a probability profile according to the pairwise distance between neurons. The profile is a normal distribution with a half-width of 200 μm, selected as a functional compromise between experimental results suggesting minimal or no distance dependence out to approximately 100 μm,[2] and results suggesting an exponential dependency at larger distances, out to about 300 μm.[3]

This distance dependence led to a natural clustering mechanism with a tendency to produce an overabundance of bidirectional connections, which competed with and overwhelmed the classical STDP-driven tendency against this. In the following sections, we will demonstrate exactly why this bias occurs, and model and explain the competition between the two competing factors. We will then go on to discuss other possible factors in the formation of bidirectional connections in the cortex.

3 TOPOLOGY, THE JENSEN INEQUALITY, AND BIDIRECTIONAL CONNECTIONS

Nonrandom connectivity in cortical microcircuits, such as the overabundance of bidirectional connections, can be expressed in the form of varying connection probabilities. In a nonrandom circuit, some pairs of neurons are more likely to be connected than others, creating a distribution of connection probabilities as opposed to a constant probability for every connection in the network. Here we analyze the effect of such a nondegenerate distribution of connection probabilities on the statistic of bidirectionally connected pairs. It turns out that under the assumption of symmetric connection probabilities within pairs, any nonrandomness in the network necessarily leads to an overrepresentation of reciprocal connections.

For this let us consider a network model of N neurons, with connection probability p_{ij} between neurons x_i and x_j for $i, j = 1, \ldots, N$. The overall connection probability P in the network, the probability for a randomly selected neuron to project to another randomly selected neuron, is just the average connection probability across all pairs,

$$P = \frac{1}{N(N-1)} \sum_{i \neq j} p_{ij}. \tag{11.1}$$

Similarly, we can calculate $P_{\mathbf{bidir}}$, the probability for a random pair of neurons to be connected bidirectionally, as

$$P_{\mathbf{bidir}} = \frac{2}{N(N-1)} \sum_{i>j} p_{ij} p_{ji}. \tag{11.2}$$

Are reciprocally connected pairs relatively over- or underrepresented in this network? To answer this question, the relative occurrence of bidirectional connections is computed as the quotient of occurrences in the network and the occurrence of reciprocal connections as expected in a random graph with connection probability P. In such an Erdős–Rényi graph, every connection is equally likely to occur with probability P, meaning in the notation here that $p_{ij} = P$ for all $i, j = 1, \ldots, N$ (excepting $p_{ii} = 0$).[17,18] The probability for a bidirectional connection P'_{bidir} in such a randomly connected network is

$$P'_{\text{bidir}} = \frac{2}{N(N-1)} \sum_{i>j} P^2 = P^2. \tag{11.3}$$

We denote the relative occurrence of bidirectional connections as σ and find with Eq. (11.3),

$$\sigma = \frac{P_{\text{bidir}}}{P^2}. \tag{11.4}$$

This relative occurrence of bidirectional connections depends on the distribution of the p_{ij}. If every connection has the same probability, $p_{ij} = P$ for $i, j = 1, \ldots, N$, there is no relative over- or underrepresentation in network, $\sigma = 1$. However, if some connections are more likely to be established than others, we say the p_{ij} follow a nondegenerate distribution, and a deviation from $\sigma = 1$ is possible. We now assume additionally that connection probabilities are symmetric in pairs, i.e., $p_{ij=}p_{ji}$ for every $i, j \in 1, \ldots, N$. This assumption holds, e.g., in the LIF-SORN,[10] in which the connection probabilities that dictate the relative likelihoods of new connection formation in structural plasticity events depend on the distance between neurons. We can now show that any nonrandomness in a network with symmetric connection probabilities necessarily results in a relative overabundance of reciprocally connected pairs. For this, we use the finite form of Jensen's inequality for convex functions[19] to give a relation between the probability for bidirectional connections in the network, P_{bidir}, and the probability P^2 for bidirectional connections in an Erdős–Rényi graph with the same overall connection probability P,

$$\begin{aligned} P^2 &= \left(\sum_{i \neq j} \tfrac{1}{N(N-1)} p_{ij} \right)^2 \\ &\leq \sum_{i \neq j} \frac{1}{N(N-1)} p_{ij}^2 = \frac{2}{N(N-1)} \sum_{i>j} p_{ij}^2 = P_{\text{bidir}}. \end{aligned} \tag{11.5}$$

Jensen's inequality could be applied as $x \to x^2$ is a strictly convex function and $\sum_{i \neq j} 1/N(N-1) = 1$. In the following equalities, the symmetry assumption $p_{ij=}p_{ji}$ was used. Recalling the definition of the relative occurrence σ, we can rewrite Eq. (11.5) as

$$\sigma \geq 1. \tag{11.6}$$

It is further known that for a strictly convex function, equality holds in Jensen's inequality, and subsequently in Eqs. (11.5) and (11.6), if and only if the p_{ij} are constant, i.e., if $p_{ij} = P$ for $i, j = 1, \ldots, N$, showing that indeed any inhomogeneity in the connection probabilities necessarily introduces a relative overrepresentation of reciprocal connections in the network as claimed. In summary, we have demonstrated that, given a network with symmetric connection probabilities, any inhomogeneity in these connection probabilities, such as those introduced by distance-dependent structural plasticity, will necessarily lead to an overrepresentation of bidirectional connections in that network.

4 A MARKOV MODEL OF COMPETING CONNECTIVITY BIASES

If we examine the development of bidirectional connections in the LIF-SORN model (which, to reiterate, has its recurrent excitatory pool imbued with an intrinsic firing threshold adaptation mechanism, classical STDP, synaptic normalization, synaptic pruning, and distance-dependent random synaptogenesis), as seen in Fig. 11.1, it is clear that there is a competition at play between the purely topological effects which drive bidirectional

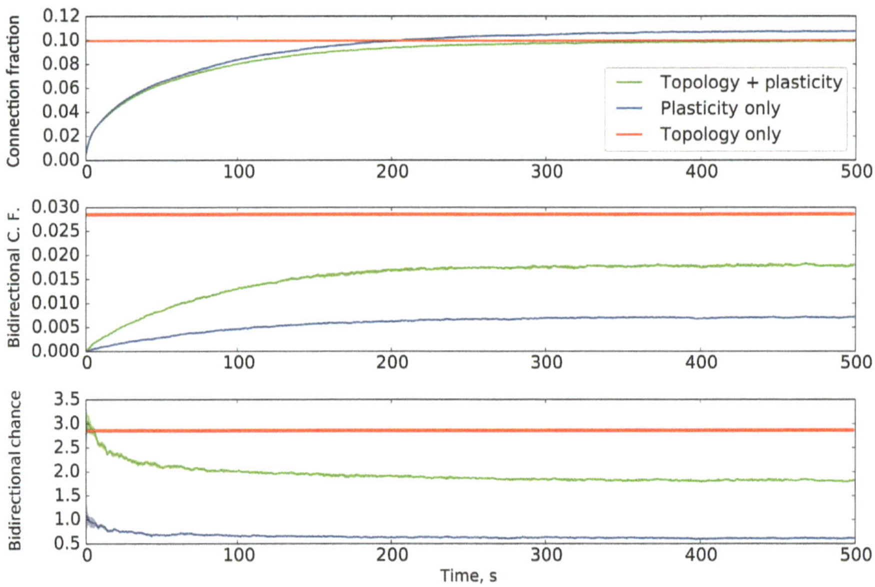

FIGURE 11.1 **Evolution of total and bidirectional connection fraction with simulation time.** Connection fraction evolution for LIF-SORN plastic networks with and without topology, as well as flat values for topology only. (*Top*) Growth and subsequent stabilization of the connection fraction of the network with simulation time. STDP operates at a biological time scale, while all other plasticity mechanisms are accelerated as much as possible while still maintaining separation of timescales in order to increase computational tractability. A 500 s simulation run might therefore approximate tens of hours or a couple of days. (*Middle*) Growth of the bidirectional connection fraction (C.F.). (*Bottom*) Evolution of the bidirectional connection fraction with time as it relates to chance level (i.e., compared to the value for an Erdős–Rényi graph with the same number of nodes and edges), i.e., Eq. (11.4). "Topology + Plasticity" refers to the complete LIF-SORN model. "Plasticity Only" refers to the complete LIF-SORN with the distance dependence removed from the synaptogenesis. "Topology Only" refers to static graphs populated to the target overall connection fraction by the distance-dependent synaptogenesis process, and therefore with all other plasticity mechanisms absent. Data averaged over 10 trials; standard deviation is shaded. Source: *From Miner D, Triesch J. Plasticity-driven self-organization under topological constraints accounts for non-random features of cortical synaptic wiring. PLoS Comput Biol. 2016;12:1–21.*

clustering, and classical STDP, which tends to naturally create a bias toward unidirectional connections.

This competition between factors driving a bias toward feedforward structures and factors driving a bias toward highly recurrent bidirectional structures can be captured and described by a simple Markov model (Fig. 11.2). In this model, each bidirectional connection pair develops independently of all the others. We may consider a single pair of excitatory neurons which has three states {U, S, D} representing that the pair of neurons is either *unconnected, singly connected,* or *doubly connected,* respectively. We may then define transition probabilities denoting the probability of transitioning from one state to another during a fixed time interval. For example, p_{US} is the probability for transitioning from the unconnected state U to the singly connected state S. The transition matrix for the Markov model may then be given by

$$T = \begin{pmatrix} p_{\mathrm{UU}} & p_{\mathrm{US}} & 0 \\ p_{\mathrm{SU}} & p_{\mathrm{SS}} & p_{\mathrm{SD}} \\ 0 & p_{\mathrm{DS}} & p_{\mathrm{DD}} \end{pmatrix}, \tag{11.7}$$

given the assumption that transitions from the unconnected state U to the doubly connected state D and vice versa occur with a negligible frequency within the considered time interval. Since the sum of the elements in each row of T has to equal one, T can be rewritten as

$$T = \begin{pmatrix} 1 - p_{\mathrm{US}} & p_{\mathrm{US}} & 0 \\ p_{\mathrm{SU}} & 1 - p_{\mathrm{SU}} - p_{\mathrm{SD}} & p_{\mathrm{SD}} \\ 0 & p_{\mathrm{DS}} & 1 - p_{\mathrm{DS}} \end{pmatrix}. \tag{11.8}$$

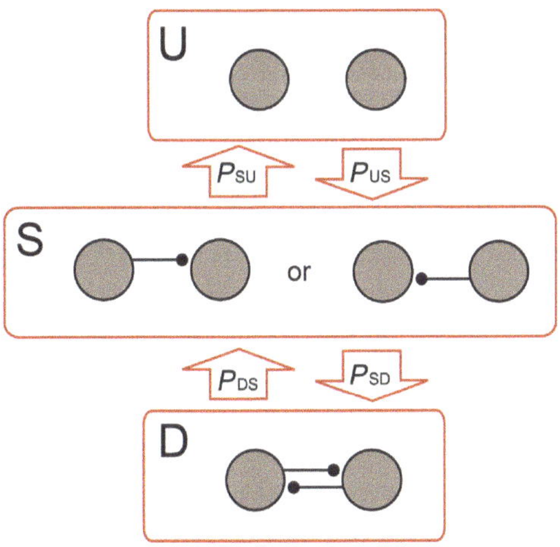

FIGURE 11.2 **A Markov model of uni- and bidirectional connectivity between pairs of neurons.** The *two circles* are a schematic of two nearby neurons. *Top*, the unconnected state U. *Middle*, the singly connected state S. *Bottom*, the doubly connected (bidirectional) state D. P_{XY} is the transition probability from state X to state Y.

It then depends on the four parameters p_{US}, p_{SU}, p_{SD}, and p_{DS}. If all of them are strictly greater than zero, then the Markov chain is regular, and we can find its stationary distribution by calculating the left eigenvector of T:

$$(u\ s\ d)\begin{pmatrix} 1-p_{US} & p_{US} & 0 \\ p_{SU} & 1-p_{SU}-p_{SD} & p_{SD} \\ 0 & p_{DS} & 1-p_{DS} \end{pmatrix} = (u\ s\ d) \tag{11.9}$$

with $u+s+d=1$. The resulting system of linear equations can be written as

$$\begin{aligned} u &= 1-s-d \\ s &= \frac{p_{US}}{p_{SU}}u \equiv \alpha u \\ d &= \frac{p_{SD}}{p_{DS}}s \equiv \beta s = \alpha\beta u, \end{aligned} \tag{11.10}$$

where we have defined $\alpha = p_{US}/p_{SU}$ and $\beta = p_{SD}/p_{DS}$. The behavior of the system then depends only on the two transition probability ratios α and β. We can express u as a function of α and β to arrive at a final solution:

$$\begin{aligned} u &= \frac{1}{1+\alpha(1+\beta)} \\ s &= \alpha u \\ d &= \alpha\beta u. \end{aligned} \tag{11.11}$$

Now, we can consider the conditions under which the model leads to an overrepresentation of bidirectional connections. The overall connection probability in the Markov model is $p = s/2 + d$. For a random graph, we would then expect:

$$\begin{aligned} u_{\text{random}} &= (1-p)^2 \\ s_{\text{random}} &= 2p(1-p) \\ d_{\text{random}} &= p^2. \end{aligned} \tag{11.12}$$

We consider an overrepresentation of bidirectional connections to be in comparison to a random graph. Therefore, using the previously defined transition ratios and a bit of algebra, we arrive at the following expression for the overrepresentation σ:

$$\sigma = \frac{d}{d_{\text{random}}} = \frac{\beta + \alpha\beta(1+\beta)}{(\alpha/4) + \alpha\beta(1+\beta)} \tag{11.13}$$

Before comparing this simple Markov model to the behavior of the full LIF-SORN model, it is instructive to consider the conditions under which the Markov model produces random graphs without any overabundance of bidirectionality.

In a random directed Erdős–Rényi graph, the presence or absence of any connection is independent of the presence or absence of any other connection. We define a probability p_c of creating a new connection per time interval and a probability p_d of destroying an already existing connection per time interval. We now express the original transition probabilities in terms of p_c and p_d, where we assume that p_c and p_d are so small that any product of the probabilities can be neglected. We have $p_{US} = 2p_c$, since there are two independent chances to grow a new connection between an unconnected neuron pair $p_{SU} = p_d$ corresponding to the probability that the only connection between the pair is removed $p_{SD} = p_c$ and $p_{DS} = 2p_d$, again because there are two independent chances of losing a connection in the D state. From this, we see that $\alpha = 2\,p_c/p_d = 2\kappa$ and $\beta = p_c/2p_d = \kappa/2$, where we have defined $\kappa = p_c/p_d$. Thus κ indicates the relative rates of creation and destruction of synapses. Also, we have $\alpha = 4\beta$. Inserting the latter relationship into Eq. (11.13), we immediately obtain $\sigma = 1$ as expected. Thus there is no overabundance of bidirectional connections under these assumptions. The overall connection probability $p = s/2 + d$ can be expressed in terms of κ using the relationships (Eq. 11.11):

$$p = \frac{\kappa + \kappa^2}{2 + 2\kappa + \kappa^2}. \tag{11.14}$$

For κ approaching zero ($p_c \ll p_d$), the graph will become increasingly unconnected. For κ approaching infinity ($p_c \gg p_d$), the graph will tend to become fully connected.

Now we return to the general case and compare the full Markov model with arbitrary α and β parameters to the full LIF-SORN model. We can empirically check it against the LIF-SORN simulation described in the 2016 study.[10] Counting and averaging connections and transitions over a section of the stable (i.e., postgrowth) phase of a simulation run of this model, we obtain $\alpha = 0.194$ and $\beta = 0.105$. This leads the Markov model to predict an overrepresentation of $\sigma = 1.80$, which is, in fact, also the approximate measured value for the average overrepresentation over the observed time period. This demonstrates that this simple Markov model, which describes the likelihoods of the ways connections change, captures the formation of the overrepresentation of bidirectional connections rather successfully.

5 BIDIRECTIONAL CONNECTIONS IN THE PRESENCE OF INHIBITORY STDP

The previous sections discuss the presence of bidirectional connections in recurrent networks, and in the presence of STDP and distance-dependent structural plasticity in the excitatory synapses. However, it has been shown that synapses of inhibitory neurons are also equipped with STDP.[20,21] A more realistic neuronal network could therefore include such plasticity mechanisms. The following section will discuss the findings in the SORN model using LIF neurons,[10] with the addition of inhibitory STDP (iSTDP) in the synapses from inhibitory to excitatory neurons, alongside the STDP and distance-dependent structural plasticity in excitatory-to-excitatory connections.

The iSTDP rule is implemented in the following way:

$$\Delta w_i = A^i_{LTP} \exp\left(-\frac{|\Delta t|}{\tau_{LTP}}\right), \tag{11.15}$$

where w_i is the inhibitory weight, A^i_{LTP} is the amplitude of the STDP window, τ_{LTP} is the time constant of the window, and $\Delta t = t_{pre} - t_{post}$ (Fig. 11.3). The iSTDP plasticity contains long-term potentiation (LTP) based on presynaptic and postsynaptic spike timings.[21,22] Like the STDP in the excitatory synapses, the iSTDP rule is additive and not weight-dependent. Additionally, inhibitory weights are bounded by the interval $[0, w_i^{max}]$.

In the current network, no short-term plasticity or intrinsic homeostasis in the excitatory neurons is applied, as the iSTDP naturally regulates the firing rates of excitatory units in the network.[22] The other plasticity mechanisms, structural rewiring, and neuronal parameters are equal to the model discussed earlier. It has been shown that inhibitory synapses are also subject to synaptic scaling.[23–25] The SORN discussed earlier already includes scaling for excitatory-to-excitatory weights. We hypothesize that synaptic scaling involves a multiplicative adjustment dependent on the sum of synaptic strengths onto a neuron, as in the excitatory synapses. Scaling of synapses is believed to occur on a longer time scale than STDP, each inhibitory weight is scaled once every 1000 ms, according to:

$$w_{i,k} \rightarrow T_{i,e} \frac{w_{i,k}}{\sum_{p=1}^{m} w_{i,p}} \quad \forall k \tag{11.16}$$

where $w_{i,k}$ is a single inhibitory weight onto the excitatory neuron and $T_{i,e}$ the sum of the inhibitory inputs onto a single excitatory neuron. Here, $T_{i,e} = w_i^{max}$. The SORN with iSTDP still displays the overrepresentation of bidirectional connections (Fig. 11.4; in the last 100 seconds of the simulation the average σ is 2.350). We conclude that the presence of iSTDP does not disturb the interaction between structural plasticity and STDP in the excitatory synapses that leads to a surplus of bidirectional connections between excitatory neuron pairs, as observed in the cortex.[2]

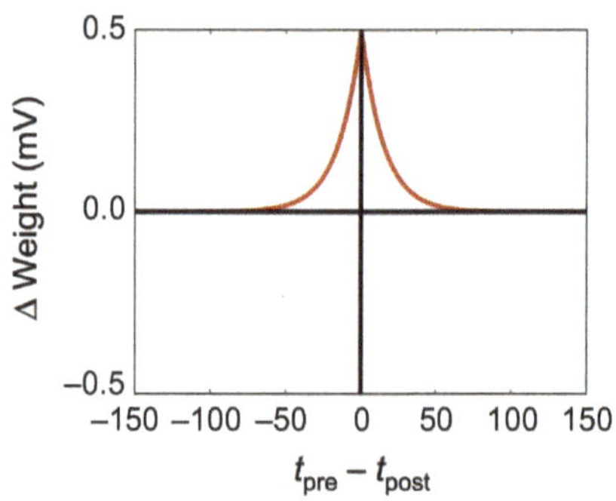

FIGURE 11.3 **Plasticity window for inhibitory STDP.** The iSTDP plasticity window with $A^i_{LTP} = 0.5$, $\tau_{LTP} = 15$ ms.

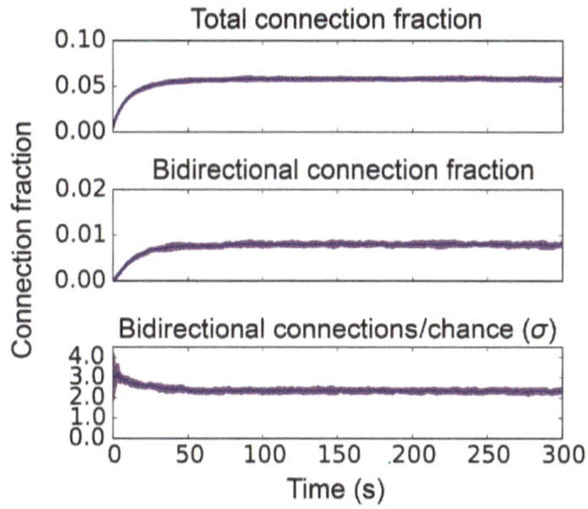

FIGURE 11.4 **Evolution of total and bidirectional connection fraction with simulation time, between excitatory neurons in the presence of structural plasticity, iSTDP and scaling for inhibitory synapses.** (*Top*) Connection fraction evolution between excitatory neurons in the SORN with iSTDP. (*Middle*) Growth of the bidirectional connection fraction between excitatory neurons. (*Bottom*) Evolution of σ, the bidirectional connection fraction as it relates to chance level (i.e., compared to the value for an Erdős–Rényi graph with the same number of nodes and edges; see Eq. (11.4)). Data averaged over 10 trials; the *shaded area* represents one standard deviation.

6 DISCUSSION

6.1 Future Experimental Studies

The models noted here make several predictions and have several implications, some of which may be tested experimentally. For example, it has been noted that for the human cortex, synaptic density at birth is approximately half the mature adult value and rapidly increases to twice the mature adult value shortly after birth, only to decay to its final adult value over the course of the latter part of childhood and throughout adolescence.[26] If we make the simplifying assumption that only changes in the synaptic growth rate lead to this effect, and that all other factors are constant, we can use the Markov model described to predict the degree of bidirectional overrepresentation as a function of developmental stage. This could then be experimentally tested via a longitudinal study over a single animal population, combing both electron microscope-based volume reconstructions and long-term spine imagining to estimate turnover and connection rates.

Rapidly advancing imaging and reconstruction techniques will allow as well for the examination of bidirectional connectivity not just within a single slice or recurrent layer, but between different cortical layers and different cell types. Studies of such specific connectivity will allow for further development of models such as the LIF-SORN, and hence for a more complete understanding of the sort of nonrandomness introduced by structural plasticity.

6.2 Future Modeling Studies

While the aforementioned models have begun to explain the overrepresentation of bidirectional connections seen in the cortex, they have not accounted for the full degree of it.

We must then ask: what other factors could contribute toward an overrepresentation of bidirectional connections in such recurrent circuits? One obvious candidate is the STDP rule itself. The classical STDP rule[27,28] relies on the relative timing of pre- and postsynaptic spikes, potentiating for pre-before-post ordering and depressing for the opposite. This necessarily leads to a preference for unidirectional connections, as in any bidirectional pair, one synapse depresses when the other potentiates. However, alternatives exist that allow for a degree of flexibility. So-called triplet and voltage rules,[29–32] which are somewhat more complicated, allow for potentiation in both directions at high rates of activity. Bursting or other high-rate activity in the cortex could then more directly support the formation and maintenance of highly bidirectionally connected cell assemblies.

In fact, the statement we have been making about classical STDP disfavoring bidirectional connections, while mostly true, is not strictly the case under all conditions and regimes. Analyses of the regimes in which this sort of bidirectional boosting of clustering could occur have been performed,[33] and demonstrated that even classical STDP, under very specific conditions, can, against common intuition, support the maintenance of bidirectional connectivity, and, as stated, that nonclassical STDP can support the formation and maintenance of such connectivity under an even wider regime.

Additionally, larger scale functional factors as well may play into the overrepresentation of bidirectional connections, as functionally similar neurons have been shown to exhibit increased connectivity between each other. Since it has been shown that pairs of neurons that share the same stimulus selectivity have a greater connection probability compared to pairs with different selectivity,[34,35] additional synchrony in firing activities due to the presentation of a stimulus can hypothetically enforce an even greater overrepresentation of bidirectionally connected neuron pairs than shown here. This may be a result of Hebbian plasticity, activity- or development-driven structural plasticity, or some mix of the aforementioned. Furthermore, ontogenetically related neurons in visual cortex have a higher chance of being connected than unrelated neurons, and share preferred orientations,[36] suggesting that ontogenesis contributes to the link between connectivity and functional selectivity. Future work is necessary to elucidate the exact interplay of coactivation of neurons due to external stimuli and structural plasticity, leading to the creation and maintenance of strong, bidirectional connectivity between neuronal pairs, as is observed in the cortex. There is also evidence for different activity-dependent subtypes of structural plasticity: Hebbian (self-enforcing) and homeostatic structural plasticity.[37] Both types of structural plasticity act on different timescales, lead to transient changes in spine densities, and have been suggested to interact in an unsupervised manner to improve the efficiency and allocation of memory in neural networks,[37] but knowledge of their effect on the overrepresentation of bidirectional connections is lacking. Exploring the role of simultaneously acting Hebbian and homeostatic structural plasticity in the formation of connectivity patterns may provide additional understanding of the emergence of bidirectional and higher order connectivity structures in the cortex.

We may also speculate on the specific functional significance of an overrepresentation of bidirectional connections. A recent theoretical study investigated the presence of bidirectional connections in recurrent networks that store memory patterns,[38] and found that networks close to the maximum capacity of memory storage are characterized by sparse connectivity and an overrepresentation of bidirectionally connected neurons (as compared to a random graph). While the exact functional role of the bidirectional connectivity patterns in the cortex is as of yet unknown, this finding suggests that they may be involved in the formation and maintenance of memory patterns, for instance by means of cell assemblies.

We have demonstrated how any directionality-blind inhomogeneity in structural plasticity processes across a local microcircuit will intrinsically lead to the formation of an overrepresentation of bidirectional connections, how there is competition between this and other factors favoring the formation of bidirectional connections, and factors acting against it (such as classical STDP in the majority of operating regimes), and what the expected outcome of such competition may be. Furthermore, we have outlined additional factors that may well contribute to an increase in bidirectional connectivity in cortical circuits. It is clear that we still have much to learn about the microstructure of the cortex and how its particular properties arise, but drawing from mathematical theory and recent computational and experimental studies, we can lay the groundwork for understanding the processes that give rise to the nonrandom features of this wiring, and, in particular, the overrepresentation of bidirectional connections.

References

1. Markram H. A network of tufted layer 5 pyramidal neurons. *Cerebral Cortex*. 1997;7:523–533.
2. Song S, Sjöström PJ, Reigl M, Nelson S, Chklovskii DB. Highly nonrandom features of synaptic connectivity in local cortical circuits. *PLoS Biol*. 2005;3:e68.
3. Perin R, Berger TK, Markram H. A synaptic organizing principle for cortical neuronal groups. *Proc Natl Acad Sci USA*. 2011;108:5419–5424.
4. Grutzendler J, Kasthuri N, Gan W-B. Long-term dendritic spine stability in the adult cortex. *Nature*. 2002;420: 812–816.
5. Yasumatsu N, Matsuzaki M, Miyazaki T, Noguchi J, Kasai H. Principles of long-term dynamics of dendritic spines. *J Neurosci*. 2008;28:13592–13608.
6. Loewenstein Y, Kuras A, Rumpel S. Multiplicative dynamics underlie the emergence of the log-normal distribution of spine sizes in the neocortex in vivo. *J Neurosci Off J Soc Neurosci*. 2011;31:9481–9488.
7. Loewenstein Y, Yanover U, Rumpel S. Predicting the dynamics of network connectivity in the neocortex. *J Neurosci*. 2015;35:12535–12544.
8. Statman A, Kaufman M, Minerbi A, Ziv NE, Brenner N. Synaptic size dynamics as an effectively stochastic process. *PLoS Computat Biol*. 2014;10:e1003846.
9. Zheng P, Dimitrakakis C, Triesch J. Network self-organization explains the statistics and dynamics of synaptic connection strengths in cortex. *PLoS Computat Biol*. 2013;9:e1002848.
10. Miner D, Triesch J. Plasticity-driven self-organization under topological constraints accounts for non-random features of cortical synaptic wiring. *PLoS Comput Biol*. 2016;12:1–21.
11. Lazar A, Pipa G, Triesch J. SORN: a self-organizing recurrent neural network. *Front Computat Neurosci*. 2009;3:23.
12. Wang Y, Markram H, Goodman PH, Berger TK, Ma J, Goldman-Rakic PS. Heterogeneity in the pyramidal network of the medial prefrontal cortex. *Nat Neurosci*. 2006;9:534–542.
13. Lefort S, Tomm C, Floyd Sarria J-C, Petersen CCH. The excitatory neuronal network of the C2 barrel column in mouse primary somatosensory cortex. *Neuron*. 2009;61:301–316.
14. Lukosevicius M, Jaeger H. Reservoir computing approaches to recurrent neural network training. *Comput Sci Rev*. 2009;3:127–149.
15. Lazar A, Pipa G, Triesch J. Emerging Bayesian priors in a self-organizing recurrent network. *Neural Networks Machine Learn*. 2011;1–8.
16. Zheng P, Triesch J. Robust development of synfire chains from multiple plasticity mechanisms. *Front Computat Neurosci*. 2014;8:1–10.
17. Erdős P, Rényi A. On random graphs, I. *Publication Math (Debrecen)*. 1959;6:290–297.
18. Gilbert E.N. Random graphs. *Ann Math Stat*. 1959;30:1141–1144.

19. Jensen JLWV. Sur les fonctions convexes et les inégalités entre les valeurs moyennes. *Acta Math.* 1906;30: 175–193.
20. Haas JS, Nowotny T, Abarbanel HDI. Spike-timing-dependent plasticity of inhibitory synapses in the entorhinal cortex. *J Neurophysiol.* 2006;96:3305–3313.
21. D'Amour J, Froemke RC. Inhibitory and excitatory spike-timing-dependent plasticity in the auditory cortex. *Neuron.* 2015;86:1–15.
22. Vogels TP, Sprekeler H, Zenke F, Clopath C, Gerstner W. Inhibitory plasticity balances excitation and inhibition in sensory pathways and memory networks. *Science.* 2011;334:1569–1573.
23. Kilman V, Rossum MCW Van, Turrigiano GG. Activity deprivation reduces miniature IPSC amplitude by decreasing the number of postsynaptic GABA A receptors clustered at neocortical synapses. *J Neurosci.* 2002;22: 1328–1337.
24. Hartman KN, Pal SK, Burrone J, Murthy VN. Activity-dependent regulation of inhibitory synaptic transmission in hippocampal neurons. *Nat Neurosci.* 2006;9:642–649.
25. Saliba RS, Michels G, Jacob TC, Pangalos MN, Moss SJ. Activity-dependent ubiquitination of GABA A receptors regulates their accumulation at synaptic sites. *J Neurosci.* 2007;27:13341–13351.
26. Huttenlocher PR. Morphometric study of human cerebral development. *Neuropsychologia.* 1990;28:517–527.
27. Bi GQ, Poo MM. Synaptic modifications in cultured hippocampal neurons: dependence on spike timing, synaptic strength, and postsynaptic cell type. *J Neurosci Off J Soc Neurosci.* 1998;18:10464–10472.
28. Froemke RC, Dan Y. Spike-timing-dependent synaptic modification induced by natural spike trains. *Nature.* 2002;416:433–438.
29. Pfister J-P, Gerstner W. Triplets of spikes in a model of spike timing-dependent plasticity. *J Neurosci Off J Soc Neurosci.* 2006;26:9673–9682.
30. Clopath C, Büsing L, Vasilaki E, Gerstner W. Connectivity reflects coding: a model of voltage-based STDP with homeostasis. *Nat Neurosci.* 2010;13:344–352.
31. Gjorgjieva J, Clopath C, Audet J, Pfister J-P. A triplet spike-timing- dependent plasticity model generalizes the Bienenstock–Cooper–Munro rule to higher-order spatiotemporal correlations. *Proc Natl Acad Sci USA.* 2011;108:19383–19388.
32. Krieg D, Triesch J. A unifying theory of synaptic long-term plasticity based on a sparse distribution of synaptic strength. *Front Synap Neurosci.* 2014;6:1–15.
33. Knoblauch A, Hauser F, Gewaltig M-O, Körner E, Palm G. Does spike-timing-dependent synaptic plasticity couple or decouple neurons firing in synchrony? *Front Computat Neurosci.* 2012;6:55.
34. Ko H, Hofer SB, Pichler B, Buchanan K a, Sjöström PJ, Mrsic-Flogel TD. Functional specificity of local synaptic connections in neocortical networks. *Nature.* 2011;473:87–91.
35. Lee W-CA, Bonin V, Reed M, Graham BJ, Hood G. Anatomy and function of an excitatory network in the visual cortex. *Nature.* 2016;532:370–374.
36. Li Y, Lu H, Cheng P, et al. Clonally related visual cortical neurons show similar stimulus feature selectivity. *Nature.* 2012;486:118–121.
37. Fauth M, Tetzlaff C. Opposing effects of neuronal activity on structural plasticity. *Front Neuroanat.* 2016;10:75.
38. Brunel N. Is cortical connectivity optimized for storing information? *Nature Neurosci.* 2016;19:749–755.

CHAPTER

12

Spike Timing-Dependent Structural Plasticity of Multicontact Synaptic Connections

Moritz Deger

University of Cologne, Cologne, Germany

OUTLINE

1 INTRODUCTION

Cortical excitatory synapses are made on dendritic spines, which are small protrusions of the dendrite toward nearby axonal boutons.[1] Dendritic spines can be observed in vivo using two-photon imaging and have been found to be highly motile structures[2] that change depending on sensory experience[3] over a range of time scales from minutes[4] to weeks.[5]

Excitatory synapses in the somatosensory barrel cortex typically consist of several distinct synaptic contacts between pairs of neurons.[6] Since these synapses are nearly

DOI: http://dx.doi.org/10.1016/B978-0-12-803784-3.00012-3

exclusively made on dendritic spines, the synaptic contacts must be made on dendritic spines. Previous modeling work suggested that the synaptic plasticity that governs the development of contacts belonging to the same synapse (i.e., connecting the same neurons) is cooperative,[7] meaning that these contacts mutually support each other. Models of the plasticity of spines should thus be able to explain the occurrence of a bimodal distribution of spine numbers across synaptic connections: either two neurons are unconnected or they have several synaptic contacts (spines).[6]

Theoretical considerations on the function of neuronal networks have addressed structural plasticity (discrete changes: removal and creation of synapses) as being distinct from synaptic plasticity (gradual changes of synaptic efficacy). It has been argued that, given a limited total number of synapses, structural plasticity can increase memory capacity of neural networks.[8–10] Moreover, even in studies where the total synapse number was unconstrained, it was shown that the capacity of the perceptron neuron model to associate input patterns with outputs is maximized if the majority of input synapses are actually zero.[11,12] These results suggest that structural synaptic plasticity might be optimized to stochastically sample the space of possible network configurations in order to optimize neural activity with respect to a given cost function.[13]

Recently structural plasticity models have been included in recurrent network simulations. First, synaptic plasticity combined with a replacement rule that exchanges weak synapses for new ones at other locations was shown to perform well in networks trained to solve association tasks.[14] Second, an artificial neural network model using several, rather abstract plasticity rules was shown to develop sparse synaptic connectivity with a lognormal distribution of synaptic weights,[15] qualitatively consistent with experimental data. In a subsequent study, this phenomenon was confirmed using a network model with spiking neurons and more realistic plasticity.[16] Third, by using a structural plasticity model designed to keep neuronal firing rates in a desired regime,[17] it was demonstrated how structural plasticity can be used to dynamically generate network structures for large-scale network simulations.[18] These previous models, however, neither considered multiple synaptic contacts and their interactions, nor stochastic synaptic transmission.

In the following sections, I will give a selective account on models that address synaptic plasticity with increasing level of biological detail. In Section 2, early models based on spike-triggered calcium release are discussed. Section 3 focuses on an approach to address cooperation of dendritic spines in the context of local synaptic plasticity models. Section 4 introduces our recent model of spike timing-dependent plasticity (STDP) of contacts, which is an attempt to merge structural and synaptic plasticity in a spike timing-dependent model. Finally, Section 5 contains conclusions and identifies future challenges in this modeling field.

2 LOCAL CORRELATION DETECTION

Since Donald Hebb postulated that neurons which fire together would wire together,[19] synaptic plasticity is generally believed to depend on the temporal relation of pre- and postsynaptic neuronal activities. Numerous attempts have been made to biophysically model the molecular signaling pathways leading to synaptic plasticity at the dendritic spine, but due to the great complexity of this system have not achieved a simple and commonly accepted consensus.[20] Here, I will focus instead on phenomenological modeling approaches without explicit biophysical details.

Most phenomenological models of synaptic plasticity are phrased in terms of neuronal firing rates.[21] In the neural circuits of the neocortex, however, neuronal interactions are mediated by action potentials (spikes). Firing rates generally have to be estimated from spike trains in one way or another, e.g., by counting the number of spikes in a time bin of certain duration, or by filtering the spike train with a smoothing kernel of some time scale. While rate-based models of plasticity have contributed substantially to our theoretical understanding of the effects of different kinds of plasticity rules,[21] they typically omit the crucial detail of how the rate terms map to the actual spiking activity of the neurons, which neglects correlations of spiking and introduces an additional filtering time scale that may change the dynamics of plasticity.

STDP models on the other hand embrace the spiking nature of cortical neurons without resorting to the abstraction of firing rates and have been shown to be particularly useful for learning tasks that require fine temporal precision.[22,23] Moreover, experiments have explicitly shown synaptic plasticity dependencies on spike timing that cannot be explained by rate-based plasticity alone.[24]

It is widely accepted and experimentally supported that synaptic plasticity strongly depends on the presence of NMDA receptors in the dendritic spine.[25] The NMDA receptor is a transmembrane ion channel protein that is both ligand- and voltage-gated: to open its pore, its ligand glutamate has to be present extracellularly, and the intracellular medium has to be slightly depolarized to release the magnesium block of the channel. Thus NMDA channels typically open if, at about the same time, presynaptic action potentials cause transmitter release and the postsynaptic dendrite is depolarized due to a back-propagating action potential[26] (Fig. 12.1). The calcium influx through the NMDA channel can thus be used as a measure of the correlation of pre- and postsynaptic spike timings.

A series of models have established that most experimental paradigms on synaptic plasticity in paired neuronal recordings in the slice can be explained on the basis of the postsynaptic calcium influx. According to the calcium control hypothesis,[27] large, transient calcium influx may cause phosphorylation of the Ca^{2+}/calmodulin-dependent protein

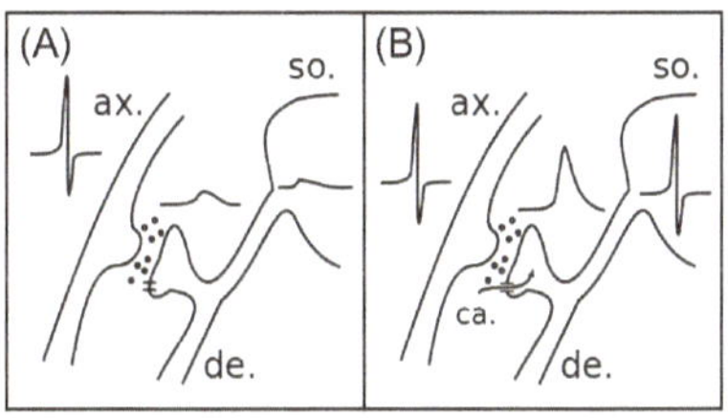

FIGURE 12.1 **Schematic of local correlation detection at dendritic spines.** (A) Presynaptic action potential (*spike*) causes neurotransmitter release (*dots*) at an axonal bouton. The postsynaptic dendritic spine and the postsynaptic soma (so.) are slightly depolarized. NMDA channels (*dashes* on spine) remain closed. (B) Coincident pre- and postsynaptic spikes cause greater depolarization at the spine, sufficient to release the magnesium block of the NMDA channels. Calcium (ca.) flows into the spine and activates signaling pathways. Calcium influx can serve as a local measure of the correlation of pre- and postsynaptic spike timing at dendritic spines. Axon (ax.), dendrite (de.).

kinase II (CAMKII), while small calcium influx may cause dephosphorylation. In turn, CAMKII is known to be central to many biomolecular processes in dendritic spines, and may well be the decisive molecule to trigger spine growth, shrinkage, AMPA receptor insertion or removal, and other genetic programs.[25] A plasticity model based on calcium influx[28] was shown to reproduce a large variety of experiments on synaptic plasticity.[29]

Apart from its influence on models of synaptic plasticity, the calcium control hypothesis was also central to the first spike-timing based model of structural plasticity.[30] In this model, dendritic spines contain a pool of CAMKII proteins which can be in an activated or inactivated state. State transitions are triggered by calcium transients, which in turn depend on pre- and postsynaptic spike timing coincidences (Fig. 12.1). If the number of activated CAMKII falls below a certain number in this model, the dendritic spine is removed. Recently this model was used to explain changes in spine stability observed in denervated dentate gyrus.[31] However, here only structural plasticity is modeled, and spines all have the same, constant efficacy, as long as they exist. Because both structural and synaptic plasticity share the same biophysical substrate (CAMKII and associated proteins), it thus remains a crucial question how the two forms of plasticity are mutually related and how they interact.

3 CONNECTIONS MADE OF MULTIPLE CONTACTS

Combined anatomical and physiological studies in the neocortex[6,32,33] have reported that, if there is an actual synaptic connection between two neurons, it typically consist of about five synaptic contacts at different locations on the dendrite. I adopt this terminology here to unambiguously distinguish individual synaptic contacts (here identified with spines) from entire synaptic connections, which can consist of several contacts.

For the modeling of structural plasticity, the bimodal distribution of synaptic contact numbers poses an additional challenge. While time-lapse imaging data provide information about the temporal dynamics of dendritic spines, the contact number distribution of a connection defines a complementary property about the steady state of network configurations in adult cortex. Fares and Stepanyants[7] realized that to achieve such bimodal distributions, synaptic contacts have to cooperate. If they formed independently from the set of all potential synaptic contacts between pairs of neurons, the contact number distributions would be unimodal instead.[7]

Most previous models of synaptic plasticity have considered synaptic connections as a single entity, commonly referred to just as a synapse.[21,23] This is justified when the quantity of interest is the compound effect of a presynaptic spike on the somatic potential of the postsynaptic neuron. Then the size of the excitatory postsynaptic potential (EPSP) can be described as a single-valued synaptic efficacy and the dynamics of this efficacy can be studied. However, in light of the well-established finding that synaptic transmission at individual synaptic contacts is prone to stochastic transmission failures,[34] it becomes more important to distinguish individual contacts from connections. The more contacts a connection consists of, the more reliably an EPSP is generated at the postsynaptic soma, even if the individual contacts may fail to transmit[35] (Fig. 12.2).

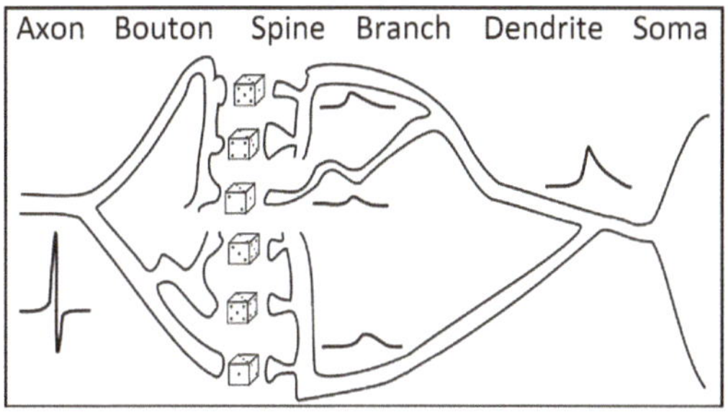

FIGURE 12.2 **Schematic of an excitatory synaptic connection made of multiple synaptic contacts.** Spikes traveling along the axon cause neurotransmitter (glutamate) release at the boutons stochastically (random transmission failures occur as indicated by *dice*). Dendritic spines contain glutamate receptors (AMPA and NMDA) and cause excitatory postsynaptic potentials (miniature EPSPs, *sketched*) in response to the transmitter release. Depolarization is greater and more reliable in dendritic branches that receive many contacts from the presynaptic neuron. Postsynaptic potentials from throughout the dendrite combine at the soma (EPSP).

My colleagues and I initially approached the question of how multiple synaptic contacts can cooperate[7] loosely based on a previously proposed spike timing-dependent model of CAMKII-dependent structural plasticity.[30] In our model,[36] we argued that cooperation between contacts arises if contact plasticity depends on pre- and postsynaptic spike-timing correlations, because each excitatory synaptic contact contributes to firing the postsynaptic neuron each time there is a presynaptic spike (see also Fig. 12.2). Hence more contacts implies more correlation. Now if the contacts have the means to measure these correlations, and plasticity processes depend on those, a bimodal contact number distribution naturally arises.[36]

However, besides the principle of cooperation of contacts through spike timing, more mechanisms are required to quantitatively explain the experimentally measured contact number distributions and rates of dendritic spine turnover. In a first step to combine structural plasticity with changes in spine size and efficacy, we considered a coarse-grained model of spine sizes[36]: spines could either be (1) unrealized (a close apposition of axon and dendrite without a synaptic contact), (2) inactive (a small spine with NMDA receptors to sense correlation, but without substantial contribution to the EPSP), or (3) active (a fully functional synaptic contact/large spine) (Fig. 12.3A). Then, to explain the contact numbers, we assumed a Markov process model where each contact could change from one state into the other through transitions called creation, maturation, shrinkage, and pruning (Fig. 12.3B).

The key to achieving cooperation of contacts in this model is to make the transitions of each contact depend on the state of the other contacts. This was implemented through maturation, shrinkage, and pruning rates that depend on the pre–post correlations (which can be measured through NMDA receptors as described earlier). We chose a sigmoidal dependence of the transition rate λ on the mean measured correlation at the contact μ (Fig. 12.3C). In spiking neuronal networks in the asynchronous irregular activity state, the spiking response of neurons to input spikes can be approximated by their linear response.[37] We therefore used a simple model for the postsynaptic neuron which responds linearly to input spikes. Now, because in this model the mean correlations μ between pre- and postsynaptic activities depend linearly on the number of active contacts, the number of active contacts modulates

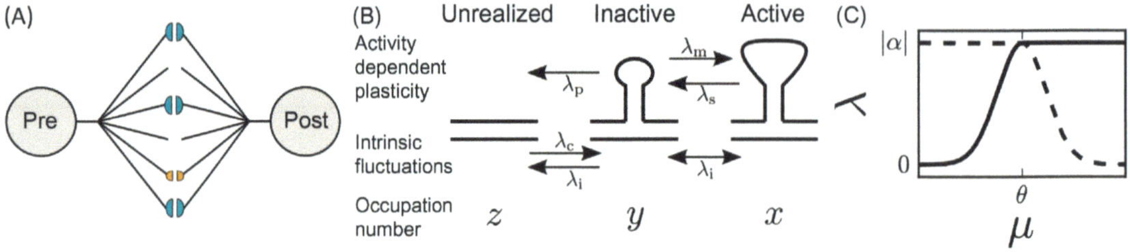

FIGURE 12.3 **Abstract model of multicontact synaptic connections.** (A) Schematic of a synapse connecting a pair of cortical neurons (pre- and postsynaptic). Individual synaptic contacts can be active (*blue*, larger; corresponding to a large dendritic spine), inactive (*orange*, smaller; corresponding to a thin dendritic spine) or unrealized. (B) State diagram of the structural plasticity model. A synaptic contact can occupy one of three states: active, inactive, or unrealized. Transitions from one state to the other are possible through either intrinsic fluctuations or activity-dependent plasticity, indicated by *arrows*. The transitions from unrealized to inactive are called "creation," from inactive to active "maturation." The transitions from active to inactive are called "shrinkage" and from inactive to unrealized "pruning." A synapse consists of several potential contact sites which may or may not host inactive or active synaptic contacts. The occupation number of the three states, active, inactive, unrealized, are called x, y, and z, respectively. (C) Assumed functional form of the dependence of the rate of activity-dependent transitions on the mean value μ of the correlation trace, which is formed at the dendritic spine. The absolute value of α sets the scale of the transition rate, while its sign flips the function from right-saturating (*solid*) to left-saturating (*dashed*). The parameter θ determines at what correlation level saturation of the transition rate is reached. Source: *Reproduced from Deger M, Helias M, Rotter S, Diesmann M. Spike-timing dependence of structural plasticity explains cooperative synapse formation in the neocortex. PLoS Comput Biol. 2012;8:e1002689.*

the transition rates $\lambda(\mu)$ of the Markov process (Fig. 12.3B). We then optimized the parameters of the correlation trace, of the transition rate functions, and the intrinsic transition rate, to achieve very good fits of the experimentally measured distribution of contact numbers in three different intracortical synaptic connection types.[6,32,33] Moreover, the expected lifetimes of inactive and active contacts in the model resembled those of the experimentally distinguished classes of transient and persistent spines.[38]

In that study,[36] we further argued that the correlation trace can be implemented by the phosphorylation state of CAMKII, and that the sigmoidal transition rate functions can be realized by events that are triggered when the (noisy) correlation trace exceeds or falls below a certain threshold. However, no actual spike timing-dependent implementation of the model was given, leaving the explicit example of such a model subject of future research. Moreover, while the three states of synaptic contacts used here (unrealized, inactive, active) made the dynamics of the plasticity model tractable in terms of a Markov process, it was still a crude approximation to the continuous dynamics of spine volume observed in time-lapse imaging of dendritic spines.[2]

The issue of describing active synaptic contacts by a fixed efficacy was partially resolved by a subsequent modeling study.[39] Here, a similar Markov process for the state of a multicontact connection (Fig. 12.3A and B) was proposed, with the following important differences: (1) contacts can only be unrealized or active, there is no inactive state; (2) the efficacy of active contacts is not fixed, but is taken to be the stable fixed point of a (rate-based) synaptic plasticity rule, obtained under the assumption that the synaptic weight converges to the fixed point much faster than structural transitions occur (by a separation of time scales argument). Analyzing this modified Markov process, the authors found conditions on the types of synaptic plasticity rules that would yield synaptic contact number distributions of the experimentally observed form. It was further demonstrated that the multicontact system shows hysteresis-like effects, which are beneficial for long-term storage of information.[39,40] More details on these studies can be found in a dedicated Chapter 16 by Fauth et al. in this book. Also this model, however, cannot fully merge the domains of structural and synaptic plasticity, due to its reliance on a separation of the time scales of the two. Recent experiments demonstrate that new spines can be formed within minutes,[4] which is as short as the typical time scale of long-term depression and potentiation experiments and models.[21] Yet a model in which synaptic weights and structural changes occur on the same time scale is beyond of what can be analyzed in the framework of Markov processes, because that generally results in a continuum of synaptic contact states.

4 STDP MODEL OF SPINE PLASTICITY AND TURNOVER

Theory of isolated multicontact synaptic connections has shown that Hebbian synaptic plasticity rules (either spike-based[36] or rate-based[39]) driving structural plasticity can explain cooperative synapse formation[7] in the neocortex. Thus far, however, no such plasticity model that can actually be implemented in simulations of spiking neurons has been proposed. Moreover, failures of synaptic transmission have not yet been considered in the context of structural plasticity. Here I would like to summarize a recently developed, explicit STDP model of structural plasticity[41,43] that addresses these shortcomings.

In this model, synaptic contacts are described by the weight $w_{i,jk}(t)$, where j identifies the presynaptic neuron and i the postsynaptic neuron. Neurons j and i may potentially be connected by $n_{i,j}$ synaptic contacts, which we number by $1 \leq k \leq n_{i,j}$. The weight of the contact $w_{i,jk}$ corresponds to the efficacy and size of the dendritic spine. The plasticity of each contact is modeled by

$$\dot{w}_{i,jk}(t) = aC_{i,jk}(t) - bC^2{}_{i,jk}(t) - cR_i^4(t) - dw_{i,jk}(t) \tag{12.1}$$

where $a, b, c, d > 0$ are constants, $C_{i,jk}(t)$ measures the correlation of pre- and postsynaptic spike trains locally at the contact, and $R_i(t)$ is a trace of the postsynaptic activity.

Presynaptic spikes from neuron j are transmitted by each contact k, but transmission failures occur independently with probability 0.5. Therefore, the local presynaptic spike train S_{jk} at the contact k is a random subset of the complete spike train S_j emitted by neuron j. The contact forms local traces r_{jk}, r_i, and R_i of the pre- (j) and postsynaptic (i) activities, which obey the differential equations:

$$\begin{aligned} \tau \dot{r}_{jk} &= -r_{jk} + S_{jk} \\ \tau \dot{r}_i &= -r_i + S_i \\ \tau_{\text{slow}} \dot{R}_i &= -R_i + S_i \end{aligned}$$

with the time constants $\tau = 20$ ms and $\tau_{\text{slow}} = 1$ min. The correlation trace $C_{i,jk}(t)$ is then computed as

$$\tau_{\text{slow}} \dot{C}_{i,jk} = - \dot{C}_{i,jk} + r_i r_{jk}, \tag{12.2}$$

and thus slowly tracks the product of the fast activity traces.

Generally, Eq. (12.1) means that synaptic contacts (k) grow if the pre- (j) and postsynaptic neurons (i) are correlated (as measured by $C_{i,jk}$) but stop growing once the correlations become large (when $aC_{i,jk} \approx bC_{i,jk}^2$). Additionally, the total input to the neuron is bounded (synaptic normalization) by the negative rate term ($-R_i^4$).[42] It can be shown analytically how the expected weight change in this model depends on the contact weight ($w_{i,jk}$), on the total weight of the connection ($w_{i,j} = w_{i,j1} + \cdots + w_{i,jn_{i,j}}$) that the contact is part of, and on the total synaptic input to the postsynaptic neuron. Using this result, the parameters of the learning rule equation (Eq. 12.1) can be adjusted to have a stable fixed point at a synaptic configuration that is consistent with the experimentally found contact numbers, turnover rates, and postsynaptic firing rates.[41]

In simulations of this model, we first considered a single postsynaptic neuron that receives synaptic connections from 1000 presynaptic neurons. All synaptic contacts obey the dynamics of Eq. (12.1), but once a contact has a weight of zero or less it is removed and its dynamics cease (Fig. 12.4A). The presynaptic neurons form multicontact synaptic connections onto the postsynaptic neuron (Fig. 12.4B). Because there is only a single postsynaptic neuron here, we drop the index i (as in $w_{i,jk}$) for convenience. Synaptic connections consist of up to 10 potential contacts, randomly drawn from the estimated distribution of potential contact numbers for layer 5 to layer 5 connections (L5–L5) in barrel cortex.[7] Each potential but inactive (zero-weight) contact can be created (set to a low, non-zero weight) at a random time, and such creations occur with a rate of 0.02 per day (for each potential contact independently). After creation, contacts remain at the creation weight for a period of grace of 15 minutes (similar as in Ref. [30]). The presynaptic neurons are modeled as Poisson spike trains with a rate of $5\,\text{s}^{-1}$ (Fig. 12.4C). Each synaptic contact transmits a random subset of the presynaptic spikes due to stochastic transmission failures. The postsynaptic neuron here is again modeled as a linear Poisson neuron, i.e., a Poisson process that increases its firing rate transiently in response to input spikes.

Simulations of this model (Fig. 12.4D) were performed using a spike-driven implementation of (Eq. 12.1) in the Neural Simulation Tool.[44] Neural dynamics proceed in discrete time steps of 1 ms. In the steady state, which is reached after about 80–100 days of simulated time, only about 10% of presynaptic neurons form non-zero connections to the postsynaptic neuron at all, and have several synaptic contacts each, consistent with the experimental data for L5–L5.[6] The turnover ratio of synaptic contacts of 0.18% per day in the model[41] is consistent with data from time-lapse spine imaging.[3] Moreover, the survival of synaptic contacts over time is qualitatively consistent with experimental data, as well as a simulated lesion experiment in which a fraction of the presynaptic neurons is manipulated. We also find that over a course of 150 days of simulation, multicontact synaptic connections are much more stable than the synaptic connections in an analogous simulation of the same model that has only one potential contact per connection. Thus our structural plasticity model[41] (1) exemplifies that synaptic contact cooperation may be mediated by STDP in an actual spiking model, (2) indicates that multiple contacts may be functionally

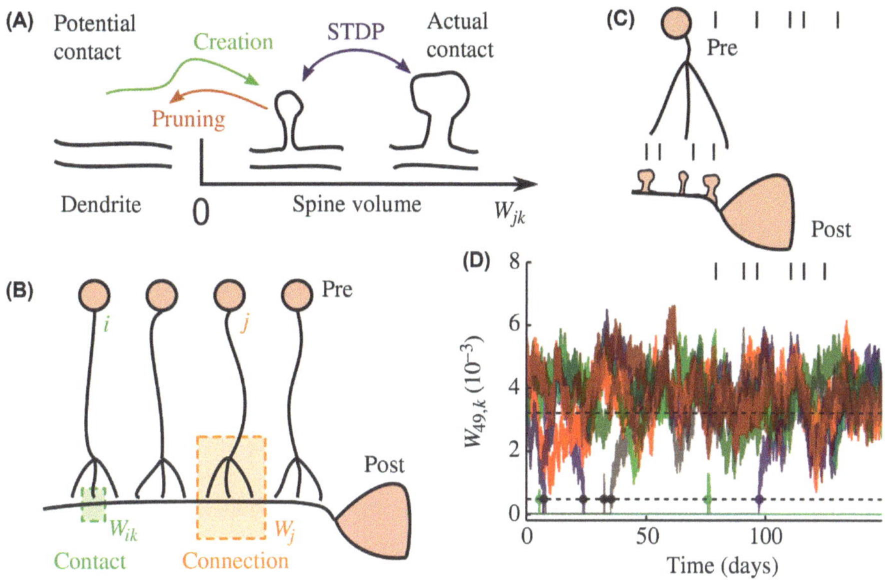

FIGURE 12.4 **Spike timing-dependent structural plasticity model with multiple contacts and synaptic transmission failures.**[41,43] (A) Schematic of synaptic contact creation (*green*), spike timing-dependent plasticity (STDP), and pruning (*orange*). Synaptic contacts correspond to dendritic spines, the weight corresponds to the spine volume or the efficacy of the contact. A weight of zero denotes a potential contact, non-zero weight an actual contact. (B) Multicontact synaptic connections from several presynaptic neurons (*i,j*) are studied. Here, the contact weight ($w_{i,jk}$ in Eq. 12.1) is denoted as w_{jk} for convenience, because there is only a single postsynaptic neuron. (C) The presynaptic spike train is transmitted at each synaptic contact, but is subject to stochastic transmission failures. Therefore each contact records slightly different activity. (D) Simulation of the multiconnection system (B) at steady state. Contact weights w_{jk} (*k*, colors) of an example connection (from presynaptic neuron $j = 49$ to post) over time. New contacts (*circles*) are created at lower *dashed line*. Persistent contact weights fluctuate around the fixed point (*upper dashed line*). Contact weight values denote the total probability of evoking a spike in the postsynaptic neuron upon successful transmission of a presynaptic spike. The synaptic connection is stable over 150 days despite ongoing contact plasticity and turnover.

important for long-term memory, and (3) provides a foundation for future studies of synaptic and structural plasticity in networks in a unified manner.

5 DISCUSSION

5.1 Functional Significance of Multiple Synaptic Contacts

Synaptic connections made of multiple contacts provide stability over very long times: the number of active contacts in a connection constitutes a degree of freedom of the state of the connection additional to the total weight, which previously has received most attention. A connection with a given total weight can be made of many or few contacts, and the number of contacts can change without any change to the total weight. These internal or hidden transitions within the synaptic connection are reminiscent of the cascades of internal states suggested in earlier models of synaptic plasticity to explain the longevity of synaptically stored memories.[45,46] Multiple contacts are further beneficial since they increase the reliability of synaptic transmission by decreasing noise due to transmission failures at individual contacts[34] in the combined EPSP at the soma[35,41,47] (see also Fig. 12.2).

5.2 Relations to Previous Models

Although previous theory[36] predicted that a dependence of structural plasticity on pre–post activity correlations can explain cooperative synapse formation, an explicit model of synaptic contacts that is able to implement such learning and reproduce the experimental data was missing until recently. In this chapter, I have summarized previous modeling approaches that have now led to a first model of STDP of dendritic spines and turnover in presence of transmission failures.[41]

Several concepts invoked in this plasticity model have previously occurred in the literature. Many variants of STDP rules and the effects of nonlinearities, including weight normalization, have been studied.[42,48] Identification of the zero-crossing of the weight with pruning of a spine was previously used in an artificial neural network model[15] and in a spiking variant of it.[16] Finally, a long-term correlation trace implemented by CAMKII at the dendritic spine was introduced in a previous spike timing-dependent structural plasticity model without plasticity of the synaptic weight.[30]

Other previous models of structural plasticity have focused on the assumption that structural plasticity homeostatically regulates the firing rates of pre- and postsynaptic neurons.[17,18,49] Indeed a mechanism that regulates the firing rate is required if structural plasticity is active, because networks that continuously change their connections can otherwise quickly develop into nonphysiological activity states. In the spike timing-dependent models summarized here, pre- and postsynaptic rates also influence the development of the connections, because both models employ a correlation trace that is not normalized but rather is the direct product of pre- and postsynaptic activities. Therefore, in the abstract model of Section 3, the correlation trace values increase when firing rates rise, leading to a greater frequency of contact shrinkage and pruning events in the parameter-optimized model.[36] In the full STDP-based model[41] of Section 4, the correlation trace similarly grows with pre- and postsynaptic rates (cf. Eq. 12.2), leading to increased synaptic depression and eventually to pruning of contacts (through Eq. 12.1). Hence a homeostasis of the firing rate is inherent to these STDP-based structural plasticity models, but here arises as a consequence of the spike timing-dependent weight dynamics.

5.3 Future Experimental Studies

The spike timing-dependent model[41] presented in Section 4 combines synaptic weight plasticity and structural plasticity through the time evolution of the efficacy of each synaptic contact (dendritic spine). This dynamics depends on the correlations of pre- and postsynaptic activities, and is thus influenced by the state of all other synaptic connections in the network. The model reproduces many experimental features of spine plasticity and contact number distributions in the steady state. Because cooperation of contacts in the model is driven by localized correlation measurements at the spine, it may also explain findings on spatial clustering of spine locations on the dendrite[50–52]: local depolarization due to presynaptic activity is strongest at those dendritic branches which receive most contacts from a presynaptic neuron (Fig. 12.2). Therefore, contacts received from a single presynaptic neuron will tend to cluster on the same dendritic branch and avoid other branches. This prediction of the model warrants further elaboration and comparison to experimental results.

5.4 Future Modeling Studies

This and other models of structural plasticity cannot, at this stage, be comprehensive models of all aspects of synaptic plasticity. It remains desirable to explain both the long-term dynamics and steady state properties reported in spine imaging experiments (as done in Ref. [41]) and the various experimental paradigms of induction of long-term depression and potentiation in a model of the dynamics of the contact weights of the type of Eq. (12.1) by modifying it and by including further terms. Furthermore, it is currently unclear how multicontact excitatory plasticity interacts with inhibitory plasticity[53] in recurrent networks, whether inhibitory synapses are made of multiple plastic contacts as well, and how silent synapses[54] fit into the picture.

Phenomenological models of synaptic plasticity as the ones discussed here hold the promise to provide an initial understanding of the principles of biological neuronal network formation and maintenance. One needs to be aware, however, that biology is complex and that phenomenological models just represent an intermediate out of many possible levels of abstraction. Ultimately the detailed biomolecular mechanisms underlying synaptic plasticity should not be neglected, and it should be made sure that any presumed dynamics of a phenomenological model is consistent with, or may be derived from, the detailed biophysics of dendritic spines.[20]

Acknowledgments

I thank Martin Nawrot for supporting this project, and Moritz Helias and Alex Seeholzer for comments on the manuscript.

References

1. Harris KM, Kater SB. Dendritic spines: cellular specializations imparting both stability and flexibility to synaptic function. *Annu Rev Neurosci*. 1994;17:341–371.
2. Yasumatsu N, Matsuzaki M, Miyazaki T, Noguchi J, Kasai H. Principles of long-term dynamics of dendritic spines. *J Neurosci*. 2008;28:13592–13608.
3. Holtmaat A, Svoboda K. Experience-dependent structural synaptic plasticity in the mammalian brain. *Nat Rev Neurosci*. 2009;10:647–658.
4. Kwon HB, Sabatini BL. Glutamate induces de novo growth of functional spines in developing cortex. *Nature*. 2011;474:100–104.
5. Xu T, Yu X, Perlik AJ, Tobin WF, Zweig JA, et al. Rapid formation and selective stabilization of synapses for enduring motor memories. *Nature*. 2009;462:915–919.

6. Markram H, Lübke J, Frotscher M, Roth A, Sakmann B. Physiology and anatomy of synaptic connections between thick tufted pyramidal neurones in the developing rat neocortex. *J Physiol*. 1997;500(Pt 2):409–440.
7. Fares T, Stepanyants A. Cooperative synapse formation in the neocortex. *Proc Natl Acad Sci USA*. 2009;106: 16463–16468.
8. Escobar G, Fares T, Stepanyants A. Structural plasticity of circuits in cortical neuropil. *J Neurosci*. 2008;28: 8477–8488.
9. Stepanyants A, Hof PR, Chklovskii DB. Geometry and structural plasticity of synaptic connectivity. *Neuron*. 2002;34:275–288.
10. Knoblauch A, Palm G, Sommer FT. Memory capacities for synaptic and structural plasticity. *Neural Comput*. 2010;22:289–341.
11. Brunel N, Hakim V, Isope P, Nadal JP, Barbour B. Optimal information storage and the distribution of synaptic weights: perceptron versus purkinje cell. *Neuron*. 2004;43:745–757.
12. Chapeton J, Fares T, LaSota D, Stepanyants A. Efficient associative memory storage in cortical circuits of inhibitory and excitatory neurons. *Proc Natl Acad Sci USA*. 2012;109:E3614–E3622.
13. Kappel D, Habenschuss S, Legenstein R, Maass W. Network plasticity as bayesian inference. *PLoS Comput Biol*. 2015;11:e1004485.
14. Bourjaily MA, Miller P. Excitatory, inhibitory, and structural plasticity produce correlated connectivity in random networks trained to solve paired-stimulus tasks. *Front Comput Neurosci*. 2011;5:37.
15. Zheng P, Dimitrakakis C, Triesch J. Network self-organization explains the statistics and dynamics of synaptic connection strengths in cortex. *PLoS Comput Biol*. 2013;9:e1002848.
16. Miner D, Triesch J. Plasticity-driven self-organization under topological constraints accounts for non-random features of cortical synaptic wiring. *PLoS Comput Biol*. 2016;12:e1004759.
17. Butz M, van Ooyen A. A simple rule for dendritic spine and axonal bouton formation can account for cortical reorganization after focal retinal lesions. *PLoS Comput Biol*. 2013;9:e1003259.
18. Diaz-Pier S, Naveau M, Butz-Ostendorf M, Morrison A. Automatic generation of connectivity for large-scale neuronal network models through structural plasticity. *Front Neuroanat*. 2016;10:57.
19. Hebb DO. *The Organization of Behavior; a Neuropsychological Theory*. New York: Wiley; 1949.
20. Kotaleski JH, Blackwell KT. Modelling the molecular mechanisms of synaptic plasticity using systems biology approaches. *Nat Rev Neurosci*. 2010;11:239–251.
21. Dayan P, Abbott LF. *Theoretical Neuroscience: Computational and Mathematical Modeling of Neural Systems*. Cambridge: MIT Press; 2001.
22. Gerstner W, Kempter R, van Hemmen JL, Wagner H. A neuronal learning rule for sub-millisecond temporal coding. *Nature*. 1996;383:76–81.
23. Gerstner W, Kistler WM. *Spiking neuron models: single neurons, populations, plasticity*. Cambridge: Cambridge University Press; 2002.
24. Bi GQ, Poo MM. Synaptic modifications in cultured hippocampal neurons: dependence on spike timing, synaptic strength, and postsynaptic cell type. *J Neurosci*. 1998;18:10464–10472.
25. Sala C, Segal M. Dendritic spines: the locus of structural and functional plasticity. *Physiol Rev*. 2014;94:141–188.
26. Markram H, Lübke J, Frotscher M, Sakmann B. Regulation of synaptic efficacy by coincidence of postsynaptic aps and epsps. *Science*. 1997;275:213–215.
27. Lisman J. A mechanism for the hebb and the anti-Hebb processes underlying learning and memory. *Proc Natl Acad Sci USA*. 1989;86:9574–9578.
28. Graupner M, Brunel N. Stdp in a bistable synapse model based on camkii and associated signaling pathways. *PLoS Comput Biol*. 2007;3:e221.
29. Graupner M, Brunel N. Calcium-based plasticity model explains sensitivity of synaptic changes to spike pattern, rate, and dendritic location. *Proc Natl Acad Sci USA*. 2012;109:3991–3996.
30. Helias M, Rotter S, Gewaltig MO, Diesmann M. Structural plasticity controlled by calcium based correlation detection. *Front Comput Neurosci*. 2008;2:7.
31. Vlachos A, Helias M, Becker D, Diesmann M, Deller T. Nmda-receptor inhibition increases spine stability of denervated mouse dentate granule cells and accelerates spine density recovery following entorhinal denervation in vitro. *Neurobiol Dis*. 2013;59:267–276.
32. Feldmeyer D, Egger V, Lübke J, Sakmann B. Reliable synaptic connections between pairs of excitatory layer 4 neurones within a single "barrel" of developing rat somatosensory cortex. *J Physiol*. 1999;521(Pt 1):169–190.

33. Feldmeyer D, Lübke J, Silver RA, Sakmann B. Synaptic connections between layer 4 spiny neurone-layer 2/3 pyramidal cell pairs in juvenile rat barrel cortex: physiology and anatomy of interlaminar signalling within a cortical column. *J Physiol*. 2002;538:803–822.
34. Hessler NA, Shirke AM, Malinow R. The probability of transmitter release at a mammalian central synapse. *Nature*. 1993;366:569–572.
35. Nawrot MP, Schnepel P, Aertsen A, Boucsein C. Precisely timed signal transmission in neocortical networks with reliable intermediate-range projections. *Front Neural Circuits*. 2009;3:1.
36. Deger M, Helias M, Rotter S, Diesmann M. Spike-timing dependence of structural plasticity explains cooperative synapse formation in the neocortex. *PLoS Comput Biol*. 2012;8:e1002689.
37. Helias M, Deger M, Rotter S, Diesmann M. Instantaneous non-linear processing by pulse-coupled threshold units. *PLoS Comput Biol*. 2010;6.
38. Holtmaat AJGD, Trachtenberg JT, Wilbrecht L, Shepherd GM, Zhang X, et al. Transient and persistent dendritic spines in the neocortex in vivo. *Neuron*. 2005;45:279–291.
39. Fauth M, Wörgötter F, Tetzlaff C. The formation of multi-synaptic connections by the interaction of synaptic and structural plasticity and their functional consequences. *PLoS Comput Biol*. 2015;11:e1004031.
40. Fauth M, Wörgötter F, Tetzlaff C. Formation and maintenance of robust long-term information storage in the presence of synaptic turnover. *PLoS Comput Biol*. 2015;11:e1004684.
41. Deger M, Seeholzer A, Gerstner W. Multi-contact synapses for stable networks: a spike-timing dependent model of dendritic spine plasticity and turnover; 2016. arXiv:1609.05730 (preprint).
42. Kempter R, Gerstner W, van Hemmen JL. Intrinsic stabilization of output rates by spike-based hebbian learning. *Neural Comput*. 2001;13:2709–2741.
43. Deger M, Seeholzer A, Gerstner W. A spike timing dependent model of dendritic spine plasticity and turnover. Bernstein Conference 2016 Abstracts; 2016.
44. Eppler JM, Pauli R, Peyser A, Ippen T, Morrison A, et al. Nest 2.8.0. Zenodo; 2015.
45. Fusi S, Drew PJ, Abbott LF. Cascade models of synaptically stored memories. *Neuron*. 2005;45:599–611.
46. Clopath C, Ziegler L, Vasilaki E, Büsing L, Gerstner W. Tag-trigger-consolidation: a model of early and late long-term-potentiation and depression. *PLoS Comput Biol*. 2008;4:e1000248.
47. Zador A. Impact of synaptic unreliability on the information transmitted by spiking neurons. *J Neurophysiol*. 1998;79:1219–1229.
48. Zenke F, Hennequin G, Gerstner W. Synaptic plasticity in neural networks needs homeostasis with a fast rate detector. *PLoS Comput Biol*. 2013;9:e1003330.
49. Butz M, Wörgotter F, Van Ooyen A. Activity-dependent structural plasticity. *Brain Res Rev*. 2009;60:287–305.
50. Yadav A, Gao YZ, Rodriguez A, Dickstein DL, Wearne SL, et al. Morphologic evidence for spatially clustered spines in apical dendrites of monkey neocortical pyramidal cells. *J Comp Neurol*. 2012;520:2888–2902.
51. Druckmann S, Feng L, Lee B, Yook C, Zhao T, et al. Structured synaptic connectivity between hippocampal regions. *Neuron*. 2014;81:629–640.
52. Weber JP, Andrásfalvy BK, Polito M, Magó Á, Ujfalussy BB, et al. Location-dependent synaptic plasticity rules by dendritic spine cooperativity. *Nat Commun*. 2016;7:11380.
53. Vogels TP, Froemke RC, Doyon N, Gilson M, Haas JS, et al. Inhibitory synaptic plasticity: spike timing-dependence and putative network function. *Front Neural Circuits*. 2013;7:119.
54. Kerchner GA, Nicoll RA. Silent synapses and the emergence of a postsynaptic mechanism for ltp. *Nat Rev Neurosci*. 2008;9:813–825.

CHAPTER

13

Selection of Synaptic Connections by Wiring Plasticity for Robust Learning by Synaptic Weight Plasticity

Naoki Hiratani[1,2] *and Tomoki Fukai*[1,3]

[1]RIKEN Brain Science Institute, Wako, Japan [2]University College London, London, United Kingdom [3]Japan Science and Technology Agency, Kawaguchi, Japan

OUTLINE

The Rewiring Brain
DOI: http://dx.doi.org/10.1016/B978-0-12-803784-3.00013-5

1 INTRODUCTION

Synaptic connections are the fundamental substrate for learning and computation in the brain. Especially, synaptic weight, which is typically defined by the amplitude of excitatory postsynaptic potential (EPSP) at the soma caused by a single presynaptic spike, is often assumed to be the primal factor of information processing in the brain.[1] It is known that EPSP amplitude has large variability from synapse to synapse ranging from 0.1–10 mV,[2,3] which provides EPSPs a large capacity for information representation. In particular, the amplitude of large EPSP is typically stable against short-term fluctuation caused by synaptic noise,[4] although rather susceptible for long-term fluctuation due to structural noise,[5] making it a reasonable candidate for information storage. Indeed, representation with synaptic weights enables better information capacity compared to binary synapses in perceptron[6] and associative memory[7] models. In addition, a synaptic weight is often highly malleable through various synaptic plasticity mechanisms,[8] such as spike timing-dependent plasticity.[9] Consequently, previous theoretical studies often focused on how to modulate EPSPs by synaptic weight plasticity for performing a given function,[10] and rewiring of synaptic connections was often considered as a source of noise[11] rather than a functional importance, even when taken into account.

Nevertheless, experimental results in the last decade strongly indicate the functional importance of wiring plasticity through elimination and creation of synaptic connections, even in the brain of adult mammalian species.[12] Firstly, imaging studies of adult rodents have found active elimination and creation of dendritic spines in sensory[13] and motor cortices,[14] as well as in hippocampus.[5] In particular, spine turnover rate is up to 15% per day in the somatosensory cortex.[13] Moreover, performances in behavioral tasks are often correlated with spine dynamics in the associated regions.[15,16] For instance, it has been shown that the performance in motor learning tasks has a tight correlation with the ratio of persistent spines to the total spines created in the early phase of learning.[16] A recent result further revealed that optical shrinkage of newly created or potentiated spines during training impairs motor performance.[17] Taken together, it is safe to claim that wiring plasticity is a crucial substrate of learning and not inevitable obstacle, even in the adult brain.

Indeed, previous modeling studies have shown that wiring plasticity helps memory storage,[18,19] especially in collaboration with nonlinear dendritic computation.[20] However, in those studies, EPSP amplitude was often assumed to be a binary variable, so it was unclear whether the proposed functional roles are peculiar to wiring plasticity or could be substituted by synaptic weight plasticity. In particular, it remains unknown what should be encoded by synaptic connection structure when synaptic weights have a rich capacity for representation through activity-dependent plasticity mechanisms. In addition, it has not yet been well understood how local creation and elimination of spines, which is noisy and stochastic,[5] can generate a functional connection structure.

To illustrate the difference between representation by synaptic connectivity and synaptic weights, let us first consider a toy example. Suppose there is a feedforward network connecting two layers that consist of M and N neurons with connectivity ρ. Here, we define connectivity ρ as the ratio in number of synaptically connected neuron pairs to all the possible pairs. We ask what the information capacity is of this network and how much of that is represented by the connection structure. From a computational perspective,[18] information capacity of the connection structure can be defined as the logarithm of the total number of possible connection structures: $I_C(\rho) \equiv \log_{MN} C_{\rho MN} \approx MN \cdot H(\rho)$ where H represents the entropy function $H(\rho) \equiv -\rho\log\rho - (1-\rho)\log(1-\rho)$, and ${}_nC_k$ is the number of k-combinations in a size-n set. Similarly, for a given connection structure between the two layers, information capacity of synaptic weights is given by $I_W(\rho; b) = \rho MN\log b$, where b is the number of distinctive synaptic states[21] (e.g., a binary synapse has $b=2$ distinctive states). Therefore, the ratio of information capacities between connections and weights is characterized by $\Gamma(\rho; b) \equiv \exp\left(\frac{1}{MN}[I_C(\rho) - I_W(\rho; b)]\right)$. As shown in Fig. 13.1A, when connectivity is sparse, information capacity of connectivity tends to be larger than the capacity of weights at a wide range of parameter b (note that the y-axis is $\log_2 b$). This result suggests potential importance of wiring plasticity and resulting connection structures for the information processing in the brain, where neural connectivity is typically sparse.[22] For instance, CA3-to-CA1 connection of mice, where connection probability is around 6%,[23] and information capacity of synaptic weight is estimated to be around 4.7 bits,[24] is arguably operated at a regime where wiring plasticity is crucial (the point in Fig. 13.1A). Expectedly, the same argument holds for neocortical networks, considering that sparse connectivity is commonly observed in the neocortex.[22] It should also be noted that fluctuation in presynaptic firing rate δr has a similar effect as the fluctuation in synaptic weight δw because $(w+\delta w)r = w(r+\delta r)$ if $\delta r = r\delta w/w$. Hence, whenever information transmission through EPSPs is degraded by noise, nonrandom connection structure potentially provides functional benefits over unstructured connectivity.

We further investigate the functional role of wiring plasticity and resulting connection structure under rich synaptic weight plasticity by considering a functional circuit model in the following. Especially, we focus on a theoretical model of an inference task that commonly appears at various stages of neural information processing.[25,26] Our main results suggest that wiring plasticity helps output neurons to learn which presynaptic neurons they should listen to, rather than how much they should listen to each presynaptic neuron. We first review the essential findings of our previous work[27] and then show novel results for learning from input variance.

2 MODEL

2.1 Neural Model of an Inference Task

We considered an inference of a hidden state s from population activity $\boldsymbol{r}_X$ of neurons in the input layer (Fig. 13.1B). We denoted firing rates of input neurons as $\boldsymbol{r}_X$, and firing rates of output neurons as $\boldsymbol{r}_Y$ for clarity. Here, we assumed that a hidden state s^t is randomly sampled from prior distribution P_s, and each neuron j shows a stochastic response following a Gaussian distribution as $r^t_{X,j} \sim N(\theta_j(s^t), \sigma_{X,j})$, where $\theta_j(s^t)$ is the mean response, and $\sigma_{X,j}$ is the noise component. Note that the framework described below is applicable for a general exponential family, although analytical consideration would be difficult for non-Gaussian models.[27] For simplicity, we assumed that there are p discrete hidden states

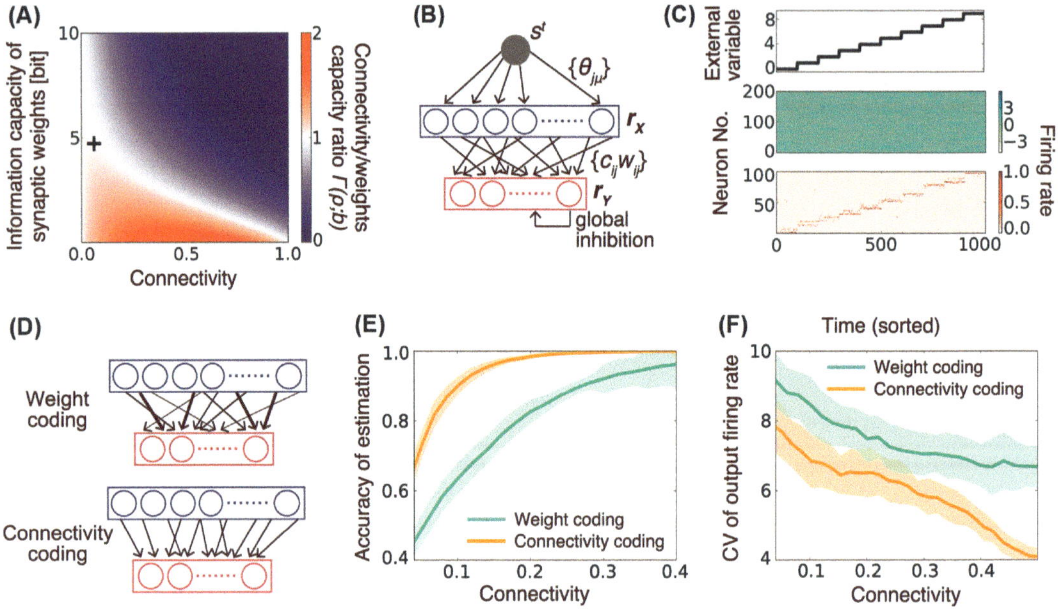

FIGURE 13.1 **Weight and connectivity coding for an inference from stochastic input activity.** (A) Information capacity ratio between synaptic connectivity and weight. *Plus* symbol represents the data point obtained from CA3-to-CA1 connections.[23,24] (B) Schematic diagram of the inference model. Here, s^t represents the external variable (*top*), from which firing rates of input neurons $\boldsymbol{r}_X$ are calculated as $r_{X,j} \sim N(\theta_{js^t}, \sigma_X)$ (*middle*). Output neurons (*bottom*) receive sparse synaptic projections $\{c_{ij}w_{ij}\}$ from input neurons and show firing activity $\boldsymbol{r}_Y$, where c_{ij} and w_{ij} represent the connectivity and weight of synaptic connection from input neuron i to output neuron j. (C) An example of the model behavior calculated at $\rho_o = 0.4$ under the weight coding. The top panel represents the external variable, which takes p discrete states ($p = 10$). Note that external variables were randomly sampled in the simulation, although here we sorted trials in ascending order for illustration purpose. The middle panel is the activity of input neurons, and the bottom panel represents the response of output neurons. Columns in the middle and the bottom panels correspond to the firing rates vector $\boldsymbol{r}_X$ and $\boldsymbol{r}_Y$, respectively. (D) Schematic representation of the weight coding and the connectivity coding. Let us say, vertical (short) connections are more informative than tilted (long) connections. In the weight coding (*top*), vertical connections are stronger (thicker) than tilted connections, whereas in the connectivity coding (*bottom*), there are more vertical connections than tilted connections. (E) Comparison of the performance between the connectivity coding and the weight coding. Throughout the chapter, the error bars represent the standard deviations over 50 independent simulations unless otherwise notified. (F) CV calculated for the firing rate of output neurons selective for the given external state. Here, we excluded low firing units ($r_{Y,i} < 10^{-6}$) from the calculation. *CV*, coefficient of variation.

($s^t = 0,1,\ldots,p-1$) and denoted the mean response for $s^t = \mu$ as $\theta_j(s^t = \mu) = \theta_{j\mu}$. In this problem setting, the posterior probability of $s^t = \mu$ is given as

$$p(s^t = \mu|\boldsymbol{r}_X) \propto p(s^t = \mu)\exp\left(-\sum_{j=1}^{M}(r^t_{X,j} - \theta_{j\mu})^2/2\sigma^2_{X,j}\right), \tag{13.1}$$

where M represents the total number of neurons in the input layer. In this study, we focused on how a layer of neurons receiving sparse projections from the input layer learns to perform this inference problem, especially on how connection structure between two layers should be modulated for achieving better decoding performance.

We defined neural dynamics with a log-linear model as

$$r^t_{Y,i} = r^o_Y \exp\left[\sum_{j=1}^M c_{ij}(w_{ij} r^t_{X,j} - h_w) - I^t_{inh}\right], I^t_{inh} = \log\left[\sum_{i=1}^N \exp\left(\sum_{j=1}^M c_{ij}[w_{ij} r^t_{X,j} - h_w]\right)\right], \quad (13.2)$$

where c_{ij} is the connectivity from input neuron j to output neuron i ($c_{ij} = 0$ or 1), w_{ij} the weight of the connection, h_w the threshold constant, $I^t{}_{inh}$ a global inhibition, N the number of neurons in the output layer, $r^t{}_{Y,i}$ the firing rate of output neuron i at time t, and $r^o{}_Y$ the maximum firing rate of output neurons. To represent the posterior distribution over p discrete states $p(s^t \mid \boldsymbol{r}_X)$ with the population activity $\boldsymbol{r}_Y$, each neuron in the output layer should represent the posterior probability of one of the states. Here, we denoted the set of output neurons that represent state $s = \mu$ as Ω_μ. For instance, if the firing activity of output neuron i follows $r^t_{Y,i} \propto p(s^t = \mu | r^t_X)$, neuron i belongs to Ω_μ ($i \in \Omega_\mu$).

If we assume $\frac{1}{M}\sum_{j=1}^M \theta^2_{j\mu}/\sigma^2_{X,j} = \alpha_o$ and $\log p(s^t = \mu)$:*const.*, then Eq. (13.1) is rewritten with respect to $\theta_{j\mu}$ as $\log p(s^t = \mu | \boldsymbol{r}_X) = \sum_{j=1}^M \theta_{j\mu} r^t_{X,j}/\sigma^2_{X,j} + const.$ Therefore, if the connection between the two layers is all-to-all, by setting the synaptic weight of a connection from input neuron j to the output neuron i as $w_{ij} = \theta_{j\mu}/\sigma^2_{X,j}$, the firing rate of the output neuron i satisfies $r^t_{Y,i} \propto p(s^t = \mu | \boldsymbol{r}_X)$ as long as $|\Omega_\mu| \approx N/p$ for all μ.

2.2 Coding Strategies for Sparsely Connected Networks

The weight and connectivity coding described in Fig. 13.1D were generated as follows: first, in both coding strategies, we randomly assigned one of the external states to each output neuron in such a way that all external states are represented by the same number of output neurons (i.e., $|\Omega_\mu| = N/p$ for all μ). In the weight coding, subsequently we set synaptic weights as $w_{ij} = \theta_{j\mu}/\rho_o \sigma^2_X$ for output neuron i representing external state μ (i.e., $i \in \Omega_\mu$) and generated a connectivity matrix such that $c_{ij} = 1$ with probability ρ_o and $c_{ij} = 0$ otherwise. Here, we denoted the mean connection probability as ρ_o. In the connectivity coding, we set synaptic weight as $w_{ij} = 1/\rho_o$, and the connections were set as $c_{ij} = 1$ with probability $\rho_{ij} = \rho_o \theta_{j\mu}/\sigma^2_X$ and $c_{ij} = 0$ otherwise. Note that in this configuration, the actual connectivity becomes smaller than ρ_o, because ρ_{ij} can be larger than 1, while the number of connections between two neurons is limited to 1 in our problem setting.

In all three coding strategies shown in Fig. 13.2, we first set the synaptic weight matrix as $w_{ij} = \theta_{j\mu}/\rho_o \sigma^2_{X,j}$, where $i \in \Omega_\mu$, because this weight configuration enables optimal inference at the dense connectivity limit (i.e., $\rho_o \to 1$) and is also easily attainable through a Hebbian learning rule. Based on these weights, in the cutoff coding strategy, for each output neuron i, we selected the $\rho_o M$ largest input synaptic weights and then connected the output neuron to the corresponding input neurons. In the dual-Hebbian coding strategy, for each output neuron i, we randomly picked up an input neuron j and then created a connection with probability $2\rho_o R_j/M$ until $\rho_o M$ connections were projected to the output neuron i, where R_j is the rank of the synaptic weight w_{ij} among all potential connections to the output neuron i. Here, we used a slightly different algorithm than the original connectivity coding explained above, to keep the input degree constant. Finally, in the random connectivity, we randomly selected input neurons, until all output neurons received $\rho_o M$ connections.

Three input configurations used in Fig. 13.2 were defined as below. Homogeneous input structure consisted of input neurons having identical noise variance $\sigma_{X,j} = \sigma_X$(fixed).

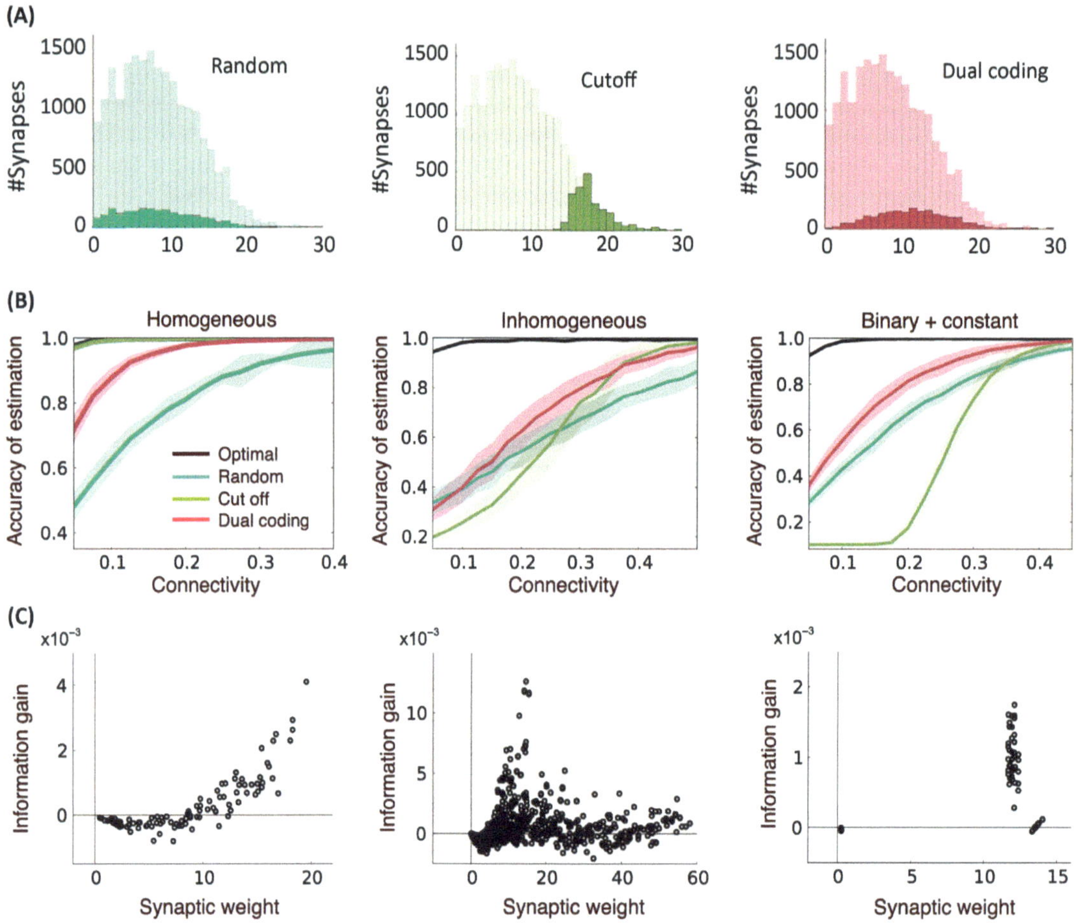

FIGURE 13.2 **Performances of different strategies for connectivity shaping.** (A) Synaptic weight distributions in random connection (*top*), cutoff (*middle*), and dual coding (*bottom*) strategies for the homogeneous input structure. Light colors represent possible connections (i.e., normalized distribution of optimal synaptic weights under all-to-all connections), whereas dark colors show the actual connections chosen under the different strategies. Connection probability was set at $\rho = 0.1$. (B) Comparisons of performance among different connection structure organizations for models with homogeneous (*left*) and inhomogeneous input variability (*middle*), and the model with binary or constant input structure (*right*; see Section 2.2 for the details of models). Note that black lines represent lower bounds for the optimal performance but not the exact optimal solutions. The error bars are the standard deviations over 100 simulations. (C) Relationships between the synaptic weight and the information gain per connection for three input configurations (see Section 2.3 for details). The middle panel shows 1000 synapses, whereas the left and the right panels depict 100 synapses.

Inhomogeneous input structure was generated by randomly setting the standard deviation of input firing rate as $\sigma_{X,j} = \bar{\sigma}_X d_\sigma e^{d_\sigma \zeta_U}/(e^{d_\sigma} - 1)$, where ζ_U is a random variable uniformly sampled from [0,1], and parameter d_σ controls the inhomogeneity of $\{\sigma_{X,j}\}$. Subsequently, the mean input responses were defined as $\tilde{\theta}_{j\mu} = \theta_o + d_\theta \sigma_{X,j}^2 \zeta_G$, and $\theta_{j\mu} = \tilde{\theta}_{j\mu} / \sqrt{\frac{1}{M}\sum_j \tilde{\theta}_{j\mu}^2 / \sigma_{X,j}^2}$, where ζ_G is a Gaussian random variable. We used $\overline{\sigma}_X = 1.0$, and $d_\sigma = 2.0$ in the simulation. In the "binary + constant" model, for randomly selected one-fourth of input neurons, we set $\tilde{\theta}_{j\mu} = 1.7\theta_o$ for all μ. The responses of the rest of the input neurons were defined as either $\tilde{\theta}_{j\mu} = 1.5\theta_o$ or $\tilde{\theta}_{j\mu} = 0.033\theta_o$ with 50% chance. The mean response variables were then normalized as $\theta_{j\mu} = \tilde{\theta}_{j\mu}/\sqrt{\frac{1}{M}\sum_j \tilde{\theta}_{j\mu}^2/\sigma_X^2}$.

2.3 Optimality of Connectivity

To evaluate information loss due to sparse connectivity, we calculated the posterior probability of the external states estimated from a sparse connection matrix C and then compared it to that from the fully connected network C_{all}. Here, we performed the comparison by considering the mean KL-divergence $I(C_{all}, C) \equiv \langle D_{KL}[p(s^t|\boldsymbol{r}_X, C_{all})||p(s^t|\boldsymbol{r}_X, C)] \rangle_{\boldsymbol{r}_X}$ over input activity $\boldsymbol{r}_X$, where KL-divergence $D_{KL}[p(x)||q(x)] \equiv \int p(x)\log[p(x)/q(x)]dx (\geq 0)$ gives a measure for similarity between two probabilistic distributions: it vanishes when two distributions $p(x)$ and $q(x)$ are identical. When the true external state is $s^t = \nu$, firing rates of input neurons are given by $r_{X,j}^t \sim N(\theta_{j\nu}, \sigma_X)$; hence the information loss $I(C_{all}, C)$ is approximately estimated as

$$
\begin{aligned}
I(C_{all}, C) &\approx \frac{1}{p}\sum_\nu \langle D_{KL}[p(s^t|\boldsymbol{r}_{X|\nu}, C_{all})||p(s^t|\boldsymbol{r}_{X|\nu}, C)] \rangle_{\boldsymbol{r}_{X|\nu}} \\
&\approx \frac{1}{p}\sum_\nu D_{KL}\left[\left\langle p\left(s^t|\left\{\theta_{\nu j} + \sigma_X\zeta_j\right\}, C_{all}\right)\right\rangle_{\{\zeta_j\}} || \left\langle p\left(s^t|\left\{\theta_{\nu j} + \sigma_X\zeta_j\right\}, C\right)\right\rangle_{\{\zeta_j\}}\right]
\end{aligned}
$$

where $\{\zeta_j\}$ are Gaussian random variables. By taking the integral over the random variables, the posterior probability is evaluated as,

$$
\begin{aligned}
&\left\langle p\left(s^t = \mu|\left\{\theta_{\nu j} + \sigma_X\zeta_j\right\}, C\right)\right\rangle_{\{\zeta_j\}} \\
&\quad = p(s^t = \mu)\left\langle \exp\left(\sum_j c_{ij}\left[2\theta_{j\mu}\left(\theta_{j\nu} + \sigma_X\zeta_j\right) - \theta_{j\mu}^2 - \left(\theta_{j\nu} + \sigma_X\zeta_j\right)^2\right]/2\sigma_X^2\right)\right\rangle_{\{\zeta_j\}} \\
&\quad \propto \left\langle \exp\left(\sum_j c_{ij}\left[2\theta_{j\mu}\theta_{j\nu} - \theta_{j\mu}^2\right]/2\sigma_X^2 + \sum_j c_{ij}\theta_{j\mu}\zeta_j/\sigma_X\right)\right\rangle_{\{\zeta_j\}} \\
&\quad \approx \exp\left(\sum_j c_{ij}\theta_{j\mu}\theta_{j\nu}/\sigma_X^2\right).
\end{aligned}
$$

Therefore, by denoting $p_\nu(s^t = \mu|C) \equiv \left\langle p(s^t = \mu|\{\theta_{\nu j} + \sigma_X\zeta_j\}, C)\right\rangle_{\{\zeta_j\}}$, the mean KL-divergence between estimations by two connection structures C_{all} and C is approximately given as

$$
I(C_{all}, C) \approx \frac{1}{p}\sum_\mu \sum_\nu p_\nu(s^t = \mu|C_{all})\frac{p_\nu(s^t = \mu|C_{all})}{p_\nu(s^t = \mu|C)}.
$$

Similarly, the information gain by a connection c_{ij} was evaluated by

$$\Delta I_{ij}(C) \equiv I(C_{all}, C) - I(C_{all}, C + \delta_{ij}),$$

where δ_{ij} is an $N \times M$ matrix in which only (i,j) element takes 1, and all other elements are 0. In Fig. 13.2B, the optimal performance was approximately evaluated through maximization of $I(C_{all}, C)$ with a hill-climbing method. In Fig. 13.2C, the mean information gain was evaluated by taking the average of $\Delta I_{ij}(C)$ over randomly generated 1000 connectivity matrices C.

2.4 Synaptic Weight Plasticity and Wiring Plasticity

To perform an accurate inference from noisy mixed inputs, synaptic weights and connectivity between the input and output layers should provide an inverse model of input neuronal activity generation. Because the accuracy of the inverse model is evaluated by $D_{KL}\left[p^*(\mathbf{r}_X^t)||p(\mathbf{r}_X^t|C, W)\right]$, unsupervised learning of synaptic weight or connectivity structure is achievable by minimizing this KL-divergence through gradient descendent methods.[28,29] The likelihood of input activity $p(\mathbf{r}_X^t|C, W)$ is given as

$$p(\mathbf{r}_X^t|C, W) \propto \sum\nolimits_{\mu} p(s^t = \mu)\exp\left[\sum\nolimits_j \left(\theta_{j\mu}^{C,W} r_{X,j}^t - \tfrac{1}{2}(\theta_{j\mu}^{C,W})^2 - \tfrac{1}{2}(r_{X,j}^t)^2\right)/\sigma_X^2\right],$$

where $\theta_{j\mu}^{C,W}$ is the estimated value of $\theta_{j\mu}$ from the weight matrix W and the connectivity matrix C. If we approximate $\theta_{j\mu}^{C,W}$ as $\theta_{j\mu}^{C,W} \approx \rho_o w_{ij}$ where $i \in \Omega_\mu$, by considering stochastic gradient on the KL-divergence, a synaptic weight plasticity rule is given as $\Delta w_{ij} \propto r_{Y,i}^t(r_{X,j}^t - \rho_o w_{ij})$. Similarly, by approximating $\theta_{j\mu}^{C,W}$ as $\theta_{j\mu}^{C,W} \approx \rho_{ij} w_o$, the update rule for connection probability is given as $\Delta \rho_{ij} \propto r_{Y,i}^t(r_{X,j}^t - \rho_{ij} w_o)$.

In the variance learning, synaptic weight plasticity was performed based on the mean variance $\overline{\sigma}_X^2$, instead of the true variance $\sigma_{X,j}^2$, so that the rule is written as

$$\Delta w_{ij}^t = (\eta_X/\rho_o)\left(r_{Y,i}^t[r_{X,j}^t - \overline{\sigma}_X^2 \rho_o w_{ij}^t] + b_h[r_Y^0/N - r_{Y,i}^t]\right),$$

where b_h is the relative strength of the homeostatic term. Because the variance of input neuron activity is approximately estimated with

$$\sigma_{X,j}^2 = \left\langle (r_{X,j}^t - \theta_{j\mu})^2 \right\rangle_{s^t=\mu} \approx \left\langle r_{Y,i}^t (r_{X,j}^t - \rho_o \overline{\sigma}_X^2 w_{ij})^2 \right\rangle_t,$$

for output neuron $i \in \Omega_\mu$, the learning rule of the variance was defined as

$$m_{ij}^{t+1} = m_{ij}^t + \eta_\rho r_{Y,i}^t[(r_{X,j}^t - \overline{\sigma}_X^2 \rho_o w_{ij})^2 - m_{ij}^t].$$

2.5 Details of Simulations

In the simulation, the external variable s^t was sampled from 10 discrete states ($p = 10$) with equal probability. The mean response probability $\{\theta_{j\mu}\}$ was generated by

$\theta_{j\mu} = \tilde{\theta}_{j\mu} / \sqrt{\frac{1}{M}\sum_j \tilde{\theta}_{j\mu}^2 / \sigma_X^2}$, where $\{\tilde{\theta}_{j\mu}\}$ are random variables sampled from a truncated Gaussian distribution $N_+(\theta_o, d_\theta)$. The threshold constant h_w was defined as $h_w = \frac{1}{pM}\sum_j \sum_\mu \theta_{j\mu}/\rho_o$. For numerical stability, we introduced a membrane potential variable $v_i^t = \sum_j c_{ij}(w_{ij} r_{X,j}^t - h_w)$ and then set the output firing rate as $r_{Y,i}^t = \exp(v_i^t - I_{inh}^t)$ if $v_i^t > \max_{i'}(v_{i'}^t - v_d)$ and $r_{Y,i}^t = \exp(-v_d - I_{inh}^t)$ otherwise. Here, v_d is the effective range of the membrane variable v_i^t: $\max_{i'}(v_{i'}^t - v_d) > -v_d$. We also excluded output neurons with $v_i^t \leq \text{maxi}(v_i^t - v_d)_i$ from the calculation of I^t_{inh}. Except for Fig. 13.4 and the inhomogeneous model in Fig. 13.2, the same noise amplitude was used for all input neurons (i.e., $\sigma_{X,j} = \sigma_X$). In all rewiring plasticity model, the weight of a newly created spine was set as $w_{ij} = (h_w + \sigma_w \zeta_G)/\sigma_X^2$ where ζ_G is a Gaussian random variable. For model parameters, we used $\theta_o = 1.0$, $d_\theta = 1.0$, $M = 200$, $N = 100$, $v_d = 10.0$, $\sigma_X = 1.0$, and for the learning parameters, we used $\eta_w = 0.01$, $\eta_\rho = 0.001$, $b_h = 0.1$, $\tau_c = 10^6$, and $\sigma_w = 0.1$, unless otherwise stated.

The accuracy of estimation was measured with a bootstrap method. To evaluate the performance around time t, we first determined the selectivity of each output neuron by using firing rates from $t - T_o \leq t' < t$, where T_o is the number of observations used for calculating selectivity. We defined that neuron i belongs to set Ω_μ at time t if activity of neuron i satisfies

$$\mu = \underset{\mu'}{\operatorname{argmax}} \sum_{\tau=t-T_o}^{t} [s^\tau = \mu']_+ r_{Y,i}^\tau / \sum_{\tau=t-T_o}^{t} [s^\tau = \mu']_+ \Big),$$

where operator $[X]_+$ returns 1 if X is true; otherwise, it returns 0. By using this selectivity, based on activity data from $t \leq t' < t + T_o$, the accuracy of inference was estimated as

$$\frac{1}{T}\sum_{\tau=t}^{t+T_o-1} \left[\frac{1}{|\Omega_{s^\tau}|} \sum_{i\in\Omega_{s^\tau}} r_{Y,i}^\tau > \max_{\mu \neq s^\tau} \frac{1}{|\Omega_\mu|} \sum_{i\in\Omega_\mu} r_{Y,i}^\tau \right]_+ .$$

In the simulation, we used $T_o = 10^3$, because this value is sufficiently short compared with weight change but long enough to suppress variability.

3 RESULTS

3.1 Coding by Synaptic Connections Enables Signal Variability Reduction

Let us consider an inference task with a feedforward network (Fig. 13.1B). Inference from noisy presynaptic activity is not only crucial for stimulus detection[26] but also an essential part of decision-making[25] or motor planning.[30] In the task, at ever time t, one hidden state s^t is sampled with equal probability from p number of external states (Fig. 13.1C, top). Neurons in the input layer show independent stochastic response $r_{X,j}^t \sim N(\theta_{j\mu}, \sigma_X)$ (Fig. 13.1C, middle), where $r^t_{X,j}$ is the firing rate of input neuron j at time t, $\theta_{j\mu}$ is the average firing rate of neuron j to the stimulus μ ($s^t = \mu$), and σ_X is the noise amplitude. Neurons in the output layer estimate the hidden variable from input neuron activity and represent the variable with population firing $\boldsymbol{r}_Y^t$ (Fig. 13.1C, bottom). This task is computationally difficult because most input neurons have mixed selectivity for several hidden inputs, and the responses of the input neurons are highly stochastic. The dynamics of output neurons obeys the log-linear model given in Eq. (13.2), in which the connection probability c_{ij} (=0 or 1) between input neuron i and output neuron j and the corresponding synaptic weight w_{ij} are treated as independent variables (see Section 2.1 for the details of the model).

In this decoding model, if the connection is all-to-all, by setting synaptic weights as$w_{ij} = \theta_{j\mu}/\sigma_X^2$, the output firing activity $\boldsymbol{r}_Y$ actually provides the optimal inference from given input $\boldsymbol{r}_X$ (i.e., the activity of output neuron i satisfies $r_{Y,i}^t \propto p(s^t = \mu|\boldsymbol{r}_X^t)$ for the corresponding external state μ; see Section 2.1 for the details). Note that, here we manually assigned one external state for each output neuron in such a way that each state is represented by N/p output neurons. On the contrary, if the connectivity between the two layers is sparse, as in most regions of the brain,[22] the optimal inference is generally unattainable because of information loss caused by the sparseness. How should we organize synaptic connectivity and their weights when the connection probability is limited to $\rho_o < 1$? Intuitively, we may randomly dilute the connections while renormalizing the synaptic weights with $w_{ij} = \theta_{j\mu}/\rho_o\sigma_X^2$ to obtain near optimal inference (see Section 2.2 for details). Let us denote this coding strategy as the weight coding, because in this strategy, all the information is carried by synaptic weights and not by synaptic connectivity (Fig. 13.1D, top). The opposite example would be the case when synaptic connectivity is employed for the representation, and synaptic weights are kept constant (Fig. 13.1D, bottom). This connectivity coding strategy can be achieved by setting the connection probability between neuron i and j as $\rho_{ij} = \rho_o\theta_{j\mu}/\sigma_X^2$, while keeping all the synaptic weights at $w_{ij} = 1/\rho_o$. We first compare these two realizations to investigate whether coding by connectivity provides any benefit over the weight coding. Note that both weight and connectivity coding strategies asymptotically converge to the optimal solution when the number of neurons in the middle layer becomes sufficiently large. However, they are not strictly optimal under given constraints.

In both coding strategies, sparse connectivity degrades the performance because the output neurons gain less information on the hidden external states under sparse connectivity (Fig. 13.1E). Here, the performance was evaluated by the accuracy of estimation through a bootstrap method (see Section 2.5). Surprisingly, under a sparse connectivity, the connectivity coding outperforms the weight coding despite its binary representation (Fig. 13.1E). This is because, in the connectivity coding, the output neurons can receive inputs from relevant input neurons, whereas in the weight coding, the selections of input neurons are random although their weights are well curated. Indeed, the coefficient of variation of output neuron activity is lower in the connectivity coding, suggesting more reliable information transmission (Fig. 13.1F). These results indicate a potential advantage of wiring plasticity over weight plasticity in the sparse connectivity.

3.2 Dual Coding by Synaptic Weights and Connections Enables Robust Inference

Results in the previous section have shown that random sparse connections disturb information transfer between the input and output layers so much, so that the connectivity coding with fixed synaptic weights outperforms the weight coding with random connectivity. Hence, both connectivity and weights should be neatly composed to achieve robust inference in the sparse regime. Ideally, we should choose the connection structure that minimizes the information loss due to sparse connectivity for a given synaptic weights matrix. However, the optimal solution for this minimization problem is typically unattainable due to combinatorial explosion[31] (see Section 2.3). Therefore, here we compare heuristic solutions in the inference task. One candidate solution is the cutoff strategy,[11,32] in which only the connections with large synaptic weights are kept, and the rest of connections are eliminated from a network with all-to-all connectivity (Fig. 13.2A, middle).

Another potential solution is to combine weight and connectivity coding introduced previously. Recent experimental results suggest that synaptic weights and connectivity often represent similar features. For instance, the EPSP size of a connection in a clustered network is typically larger than the average EPSP size,[4,33] and a similar tendency is also observed in interlayer connections.[34,35] Hence, dual coding by weight and connectivity structure may provide some benefits over the cutoff strategy. In this dual coding strategy, the synaptic weight distribution of surviving connections show a broad distribution as observed in experiments,[4,3] in contrast to the sharp synaptic weight distribution in the cutoff strategy (Fig. 13.2A, bottom and middle).

We further investigate how networks organized by these two strategies, as well as a network with random connectivity, perform the task in various input configurations. When the noise level of input neurons is homogeneous (i.e., $\sigma_{X,j} = \sigma_X$: *const.*) as in the previous section, networks with the cutoff coding strategy perform almost optimally, although the dual coding also improves the performance compared to networks with the random connectivity (Fig. 13.2B, left). By contrast, when the noise level is inhomogeneous, and the response properties of the input neurons are positively correlated with their noise level (see Section 2.2 for details), the performance of the cutoff strategy is worse than that of the random connectivity at the sparse regime, whereas the dual coding strategy constantly outperforms the rest (Fig. 13.2B, middle). To further illustrate this phenomenon, we next consider the case where one-fourth of input neurons show a constant high response to all of the external states, while the rest of the input neurons show high response only to randomly selected half of the external states (we denote this as "binary + constant" model). In this example, although the cutoff strategy shows a near-optimal performance under a dense connectivity, it fails to perform the inference at the sparse regime (Fig. 13.2B, right). In contrast, the performance of the dual coding stays at a relatively high level even at sparse connectivity.

To get insight into the observed robust performance of the dual coding, we next study the relationship between synaptic weight w_{ij} and the information gain caused by that synaptic connection ΔI_{ij} (see Section 2.3 for the definition). When the noise level in the input layer is homogeneous, information gain grows approximately linearly with synaptic weights (Fig. 13.2C, left). Thus, the maximum information gain is achieved by keeping connections with large synaptic weights, while removing the rest of connections. Indeed, the performance of the cutoff strategy is almost optimal for the homogeneous input structure. However, under the inhomogeneous noise, connections with large synaptic weights tend to have small information gains, whereas some connections with small synaptic weights attain large information gains (Fig. 13.2C, middle). Hence, the performance is better when connections with various weights are sampled through the dual coding strategy. Similarly, when connections with large weights have zero information gains (Fig. 13.2C right), the dual coding has an advantage over other strategies. These results suggest that the dual coding strategy by synaptic weights and connectivity enables robust inference for various input structure because of robust information gain.

3.3 Hebbian Wiring Plasticity Prevents Information Loss by Spine Elimination

Thus, combining the weight coding and connectivity coding enables the network to robustly perform inference from input with various structures. However, in the previous section, specific connection and weight structures were not achieved by the remodeling of neural circuits by weight and wiring plasticity but given a priori. Thus, we next consider how local spine creation and elimination generate a functional wiring plasticity.

Let us first consider a local learning rule of synaptic weights. Suppose, output neuron i represents state μ, then synaptic weight w_{ij} should converge to $w_{ij} \to \theta_{j\mu}/(\sigma_X^2 \rho_o) = \left\langle r_{X,j}^t r_{Y,i}^t / \left(\sigma_X^2 \rho_o r_{Y,i}^t\right)\right\rangle$ (see Section 2.4 for details). Therefore, a synaptic weight change rule is given as

$$\Delta w_{ij}^t = (\eta_X/\rho_o)(r_{Y,i}^t[r_{X,j}^t - \sigma_X^2 \rho_o w_{ij}^t] + b_h[r_Y^0/N - r_{Y,i}^t]), \tag{13.3}$$

where $\Delta w_{ij}^t = w_{ij}^{t+1} - w_{ij}^t$. Here, the homeostatic term was added to constrain the average firing rates of output neurons. This weight-learning rule alone can indeed improve the decoding performance in a sparsely connected network up to some extent (cyan line in Fig. 13.3B).

We next investigate the learning of connection structures by synaptic wiring plasticity. As expected, if we randomly perform creation and elimination of synaptic connections, the performance gets worse compared to the fixed sparse connection, because random rewiring causes instability in representation by output neurons (orange line vs cyan line in Fig. 13.3B). Instead, here we introduce the connection probability variable ρ_{ij} for all potential connections and then perform rewiring based on this value[36] (e.g., in random rewiring, $\rho_{ij}=\rho_o$: fixed). If we stochastically create a connection between neurons (i,j) with probability ρ_{ij}/τ_c and eliminate the connection with probability $(1-\rho_{ij})/\tau_c$, as long as c_{ij} is bounded by 1, the average connection probability stays at ρ_{ij}. To achieve connectivity coding with this rewiring strategy, the connection probability should be modified as

$$\Delta \rho_{ij}^t = \eta_\rho r_{Y,i}^t[r_{X,j}^t - \sigma_X^2 \rho_{ij}^t w_o], \tag{13.4}$$

where $\Delta \rho_{ij}^t \equiv \rho_{ij}^{t+1} - \rho_{ij}^t$, and w_o is the expected mean synaptic weight (see Section 2.4 for details).

Fig. 13.3A shows the typical behavior of ρ_{ij} and w_{ij} under the combination of this rewiring rule and the weight update rule (i.e., Eqs. (13.3) and (13.4); we denote this combination as "dual-Hebbian" below). When the connection probability is low, connections between two neurons are rare, and even when a spine is created stochastically, the created spine typically undergoes rapid elimination (Fig. 13.3A, top). In the moderate connection probability, spine creation is more frequent, and created spine survives longer (Fig. 13.3A, middle).

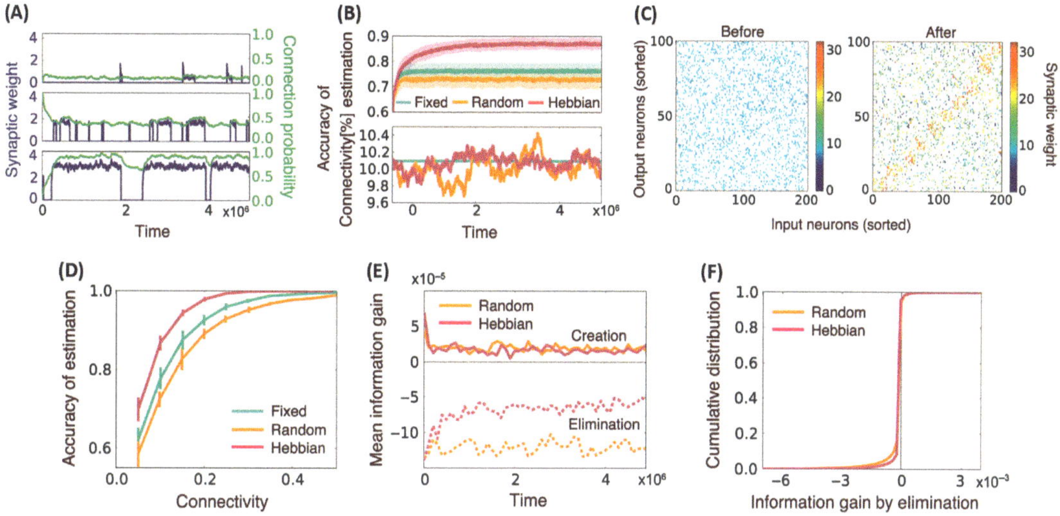

FIGURE 13.3 **Dynamics under Hebbian weight and wiring plasticity.** (A) Examples of spine creation and elimination. In all three panels, *blue lines* show synaptic weights w and *green lines* are connection probability ρ. When there is not a synaptic connection between two neurons, the synaptic weight becomes zero, but the connection probability can take a non-zero value. Simulation was calculated at $\rho = 0.5$, and $\tau_c = 10^5$. (B) Development of performance (*top*) and the connectivity (*bottom*) under different rewiring strategies. In all three models, synaptic weights were modulated by Eq. (13.3). (C) Synaptic weight matrices before (*left*) and after (*right*) learning. Both x- and y-axes were sorted based on their preferred external states. (D) Comparison of performance after the learning at various degree of connectivity. (E) Mean information gain provided by creation of a synaptic connection (*solid lines*) and by elimination of a connection (*dotted lines*) under random and Hebbian rewiring. Note that spine elimination typically causes information loss, so that the gain is on an average negative (see Section 2.3 for details). (F) Cumulative distribution of information gain by spine elimination in the two rewiring strategies.

When the connection probability is high enough, there is almost always a connection between the two neurons and the synaptic weight of the connection is large, because here synaptic weight dynamics also follow a similar Hebbian rule (Fig. 13.3A, bottom).

This rewiring rule indeed performs better than the learning with weight plasticity in a fixed connection unlike random rewiring (Fig. 13.3B,D). After the learning process, informative connections (diagonal components of the sorted weight matrix in Fig. 13.3C right) develop relatively large synaptic weights, at the same time, denser connectivity than other connections. Comparing the Hebbian rewiring rule with the random rewiring, information gain by a newly created connection is nearly the same in the two rules except for the very beginning of learning (solid lines in Fig. 13.3E). However, information loss (i.e., negative information gain) by elimination is smaller in the Hebbian rewiring than in the random rewiring even in the steady state (dotted lines in Fig. 13.3E). About 20% of the spine elimination in the random rewiring causes information loss, whereas less than 10% of spine elimination affects the information transfer in the Hebbian rewiring (Fig. 13.3F). These results suggest that selective elimination is crucial for achieving and sustaining functional connectivity.

3.4 Variance Learning With Synaptic Wiring Plasticity

In the arguments above, we assumed that all input neurons have the same variance σ_X^2, and the output neurons know the exact value of σ_X^2. However, in the brain, the stochasticity of firing is possibly different from neuron to neuron (i.e., $\sigma_{X,j} \neq \sigma_{X,m}$) because the individual neurons may represent many different features that appear as the noise in the decoding of one feature.[37,38] In addition, presumably the output neurons do not know the variances of presynaptic firings a priori. How should synaptic rewiring be performed under such variability?

Fig. 13.4A illustrates the effect of such variability. Here, neurons have smaller (larger) variances than those in lower (upper) panels. In this case, the output neurons should receive connections from the input neurons with the smallest variability (in Fig. 13.4A, neurons on the top) to suppress the noise, regardless of the selectivity of these input neurons. However, how can a synapse learn the variance? One possible way is to use the following two-step learning rule for connectivity:

$$m_{ij}^{t+1} = m_{ij}^t + \eta_\rho r_{Y,i}^t[(r_{X,j}^t - \overline{\sigma}_X^2 \rho_o w_{ij})^2 - m_{ij}^t],\ \rho_{ij}^{t+1} = \rho_o \overline{\sigma}_X^2 / m_{ij}^{t+1}, \tag{13.5}$$

where m_{ij} is an online estimation of the variance of input neuron j, and $\overline{\sigma}_X$ is the mean standard deviation of input firing activity (see Section 2.4 for details). Below, we denote the combination of Eqs. (13.3) and (13.5) as "variance learning." Under this variance learning, connection probability develops a negative correlation with the variance of input neurons, so connections are preferably made from less noisy input neurons (Fig. 13.4B). As a result, the variance learning eventually outperforms synaptic weight learning with fixed connectivity when compared at the same connection probability (Fig. 13.4C).

This variance rule is especially useful when the input structure $\{\theta_{j\mu}\}$ changes completely at a certain interval, while the noise levels of input neurons $\{\sigma_{X,j}\}$ stay the same. In this case, the network with the variance-learning rule slowly develops preference for the input neurons with smaller variances, although at each change in the input structure, performance goes back to a baseline level (Fig. 13.4D). As a result, the network shows faster convergence to better performance for each input structure (Fig. 13.4E: note that here the timescale of x-axis is not 10^6 but 10^4). As observed in Fig. 13.4B, synaptic connectivity shows a clear preference for input neurons with low input noise (green line in Fig. 13.4F) but not for input neurons with strong selective responses (green line in Fig. 13.4G). However, the learned synaptic weights are somewhat positively correlated with input noise (blue line in Fig. 13.4F) while showing a clear preference for input neurons with strong selective responses (i.e., $w_{ij} \propto \theta_{j\mu}$ if $i \in \Omega_\mu$; blue line in Fig. 13.4G). These results indicate that in the variance-learning rule, wiring plasticity generates a noise-resilient connection structure to help learning with synaptic weight plasticity.

4 DISCUSSION

In this chapter, we have investigated some potential mechanisms in which synaptic wiring plasticity cooperates with synaptic weight plasticity in learning of the structure of

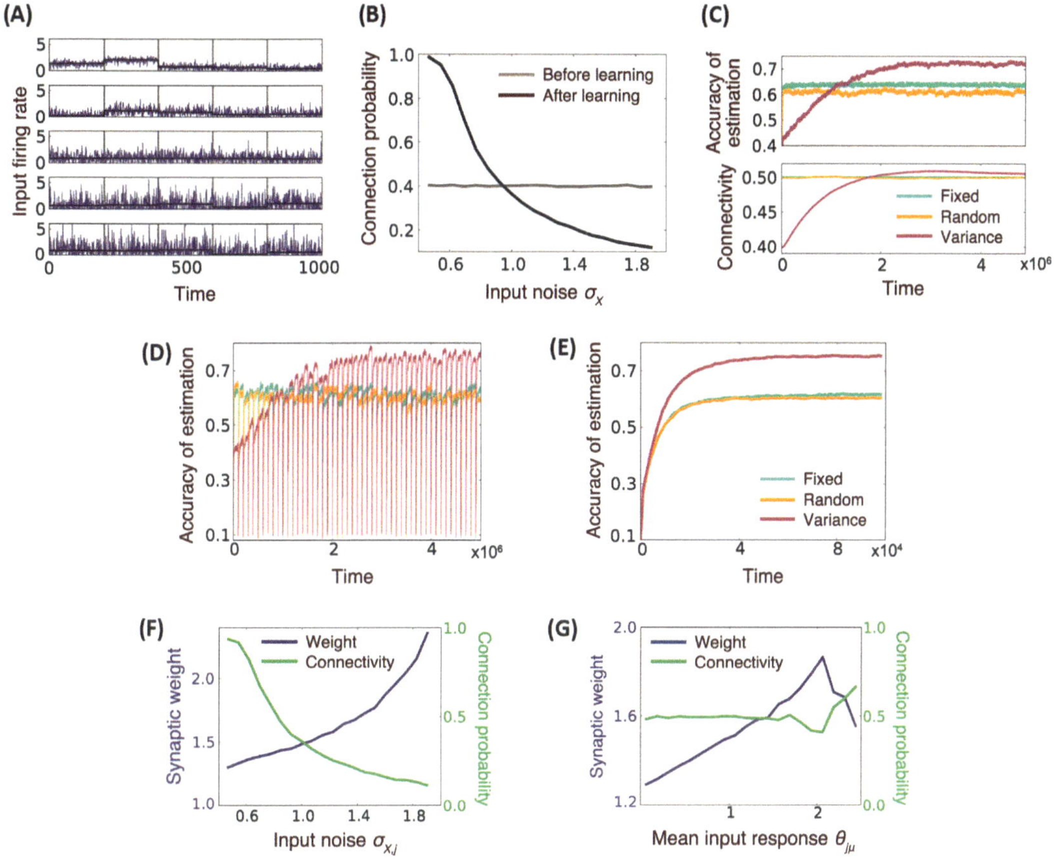

FIGURE 13.4 **Variance-based wiring plasticity enables synaptic weight learning from reliable presynaptic neurons.** (A) Examples of input neuron activity for five different external states. Gray vertical bars separate responses for different external states. Input neurons were sorted in such a way that the neuron on the top shows the smallest variance. (B) Connection probability from input neurons with various input noise levels before and after learning. (C) Development of performance and the connectivity. We set $\rho_o = 0.5$ for the networks with fixed connectivity and random rewiring, and $\rho_o = 0.4$ for the network with variance learning. (D, E) Learning curves of the three learning rules when the whole input structure $\{\theta_{j\mu}\}$ changes every 10^5 time steps, while the noise levels $\{\sigma_{X,j}\}$ are kept constant. Panel (D) describes the entire learning curves, while panel (E) describes the average behavior for a new input structure after 3×10^6 steps of learning. (F, G) Dependence of synaptic weights and connection probability on the noise level (F) and the mean response (G) of input neurons.

external stimuli. We have first demonstrated that when connectivity is sparse, representation by synaptic weights is far from optimal due to enhanced signal variability (Fig. 13.1); we have then shown that incorporation of representation by connectivity structure enables robust inference against inhomogeneity in the input layer (Fig. 13.2). Subsequently, our model has revealed that Hebbian wiring plasticity prevents information loss by stochastic synaptic elimination (Fig. 13.3). We have also shown that wiring plasticity enables learning with weight plasticity from reliable input neurons, when synaptic rewiring is performed based on the variance of input neuron activity (Fig. 13.4).

4.1 Future Experimental Studies

Regarding synaptic elimination, it is known that small spines[5] and spines that recently experienced LTD[39] are more likely to be eliminated. By contrast, exact conditions for the creation of new synapses are still not yet well understood. Our model suggests that spine creation may not necessarily be selective, as long as spine elimination is activity-dependent (Fig. 13.3). In addition, our model indicates that spine elimination based on stochasticity of presynaptic activity is functionally beneficial especially in a rapidly changing environment (Fig. 13.4). Correspondingly, it is suggested that a synapse with low presynaptic release probability is more susceptible to elimination after LTD.[39] Further studies are awaited on the presynaptic contribution for the elimination and creation of a synaptic contact.

4.2 Future Theoretical Studies

In a recent connectomics study,[40] it was revealed that there are many multiple connections between identified axon–dendrite pairs even in a small portion of a dendritic tree in the somatosensory cortex. Moreover, from computational reconstruction,[41] Markram and colleagues estimated that in the barrel cortex most interneuronal connections are actually realized by multiple synapses, and the mean number of synapses per connection is around 10. These results suggest that synaptic connections are much more redundant than we previously thought they were. However, little is known about their functional roles. From the information theoretic perspective discussed in this chapter, multiple connections do not seem to provide much advantage. In particular, if synaptic transmission is reliable (i.e., if synaptic failure is rare), the amount of information a postsynaptic neuron can obtain remains mostly the same regardless of the number of total synaptic connections. Further theoretical investigation should be performed on this issue.

Acknowledgment

This work was partly supported by Grants-in-Aid for Scientific Research (KAKENHI) from MEXT (no 15H04265 and 16H01289 to TF) and CREST grant (no JPMJCR13W1) from Japan Science and Technology Agency (JST).

References

1. Bliss TV, Collingridge GL. A synaptic model of memory: long-term potentiation in the hippocampus. *Nature.* 1993;361(6407):31–39.
2. Song S, Sjöström PJ, Reigl M, Nelson S, Chklovskii DB. Highly nonrandom features of synaptic connectivity in local cortical circuits. *PLoS Biol.* 2005;3(3):e68.
3. Ikegaya Y, Sasaki T, Ishikawa D, et al. Interpyramid spike transmission stabilizes the sparseness of recurrent network activity. *Cereb Cortex.* 2013;23(2):293–304.

4. Lefort S, Tomm C, Floyd Sarria J-C, Petersen CCH. The excitatory neuronal network of the C2 barrel column in mouse primary somatosensory cortex. *Neuron*. 2009;61(2):301–316.
5. Yasumatsu N, Matsuzaki M, Miyazaki T, Noguchi J, Kasai H. Principles of long-term dynamics of dendritic spines. *J Neurosci Off J Soc Neurosci*. 2008;28(50):13592–13608.
6. Brunel N, Hakim V, Isope P, Nadal J-P, Barbour B. Optimal information storage and the distribution of synaptic weights: perceptron versus Purkinje cell. *Neuron*. 2004;43(5):745–757.
7. Hiratani N, Teramae J-N, Fukai T. Associative memory model with long-tail-distributed Hebbian synaptic connections. *Front Comput Neurosci*. 2013;6:102.
8. Feldman DE. Synaptic mechanisms for plasticity in neocortex. *Annu Rev Neurosci*. 2009;32:33–55.
9. Caporale N, Dan Y. Spike timing–dependent plasticity: a Hebbian learning rule. *Annu Rev Neurosci*. 2008;31(1):25–46.
10. Brea J, Gerstner W. Does computational neuroscience need new synaptic learning paradigms? *Curr Opin Behav Sci*. 2016;11:61–66.
11. Chechik G, Meilijson I, Ruppin E. Synaptic pruning in development: a computational account. *Neural Comput*. 1998;10(7):1759–1777.
12. Holtmaat A, Svoboda K. Experience-dependent structural synaptic plasticity in the mammalian brain. *Nat Rev Neurosci*. 2009;10(9):647–658.
13. Holtmaat AJGD, Trachtenberg JT, Wilbrecht L, et al. Transient and persistent dendritic spines in the neocortex in vivo. *Neuron*. 2005;45(2):279–291.
14. Zuo Y, Lin A, Chang P, Gan W-B. Development of long-term dendritic spine stability in diverse regions of cerebral cortex. *Neuron*. 2005;46(2):181–189.
15. Xu T, Yu X, Perlik AJ, et al. Rapid formation and selective stabilization of synapses for enduring motor memories. *Nature*. 2009;462(7275):915–919.
16. Yang G, Pan F, Gan W-B. Stably maintained dendritic spines are associated with lifelong memories. *Nature*. 2009;462(7275):920–924.
17. Hayashi-Takagi A, Yagishita S, Nakamura M, et al. Labelling and optical erasure of synaptic memory traces in the motor cortex. *Nature*. 2015;525(7569):333–338.
18. Stepanyants A, Hof PR, Chklovskii DB. Geometry and structural plasticity of synaptic connectivity. *Neuron*. 2002;34(2):275–288.
19. Knoblauch A, Palm G, Sommer FT. Memory capacities for synaptic and structural plasticity. *Neural Comput*. 2010;22(2):289–341.
20. Poirazi P, Mel BW. Impact of active dendrites and structural plasticity on the memory capacity of neural tissue. *Neuron*. 2001;29(3):779–796.
21. Varshney LR, Sjöström PJ, Chklovskii DB. Optimal information storage in noisy synapses under resource constraints. *Neuron*. 2006;52(3):409–423.
22. Potjans TC, Diesmann M. The cell-type specific cortical microcircuit: relating structure and activity in a full-scale spiking network model. *Cereb Cortex*. 2014;24(3):785–806.
23. Sayer RJ, Friedlander MJ, Redman SJ. The time course and amplitude of EPSPs evoked at synapses between pairs of CA3/CA1 neurons in the hippocampal slice. *J Neurosci*. 1990;10(3):826–836.
24. Bartol TM, Bromer C, Kinney JP, et al. Nanoconnectomic upper bound on the variability of synaptic plasticity. *eLife*. November 2015;:e10778.
25. Beck JM, Ma WJ, Kiani R, et al. Probabilistic population codes for Bayesian decision making. *Neuron*. 2008;60(6):1142–1152.
26. Lochmann T, Deneve S. Neural processing as causal inference. *Curr Opin Neurobiol*. 2011;21(5):774–781.
27. Hiratani N, Fukai T. Hebbian wiring plasticity generates efficient network structures for robust inference with synaptic weight plasticity. *Front Neural Circuits*. 2016;10:41.
28. Dayan P, Hinton GE, Neal RM, Zemel RS. The Helmholtz machine. *Neural Comput*. 1995;7(5):889–904.
29. Nessler B, Pfeiffer M, Buesing L, Maass W. Bayesian computation emerges in generic cortical microcircuits through spike-timing-dependent plasticity. *PLoS Comput Biol*. 2013;9(4):e1003037.
30. Körding KP, Wolpert DM. Bayesian decision theory in sensorimotor control. *Trends Cogn Sci*. 2006;10(7):319–326.
31. Donoho DL. Compressed sensing. *IEEE Trans Inf Theory*. 2006;52(4):1289–1306.
32. Navlakha S, Barth AL, Bar-Joseph Z. Decreasing-rate pruning optimizes the construction of efficient and robust distributed networks. *PLoS Comput Biol*. 2015;11(7):e1004347.

33. Perin R, Berger TK, Markram H. A synaptic organizing principle for cortical neuronal groups. *Proc Natl Acad Sci*. 2011;108(13):5419–5424.
34. Yoshimura Y, Dantzker JLM, Callaway EM. Excitatory cortical neurons form fine-scale functional networks. *Nature*. 2005;433(7028):868–873.
35. Ryan TJ, Roy DS, Pignatelli M, Arons A, Tonegawa S. Engram cells retain memory under retrograde amnesia. *Science*. 2015;348(6238):1007–1013.
36. Fauth M, Wörgötter F, Tetzlaff C. The formation of multi-synaptic connections by the interaction of synaptic and structural plasticity and their functional consequences. *PLoS Comput Biol*. 2015;11(1):e1004031.
37. Rigotti M, Barak O, Warden MR, et al. The importance of mixed selectivity in complex cognitive tasks. *Nature*. 2013;497(7451):585–590.
38. Goris RLT, Movshon JA, Simoncelli EP. Partitioning neuronal variability. *Nat Neurosci*. 2014;17(6):858–865.
39. Wiegert JS, Oertner TG. Long-term depression triggers the selective elimination of weakly integrated synapses. *Proc Natl Acad Sci U S A*. 2013;110(47):E4510–E4519.
40. Kasthuri N, Hayworth KJ, Berger DR, et al. Saturated reconstruction of a volume of neocortex. *Cell*. 2015;162(3):648–661.
41. Markram H, Muller E, Ramaswamy S, et al. Reconstruction and simulation of neocortical microcircuitry. *Cell*. 2015;163(2):456–492.

SECTION IV

STRUCTURAL PLASTICITY AND LEARNING AND MEMORY

CHAPTER

14

Within a Spine's Reach

Will DeBello and Karen Zito

University of California, Davis, CA, United States

OUTLINE

1 INTRODUCTION

One of the most striking features of the brain is its ability to produce adaptive behaviors on the basis of individual experience. From development of human language to problem solving by cephalopods, the diversity of sensory, motor, and cognitive skills displayed across the animal kingdom is staggering. Although task-related specializations are evident

The Rewiring Brain
DOI: http://dx.doi.org/10.1016/B978-0-12-803784-3.00014-7

across species and at all levels of the nervous system, the seeming ease by which animals learn new tasks and acquire lasting memories raises the question of what, if any, common principles are at work.

The foundation for answering this question was laid in the 20th century with three fundamental advances: establishment of the neuron doctrine,[1] demonstration of chemical synaptic transmission,[2–6] and elucidation of the ionic basis of the action potential.[7–11] Increasingly precise tools for delving into the structure and function of synapses, and into the electrical signaling properties of neurons, have provided proof that the brain is a network. Contemporary with this insight was development of artificial neural networks, beginning a history of cross-pollination between the fields of neuroscience and computer science.

Information processing and storage within any network, biological or artificial, can be achieved by modifications in three parameters:

- The weights of each connection.
- The pattern of connections between nodes.
- The rules of integration that occurs at each node.

Interestingly, the dominant theory for learning in neuronal circuits from the 1950s to around 2000 focused mainly on the first parameter: weight changes induced at synapses according to their history of electrical activity. Perhaps, that narrow focus resulted either from the compelling logic (and truth) of Hebb's postulate of temporal precedence,[12] now fully described by rules of spike timing-dependent plasticity,[13–16] or from the discovery of long-term potentiation[17] (LTP) and long-term depression[18] in the 1970s–80s, which provided a path toward unraveling the molecular mechanisms of learning and memory[19–21] and the associated hope of clinical applications.

More recent theoretical and experimental results, however, have demonstrated the crucial role of changes in two other parameters: the pattern of connections (rewiring) and the rules of integration performed by each neuron or dendritic branch. This chapter focuses on developments of approximately the past 15 years beginning with discovery of microstructural dynamics in the living brain. We consider how, in principle, these changes might be harnessed to fine-tune information processing and storage and critically review empirical tests of the input clustering model[22–25] developed by Poirazi and Mel (Fig. 14.1). We identify gaps in current understanding and use potential experiments to fill those gaps. Finally, we speculate on the implications of this emerging understanding for neurorealistic simulation and neuromorphic computing.

2 MICROSTRUCTURAL PLASTICITY: SPINE DYNAMICS AND THE MAKING AND BREAKING OF SYNAPTIC CONNECTIONS

Macroanatomical changes in brain structure have long been known to occur during development and in postdevelopmental circuits following injury, neurodegeneration, or neurogenesis. In contrast, the notion that ongoing changes at the microscale level might provide a substrate for learning and memory was still not widely accepted in 1982 with the publication of Francis Crick's commentary "Do dendritic spines twitch?"[26] The

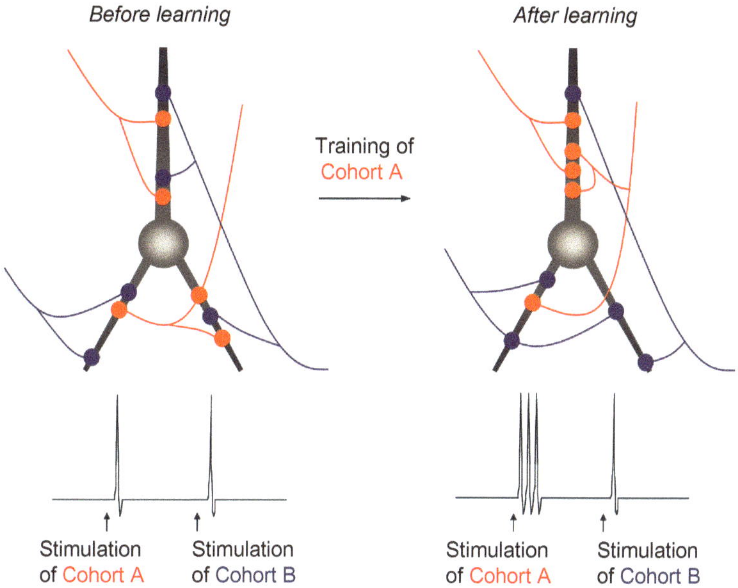

FIGURE 14.1 **The input clustering hypothesis.** Two cohorts of afferent input are depicted in *red* and *blue*. Activity evoked in each cohort is asynchronous with the other. Before learning, synapses are distributed randomly across the dendritic field of the postsynaptic neuron (*gray*). Activation of either cohort evokes a similar response from the soma. During a training period in which cohort A is repeatedly activated, synapses are made with all axons in the local neighborhood. Synapses from cohort A are selectively preserved only if they are located near one another on the same branch of dendrite—an input cluster. Activation of the cluster drives supralinear summation and enhances the response from the soma. That occurs even with the same total number and strengths of synapses present before training. The storage capacity of circuits is predicted to be greater when input clustering is employed than when plasticity is limited to changes in synaptic weight.

postulate was that cytoskeletal-driven changes in the morphology of dendritic spines would alter their electrical properties, thereby effecting synapse-specific weight change that encoded memories. Indeed this "twitching" would be experimentally verified *in vivo* two decades later (discussed in the following paragraph) and the proposed actin-based mechanism confirmed.[27–29] Therefore, compelling was the notion at the time that synaptic wiring was static, even Crick's prescient article did not include speculation that spine twitching might also lead to the formation of new synaptic connections.

The first observations of chronic dendritic spine dynamics in the living brain were reported in 2002 by two research groups using two-photon microscopy to acquire days to weeks of time-lapse images of dendrites of fluorescently labeled neurons in rodent cortex.[30,31] They found that dendritic spines could completely extend or retract in the adult rodent brain. The structure of dendritic arbors, however, was remarkably stable. In most cases the upper range of spine extension was ~2 μm. The groups disagreed on the prevalence and frequency of spine dynamics, but subsequent work by many others has confirmed the basic observation that spines are extending and retracting in the living brain,[32–35] charted the behavioral and sensory paradigms that affect motility,[34,36–44] determined the relationship between spine growth and retraction and the formation and elimination of synapses,[36,45–50,51] begun to define the cell types,[36,37,52,53] the cellular, and molecular mechanisms,[54–59] and, in a recent *tour de force*, demonstrated that spine growth is necessary for learning.[60] As this topic receives detailed treatment in a separate chapter, we summarize the relevant findings here:

- Dendritic and axonal branching patterns in postdevelopmental circuits are stable, absent injury, neurogenesis, or neurodegeneration.[30]
- Spine motility occurs throughout life.[32,33,61] Presynaptically, bouton turnover and microstructural extensions of axonal branchlets and terminal boutons have also been observed.[62–65]
- Spine motility is most prevalent in juveniles and declines in adults where most connections are stable over weeks to months, although a few (~5%) continue to be dynamic.[32,33]
- In both juveniles and adults, newly sprouted spines that eventually stabilize are associated with newly formed synapses.[36,46] Retraction of previously stable spines is associated with synapse elimination.[30,50,51]
- In conclusion, spine motility provides a substrate for sampling the local neighborhood for new synaptic partners. Because the dendritic and axonal branches are stable, the spatial range of remodeling is limited to approximately one spine's reach.

3 WHAT IS WITHIN A SPINE'S REACH?

Neuropil is a densely packed forest of cellular processes and extracellular matrix. What could a single spine's extension within this milieu realistically hope to achieve?

This question was first addressed using geometric modeling and bulk neuropil statistics.[66,67] Informed by the vital imaging studies described above, a cylindrical zone extending one spine's reach from the surface of each dendrite was identified as the spatial realm of remodeling. The number of actual synapses made onto the dendrite was calculated, and the number of potential synapses estimated from total length of axons positioned within the zone. Filling fraction—the ratio of actual to potential synapses—was calculated from anatomical measurements made independently in seven local circuits populating mouse neocortex, macaque neocortex, and rat hippocampus: pyramidal neuron density (range $2–22 \times 10^4$ mm^{-3}), dendritic length/neuron (1.4–12.3 mm), interbouton interval (3.7–6.2 μm), and average spine length (1.8–2.6 μm). The calculated filling fractions ranged from 0.12 to 0.34. Thus, the majority of potential synaptic partners coursing through the remodeling zone were unconnected with the target dendrite.

These low filling fractions imply a high capacity for information storage. Indeed, the storage capacity per synapse was predicted to be 3–4 bits, exceeding the naive value of 1 bit because of the latent potential connections. Thus, one answer to the question of what is within a spine's reach is a large number of unconnected axons providing opportunity for a very large number of distinct wiring diagrams.

A related model explored whether radially random axonal outgrowth could generate the observed filling fractions.[68] In this simulation, as neuronal density increased from a few neurons to thousands (an anatomically realistic number), the filling fraction fell from

1 and stabilized at ~0.26, in the center of the range calculated by the previous study. In addition, the incidence of unconnected neighboring neurons increased, consistent with experimental data. These modeling results demonstrate that a simple geometric rule for outgrowth, in combination with Hebbian-type synaptic competition, can explain key features of cortical connectivity.

Both modeling studies highlight the need for more precise connectivity measurements. With the advent of tools for structural and functional connectome reconstruction, the desired data are beginning to emerge. One project employed multiscale imaging to analyze a 0.13 mm^3 volume of somatosensory cortex from a young adult mouse.[69] An $80 \times 10^3\ \mu m^3$ subvolume was imaged at 3 nm per pixel resolution using scanning electron microscopy (EM), and within it, all neuropil elements (axons, dendrites, spines, and synapses) were reconstructed. This smaller "saturated" volume contained two apical dendrites of pyramidal cells and all of their synaptic inputs. This is the type of primary data needed to definitely address questions of connectivity at the microscale. The first application was to test Peters' rule.

Peters' rule, as formalized by Valentino Braitenberg,[70] postulates that the number of synaptic connections between two neurons can be predicted from geometric overlap of their axonal and dendritic arbors.[71] In this model, axodendritic "touches" are assigned as synapses and the entire population winnowed to match actual bulk synapse density. To test Peters' rule using the saturated volume of neocortex, a geometric bootstrap analysis was performed on all 1037 synapses, 916 axons, 1036 dendritic spines, and 7505 spine touches. Redundant synapses were identified when a single axon made more than one synapse with the target dendrite. In the simulation, redundant synapses were calculated from axospinous touches in each of 80,000 randomization trials of axonal positions. The median number of redundant synapses in these trials, 52, was far lower than the actual number, 78, ($P<0.00001$), indicating that axodendritic overlap is not sufficient to explain connectivity in adult somatosensory cortex.[69] The conclusion was that some factor—a molecular tag or activity-based signal—endowed certain axons with privileged access to the dendrite.

These results, and more recent ones employing unbiased large-scale methods to uncover connectivity principles between finely parsed cell type populations in adult neocortex,[72] lend support to the notion that microscale connectivity is nonrandom. Still, the focus of many studies so far has been to map connectivity by quantifying the strength of the bulk connection (including redundant synapses) between two identified neurons. This can be measured functionally using paired recordings or estimated from light or EM images by summing the number of synapses made between neurons A and B after differentially weighting individual synapses on the basis of proxy measures such as volume, number of synaptic vesicles, or protein content. These methods of quantifying bulk connection strength provide valuable information yet do not account for one potentially crucial aspect of postsynaptic integration, which is not just the number and strength of individual synapses but their dendritic address.

4 DISTRIBUTED VERSUS CLUSTERED INPUTS

Neurons can integrate up to 10,000 synaptic inputs arising from hundreds of input sources. Dendritic fields performing this operation exhibit diverse morphologies speckled with passive and active properties that shape the integration. To identify the summation rules for excitatory synapses, theoretical studies[24,73–77] have used compartmental models of anatomically and biophysically realistic neurons, and experimental studies[78–85] have employed in vitro slice preparation, whole-cell recording, and stimulation of multiple input locations using glutamate iontophoresis or uncaging. A few common principles have emerged.[1] For excitatory postsynaptic potentials (EPSPs) delivered simultaneously to a: (1) small number of different shaft locations on the same branch, summation is sublinear due to local reductions in driving force; (2) small number of spine synapses on the same dendrite, summation is linear, as the high neck resistance restores linearity by electrically isolating each spine head from local reductions in driving force; (3) larger number of spine synapses clustered on a subregion of dendrite, summation is supralinear because the summed EPSP crosses a local threshold for activation of voltage-dependent channels (sodium channels, calcium channels, or NMDA receptors). Whether this supralinearity results in a dendritic spike,[86–88] the activation of clustered inputs results in strong depolarization at the soma.

How these summation rules interact with weight and wiring changes is the key distinction between two competing cellular models for information processing and storage: distributed versus clustered inputs. The distributed model proposes that excitatory spine synapses originating at different afferent sources are distributed randomly along the dendritic tree.[2] Narrow spine necks isolate each spine from dendritic depolarizations, active conductances serve to amplify more distal EPSPs, and, as a result, summation is linear with each synapse contributing to the neuronal computation according to its weight, not location.[90] The output of this global integration would be influenced by adaptable weight changes and plasticity of intrinsic properties but not by readdressing of inputs already present. The computational power of the network would arise both from these cellular properties and emergent features of distributed processing at higher levels.

In contrast, the clustered model proposes that synaptic inputs bringing synergistic afferent information—for example, independent sensory features needed to construct the postsynaptic neurons' receptive field—would be most effectively positioned as dendritic neighbors. This would increase their chance of crossing the local threshold for amplification and thus increase the capacity of the neuron for pattern detection. Indeed, any position-dependent nonlinearity is predicted to add to the repertoire of neuronal computations.[91] In summary, dendritic readdressing even of preexisting inputs, in combination

[1]Cell-type specific variations in dendritic processing can be striking. For example, the dendrites of cerebellar interneurons typically operate in sublinear mode.[89] The discussion below refers primarily, although not exclusively, to pyramidal neurons from the mammalian neocortex or hippocampus and emergent common principles. Cell-type specific exceptions, when relevant, are noted.

[2]The question of dendritic addressing of different afferent sources refers to sources carrying the same type of input, i.e., excitatory glutamatergic synapses encoding the value of a graded variable such as location of a sensory stimulus. The dendritic addressing of these inputs is believed to be sculpted by stochastic mechanisms and activity-dependent feedback. In contrast, dendritic addressing of afferent sources carrying different types of input (neurochemical or source) is in many cases directed by genomic programs. This chapter focuses on the former, not the latter.

with adaptable weight changes and plasticity of intrinsic properties, is predicted to enhance the computational power of the network.

This input clustering hypothesis predicts (1) that clusters of coactive synapses exist and that the number, strength, and spacing of these *dendritic input clusters* is appropriate to drive supralinear summation on dendrites in response to natural stimuli and (2) that experience-dependent remodeling will drive formation of dendritic input clusters from an initially global distribution (Fig. 14.1).

Regarding the first prediction, cluster composition is expected to depend on synaptic weights and kinetics, dendritic dimensions, and membrane properties and thus will vary from cell to cell, dendrite to dendrite, and across a cell's lifetime as synaptic and intrinsic properties are modified. Using biophysically realistic models of pyramidal and hippocampal neurons, Poirazi and Mel's model predicted supralinear interactions resulting from as few as two moderately strong spine synapses located $<40\,\mu m$ apart on a dendrite. Physiological studies informed by these predictions, across a wider range of cell types and conditions, have found supralinear summation resulting from coactivation of 7–10[82] or ~20[83] nearby synapses or as few as two stronger ones.[80] In accordance with model predictions, experimental results confirm that fewer/weaker input patterns result in linear summation.[78,80,81] Thus, functional clusters occupy a niche within the larger parameter space of spatiotemporal input patterns.

There have been more than 20 tests of the two predictions of the input clustering hypothesis since 2008. These studies were performed in a wide array of species, analyzed diverse circuits, and used different tools for functional (calcium imaging, electrophysiology) or anatomical (confocal and two-photon imaging, EM, array tomography) mapping. In some cases they reached opposing conclusions. We highlight many of the significant findings in the subsequent sections.

5 EXPERIMENTAL TESTS OF THE PREDICTIONS OF THE INPUT CLUSTERING HYPOTHESIS

5.1 Do Dendritic Input Clusters Exist?

Within a few years of publication of Poirazi and Mel's modeling predictions, several independent studies established that coactivation of clustered inputs on dendrites *can* drive supralinear summation[80,82,83]; however, determining whether clustered dendritic inputs are *naturally* coactive and whether they *do* drive supralinear summation required experiments that utilized natural activity patterns, such as spontaneous or sensory-evoked activity. Such studies began to appear in 2010–12.[92–96]

Initial support for the existence of naturally coactive input clusters came from Kleindienst et al. (2011) and Takahashi et al. (2012). Using calcium imaging to monitor spontaneous calcium transients in CA3 pyramidal neurons in organotypic hippocampal slice cultures, both studies found that calcium transients arising from neighboring locations on the dendrite were more likely to be coincident than those arising from distant locations; these local clusters of coactive inputs spanned $16\,\mu m$ in early postnatal slices[95] and $8\,\mu m$ in older slices that had been an additional few weeks in culture.[96] Comparable results were obtained from dendrites of spontaneously active L2/3 pyramidal neurons *in vivo* in the somatosensory cortex of young adult mice, where local clusters of coactive spines were observed within a 6 μm span of dendrite.[96]

In contrast, Konnerth and colleagues reported no evidence for clustering of coactive inputs on dendrites of L2/3 pyramidal neurons *in vivo* in the sensory cortices of young adult mice in response to sensory stimulation.[92–94] Notably, these pioneering studies provided the first observations of sensory-evoked single-spine activity in mammalian cortical neurons *in vivo*. Using two-photon imaging to monitor dendritic calcium transients in L2/3 neurons of the mouse visual cortex,[92] auditory cortex,[94] or somatosensory cortex,[93] the authors observed that calcium transients evoked by closely related sensory stimuli appeared widely distributed on the dendritic tree, and that adjacent inputs often represented signals of broadly different sensory feature values. The authors concluded that the results of their studies are most consistent with the global integration model in which neurons compute their output by integrating spatially distributed synaptic inputs.

While the data from Konnerth and colleagues did not provide support for the preferential activation of clustered inputs, they did not rule out that certain aspects of receptive field arise due to dendritic input clusters. Indeed, a recent study examining a broader array of stimulus characteristics demonstrated just that. Wilson et al. used *in vivo* two-photon calcium imaging of individual spines on dendrites of L2/3 cells to examine orientation preference and orientation selectivity in the visual cortex of young adult mice.[97] They showed that while reliable in predicting orientation preference, summed synaptic input to individual neurons did not predict orientation selectivity. Instead, orientation selectivity strongly correlated with the spatial clustering of cotuned synaptic inputs. Furthermore, those dendritic branches with more cotuned clusters showed a greater rate of local dendritic calcium events, suggesting that functional clustering of synaptic inputs plays an important role in the dendritic nonlinearities that shape orientation selectivity.[97]

Two recent functional studies provide further support for the existence of dendritic input clusters *in vivo*. Winnubst et al. quantified spontaneous calcium transients *in vivo* in L2/3 cells of the visual cortex of young postnatal mice, demonstrating that, even at these early ages, nearby synapses (<12 μm) were more coactive than synapses far apart.[98] Gokce et al. combined optogenetics and two-photon calcium imaging to map the spatial organization of glutamatergic synapses between L5 pyramidal neurons in mouse neocortical slices, finding that synapses of intralaminar inputs form clusters (4–14 synapses) spanning 30 μm on the basal dendrites of L5 pyramidal neurons.[99]

Anatomical studies have provided complementary data. No evidence for clustering was reported in experiments using correlative light and EM to study focally labeled thalamocortical synapses onto L4 cells.[100] Because of the topographic organization of the visual system, focal labeling should target neurons, and thus axons, that are likely to be coactive. Yet evidence for clustering of labeled inputs onto dendrites was not observed for LM-identified contacts nor for EM-identified synapses.[100] In contrast, a recent study examining thalamocortical synapses onto excitatory L4 neurons showed a significant clustering of inputs onto barrel star pyramid neurons and a trend toward input clustering onto a septal star pyramid neuron but no evidence for clustering on barrel spiny stellate neurons.[101] The authors argue that this clustering could help explain the surprisingly potent ability of the thalamus to drive cortical activity despite that corticocortical synapses outnumber thalamocortical synapses in a range of 10 to 1.

High-throughput anatomical approaches provide further support for dendritic input clusters. Rah and colleagues used array tomography to examine the spatial distribution of thalamocortical synapses onto L5 neurons in the mouse somatosensory cortex,[102] finding within-branch clustering (5–15 μm) more often than predicted from a random distribution. In another study, Druckmann and colleagues used mGRASP to visualize inputs of presynaptic CA3 neurons to postsynaptic CA1 neurons in the hippocampus.[103] The authors analyzed within branch clustering and found significant deviation from random distribution in 17 of 27 neurons: all exhibited an overabundance of the shorter distances, supporting clusters. Furthermore, by labeling only neurons that were born in a similar time window, input clustering (<1.5 μm) was significantly enhanced, suggesting that neurons born at the same time could be functionally related and signaling together.

5.2 Does Experience-Dependent Remodeling Drive Formation of Dendritic Input Clusters?

The second major prediction of the hypothesis is that learning should drive the redistribution of synapses from a dispersed into a more clustered state.[98,104] Indeed, experience-driven changes in clustering have been observed using naturalistic learning paradigms in owl and mouse, and in both juvenile and adult animals.

The first study to show clustering of inputs on dendrites following a learning paradigm was McBride et al. using juvenile barn owls.[105] When owls are reared wearing prism glasses that shift the frontal visual field, new axonal growth in the projection from the central nucleus of the inferior colliculus (ICC) to the external nucleus (ICX) serves to realign the auditory and visual space maps and drive adaptive localization behavior. Notably, the functionally dormant normal circuit is anatomically preserved alongside the active learned circuit,[106] providing an internal control for cluster analysis. Because of the topographic nature of the auditory projections, focal labeling of neurons in the ICC identifies coactive neurons and thus coactive axonal inputs to the ICX. Measurements of the inter-contact distance (ICD) of labeled axonal inputs to the nearest neighbors measured from the dormant normal circuit in prism-adapted owls were on average <10 μm; however, a significant fraction were >20 μm. In comparison, ICDs >20 μm were almost never observed in the active learned circuit; nearly all inputs resided in clusters. When prisms were removed in adulthood, increased clustering was observed in the now reactivated normal circuit.[107] Interestingly, even the synapses in functionally weak circuits in adult owls resided in clusters. This could represent a trace of learning in the brain—an efficient way of storing a disused input pattern that could be needed later in life. Altogether, these results demonstrate that behaviorally relevant experience can drive the formation or disassembly of dendritic input clusters.

A subsequent study in young adult rodents by Makino and Malinow demonstrated experience-dependent input clustering on dendrites of L2/3 cells in acute slices of somatosensory cortex of mice.[108] Using an optical approach to track the delivery of fluorescently tagged AMPA receptors into spines, the authors were able to monitor single-synapse plasticity. Normal sensory experience preferentially produced synaptic potentiation onto nearby dendritic synapses (28/95 dendrites), consistent with the input clustering hypothesis. In contrast, sensory deprivation via whisker trimming showed much less correlation (5/68 dendrites) and instead led to a global homeostatic enhancement.

Further evidence supporting learning-induced input clustering on dendrites was first reported in rodents by Fu and colleagues.[41] Using two-photon imaging of GFP-labeled L5 pyramidal neurons in the motor cortex following training on a novel forelimb task, Fu et al. showed that one-third of new spines appeared in clusters (adjacent to existing spines, $<5\ \mu m$), and that these were more resistant to elimination than nonclustered new spines. Cross-training of the mice on a new task as adults resulted in the emergence of new clusters that were largely separated from those associated with the first task. Notably, formation of new spine clusters required repetition of the same motor task; thus, the authors show that clustering of new synapses along dendrites is induced by repetitive activation of cortical circuitry during learning.[41] Reiner and colleagues[109] confirmed these studies in wild-type animals and further showed that FMR1 knockout animals have deficits in learning-induced spine clustering.

Finally, two behavioral studies utilized EM analyses to demonstrate learning-induced synaptic clustering.[110,111] In the first study, Lee et al. trained adult rats in a motor skills task. EM analysis of parallel fiber to Purkinje cell synapses showed that motor skill training promoted the formation of parallel fiber multiple synapse boutons (MSBs) contacting two dendritic spines from the same dendrite of Purkinje cells in the rat cerebellum.[110] Notably, excitatory synapses near MSBs were smaller in motor-trained animals, suggesting a compensatory depression of MSB-neighbor synapses.[110] A second study by Pereira and colleagues set out to define the effects of riluzole, a drug that decreases glutamate release and facilitates astrocytic glutamate uptake, on age-related cognitive decline. They measured both synaptic structure (EM) and memory performance and showed that riluzole-treated rats were protected against age-related cognitive decline and they exhibited a correlated increase in the density and clustering of thin spines.[111]

In summary:

- A diverse array of species (owl,[105,107] rat,[95,96,108,110–112] mouse,[41,92–94,98,99,102,103] cat,[100] ferret[97]) and circuits (inferior colliculus,[105,107] hippocampus,[95,96,103,111,112] primary sensory cortices,[92–94,96–100,102,108] motor cortex,[41] cerebellum[110]) has been studied.
- Most researchers analyzed intact circuitry in vivo, ex vivo, or fixed.
- Nearly all studies focused on excitatory synapses convergent on pyramidal neurons with large dendritic fields.
- Coactive synapses were identified by electrical or optical recordings of spontaneous activity, activity evoked by sensory presentation, or by artificial stimulation. Their dendritic locations were apparent from imaging fields with wide coverage of the postsynaptic neuron.
- *Most, not all, studies found a correlation between co-activity and dendritic location.* Specifically, coactive synapses were more likely to occur within $<20\ \mu m$ of one another, with most results indicating $<10\ \mu m$ and several identifying immediate neighbors as part of the cluster. This empirically determined spatial window is somewhat smaller than predicted by computational modeling.

- Activity-dependent cluster formation has been observed in several systems.[98,112–116] Experience-driven changes in clustering have been observed using naturalistic learning paradigms in owl[105,107] and rodents,[41,108–110] and in both juvenile and adult animals.
- Clusters are not composed exclusively of coactive synapses but are often interdigitated with other input cohorts. As a purely random distribution of afferent cohorts is expected to produce some clusters by chance, convincing demonstration of true clustering requires data sets with near saturating coverage and bootstrap analysis.

6 CELLULAR AND MOLECULAR MECHANISMS DRIVING INPUT CLUSTER FORMATION

The formation of dendritic input clusters could result from the appearance of new coactive neighboring synapses and/or from the elimination of neighboring inactive synapses. Several studies, outlined below, have identified potential molecular and cellular mechanisms.

Selective outgrowth and/or stabilization of new coactive neighboring synapses could be accomplished through local cross talk, as reported by Harvey and colleagues in 2007. This study demonstrated that activation of a single spine in a pattern that induces input-specific LTP of synaptic structure and function[113] also leads to the spread of a signal that converts a subthreshold LTP stimulus into stable LTP at neighboring synapses. This cross talk between synapses was due to an intracellular diffusible molecule that acts on a spatial scale of 10 μm and a temporal scale of 10 min. Later studies identified Ras,[117] Rac1, and RhoA as intracellular diffusible signals[59] that support cross talk and thus selective strengthening of nearby synapses. Such a cross talk mechanism could be responsible for the selective outgrowth of new spines nearby active spines observed in 2008 by De Roo and colleagues.[112]

Alternatively, selective outgrowth and/or stabilization could be accomplished through a synaptic tagging and capture mechanism, as reported by Govindarajan and colleagues in 2011.[118] In this model, a stimulus that drives synaptic strengthening at strongly activated synapses also leads to the protein synthesis-dependent production of PrPs (plasticity-related protein products), which are captured at neighboring weakly activated, tagged synapses. If successful in capturing PrPs, the weak, tagged synapse will undergo long-lasting late LTP. This mechanism is similar to cross talk; however, it operates at a more global spatial scale of 70 μm (vs 10 μm) and temporal scale of 90 min (vs 10 min). Furthermore, synaptic tagging and capture requires protein synthesis, whereas cross talk does not.

Finally, selective outgrowth and/or stabilization could occur through location-dependent plasticity mechanisms that rely upon local dendritic excitability and integrative properties.[116] Specifically, Weber and colleagues report a proximodistally increasing gradient of nonlinear NMDA receptor-mediated amplification of spine Ca^{2+} signals; the enhanced synaptic cooperativity at distal dendritic compartments could promote the formation of input clusters on distal dendrites.

In addition to selective outgrowth and/or stabilization, formation of dendritic input clusters could also occur via the selective weakening and/or elimination of nearby inactive inputs. One such mechanism was reported by Oh and colleagues at synapses of hippocampal pyramidal neurons in slice culture.[115] In that study, the authors observed that robust stimulation, leading to the growth and strengthening of several inputs on a single dendrite, also led to the selective weakening of nearby ($<3.5\,\mu m$) inactive synapses. Heterosynaptic shrinkage of inactive spines required activation of calcineurin, metabotropic glutamate receptors, and IP_3 receptors,[115] yet the specific mechanism has not yet been determined. A recent report identified activity-induced competition for beta-catenin as critical for mediating interspine competition during heterosynaptic spine pruning.[119]

An alternative mechanism for weakening of nearby inactive inputs was reported by Winnubst et al. The authors used whole-cell patch clamp to monitor naturally occurring changes in spontaneous activity at individual synapses in L2/3 neurons in the mouse visual cortex in vivo and in hippocampal pyramidal neurons in slice culture.[120] They found that synapses that exhibit low synchronicity with neighbors ($<12\,\mu m$) become depressed in their transmission frequency, but not in their amplitude, suggesting a retrograde signaling mechanism that influences on presynaptic function. Experimentally increasing local synchronicity stabilized synaptic transmission. Signaling through the high-affinity pro-BDNF receptor $p75^{NTR}$ was required for depression of asynchronously stimulated synapses.[120]

Finally, the formation of dendritic input clusters could rely upon the coordinated local plasticity of both excitatory and inhibitory inputs, as observed by Chen and colleagues.[114] Using Teal-Gephyrin as a marker for inhibitory synapses, Chen et al. were able to simultaneously image putative inhibitory synapses and putative excitatory synapses (dendritic spines) in the visual cortex in vivo. They found that inhibitory synapse and dendritic spine remodeling were spatially clustered, suggesting local coordination of inhibitory and excitatory synaptic rearrangements.[114] Such coordinated rearrangements could play an important role in the establishment of coactive clustered inputs on dendrites.

7 FUTURE EXPERIMENTAL STUDIES

Below we outline four potential areas for future work. The first describes a stringent experiment (currently just outside the realm of feasibility) regarding the functional significance of input clustering. The second poses a new question—how microstructural dynamics drive cluster formation. The final two topics are more open-ended.

7.1 What Is the Functional Significance of Clustering?

The empirical studies summarized earlier derive from an impressive arsenal of technologies. They provide compelling support for the existence of clusters and for the capacity of dendrites to act as supralinear integrators in certain circumstances. Yet direct demonstration of whether naturally occurring input patterns drive supralinear summation through clustered inputs is lacking, primarily because the natural input patterns are unknown. Testing the model with natural patterns as opposed to the artificial patterns used in previous studies would be an important advance because postsynaptic integration—known to depend on timing, number, and strengths of inputs—could differ markedly across input regimes. Moreover, naturalistic inputs could cause state changes[121]

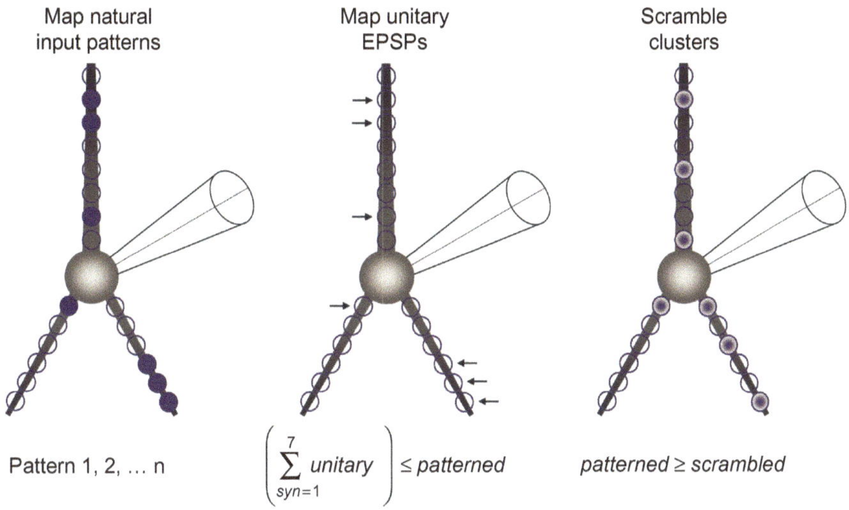

FIGURE 14.2 **Hypothetical experiment.** (A) Map natural input patterns *in vivo*. This could be accomplished via two-photon imaging of calcium transients evoked by natural sensory stimuli. One challenge is developing an imaging system with single synapse resolution and total neuronal coverage (thousands of synapses). Each stimulus would be expected to evoke a different pattern of activation, illustrated for one stimulus by *solid fill* in 7 of 20 synapses. For each pattern, the spatial organization of synchronous versus asynchronous synapses could be analyzed using bootstrap methods to determine deviations from a random pattern. The model predicts different input clusters uniquely associated with different stimuli. (B) Map unitary EPSPs. Whole-cell recordings could be made from the same neuron. Glutamate uncaging could be used to measure EPSPs evoked by individual synapses (*arrows*). Input clustering predicts that patterned activation would produce a stronger response than the sum of unitary activations. (C) Scramble clusters. Multispot photoactivation could be used to simultaneously activate a spatially randomized cohort of synapses (*radial fill*), matching the patterned cohort in its sum of unitary EPSPs. Scrambled activation is predicted to produce a weaker response than patterned activation.

in the neuron via longer lasting biochemical modifications.[117] We propose an experiment with three components (Fig. 14.2).

The first is to determine the spatiotemporal patterns of synaptic input onto the entire dendritic tree of a neuron in vivo. This is not a small task. The challenge lies in visualizing the activations of thousands of afferent synapses because a diverse array of naturalistic sensory stimuli is delivered to an awake animal. Simultaneous optical recording of each synapse is theoretically feasible, and recent unpublished reports suggest that current platforms have achieved the goal of complete dendritic input coverage for some cell types, yet confidence is low that all synaptic events are reliably detected. Meeting this challenge likely depends on the development of higher sensitivity calcium or voltage-based dyes and further tools for high-speed multiphoton imaging, explicit goals of the BRAIN Initiative. Failing complete neuronal coverage, mapping the input pattern to an entire dendrite should suffice, as current evidence supports the dendritic branch as an independent computational subunit.

The second is to measure EPSPs evoked both by each individual synapse and by the naturally occurring input patterns. This could be achieved using multisport photoactivation and electrical or optical recordings, first to quantify (or estimate) the magnitude of

EPSPs resulting from activation of each synapse and second to recapitulate the natural input patterns. Retrospective analysis would determine whether native input clusters are *sufficient* to drive supralinear summation and also should reveal motifs that most effectively drive the neuron. With a diverse array of naturalistic patterns, one might also be able to estimate the neurons capacity for pattern recognition and/or storage.

The third is to manipulate the multispot pattern in artificial ways that probe the *necessity* of clusters. The number, strength, and relative timing of synapses within the effective naturalistic cohort would be preserved but the spatial structure degraded. This would disambiguate the roles of the three parameters that govern function of the synaptic network—location, weight, and integration rules—and provide important new data for computational modeling. Close agreement of the model output and experimental results would provide strong proof of input hypothesis and elevate the hypothesis to paradigm status.

7.2 How Do Clusters Form?

Spine motility is a necessary substrate for cluster formation but in itself does not explain structured dendritic addressing. One possibility is that spines extend randomly into the milieu of their potential inputs and form "tester" synapses with each axon they touch. Those synapses that, by chance, are positioned within an effective cohort receive a reward or consolidation signal, and the synapses would be stabilized. Spines that have formed in the vicinity of asynchronously active synapses would destabilize and be eliminated coincident with spine retraction. This is a straightforward Hebbian model.

The alternative is a "Directed model" in which only coactive dendrites are recruited by an axon.[122] This could be achieved by the release of diffusible factors that guide spine outgrowth in the direction of the axons or by cell surface expression of receptors that signal a match. Activity-based rules would still be needed to confirm specificity and fine-tune the synaptic weight. This Directed model might be expected to rely on fewer tester synapses and thus produce more efficient or faster remodeling. However, it would require a panel of molecular tags (diffusible or cell surface), each encoding a separate activity pattern, to provide target specificity within the milieu and therefore is mechanistically more complex. Occam's razor favors the Hebbian model.

Still, if directed outgrowth were found, the translational significance would be high. Before a search for molecular tags is launched, a simple experiment would be to systematically manipulate the correlative activity between a postsynaptic dendrite and its neighboring axons while monitoring spine dynamics over an extended time period. Is spine outgrowth geometrically random or targeted to processes with registered activity?

7.3 Does Clustering Occur at Higher Levels in the Network?

Lost in the discussion about whether the dendritic branch or neuronal cell body is the fundamental computational subunit is the notion of recurrent structure across scales. This chapter has focused on dendritic input clustering as a mechanism to boost neuronal storage capacity. One open question is whether clustering is an organizing principle at the level of neurons, circuits, and systems.[123] Recordings using high-density multielectrode arrays (MEAs) have shown that a subset of neurons in mouse somatosensory cortex transfer and receive far more information than others.[124] These "hub neurons" may route as much as 70% or cortical traffic despite representing just 20% of all neurons. A separate theoretical study indicates that neuronal hubs could have an outsized effect on network processing, in particular in orchestrating state changes.[125] Analogous results at the level of macroscale connectomics, based on diffusion tensor imaging tractography with a resolution limited to voxels containing millions of circuit elements, have shown the brain to be a small-world network with hub regions that make and receive significantly more connections than others.[126] It would be interesting to compare the organization of these nonrandom features that appear at markedly different spatial scales

7.4 Did Clustering Evolve in Circuits With Higher Demands for Adaptive Pattern Recognition?

The experiments and models described in this chapter provide good (not conclusive) support of the input clustering model—at least in certain cell types and conditions. However, global integration is hardly ruled out. One could speculate that both have a role in microscale operation.

Few circuits may require the extra capacity afforded by clustering to achieve their purpose. Moreover, many are best left consolidated as they were at the tail end of development. Absent injury, degeneration, or neurogenesis, both the information, input and computed output of these types of circuits, is stationary over the adult lifetime. Building in a seldom-needed capacity for structural plasticity may simply not be worth the metabolic cost. Thus, a comparative analysis of input clustering in circuits that evolved for adaptive pattern recognition would be informative.

This search should not be regarded as a one-way trip through the phylogenetic tree nor up the neuroaxis from spinal cord to cortex. For example, primary visual cortex in mammals—whose capacity for invariant feature detection provides a reliable substrate from which association cortex extracts complex percepts—may be less in need of adaptive pattern recognition than, for example, the inferior colliculus in barn owls, whose role is lifelong adaptation to slight changes in registration of auditory and visual cues necessary for survival. Consistent with this notion, computational estimates of dendritic input discrimination capability (M) across morphologically defined cell types found in fish, insects, amphibians, birds, and mammals revealed large neuron-to-neuron variability[127] that did not correlate with evolutionary complexity. Predicting which circuits employ input clustering demands careful evaluation of each input–output transformation and whether it is stationary over long timescales.

8 FUTURE MODELING STUDIES

As experimental studies move forward to test the sufficiency and necessity of structured dendritic addressing in a variety of circuits—a journey that in our view is more likely to reveal a mixed bag of processing strategies than a canonical one—we consider the implications of distributed versus clustered processing for two "big science" projects at the intersection of neuroscience and computer science: neurorealistic simulation and neuromorphic computing.

8.1 Neurorealistic Simulation

Quantitative models are ultimately required to demonstrate understanding of complex systems such as Earth's climate or mammalian brains. Simulating *in silico* the biophysical operation of the human brain at the level of individual molecules is infeasible and unnecessary. The question is what level of biophysical detail is required to produce a reasonable emulation of even one semiautonomous neural circuit such as a cortical column. There are at least three schools of thought.

The *structural connectome approach* uses networks built of $\sim 10^3 - 10^6$ s of anatomically and biophysically realistic neurons represented by compartmental models or computationally simplified two-stage processors. There are two distinct strategies based on *predicted connectivity*[128–136] or *actual connectivity*. One example of the former is the Human Brain Project co-funded by the European Union. To simulate a microcircuit in rat somatosensory cortex, a near complete catalog of resident cell types was assembled from morphological and electrophysiological data gathered from hundreds of independent experiments. The neurons ($\sim$31,000) were superimposed digitally according to laminar landmarks, and connectivity was calculated from axodendritic overlap.[137] The number of synapses ($\sim$37 million) was scaled to match data, and synaptic weights were assigned. Remarkably, this network built of predicted connectivity, as opposed to actual microscale wiring, was able to reproduce an array of functional data including some in vivo features, all without parameter tuning.[138] Such a result is consistent with the distributed model of neural processing.

The general applicability of this strategy, which does not incorporate empirically determined dendritic addresses, is not clear. Many higher order aspects of cortical function, including capacity for learning and memory, have yet to be explored. Indeed, the value of input clustering is maximal when networks are pushed to perform sophisticated pattern recognition and adaptive processing. In this second school of thought, achieving high performance from a neurorealistic simulation will depend on incorporation of *actual connectivity*. Ultimately this may require dense reconstruction of a serial EM volume coregistered with prospective functional mapping.[139] Several such projects are underway but will likely take years to decades to complete.

In the interim, neurorealistic models could be infused with plasticity rules that produce biologically observed connection motifs such as input clusters, sparse connectivity, and other nonrandom features. Toward this goal, an abstract neural network (i.e., not neurorealistic) was constructed to solve a hidden variable estimation task,[140] a type of inference that occurs in the brain. Robust computations were achieved only when both synaptic weights and connectivity were left as adjustable parameters. This result was more striking for sparsely connected networks—of which the brain is one example—than densely connected ones. The network was also trained to mimic a motor learning task under conditions where wiring plasticity was delimited by empirical measurements of the rates and scope of spine dynamics. Collectively, these theoretical results support the notion that simulation of microstructural dynamics may be essential for maximizing the performance of neurorealistic models.

In a third school of thought, the *functional connectome approach* dispenses with the notion of simulating synaptic signaling and neuronal integration and instead focuses on the real-time activity pattern at millions of neuronal cell bodies.[141] The raw data that would feed such a model is expected to come from advances in high-throughput recording technology, either electrical or optical.[142,143] With days to weeks of recording naturally driven patterns of activity with single neuron and single action potential resolution and circuit-wide coverage, the algorithms performed by canonical circuits might be elucidated. One emerging example of this approach is neural control of prosthetic limbs[144] via real-time analysis of MEA signals implanted in motor cortex of primates.[145] With enhanced coverage and resolution, one could imagine the ability to "readout" of increasingly sophisticated cognitive processes[146] and eventually to piece together holistic emulation of natural behavior.

8.2 Neuromorphic Computing

One other potential application of understanding distributed versus clustered processing is to the field of neuromorphic computing. For more than 80 years, computer cores have employed Von Neumann type architecture in which the sites of processing and memory storage are physically separate. The simple existence of the digital revolution indicates the power of this platform. Yet the design is quite unlike the brain in which the sites of processing and storage are colocated at synapses. Neuromorphic chips, in contrast, loosely mimic real synaptic networks. Prototype designs developed by academic groups at MIT, Stanford, and Zhejiang University (China), and commercial efforts at IBM (True North), Qualcomm (Zeroth NPU), Numenta (NuPIC), and Knowm (memristors) already demonstrate one advantage of such an approach—radically lower power consumption compared with standard processors.[147] This feature alone promises enormous practical advantages for mobile computing and remote sensing applications.

Yet another feature of neuromorphic architecture holds added potential. Recent advances in "deep learning" have produced computer algorithms that, for the first time in history, rival or exceed human performance in complex visual tasks such as object or face recognition.[148] Running these biologically inspired algorithms on Von Neumann architecture, however, is viewed as inefficient. It is believed that embedding deep learning algorithms in neuromorphic chips could maximize the utility of using synaptic networks. The most futuristic musings on this topic predict a full 80 billion neuron simulation by 2025 and human-like artificial intelligence on the near horizon.

Putting aside fantastic claims, unraveling the brain's microscale structure could concretely inform the next generation of neuromorphic designs. This is because current designs are based on a preconnectomics era understanding of synaptic networks and therefore do not implement microscale connection motifs. As neuroscience continues to discover motifs at an accelerating pace—and demonstrate their role in diverse information processing and storage functions—it could provide a deep well of biological inspiration for hardware engineers.

References

1. Ramon y, Cajal S. *Neuron Theory or Reticular Theory?: Objective Evidence of the Anatomical Unity of Nerve Cells.* Madrid: Consejo Superior de Investigaciones Cientificas, Instituo Ramon y Cajal; 1954.
2. Sherrington CS. *The integrative action of the nervous system.* New Haven, CT: Yale University Press; 1920.
3. Loewi O. Über humorale Übertragbarkeit der Herznervenwirkung. *Pflügers Arch Ges Physiol.* 1921;189:239–242.
4. Dale HH. The chemical transmission of nerve impulses. *Science.* 1934;80(2081):450.
5. De Robertis ED, Bennett HS. Some features of the submicroscopic morphology of synapses in frog and earthworm. *J Biophys Biochem Cytol.* 1955;1(1):47–58.
6. Palay SL, Palade GE. The fine structure of neurons. *J Biophys Biochem Cytol.* 1955;1(1):69–88.
7. Hodgkin AL, Huxley AF, Katz B. Measurement of current-voltage relations in the membrane of the giant axon of Loligo. *J Physiol.* 1952;116(4):424–448.
8. Hodgkin AL, Huxley AF. Currents carried by sodium and potassium ions through the membrane of the giant axon of Loligo. *J Physiol.* 1952;116(4):449–472.
9. Hodgkin AL, Huxley AF. The components of membrane conductance in the giant axon of Loligo. *J Physiol.* 1952;116(4):473–496.
10. Hodgkin AL, Huxley AF. The dual effect of membrane potential on sodium conductance in the giant axon of Loligo. *J Physiol.* 1952;116(4):497–506.
11. Hodgkin AL, Huxley AF. A quantitative description of membrane current and its application to conduction and excitation in nerve. *J Physiol.* 1952;117(4):500–544.
12. Hebb DO. *Organization of Behavior.* New York, NY: Wiley; 1949.
13. van Rossum MC, Bi GQ, Turrigiano GG. Stable Hebbian learning from spike timing-dependent plasticity. *J Neurosci.* 2000;20(23):8812–8821.
14. Song S, Miller KD, Abbott LF. Competitive Hebbian learning through spike-timing-dependent synaptic plasticity. *Nat Neurosci.* 2000;3(9):919–926.
15. Bi G, Poo M. Synaptic modification by correlated activity: Hebb's postulate revisited. *Annu Rev Neurosci.* 2001;24:139–166.
16. Abbott LF, Nelson SB. Synaptic plasticity: taming the beast. *Nat Neurosci.* 2000;3(Suppl:):1178–1183.
17. Bliss TV, Lomo T. Long-lasting potentiation of synaptic transmission in the dentate area of the anaesthetized rabbit following stimulation of the perforant path. *J Physiol.* 1973;232(2):331–356.
18. Ito M, Sakurai M, Tongroach P. Climbing fibre induced depression of both mossy fibre responsiveness and glutamate sensitivity of cerebellar Purkinje cells. *J Physiol.* 1982;324:113–134.
19. Malinow R, Madison DV, Tsien RW. Persistent protein kinase activity underlying long-term potentiation. *Nature.* 1988;335(6193):820–824.
20. Kandel ER. The molecular biology of memory storage: a dialog between genes and synapses. *Biosci Rep.* 2001;21(5):565–611.
21. Malinow R, Malenka RC. AMPA receptor trafficking and synaptic plasticity. *Annu Rev Neurosci.* 2002;25:103–126.
22. Mel BW. The clusteron: toward a simple abstraction for a complex neuron. In: Moody J, Hanson S, Lippmann R, eds. *Advances in Neural Information Processing Systems.* Vol 4. San Mateo, CA: Morgan Kaufmann; 1992:35–42.
23. Poirazi P, Mel BW. Impact of active dendrites and structural plasticity on the memory capacity of neural tissue. *Neuron.* 2001;29(3):779–796.
24. Poirazi P, Brannon T, Mel BW. Arithmetic of subthreshold synaptic summation in a model CA1 pyramidal cell. *Neuron.* 2003;37(6):977–987.
25. Poirazi P, Brannon T, Mel BW. Pyramidal neuron as two-layer neural network. *Neuron.* 2003;37(6):989–999.
26. Crick F. Do dendritic spines twitch? *Trends in Neurosciences.* 1982;5:44–46.
27. Fischer M, Kaech S, Knutti D, Matus A. Rapid actin-based plasticity in dendritic spines. *Neuron.* 1998;20(5):847–854.
28. Dunaevsky A, Tashiro A, Majewska A, Mason C, Yuste R. Developmental regulation of spine motility in the mammalian central nervous system. *Proc Natl Acad Sci U S A.* 1999;96(23):13438–13443.
29. Halpain S, Hipolito A, Saffer L. Regulation of F-actin stability in dendritic spines by glutamate receptors and calcineurin. *J Neurosci.* 1998;18(23):9835–9844.

30. Trachtenberg JT, Chen BE, Knott GW, et al. Long-term in vivo imaging of experience-dependent synaptic plasticity in adult cortex. *Nature*. 2002;420(6917):788–794.
31. Grutzendler J, Kasthuri N, Gan WB. Long-term dendritic spine stability in the adult cortex. *Nature*. 2002;420 (6917):812–816.
32. Holtmaat AJ, Trachtenberg JT, Wilbrecht L, et al. Transient and persistent dendritic spines in the neocortex in vivo. *Neuron*. 2005;45(2):279–291.
33. Zuo Y, Lin A, Chang P, Gan WB. Development of long-term dendritic spine stability in diverse regions of cerebral cortex. *Neuron*. 2005;46(2):181–189.
34. Roberts TF, Tschida KA, Klein ME, Mooney R. Rapid spine stabilization and synaptic enhancement at the onset of behavioural learning. *Nature*. 2010;463(7283):948–952.
35. Loewenstein Y, Yanover U, Rumpel S. Predicting the dynamics of network connectivity in the neocortex. *J Neurosci*. 2015;35(36):12535–12544.
36. Holtmaat A, Wilbrecht L, Knott GW, Welker E, Svoboda K. Experience-dependent and cell-type-specific spine growth in the neocortex. *Nature*. 2006;441(7096):979–983.
37. Hofer SB, Mrsic-Flogel TD, Bonhoeffer T, Hubener M. Experience leaves a lasting structural trace in cortical circuits. *Nature*. 2009;457(7227):313–317.
38. Yang G, Pan F, Gan WB. Stably maintained dendritic spines are associated with lifelong memories. *Nature*. 2009;462(7275):920–924.
39. Xu T, Yu X, Perlik AJ, et al. Rapid formation and selective stabilization of synapses for enduring motor memories. *Nature*. 2009;462(7275):915–919.
40. Lai CS, Franke TF, Gan WB. Opposite effects of fear conditioning and extinction on dendritic spine remodelling. *Nature*. 2012;483(7387):87–91.
41. Fu M, Yu X, Lu J, Zuo Y. Repetitive motor learning induces coordinated formation of clustered dendritic spines in vivo. *Nature*. 2012;483(7387):92–95.
42. Yang G, Gan WB. Sleep contributes to dendritic spine formation and elimination in the developing mouse somatosensory cortex. *Developmental neurobiology*. 2012;72(11):1391–1398.
43. Moczulska KE, Tinter-Thiede J, Peter M, et al. Dynamics of dendritic spines in the mouse auditory cortex during memory formation and memory recall. *Proc Natl Acad Sci U S A*. 2013;110(45):18315–18320.
44. Yang G, Lai CS, Cichon J, Ma L, Li W, Gan WB. Sleep promotes branch-specific formation of dendritic spines after learning. *Science*. 2014;344(6188):1173–1178.
45. Zito K, Knott G, Shepherd GM, Shenolikar S, Svoboda K. Induction of spine growth and synapse formation by regulation of the spine actin cytoskeleton. *Neuron*. 2004;44(2):321–334.
46. Knott GW, Holtmaat A, Wilbrecht L, Welker E, Svoboda K. Spine growth precedes synapse formation in the adult neocortex in vivo. *Nat Neurosci*. 2006;9(9):1117–1124.
47. Nagerl UV, Kostinger G, Anderson JC, Martin KA, Bonhoeffer T. Protracted synaptogenesis after activity-dependent spinogenesis in hippocampal neurons. *J Neurosci*. 2007;27(30):8149–8156.
48. Zito K, Scheuss V, Knott G, Hill T, Svoboda K. Rapid functional maturation of nascent dendritic spines. *Neuron*. 2009;61(2):247–258.
49. Kwon HB, Sabatini BL. Glutamate induces de novo growth of functional spines in developing cortex. *Nature*. 2011;474(7349):100–104.
50. Woods GF, Oh WC, Boudewyn LC, Mikula SK, Zito K. Loss of PSD-95 enrichment is not a prerequisite for spine retraction. *J Neurosci*. 2011;31(34):12129–12138.
51. Cane M, Maco B, Knott G, Holtmaat A. The relationship between PSD-95 clustering and spine stability in vivo. *J Neurosci*. 2014;34(6):2075–2086.
52. Ma L, Qiao Q, Tsai JW, Yang G, Li W, Gan WB. Experience-dependent plasticity of dendritic spines of layer 2/3 pyramidal neurons in the mouse cortex. *Developmental neurobiology*. 2016;76(3):277–286.
53. Miquelajauregui A, Kribakaran S, Mostany R, Badaloni A, Consalez GG, Portera-Cailliau C. Layer 4 pyramidal neurons exhibit robust dendritic spine plasticity in vivo after input deprivation. *J Neurosci*. 2015;35 (18):7287–7294.
54. Tashiro A, Dunaevsky A, Blazeski R, Mason CA, Yuste R. Bidirectional regulation of hippocampal mossy fiber filopodial motility by kainate receptors: a two-step model of synaptogenesis. *Neuron*. 2003;38(5):773–784.
55. Tanaka J, Horiike Y, Matsuzaki M, Miyazaki T, Ellis-Davies GC, Kasai H. Protein synthesis and neurotrophin-dependent structural plasticity of single dendritic spines. *Science*. 2008;319(5870):1683–1687.

56. Penzes P, Jones KA. Dendritic spine dynamics—a key role for kalirin-7. *Trends Neurosci*. 2008;31(8):419–427.
57. Saneyoshi T, Hayashi Y. The Ca2 + and Rho GTPase signaling pathways underlying activity-dependent actin remodeling at dendritic spines. *Cytoskeleton*. 2012;69(8):545–554.
58. Harward SC, Hedrick NG, Hall CE, et al. Autocrine BDNF-TrkB signalling within a single dendritic spine. *Nature*. 2016;538(7623):99–103.
59. Hedrick NG, Harward SC, Hall CE, Murakoshi H, McNamara JO, Yasuda R. Rho GTPase complementation underlies BDNF-dependent homo- and heterosynaptic plasticity. *Nature*. 2016;538(7623):104–108.
60. Hayashi-Takagi A, Yagishita S, Nakamura M, et al. Labelling and optical erasure of synaptic memory traces in the motor cortex. *Nature*. 2015;525(7569):333–338.
61. Mostany R, Anstey JE, Crump KL, Maco B, Knott G, Portera-Cailliau C. Altered synaptic dynamics during normal brain aging. *J Neurosci*. 2013;33(9):4094–4104.
62. De Paola V, Holtmaat A, Knott G, et al. Cell type-specific structural plasticity of axonal branches and boutons in the adult neocortex. *Neuron*. 2006;49(6):861–875.
63. Stettler DD, Yamahachi H, Li W, Denk W, Gilbert CD. Axons and synaptic boutons are highly dynamic in adult visual cortex. *Neuron*. 2006;49(6):877–887.
64. Grillo FW, Song S, Teles-Grilo Ruivo LM, et al. Increased axonal bouton dynamics in the aging mouse cortex. *Proc Natl Acad Sci U S A*. 2013;110(16):E1514–E1523.
65. Johnson CM, Peckler H, Tai LH, Wilbrecht L. Rule learning enhances structural plasticity of long-range axons in frontal cortex. *Nat Commun*. 2016;7:10785.
66. Stepanyants A, Hof PR, Chklovskii DB. Geometry and structural plasticity of synaptic connectivity. *Neuron*. 2002;34(2):275–288.
67. Stepanyants A, Chklovskii DB. Neurogeometry and potential synaptic connectivity. *Trends Neurosci*. 2005;28(7):387–394.
68. Kaiser M, Hilgetag CC, van Ooyen A. A simple rule for axon outgrowth and synaptic competition generates realistic connection lengths and filling fractions. *Cerebral cortex*. 2009;19(12):3001–3010.
69. Kasthuri N, Hayworth KJ, Berger DR, et al. Saturated reconstruction of a volume of neocortex. *Cell*. 2015;162(3):648–661.
70. Braitenberg VaS A. *Peters' rule and White's exception. Anatomy of the Cortex*. New York, NY: Springer-Verlag Publishing; 1991:109–112.
71. Peters A, Feldman M, Saldanha J. The projection of the lateral geniculate nucleus to area 17 of the rat cerebral cortex. II. Terminations upon neuronal perikarya and dendritic shafts. *J Neurocytol*. 1976;5(1):85–107.
72. Jiang X, Shen S, Cadwell CR, et al. Principles of connectivity among morphologically defined cell types in adult neocortex. *Science*. 2015;350(6264):aac9462.
73. Llinas RaH DE. Physiological and morpholigical organization of the cerebellar circuits in various vertebrates. In: Llinas R, ed. *Neurobiology of Cerebellar Evolution and Development*. Chicago, IL: American Medical Association; 1969:43–73.
74. Rall W. Dynamic patterns of brain cell assemblies. II. Concept of dynamic patterns. Nonequilibrium steady states and nerve membrane biophysics. *Neurosci Res Program Bull*. 1974;12(1):27–30.
75. Rinzel J, Rall W. Transient response in a dendritic neuron model for current injected at one branch. *Biophys J*. 1974;14(10):759–790.
76. Gomez Gonzalez JF, Mel BW, Poirazi P. Distinguishing linear vs. non-linear integration in CA1 radial oblique dendrites: it's about time. *Front Comput Neurosci*. 2011;5:44.
77. Behabadi BF, Polsky A, Jadi M, Schiller J, Mel BW. Location-dependent excitatory synaptic interactions in pyramidal neuron dendrites. *PLoS Comput Biol*. 2012;8(7):e1002599.
78. Cash S, Yuste R. Linear summation of excitatory inputs by CA1 pyramidal neurons. *Neuron*. 1999;22(2):383–394.
79. Tamas G, Szabadics J, Somogyi P. Cell type- and subcellular position-dependent summation of unitary postsynaptic potentials in neocortical neurons. *J Neurosci*. 2002;22(3):740–747.
80. Polsky A, Mel BW, Schiller J. Computational subunits in thin dendrites of pyramidal cells. *Nat Neurosci*. 2004;7(6):621–627.
81. Araya R, Eisenthal KB, Yuste R. Dendritic spines linearize the summation of excitatory potentials. *Proc Natl Acad Sci U S A*. 2006;103(49):18799–18804.
82. Gasparini S, Magee JC. State-dependent dendritic computation in hippocampal CA1 pyramidal neurons. *J Neurosci*. 2006;26(7):2088–2100.

83. Losonczy A, Magee JC. Integrative properties of radial oblique dendrites in hippocampal CA1 pyramidal neurons. *Neuron*. 2006;50(2):291–307.
84. Carter AG, Soler-Llavina GJ, Sabatini BL. Timing and location of synaptic inputs determine modes of subthreshold integration in striatal medium spiny neurons. *J Neurosci*. 2007;27(33):8967–8977.
85. Harnett MT, Makara JK, Spruston N, Kath WL, Magee JC. Synaptic amplification by dendritic spines enhances input cooperativity. *Nature*. 2012;491(7425):599–602.
86. Major G, Polsky A, Denk W, Schiller J, Tank DW. Spatiotemporally graded NMDA spike/plateau potentials in basal dendrites of neocortical pyramidal neurons. *J Neurophysiol*. 2008;99(5):2584–2601.
87. Larkum ME, Nevian T, Sandler M, Polsky A, Schiller J. Synaptic integration in tuft dendrites of layer 5 pyramidal neurons: a new unifying principle. *Science*. 2009;325(5941):756–760.
88. Polsky A, Mel B, Schiller J. Encoding and decoding bursts by NMDA spikes in basal dendrites of layer 5 pyramidal neurons. *J Neurosci*. 2009;29(38):11891–11903.
89. Tran-Van-Minh A, Abrahamsson T, Cathala L, DiGregorio DA. Differential dendritic integration of synaptic potentials and calcium in cerebellar interneurons. *Neuron*. 2016;91(4):837–850.
90. Yuste R. Dendritic spines and distributed circuits. *Neuron*. 2011;71(5):772–781.
91. Tran-Van-Minh A, Caze RD, Abrahamsson T, Cathala L, Gutkin BS, DiGregorio DA. Contribution of sublinear and supralinear dendritic integration to neuronal computations. *Frontiers in cellular neuroscience*. 2015;9:67.
92. Jia H, Rochefort NL, Chen X, Konnerth A. Dendritic organization of sensory input to cortical neurons in vivo. *Nature*. 2010;464(7293):1307–1312.
93. Varga Z, Jia H, Sakmann B, Konnerth A. Dendritic coding of multiple sensory inputs in single cortical neurons in vivo. *Proc Natl Acad Sci U S A*. 2011;108(37):15420–15425.
94. Chen X, Leischner U, Rochefort NL, Nelken I, Konnerth A. Functional mapping of single spines in cortical neurons in vivo. *Nature*. 2011;475(7357):501–505.
95. Kleindienst T, Winnubst J, Roth-Alpermann C, Bonhoeffer T, Lohmann C. Activity-dependent clustering of functional synaptic inputs on developing hippocampal dendrites. *Neuron*. 2011;72(6):1012–1024.
96. Takahashi N, Kitamura K, Matsuo N, et al. Locally synchronized synaptic inputs. *Science*. 2012;335 (6066):353–356.
97. Wilson DE, Whitney DE, Scholl B, Fitzpatrick D. Orientation selectivity and the functional clustering of synaptic inputs in primary visual cortex. *Nat Neurosci*. 2016;19(8):1003–1009.
98. Winnubst J, Lohmann C. Synaptic clustering during development and learning: the why, when, and how. *Front Mol Neurosci*. 2012;5:70.
99. Gokce O, Bonhoeffer T, Scheuss V. Clusters of synaptic inputs on dendrites of layer 5 pyramidal cells in mouse visual cortex. *eLife*. 2016;19:5.
100. da Costa NM, Martin KA. How thalamus connects to spiny stellate cells in the cat's visual cortex. *J Neurosci*. 2011;31(8):2925–2937.
101. Schoonover CE, Tapia JC, Schilling VC, et al. Comparative strength and dendritic organization of thalamocortical and corticocortical synapses onto excitatory layer 4 neurons. *J Neurosci*. 2014;34(20):6746–6758.
102. Rah JC, Bas E, Colonell J, et al. Thalamocortical input onto layer 5 pyramidal neurons measured using quantitative large-scale array tomography. *Front Neural Circuits*. 2013;7:177.
103. Druckmann S, Feng L, Lee B, et al. Structured synaptic connectivity between hippocampal regions. *Neuron*. 2014;81(3):629–640.
104. Larkum ME, Nevian T. Synaptic clustering by dendritic signalling mechanisms. *Curr Opin Neurobiol*. Jun 2008;18(3):321–331.
105. McBride TJ, Rodriguez-Contreras A, Trinh A, Bailey R, Debello WM. Learning drives differential clustering of axodendritic contacts in the barn owl auditory system. *J Neurosci*. 2008;28(27):6960–6973.
106. Knudsen EI. Instructed learning in the auditory localization pathway of the barn owl. *Nature*. 2002;417 (6886):322–328.
107. McBride TJ, DeBello WM. Input clustering in the normal and learned circuits of adult barn owls. *Neurobiol Learn Mem*. 2015;121:39–51.
108. Makino H, Malinow R. Compartmentalized versus global synaptic plasticity on dendrites controlled by experience. *Neuron*. 2011;72(6):1001–1011.
109. Reiner BC, Dunaevsky A. Deficit in motor training-induced clustering, but not stabilization, of new dendritic spines in FMR1 knock-out mice. *PloS one*. 2015;10(5):e0126572.

110. Lee KJ, Park IS, Kim H, Greenough WT, Pak DT, Rhyu IJ. Motor skill training induces coordinated strengthening and weakening between neighboring synapses. *J Neurosci*. 2013;33(23):9794–9799.
111. Pereira AC, Lambert HK, Grossman YS, et al. Glutamatergic regulation prevents hippocampal-dependent age-related cognitive decline through dendritic spine clustering. *Proc Natl Acad Sci U S A*. 2014;111(52):18733–18738.
112. De Roo M, Klauser P, Muller D. LTP promotes a selective long-term stabilization and clustering of dendritic spines. *PLoS Biol*. 2008;6(9):e219.
113. Harvey CD, Svoboda K. Locally dynamic synaptic learning rules in pyramidal neuron dendrites. *Nature*. 2007;450(7173):1195–1200.
114. Chen JL, Villa KL, Cha JW, So PT, Kubota Y, Nedivi E. Clustered dynamics of inhibitory synapses and dendritic spines in the adult neocortex. *Neuron*. 2012;74(2):361–373.
115. Oh WC, Parajuli LK, Zito K. Heterosynaptic structural plasticity on local dendritic segments of hippocampal CA1 neurons. *Cell Rep*. 2015;10(2):162–169.
116. Weber JP, Andrasfalvy BK, Polito M, Mago A, Ujfalussy BB, Makara JK. Location-dependent synaptic plasticity rules by dendritic spine cooperativity. *Nat Commun*. 2016;7:11380.
117. Harvey CD, Yasuda R, Zhong H, Svoboda K. The spread of Ras activity triggered by activation of a single dendritic spine. *Science*. 2008;321(5885):136–140.
118. Govindarajan A, Kelleher RJ, Tonegawa S. A clustered plasticity model of long-term memory engrams. *Nat Rev Neurosci*. 2006;7(7):575–583.
119. Bian WJ, Miao WY, He SJ, Qiu Z, Yu X. Coordinated spine pruning and maturation mediated by inter-spine competition for cadherin/catenin complexes. *Cell*. 2015;162(4):808–822.
120. Winnubst J, Cheyne JE, Niculescu D, Lohmann C. Spontaneous activity drives local synaptic plasticity in vivo. *Neuron*. 2015;87(2):399–410.
121. Muller C, Beck H, Coulter D, Remy S. Inhibitory control of linear and supralinear dendritic excitation in CA1 pyramidal neurons. *Neuron*. 2012;75(5):851–864.
122. DeBello WM. Micro-rewiring as a substrate for learning. *Trends Neurosci*. 2008;31(11):577–584.
123. Zippo AG, Gelsomino G, Van Duin P, et al. Small-world networks in neuronal populations: a computational perspective. *Neural Netw*. 2013;44:143–156.
124. Nigam S, Shimono M, Ito S, et al. Rich-club organization in effective connectivity among cortical neurons. *J Neurosci*. 2016;36(3):670–684.
125. Gutierrez GJ, Marder E. Modulation of a single neuron has state-dependent actions on circuit dynamics. *eNeuro*. 2014;1:1.
126. van den Heuvel MP, Sporns O. Rich-club organization of the human connectome. *J Neurosci*. 2011;31 (44):15775–15786.
127. Zippo AG, Biella GE. Quantifying the number of discriminable coincident dendritic input patterns through dendritic tree morphology. *Sci Rep*. 2015;5:11543.
128. Binzegger T, Douglas RJ, Martin KA. A quantitative map of the circuit of cat primary visual cortex. *J Neurosci*. 2004;24(39):8441–8453.
129. Kalisman N, Silberberg G, Markram H. The neocortical microcircuit as a tabula rasa. *Proc Natl Acad Sci U S A*. 2005;102(3):880–885.
130. Amirikian B. A phenomenological theory of spatially structured local synaptic connectivity. *PLoS computational biology*. 2005;1(1):e11.
131. Lang S, Drouvelis P, Tafaj E, Bastian P, Sakmann B. Fast extraction of neuron morphologies from large-scale SBFSEM image stacks. *J Comput Neurosci*. 2011;31(3):533–545.
132. Oberlaender M, de Kock CP, Bruno RM, et al. Cell type-specific three-dimensional structure of thalamocortical circuits in a column of rat vibrissal cortex. *Cerebral cortex*. 2012;22(10):2375–2391.
133. Hill SL, Wang Y, Riachi I, Schurmann F, Markram H. Statistical connectivity provides a sufficient foundation for specific functional connectivity in neocortical neural microcircuits. *Proc Natl Acad Sci U S A*. 2012;109(42): E2885–E2894.
134. Ramaswamy S, Hill SL, King JG, Schurmann F, Wang Y, Markram H. Intrinsic morphological diversity of thick-tufted layer 5 pyramidal neurons ensures robust and invariant properties of in silico synaptic connections. *J Physiol*. 2012;590(4):737–752.
135. Ascoli GA. Potential connectomics complements the endeavour of 'no synapse left behind' in the cortex. *J Physiol*. 2012;590(4):651–652.

136. da Costa NM, Martin KA. Sparse reconstruction of brain circuits: or, how to survive without a microscopic connectome. *NeuroImage*. 2013;80:27–36.
137. Reimann MW, King JG, Muller EB, Ramaswamy S, Markram H. An algorithm to predict the connectome of neural microcircuits. *Front Comput Neurosci*. 2015;9:120.
138. Markram H, Muller E, Ramaswamy S, et al. Reconstruction and Simulation of Neocortical Microcircuitry. *Cell*. 2015;163(2):456–492.
139. Bock DD, Lee WC, Kerlin AM, et al. Network anatomy and in vivo physiology of visual cortical neurons. *Nature*. 2011;471(7337):177–182.
140. Hiratani N, Fukai T. Hebbian wiring plasticity generates efficient network structures for robust inference with synaptic weight plasticity. *Front Neural Circuits*. 2016;10:41.
141. Alivisatos AP, Chun M, Church GM, Greenspan RJ, Roukes ML, Yuste R. The brain activity map project and the challenge of functional connectomics. *Neuron*. 2012;74(6):970–974.
142. Alivisatos AP, Chun M, Church GM, et al. Neuroscience. The brain activity map. *Science*. 2013;339 (6125):1284–1285.
143. Alivisatos AP, Andrews AM, Boyden ES, et al. Nanotools for neuroscience and brain activity mapping. *ACS Nano*. 2013;7(3):1850–1866.
144. Ifft PJ, Shokur S, Li Z, Lebedev MA, Nicolelis MA. A brain-machine interface enables bimanual arm movements in monkeys. *Sci Transl Med*. 2013;5(210):210ra154.
145. Schwarz DA, Lebedev MA, Hanson TL, et al. Chronic, wireless recordings of large-scale brain activity in freely moving rhesus monkeys. *Nat Methods*. 2014;11(6):670–676.
146. Nicolelis MA. Brain-to-Brain Interfaces: When Reality Meets Science Fiction. *Cerebrum*. 2014;2014:13.
147. Merolla PA, Arthur JV, Alvarez-Icaza R, et al. Artificial brains. A million spiking-neuron integrated circuit with a scalable communication network and interface. *Science*. 2014;345(6197):668–673.
148. Taigman Y, Yang M, Ranzato M, Wolf L. Deepface: Closing the gap to human-level performance in face verification. In: Proceedings of the 2014 IEEE Conference on Computer Vision and Pattern Recognition (CVPR). IEEE: Washington, DC; 2014:1701–1708.

CHAPTER

15

Impact of Structural Plasticity on Memory Capacity

Panagiotis Bozelos[1,2] *and Panayiota Poirazi*[1]

[1]Foundation for Research and Technology-Hellas (FORTH), Crete, Greece

[2]Democritus University of Thrace, Alexandroupolis, Greece

OUTLINE

The Rewiring Brain
DOI: http://dx.doi.org/10.1016/B978-0-12-803784-3.00015-9

1 INTRODUCTION

After years of intensive research and according to a large body of experimental evidence, memories are thought to be encoded in the brain as tiny chunks of information distributed among sparse neural networks.[1,2] The encoding process begins immediately after learning, with the initiation of a well-orchestrated, temporally evolving cascade of biochemical and structural alterations in the underlying neural circuits.[3,4] Intriguingly, this diverse set of processes exploits different levels of neural and biochemical organization and is increasingly thought to be coordinated by the concurrent activation and interaction between large brain structures.[5,6] The resultant outcome of reorganizational activity decides the fate of potential memories and reflects multiple functional aspects of adaptive plasticity in regard to learning and memory capacity. Nonetheless, the effective range and scope of structural plasticity mechanisms in relation to memory is not yet quite understood.

1.1 Chapter Organization and Content

In the first part of this chapter, we provide a brief overview of some landmark experimental studies that shed light on various forms of structural plasticity. As we navigate through the numerous examples of well-studied paradigms, we put a strong emphasis on the structural alterations that happen on the microscale, such as the synapses, dendrites, and axons. In parallel, we attempt to sort through the possible implications for memory capacity and to provide a tangible link with the current state of understanding in the field.

Despite the abundance of experimental data, a comprehensive theoretical framework that binds the various aspects of structural plasticity with regard to learning and memory is still subject to investigation. One problem is associated with the largely unknown network of interdependent factors that are involved in the processes underlying learning-induced structural plasticity.[7] The complex nature of the underlying phenomena makes it difficult to dissect and isolate them for any subsequent integration efforts into a unifying theoretical model. Another difficulty traces back to the lack of suitable memory metrics to capture the full extent of learning phenomena in the activated neural networks, beyond their obvious behavioral manifestation.

In this context, we discuss in the second part of this chapter a number of computational approaches that simulate learning paradigms in model neurons and neural networks, while accounting for structural plasticity and its effects on information storage. While the big, unifying picture is still missing, the theoretical approaches provide an invaluable toolset toward exploring the potential advantages of structural plasticity for neuronal function, learning, and memory. We conclude by proposing novel hypotheses and future experiments that may help further our understanding as to how structural plasticity contributes to the brain's capacity to form, store, and recall new memories.

2 FORMS OF STRUCTURAL PLASTICITY

2.1 Synaptogenesis

During the early stages of life, there is an overproduction (and turnover) of excitatory synapses that continuously form onto (and die off) specialized protrusions of dendrites called dendritic spines.[8,9] Over the course of this dynamic process, which is generally called synaptogenesis, a heuristic exploration of the potential neural contact space is accomplished through the continuous creation, fixation, or elimination of innumerable synapses. Synaptogenesis is largely dependent on the spontaneous, but still patterned, electrophysiological activity of developing neurons[10,11] and obeys a diverse set of potentiation and depression rules.[12,13] Among other biological effects, synaptogenesis actively configures the geometry, topology, and connectivity of the developing neural circuitry.[14,15] Following development, there is a period of marked selective synaptic pruning, speculated to deliver an optimal implementation of neural function and performance.[16]

The intense manifestation of structural plasticity is partly ceased in later stages of life; however, synaptogenesis continues to mold finer scales of the neural tissue, as the organism reacts and adapts to new experiences. Experiments in vertebrate and invertebrate animal models indicate that, during learning, reversible anatomical and biophysical changes occur in the nervous system, and that their stabilization is required for memory consolidation.[17,18] In particular, short-term plasticity has been correlated with rapid molecular changes in cytoskeletal and adhesion components, while the persistence of memory appears to be dependent on de novo protein synthesis.[19,20] These kinds of alterations are triggered by signaling cascades and act upon multiple scales of neural structures that range from as small as lipid nanodomains[21] of the postsynaptic density, to as large as the entire dendritic arbor.[22]

2.2 Filopodia Dynamics

A considerable amount of research has been directed toward the investigation of filopodia dynamics and so far has revealed that numerous protrusions of the dendritic shaft emerge and retract on time scales of seconds to minutes.[23] These small dendritic specializations, which are much longer than spines and without an apparent bulbous head,[24] are considered to be premature forms of spines, ready to participate in synaptogenesis if contacted by an axon and stabilized by the right pattern of incoming stimuli.[25,26] During the early postnatal period in mice, even when filopodia dynamics are significantly increased,[8,27] only 25% of them grow to become a mature spine with a synapse.[24]

Interestingly, however, in the adult neuropil, nonsynaptic filopodia are less likely to contract back to the dendritic shaft[28] and often convert directly into spines with bulbous heads.[29] This can be explained by the enriched local environment of mature neurons that can supply a filopodium with abundant resources of proteins, mRNA, and organelles to support its maturation.[30] Additionally, Yuste et al. have suggested that there are two functionally distinct populations of filopodia, relying on different stabilization mechanisms.[31] The first one emerges on growth cones of dendritic branches and is stabilized in an activity-independent manner and the latter one is found on shafts and proves to be responsible for activity-dependent synaptogenesis.

Not surprisingly, the dynamic nature of filopodia is stimulated under conditions of increased synaptic activity.[32,33] Work in organotypic hippocampal slices, conducted by Maletic-Savatic and coworkers, revealed a substantial increase in filopodia dynamics after dendritic stimulation with an electrode.[34] Researchers observed under the two-photon laser scanning microscope an enhanced growth of existing filopodia and an emergence of new ones within minutes of stimulation and in the close proximity of the target stimulation site. Interestingly, increased dynamics were apparent for at least 30 minutes after stimulation, but it was not further assessed if any activity-induced filopodia were potentiated to form stable spines. Nevertheless, increased filopodia dynamics due to ongoing neural activity provide a window of opportunity to establish connections with recently active axon terminals and thus suggest a viable mechanism for Hebbian synaptic plasticity.

More recently, Lohman et al. attempted to shed some light on the mechanisms that influence filopodia dynamics, and more specifically their emergence, stabilization, or elimination. They investigated the role of calcium signaling by simultaneously recording filopodia outgrowth and calcium transients in rat hippocampal slice cultures.[35] They observed a high correlation between low frequency of calcium signaling events and emergence of filopodia, whereas increased frequency was accompanied by stabilization of pre-existing filopodia and rare emergence of new ones. Accordingly, they replicated this finding by experimentally manipulating the frequency of calcium transient and found the same correlations. In a later study, they built up on their results by probing how calcium signaling is exploited by filopodia to either select a synaptic partner or perish.[36] It was found that early axonal contact, followed by high calcium transient frequency, is sufficient to determine the survival of a filopodium, even before it grows into a mature spine. It becomes apparent that dynamic filopodia function not only as precursors to isolated spine-like biochemical compartments, but can readily integrate the synaptic activation effects with local dendritic processing, to ensure the encoding of new information via structural plasticity.

2.3 Ca^{2+} Influx as a Key Determinant of Synaptic Plasticity and Structural Rearrangements

One key associative learning mechanism is Hebbian plasticity, which postulates that coincident activation of pre- and postsynaptic neurons enhances the intersynaptic transmission efficacy by inducing a set of durable structural modifications.[37] It is believed that the initiatory signal of Hebbian plasticity constitutes of calcium influx into the postsynaptic neurons via the excitatory *N*-methyl-D-aspartate (NMDA) receptors and the voltage-gated calcium channels (VGCC).[38] Intriguingly, while calcium conductance through NMDA receptors is reversibly blocked by magnesium at a resting voltage, the inhibition is alleviated when postsynaptic depolarization occurs by spike-triggering events at other synapses.[39] Besides the emerging synaptic cooperativity property, calcium influx is sensitive to the temporal coincidence of presynaptic glutamate release events and the postsynaptic depolarization activity.[40] Thus the NMDA receptor acts as a coincidence detector. Additionally, when the plasma membrane is depolarized, either due to recent excitatory postsynaptic potentials (EPSPs) or backpropagating action potentials (BAPs), VGCC channels are activated and allow calcium to flow into the cell.[41] BAPs are considered to communicate a signal integration event at the axon hillock compartment, thus providing a positive feedback to remote dendrites,[42] which is incorporated in the widely used spike-timing-dependent plasticity (STDP) rule.[43]

Essentially, the sufficient elevation of intracellular calcium concentration facilitates synaptic plasticity by triggering a variety of signal transduction pathways. A significant molecular mediator to these biochemical processes is the calmodulin-dependent phosphorylation of constitutively present calcium/calmodulin-dependent protein kinases (CaMKs) and protein kinase C, leading to an increased accumulation of AMPA receptors onto the postsynaptic density (PSD) and to rapid actin polymerization.[44,45] This range of immediate structural rearrangements can support memory persistence without the implication of nuclear activity-dependent gene expression and dendritic protein synthesis.[46]

2.4 Morphological Changes of the Dendritic Spines

In addition to filopodia, emerging and retracting, other, more persistent dendritic protrusions, the so-called spines, are characterized by a bulbous head connected to the dendritic shaft via a thin neck.[9] Not surprisingly, the isolated dendritic spine compartments are most apt for rapid changes to occur in response to enhanced neural activity and novel incoming stimuli.[47] Long-term potentiation (LTP) and depression (LTD) are the most widely used protocols for inducing persistent spine modifications (strengthening/weakening), when specific electrical input patterns are applied to the cell. LTP was first described by Bliss and Lomo, who reported that high frequency stimulation training to the perforant pathway can increase the transmission efficacy in the granule cells of the dentate gyrus.[48] Around 1980s, LTD was also discovered[49] and since then it has been extensively documented.[50] LTP and LTD have been suggested to serve as the underlying cellular manifestations of learning by which novel experiences modulate the neural transmission efficacy.[51,52] Experimentalists can enforce the potentiation of dendritic spines through high frequency stimulation of axonal afferents by 100 Hz or by applying "theta burst" of 3–4 pulses of stimuli at 100 Hz, each separated by time intervals of 200 ms. Conversely, LTD can be induced by applying low frequency presynaptic stimulation (900 or 1800 stimuli, 1 Hz) without triggering pronounced postsynaptic activity events.

Since the discovery of LTP, a substantial number of studies have attempted to provide valuable insights into the structural reorganization of synaptic architecture, mainly by measuring the number or shape of spines. One of the first studies reporting a statistical correlation between morphological changes of dendritic spines and synaptic activity enhancement was conducted by Van Harreveld and Fifkova.[53] The researchers employed an LTP stimulation protocol near the distal dendritic region of granule cells and subsequently observed under the electron microscope (EM) a significant increase of 40%–53% of the spine area spine in comparison with an unstimulated control area. Additionally, they reported a shortening and widening of the neck of the spines. An even more accentuated effect of LTP was reported by a similar study conducted by the Andersen group.[54] The investigators used a serial reconstruction EM after applying in vivo LTP to detect a 30%–50% increase of the spine number. It's worth noting that the technical implementation of the aforementioned approaches couldn't really point out structural changes of particular stimulated spines. However, later studies addressed this shortcoming by marking the calcium precipitate in the stimulated spines.[55]

In more recent studies, the use of advanced live single- or two-photon microscopy allowed a more direct observation of the actual structural changes happening at stimulated synapses. Typically, the researchers prepare acute or organotypic slices taken from mice that are transfected with a fluorescent protein marker in various experimental ways, and more recently have managed to put into effect in vivo manipulations.[9] Hosokawa et al. described for the first time an extension of spine length and signs of increased spine motility when hippocampal pyramidal neurons of rats were potentiated by LTP.[56] Counterintuitively, new spines were not detected. This finding was reported in live preparations by Maletic-Savatic and coworkers, by using organotypic hippocampal slices trained with intense synaptic stimulation.[34] The researchers reported the generation of filopodia, along with the extension of preexisting ones, but didn't confirm any functional enhancement of transmission efficacy. Furthermore, several studies have shown that dendritic spines change in volume after synaptic potentiation. Matsuzaki et al. reported a significant increase of more than 50% in spine volume within 1–5 minutes after high frequency training stimulation and observed that smaller spines were able to hold the change for longer time than larger ones.[57] This study also confirmed that the structural changes are AMPA receptor mediated and dependent on NMDA receptor activation.

Taken together, the advent of modern imaging instrumentation made it feasible to follow and describe the various structural changes that are associated with LTP; however, the overall impact of structural plasticity events is difficult to isolate and interpret without the use of model abstractions, which we will later elaborate on.

2.5 Learning-Induced Synapse Clustering

Apart from LTP/LTD-induced alterations, behavioral paradigms have also been used to study learning-induced structural modifications in the *arrangement* of synaptic contacts. Barn owls reared with prismatic spectacles were shown to have increased "clustering" (namely the formation of synaptic contacts located within a short distance, in this case <20 microns). These axodendritic contacts (potential synapses) formed clusters in the absence of obvious changes in the total number of synapses,[58] indicating that synaptic contacts were both created and eliminated. By imaging the formation of spines in motor cortex dendrites during the learning of a repeated motor task, Fu et al. showed that the majority of newly formed spines were located closer (within distances <5 μm) to preexisting synapses than control spines.[59] This study showed that newly formed spines are highly likely to be added to the existing, stable clusters, thus contributing to the refinement and stability of motor learning. Increased anatomical clustering of potentiated synapses has also been observed in an in vitro study which simulated spaced learning in the hippocampus.[60] Makino and Malinow visualized the trafficking of AMPA receptors during normal sensory experience as well as during sensory deprivation (whisker removal) in mice.[61] The authors showed that clustering of GluR1 subunits to nearby synapses on the dendrites of somatosensory neurons was evident in control but not sensory-deprived animals, indicating that rich sensory experience results in synapse clustering. Finally, Yang et al. showed that different learning tasks cause spine formation on different sets of dendritic branches after sleep consolidation.[62] The formation of these spines was dependent on the reactivation during sleep of the same population of neurons that were active during learning. It is important to note that spatial clustering of synapses facilitates the activation of voltage-gated conductances, such as the NMDA receptors, thus aiding the generation of dendritic spikes, an important hallmark of successful neuronal signaling and initiator of many plasticity mechanisms.

2.6 Dendritic and Axonal Remodeling

Other forms of learning-induced structural plasticity include dendritic and axonal remodeling. In previous studies, it had been shown that dendritic development and maturation is dependent on presynaptic innervation, which is considered to act like a stabilization factor regulating selective dendritic maturation.[8] Recently, Tronel et al. have shown that increased learning demands due to spatial exploration can induce dendritic arbor alterations in rat hippocampal dentate gyrus (DG) neurons that are dependent on NMDA receptor activation.[22] In more detail, they used doublecortin (Dcx) and retroviral labeling to elucidate the effects of Morris water navigation task on the neo-dendritic maturation of adult-born neurons. The spatial learning task was shown to induce a marked increase in dendritic complexity, including increases in the total length, the number of branch points and the number of terminal points. Interestingly, the observed changes in dendritic arborization were specific to adult-born neurons.

Apart from explicit training protocols, environmental enrichment has also been found to induce structural modifications. Leggio et al. reared Wistar rats in cages that promoted exploration, manipulation of objects (toys, constructions, etc.), and social interactions and then proceeded to a morphological comparison analysis of layer III pyramidal neurons of parietal cortex with a control group of rats reared in standard conditions (two rats in a cage without objects).[63] The researchers demonstrated a broad range of statistically significant structural alterations, including increased dendritic arborization, as well as increased spine density of both the basal and apical regions of the tree. The researchers proposed that the observed morphological changes may be the result of putative plasticity mechanisms that lead to functional reorganization of the neural structure, to help the animal cope with the increased sensory complexity of the environment.

Complementary to dendritic remodeling, there is now persuading evidence for presynaptic axonal bouton and side-branches structural plasticity. Axonal remodeling might be substantially pervasive due to the extensive ramifications of axons, which usually sample relatively large neural volumes and, thus, correlate well with learning-induced rewiring. In a series of two studies conducted by Ramírez-Amaya in 1999 and 2001, it was shown that mossy fibers of rats greatly expand following a water maze task, inducing synaptogenesis in the CA3 stratum oriens (SO) field[64] and, subsequently, that the performance of the animals was linked to the extent of mossy fiber innervation to CA3.[65] The researchers theorized that increased synaptic activity promotes the structural remodeling of presynaptic terminal boutons, in order to convert novel environmental stimuli into network connectivity changes.

2.7 Structural Plasticity and Dendritic Nonlinearities

The effects of structural modifications in synaptic and dendritic properties on neural output are greatly enhanced by the presence of dendritic nonlinearities. Dendrites of pyramidal neurons are furnished with a variety of biophysical mechanisms (for extensive reviews, see Refs.[66–70]) that allow them to integrate synaptic inputs in sublinear, linear, or supralinear ways.[67,70–76] Coupled with different plasticity rules, like spike-timing-dependent plasticity (STDP),[43,77] synaptic tagging and capture (STC), calcium-dependent Hebbian LTP/LTD, homeostatic plasticity, branch strength potentiation,[78] plasticity of inhibition and intrinsic excitability, these nonlinearities enable individual neurons as well as neural networks to acquire robust learning capabilities. These capabilities have been investigated via the use of computational and theoretical approaches.

3 COMPUTATIONAL MODELS OF STRUCTURAL PLASTICITY AND MEMORY FORMATION

Theoretical models of structural plasticity can generally be categorized into three major classes: (1) models that account for synaptic weight plasticity, whereby the strength of individual synapses is modified but de novo generation of synapses is not implemented, (2) models that simulate both weight and rewiring changes via the implementation of synapse formation/elimination, and (3) models that account for the remodeling of dendritic arbors. While all of these models are based on rather simplified notions about the mechanisms of structural plasticity, either on fast and or slower time scales, they provide a valuable toolset toward a better understanding of the workings of memory at the cellular and dendritic levels.

Plasticity models limited to synaptic weight modifications are numerous and widely studied. For an in-depth analysis, please see Refs.[79–81] In this chapter, we focus on computational models that account for rewiring plasticity[82] (namely synapse formation/elimination that changes the connectivity diagram between neurons) and/or plasticity of dendritic arbors and their effects on information storage. The following paragraphs provide brief descriptions of the most representative of these models.

In 2001, Poirazi and Mel introduced a simplified model of a neuron with active or passive dendrites in order to investigate whether changes in the wiring diagram, namely the formation and elimination of synapses at different locations within a neuron, guided by local plasticity rules that rely on dendritic nonlinearities, may influence the storage capacity of neural tissue.[83] Toward this goal, the authors developed a simplified model of a stellate-like neuron with m identical dendritic branches connected directly to a cell body (graphically depicted in Fig. 15.1). Each dendritic branch received k excitatory synaptic contacts originating from d presynaptic neurons. The total number of synaptic contacts onto a model neuron (s) was equal to k^*m. Dendritic outputs were modeled either as a linear sum of their weighted synaptic inputs (linear model) or as a weighted sum passed through a static branch nonlinearity b (nonlinear model). Weights were only allowed to take small, positive integer values ($0 < w < l$, where l takes the values of 1, 2, 3, or 4 and represents the number of states a synapse can be found in; binary synapses have an $l = 1$). Given that synaptic weights (w) and inputs (x) are not allowed to have negative values, in order to account for inhibitory signals the output of each model cell ($y_L(x)$ or $y_N(x)$) was determined by the subtraction of the output of an inhibitory ($a_L^-(x)$ or $a_N^-(x)$) neuron from the output of an excitatory ($a_L^+(x)$ or $a_N^+(x)$) neuron. As a result, neuronal outputs

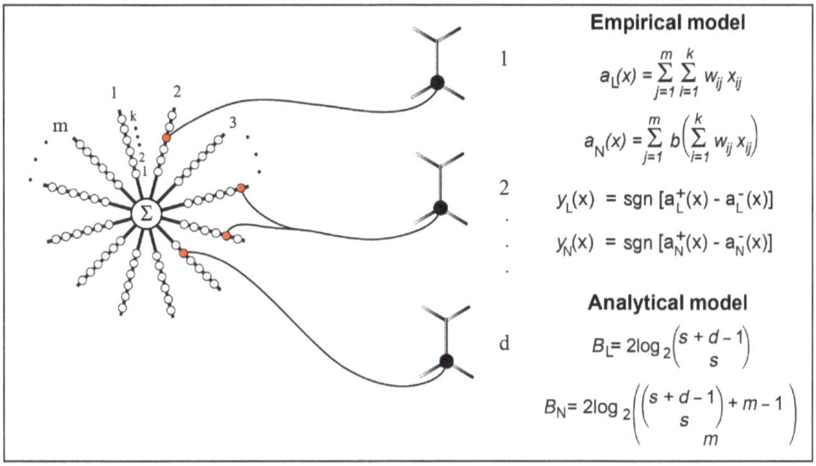

FIGURE 15.1 **Schematic of the simplified neuron model implemented by Poirazi and Mel[83] (*left*) and the respective empirical and analytical equations (*right*).** In more detail, $a_L(x)$ and $a_N(x)$ are the cell's activation levels in the linear and nonlinear models, respectively, before any output thresholding. These activation levels can correspond to either an excitatory ($a_{L,N}{}^{+}(x)$) or an inhibitory ($a_{L,N}{}^{-}(x)$) neuron. $y_L(x)$ and $y_N(x)$, respectively, are the thresholded outputs of each neuron pair and correspond to binary-valued "discriminant functions" that if positive indicates a preference for the x pattern and, likewise, if negative a nonpreference for the particular pattern. The variable w_{ij} denotes the weight of the ith synapse on the jth branch, and x_{ij} represents the activity arriving at the ith synapse on the jth branch. B_L calculates the number of different ways of assigning s synapses to the d presynaptic neurons in the linear model, and B_N gives the same for the nonlinear model. Source: *Adapted with permission from Poirazi P, Mel BW. Impact of active dendrites and structural plasticity on the memory capacity of neural tissue. Neuron. 2001;29(3):779–796.*

$y(x)$ could take both positive and negative values and specifically +1 or −1 due to the use of the sgn[] operator. Empirical equations for the two model types are shown in Fig. 15.1, top right.

To estimate analytically the storage capacity of the linear and nonlinear model neurons, the authors derived combinatorial expressions that counted the number of all distinct input–output configurations realizable by the two models (depicted in Fig. 15.1, bottom right). These expressions measured storage capacity in *bits* and predicted that a linear model cell would be blind to changes in the location (but not the number) of incoming contacts. This is explained better by looking at the example in Fig. 15.2.

Based on the analytical expressions, the authors predicted that neurons with a fixed number of presynaptic partners would have to implement different types of structural modifications (growing more, longer, or shorter dendrites) in order to maximize their storage capacity. However, in all cases, the nonlinear model gained massively more in terms of storage capacity compared to the linear model.

To validate the analytical model, the empirical model neurons were trained to discriminate between two sets of random input patterns, the "preferred" and "nonpreferred" ones, by responding positively to the first and negatively to the latter. The authors implemented a plasticity rule (stochastic gradient descent) according to which new synapses were formed at random locations and were stabilized only if consistently active with other

		Linear cell	Nonlinear cell
	○ d_1 ◐ d_2 ● d_3 $d,m,k = 3$	$a_L(x)$	$a_N(x)$
Wiring configurations	(1) 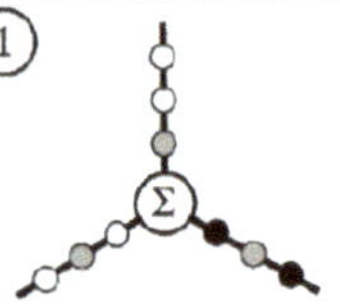	$4x_1 + 3x_2 + 2x_3$	$b(2x_1 + x_2) +$ $b(2x_1 + x_2) +$ $b(x_2 + 2x_3)$
Wiring configurations	(2) 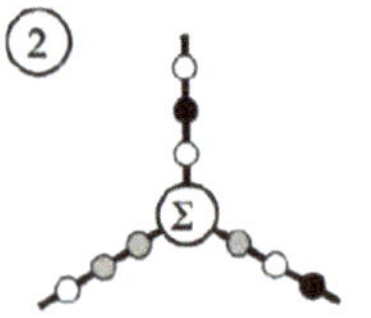	$4x_1 + 3x_2 + 2x_3$	$b(2x_1 + x_3) +$ $b(x_1 + 2x_2) +$ $b(x_1 + x_2 + x_3)$
	Total number of distinct i/o functions	110	220

FIGURE 15.2 **Two different spatial configurations of inputs to a cell with $m = 3$ branches, each containing $k = 3$ synaptic sites with access to $d = 3$ presynaptic axons firing at x_1, x_2, and x_3 Hz.** Synaptic weights are set to 1. The linear cell cannot detect the differences in the two configurations, resulting in identical output functions. The total number of distinct i/o functions for the two types of cells are estimated using the B_L and B_N equations of Fig. 15.1. Source: *Adapted with permission from Poirazi P, Mel BW. Impact of active dendrites and structural plasticity on the memory capacity of neural tissue. Neuron. 2001;29(3):779–796.*

synapses on their branch and aligned with the desired somatic response to each input. The authors found a close match between analytical models and empirically trained models and predicted that for a cell with a fixed number of presynaptic partners, maximization of storage capacity is achieved when the cell has many short nonlinear dendritic branches. In other words, when the number of nonlinear branches is large, each branch contains only a few synaptic sites (see Fig. 15.3A). Moreover, the authors showed that this maximized capacity is massively larger than the one realizable by the respective linear model cell, at both the single cell and the population levels, even for synapses with multiple activity states (see Fig. 15.3Aand B).

These predictions suggest that storage capacity is effectively maximized in a cell with many nonlinear subunits each of which processes only a few active sites that are correlated in their firing. In other words, this work predicted for the first time that the spatial clustering of correlated inputs within nonlinear branches can maximize the storage capacity of neurons as well as networks of neurons equipped with dendritic nonlinearities. This result came in stark contrast with the previous held notion by the neuroscience community, that memories are encoded in neural networks via synaptic weight plasticity alone. The proposed, biologically plausible, structural plasticity learning rule was able to

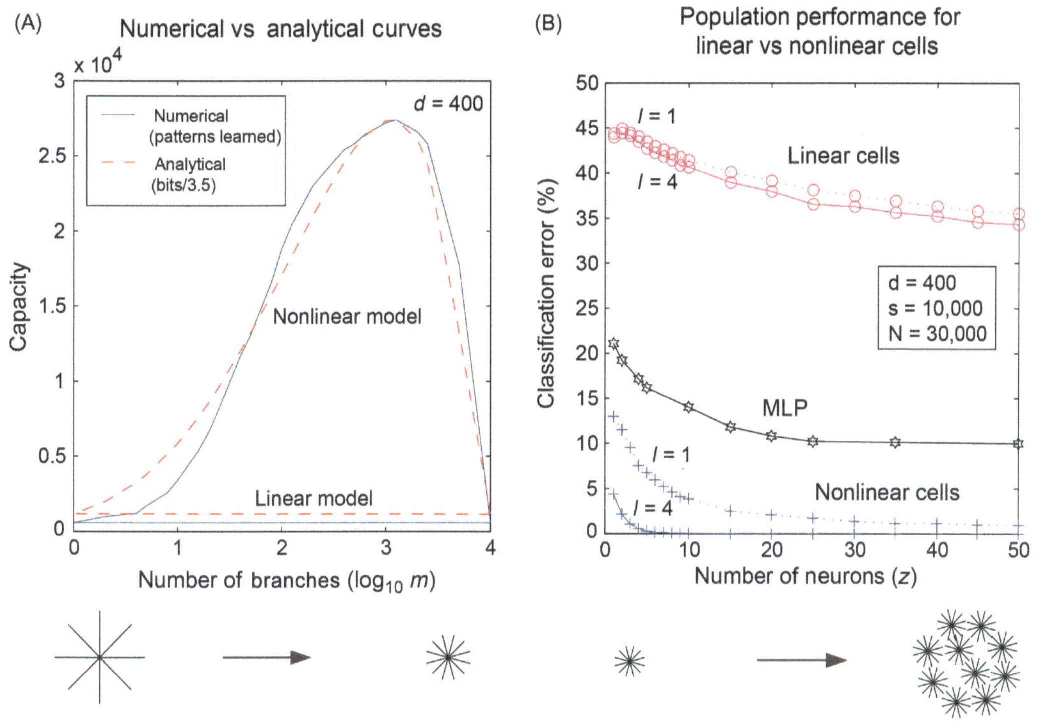

FIGURE 15.3 **Comparison of memory capacity predicted by analysis with that found empirically in random memorization problems.** (A) *Dashed lines* show analytical curves for linear and nonlinear cells. *Solid curves* show capacity measured empirically at 2% error criterion, using a subunit nonlinearity $b(x) = x^{10}$. (B) Comparison of error rates for populations of linear vs nonlinear cells and a multilayer perceptron (MLP) classifier trained on the same problem. Error rates are plotted for populations ranging from 1 to 50 cells. All cells were trained with $m = 400$, $k = 25$. Runs involving cells with unitary weights are indicated by $l = 1$, those trained with 4-level weights are indicated by $l = 4$. Source: *Adapted with permission from Poirazi P, Mel BW. Impact of active dendrites and structural plasticity on the memory capacity of neural tissue. Neuron. 2001;29(3):779–796.*

discriminate among different memories by addressing particular synaptic contacts onto specific dendritic subunits. The authors proposed that neurons are likely to exploit a fast-loading but low-capacity storage reservoir by regulating the synaptic weight values by conventional activity-dependent learning rules. In addition to this fast-loading reservoir, activity-dependent rewiring of presynaptic contacts provides the cell with an additional, much higher but slow-loading, reservoir for information storage.

In 2008, Kelleher et al. further researched the implications of clusteron,[84] a neuronal model of learning developed by Bartlett Mel, which is based on structural remodeling and spatial patterning of synapses rather than synaptic weight plasticity rules.[85] Its primary feature lies in the relative dependence of postsynaptic depolarization activity on the nonlinear integration carried out by the total number of active synapses in a particular vicinity. Due to the immutable nature of synaptic weights, there is only one way to differentiate the response to a particular stimulus pattern: to redistribute the synaptic contacts across the dendritic subunits, thus changing their spatial distribution. The investigators used two versions of the model: (1) the first one was similar to the original clusteron[85] composed of a cell body and a single dendrite and (2) the other one was a more realistic "branched" model, considered to be a reductionist approach to the two-layer model of dendritic integration proposed by Poirazi et al.[75] With the goal of assessing the degree of synaptic adaptation for both models, a supervised machine learning approach was applied after the presentation of each input pattern to stabilize the most highly active synapses and force the most poorly performing ones to reallocate. Two main findings were derived from this computational study. The first finding is that there is a direct relationship between the strength and speed of associating memories with the learning threshold, which represents the degree of local activation needed to induce LTP. In the case of a low value of the learning threshold, the learning is rapid but leads to small increases in somatic responses. Alternatively, a high value leads to optimal learning of the associative memory, but over longer time scales. The second finding is that there is a positive effect of the number of participating synapses on the strength of the memory, measured as the increase in the model response compared to baseline. Afterwards, the researchers extended the model to perform novel learning tasks, such as feature detection, and learning new spatiotemporal patterns. Once again, the model reflects the experimental observation of network reorganization as a consequence of learning.[86]

A different theoretical approach toward determining the effect of dendritic arborization patterns on memory storage capacity was employed by de Sousa et al.[87] The researchers used an algorithm to exhaustively generate all possible binary trees terminating in 22 points, thus creating a fully explorable space of different dendritic topologies on which they based the construction of their multicompartmental models. This strategy mimics to some extent the developmental structural plasticity, which is considered to substantially enrich the overall storage capacity of the brain's memory systems. To explore the potential of this speculation, the researchers sought to assess the computational performance of neuron models by quantifying their ability to discriminate between learnt and novel input feature patterns. Specifically, they assessed the performance of a passive neuron model and one furnished with voltage-gated ion channels. Passive and active properties were according to Duijnhouwer et al.[88] Synaptic excitation was simulated through AMPA receptor-type time-varying conductances. In order to distinguish between different topologies, de Sousa et al. used four morphological metrics: asymmetry index, mean depth, and mean and variance of electrotonic path length. They found that dendritic morphology has a considerable impact on the pattern recognition capabilities of the model cells. They compared the results with a representative selection of larger trees with 128 terminal points and found that fully symmetric morphologies had a better overall performance and that there was an inverse correlation with the mean value of the branch centrifugal order. When dendritic tapering was introduced to the models, the mean and variance of the electrotonic distance were the best predictors for the success of the pattern recognition task. Importantly, the study predicted that passive neurons had to linearly integrate the weight contribution of each synapse activated by the input pattern, and that active models had to compare the sum to the spike thresholds.

While the models of synaptic plasticity have shed some light on the basic mechanistic rules of learning and memory, there are still several intriguing questions that remain largely unsolved. In a recently published study, Fauth et al. employed an LTP stimulation protocol to address the challenging problem of long-term memory storage despite the largely volatile neural substrate that supports it.[89] By using a generic and simple probabilistic model for the simulation of synaptic turnover, they found that information can be held for months when exploiting multisynaptic intrinsic dynamics. Specifically, they modeled the collective dynamics by relating synapse stability to the total number of synapses. This theoretical model provided evidence that the actual number of synapses is fluctuating within a certain range to ensure flexibility. It was also shown that when the model follows the experimentally validated bimodal distribution, the storage duration of the neuron extends to orders of magnitude larger time scales than that of the lifetime of a single synapse. The resulting model dynamics provided support for a scheme of fast learning and slow forgetting, due to the high volatility of single spines with the memory stability being represented by the compound state of the synapses.

In another research article published by Stepanyants et al. in 2002, spine remodeling was studied in relation to its potential effect on expanding memory capacity.[90] The researchers postulated that since structural plasticity poses a significant effect on learning and memory, the quantitative potential of exploring all the different connectivity patterns ultimately determines the limit of memory capacity associated with spine remodeling. In more detail, they attempted to make an estimation of all the different interneuronal connectivity patterns that are feasible without remodeling the dendritic or axonal structures. It is critical to mention that this number is related to ratio of the number of potential spines along a dendrite to the number of axons that pass within a spine length close to that dendrite. The researchers named this ratio filling fraction and proceeded to make calculations based on data from anatomical and geometrical analyses, thus publishing interesting evaluations of the upper bound of memory capacity for various cortical areas and species.

An interesting study was published in 2014 by Knoblauch, presenting a parsimonious computational model for learning, by putting the emphasis on spine remodeling.[91] The model was based on the concept that potential synapses can grow at any location along a modeled network and that this property is subject to plasticity. In order to assess the contribution of this type of plasticity, the researchers defined a metric named effectual connectivity as the fraction of the existing synapses to the total number of synapses required for a particular computation. Activity-dependent structural plasticity can improve effectual connectivity by "migrating" synapses to the most suitable dendritic locations for learning particular input patterns. The researchers implemented models with well-known Hopfield- or Willshaw-type rules for memory storage and retrieval, thus accounting for structural plasticity on a slow time-scale and weight plasticity on a faster time scale. Surprisingly, the simulation results substantiated the notion that structural plasticity provides largely increased performance in comparison with weight plasticity alone. Furthermore, the model explains the interesting cognitive phenomena of graded amnesia, catastrophic forgetting, and the spacing effect as it produces gradients of effectual connectivity in the course of learning. One of the advantages of this model is that it can generalize well already existing models of memory storage and retrieval using weight plasticity as the main rule for learning and memory.

3.1 Toward Hardware Implementations of Structural Plasticity Models

In recent years, the scientific community has witnessed a surge in interdisciplinary research regarding neuromorphic chip design in the pursuit of developing hardware devices that execute complex tasks using brain-inspired approaches in real-time, with low power consumption.[92–95] The current designs are exploiting some basic event-driven principles adopted from neuroscience research, such as sparse network connectivity, spike-based learning rules, analog synaptic weight values, and nonlinear integration properties. However, the very nature of hybrid analog–digital devices, such as very large-scale integration (VLSI) technology, is limited by low precision accuracy of analog synaptic weights[96] in comparison with software simulations, and the benefits tend to diminish all along the miniaturization trend. Hence, the use of analog synaptic weights leads to unreliable performance with respect to the implemented function, e.g., the encoding and recall of complex sensory stimuli. In order to bypass these issues, researchers have moved to the direction of increasing the total number of neurons comprising the neural network and/or using highly recurrent synaptic connections, albeit at the expense of larger chip area and higher energy demands. Interestingly, Hussain and Basu[97] opted to follow a different approach by including structural plasticity (as introduced by Poirazi and Mel[83]) in a model with nonlinear dendrites and delivered proof that it can alleviate the aforementioned restrictions and boost the memory capacity in hardware system implementations.

Specifically, the authors developed a network of adaptive (integrate-and-fire) neurons with nonlinear single-compartment dendrites that are connected through a sparse binary synaptic scheme (see Fig. 15.4) and attempted to sort out a standard multiclass classification problem. Specifically, they set to tackle the problem of algorithmically discriminating the handwritten digits contained in the benchmark MNIST dataset.[98] At first, they employed an adaptive approach of gradually adding new dendrites with a fixed number of synapses to each neural classifier as the complexity of the task increased. This adaptive learning method is used to determine the optimal size of each neuron for the specific task learnt (e.g., a particular handwritten digit), by increasing its number of dendrites, m. While the adaptation is under way, the number of synapses per branch, k, is kept constant for all neurons (see Fig. 15.5). Next, they enforced a learning procedure based on structural plasticity by forming new synaptic contacts within dendrites and eliminating the worst performing ones, via the use of "silent synapses" along the lines described by Poirazi and Mel.[83] As a result, each of the N_c neural classifiers acquired an optimal topology and connectivity configuration via which it reached a high performance (% of correct predictions) for the specific problem assigned.

Two main conclusions were derived from this approach: (1) the best performing ensembles of classifiers are those with moderate numbers of dendrites, i.e., classifiers that are neither perceptrons nor have highly complex tree structures, (2) the structural plasticity rule greatly improves the memory capacity of a hardware network without the use of continuous synaptic weights. Importantly, the researchers noted that the attained performance compares with the highest classification scores reported in the literature while using no more than 7% of synaptic resources. A major disadvantage of this implementation consists in the lack of any spatial information along the dendrites, which are modeled as single compartments. As a consequence, the synapses are lumped together without any time lag

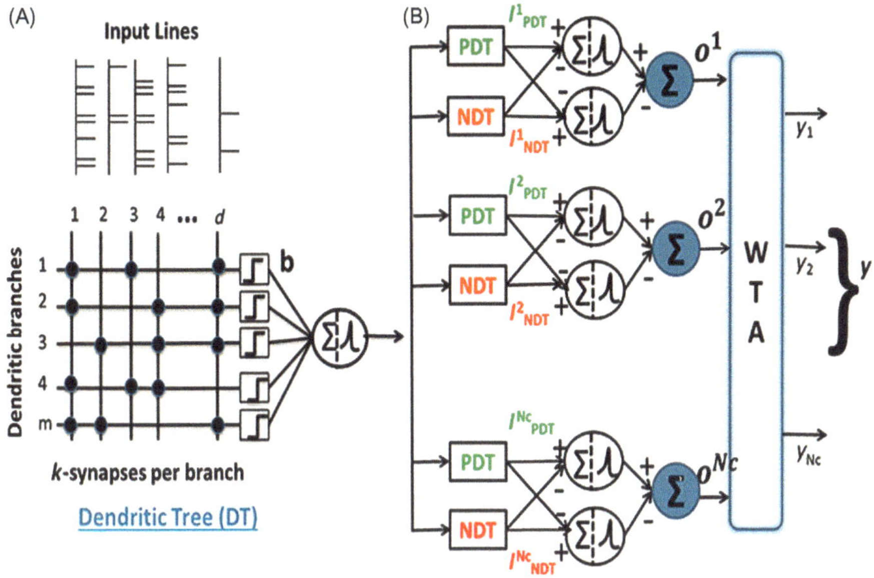

FIGURE 15.4 (A) Schematic diagram of the model by Hussain and Basu[97] showing a neuron with nonlinear dendrites implementing lumped nonlinear functions for each dendritic branch. A number of k binary synapses are formed by d input afferents on m dendrites, where b indicates the branch nonlinearity. (B) The architecture of the multiclass pattern classifier is made up of nonlinear neurons that receive input (I^1, I^2, . . ., I^{Nc}) from both positive (excitatory) and negative (inhibitory) dendritic trees (PDT and NDT, respectively), while the neurons relay their output (O^1, O^2, . . ., O^{Nc}) to a winner-take-all (WTA) to compare the outputs of the Nc classifiers. Source: *Adapted with permission from Hussain S, Basu A. Multiclass classification by adaptive network of dendritic neurons with binary synapses using structural plasticity. Front Neurosci. 2016;10.*

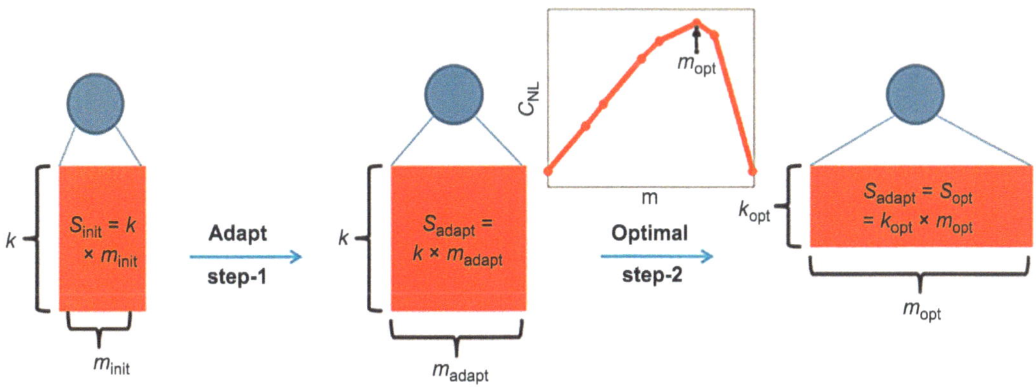

FIGURE 15.5 **A neural classifier is adaptively trained by the addition of dendrites (m) (step-1), while keeping the number of synapses/dendrite (k) constant.** In this way, the neuron acquires a certain total number of synapses (S) and the corresponding optimal topology is resolved (step-2). C_{NL} denotes the memory capacity of neurons with nonlinear dendrites in bits. Source: *Adapted with permission from Hussain S, Basu A. Multiclass classification by adaptive network of dendritic neurons with binary synapses using structural plasticity. Front Neurosci. 2016;10.*

interfering in the signal propagation between them. If multiple dendritic compartments were introduced, the structural learning rule would likely yield even better performance.

4 DISCUSSION

A growing body of experimental evidence demonstrates that structural plasticity is a key determinant of the brain's ability to learn and has a large impact on its storage capacity. However, the exact mechanistic details, as well as the emerging functional consequences of structural plasticity, remain elusive. In this chapter, we provided a brief overview of the experimental evidence regarding structural plasticity in the brain and presented some of the computational models that link this type of plasticity to the storage and/or information processing capacity of the neural tissue and of emerging hardware devices. Although certainly not a complete coverage of the existing literature, the studies presented in this chapter challenge the old view whereby synapses are considered as independent and elementary units of plasticity. The activity-dependent formation of synaptic clusters within dendritic branches as suggested by both experimental and modeling studies[10,59,83,99] implies that synapses may act in groups in order to maximize information storage. The emerging picture suggests that clusters of functionally related synapses, which are formed within dendritic branches under the influence of structural plasticity mechanisms, are likely to serve as a key computational and memory storage unit in the brain.

4.1 Future Experimental Studies

Over the past couple of decades, theoretical studies of structural plasticity and its implications for neural functions have generated an array of predictions that remain subject to experimental verification. What are some wet-lab approaches that could possibly aid this effort? A number of recent studies have already demonstrated rigorous protocols that ensure successful labeling and manipulation of the neural ensembles recruited for encoding certain memories.[100,101] This is usually achieved by coupling the promoters of immediate early genes, which are induced by recent neural activity,[102] with the expression of genetically encoded calcium indicators.[103] Afterwards, two- or multiphoton excited laser scanning microscopy can be employed to monitor the deep tissue in vivo. In future studies, it may be possible to register all structural changes happening on both the short and longer time scales of distributed and sparsely activated neuronal populations. Beyond just following the structural changes to formulate hypotheses about plasticity mechanisms and their potential contribution to memory capacity, researchers can also induce these changes. For example, glutamate uncaging experiments are known to influence spine dynamics[104] and other advanced techniques, such as femtosecond laser dendrotomy, can prune targeted regions of the tree.[105] Interestingly, treating animal models with biochemical agents, like corticosteroids, can also induce structural reorganization in multiple brain regions such as the hippocampus and the amygdala.[106] Tracking the effect of structural plasticity onto the activity of neural cells could be achieved via the use of multielectrode arrays and/or calcium imaging. A major restriction for the assessment of memory capacity at the single cell or neural-circuit level is the lack of relevant measures/tests. Thus it is imperative to formulate and test theoretical assumptions about how memory processes are reflected in electrophysiological recordings.

In the pursuit of experimental support for the theoretically derived ramifications presumably exerted by structural plasticity, Poirazi and Mel proposed, back in 2001,[83] an elegant experiment—technically challenging at the time, but perhaps feasible today. As shown in the schematic of Fig. 15.6, the experimenter should be able to stimulate and record from five neuronal ensembles Axons of neurons in each of the four "surround" ensembles (labeled A, B, C, and D) should have equal access to the dendrites of the "center" ensemble (labeled X). During a period of increased synaptogenesis, ensembles A, X, D and B, X, C should be stimulated in alternating fashion, so as to establish strong positive correlations between ensembles A and D or B and C, and strong negative correlations between ensembles A and B or C and D. Note that ensemble X may not need to be stimulated if stimulation of A and D (or B and C) is sufficient to generate dendritic spikes in the ensemble X. After training, neurons in correlated (e.g., A and D) or uncorrelated ensembles (e.g., A and B) should be filled with dyes of different color, allowing their synaptic

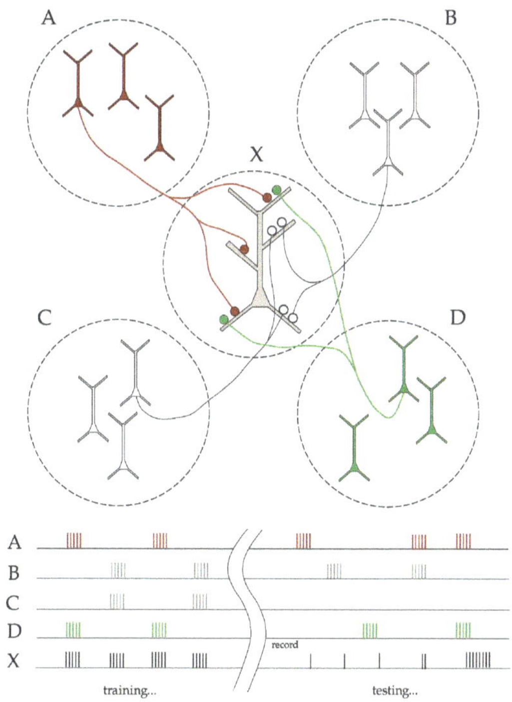

FIGURE 15.6 **Schematic of a conceptual experiment designed to test the role of structural plasticity in neural learning.** Temporally correlated activation of populations A and D, alternating with correlated activation of populations B and C, hypothetically would lead to a more frequent stabilization of synapses arising from neurons in ensembles A and D on the same dendrites or minor dendritic branches (likewise for synapses from ensembles B and C). Source: *Adapted with permission from Poirazi P, Mel BW. Impact of active dendrites and structural plasticity on the memory capacity of neural tissue. Neuron. 2001;29(3):779–796.*

contacts onto the cells of ensemble X (also stained) to be visualized and distinguished. If the theoretical predictions of Poirazi and Mel are correct, synapses arising from cells in ensemble A should be more frequently colocalized with those of ensemble D (i.e., found in close proximity within a dendrite) than with those of ensemble B. Moreover, costimulation of ensembles A and D should induce stronger responses of cells in ensemble X than costimulation of ensembles A and B, and this boosting effect should depend on excitatory voltage-dependent channels within individual cells in ensemble X (including NMDA channels and various types of voltage-dependent Na^+ and Ca^{2+} channels). Thus the blockade of such channels via, e.g., strong hyperpolarization or depolarization, or injection of intracellular channel blockers, should lead to a reduction or elimination of the cell's integrative nonlinearity.

As proposed by the authors, this experiment could be carried out in a developing neuronal culture grown on a microelectrode array, which makes possible long-term stimulation and recording from known subpopulations of cells. Alternatively, in the intact animal, the five neuronal ensembles could consist of nearby neuron groups in, e.g., the auditory cortex, activated by playing appropriate combinations of tones over several days. The "chords" should consist of tones that drive columns of cells whose horizontal connections are symmetrically distributed and within reach of each other in the tangential plane of the cortex. During the test phase, recordings from neurons in ensemble X would be carried out while playing chords to the animal through the intact auditory system.

4.2 Future Modeling Studies

Despite the overall progress in our understanding of dendritic function, there are many steps to be taken in order to untangle the mysteries of structural plasticity in relation to memory, especially on expanded spatiotemporal scales. First, we need to gain a more comprehensive insight into the intricate relationship between dendritic geometry and computational function. Databases such as NeuroMorpho.org[107] and projects like BigNeuron[108] are currently the major contributors to the ever-increasing availability of virtual realistic neuronal morphologies. In conjunction with the implementation of elegant and simplified mathematical models, the defining principles of the vast structural diversity and organization of the neural tissue could be clearly assessed and revealed. Although many studies have researched this issue (for an extensive review, see Ref.[109]), more investigation is certainly needed. Secondly, there is a pressing need for modeling approaches that account for the adaptive remodeling of whole dendritic and axonal arbors in response to well-established paradigms of electrophysiological activity patterns.[110] In this way, we'll be able to explore the implications of structural plasticity on slower time scales than what current approaches have allowed. Importantly, the advent of modern imaging instrumentation provides us with the opportunity to monitor and mathematically characterize the aforementioned plasticity changes, thus feeding the abstract models with all the necessary parameters and plasticity principles. Lastly, the researchers could exploit and expand the existing software tools[111–113] so as to enforce morphological manipulations on single cells and neural networks and investigate the possible causal relationships governing the adaptive function of the neural tissue in response to various incoming stimuli. We look forward to the next generation of experimental, modeling, and theoretical approaches to see success in unraveling the mysteries of memory by accounting for structural plasticity of the ever-evolving neural tissue on all accessible spatiotemporal scales.

Acknowledgment

This work was supported by the ERC Starting Grant dEMORY (ERC-2012-StG-311435).

References

1. Mathy F, Feldman J. What's magic about magic numbers? Chunking and data compression in short-term memory. *Cognition.* 2012;122(3):346–362.
2. Tonegawa S, Pignatelli M, Roy DS, Ryan TJ. Memory engram storage and retrieval. *Curr Opin Neurobiol.* 2015;35:101–109.
3. Stock JB, Zhang S. The biochemistry of memory. *Curr Biol.* 2013;23(17).
4. Bailey CH, Kandel ER, Harris KM. *Structural Components of Synaptic Plasticity and Memory Consolidation.* 2015;1–30.
5. Benchenane K, Peyrache A, Khamassi M, et al. Coherent theta oscillations and reorganization of spike timing in the hippocampal- prefrontal network upon learning. *Neuron.* 2010;66(6):921–936.
6. Newman EL, Gillet SN, Climer JR, Hasselmo ME. Cholinergic blockade reduces theta-gamma phase amplitude coupling and speed modulation of theta frequency consistent with behavioral effects on encoding. *J Neurosci.* 2013;33(50):19635–19646.
7. Polleux F, Ghosh A. Molecular determinants of dendrite and spine development. *Dendrites.* 2012.
8. Cline HT. Dendritic arbor development and synaptogenesis. *Curr Opin Neurobiol.* 2001;11(1):118–126.
9. Harris KM. *Dendritic spines. eLS.* John Wiley & Sons, Ltd; 2001.
10. Kleindienst T, Winnubst J, Roth-Alpermann C, Bonhoeffer T, Lohmann C. Activity-dependent clustering of functional synaptic inputs on developing hippocampal dendrites. *Neuron.* 2011;72(6):1012–1024.
11. Mennerick S, Zorumski CF. Neural activity and survival in the developing nervous system. *Mol Neurobiol.* 2001;22(1–3):41–54.
12. Feldman DE. Synaptic mechanisms for plasticity in neocortex. *Annu Rev Neurosci.* 2009;32:33–55.
13. Kasai H, Hayama T, Ishikawa M, Watanabe S, Yagishita S, Noguchi J. Learning rules and persistence of dendritic spines. *Eur J Neurosci.* 2010;32(2):241–249.
14. Choquet D, Triller A. The dynamic synapse. *Neuron.* 2013;80(3):691–703.
15. Bhatt DH, Zhang S, Gan W-B. Dendritic spine dynamics. *Annu Rev Physiol.* 2009;71:261–282.
16. Hua JY, Smith SJ. Neural activity and the dynamics of central nervous system development. *Nat Neurosci.* 2004;7(4):327–332.
17. Dudai Y. Consolidation: fragility on the road to the engram. *Neuron.* 1996;17(3):367–370.
18. McGaugh JL. Neuroscience—Memory—a century of consolidation. *Science (80-).* 2000;287(5451):248–251.
19. Govindarajan A, Kelleher RJ, Tonegawa S. A clustered plasticity model of long-term memory engrams. *Nat Rev Neurosci.* 2006;7(7):575–583.
20. Kandel ER. The molecular biology of memory storage: a dialogue between genes and synapses. *Science (80-).* 2001;294(5544):1030–1038.
21. Aureli M, Grassi S, Prioni S, Sonnino S, Prinetti A. Lipid membrane domains in the brain. *Biochim Biophys Acta.* 2015;1851(8):1006–1016.
22. Tronel S, Fabre A, Charrier V, Oliet SHR, Gage FH, Abrous DN. Spatial learning sculpts the dendritic arbor of adult-born hippocampal neurons. *Proc Natl Acad Sci USA.* 2010;107(17):7963–7968.
23. Dailey ME, Smith SJ. The dynamics of dendritic structure in developing hippocampal slices. *J Neurosci.* 1996;16(9):2983–2994.
24. Fiala JC, Feinberg M, Popov V, Harris KM. Synaptogenesis via dendritic filopodia in developing hippocampal area CA1. *J Neurosci.* 1998;18(21):8900–8911.
25. Petrak LJ, Harris KM, Kirov SA. Synaptogenesis on mature hippocampal dendrites occurs via filopodia and immature spines during blocked synaptic transmission. *J Comp Neurol.* 2005;484(2):183–190.

26. Ziv NE, Smith SJ. Evidence for a role of dendritic filopodia in synaptogenesis and spine formation. *Neuron*. 1996;17(1):91–102.
27. Lendvai B, Stern E a, Chen B, Svoboda K. Experience-dependent plasticity of dendritic spines in the developing rat barrel cortex in vivo. *Nature*. 2000;404(6780):876–881.
28. Holtmaat AJGD, Trachtenberg JT, Wilbrecht L, et al. Transient and persistent dendritic spines in the neocortex in vivo. *Neuron*. 2005;45(2):279–291.
29. Matus A. Growth of dendritic spines: a continuing story. *Curr Opin Neurobiol*. 2005;15(1):67–72.
30. Marchenko O, Wolgemuth C, Loew L. Actomyosin contractility and actin polymerization explain dynamic behavior of dendritic filopodia. *Biophys J*. 2016;104(2):317a.
31. Portera-Cailliau C, Pan DT, Yuste R. Activity-regulated dynamic behavior of early dendritic protrusions: evidence for different types of dendritic filopodia. *J Neurosci Off J Soc Neurosci*. 2003;23(18):7129–7142.
32. Zito K, Svoboda K. Activity-dependent synaptogenesis in the adult mammalian cortex. *Neuron*. 2002;35 (6):1015–1017.
33. Knott GW, Quairiaux C, Genoud C, Welker E. Formation of dendritic spines with GABAergic synapses induced by whisker stimulation in adult mice. *Neuron*. 2002;34(2):265–273.
34. Maletic-Savatic M. Rapid dendritic morphogenesis in CA1 hippocampal dendrites induced by synaptic activity. *Science (80-)* 1999;283(5409):1923–1927.
35. Lohmann C, Finski A, Bonhoeffer T. Local calcium transients regulate the spontaneous motility of dendritic filopodia. *Nat Neurosci*. 2005;8(3):305–312.
36. Lohmann C, Bonhoeffer T. A role for local calcium signaling in rapid synaptic partner selection by dendritic filopodia. *Neuron*. 2008;59(2):253–260.
37. Hebb DO. The organization of behavior: a neuropsychological approach. *Am J Psychol*. 1949;63(4):633.
38. Bauer EP, Schafe GE, LeDoux JE. NMDA receptors and L-type voltage-gated calcium channels contribute to long-term potentiation and different components of fear memory formation in the lateral amygdala. *J Neurosci*. 2002;22(12):5239–5249.
39. Mayer ML, Westbrook GL, Guthrie PB. Voltage-dependent block by Mg^{2+} of NMDA responses in spinal cord neurones. *Nature*. 1984;309(5965):261–263.
40. Yuste R, Majewska A, Cash SS, Denk W. Mechanisms of calcium influx into hippocampal spines: heterogeneity among spines, coincidence detection by NMDA receptors, and optical quantal analysis. *J Neurosci*. 1999;19 (6):1976–1987.
41. Miyakawa H, Ross WN, Jaffe D, et al. Synaptically activated increases in Ca^{2+} concentration in hippocampal CA1 pyramidal cells are primarily due to voltage-gated Ca^{2+} channels. *Neuron*. 1992;9(6):1163–1173.
42. Fuenzalida M, Fernández de Sevilla D, Couve A, Buño W. Role of AMPA and NMDA receptors and back-propagating action potentials in spike timing-dependent plasticity. *J Neurophysiol*. 2010;103(1):47–54.
43. Markram H, Lubke J, Frotscher M, Sakmann B. Regulation of synaptic efficacy by coincidence of postsynaptic APs and EPSPs. *Science (80-)*. 1997;275(5297):213–215.
44. Lisman J, Schulman H, Cline H. The molecular basis of CaMKII function in synaptic and behavioural memory. *Nat Rev Neurosci*. 2002;3(3):175–190.
45. Sanhueza M, Fernandez-Villalobos G, Stein IS, et al. Role of the CaMKII/NMDA receptor complex in the maintenance of synaptic strength. *J Neurosci*. 2011;31(25):9170–9178.
46. Lamprecht R, LeDoux J. Structural plasticity and memory. *Nat Rev*. 2004;5:45–54.
47. Yuste R, Denk W. Dendritic spines as basic functional units of neuronal integration. *Nature*. 1995;375 (6533):682–684.
48. Bliss TVP, Lomo T. Long-lasting potentiation of synaptic transmission in the dentate area of the anaesthetized rabbit following stimulation of the preforant path. *J Physiol*. 1973;232:331–356:4727084.
49. Ito M, Sakurai M, Tongroach P. Climbing fibre induced depression of both mossy fibre responsiveness and glutamate sensitivity of cerebellar Purkinje cells. *J Physiol*. 1982;324:113–134:VL-324.
50. Ekerot CF, Kano M. Long-term depression of parallel fibre synapses following stimulation of climbing fibres. *Brain Res*. 1985;342(2):357–360.
51. Gold MG. A frontier in the understanding of synaptic plasticity: solving the structure of the postsynaptic density. *Bioessays*. 2012;34(7):599–608.
52. Nabavi S, Fox R, Proulx CD, Lin JY, Tsien RY, Malinow R. Engineering a memory with LTD and LTP. *Nature*. 2014.

53. Van Harreveld A, Fifkova E. Swelling of dendritic spines in the fascia dentata after stimulation of the perforant fibers as a mechanism of post-tetanic potentiation. *Exp Neurol.* 1975;49(3):736–749.
54. Trommald M, Hulleberg G, Andersen P. Long-term potentiation is associated with new excitatory spine synapses on rat dentate granule cells. *Learn Mem.* 1996;3(2-3):218–228.
55. Toni N, Buchs P a, Nikonenko I, Bron CR, Muller D. LTP promotes formation of multiple spine synapses between a single axon terminal and a dendrite. *Nature.* 1999;402(6760):421–425.
56. Hosokawa T, Rusakov D a, Bliss TV. Fine a. Repeated confocal imaging of individual dendritic spines in the living hippocampal slice: evidence for changes in length and orientation associated with chemically induced LTP. *J Neurosci.* 1995;15(8):5560–5573:0270-6474/95/155560-14.
57. Matsuzaki M, Honkura N, Ellis-Davies GCR, Kasai H. Structural basis of long-term potentiation in single dendritic spines. *Nature.* 2004;429(6993):761–766.
58. McBride TJ, Rodriguez-Contreras A, Trinh A, Bailey R, Debello WM. Learning drives differential clustering of axodendritic contacts in the barn owl auditory system. *J Neurosci.* 2008;28(27):6960–6973.
59. Fu M, Yu X, Lu J, Zuo Y. Repetitive motor learning induces coordinated formation of clustered dendritic spines in vivo. *Nature.* 2012;483(7387):92–95.
60. Kramár E a, Babayan AH, Gavin CF, et al. Synaptic evidence for the efficacy of spaced learning. *Proc Natl Acad Sci USA.* 2012;109(13):5121–5126.
61. Makino H, Malinow R. Compartmentalized versus global synaptic plasticity on dendrites controlled by experience. *Neuron.* 2011;72(6):1001–1011.
62. Yang G, Lai CSW, Cichon J, Ma L, Li W, Gan W-B. Sleep promotes branch-specific formation of dendritic spines after learning. *Science.* 2014;344(6188):1173–1178.
63. Leggio MG, Mandolesi L, Federico F, et al. Environmental enrichment promotes improved spatial abilities and enhanced dendritic growth in the rat. *Behav Brain Res.* 2005;163(1):78–90.
64. Ramírez-Amaya V, Escobar ML, Chao V, Bermúdez-Rattoni F. Synaptogenesis of messy fibers induced by spatial water maze overtraining. *Hippocampus.* 1999;9(6):631–636.
65. Ramírez-Amaya V, Balderas I, Sandoval J, Escobar ML, Bermúdez-Rattoni F. Spatial long-term memory is related to mossy fiber synaptogenesis. *J Neurosci.* 2001;21(18):7340–7348.
66. Branco T, Clark B a, Häusser M. Dendritic discrimination of temporal input sequences in cortical neurons. *Science.* 2010;329(5999):1671–1675.
67. Häusser M, Mel B. Dendrites: bug or feature? *Curr Opin Neurobiol.* 2003;13(3):372–383.
68. Major G, Larkum ME, Schiller J. Active properties of neocortical pyramidal neuron dendrites. *Annu Rev Neurosci.* 2013;36.
69. Segev I, London M. Untangling dendrites with quantitative models. *Science.* 2000;290(5492):744–750.
70. Silver RA. Neuronal arithmetic. *Nat Rev Neurosci.* 2010;11(7):474–489.
71. Ariav G, Polsky A, Schiller J. Submillisecond precision of the input-output transformation function mediated by fast sodium dendritic spikes in basal dendrites of CA1 pyramidal neurons. *J Neurosci.*. 2003;23 (21):7750–7758.
72. Branco T, H??usser M. Synaptic integration gradients in single cortical pyramidal cell dendrites. *Neuron.* 2011;69(5):885–892.
73. Longordo F, To M-S, Ikeda K, Stuart GJ. Sublinear integration underlies binocular processing in primary visual cortex. *Nat Neurosci.* 2013;16(6):714–723.
74. Losonczy A, Magee JC. Integrative properties of radial oblique dendrites in hippocampal CA1 pyramidal neurons. *Neuron.* 2006;50(2):291–307.
75. Poirazi P, Brannon T, Mel BW. Pyramidal neuron as two-layer neural network. *Neuron.* 2003;37(6):989–999.
76. Yuste R. Dendritic spines and distributed circuits. *Neuron.* 2011;71(5):772–781.
77. Bi GQ, Poo MM. Synaptic modifications in cultured hippocampal neurons: dependence on spike timing, synaptic strength, and postsynaptic cell type. *J Neurosci.* 1998;18(24):10464–10472.
78. Legenstein R, Maass W. Branch-specific plasticity enables self-organization of nonlinear computation in single neurons. *J Neurosci.* 2011;31(30):10787–10802.
79. Castellani GC, Quinlan EM, Bersani F, Cooper LN, Shouval HZ. A model of bidirectional synaptic plasticity: from signaling network to channel conductance. *Learn Mem.* 2005;12(4):423–432.
80. Dayan P, Abbott LF. Theoretical neuroscience: computational and mathematical modeling of neural systems. *Comput Math Model Neural.* 2001;480.

81. Kitajima T, Hara K. A generalized Hebbian rule for activity-dependent synaptic modifications. *Neural Networks*. 2000;13(4-5):445–454.
82. Chklovskii DB, Mel BW, Svoboda K. Cortical rewiring and information storage. *Nature*. 2004;431 (October):782–788.
83. Poirazi P, Mel BW. Impact of active dendrites and structural plasticity on the memory capacity of neural tissue. *Neuron*. 2001;29(3):779–796.
84. Kelleher KJ, Hajdik V, Colbert CM, Josié K. Learning by structural remodeling in a class of single cell models. *J Comput Neurosci*. 2008;25(2):282–295.
85. Mel BW. The clusteron: toward a simple abstraction for a complex neuron. *Nips*. 1991;35–42.
86. Doyon J, Benali H. Reorganization and plasticity in the adult brain during learning of motor skills. *Curr Opin Neurobiol*. 2005;15(2):161–167.
87. de Sousa G, Maex R, Adams R, Davey N, Steuber V. Dendritic morphology predicts pattern recognition performance in multi-compartmental model neurons with and without active conductances. *J Comput Neurosci*. 2014;221–234.
88. Duijnhouwer J, Remme MWH, Van Ooyen A, Van Pelt J. Influence of dendritic topology on firing patterns in model neurons. *Neurocomputing*. 2001;38-40:183–189.
89. Fauth M, Wörgötter F, Tetzlaff C. Formation and maintenance of robust long-term information storage in the presence of synaptic turnover. *PLoS Comput Biol*. 2015;11(12):1–22.
90. Stepanyants A, Hof PR, Chklovskii DB. Geometry and structural plasticity of synaptic connectivity. *Neuron*. 2002;34(2):275–288.
91. Knoblauch A, Körner E, Körner U, Sommer FT. Structural synaptic plasticity has high memory capacity and can explain graded amnesia, catastrophic forgetting, and the spacing effect. *PLoS One*.. 2014;9(5).
92. Serrano-Gotarredona R, Oster M, Lichtsteiner P, et al. CAVIAR: A 45k neuron, 5M synapse, 12G connects/s AER hardware sensory-processing-learning-actuating system for high-speed visual object recognition and tracking. *IEEE Trans Neural Networks*. 2009;20(9):1417–1438.
93. Liu SC, Van Schaik A, Minch BA, Delbruck T. Asynchronous binaural spatial audition sensor with 2??64??4 Channel output. *IEEE Trans Biomed Circuits Syst*. 2014;8(4):453–464.
94. Painkras E, Plana LA, Garside J, et al. SpiNNaker: A 1-W 18-core system-on-chip for massively-parallel neural network simulation. *IEEE J Solid-State Circuits*. 2013;48(8):1943–1953.
95. Merolla P a, Arthur JV, Alvarez-Icaza R, et al. A million spiking-neuron integrated circuit with a scalable communication network and interface. *Science (80-)*. 2014;345(6197):668–673.
96. Alvarez C, Corbal J, Valero M. Fuzzy memoization for floating-point multimedia applications. *IEEE Trans Comput*. 2005;54(7):922–927.
97. Hussain S, Basu A. Multiclass classification by adaptive network of dendritic neurons with binary synapses using structural plasticity. *Front Neurosci*. 2016;10.
98. Deng L. The MNIST database of handwritten digit images for machine learning research. *IEEE Signal Process Mag*. 2012;29(6):141–142.
99. Takahashi N, Kitamura K, Matsuo N, et al. Locally synchronized synaptic inputs. *Science (80-)*. 2012;335 (6066):353–356.
100. Liu X, Ramirez S, Tonegawa S. Inception of a false memory by optogenetic manipulation of a hippocampal memory engram. *Philos Trans R Soc Lond B Biol Sci*. 2014;369(1633):20130142.
101. Mayford M. The search for a hippocampal engram. *Philos Trans R Soc Lond B Biol Sci*. 2014;369 (1633):20130161.
102. Kubik S, Miyashita T, Guzowski JF. Using immediate-early genes to map hippocampal subregional functions. *Learn Mem*. 2007;14(11):758–770.
103. Mank M, Santos AF, Direnberger S, et al. A genetically encoded calcium indicator for chronic in vivo two-photon imaging. *Nat Methods*. 2008;5(9):805–811.
104. Alvarez V a, Sabatini BL. Anatomical and physiological plasticity of dendritic spines. *Annu Rev Neurosci*. 2007;30:79–97.
105. Bekkers JM, Häusser M. Targeted dendrotomy reveals active and passive contributions of the dendritic tree to synaptic integration and neuronal output. *Proc Natl Acad Sci USA*. 2007;104(27):11447–11452.
106. Vyas A, Mitra R, Rao BSS, Chattarji S. Chronic stress induces contrasting patterns of dendritic remodeling in hippocampal and amygdaloid neurons. *J Neurosci*. 2002;22(15):6810–6818.

107. Ascoli G a, Donohue DE, Halavi M. NeuroMorpho.Org: a central resource for neuronal morphologies. *J Neurosci*. 2007;27(35):9247–9251.
108. Peng H, Hawrylycz M, Roskams J, et al. BigNeuron: large-scale 3D neuron reconstruction from optical microscopy images. *Neuron*. 2015;87(2):252–256.
109. Butz M, Wörgötter F, van Ooyen A. Activity-dependent structural plasticity. *Brain Res Rev*. 2009;60 (2):287–305.
110. Christian KM, Miracle AD, Wellman CL, Nakazawa K. Chronic stress-induced hippocampal dendritic retraction requires CA3 NMDA receptors. *Neuroscience*. 2011;174:26–36.
111. Cuntz H, Forstner F, Borst A, Häusser M. The TREES toolbox-probing the basis of axonal and dendritic branching. *Neuroinformatics*. 2011;9(1):91–96.
112. Bozelos P, Stamatiadis SS, Bouloukakis G, Melachrinos C, Poirazi P. REMOD: a tool for analyzing and remodeling the dendritic architecture of neural cells. *Front Neuroanat*. 2015;9(January):1–11.
113. Koene R a, Tijms B, van Hees P, et al. NETMORPH: a framework for the stochastic generation of large scale neuronal networks with realistic neuron morphologies. *Neuroinformatics*. 2009;7(3):195–210.

CHAPTER

16

Long-Term Information Storage by the Interaction of Synaptic and Structural Plasticity

Michael Fauth[1], *Florentin Wörgötter*[1] *and Christian Tetzlaff*[2]

[1]University of Göttingen, Göttingen, Germany

[2]Max Planck Institute for Dynamics and Self-Organization, Göttingen, Germany

OUTLINE

The Rewiring Brain
DOI: http://dx.doi.org/10.1016/B978-0-12-803784-3.00016-0

1 INTRODUCTION

One of the most fundamental skills enabling humans and animals to survive is their ability to adapt to complex, changing environments. This adaptation, in turn, relies on the ability to store and retrieve information of past experiences, commonly known as learning and memory. In humans and higher animals, learning is associated with changes of the neuronal networks in memory-related brain areas as the cortex, while memory is associated with the persistence of these changes. Yet, the interaction of the underlying neuronal processes leading to such network changes and how this interaction induces information storage for timespans of long-term memory (months–decades) is still not completely understood.[1,2]

It is commonly assumed that experiences which trigger learning are "observed" by changes in the stimuli the brain perceives. Thus the inputs to the adaptive, memory-related neuronal network are altered. These changes in the inputs, in turn, trigger alterations in the neuronal activities, which lead to adjustments of the connections between the neurons.[3–6] Accordingly, memory retention is hypothesized to be linked to the persistence of these elicited changes in the connectivity.[4,7–9] This hypothesis is, however, strongly challenged by the dynamics of the small protrusions of the dendrite, on which the majority of the cortical synapses resides, the so-called dendritic spines. Dendritic spines are highly dynamic and they are created and removed on a daily basis.[7,9–15] For example, in the adult mouse auditory cortex, more than half of the spine population is removed in a time interval of only 3 weeks.[13] Accordingly, also the synapses which have been changed or created upon learning or memory formation are deleted on similar timescales. In the mouse motor cortex, for example, only 5% of the spines formed during motor learning survive longer than 3 months.[9,15] It has been proposed that this small number of long-term stable spines form the morphological basis of memory.[9,15] However, considering the strong correlation between the number of spines formed by learning and behavioral performance,[9,15] it is unlikely that these few persisting spines are sufficient to successfully trigger the learned behavior a long time after its encoding. We therefore conclude that the daily turnover of dendritic spines (or synapses) strongly contradicts the retention of information by individual synapses or their individual properties as their synaptic transmission efficacies.

In this chapter, we will argue that this contradiction can be solved when, instead of storing information by the persistence of single synapses, information is stored and retained by the collective dynamics of multiple synapses.

First, to understand this collective dynamics of synapses, we will consider the underlying plasticity processes which change the connectivity, i.e., remove, alter, or create synapses according to neuronal activities. Essentially, there are two main classes of activity-dependent plasticity processes associated to learning and memory: synaptic and structural plasticity.

Activity-dependent (long-term) synaptic plasticity implies processes which change the transmission efficacies—or synaptic weights—of existing synapses. These processes are subdivided into Hebbian synaptic plasticity and homeostatic synaptic plasticity. Hebbian synaptic plasticity strengthens the synaptic efficacy between pre- and postsynaptic neurons, if they are strongly active together, and weakens it otherwise (long-term potentiation and depression[16–18]; for reviews, see Refs.[19,20]). Thereby, Hebbian synaptic plasticity enforces a positive feedback loop between neuronal activity and connectivity.[4] On the other hand, homeostatic synaptic plasticity implies processes regulating the neuronal activity to a target value or regime by adapting the incoming synaptic weights accordingly. This implements a negative feedback loop between neuronal activity and connectivity. For example, synaptic scaling decreases (increases) synaptic weights of all incoming synapses connecting to a postsynaptic neuron if its activity is quite high (low) for a long period of time.[21,22] Note that the Hebbian processes, acting on a timescale of seconds to minutes, are typically faster than the homeostatic processes which operate on a timescale of hours to days.[2]

The second class of processes changing the neuronal connectivity in an activity-dependent manner is structural plasticity, which describes direct changes of the network structure—namely the creation and removal of synapses. For excitatory synapses, on which we will focus here, such changes can be monitored by imaging the dynamics of the dendritic spines.[23] Similar to activity-dependent synaptic plasticity described earlier, structural plasticity can also be subdivided into Hebbian and homeostatic (structural plasticity) processes.[24] Homeostatic structural plasticity creates new synapses, when neuronal activity is low, and removes synapses, when it is high. Like homeostatic synaptic plasticity, homeostatic structural plasticity enforces a negative feedback loop between neuronal activity and connectivity. These processes typically occur on a timescale of several days to weeks.[12,25–27] On the other hand, Hebbian structural plasticity processes create synapses during phases of high neuronal activity (when neurons tend to fire together) and, thus, also implement a positive feedback loop between neuronal activity and connectivity. These Hebbian changes occur already after several hours or a few days.[9,15,28–30] As the structural changes related to motor and auditory learning[9,15,31] typically take place on the (fast) timescales of Hebbian structural plasticity, in this chapter, we will focus on these processes.

Interestingly, the processes of synaptic and structural plasticity are strongly interwoven. For example, the synaptic weight is correlated to the volume of the dendritic spine[32] of the corresponding synapse. The spine volume, in turn, is linked to the lifetime or removal of the synapse or spine.[13,33] This directly links the dynamics of synaptic plasticity (synaptic weight) with structural plasticity (synapse removal) implying that the activity-dependent changes of the connectivity related to learning and memory can only be understood by investigating the interaction of both processes.

Our hypothesis is that the collective dynamics of multiple synapses can store information for long durations despite synaptic turnover. Thus we have to show that the interaction of activity-dependent synaptic and structural plasticity described earlier yields the formation of multiple synapses with collective dynamics. For simplicity, in contrast to whole network analysis, here we will focus our analysis on investigating the dynamics of the interaction at one connection between two pyramidal neurons in the cortex.

Experiments assessing the number of synapses between two neurons found that different pairs of neurons are connected with different numbers of synapses. However, this number is not random. Summarizing the individual results leads to a specific probability distribution of the number of synapses between two neurons: this distribution is bimodal

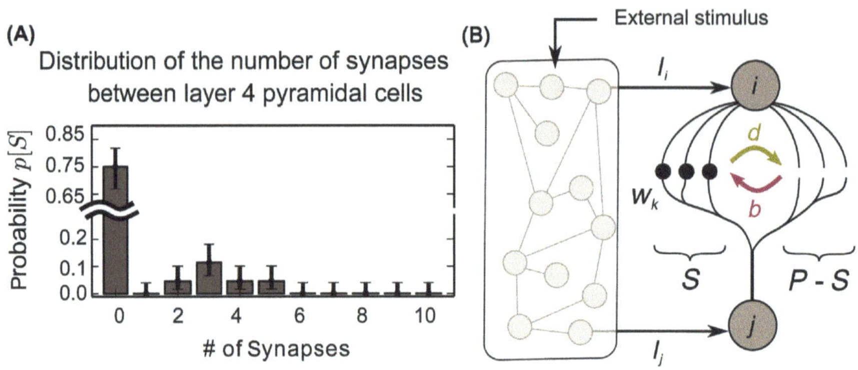

FIGURE 16.1 **Bimodal distributions of the number of synapses between cortical neurons and scheme of the model used in this chapter.** (A) Experimentally measured probability distribution of the number of synapses between two-layer four pyramidal neurons.[34] Error bars indicate 65% confidence intervals for the corresponding probability of each individual number of synapses from statistical bootstrapping (replotted from Ref. [43]). (B) Schematic of the model of structural plasticity used in this chapter. Each connection from neuron j to i has P potential synaptic locations. Out of these P locations, only S actually hosts a functional synapse. These functional synapses have individual weights w_k, which are adapted by activity-dependent synaptic plasticity. Hereby, neuronal activities are determined by the external stimulation I_i and I_j and the current transmitted by the existing synapses. In each time step, synapses are created with a probability b and removed with a probability d, which can depend on the current state of the connection (see main text).

with a high probability around zero synapses and another high probability around three to five synapses (Fig. 16.1A, Ref. [34–38]). Connections consisting of one or two synapses are only very rarely observed and, thus, have a low probability. In contrast, the probability distribution of the number of the anatomically possible synapses—the so-called potential synapses (positions at which a synapse could bridge the spatial gap between axon and dendrite)[39–41]—just decays monotonically.[42] These completely different forms of the two distributions entail that the bimodal distribution of the number of existing or realized synapses cannot emerge from random sampling from the monotonically decaying distribution of potential synapses. This indicates that the number of existing synapses depends on a collective dynamics of all synapses of a connection.[42] In other words, every realized synapse between two neurons has to "know" about the existence of other synapses between these neurons.

In the following, we show with theoretical models (see Sections 2 and 6) that the bimodal distribution of the number of synapses and the associated collective dynamics emerge from the interaction of activity-dependent synaptic and structural plasticity (see Sections 3 and 4). Furthermore, we show that, to a certain degree, external stimuli influence the emerging collective dynamics (see Section 5). We, then, demonstrate that this collective dynamics of multiple synapses on a single connection between two neurons can indeed retain information for long times despite faster synaptic turnover (see Section 7). We report that, given the stimulation dependency of the collective dynamics, information conveyed by external stimuli can be stored orders of magnitude faster than it decays (see Sections 8 and 9).

2 MODEL

As discussed earlier, the activity-dependent changes of connectivity resulting in learning and memory can particularly be understood from investigating the interaction of activity-dependent synaptic and structural plasticity. However, due to the different timescales of synaptic and structural plasticity, this interaction is hard to tackle experimentally. Therefore, in the following, we will derive a theoretical model of the interaction of synaptic and structural plasticity (Fig. 16.1B) to analyze the dynamics of the connectivity between two cortical neurons.

In this model, we assume that each pair of pre- and postsynaptic neurons has P potential or anatomically possible synapses. Out of these P potential synapses, only S synapses are actually realized and host a functional synapse with a certain synaptic transmission efficacy or weight. These weights are adapted by the processes of activity-dependent synaptic plasticity depending on the pre- and postsynaptic neuronal activities or firing rates u and v. These firing rates are, in turn, influenced by external stimuli I_j (for presynaptic neuron j) and I_i (for postsynaptic neuron i) transmitting information to the system during learning or memory formation.

The stochastic appearance and disappearance of dendritic spines is included in the model of structural plasticity by using probabilities: at each time step, new synapses are created with a constant build-up probability b and old synapses are removed with synapse-dependent deletion probability d_k (with synapse index $k = 1, \ldots, S$). Note, the interaction between synaptic and structural plasticity results from the fact that the synaptic weights and the number of existing synapses, both influence the (postsynaptic) neuronal activity and that the deletion probability can depend on the synaptic weight w_k. The influence of this interaction on the connectivity will be analyzed next.

3 CONNECTIVITY EMERGING FROM THE INTERACTION OF SYNAPTIC AND STRUCTURAL PLASTICITY

In a first step, to investigate the dynamics emerging from the interaction of synaptic and structural plasticity, we analyze under which conditions this interaction leads to a bimodal distribution of the number of synapses as observed in the experiments.[34–38] In other words, we have to analyze the long-term equilibrium connectivity of the model corresponding to the stationary probability distribution of the number of synapses. In the following, we will briefly introduce the ansatz to derive this stationary probability distribution.

For this, we have to consider the link between synaptic and structural plasticity in more detail. As discussed earlier, a strong interaction between synaptic and structural plasticity is indicated by (1) a correlation between the synaptic weight and the volume of the corresponding dendritic spine[32] and by (2) a link between the spine volume and the lifetime of the spine or synapse.[33] These two relations (1) and (2) can be combined and understood as a direct link between the synaptic weight and the synapse stability. Here we model this link by a weight-dependent synapse-deletion probability$d_k := \tilde{d}[w_k]$. To match experimental findings,[13,32,33,44] this weight-dependent deletion probability has to decrease with increasing synaptic weight, i.e., synapses with a small synaptic weight have a higher removal probability than synapses with larger weights.

As described earlier, the synaptic weights are determined by an activity-dependent synaptic plasticity rule.[45] As we investigate phenomena on the timescale of structural plasticity (hours to days), we focus on rate-based synaptic plasticity rules with time-dependent synaptic changes of the form

$$\frac{\mathrm{d}w_k}{\mathrm{d}t} = f(u, v, w_k) \tag{16.1}$$

where f is a function depending on the pre- and postsynaptic firing rates u and v, and the synaptic weight w_k of the kth synapse between the pre- and postsynaptic neurons. Note, the synaptic plasticity rules considered here are local, i.e., they only depend on quantities which are accessible directly to the individual synapse.[45] Furthermore, we assume that the synaptic plasticity rules are inherently stable, such that in general, for constant pre- and postsynaptic activities, the synaptic weights converge to a stable (activity-dependent) fixed point (denoted by *)

$$w^*(u, v) \tag{16.2}$$

For constant pre- and postsynaptic stimuli I_j and I_i, the postsynaptic activity converges to a fixed value dependent on the number of realized synapses S. As all synapses on a multisynaptic connection experience the same pre- and postsynaptic activities, this implies that all synapses converge to the same fixed weight

$$w^*(u, v^*(S)) \tag{16.3}$$

depending on the (constant) activities u and $v^*(S)$ emerging for the given stimuli and number of synapses.

To investigate the long-term stationary distribution emerging from the interaction of these inherently stable synaptic plasticity rules with the weight-dependent structural plasticity described earlier, we make two approximations: first, the dynamics of synaptic plasticity is fast compared to the dynamics of structural plasticity. Thus, on the timescale of structural plasticity, we can assume that the dynamics of synaptic plasticity is already converged to its fixed point w^* (adiabatic approximation), which holds for all synapses. As discussed earlier, for constant pre- and postsynaptic stimuli I_j and I_i, there is a unique fixed weight for each number of synapses (see Eq. 16.3) to which all of the synapses converge. In combination with our assumption, this entails that the weight-dependent deletion probability $\tilde{d}[w_k]$ only changes dependent on the number of synapses:

$$d[S] := \tilde{d}[w^*(u, v^*(S))] \tag{16.4}$$

Second, we assume that per time step maximally one synapse is either formed or removed at a connection between two neurons (first-step approximation, see Refs.[43,46]). Then, together with the adiabatic approximation, the following relation holds for the stationary long-term equilibrium probability distribution of the number of realized synapses[43,46,47]:

$$\frac{p[S]}{p[S-1]} = \frac{P - S + 1}{S} \frac{b}{d[S]} \tag{16.5}$$

with $p[S]$ being the long-term equilibrium probability of finding a pair of neurons connected by S realized synapses. From this relation (see Eq. 16.5), the whole stationary probability distribution can be recursively calculated by starting with a fixed value for $p[0]$ and calculating $p[1]$ and repeatedly applying Eq. (16.5) until all probabilities up to $p[P]$ are calculated. Then, the overall distribution has to be normalized to a sum of one, i.e., $\sum_{S=0}^{S=P} p[S] = 1$. Thus, from Eq. (16.5) we can directly derive the stationary probability distribution required to analyze the interaction between activity-dependent synaptic and structural plasticity.

4 NECESSARY CONDITIONS TO YIELD A BIMODAL DISTRIBUTION OF THE NUMBER OF SYNAPSES

Eq. (16.5) can be used to directly investigate whether the bimodal probability distribution of the number of synapses can be reproduced by the interaction of synaptic and structural plasticity.[43] To get such a bimodal distribution, the relation $p[S]/p[S-1]$ (see Eq. 16.5) has to change in a specific way with increasing number of synapses S. As described earlier (see also Fig. 16.1), for zero synapses the system has a high probability. This probability $p[S]$ decreases with increasing number of synapses S, thus $p[S]/p[S-1] < 1$. However, the system has again a high probability to form four synapses. This implies that, in between zero and four synapses, after decreasing, the probabilities have to increase with increasing number of synapses and, thus, $p[S]/p[S-1] > 1$. Thus, in a certain regime, the left-hand side of Eq. (16.5) has to grow from below one to above one with increasing S. However, the first term on the right-hand side decreases monotonically with S. Hence, a bimodal probability distribution can only emerge, when, in the second term on the right-hand side, the ratio between build-up probability b and deletion probability $d[S]$ increases with the number of synapses S. This implies that the deletion probability has to decrease when the number of synapses is increased. In other words, in the long-term equilibrium, a bimodal probability distribution of the number of synapses between two cortical neurons can only emerge, when synapses become more stable (decrease in deletion probability) with more realized synapses or rather when multiple synapses mutually stabilize each other. Thus these multiple synapses need to have collective dynamics.

As described in Section 3, the weight-dependent deletion probability $\tilde{d}[w]$ decreases with increasing synaptic weight. Thus, as deletion probability also has to decrease with increasing numbers of realized synapses, a bimodal probability distribution can only emerge if the (fixed) synaptic weights grow with increased number of synapses. Yet, the biologically plausible locality of the synaptic plasticity rules (see Eq. 16.1) does not give an individual synaptic weight direct "access" to the overall number of synapses on a dendrite. The only possibility to "share" information across synapses connected to the same postsynaptic neuron is via the postsynaptic activity. As the synaptic weights should increase with increasing number of synapses, and as obviously the postsynaptic activity increases with increasing number of (excitatory) synapses, the synaptic weights also have to increase with increased postsynaptic activity. This condition constrains the possible activity-dependent synaptic plasticity rules that will yield, in conjunction with structural plasticity, the bimodal probability distribution of the number of synapses as observed in the cortex.[34–38]

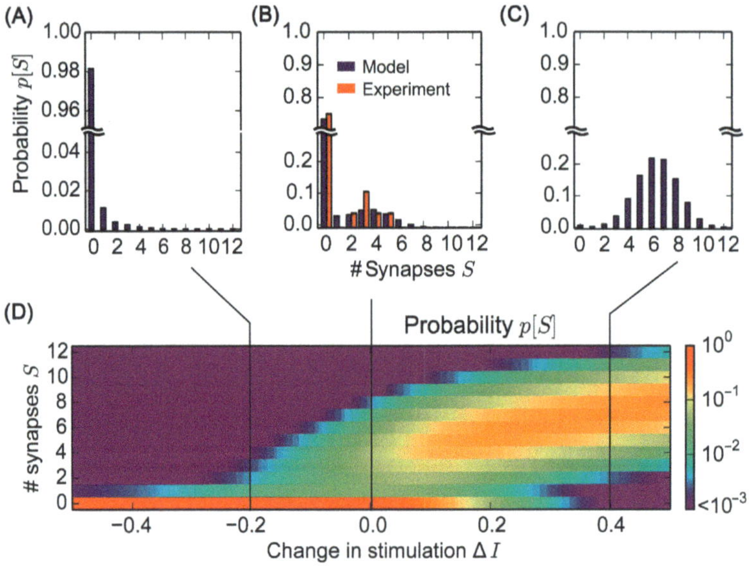

FIGURE 16.2 **Interaction of synaptic and structural plasticity yields a collective dynamics which can be altered by external stimulations and account for experimental data.** (A–C) The long-term equilibrium probability distribution of the number of synapses between two neurons from the theoretical model (*blue*) at different strengths of external stimulation (replotted from Ref. [43]). (A) At low stimulation levels, two neurons are in general unconnected ($S = 0$) or connected with only a few synapses. (B) At intermediate stimulation levels, the model matches the experimentally observed bimodal distribution (*red bars*). (C) At high stimulation levels, two neurons are most likely connected with each other by multiple synapses. (D) Long-term equilibrium probability distribution of the number of synapses (y-axis) for different levels of external stimulation (x-axis). Color code shows the logarithm of the probabilities $p[S]$. Note that one of the high probability peaks at zero and multiple synapses vanishes for sufficiently high or low stimulation levels, respectively.

We find that a large class of biologically plausible synaptic plasticity rules fulfills this constraint, although some of the commonly used synaptic plasticity rules do not.[43] For instance, the BCM rule[48] (without sliding threshold) in combination with weight-dependent synaptic scaling[49] fulfills this constraint[43] and leads to the formation of a bimodal probability distribution similar to experimental findings (Fig. 16.2B and D, see Ref. [43] for detailed methods and parameters).

These results indicate that, under reasonable constraints, the interaction of activity-dependent synaptic and structural plasticity leads to a bimodal distribution of the number of realized synapses. As already indicated before,[42] this can only be achieved by considering the collective dynamics of multiple synapses. Note, although the here presented results have been derived by using rate-based synaptic plasticity rules, similar results are also obtained by using spike-timing-dependent synaptic or structural plasticity rules.[43,46]

5 STIMULATION-DEPENDENT CHANGES OF THE STATIONARY CONNECTIVITY

In order to store and retain information transmitted by external stimuli, the collective dynamics of multiple synapses and the emerging connectivity has to be responsive to alterations in external stimulation. To investigate how far the system is sensitive to such alterations, first, we define the stimulation level, at which the experimentally verified bimodal probability distribution of the number of realized synapses is generated, as the biological working point of the system. Then, starting from this working point, we varied the stimuli transmitted to the pre- and postsynaptic neurons by the same amount ΔI and evaluated the resulting long-term equilibrium probability distributions of the number of synapses (Fig. 16.2D). By this procedure, we find that the external stimuli, indeed, have a strong influence on the resulting probability distributions. At intermediate stimulation levels, the model can account for the experimentally observed bimodal probability distribution (Fig. 16.2B). When the external stimulation is decreased, e.g., by stronger inhibitory inputs, the collective dynamics leads to the disconnection of the two neurons, i.e., the probability distribution only shows high probabilities around zero synapses (Fig. 16.2A). On the other hand, when the system receives high external stimulation levels, which could result from increased excitatory inputs, the two neurons have a high chance to be connected with each other by multiple synapses. Thus numbers of synapses around zero have a low probability to be realized, while multiple synapses have a high probability level (Fig. 16.2C).

As a consequence, the collective dynamics, emerging at low or high stimulation levels, can be used to drive two neurons to be unconnected or to be connected (by multiple synapses), respectively. Yet, considering that changing the connectivity by altered stimulation conditions represents learning, it still remains unclear whether these changes can also be retained for long periods of time, during which the system receives only stimulations on the control (working point) level. Thus, in the following, we will investigate the information retention properties of the system.

6 GENERALIZATION OF THE MODEL FOR INVESTIGATING DYNAMICS

So far, we have studied the influence of different stimuli on the connectivity between two neurons in the long-term equilibrium. However, to investigate the storage and retention of information, we also need to investigate the temporal development of the connectivity.

In the previous sections, we had demonstrated that changes in the external stimulations influencing neuronal activities alter the collective dynamics emerging from the interaction of synaptic and structural plasticity, such that different equilibrium distributions of the number of synapses emerge. As our hypothesis is that the collective dynamics are important to store and retain information, in the following, we will focus on this collective dynamics. For this, we will consider the parts of the model which are required to implement the collective dynamics of multiple synapses omitting the details of how these dynamics emerge (which has been shown before). Given that Eq. (16.5) describes the dependency between the equilibrium probability distribution of the number of synapses $p[S]$ and the probabilities of synapse deletion $d[S]$ and synapse formation b, which is constant, we can calculate the deletion probabilities leading to different equilibrium distributions. Then, in turn, we can simulate a collective dynamics yielding a certain distribution by directly using or setting the set of corresponding deletion probabilities $d[S]$ (with S from zero to P). In the following, we will do this for the three qualitatively different probability distributions observed in Fig. 16.2: (1) at low stimulation with high probabilities only for numbers of synapses around zero, (2) at intermediate or working-point stimulation with a bimodal distribution of the number of synapses as observed in experiments, and (3) at high stimulations with high probabilities only around multiple synapses. Hereby, we summarize distributions of similar shapes and we approximate the different probability distributions as follows. The probability distribution of the number of synapses resulting at low stimulations is abstracted by a (truncated) Poisson distribution. The probability distribution at high stimulations is abstracted by a (truncated) Gaussian distribution. Finally, bimodal probability distributions, emerging at intermediate stimulation levels, are abstracted by a weighted sum of the Poisson and Gaussian distributions (for details, see Ref. [47]).

The usage of this abstracted model including only the basic dynamics also has the advantage that all following results hold for all neuron models and synaptic plasticity rules leading to these kinds of collective dynamics of multiple synapses.

7 INFORMATION RETENTION AT THE BIOLOGICAL WORKING POINT

Given the abstracted model, in the following, we will investigate whether the collective dynamics revealed above will store and retain information despite constant turnover of individual synapses.

First, we will investigate how well multisynaptic connections retain the information about their initial state and how this information retention depends on the collective dynamics of the synapses of such a connection. Here we are especially interested in the dynamics at intermediate stimulation levels, which leads to the experimentally observed probability distribution of the number of synapses (Fig. 16.2B), considered to be the biological working point of the system.

Given a system in this working-point state, the number of synapses fluctuates on different timescales (Fig. 16.3A). If a connection hosts multiple synapses (here five synapses), the number of synapses starts to fluctuate strongly after a short time span (up to 10^8 time steps), but the connection always maintains multiple synapses. Only at much slower timescales, hence after a long duration, also transitions to zero synapses and back occur. Similarly, if a connection initially has zero realized synapses, it only fluctuates around zero synapses at fast timescales and may switch to multiple synapses only at much longer time spans. This indicates that, although the number of synapses is fluctuating due to the continuous synaptic turnover, the information, whether a connection was initialized at zero or multiple synapses, is maintained for long timescales.

To quantify this hypothesis, we analyzed a population of connections given stimuli at working-point level and evaluated the maintenance of information by using the statistical measure of mutual information. More precisely, we calculated the mutual information between the number of synapses at time point t and the initial number of synapses at time point zero.

If the population of connections is initialized from the bimodal probability distribution (Fig. 16.3Bi), with ongoing time, the probability distribution of the number of synapses

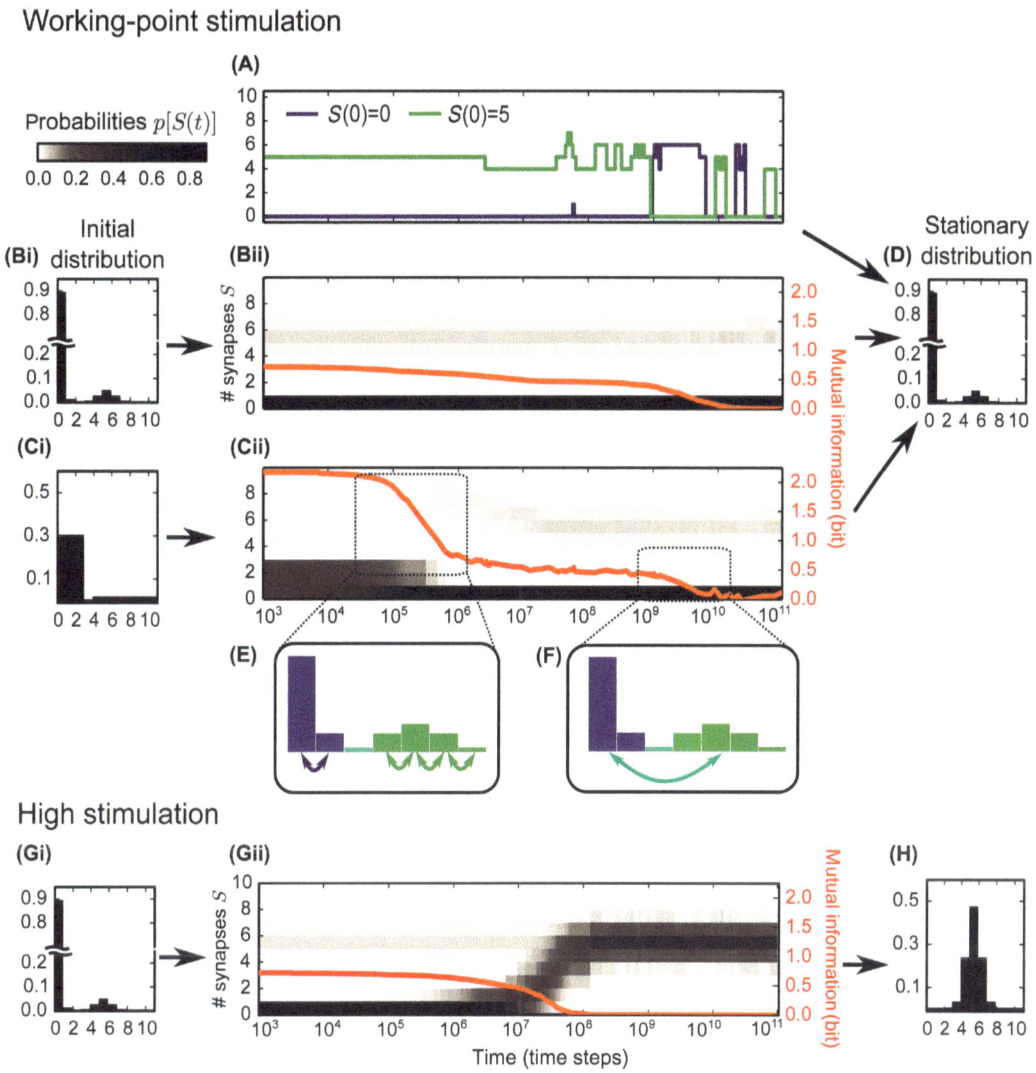

FIGURE 16.3 **Information retention at working point and high stimulation levels.** (A) Temporal development of the number of synapses for two examples of multisynaptic connections initialized at zero (*blue*) and five (*green*) synapses. (Bii, Cii) Temporal development of the probability distribution of the number of synapses in a population of multisynaptic connections (gray scale, logarithmic) initialized with numbers of synapses (Bi) in a bimodal probability distribution and (Ci) in a probability distribution with two uniform peaks. The system receives working-point stimulation levels and, thus, converges to the working-point probability distribution shown in (D). Underlying this temporal development of the number of synapses, the connections retain information about their initial conditions as measured by the mutual information between the number of synapses at time point 0 and time point t (*red curve*). (E) During the first decay phase of mutual information, at fast timescales, connections mostly fluctuate within the peak of the bimodal probability distribution, in which they have been initialized. Transitions between peaks (low to high and vice versa) occur only rarely, such that the connection retains the information about the peak in which it was initialized (zero or multiple synapses) until the second decay phase sets in. (F) During the second decay phase at much slower timescales, also transitions between the peaks set in, such that the number of synapses reaches its long-term equilibrium probability distribution and all information about the initial information is lost. (Gii) Same as panel Bii but for a population of connections receiving high stimulation levels yielding a (H) unimodal equilibrium distribution. (Gi) Initial distribution for panel Gii. Note, the information about the initial conditions decays completely on a fast timescale (i.e., during the first decay phase observed in panel c). Panels Bii, Cii, and Gii have been replotted based on simulations in Ref. [47]

(logarithmic grayscale in Fig. 16.3Bii) does not change significantly. However, the individual connections undergo fluctuations due to synaptic turnover (Fig. 16.3A) and "forget" the information about their initial conditions. Accordingly, for the whole population of connections, the mutual information between initial condition and the number of synapses at time point t decays over time (red curve, Fig. 16.3Bii). However, a large proportion of information about the initial condition is preserved orders of magnitude longer than the typical timescale of structural changes (given by the inverse synapse creation probability, $b^{-1} = 10^8$). This prolongation of information storage is also given for other initial conditions (Fig. 16.3C). Furthermore, the time course of mutual information indicates that the information about the initial conditions decays in two distinct phases—each corresponding to a different type of fluctuation of multisynaptic connections (Fig. 16.3D and E, see Ref. [47]). On fast timescales, multisynaptic connections following the dynamics at working-point stimulation level typically stay in the peak (zero or multiple synapses) of the bimodal probability distribution in which they have been initialized. Thus the first decay in mutual information occurs due to fluctuation within the peaks (Fig. 16.3D, see Ref. [47] for a detailed analysis of the fluctuations and timescales). This also means that, on this timescale, the information about the initial peak (zero or multiple synapses) is preserved such that the time course of mutual information decay exhibits a plateau (see Fig. 16.3Bii and Cii). Only at much longer time spans and, thus, on slower timescales, transitions between peaks (low to high and vice versa) occur and the information about the initial condition is completely lost (Fig. 16.3E, see also Ref. [47]). This is due to the fact that, for the bimodal probability distribution, connections have a very low probability to host certain number of synapses. These are typically the number of synapses between both peaks. Thus, to make a transition from one peak to the other, the connection has to transit these number of synapses, which have a low probability to be realized by the fluctuations. This, in turn, implies that the probability of transitions from one peak to the other is also quite low and, therefore, transitions are very rare.[47] Thus the timescale of the second decay of mutual information is orders of magnitude larger than the typical timescale of structural changes.

Hence, multisynaptic connections following the collective dynamics at the intermediate working-point stimulation level retain the information, whether they were initialized around zero or at multiple synapses, in spite of the fluctuations of the number of synapses due to synaptic turnover. Interestingly, these two states (zero or multiple synapses) correspond to the states to which a connection can be driven by high or low stimulation levels. This indicates that the dynamics at altered, nonworking-point stimulation levels is suitable for storing information on multisynaptic connections. However, at this point it is still unclear, whether the probability distribution under altered stimulation is reached sufficiently fast, i.e., whether the timescale of information storage is faster than the timescale of information retention. Only then the here-observed characteristic mechanism might indeed be useful for a memory system.

8 INFORMATION RETENTION AT ALTERED STIMULATION LEVELS

In a first approach to clarify how the collective dynamics at altered stimulation levels can be used to store information, we investigate the information retention of multisynaptic connections following the collective dynamics at high or low stimulation levels (see above). The rationale is that, if the system maintains information under these stimuli for only a short period, the system converges fast to a new state. Thus these stimuli can be used for fast learning. As an example, we repeat the same experiment as in Fig. 16.3B but with high stimulation levels (Fig. 16.3H). Interestingly, multisynaptic connections under high (or low[47]) stimulation levels do not exhibit a prolonged information retention, different from the result found under working-point stimulation (Fig. 16.3Gii). Instead, the mutual information quickly decreases to zero indicating a complete loss of information about the initial condition at fast timescales. This decay is comparable to the first decay phase observed under working-point stimulation (Fig. 16.3Bii and Cii). It can also be seen that, under high or low stimulation levels, the connections converge to their new equilibrium distribution quickly and, thus, cannot retain information about their past for very long durations. This also implies that the dynamics of the multisynaptic connections is only determined by these stimuli (and not by their history) such that high or low stimulation levels can be used to actively and quickly store information on multisynaptic connections. Comparing the time courses of mutual information to the working-point stimulation level, it can indeed be expected that these processes of information storage would be much faster than information retention, which will be verified in the following.

9 INFORMATION CAN BE STORED FASTER THAN IT DECAYS

To test whether the stimulation-dependent collective dynamics can really be used to store information faster than the information decays afterwards, in the next step, we use a population of multisynaptic connections to store and retain a 4-bit grayscale image (Fig. 16.4A; adapted from Ref. [50]). More precisely, for each pixel of the image, we have 16 connections.

In the learning phase, a fraction of these 16 connections corresponding to the gray value of the pixel follows the dynamics observed for high-level stimulations, whereas the rest follows the dynamics observed for low-level stimulations. In the following retention phase, all connections are set back to the collective dynamics observed for working-point stimulation levels. The image, which is stored by the model connections, is read out by summing over the numbers of synapses of all 16 multisynaptic connections for each pixel (Fig. 16.4B–D and Ei–Hi). To assess the quality of information storage and retention, we also calculate the correlation between the original image and the image stored by the population of connections over time (Fig. 16.4J).

As expected from the different time courses of information retention (Fig. 16.3Bii, Cii, Gii), the image can be learned quite fast. This is indicated by a sharp rise in the correlation between original and encoded images (Fig. 16.4J). Accordingly, the stored image is already quite similar to the original image at 154×10^6 time steps (Fig. 16.4C) and almost perfectly correlated at the end of the learning phase at 300×10^6 time steps (Fig. 16.4D). On the other hand, during the retention phase, the correlation decays very slowly as compared to the learning phase. The stored image is still recognizable at the double (Fig. 16.4Fi) or even the 10-fold duration of the learning phase (Fig. 16.4Gi). Only after the 50-fold duration of the learning phase (Fig. 16.4Hi), the correlation has dropped significantly and the

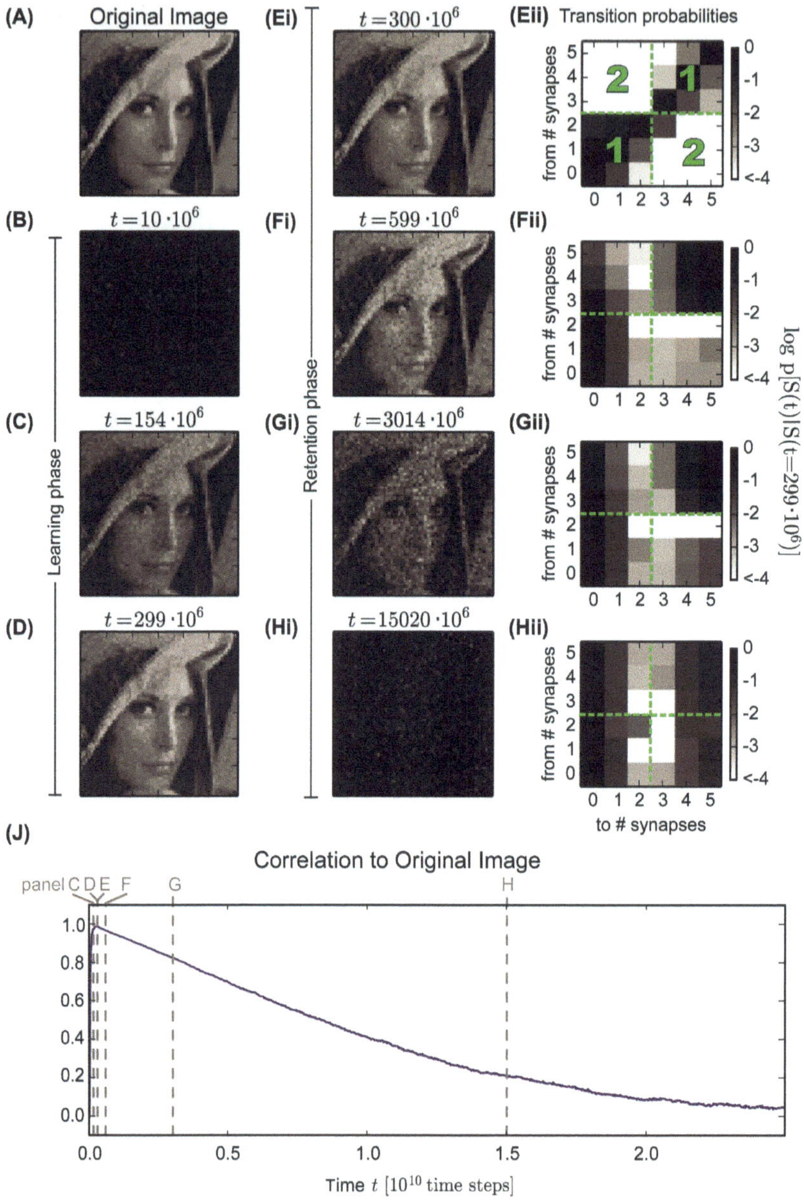

FIGURE 16.4 **A population of multisynaptic connections stores and retains an image.** (A) Original image *(adapted from Viterbi School of Engeneering. The USC-SIPI Image Database 1981)*. (B–D) Different snapshots of the temporal development of the encoded image by the population of multisynaptic connections during the learning phase. The gray value of each pixel is determined by the sum of the number of synapses over all 16 connections corresponding to the pixel. (Ei–Hi) Same for retention phase. (Eii–Hii) Gray-scale represents the logarithmic transition probabilities between the number of synapses after learning ($t = 299 \times 10^6$ time steps) and the number of synapses at later time points indicated above the images on the left (panels Ei–Hi). The *green dotted lines* in panel Eii separate the two peaks of the bimodal probability distribution. Regions marked with "1" correspond to transitions within the peak, whereas transitions between the peaks are marked with "2." Initially (Eii–Gii), the majority of transitions occur within the peaks. Later (Hii), also transitions between the peaks take place and the image becomes unrecognizable. (J) Time course of the correlation between original image and the image encoded by the population of multisynaptic connections. The sharp rise and slow decay of correlation indicate fast learning and slow forgetting (for methods refer to Ref. [47], $P = 5, \mu = 5, \sigma = 1, \lambda = 0.05, C = 0.1$).

image becomes unrecognizable. These results confirm that, due to the changes in the collective dynamics at different external stimulation levels, information can be retained orders of magnitude longer than it took to store it.

Therefore, we used this simulation to directly verify whether this long-term information retention occurs due the fact that transitions between the peaks of the bimodal equilibrium distribution only occur on long timescales (as proposed in Section 8). For this, we evaluated the transition probabilities between the number of synapses at the end of the learning phase (299×10^6 time steps) and the number of synapses at later time points (Fig. 16.4Eii–Hii, logarithmical transition probabilities in grayscale, time points are indicated above the image panels on the left). We find that, as long as the image can be retained reasonably well (Fig. 16.4Eii–Gii), the majority of the transitions occur within the peaks of the bimodal distribution (lower left and upper right quadrant marked with "1" in Fig. 16.4Eii). Only after longer times, the correlation between original and stored image drops and transitions between both peaks of the bimodal distribution take place. These results demonstrate that long-term information retention indeed relies on the fact that transitions between the peaks of the bimodal probability distribution are initially rare and only occur on very long timescales.

10 DISCUSSION

In summary, we showed that the interaction of synaptic and structural plasticity leads to collective dynamics of multiple synapses, which can be influenced by external stimulations. These collective dynamics can, in turn, retain information orders of magnitude longer than the typical timescale of structural changes implying that the collective dynamics of multiple synapses allows information retention despite synaptic turnover. Furthermore, the stimulation-dependent changes in the collective dynamics can be used to store information much faster than it decays, which renders the collective dynamics of multiple synapses as a candidate system for long-term memory storage.

Along these lines, when adapting the collective dynamics of synapses of our model to experimental data on dendritic spine creation and removal during motor learning, the difference in timescales in learning and forgetting observed in our model allows us to explain how information, which is stored in a few days, can be retained over several months.[47] Thus the collective dynamics of multiple synapses between two neurons can indeed account for information storage on the timescale of months and probably also years.[47] Therefore, to our knowledge, these collective dynamics provide the first biological plausible explanation for information storage on the timescale of long-term memory under ongoing synaptic turnover that does not rely on plasticity processes at the same timescale as memory[51] or the periodical occurrence of consolidation stimuli.[52]

However, this prolongation in information retention is accompanied by a decrease in storage capacity per synapse. On long timescales, the collective dynamics of multisynaptic connections can only retain whether two neurons are unconnected or connected by multiple synapses. Thus, on these timescales, all synapses of a multisynaptic connection together store at most one bit of information and the transmission efficacy between two neurons becomes effectively binary (connected or unconnected). Thus the long-term memory capacity of a neuronal network consisting of multisynaptic connections would be best described by models using binary synapses.[53–58] Yet, on shorter timescales, the multisynaptic connections also retain information about their precise number of synapses and corresponding synaptic weights. This (transient) memory storage might be essential to bridge the gap between fast changes of stimuli and activities, which convey information, and the here-described slow changes of the network structure, which store and retain the information on much longer timescales.

10.1 Future Modeling Studies

Although the collective dynamics of multisynaptic connections provides a possible solution for the stable information storage and retention despite synaptic turnover, this dynamics has only been studied on feedforward connections between two neurons. Apparently, many connections in cortical networks are recurrent. We expect that such recurrent structures influence, in turn, the emerging dynamics of multisynaptic connections. Thus in the next step to investigate the dynamics of structural plasticity on multisynaptic connections, their emergence and functional impact in recurrent networks has to be investigated in future modeling studies.

10.2 Future Experimental Studies

Based on the results of our modeling studies shown here, we are able to provide several experimentally testable predictions. The most important prediction from the above-described collective dynamics is that the number of synapses between two neurons influences the lifetime of each synapse. Thus a synapse on a connection with multiple other realized synapses has a longer lifetime than a synapse on a connection with no or just a small number of other synapses. To test this prediction, the lifetime of synapses has to be measured together with the number of realized synapses between the same pre- and postsynaptic neurons. In this way, the relation between lifetime and the number of synapses can be evaluated. However, to obtain these data, a method to identify synapses between the same pair of neurons in living tissue has to be derived.

On the behavioral level, from our model we can predict that memory retention does not rely on the fixed set of synapses, which has been manipulated upon memory storage, but rather on the overall strength of connectivity between the involved neurons. To test this, the number of synapses between neurons has to be repeatedly determined during behavioral experiments investigating memory retention, such that the correlation between connectivity and memory retention can be assessed.

Yet, to test these predictions, an in vivo method to determine the number of synapses between the same pre- and postsynaptic neurons is needed. One possible approach would be to find synapses receiving the same pre- and postsynaptic activities. For this, one could compare the correlations in the calcium signaling between different synapses hosted on the same dendrite.[59] So far, this approach has only been applied for parts of the dendrite. Evaluating these correlations over the whole dendritic tree could be one possibility to find the here-proposed neuronal substrate of long-term memory.

References

1. Lisman J. The challenge of understanding the brain: where we stand in 2015. *Neuron*. 2015;86:864–882.
2. Tetzlaff C, Kolodziejski C, Markelic I, Wörgötter F. Time scales of memory, learning, and plasticity. *Biol Cybern*. 2012;106:715–726.
3. Cajal SRY. The Croonian Lecture: La Fine Structure des Centres Nerveux. *Proc Royal Soc Lond*. 1894;55:444–468.
4. Hebb D.O. *The organization of behavior: a neuropsychological theory*. Wiley Book in Clinical psychology. Wiley 1949.
5. Kandel ER. Nobel lecture: the molecular biology of memory storage: a dialog between genes and synapses. *Biosci Rep*. 2001;21:565–611.
6. Konorski J. *Conditioned Reflexes and Neuron Organization*. Cambridge University Press; 1948.
7. Hofer SB, Mrsic-Flogel TD, Bonhoeffer T, Hübener M. Experience leaves a lasting structural trace in cortical circuits. *Nature*. 2009;457:313–317.
8. Rolls ET. Attractor networks. *Wiley Interdiscip Rev Cogn Sci*. 2010;1:119–134.
9. Yang G, Pan F, Gan W-B. Stably maintained dendritic spines are associated with lifelong memories. *Nature*. 2009;462:920–924.
10. Attardo A, Fitzgerald JE, Schnitzer MJ. Impermanence of dendritic spines in live adult CA1 hippocampus. *Nature*. 2015;
11. Holtmaat AJGD, Trachtenberg JT, Wilbrecht L, Shepherd GM, Zhang X, Knott GW, et al. Transient and persistent dendritic spines in the neocortex in vivo. *Neuron*. 2005;45:279–291.
12. Keck T, Mrsic-Flogel TD, Vaz AM, Eysel UT, Bonhoeffer T, Hubener M. Massive restructuring of neuronal circuits during functional reorganization of adult visual cortex. *Nat Neurosci*. 2008;11:1162–1167.
13. Loewenstein Y, Yanover U, Rumpel S. Predicting the dynamics of network connectivity in the neocortex. *J Neurosci*. 2015;35:12535–12544.
14. Trachtenberg JT, Chen BE, Knott GW, et al. Long-term in vivo imaging of experience-dependent synaptic plasticity in adult cortex. *Nature*. 2002;420:788–794.
15. Xu T, Yu X, Perlik AJ, Tobin WF, Zweig JA, Tennant K, et al. Rapid formation and selective stabilization of synapses for enduring motor memories. *Nature*. 2009;462:915–919.
16. Bliss TV, Lomo T. Long-lasting potentiation of synaptic transmission in the dentate area of the anaesthetized rabbit following stimulation of the perforant path. *J Physiol*. 1973;232:331–356.
17. Dudek SM, Bear MF. Homosynaptic long-term depression in area CA1 of hippocampus and effects of *N*-methyl-D-aspartate receptor blockade. *Proc Natl Acad Sci USA*. 1992;89:4363–4367.
18. Ito M, Kano M. Long-lasting depression of parallel fiber-Purkinje cell transmission induced by conjunctive stimulation of parallel fibers and climbing fibers in the cerebellar cortex. *Neurosci Lett*. 1982;33:253–258.
19. Collingridge GL, Peineau S, Howland JG, Wang YT. Long-term depression in the CNS. *Nat Rev Neurosci*. 2010;11:459–473.
20. Feldman DE. Synaptic mechanisms for plasticity in neocortex. *Annu Rev Neurosci*. 2009;32:33–55.
21. Turrigiano GG, Leslie KR, Desai NS, Rutherford LC, Nelson SB. Activity-dependent scaling of quantal amplitude in neocortical neurons. *Nature*. 1998;391:892–896.
22. Turrigiano GG, Nelson SB. Homeostatic plasticity in the developing nervous system. *Nat Rev Neurosci*. 2004;5:97–107.
23. Yuste R. *Dendritic Spines*. Cambridge, MA: The MIT Press; 2010.
24. Fauth M, Tetzlaff C. Opposing effects of neuronal activity on structural plasticity. *Front Neuroanat*. 2016;10:75.
25. Isokawa M. Remodeling dendritic spines in the rat pilocarpine model of temporal lobe epilepsy. *Neurosci Lett*. 1998;258:73–76.
26. Paul LA, Scheibel AB. Structural substrates of epilepsy. *Adv Neurol*. 1986;44:775–786.
27. Wolff JR, Joó F, Dames W. Plasticity in dendrites shown by continuous GABA administration in superior cervical ganglion of adult rat. *Nature*. 1978;274:72–74.
28. Maletic-Savatic M, Malinow R, Svoboda K. Rapid dendritic morphogenesis in CA1 hippocampal dendrites induced by synaptic activity. *Science*. 1999;283:1923–1927.
29. Nägerl UV, Eberhorn N, Cambridge SB, Bonhoeffer T. Bidirectional activity-dependent morphological plasticity in hippocampal neurons. *Neuron*. 2004;44:759–767.
30. Nägerl UV, Köstinger G, Anderson JC, Martin KAC, Bonhoeffer T. Protracted synaptogenesis after activity-dependent spinogenesis in hippocampal neurons. *J Neurosci*. 2007;27:8149–8156.

31. Moczulska KE, Tinter-Thiede J, Peter M, Ushakova L, Wernie T, Bathellier B, et al. Dynamics of dendritic spines in the mouse auditory cortex during memory formation and memory recall. *Proc Natl Acad Sci USA*. 2013;110:18315–18320.
32. Matsuzaki M, Ellis-Davies GC, Nemoto T, Miyashita Y, Iino M, Kasai H. Dendritic spine geometry is critical for AMPA receptor expression in hippocampal CA1 pyramidal neurons. *Nat Neurosci*. 2001;4:1086–1092.
33. Yasumatsu N, Matsuzaki M, Miyazaki T, Noguchi J, Kasai H. Principles of long-term dynamics of dendritic spines. *J Neurosci*. 2008;28:13592–13608.
34. Feldmeyer D, Egger V, Lübke J, Sakmann B. Reliable synaptic connections between pairs of excitatory layer 4 neurones within a single "barrel" of developing rat somatosensory cortex. *J Physiol*. 1999;521(Pt 1):169–190.
35. Feldmeyer D, Lübke J, Sakmann B. Efficacy and connectivity of intracolumnar pairs of layer 2/3 pyramidal cells in the barrel cortex of juvenile rats. *J Physiol*. 2006;575:583–602.
36. Feldmeyer D, Lübke J, Silver RA, Sakmann B. Synaptic connections between layer 4 spiny neurone-layer 2/3 pyramidal cell pairs in juvenile rat barrel cortex: physiology and anatomy of interlaminar signalling within a cortical column. *J Physiol*. 2002;538:803–822.
37. Hardingham NR, Read Jenny CA, Trevelyan AJ, Nelson JC, Jack JJB, Bannister NJ. Quantal analysis reveals a functional correlation between presynaptic and postsynaptic efficacy in excitatory connections from rat neocortex. *J Neurosci*. 2010;30:1441–1451.
38. Markram H, Lübke J, Frotscher M, Roth A, Sakmann B. Physiology and anatomy of synaptic connections between thick tufted pyramidal neurones in the developing rat neocortex. *J Physiol*. 1997;500(Pt 2):409–440.
39. Chklovskii DB, Mel BW, Svoboda K. Cortical rewiring and information storage. *Nature*. 2004;431:782–788.
40. Stepanyants A, Hirsch JA, Martinez LM, Kisvárday ZF, Ferecskó AS, Chklovskii DB. Local potential connectivity in cat primary visual cortex. *Cereb Cortex*. 2008;18:13–28.
41. Stepanyants A, Hof PR, Chklovskii DB. Geometry and structural plasticity of synaptic connectivity. *Neuron*. 2002;34:275–288.
42. Fares T, Stepanyants A. Cooperative synapse formation in the neocortex. *Proc Natl Acad Sci*. 2009;106:16463–16468.
43. Fauth M, Wörgötter F, Tetzlaff C. The formation of multi-synaptic connections by the interaction of synaptic and structural plasticity and their functional consequences. *PLoS Comput Biol*. 2015;11:e1004031.
44. Le Bé J-V, Markram H. Spontaneous and evoked synaptic rewiring in the neonatal neocortex. *Proc Natl Acad Sci USA*. 2006;103:13214–13219.
45. Gerstner W, Kistler WM. Mathematical formulations of Hebbian learning. *Biol Cybern*. 2002;87:404–415.
46. Deger M, Helias M, Rotter S, Diesmann M. Spike-timing dependence of structural plasticity explains cooperative synapse formation in the neocortex. *PLoS Comput Biol*. 2012;8:e1002689.
47. Fauth M, Wörgötter F, Tetzlaff C. Formation and maintenance of robust long-term information storage in the presence of synaptic turnover. *PLoS Comput Biol*. 2015;11:e1004684.
48. Bienenstock EL, Cooper LN, Munro PW. Theory for the development of neuron selectivity: orientation specificity and binocular interaction in visual cortex. *J Neurosci*. 1982;2:32–48.
49. Tetzlaff C, Kolodziejski C, Timme M, Wörgötter F. Synaptic scaling in combination with many generic plasticity mechanisms stabilizes circuit connectivity. *Front Comp Neurosci*. 2011;5:47.
50. Viterbi School of Engeneering. The USC-SIPI Image Database 1981.
51. Fusi S, Drew PJ, Abbott LF. Cascade models of synaptically stored memories. *Neuron*. 2005;45:599–611.
52. Tetzlaff C, Kolodziejski C, Timme M, Tsodyks M, Wörgötter Florentin. Synaptic scaling enables dynamically distinct short- and long-term memory formation. *PLoS Comput Biol*. 2013;9:e1003307.
53. Amit DJ, Gutfreund H, Sompolinsky H. Spin-glass models of neural networks. *Phys Rev A*. 1985;32:1007–1018.
54. Amit DJ, Gutfreund H, Sompolinsky H. Information storage in neural networks with low levels of activity. *Phys Rev A*. 1987;35:2293–2303.
55. Dubreuil AM, Amit Y, Brunel N. Memory capacity of networks with stochastic binary synapses. *PLoS Comput Biol*. 2014;10:e1003727.
56. Hopfield JJ. Neural networks and physical systems with emergent collective computational abilities. *Proc Natl Acad Sci USA*. 1982;79:2554–2558.
57. Steinbuch K. *Die Lernmatrix Kybernetik*. 1961;1:36–45.
58. Willshaw DJ, Buneman OP, Longuet-Higgins HC. Non-holographic associative memory. *Nature*. 1969;222:960–962.
59. Winnubst J, Cheyne JE, Niculescu D, Lohmann C. Spontaneous activity drives local synaptic plasticity in vivo. *Neuron*. 2015;87:399–410.

CHAPTER

17

Impact of Structural Plasticity on Memory Formation and Decline

Andreas Knoblauch

University of California, Davis, CA, United States

OUTLINE

1 INTRODUCTION

Traditional theories attribute adult learning and memory to Hebbian modification of synaptic weights[1–4] in neural networks with a fixed structure. More recent evidence suggests that network rewiring by structural plasticity has a significant contribution to

DOI: http://dx.doi.org/10.1016/B978-0-12-803784-3.00017-2

adult learning as well.[5–14] Such structural network modification includes generation and elimination of synapses, growth and retraction of spines, and remodeling of dendritic and axonal branches,[15–18] which typically acts on a much longer time scale than synaptic weight modification.[19]

One likely function of structural plasticity is to increase information storage efficiency in terms of space and energy demands.[8,20–22] Indeed, due to space limitations, neural networks in the brain are only sparsely connected, even on a local scale.[23–25] Moreover, it is believed that the energy consumption of the brain is dominated by the number of postsynaptic potentials or, equivalently, the number of functional *non-silent* synapses.[26–28] Together, this implies a pressure to minimize the number and density of functional non-silent synapses. The major function of structural plasticity may therefore be to "move" the rare expensive synapses to the most useful locations, while keeping the mean synaptic connectivity on a constant low level.[22,29] By this, sparsely connected networks may be computationally equivalent to densely connected networks.

For example, it is well known from earlier theories on neural networks without structural plasticity[30–34] that memory storage capacity scales in proportion to synapse density, such that densely connected networks can store many more memories than sparsely connected networks.[35–37] This implies that sparsely connected networks *with* structural plasticity may store the same large number of memories as densely connected networks without structural plasticity, at least, if the number of "required" (non-silent) synapses in the dense network is not much larger than the actual synapse number in the corresponding sparse network.[21,22,38]

By considering such equivalences between sparse and dense networks, it is possible to generalize the earlier network theories to assess the impact of structural plasticity for memory. One simple strategy is to analyze pruning of irrelevant synapses *after* learning in densely connected networks during a developmental phase,[21,38,39] where synapse density is known to be significantly higher than in adulthood.[40–43] These results can then be generalized further to adult learning in sparsely connected networks with ongoing structural plasticity if considering additional space and time constraints.[22,29,44] In particular for the latter case, it is then necessary to distinguish between different connectivity measures, for example, relating to the actual number of anatomical synapses per neuron versus the number of potential synapses[8,45] per neuron or the number of "effectual" synapses[22,29] per neuron of a Hebbian cell assembly that represents a particular memory.[1,46,47]

By analysis and model simulation, one can estimate the increase in storage efficiency by structural plasticity and the relevance of learning parameters such as the resolution of synaptic weights. Interestingly, it turns out that, for plausible network parameters, structural plasticity can increase storage efficiency per actual synapse by at least one order of magnitude, whereas weight resolution appears to have only a minor role.[21,22,38] Moreover, it is possible to simulate long-term memory storage and consolidation[22,29] and compare the results to various known memory phenomena such as graded retrograde amnesia,[48–50] catastrophic forgetting (CF),[51,52] and the spacing effect.[53–55]

These previous theoretical works on memory storage by structural plasticity have still limitations: Some works did not consider any particular model for memory retrieval at all but just counted the number of synaptic configurations to assess information storage capacity.[8,20,45] Other approaches used plausible network models for memory retrieval but considered only synaptic pruning after developmental learning in densely connected networks.[21,39] And models for memory storage and retrieval with ongoing structural plasticity in sparsely connected networks[22,29] relied on a very simple binary model of associative memory, the so-called Steinbuch or Willshaw model with extremely simple clipped Hebbian learning of binary synapses.[21,30,31,33] In such Willshaw-type models a binary synapse is typically potentiated and/or consolidated already after a *single* pre/postsynaptic coincidence. Unless neural activity is extremely sparse, this entails the problem that *all* synapses will be potentiated and consolidated already after learning a few memories. In networks without structural plasticity, this effectively extincts all stored information corresponding to CF. Previous works have shown that structural plasticity can prevent such CF if network connectivity is sufficiently sparse.[22,29] However, stable storage with sparse connectivity requires relatively large cell assemblies that are expensive in terms of space requirements such that only a relatively small number of memories could be stored in a neural network of a given size.

This work summarizes, reevaluates, and extends the previous findings in the context of a more general model of associative memory with structural plasticity that attempts to avoid some limitations of the previous models.[38,44] It is argued that structural plasticity in the general model still gives very natural explanations to memory phenomena such as high storage capacity, graded retrograde amnesia, and the spacing effect. In addition, it offers novel explanations for sharpening of neural activity during consolidation,[56–59] and contributes novel aspects to understand the absence of CF and the stability–plasticity dilemma under normal learning conditions.

2 MODELING FRAMEWORK

2.1 Memories and Cell Assemblies

Memories are commonly identified with patterns of neural activity that can be evoked and stabilized by appropriately modified synaptic connections.[1,2,24,60–62] In the simplest case a memory corresponds to a group of neurons that fire at the same time, and, according to the Hebbian hypothesis "what fires together wires together" develop strong mutual synaptic connections.[1,63] Such groups of strongly connected neurons are called cell assemblies[1,47,64] and have a number of properties that suggest a function for associative memory[31–34,37] or more complex state-based computation.[65,66] For example, if a stimulus activates a part of a cell assembly, the mutual synaptic connections will quickly activate the whole cell assembly, which is believed to correspond to the retrieval or completion of a memory pattern. Similarly, a cell assembly in one brain area u can activate another cell assembly v in a distant brain area via associative synaptic connections. By interconnecting multiple areas by associative synaptic networks, it is possible to realize complex cognitive systems, for example, for language understanding or to control a robot.[65–67]

2.2 Memory Retrieval, Storage Capacity, and Output Noise

After storing a set of M memory associations represented by activity patterns u^t and v^t at times $t = 1,2,\ldots,M$ in two neuron populations u and v (which may be identical for auto-association in a recurrent network), we can simulate *memory retrieval* by stimulating the network with a noisy input pattern $u^{t'}$, which will evoke a corresponding output pattern $v^{t'}$. In the simplest case, the output pattern $v^{t'}$ can be obtained from a *one-step retrieval* procedure applying a threshold Θ_j on the membrane potentials of each output unit v_j,

$$v^{t'}_j = H\left(\sum_{i=1}^{n} W_{ij} u^{t'}_i - \Theta_j\right) \tag{17.1}$$

where W_{ij} are synaptic weights (of the connection from u_i to v_j), and H is a sigmoid transfer function, e.g., for binary patterns the Heaviside function ($H(x) = 1$ for $x \geq 0$ and 0 for $x < 0$). If cell assemblies have a fixed size k, simulations will sometimes assume a k-winners-take-all (WTA) recurrent inhibition selecting the k neurons with largest dendritic potentials (for Figs.17.7, 17.8). Instead of regulating the number of active neurons, inhibition may also regulate total activity of the population (for Fig. 17.9). To mimic auto-associative attractor dynamics of more realistic spiking network models,[68] we sometimes also simulate multistep *iterative retrieval*[69] (for Figs. 17.7–17.9), where the output pattern from step $r-1$ is used as input pattern for step r. If cell assemblies are sufficiently stable, then activity typically converges within a few steps (e.g., 3–5), completely activating one of the cell assemblies.

To measure the performance of an associative memory, we can compute the stored Shannon information[70,71] or transinformation between the original set of M activity patterns v^t and the corresponding output patterns $v^{t'}$ that get activated during retrieval ($t = 1,2, \ldots, M$). For example, normalizing this value by the number of synapses yields the *storage capacity* C in bit per synapse.[21] As C is often maximal for low retrieval quality,[72,73] we will more often consider *output noise*.[21,37]

$$\hat{\varepsilon} = \frac{\sum_j \left| v^{t'}_j - v^t_j \right|}{k} \tag{17.2}$$

which is for binary activity patterns defined as the mean Hamming distance between v^t and $v^{t'}$ normalized to the mean size k of a cell assembly. For example, output noise $\varepsilon = 0.01$ means that 1% of active units differ between original activity pattern v^t and retrieval output $v^{t'}$.

2.3 Synaptic Weight Plasticity

Cell assemblies are believed to develop by simple *local* learning rules,[74,75] where changes of synaptic weights W_{ij} depend only on local information, for example, pre and postsynaptic activities u_i and v_j. For episodic memory, a learning rule should also be *immediate*[76] such that a corresponding sensory configuration can be stored in a *single shot*. Moreover, it is often desirable that a learning rule should be *incremental*, meaning that storing a novel memory need not refer to previous memories. It is possible to infer an optimal Bayesian rule for neural associative memory with local, immediate, and incremental learning.[37] If the cell assembly size k scales linearly with the number of neurons n, then the Bayesian rule becomes equivalent to a biologically more plausible covariance-type linear learning rule,[77]

$$W_{ij}^{M} = \sum_{t=1}^{M} R\left(u_i^t, v_j^t\right) \tag{17.3}$$

where each of the M memory associations between activity patterns u^t and v^t contributes equally according to the linear learning rule R. For binary activity patterns with $u^t{}_i, v^t{}_j \in \{0,1\}$ and mean activity $p = k/n$, the rule R can be specified by only four parameters. For example, if neuron populations u and v have identical sizes and activities, the optimal *covariance rule* is $R(0,0) = p^2$, $R(0,1) = R(1,0) = -p(1-p)$, and $R(1,1) = (1-p)^2$. Interestingly, for learning with structural plasticity, even simpler rules such as the *homosynaptic rule* with $R(0,0) = R(0,1) = 0$, $R(1,0) = -p$, and $R(1,1) = 1 - p$ turn out to be optimal as well.[38] For very sparse neural activity[78] with $p \to 0$, both rules become equivalent to the *Hebbian rule* $R(0,0) = R(0,1) = R(1,0) = 0$, $R(1,1) = 1$ and, if learning events are sparse themselves, even equivalent to clipped Hebbian learning according to the binary *Steinbuch* or *Willshaw* model,[17,26,29,49]

$$W_{ij}^{t+1} = max\left(W_{ij}^t, u_i^t \cdot v_j^t\right) \quad \in \{0, 1\} \tag{17.4}$$

where initially silent synapses ($W_{ij}{}^0 = 0$) get potentiated and/or consolidated ($W_{ij}{}^0 = 1$) after a single pre/postsynaptic coincidence. Although extremely simple, the Willshaw model, Eq. (17.4), has a number of desirable properties: First, it reaches almost the same high storage capacity as optimal Bayesian learning (0.69 vs 0.72 bit/synapse), however, only for extremely "log"-sparse activity, where assembly size k scales with log(n). Second, being a binary rule, it can easily be combined with simple models of structural plasticity[22,29] where the silent synapses (weight 0) take part in synaptic turnover, whereas the potentiated synapses (weight 1) correspond to consolidated stable synapses. Third, in addition to being local, incremental, and immediate, the Willshaw model is the only learning model that has a property that I call *"final"*: Once a synapse gets potentiated, it will remain potentiated forever, independent of possible future memories. I will discuss further on that, together with structural plasticity, this property may contribute to solve the stability–plasticity dilemma and prevent CF as often observed in artificial neural networks.[51,52]

Unfortunately, the Willshaw model performs less favorably for larger cell assemblies with $k >> \log(n)$ that are considered more realistic and that are required for stable memory storage in sparsely connected networks.[79] To still have an efficient weight learning method for this case that can be combined with structural plasticity, I have recently proposed a further learning rule[38] that optimally discretizes the real-valued weights of the linear learning rule, Eq. (17.3). As this strongly reduces entropy of synaptic weights and, thus, computing costs on both digital and neuromorphic hardware, I refer to this model briefly as *zip net rule*. In the simplest case, the *binary zip rule*[38,80]

$$W_{ij}^{M} = H\left(\sum_{t=1}^{M} R\left(u_i^t, v_j^t\right) - \theta_{ij}\right) \tag{17.5}$$

applies a threshold θ_{ij} to the real-valued weights (also called *synaptic potentials*), where H is again the Heaviside function. To maximize efficiency and flexibility, the threshold may be regulated homeostatically such that always a desired fraction p_1 of a neuron's synapses are in the potentiated state.[22] Similar to the Bayesian and linear learning rules, zip nets can store close to n^2/k cell assemblies.[38] For large $k >> \log n$, this is much more than n^2/k^2 for the Willshaw model.[21]

2.4 Structural Plasticity

While weight plasticity modifies the weights of *existing* synapses, structural plasticity rather modifies the ontological state of *potential synapses*. Here a potential synapse ij^{ν} may be defined, for example, by a cortical location where axonal and dendritic branches of neurons i and j are close enough to allow the formation of a novel connection, for example, by spine growth and synaptogenesis.[45] In general, as dendrites and axons may closely overlap at multiple locations, there may be multiple potential synapses ($\nu = 1,2, \ldots$) between a neuron pair ij. Most of the following simulation experiments use a minimal model of structural plasticity (Fig. 17.1A) with only three states:[22,29] A synapse can be potential but not yet realized (state π), realized but unstable (state 0), or realized and consolidated (state C). Real synapses are also called actual synapses. For the binary weights from plasticity rules of Eqs. (17.4) and (17.5), the states 0 and C correspond to weights 0 and 1, respectively.

State transitions occur with probabilities p_g, $p_{e|s}$, $p_{c|s}$, and $p_{d|s}$ as illustrated in Fig. 17.1. These transitions may depend on a binary learning signal $s = S_{ij}$ which is here identified with synaptic tagging[81] by the Willshaw rule, Eq. (17.4), i.e., $S_{ij} = 1$ if in the current learning epoch, there is at least one coincidence of pre/postsynaptic activity (and $S_{ij} = 0$ otherwise). Typically, $S_{ij} = 1$ supports consolidation ($p_{c|1} >> p_{c|0}$) and prevents deconsolidation ($p_{d|1} << p_{d|0}$) and elimination ($p_{e|1} << p_{e|0}$) of a synapse ij. In all the following simulations, synapse generation (by p_g) and elimination ($p_{e|s}$) is assumed in homeostatic balance. That is, in each simulation step the number of generated synapses equals the number of eliminated synapses, thereby maintaining a constant anatomical connectivity P. As the number of synapses of a connected neuron pair ij seems narrowly distributed around a small number (e.g., 4),[82] most of the following simulations use a single synapse per connected neuron pair ij, where each model synapse may be believed of representing a compound connection of multiple synapses[83] (but see Discussion section).

The extended model of Fig. 17.1B has two potentiated states 1 and C (both with weight 1) to account for different levels of stability, similar as previously suggested in cascade models

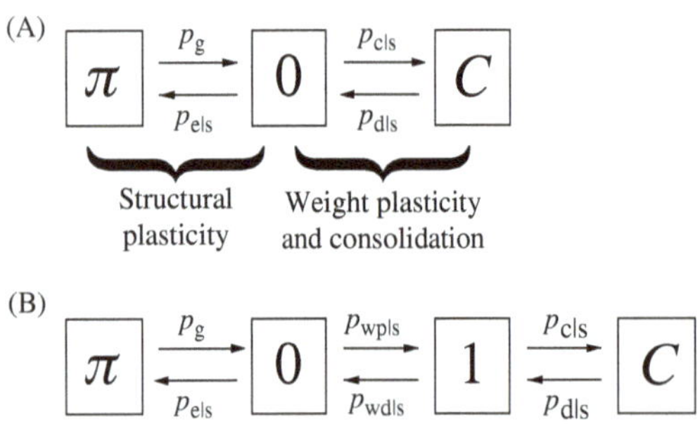

FIGURE 17.1 **Synapse model.** Most simulations use a three-state model (panel A) having states π, 0, and C to represent potential, silent, and consolidated synapses. Transitions between states occur with probabilities p_g, $p_{e|s}$, $p_{c|s}$, and $p_{d|s}$ that may depend on a learning signal s. Some simulations use an extended four-state model (panel B) having additionally a potentiated but unconsolidated state 1 and corresponding transition probabilities $p_{wp|s}$, $p_{wd|s}$ for weight potentiation and depression.

of synaptic learning.[84] This model variant is more useful for simulating sparsification or sharpening of cell assembly activity during memory consolidation.

2.5 Effectual Connectivity

Network connectivity is an important parameter for neural associative memory because storage capacity scales typically in proportion to or even supralinearly with the number of synapses per neuron.[22,37] If including structural plasticity in a memory theory, we have to distinguish between different types of connectivity (Fig. 17.2). First, *anatomical connectivity* P is the fraction of neuron pairs that are connected by at least one real actual synapse. Second, *potential connectivity* P_{pot} is the fraction of neuron pairs that have at least one potential synapse (Fig. 17.2A). Finally, we define effectual connectivity P_{eff} as the fraction of required synapses (specified by the learning Signal S_{ij}) that have already been realized (Fig. 17.2B).[22,29] For memory storage and cell assemblies, this means that P_{eff} is the average anatomical connectivity *within* a cell assembly. For example, $P_{eff} = 0.5$ means that half of the neuron pairs of a cell assembly are connected by at least one actual synapse.

The main hypothesis of our memory theory is that structural plasticity increases P_{eff} to increase storage efficiency of sparsely connected networks with low anatomical connectivity P. Thus, after ongoing learning by structural plasticity (Fig. 17.3), P_{eff} may be high (close to P_{pot}), whereas anatomical connectivity P remains low. This is illustrated in

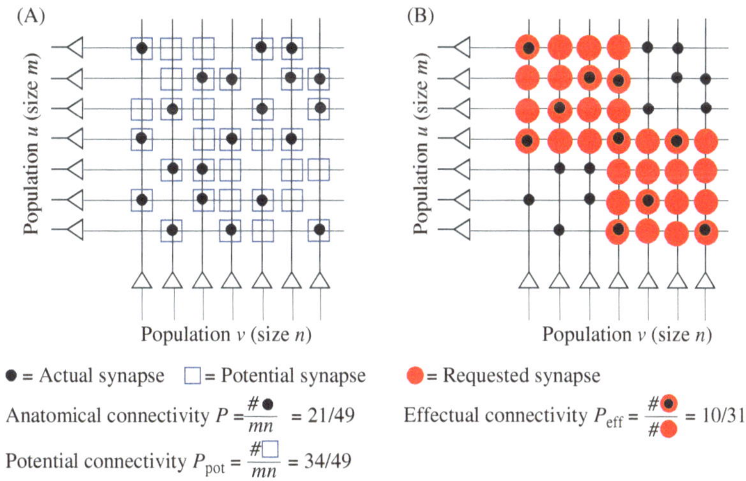

FIGURE 17.2 **Different types of "connectivity" of two neuron populations u and v.** (A) Anatomical connectivity P is the fraction of neuron pairs connected by an actual synapse (*black circles*). Potential connectivity P_{pot} is the fraction of neuron pairs having a potential synapse (*blue rectangles*). (B) Effectual connectivity P_{eff} is the fraction of requested synapses that have been realized and potentiated. Requested synapses are defined by a binary learning signal S_{ij} and correspond to synapses that support storage of a particular set of memories (*red circles*; see text for details). Source: *Modified from Knoblauch, A, Körner, E, Körner, U, Sommer, FT. Structural plasticity has high memory capacity and can explain graded amnesia, catastrophic forgetting, and the spacing effect. PLoS One. 2014;9(5): e96485, 1–19.*

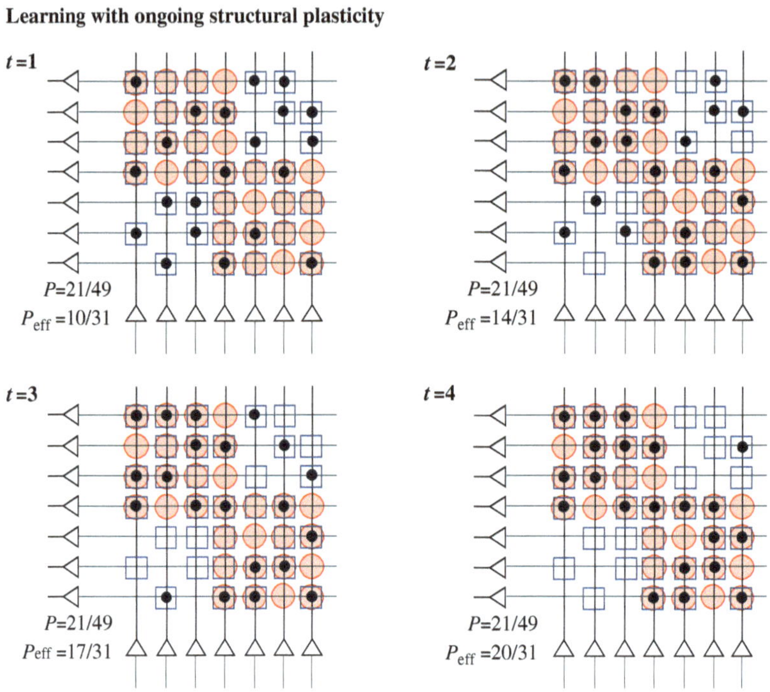

FIGURE 17.3 **Increase of effectual connectivity P_{eff} during a simulation of ongoing structural plasticity.** The initial random connectivity ($t=1$; same as in Fig. 17.2) is adapted to the ensemble S of requested synapses (*red circles*) for storing a given set of two memories (represented by neurons 1–4 and 4–7, respectively, in both populations). This happens by blind elimination ($p_{\text{e}|0}$) and regeneration of silent synapses (p_{g}) at random locations, where new synapses can be consolidated depending on a Hebbian learning signal ($p_{\text{c}|\text{s}}$), thereby escaping elimination. Although anatomical connectivity P is regulated to remain constant, effectual connectivity P_{eff} can slowly increase to a level close to potential connectivity P_{pot}. Source: *Modified from Knoblauch, A. & Sommer, F.T. Structural plasticity, effectual connectivity and memory in cortex. Front Neuroanat. 2016;10:63.*

Fig. 17.3 showing the simulated evolution of a toy network connecting two populations each consisting of $n=7$ neurons. The learning signal S specifies or tags the required synapses to be consolidated (red circles) corresponding to two overlapping cell assemblies of size $k=4$ (cells 1–4 and 4–7 in both neuron populations). At the beginning, P_{eff} is similarly low as anatomical connectivity P. In each time step, silent synapses ij that are required for stabilizing the cell assemblies ($S_{ij}=1$) are potentiated and consolidated and thus escape further synaptic turnover. By contrast, synapses that are not required ($S_{ij}=0$) are replaced by new synapses (black circles) growing randomly at the location of potential synapses (blue squares). With time, the required synapses accumulate within the cell assemblies and, thus, effectual connectivity P_{eff} increases to the level of potential connectivity, whereas anatomical connectivity P remains constantly low. In the following section, we point out the impact of such ongoing learning by structural plasticity for memory formation and decline.

3 RESULTS

3.1 Structural Plasticity, Learning Load, and Distribution of Synaptic Weights and Numbers

To understand the basic effects and predictions of our model on the distribution of synaptic weights and numbers, we can repeat the toy simulation of Fig. 17.3 for larger networks and cell assemblies. Fig. 17.4 shows such simulations for a network of $n = 1000$ neurons and cell assemblies of size $k = 50$ (left panels) with anatomical connectivity $P = 0.1$. Similarly as before, effectual connectivity P_{eff} increases with time toward potential connectivity ($P_{\text{pot}} = 1$). Correspondingly, the fraction P_1 of neuron pairs having a potentiated or consolidated synapse increases, whereas the fraction P_0 of unstable silent synapses decreases. This increase should depend on the *learning load* p_{1S} defined as the fraction of tagged neuron pairs with $S_{ij} = 1$ corresponding to the number M of novel memories. For a low learning load, a lower number of synapses will be consolidated (e.g., $P_1/P = 0.3$ in panel A for $M = 12$ and $p_{1S} = 0.46$), whereas a high learning load requires more consolidated synapses (e.g., $P_1/P = 0.85$ in panel B for $M = 36$ and $p_{1S} = 0.84$). As the effect of structural plasticity will depend on the number of unstable synapses taking part in synaptic turnover, one prediction of our basic model is thus an increase of stable synapses and a decrease of synaptic turnover with lifetime or number of stored memories, at least in those brain areas that are responsible for permanent storage of long-term memories. We will come back to this prediction later when discussing CF and the stability–plasticity dilemma.

Fig. 17.4 also shows some simulations with multisynaptic compound connections, where one neuron pair *ij* can have multiple potential synapses (dashed curves, binomial distribution). The qualitative behavior of this model variant is similar as for single synapses per neuron pair (solid curves), although the increase of P_{eff} is slower (panel A) and, for higher learning loads, may converge to lower values (panel B). The reason for this is an accumulation of multiple consolidated synapses between those neuron pairs *ij* that are tagged by the learning signal ($S_{ij} = 1$). As the total number of synapses is regulated to be constant, the number of unconnected neuron pairs increases, and, thus, the net anatomical connectivity P decreases with time (cyan dashed lines). Correspondingly, the distribution of actual synapses per neuron after learning by structural plasticity can significantly deviate from the distribution of potential synapses (panels B and D). Typically, most neuron pairs are unconnected or have only one or two silent synapses, whereas, depending on the learning load, a smaller number of neuron pairs (those tagged by $S_{ij} = 1$) can have multiple stable synapses. However, reproducing bimodal distributions of synapse numbers resembling those reported experimentally for closely neighbored cells[82] is not easy with our model. This would require particularly tuned learning parameters (e.g., p_{1S}) and already a bell-shaped distribution of potential synapses (black dashed curves in panels B and D), whereas realistic distributions have a more skewed Poisson-type shape.[82] I will argue in the discussion that, due to the very small distance of analyzed neurons (<50 μm), the experimentally reported synapse distributions may rather reflect minicolumnar structure,[85,86] whereas neural network models of associative memory with local connectivity ($P = 0.1$) typically are believed to correspond to much larger structures such as cortical macrocolumns (size 1 mm^3).[25,87]

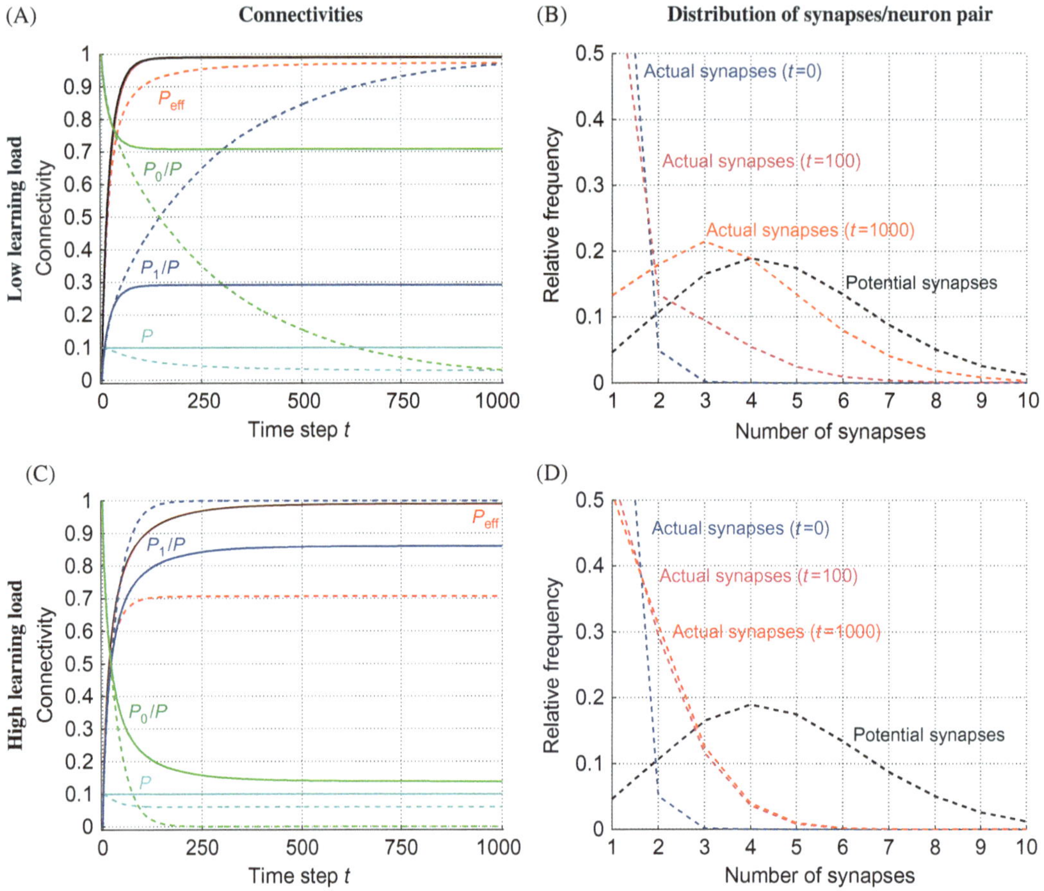

FIGURE 17.4 **Temporal evolution of connectivity** during ongoing learning of memories ($k/n = 50/1000$) with structural plasticity ($p_{e|0} = 0.5$, $p_{e|1} = p_{d|s} = 0$, constant $P = 0.1$). Left panels (A and C) show effectual connectivity P_{eff} (*red*), anatomical connectivity P (*cyan*), fraction of silent synapses P_0/P (*green*), and fraction of consolidated synapses P_1/P (*blue*) as function of time t for potential connectivity $P_{pot} = 1$. Curves correspond to model variants with single synapses per neuron pair (*solid*) and multiple synapses per neuron pair (*dashed*). Right panels (B and D) show distribution of synapse number per neuron pair in the multisynapse model variant for time $t = 0$ (*blue*), $t = 100$ (*magenta*), and $t = 1000$ (*red*). *Black curves* correspond to distribution of potential synapses (binomial, $P_{pot} = 0.99$). Top panels (A and B) correspond to relatively low learning load ($M = 12$, $p_{1S} = 0.46$), bottom panels (C and D) to high learning load ($M = 36$, $p_{1S} = 0.84$).

3.2 Impact of Increasing P_{eff} on Retrieval Quality and Storage Efficiency

The increase of effectual connectivity P_{eff} by structural plasticity as shown in Fig. 17.4 has considerable functional implications for memory formation and retrieval. This can be seen in Fig. 17.5 showing stored (trans-)information C and output noise ε as functions of P_{eff} for a macrocolumn-size network of $n = 10^5$ neurons storing a larger number M of cell assemblies having sizes $k = 50$ (panel A) or $k = 500$ (panel B). It can be observed that both C and ε are monotonically related to P_{eff}. For example, the logarithm of output noise

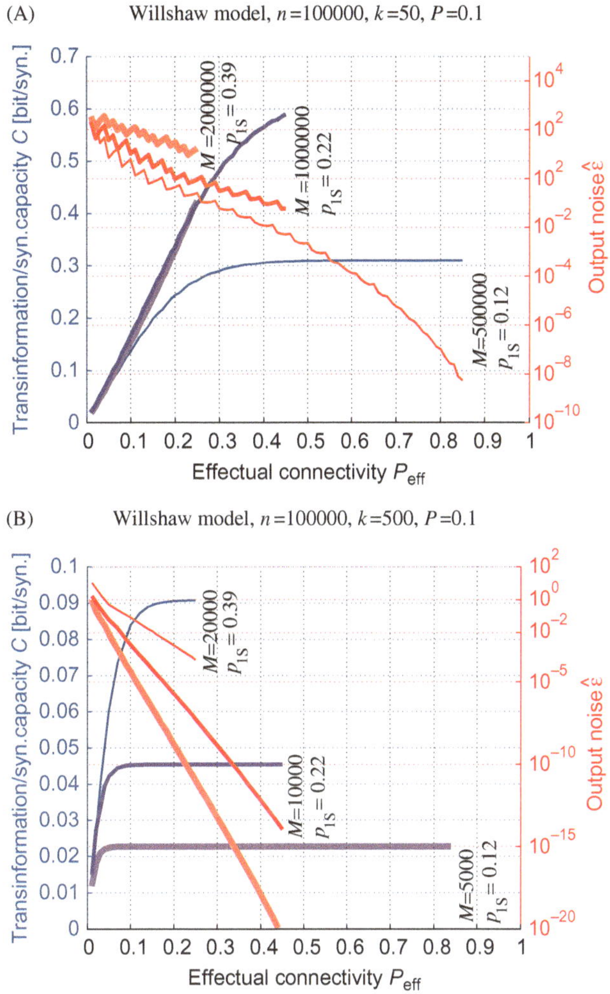

FIGURE 17.5 **Transinformation and output noise depend monotonically on P_{eff}.** An increase in effectual connectivity P_{eff} corresponds to an increase in stored (trans)information C (*blue*; in bit per actual synapse) and to a decrease in output noise ε (*red*). Panel (A) shows data corresponding to a Willshaw-type associative memory of $n = 100{,}000$ neurons, cell assembly size $k = 50$, and fixed anatomical connectivity $P = 0.1$. Curves correspond to low learning load (*thin lines*, $M = 0.5 \times 10^6$, $p_{1S} = 0.12$), medium learning load (*medium lines*, $M = 10^6$, $p_{1S} = 0.22$), and high learning load (*thick lines*, $M = 2 \times 10^6$, $p_{1S} = 0.39$). Panel (B) shows similar data but for larger cell assemblies ($k = 500$) and less memories ($M = 500$, 1000, 2000). Note that P_{eff} cannot exceed P/p_{1S}. Data have been computed from the exact theory[44,112] of the Willshaw model for zero input noise and choosing optimal firing thresholds Θ to maximize C.

decreases linearly with effectual connectivity P_{eff}. This means that the longer a memory is consolidated, the higher becomes retrieval quality. When storing novel memories, P_{eff} of the corresponding cell assemblies is low at the level of anatomical connectivity ($P = 0.1$ within a macrocolumn) and, depending on the learning load, output noise ε may be high, corresponding to low-quality retrieval. By structural plasticity and synaptic consolidation, P_{eff} increases and output noise ε decreases with time, corresponding to high-quality memory retrieval. Similarly, stored information C increases linearly with P_{eff} until output noise is close to zero where C saturates at a maximal value.

We have shown elsewhere[44] that, for a large range of activity parameters, transinformation C is basically equivalent to transfer entropy[88] being a measure commonly used to analyze "effective brain connectivity" from functional brain imaging data. Together with the demonstrated linear relationship between effectual connectivity P_{eff} and C, this means that P_{eff} is also an equivalent measure of "effective brain connectivity" but, unlike transfer entropy, P_{eff} is directly grounded in the neuroanatomy of cell assemblies.

3.3 Impact of Increasing P_{eff} on the Memory Capacity of a Cortical Macrocolumn

Fig. 17.5 shows that, for *fixed* sets of M memories, the stored information C increases in proportion to P_{eff}. As structural plasticity is believed to increase P_{eff} from anatomical connectivity P up to potential connectivity P_{pot}, it is suggested that the increase in storage efficiency may be limited by a factor P_{pot}/P, for example $0.5/0.1 = 5$ within a cortical macrocolumn.[25,45,87] However, after reducing output noise for the fixed set of memories, additional memories may be stored and consolidated by structural plasticity, until the minimal output noise ε at the upper limit of $P_{eff} = P_{pot}$ can no longer satisfy some required retrieval quality. Fig. 17.6 (top panels A and C) therefore computes the *maximal* number M_ε of random cell assemblies that can be stored at a certain output noise level ($\varepsilon = 0.01$) in a cortical macrocolumn ($n = 100{,}000$, $P = 0.1$) as a function of effectual connectivity P_{eff} and cell assembly size k. Correspondingly, bottom panels (B and D) show the total synaptic storage capacity C^{tot}_{ε}, that is, the information that can then be stored per *stable* synapse at P_{eff}.[22] The left panels (A and B) correspond to synaptic learning according to the Willshaw model, Eq. (17.4), whereas the right panels (C and D) correspond to binary zip nets, Eq. (17.5).

It is visible, in particular for the Willshaw model and small cell assemblies (e.g., $k < 100$), that increasing P_{eff} by structural plasticity (from P to P_{pot}) can strongly increase M_ε by several orders of magnitudes. The reason for this is the fact that small cell assemblies cannot be stabilized in sparsely connected networks (see Section 2).[79] Therefore, without structural plasticity ($P_{eff} = P = 0.1$) storage capacity is close to zero, and only an increase of P_{eff} by structural plasticity (e.g., up to $P_{eff} = 0.5$) can lift the working point of the network into a regime where efficient memory storage is possible. For the lower limit of experimentally estimated cell assembly sizes,[78] $p = 0.0005$ corresponding to $k = 50$, this means that structural plasticity may increase the number of cell assemblies from about 10 for $P_{eff} = 0.1$ up to almost 10^6 for $P_{eff} = 0.5$. For larger cell assemblies corresponding to the upper limit of experimentally determined sizes (e.g., $k > 500$), the capacity increase is not as impressive but still confirms the linear P_{pot}/P rule (e.g., factor 5 for an increase from $P_{eff} = 0.1$ to $P_{eff} = P_{pot} = 0.5$).

As mentioned in Section 2, the Willshaw learning rule, Eq. (17.4) is Bayesian-optimal for very small cell assemblies (e.g., $k < 20$) but performs poorly for large cell assemblies

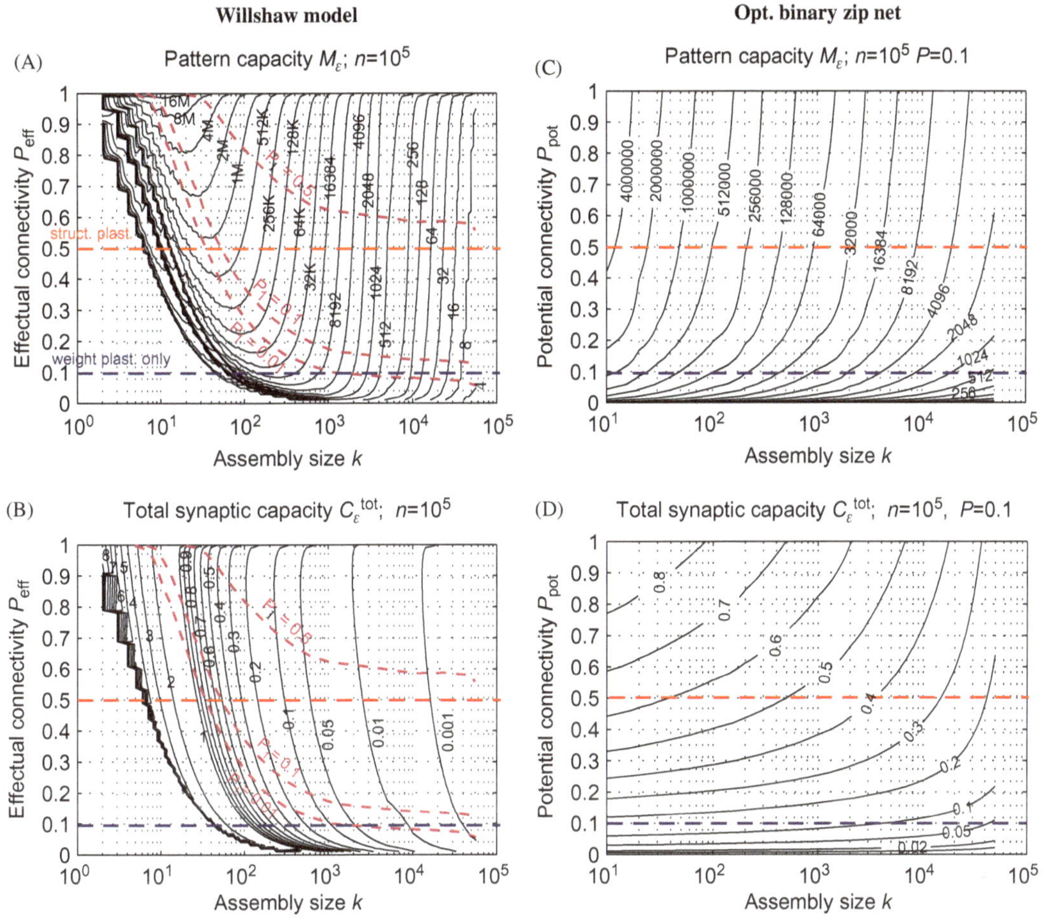

FIGURE 17.6 **Maximal storage capacity (M_ε, C^{tot}_ε) of a cortical macrocolumn** (size $n = 100{,}000$, connectivity $P = 0.1$) as a function of cell assembly size k and effectual connectivity P_{eff} (or potential connectivity P_{pot}) at output noise level $\varepsilon = 0.01$ for zero input noise. Top panels (A and C) show contour plots of pattern capacity M_ε (maximal number of cell assemblies). Bottom panels (B and D) show the corresponding total synaptic capacity C^{tot}_ε (stored information in bits per non-silent synapse). Left panels (A and B) correspond to exact values for the Willshaw model[44,112] (which is optimal for very small k). Note that the Willshaw model is consistent with an anatomical connectivity $P = 0.1$ only for values of k, P_{eff} where $P_1 < P$ (*magenta dashed line* indicate contours $P_1 = 0.01$, $P_1 = 0.1$, $P_1 = 0.5$; for details, see references[22,44]). Right panels (C and D) correspond to the approximation theory of the optimal binary zip net model (which is optimal for larger k) by using the optimal covariance/homosynaptic learning rule,[38,44] where $P = 0.1$ is fixed and P_{eff} is bounded by P_{pot}. Here, M_ε and C^{tot}_ε are given as functions of P_{pot} as they cannot easily be computed from P_{eff}.[38,44] Source: *Modified from Knoblauch, A, Sommer, FT. Structural plasticity, effectual connectivity and memory in cortex. Front Neuroanat. (2016);10:63.*

(e.g., $k > 1000$), where the binary zip net rule, Eq. (17.5) perform much better (see panels C and D). However, unlike for the Willshaw model, there is no exact theory to compute storage capacity for zip nets. In fact the approximated data of panels C and D hold only if $M (k/n)^2 >> 1$ is large,[38] which is not fulfilled for small assemblies or low P_{eff} (lower left corners of panels C and D). Preliminary simulations (not shown) rather indicate that, zip nets have a similarly low performance as the Willshaw model, confirming a supralinear performance increase by structural plasticity.

Another major difference between zip nets and the Willshaw model is that the fraction P_1 of neuron pairs having a potentiated synapse increases with M in the Willshaw model, Eq. (17.4), but is assumed as a fixed parameter in zip nets (see Section 2). For the Willshaw model, this means that a plausible parameter range of P_{eff} and k must be consistent with anatomical connectivity ($P_1 < P = 0.1$) and fraction of potentiated synapses reported in experiments.[89,90] Therefore, the Willshaw model may be consistent with experiments only in a narrow band of the P_{eff}-k-plane (perhaps for P_1 between 0.01 and 0.1; see magenta dashed lines in panels A and B). It is interesting that this band follows closely the ridge-line of maximal memory number M_ε in panel A. Similarly interesting, this band intersects with the line $P_{\mathrm{eff}} = 0.5$ of plausible potential connectivity in cortical macrocolumns[45] at the lower end of experimentally reported cell assembly sizes ($k = 50$), and with $P_{\mathrm{eff}} = P = 0.1$ at the upper end ($k > 500$). Together this supports the idea that novel memories may initially be represented by large cell assemblies ($k > 500$) that can easily be stabilized by zip-net-type weight plasticity in sparsely connected networks, whereas a subsequent Willshaw-type consolidation procedure by structural plasticity not only increases P_{eff} but may also decrease cell assembly size (down to $k = 50$) to maximize storage efficiency.[56–58,78] We will rediscuss this hypothesis later in light of the following simulations of memory phenomena like the spacing effect, graded amnesia, and, in particular, CF.

3.4 Structural Synaptic Plasticity and the Cellular Basis of the Spacing Effect

It is well known that learning is more effective if rehearsal is spaced over time compared to single block rehearsal.[53–55,91] For example, learning a list of vocabularies in two sessions each lasting 10 min turns out to be more effective than learning in a single session lasting 20 min. This so-called spacing effect is remarkably robust and occurs in many explicit and implicit memory tasks in humans and many animals being effective over many time scales from single days to months. Interestingly, the spacing effect is also an inherent property of our model of structural plasticity.[22,44,80,92,93] Fig. 17.7 (panels A and B) shows a simulation of storing memory associations between activity patterns u^t and v^t (e.g., corresponding to vocabulary of two languages) in a heteroassociative synaptic connection between two neuron populations u and v. In a first simulation (blue) the memories are rehearsed in a single long time block (20 time steps), whereas in the second simulation (red) rehearsal is spaced over several shorter blocks (4×5 time steps), such that the total rehearsal time is equal in both simulations. However, for spaced rehearsal the resulting effectual connectivity P_{eff} of the memories turns out to be much higher and, correspondingly, the output noise much lower than for single block rehearsal.

This effect is very stable and occurs for virtually all plausible parameter settings, in particular, if structural plasticity is slower than weight plasticity and consolidated synapses are more stable than silent synapses, as consistently reported in experiments.[19] Then brief rehearsal sessions are sufficient for fast weight plasticity to quickly potentiate and consolidate

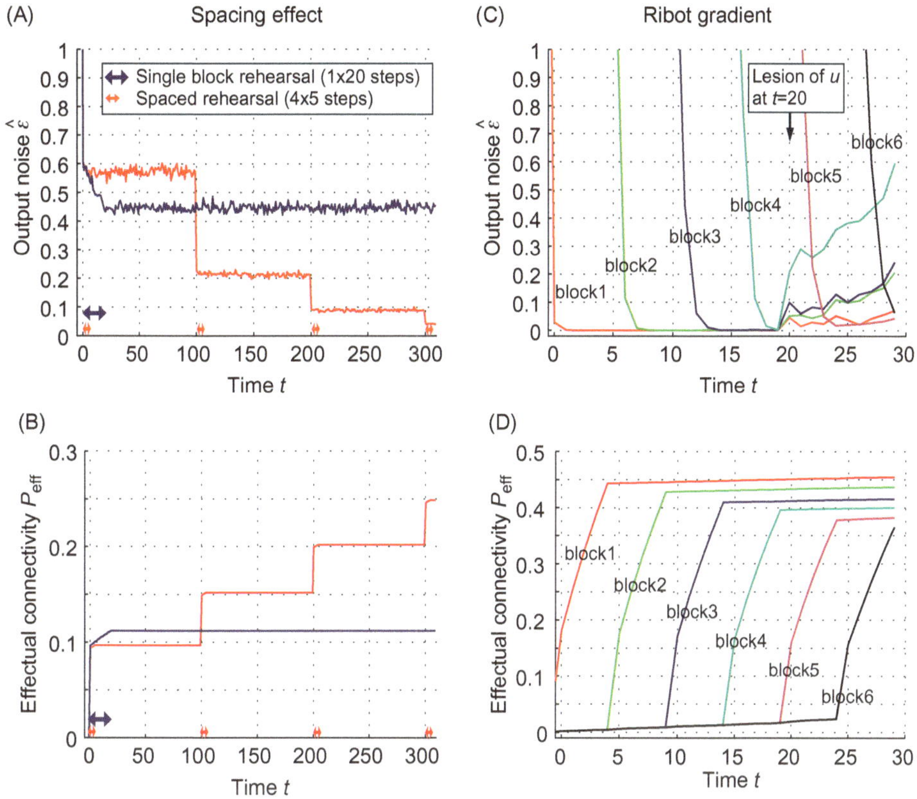

FIGURE 17.7 **Simulation of the spacing effect and Ribot gradient.** Left panels (A and B) show results from two simulations training a single memory set in either a single long rehearsal session (20 steps, *blue arrow*) or in four spaced rehearsal sessions (each 5 steps, *small red arrows*) of the same total duration. Spaced rehearsal (*red curves*) is much more effective in increasing effectual connectivity P_{eff} (B) and thus reducing output noise $\hat{\varepsilon}$ (A) than single block rehearsal (*blue curves*). Similarly, right panels (C and D) show P_{eff} and $\hat{\varepsilon}$ for sequential rehearsal (or replay) of several memory sets (block1, block2, ...) each rehearsed for five steps (beginning at times $t = 0, 5, \ldots$). This results in a gradient of P_{eff} where earlier memories reach higher P_{eff} than recent memories (D). At time $t = 20$, a lesion is simulated by inactivating half the neurons of the input population u. This increases output noise resembling retrograde amnesia (C): Recent memories may get lost (e.g., block 4), whereas earlier memories can be preserved (e.g., block 1). Model parameters: $n = 1000$, $k = 50$, $M = 20$ (A and B) or $M = 4$ per block (C and D); $p_{e|1} = 0$ and $p_{e|0} = 0.01$ (A and B), or $p_{e|0} = 1$ (C and D); $p_{d|s} = 0$; $p_{c|s} = $ s. Source: *Adapted from Knoblauch, A, Körner, E, Körner, U, Sommer, FT. Structural plasticity has high memory capacity and can explain graded amnesia, catastrophic forgetting, and the spacing effect. PLoS One. 2014;9(5):e96485, 1–19.*

all available synapses that are useful to store the new memories. After that, instead of continuing a rehearsal session for a longer time, it is more effective to wait until slower structural plasticity has grown additional useful synapses that can then be consolidated in another (brief) rehearsal session. Consequently, our simulations suggest that spacing effects occur whenever learning in the brain depends on structural plasticity. As structural plasticity is considered a regular feature in both developmental and adult learning, this suggests that structural plasticity may be the common cellular basis of spacing effects.

3.5 Gradients in P_{eff} and Graded Amnesia

Another stable effect of our model is that the number of consolidated synapses increases with time (see the evolution of P_1 in Fig. 17.4). As a consequence, the number of unstable synapses that take part in synaptic turnover will decrease such that structural plasticity will become slower with lifetime. For example, Fig. 17.7 (panels C and D) show simulations where several memory sets (block1, block2, ...) are learned sequentially one after the other. All memory blocks are rehearsed (or consolidated, e.g., assuming hippocampal replay[29,94]) with time intervals of identical lengths (five time steps). However, as structural plasticity becomes slower with lifetime, this leads to a gradient in effectual connectivity P_{eff}, where early memories can reach a higher final P_{eff} than recent memories (panel D). Under normal conditions, all memories can be retrieved with similar high quality ($\hat{\varepsilon}$ close to zero) for long lifetimes. However, after a simulated network lesion (at $t = 20$), the effect of gradients in P_{eff} is magnified such that older memories with high P_{eff} are preserved, whereas recent memories with low P_{eff} get lost (see graded increase of output noise for memory blocks 1–4).

3.6 Memory Decline, CF, and the Stability–Plasticity Dilemma

The stability–plasticity dilemma refers to the general problem of learning new while preserving old contents in neural networks.[51,95–97] For example, optimizing synaptic weights to store a new set of memories will deteriorate or even destroy earlier memories. Vice versa, freezing synaptic weights to preserve old memories will prevent acquiring new memories. A virtue of associative memory models as described in Section 2 is that they suffer to a lesser degree from this dilemma because they typically use *incremental* learning methods where learning a new item is independent of previous items. However, as their storage capacity is not unlimited, they suffer from a related phenomenon called CF that is illustrated by Fig. 17.8 for learning without structural plasticity (panel A). There, a large number of memories can be sequentially stored and retrieved with high quality. However, when the number of stored memories approaches the capacity limit of the network (M_ε), storing only a few further memories will exponentially increase output noise and, therefore, finally extinct *all* stored information.[98] As such a destructive behavior does not normally occur in our brains, this poses a problem to such memory models based on weight plasticity only.

Interestingly, memory models with structural plasticity may much more easily explain the absence of CF (panel B).[22,29] There, the gradients in effectual connectivity described in the previous section cause a gradual decline of the ability to store new memories. Again, the network can store many memories at high quality (P_{eff} high, ε close to zero). However, as the number of consolidated synapses increases and structural plasticity becomes slower with lifetime, novel memories will be stored at progressively lower P_{eff} unless rehearsal or consolidation time would be increased substantially. In the end, if all unstable synapses were depleted and synaptic turnover had stopped, the network would have gained its final stable structure and learning new memories would be impossible.

The latter implies that such learning by structural plasticity approaches the stability–plasticity dilemma rather from the stability side. Of course, structural plasticity cannot

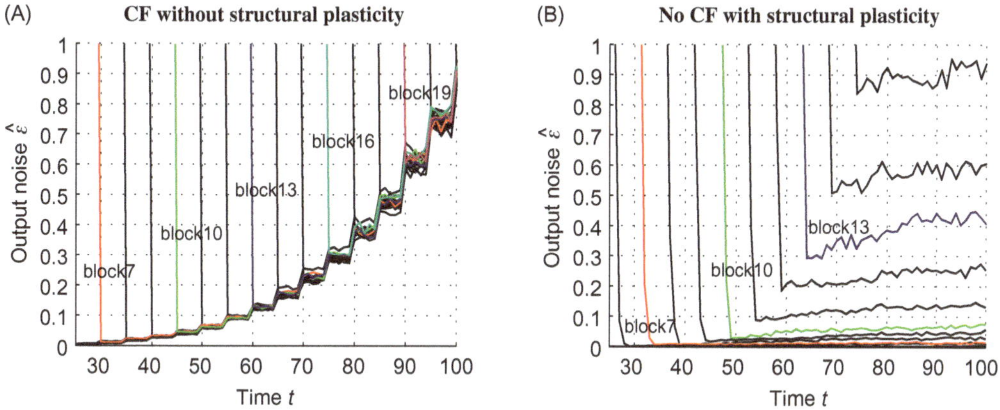

FIGURE 17.8 **Simulation of catastrophic forgetting.** Similar simulation experiment as in Fig 17.7C and 17.7D, but without a lesion, only showing consolidation of the later memory blocks (7–21). In panel (A), structural plasticity was switched off from the start ($p_{e|s} = 0$) leading to catastrophic forgetting (*CF*): If the capacity limit of the network is approached, output noise ε increases strongly and all memories get finally extinct (high ε for all memory blocks). In panel (B) with structural plasticity ($p_{e|s} = s$), there is no catastrophic forgetting. Although the ability to store new memories at low output noise decreases with time (because structural plasticity becomes slower as more and more consolidated synapses escape turnover), the remote memories can still be retrieved with low output noise. Memory blocks 16–21 are not visible because their output noise is too large ($\varepsilon > 1$). Source: *Adapted from Knoblauch, A, Körner, E, Körner, U, Sommer, FT. Structural plasticity has high memory capacity and can explain graded amnesia, catastrophic forgetting, and the spacing effect. PLoS One. 2014;9(5):e96485, 1–19.*

really solve the dilemma (as it is generally impossible to stably store an infinite memory number in a finite neural network), but, unlike alternative learning methods, it can at least balance between the two sides as long as possible.

Obviously, the stability of learning by structural plasticity in our simple model depends on stable synapses with zero deconsolidation probability. For positive $p_{d|s} > 0$, it may rather work as a palimpsests-type short-term or mid-term memory.[99,100] Then consolidated synapses may be eliminated again, and the corresponding memories may get lost, unless memories are reactivated regularly, for example, by memory replay, attractor-hopping by ongoing random activity, or pseudo-rehearsal.[98] While memory replay just shifts the stability problem by presuming another brain area where the memories are "stably" stored, attractor-hopping or pseudo-rehearsal can substantially alleviate CF, for example, in gradient-descent based learning models, but cannot prevent a slow drifting and, on the long run, forgetting of memories, in particular, in networks operating close to the capacity limit. Thus, it is at least questionable if such approaches alone can account for long-term stability of human memory.

However, it is difficult to test experimentally the long-term stability of individual synapses for durations beyond some months.[19] Moreover, an irreversible consolidation with $p_{d|s} = 0$ bears the risk of wasting resources by consolidating synapses that turn out later to be dispensable. To exclude this risk, plasticity rules selecting synapses to be consolidated should be *final*, i.e., they should be able to decide at the time of memory formation which synapses will be required on the long run. In Section 2, we have argued that the Willshaw rule, Eq. (17.4), has this property but works efficiently only for very sparse neural activity or very small cell assemblies (Fig. 17.6). As such sparseness of neural activity is inconsistent with a stable initial storage in networks having a plausible anatomical connectivity ($P = 0.1$), this suggests a role for sparsification or sharpening of activity during consolidation.[56–58]

We have tested this idea in an extended model of three neuron populations as illustrated in Fig. 17.9 (panel A) by using the extended synapse model of Fig. 17.1B. Similarly as before, the general task is to link given activity patterns of a population u to another population v, but here this link is via a third population w. Whenever storing a novel association from u^t to v^t, a sufficiently large random activity pattern w^t is activated and initially stabilized by weight plasticity of the synapses from u to w, w to v, and within w. Importantly, activity within w is not fixed but controlled by recurrent inhibition regulating the total population activity to a constant level. Such a regulation allows firing of either many neurons at low frequency or few neurons at high frequency. For example, when storing a novel memory association $u^t \to v^t$, stimulation of w by u^t is unspecific, thereby activating a large random pattern that can be stabilized by reversible weight plasticity

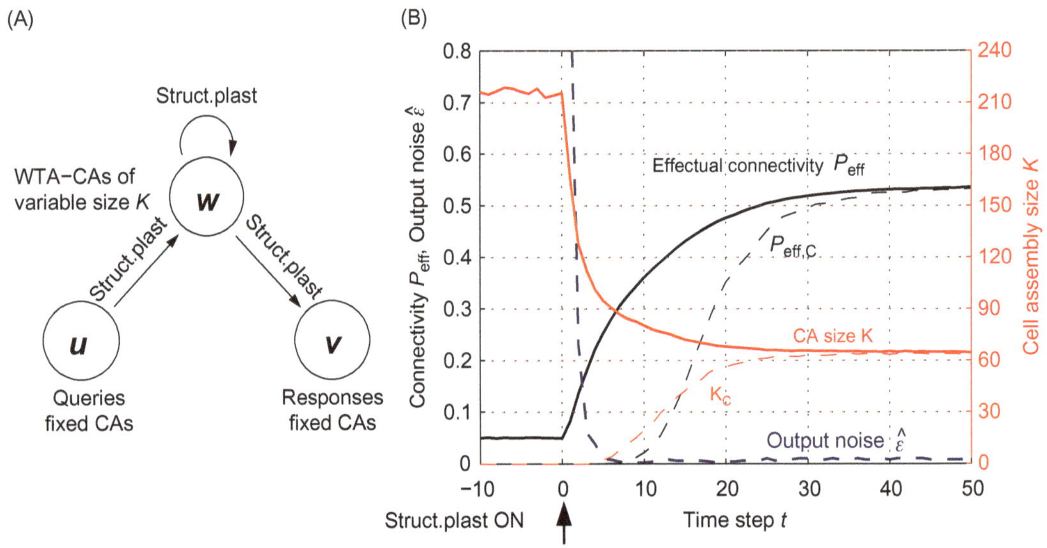

FIGURE 17.9 **Sharpening of cell assembly activity during consolidation.** Simulation of consolidating a memory set by structural plasticity using the synaptic cascade model (Fig. 17.1B) in a three-population network (panel A): Here, associations between fixed cell assemblies (*CAs*) u^t and v^t is mediated via an intermediary population w linked with three structurally plastic synaptic projections (*arrows*). CAs in w are not fixed but defined by a recurrent winners-take-all (*WTA*) inhibition regulating total activity within w to a constant value (see text for details). (Panel B) Neural activity in w shows a significant sharpening during consolidation, where average size K of CAs decreases with time (*red solid curve*) as effectual connectivity P_{eff} increases. For larger P_{eff}, firing activity within CAs increases and eventually reaches the threshold required for synaptic consolidation (transition from synaptic state 1 to state C; see Fig. 17.1B). As a consequence, there is a lagged increase of effectual connectivity $P_{eff,C}$ due to consolidated synapses (*black dashed*). Correspondingly, the number K_C of CA neurons exceeding the consolidation threshold (*red dashed*) converges to the CA size K. Output noise ε (*blue dashed*) is measured in output population v following stimulation of u.

(transition from state 0 to 1) even for sparse connectivity. If the memory is continuously rehearsed or replayed, effectual connectivity P_{eff} of the corresponding cell assembly will increase (see Fig. 17.9B, black solid line). This will increase dendritic potentials and, thus, firing rates of the involved neurons. Then recurrent inhibitory control will select fewer neurons firing at a higher rate. This means that cell assembly size K in population w decreases with time (red solid line). At some point, the cell assembly is small enough and the firing rates of an increasing neuron number K_C (red dashed line) exceeds the threshold required for synapses reaching the irreversible consolidated state (transition from state 1 to C). Correspondingly, the increase of effectual connectivity $P_{\mathrm{eff},C}$ of C synapses (dashed black line) is similar but lagged to P_{eff} of all potentiated synapses. Besides avoiding CF and consolidation of dispensable synapses, the described process would also be consistent with experimentally described sparsening or sharpening of neural activity.[56–58] Moreover, if parameters are adequately tuned, consolidation may follow the ridgeline of maximal storage capacity mentioned earlier (magenta dashed line for $P_1 = 0.1$ in Fig. 17.6A).

4 DISCUSSION

The simulations presented in this work confirm the hypothesis that a major function of structural plasticity is to increase storage efficiency by "moving" the rare synapses in sparsely connected networks to the locations that are most beneficial, for example, for storing a certain set of memories at high quality. Earlier works have established this hypothesis by counting and evaluating the Shannon information of possible synaptic configurations selected from potential connectivity.[8,20,45] Such approaches predict that structural plasticity can increase storage capacity up to a factor P_{pot}/P that is the inverse of the filling fraction.[45] For example, for a prototypical cortical macrocolumn of 1 mm^3 containing about $n = 10^6$ neurons in a network with anatomical connectivity $P = 0.1$ and potential connectivity $P_{\mathrm{pot}} = 0.5$ (corresponding to a filling fraction $P/P_{\mathrm{pot}} = 0.2$),[25,45,87] the increase would be about factor 5.

However, as neural networks cannot read out synaptic configurations directly, it is necessary to consider plausible retrieval procedures in associative memory models to estimate the impact of structural plasticity on memory storage capacity. We have therefore analyzed and simulated common models of neural associative memory[21,31,34] and Hebbian cell assemblies.[1,64] By introducing effectual connectivity P_{eff} to measure the progress of learning or consolidation of memories with time, our model shows that the dependence of storage capacity on connectivity can be strongly nonlinear. Most interestingly, for sparse neural activity as measured experimentally,[78] structural plasticity can actually boost storage capacity by several orders of magnitude. Here it is known that the corresponding small cell assemblies (e.g., with sizes $k = 50–500$) cannot be stably stored in sparsely connected network without structural plasticity.[79,101] But by increasing P_{eff} from P to P_{pot}, structural plasticity can lift the network to a processing mode where small cell assemblies become stable and storage capacity is known to be maximal (Fig. 17.6).

Besides increasing storage efficiency, structural plasticity can explain a number of memory phenomena more easily than models based on weight plasticity only. For example, the spacing effect[53–55] is an inherent property of learning by structural plasticity and relies only on the well-supported assumption that structural plasticity is slower than weight plasticity.[19] By contrast, many alternative explanations of the spacing effect have to assume additional cognitive high-level memory structures[53,54,91] or can explain only spacing effects on relatively small time scales.[102] By a plausible choice of plasticity parameters (e.g., $p_{c|s}$, $p_{d|s}$ which may differ depending on brain area), our model can also explain long-term spacing effects observed over months to years.[93,103] These results suggest that structural plasticity is the common cellular mechanism of various spacing effects described in experiments.

Another memory phenomenon is retrograde amnesia where, following brain lesions, remote memories are typically better preserved than recent memories.[48–50] Our model of structural plasticity can easily explain such "Ribot gradients" by gradients in effectual connectivity P_{eff}: As it predicts structural plasticity to become slower with lifetime (as more and more consolidated synapses escape synaptic turnover), recent memories stored at a later time cannot reach the same high P_{eff} as remote memories stored at an earlier time and are therefore more vulnerable against lesions. Unlike alternative explanations based on weight plasticity,[104] our model does not require the assumption of a lifelong consolidation, for example, by replay from hippocampus to cortex, where remote memories accumulate a longer total consolidation time. Instead, our model is consistent with the common idea that short-term memory systems such as hippocampus have only a time-limited role in consolidation of long-term memory.[48,49,105] Moreover, our model does not need to assume a continuous increase of synaptic weights (beyond optimal values) to obtain such gradients.

Note that our hypothesis of the function of structural plasticity depends crucially on potential and effectual connectivity being significantly larger than anatomical connectivity. It may be challenged by some recent experimental and theoretical work on the distribution of synapse number per neuron pair suggesting that cortical neuron pairs would be typically connected with multiple synapses (e.g., synapse number per pair being narrowly distributed around four to five), whereas single synapses could not stably exist.[82,83,106,107] Our current model cannot easily produce the experimental distributions of synapse number (Fig. 17.4) unless introducing further mechanisms to regulate synapse number. More serious, our model cannot easily be reconciled with the assumption of single synapses being generally unstable: As most potentially connected neuron pairs seem to have less than four potential synapses, this assumption would mean that most potential synapses could not become stable actual synapses as required by our model to increase effectual connectivity P_{eff}. In that case the actual potential connectivity (P_{eff} could be increased to) would not be much larger than the anatomical connectivity P, and the predictions of our model on storage efficiency would turn out to be invalid. As an alternative hypothesis for the function of structural plasticity, it has therefore been suggested that structural plasticity may rather serve to increase robustness of long-term information storage,[83] thereby realizing the function of synaptic cascade models based on weight plasticity.[84] However, this idea relies on experiments that have analyzed synaptic connectivity only for very closely neighbored neurons having distances less than 50 μm along cortical surface[82] and may therefore rather reflect minicolumnar structure of cortex.[76,77] By contrast, our model rather assumes processing within locally connected macrocolumns (of size 1 mm^3) where neuron distances may be factor 10–20 larger,[25,87] and only a tiny fraction of neuron pairs

(<1% to 1 per mill) would have distances <50 μm. For most other connected neuron pairs within a macrocolumn, it seems unlikely that they would have more than one or two actual synapses. However, this prediction has to be tested in future neurophysiological experiments (see Section 4.1).

In any case, our model is not inconsistent with the idea that another major function of structural plasticity is to increase stability of long-term memory. Indeed, we have argued before [22,29] that long-term memory models using structural plasticity are more resilient against CF than models using only weight plasticity [98] and can also better explain aging effects of human learning and memory.[108] There are three reasons for these desirable properties: First, by increasing storage efficiency, the capacity limit where CF would set in is reached much later. Second, even if the capacity limit is reached, synaptic turnover has become so slow (by depleting the unstable silent synapses) that learning will rather stop before CF occurs. Third, unlike many gradient-based methods, our model presumes only a simple Willshaw-type local, incremental, immediate, and final learning rule where, once a high effectual connectivity is reached, memories do not require a further lifelong rehearsal to prevent CF.

The classical Willshaw rule, Eq. (17.4), of binary associative learning has actually often been criticized as being inefficient and implausible for the following reasons: First, it has a very low storage capacity unless cell assemblies are very small (ideally with log-sparse activity $k \sim \log n$) and network connectivity P is high (Fig. 17.6). Moreover, the binary weights may be considered implausible as real synapses have continuous weight distributions.[109] However, our results suggest not to underestimate the Willshaw model: First, there exist memory-related cortical areas that seem to (almost) match the log-sparsity requirements of the Willshaw model.[78] Second, by structural plasticity increasing P_{eff} from P to P_{pot}, cortical networks may well provide the high effectual connectivity required for the Willshaw model to operate almost equivalent to optimal Bayesian learning.[37] Third, in physiological experiments, synaptic learning has often been described to have discrete or even binary switch-like properties.[89,90,110] Moreover, theoretical works suggest that the location of a synapse can store much more information than its weight,[22] and that binary weights are actually sufficient to reach maximal storage capacity.[38] Correspondingly, the strength of a synapse may rather reflect the stability of a connection.[83,111] Fourth, besides being "local," "incremental," and "immediate,"[76] the Steinbuch/Willshaw learning rule, Eq. (17.4), is also "final" and therefore has an invaluable advantage over other known learning models. Here "final" means that storing a sparse activity pattern *finally* determines those synapses that are required for storage of a memory. This property may provide a solution to the stability–plasticity dilemma and prevent CF as argued earlier.

4.1 Future Experimental Studies

When regarding the capacity landscape of neural networks (Fig. 17.6), the latter hypothesis, together with experiments reporting a sharpening or sparsification of neural activity during learning and consolidation,[56–58] may lead finally to a consistent theory for the formation of a cell assembly representing a memory by structural plasticity. This theory would imply the following assumptions and predictions, some of which may also be tested experimentally:

First, to allow stable activation of cell assemblies having different sizes K and different connectivities P_{eff}, recurrent inhibitory neurons should serve to maintain a constant activity level in a certain area, e.g., to keep the total number of spikes in a certain time interval constant. This differs from a k-WTA inhibition or fixed firing thresholds as assumed in many previous works.[38,44]

Second, we have to assume an extended learning model (Fig. 17.1B) with a cascade of at least two "potentiated" synaptic states, for example, a labile unconsolidated state "1" and further a stable consolidated state "C," which can only be reached by a coincidence of strong pre/postsynaptic activity.

Third, initially, a novel stimulus or memory is represented by a relatively broad neural activity distribution, where many excitatory neurons fire at relatively low frequencies. Such large activity patterns (e.g., $k > 500$) can easily be stored temporally in networks having a low anatomical connectivity (Fig. 17.6). Thus, the memory will initially be represented by a large cell assembly having low effectual connectivity P_{eff} and low firing rate with most synapses being in the potentiated but unstable state "1."

Fourth, by ongoing reactivation of the cell assembly (e.g., by repetitive rehearsal, hippocampal replay, or ongoing random "attractor-hopping"), structural plasticity will slowly increase P_{eff}. By that, individual neurons of the cell assembly will fire more strongly, and recurrent inhibition will therefore select fewer neurons to keep total activity constant. As a consequence, the size of the cell assembly, k, will decrease with time, whereas the remaining neurons of the cell assembly will increase their firing rates.

Fifth, if consolidation is continued, at some point P_{eff} is large enough (corresponding to a certain small k) that firing rates exceed the threshold required for synaptic consolidation. Then the synapses of the cell assembly can finally reach their consolidated state "C."

Sixth, unless $P_{\text{eff}} = P_{\text{pot}}$ is reached, consolidation may continue to increase P_{eff}, whereas inhibition may allow activation of only parts of the whole cell assembly at a time. Alternatively, if the threshold of consolidation is coordinated with P_{pot}, consolidation will finally stop with all synapses of the cell assembly being in state "C." Then no further reactivation of the memory is necessary to keep the memory stable. After consolidation of all memories, the network will be virtually equivalent to the Willshaw model.

Last, another important prediction of our model that may be tested is, as discussed earlier, distant neurons can be connected by single stable synapses rather than compounds of 4–5 synapses as reported for very close neurons. Otherwise the effect of structural plasticity in increasing P_{eff}, sharpening of cell assemblies, and thus improving storage efficiency would be strongly limited.

4.2 Future Modeling Studies

A preliminary simulation experiment by using the extended synaptic model (Fig. 17.1B) showed already such consolidation and sharpening of neural activity as proposed earlier (Fig. 17.9). However, one problem of the current model is to produce sufficiently sparse cell assemblies that depend on the particular form of inhibitory feedback. In particular, a simple inhibition that depends linearly on the excitation in the neural population turned out insufficient for reaching experimentally reported levels of sparsity (e.g., from $k = 1000$ to $k = 50$).

Another issue is to test the stability of fully plastic cell assemblies that are not constrained by fixed activity patterns (such as in population w in Fig. 17.9) when consolidating multiple memory sets sequentially (as in Figs. 17.7C, 17.7D, and 17.8). Then severe cross talk effects are expected (e.g., fusion of similar cell assemblies), and it may be necessary to identify further plasticity mechanisms to maintain efficient memory storage. Future work should therefore more closely investigate the precise conditions under which structural plasticity and inhibitory feedback support realistic sparsification of cell assemblies to maximize storage efficiency, prevent CF, and solve the stability–plasticity dilemma.

References

1. Hebb DO. *The Organization of Behavior: A Neuropsychological Theory*. New York, NY: John Wiley and Sons, Inc.; 1949.
2. Bliss TVP, Collingridge GL. A synaptic model of memory: long-term potentiation in the hippocampus. *Nature*. 1993;361:31–39.
3. Paulsen O, Sejnowski TJ. Natural patterns of activity and long-term synaptic plasticity. *Curr Opin Neurobiol*. 2000;10:172–179.
4. Song S, Miller KD, Abbott LF. Competitive Hebbian learning through spike-timing-dependent synaptic plasticity. *Nat Neurosci*. 2000;3(9):919–926.
5. Raisman G. Neuronal plasticity in the septal nuclei of the adult rat. *Brain Res*. 1969;14:25–48.
6. Engert F, Bonhoeffer T. Dendritic spine changes associated with hippocampal long-term synaptic plasticity. *Nature*. 1999;399:66–70.
7. Witte S, Stier H, Cline HT. In vivo observations of timecourse and distribution of morphological dynamics in Xenopus retinotectal axon arbors. *J Neurobiol*. 1996;31:219–234.
8. Chklovskii DB, Mel BW, Svoboda K. Cortical rewiring and information storage. *Nature*. 2004;431:782–788.
9. Butz M, Wörgötter F, van Ooyen A. Activity-dependent structural plasticity. *Brain Res Rev*. 2009;60(2):287–305.
10. Holtmaat A, Svoboda K. Experience-dependent structural synaptic plasticity in the mammalian brain. *Nat Rev Neurosci*. 2009;10:647–658.
11. Yang G, Pan F, Gan W-B. Stably maintained dendritic spines are associated with lifelong memories. *Nature*. 2009;462:920–924.
12. Xu T, et al. Rapid formation and selective stabilization of synapses for enduring motor memories. *Nature*. 2009;462:915–919.
13. Yu X, Zuo Y. Spine plasticity in the motor cortex. *Curr Opin Neurobiol*. 2011;21(1):169–174.
14. Fu M, Zuo Y. Experience-dependent structural plasticity in the cortex. *Trends Neurosci*. 2011;34(4):177–187.
15. Adelsberger-Mangan DM, Levy WB. The influence of limited presynaptic growth and synapse removal on adaptive synaptogenesis. *Biol Cybern*. 1994;71:461–468.
16. van Ooyen A, van Pelt J. Activity-dependent outgrowth of neurons and overshoot phenomena in developing neural networks. *J Theor Biol*. 1994;167:27–43.
17. Butz M, van Ooyen A, Wörgötter F. A model for cortical rewiring following deafferentation and focal stroke. *Front Comput Neurosci*. 2009;3:1–15.
18. Butz M, van Ooyen A. A simple rule for dendritic spine and axonal bouton formation can account for cortical reorganization after focal retinal lesions. *PLoS Comput Biol*. 2013;9(10):e1003259.
19. Tetzlaff C, Kolodziejski C, Markelic I, Wörgötter F. Time scales of memory, learning, and plasticity. *Biol Cybern*. 2012;106:715–726.
20. Poirazi P, Mel BW. Impact of active dendrites and structural plasticity on the memory capacity of neural tissue. *Neuron*. 2001;29:779–796.
21. Knoblauch A, Palm G, Sommer FT. Memory capacities for synaptic and structural plasticity. *Neural Comput*. 2010;22(2):289–341.

22. Knoblauch A, Körner E, Körner U, Sommer FT. Structural plasticity has high memory capacity and can explain graded amnesia, catastrophic forgetting, and the spacing effect. *PLoS One*. 2014;9(5):e96485:1–19.
23. Braitenberg V, Schüz A. *Anatomy of the Cortex: Statistics and Geometry*. Berlin, Germany: Springer-Verlag; 1991.
24. Abeles M. *Corticonics: Neural Circuits of the Cerebral Cortex*. New York: Cambridge University Press; 1991.
25. Hellwig B. A quantitative analysis of the local connectivity between pyramidal neurons in layers 2/3 of the rat visual cortex. *Biol Cybern*. 2000;82:111–121.
26. Lennie P. The cost of cortical computation. *Curr Biol*. 2003;13:493–497.
27. Laughlin SB, Sejnowski TJ. Communication in neuronal networks. *Science*. 2003;301:1870–1874.
28. Attwell D, Laughlin SB. An energy budget for signaling in the grey matter of the brain. *J Cereb Blood Flow Metab*. 2001;21:1133–1145.
29. Knoblauch A. The role of structural plasticity and synaptic consolidation for memory and amnesia in a model of cortico-hippocampal interplay. In: Mayor J, Ruh N, Plunkett K, eds. Connectionist Models of Behavior and Cognition II: Proceedings of the 11th Neural Computation and *Psychology Workshop*. World Scientific Publishing; 2009:79–90.
30. Steinbuch K. Die lernmatrix. *Kybernetik*. 1961;1:36–45.
31. Willshaw DJ, Buneman OP, Longuet-Higgins HC. Non-holographic associative memory. *Nature*. 1969;222:960–962.
32. Marr D. Simple memory: a theory for archicortex. *Philos Trans R Soc Lond Ser B*. 1971;262:24–81.
33. Palm G. On associative memories. *Biol Cybern*. 1980;36:19–31.
34. Hopfield JJ. Neural networks and physical systems with emergent collective computational abilities. *Proc Natl Acad Sci USA*. 1982;79:2554–2558.
35. Buckingham JT, Willshaw DJ. On setting unit thresholds in an incompletely connected associative net. *Netw Comput Neural Syst*. 1993;4:441–459.
36. Bosch H, Kurfess F. Information storage capacity of incompletely connected associative memories. *Neural Netw*. 1998;11(5):869–876.
37. Knoblauch A. Neural associative memory with optimal Bayesian learning. *Neural Comput*. 2011;23(6):1393–1451.
38. Knoblauch A. Efficient associative computation with discrete synapses. *Neural Comput*. 2016;28(1):118–186.
39. Chechik G, Meilijson I, Ruppin E. Synaptic pruning in development: a computational account. *Neural Comput*. 1998;10(7):1759–1777.
40. Huttenlocher PR. Synaptic density in human frontal cortex—developmental changes and effects of aging. *Brain Res*. 1979;163(2):195–205.
41. Huttenlocher PR, De Courten C, Garey LJ, van der Loos H. Synaptogenesis in human visual cortex – evidence for synapse elimination during normal development. *Neurosci Lett*. 1982;33:247–252.
42. Huttenlocher PR, Dabholkar AS. Regional differences in synaptogenesis in human cerebral cortex. *J Comp Neurol*. 1997;387:167–178.
43. Elston GN, Oga T, Fujita I. Spinogenesis and pruning scales across functional hierarchies. *J Neurosci*. 2009;29 (10):3271–3275.
44. Knoblauch A, Sommer FT. Structural plasticity, effectual connectivity and memory in cortex. *Front Neuroanat*. 2016;10:63.
45. Stepanyants A, Hof PR, Chklovskii DB. Geometry and structural plasticity of synaptic connectivity. *Neuron*. 2002;34:275–288.
46. Braitenberg V. Cell assemblies in the cerebral cortex. In: Heim R, Palm G, eds. *Lecture Notes in Biomathematics (21) Theoretical Approaches to Complex Systems*. Springer-Verlag; 1978:171–188.
47. Palm G. *Neural Assemblies. An Alternative Approach to Artificial Intelligence*. Berlin, Germany: Springer; 1982.
48. Squire LR, Bayley PJ. The neuroscience of remote memory. *Curr Opin Neurobiol*. 2007;17:185–196.
49. Ross RS, Eichenbaum H. Dynamics of hippocampal and cortical activation during consolidation of a nonspatial memory. *J Neurosci*. 2006;26(18):4852–4859.
50. Ribot T. *Les maladies de la memoire*. (Germer Baillare; 1881.
51. French RM. Catastrophic forgetting in connectionist networks: causes, consequences and solutions. *Trends Cogn Sci*. 1999;3(4):128–135.
52. Grossberg S. Competitive learning: from interactive activation to adaptive resonance. *Cogn Sci*. 1987;11:23–63.
53. Crowder RG. *Principles of Learning and Memory*. Hillsdale, NJ: Lawrence Erlbaum; 1976.
54. Greene RL. Spacing effects in memory: evidence for a two-process account. *J Exp Psychol Learn Mem Cogn*. 1989;15(3):371–377.

55. Ebbinghaus, H Über das Gedächtnis: Untersuchungen zur experimentellen Psychologie. (Duncker & Humblot, 1885).
56. Freedman DJ, Riesenhuber M, Poggio T, Miller EK. Experience-dependent sharpening of visual shape selectivity in inferior temporal cortex. *Cereb Cortex*. 2006;16(11):1631–1644.
57. McNamara A, et al. Neural dynamics of learning sound-action associations. *PLoS One*. 2008;3(12):e3845.
58. Weinberger NM, Miasnikov AA, Chen JC. Sensory memory consolidation observed: Increased specificity of detail over days. *Neurobiol Learn Mem*. 2009;91:273–286.
59. Senn W, Sacramento J. Backward reasoning the formation rule. *Nat Neurosci*. 2001;18(12):1705–1706.
60. Martin SJ, Grimwood PD, Morris RG. Synaptic plasticity and memory: an evaluation of the hypothesis. *Annu Rev Neurosci*. 2000;23:649–711.
61. Arshavsky YI. The seven sins' of the Hebbian synapse: Can the hypothesis of synaptic plasticity explain long-term memory consolidation? *Prog Neurobiol*. 2006;80:99–113.
62. Bienenstock E. A model of neocortex. *Netw Comput Neural Syst*. 1994;6:179–224.
63. Knoblauch A, Hauser F, Gewaltig M-O, Körner E, Palm G. Does spike-timing-dependent synaptic plasticity couple or decouple neurons firing in synchrony? *Front Comput Neurosci*. 2012;6:1–27.
64. Palm G, Knoblauch A, Hauser F, Schüz A. Cell assemblies in the cerebral cortex. *Biol Cybern*. 2014;108(5):559–572.
65. Fay R, Kaufmann U, Knoblauch A, Markert H, Palm G. Combining visual attention, object recognition and associative information processing in a neurobotic system. In: Wermter S, Palm G, Elshaw M, eds. Biomimetic Neural Learning for Intelligent Robots, Lecture Notes in Artificial Intelligence, *LNAI* 3575. Springer-Verlag; 2005:118–143.
66. Markert H, Knoblauch A, Palm G. Modelling of syntactical processing in the cortex. *BioSystems*. 2007;89(1-3):300–315.
67. Markert H, Knoblauch A, Palm G. Detecting sequences and understanding language with neural associative memories and cell assemblies. In: Wermter S, Palm G, Elshaw M, eds. Biomimetic Neural Learning for Intelligent Robots, Lecture Notes in Artificial Intelligence, *LNAI* 3575. Berlin, Germany: Springer-Verlag; 2005:107–117.
68. Knoblauch A, Palm G. Pattern separation and synchronization in spiking associative memories and visual areas. *Neural Netw*. 2001;14:763–780.
69. Schwenker F, Sommer FT, Palm G. Iterative retrieval of sparsely coded associative memory patterns. *Neural Netw*. 1996;9:445–455.
70. Shannon CE, Weaver W. *The Mathematical Theory of Communication*. Champaign, IL: University of Illinois Press; 1949.
71. Cover TM, Thomas JA. *Elements of Information Theory*. Hoboken, NJ: John Wiley & Sons, Inc.; 1991.
72. Nadal J-P, Toulouse G. Information storage in sparsely coded memory nets. *Netw Comput Neural Syst*. 1990;1:61–74.
73. Buckingham JT, Willshaw DJ. Performance characteristics of the associative net. *Netw Comput Neural Syst*. 1992;3:407–414.
74. Diederich S, Opper M. Learning of correlated patterns in spin-glass networks by local learning rules. *Phys Rev Lett*. 1987;58(9):949–952.
75. Palm G. On the information storage capacity of local learning rules. *Neural Comput*. 1992;4(5):703–711.
76. Storkey, A *Artificial Neural Networks—ICANN 1997*. 451–456 (Springer Verlag, 1997).
77. Dayan P, Willshaw DJ. Optimising synaptic learning rules in linear associative memory. *Biol Cybern*. 1991;65:253–265.
78. Waydo S, Kraskov A, Quiroga RQ, Fried I, Koch C. Sparse representation in the human medial temporal lobe. *J Neurosci*. 2006;26(40):10232–10234.
79. Latham PE, Nirenberg S. Computing and stability in cortical networks. *Neural Comput*. 2004;16(7):1385–1412.
80. Knoblauch, A. Zip nets: Efficient associative computation with binary synapses. In: *Proceedings of the International Joint Conference on Neural Networks (IJCNN)* 4271–4278. IEEE World Congress on Computational Intelligence (WCCI), 2010.
81. Frey U, Morris RGM. Synaptic tagging and long-term potentiation. *Nature*. 1997;385:533–536.
82. Fares T, Stepanyants A. Cooperative synapse formation in the neocortex. *Proc Natl Acad Sci USA*. 2009;106(38):16463–16468.
83. Fauth M, Wörgötter F, Tetzlaff C. Formation and maintenance of robust long-term information storage in the presence of synaptic turnover. *PLoS Comput Biol*. 2015;11(12):e1004684.

84. Fusi S, Drew PJ, Abbott LF. Cascade models of synaptically stored memories. *Neuron*. 2005;45:599–611.
85. Mountcastle VB. The columnar organization of the neocortex. *Brain*. 1997;120(4):701–722.
86. Sporns O, Tononi G, Kötter R. The Human connectome: a structural description of the human brain. *PLoS Comput Biol*. 2005;1(4):e42.
87. Braitenberg V, Schüz A. *Anatomy of the Cortex: Statistics and Geometry*. Berlin, Germany: Springer-Verlag; 1991.
88. Schreiber T. Measuring information transfer. *Phys Rev Lett*. 2000;85:461–464.
89. Petersen CCH, Malenka RC, Nicoll RA, Hopfield JJ. All-or-none potentiation at CA3-CA1 synapses. *Proc Natl Acad Sci USA*. 1998;95:4732–4737.
90. O'Connor DH, Wittenberg GM, Wang SS-H. Graded bidirectional synaptic plasticity is composed of switch-like unitary events. *Proc Natl Acad Sci USA*. 2005;102(27):9679–9684.
91. Cepeda NJ, Pashler H, Vul E, Wixted JT, Rohrer D. Distributed practice in verbal recall tasks: A review and quantitative synthesis. *Psychol Bull*. 2006;132(3):354–380.
92. Knoblauch A. Structural plasticity, cortical memory, and the spacing effect. *BMC Neurosci*. 2009;10(Suppl 1): O16.
93. Knoblauch A. Bimodal structural plasticity can explain the spacing effect in long-term memory tasks. *Front Syst Neurosci Conf Abstr Comput Syst Neurosci*. 2010;
94. Buhry L, Azizi AH, Cheng S. Reactivation, replay, and preplay: how it might all fit together. *Neural Plast*. 2011;2011(203462).
95. Grossberg S. How does a brain build a cognitive code. *Psychol Rev*. 1980;87:1–51.
96. McCloskey M, Cohen N. Catastrophic interference in connectionist networks: The sequential learning problem. In: Bower GH, ed. *Psychology of Learning and Motivation*. Vol. 24. New York: Academic Press; 1989:109–164.
97. Abraham C, Robins A. Memory retention—the synaptic stability versus plasticity dilemma. *Trends Neurosci*. 2005;28(2):73–78.
98. Robins A, McCallum S. Catastrophic forgetting and the pseudorehearsal solution in Hopfield type networks. *Connect Sci*. 1998;7:121–135.
99. Nadal J-P, Toulouse G, Changeux JP, Dehaene S. Networks of formal neurons and memory palimpsests. *Europhys Lett*. 1986;1(10):535–542.
100. Sandberg A, Lansner A, Petersson KM, Ekeberg Ö. A palimpsest memory based on an incremental Bayesian learning rule. *Neurocomputing*. 2000;32-33:987–994.
101. Aviel Y, Horn D, Abeles M. Memory capacity of balanced networks. *Neural Comput*. 2005;17:691–713.
102. Pagani MR, Oishi K, Gelb BD, Zhong Y. The phosphatase SHP2 regulates the spacing effect for long-term memory induction. *Cell*. 2009;139:1–13.
103. Cepeda NJ, Vul E, Rohrer D, Wixted JT, Pashler H. Spacing effects in learning: A temporal ridgeline of optimal retention. *Psychol Sci*. 2008;19(11):1095–1102.
104. Meeter M, Murre JMJ. TraceLink: a model of consolidation and amnesia. *Cogn Neuropsychol*. 2005;22 (5):559–587.
105. Ji D, Wilson MA. Coordinated memory replay in the visual cortex and hippocampus during sleep. *Nat Neurosci*. 2007;10(1):100–107.
106. Deger M, Helias M, Rotter S, Diesmann M. Spike-timing dependence of structural plasticity explains cooperative synapse formation in the neocortex. *PLoS Comput Biol*. 2012;8(9):e1002689.
107. Fauth M, Wörgötter F, Tetzlaff C. The formation of multi-synaptic connections by the interaction of synaptic and structural plasticity and their functional consequences. *PLoS Comput Biol*. 2015;11(1):e1004031.
108. Hedden T, Gabrieli JDE. Insights into the ageing mind: a view from cognitive neuroscience. *Nat Rev Neurosci*. 2004;5:87–96.
109. Song S, Sjöström PJ, Reigl M, Nelson S, Chklovskii DB. Highly nonrandom features of synaptic connectivity in local cortical circuits. *PLoS Biol*. 2005;3(3):507–519.
110. Montgomery JM, Madison DV. Discrete synaptic states define a major mechanism of synapse plasticity. *Trends Neurosci*. 2004;27(12):744–750.
111. Matsuzaki M, et al. Dendritic spine geometry is critical for AMPA receptor expression in hippocampal CA1 pyramidal neurons. *Nat Neurosci*. 2001;4:1086–1092.
112. Knoblauch A. Neural associative memory and the Willshaw-Palm probability distribution. *SIAM J Appl Math*. 2008;69(1):169–196.

SECTION V

NEUROGENESIS-RELATED STRUCTURAL PLASTICITY

CHAPTER

18

Adult Neurogenesis and Synaptic Rewiring in the Hippocampal Dentate Gyrus

Arjen van Ooyen[1], *Gertraud Teuchert-Noodt*[2], *Keren Grafen*[3] *and Markus Butz-Ostendorf*[4]

[1]VU University Amsterdam, Amsterdam, The Netherlands

[2]University of Bielefeld, Bielefeld, Germany

[3]Office of Biological Medicine, Bielefeld, Germany

[4]Biomax Informatics AG, Planegg, Germany

OUTLINE

The Rewiring Brain
DOI: http://dx.doi.org/10.1016/B978-0-12-803784-3.00018-4

1 INTRODUCTION

In higher mammals, including humans, cell proliferation (CP) of neuronal precursors continues to add new neurons to the hippocampal dentate gyrus (DG) throughout life.[1,2] Some of these neurons persist and develop into synaptically integrated, functional granule cells.[3,4] Initially, progenitor cells migrate from the germinative zone of the subgranular layer (sGL) of the DG into the granular layer (GL) of the DG,[1,5] where they form transient synapses and send out axons to the CA3 region of the hippocampus.[6,7] For at least two months after mitosis, young granule cells keep remodeling their synaptic connectivity.[8,9] Young granule cells extend their dendrites into the inner molecular layer (iML) of the DG,[10,11] expressing neurotrophic factors and competing with neighboring cells for presynaptic contacts.[10,11] The survival of young cells strongly depends on afferent support from other neurons[12] as well as on a balanced level of excitatory and inhibitory input.[13,14] Cells that fail to secure sufficient synaptic input undergo death by apoptosis.[15]

The need of young cells to become fully incorporated in the DG raises the question to what extent young granule cells induce structural changes in preexisting dentate networks. In particular, to what extent do young cells reorganize synaptic connectivity, in the sense of breaking preexisting connections that subsequently become rewired to other targets (synaptic rewiring)? This chapter describes the experimental and computational work previously reported on in Butz et al.[16] that we carried out to address these questions.

To examine the impact of CP on structural changes in synaptic connectivity, we used young adult gerbils (*Meriones unguiculatus*) that were raised under different conditions or subjected to treatment with methamphetamine (MA). Rearing conditions and MA treatment affect CP rate, so that we can investigate synaptic rewiring in animal groups with different CP rates.[17–19] Synaptic rewiring in the DG was measured by counting lysosomal accumulations (LAs), which appear in axon terminals that undergo autophagic degradation, as part of the synaptic remodeling process.[17,20,21] Surprisingly, we found that higher CP rates led to less synaptic rewiring.[16,18,19] In seeking to explain this seemingly counterintuitive finding, we used a novel neuronal network model that incorporates key structural processes in the DG, including synaptogenesis, neurogenesis, apoptosis, and synapse breakup.[22] Furthermore, each neuron in the model establishes its afferent excitatory and inhibitory connections so as to reach and maintain a desired level of electrical activity (homeostasis).[23,24] The results of the model revealed that this homeostatic regulation of activity may underlie the observed inverse relationship between CP rate and synaptic rewiring.[16]

In this chapter, we first briefly describe the experimental study and its results, followed by a detailed account of the structural plasticity model and its main findings. We discuss the experimental and simulation results and provide some perspectives on future experimental and modeling studies.

2 EXPERIMENTAL STUDY

To study the relationship between synaptic rewiring and CP in the adult DG, we used three groups of young adult gerbils (*Meriones unguiculatus*) with different CP rates. Synaptic rewiring was assessed by visualizing and counting LAs. The number of LAs served as a measure for the number of axon terminals undergoing degradation and remodeling.

2.1 Design of Study

We used a total of 24 male gerbils, divided into three different groups. In the first or control group, the animals were reared under (semi) natural conditions (NRC group). That is, after weaning, they were brought up in a social group and housed in large cages (1 m × 1 m) with tubes and boxes to hide in and wooden branches to play with. Such conditions lead to normal behavior and brain development.[25]

In the second group, the animals were instead kept in small, standard Macrolon cages (30 cm × 40 cm) under isolated and impoverished conditions (IRC group). Gerbils reared under such conditions develop anxiety and stereotypic behaviors.[25] Compared with NRC-reared animals, they also have strongly increased CP.[17,18]

In the third group, the animals were treated as in the IRC group, but on postnatal day 14 (p14) they were administered a single dose of MA (IRC/MA group). MA acutely suppresses CP in adult animals and juveniles.[18,19]

Thus, as regards CP rates, the NRC group had normal CP rates, the IRC group had increased CP rates, whereas in the IRC/MA group, the high CP rates of the IRC group were downregulated by MA.

2.2 Measuring Synaptic Rewiring

In gerbils and other species, synaptic rewiring begins by autophagic degradation of axon terminals,[17,20] in which transient LAs start to appear.[21] Autophagic degradation by LAs is a fully reversible and functional process and goes along with rewiring of axon terminals rather than a mere loss of synapses. To quantify LAs by light microscopy, we used Gallyas silver impregnation staining[26] (Fig. 18.1), which has proved to be a suitable method for highlighting LAs and characterizing periods of synaptic rewiring.[27]

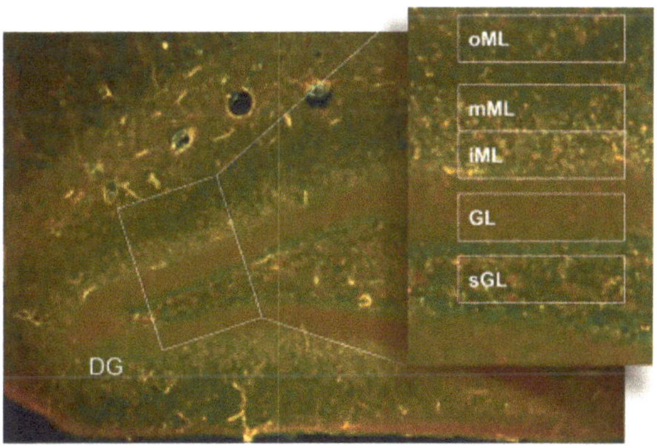

FIGURE 18.1 **Dark field light microscopy image of the hippocampal dentate gyrus (*DG*), with silver stained (Gallyas technique[26]) lysosomal accumulations appearing as bright granules.** As shown in the inset, measuring fields were positioned in the outer molecular layer (*oML*), the intermediate molecular layer (*mML*), the inner molecular layer (*iML*), the granular layer (*GL*), and the subgranular layer (*sGL*) of both the suprapyramidal and the infrapyramidal blade of the DG. Source: *Reproduced, with permission, from Butz M, Teuchert-Noodt G, Grafen K, van Ooyen A. Inverse relationship between adult hippocampal cell proliferation and synaptic rewiring in the dentate gyrus. Hippocampus. 2008;18(9):879–898, John Wiley & Sons.*

On p90, eight left and six right hemispheres from NRC animals, seven left and five right hemispheres from IRC animals, and seven left and eight right hemispheres from IRC/MA animals were successfully stained and quantified. Brain sections were viewed in dark field with a light microscope at 125 × magnification. Equal measuring fields were positioned in the iML, the intermediate molecular layer (mML), the outer molecular later (oML), the GL, and the sGL of the DG (Fig. 18.1). All measurements were carried out for both the suprapyramidal and the infrapyramidal blade of the DG in both hemispheres. The parietal cortex (PC) was chosen as a reference area.

To count the number of LAs automatically, we developed a Matlab algorithm for distinguishing the light silver grains from the dark background.[16] Analysis of variance was used to compare the LA means of the three different groups for each laminar measuring field. For those laminae with significantly different means between groups, a post hoc test was performed: a least significance difference test,[28] in case of equal variances in all groups, and a Tamhane test,[29] in case of unequal variances.

3 RESULTS OF EXPERIMENTAL STUDY

In the NRC group, the iML and mML had the highest number of LAs (Fig. 18.2), indicating pronounced synaptic rewiring in these layers. The oML and the sGL, as well as the PC reference area, showed only sparsely scattered staining of LAs, defining a baseline of synaptic reorganization within the hippocampus and cortex, respectively. The GL had the lowest LA count. This distribution of LAs across layers was present in both the suprapyramidal and the infrapyramidal blade of the DG and in both hemispheres.

In the IRC group, which had increased CP,[17,18] synaptic rewiring was reduced, as shown by the overall lower LA counts. Significant changes occurred in the iML and mML in both the suprapyramidal and the infrapyramidal blade of the left hemisphere. Other layers, including the PC, also exhibited significantly reduced LA counts, especially in the left hemisphere.

In the IRC/MA group, in which CP was suppressed by MA administration,[18,19] synaptic rewiring was increased compared with the IRC group, as shown by the elevated LA counts. The increase in LA counts in the IRC/MA group, occurring across all layers, was significant only in the left hemisphere. This is consistent with the observation that the CP-suppressing effect of MA is present in both hemispheres but most pronounced in the left hemisphere.[18]

Together, these results revealed an inverse relationship between CP and synaptic rewiring. Higher CP (IRC group) resulted in lower synaptic rewiring, whereas lower CP, either obtained by rearing conditions (NRC group) or by MA downregulation of CP in IRC animals (IRC/MA group), yielded higher synaptic rewiring.

4 COMPUTATIONAL MODEL

In an attempt to account for the experimental observation of an inverse relationship between CP and synaptic rewiring, we used a novel model of structural plasticity.[22] The

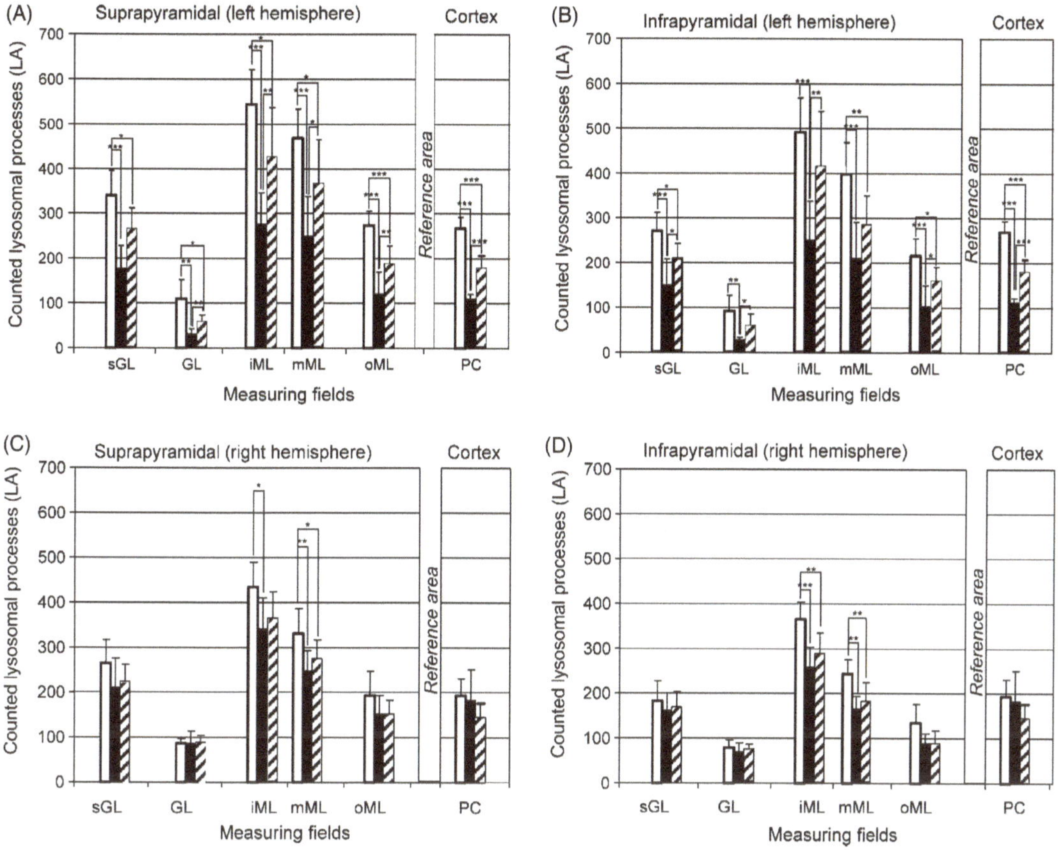

FIGURE 18.2 **Synaptic rewiring in the dentate gyrus (DG) of young adult gerbils. Synaptic rewiring was measured by counting lysosomal accumulations (*LAs*).** Each bar shows the average number of LAs in the DG of three different groups of gerbils: the NRC group (normal rearing conditions; *white bars*), the IRC group (impoverished rearing conditions; *black bars*) and the IRC/MA group (impoverished rearing conditions plus treatment with methamphetamine, MA; *shaded bars*). From the literature,[17–19] we know that the NRC group had normal cell proliferation (CP) rates, the IRC group increased CP rates, and the IRC/MA CP rates that were downregulated by MA. Results are shown from the inner molecular layer (*iML*), the intermediate molecular layer (*mML*), the outer molecular later (*oML*), the granular layer (*GL*), and the subgranular layer (*sGL*) of the DG in the suprapyramidal and infrapyramidal blades of both hemispheres. Also shown are LA counts in the parietal cortex (*PC*). The IRC group had lower LA counts than the NRC group, indicating reduced synaptic rewiring. In the IRC/MA group, as compared with the IRC group, the LA counts were increased again, indicating elevated synaptic rewiring. The changes were most pronounced in the left hemisphere. Source: *Reproduced, with permission, from Butz M, Teuchert-Noodt G, Grafen K, van Ooyen A. Inverse relationship between adult hippocampal cell proliferation and synaptic rewiring in the dentate gyrus. Hippocampus. 2008;18(9):879–898, John Wiley & Sons.*

model implements the structural processes of synaptogenesis, neurogenesis, apoptosis, and synapse breakup. In the model, each neuron has a variable number of axonal elements (representing axon terminals) and dendritic elements (representing dendritic spines) that can combine to form synapses. Importantly, the number of elements (and hence synaptic connectivity) is controlled by a neuron's own activity level so as to keep the average electrical activity of the neuron at a desired or set point level (homeostasis[23,24,30]). Rather than trying to produce an a priori defined behavior, we used the model in a bottom-up fashion to study the consequences of the interplay between neurogenesis, synaptogenesis, and neuronal activity.

The model network contains N neurons, NE excitatory neurons and $N - NE$ inhibitory neurons. In the Matlab computer program, the neurons are ordered so that cell indices $1 \leq i \leq NE$ denote the excitatory neurons, and cell indices $NE + 1 \leq i \leq N$ denote the inhibitory neurons. Because cells can be added or deleted, N and NE refer to the current number of neurons, whereas N^0 and NE^0 indicate the initial number of neurons. The excitatory neurons represent glutamatergic granule and mossy cells, and the inhibitory cells GABAergic basket cells (Fig. 18.3). The network is initialized with nonproliferating neurons, i.e., mossy and basket cells. Granule cells are added during the simulation. The initial number of neuron, N^0, is set to 30, of which 10% are inhibitory. A simple spiking neuron model is used to model neuronal electrical activity.

Each neuron contains A_i axonal elements (representing axon terminals or varicosities) and D_i dendritic elements (representing dendritic spines) (Fig. 18.4). Axonal and dendritic elements, collectively called synaptic elements, are labeled as either excitatory or inhibitory. Excitatory neurons carry exclusively excitatory axonal elements, and inhibitory neurons carry exclusively inhibitory axonal elements. Thus, A_i for $1 \leq i \leq NE$ refers to excitatory axonal elements, and A_i for $NE + 1 \leq i \leq N$ refers to inhibitory axonal elements. Every neuron, regardless of whether it is excitatory or inhibitory, has D_i^{exc} excitatory and D_i^{inh} inhibitory dendritic elements. The numbers of synaptic elements are treated as real-valued variables, but when synaptic elements are deleted (iteration Step 2, below) or used for synapse formation (iteration Step 3, below), the real-valued numbers are rounded off to their smallest integer values.

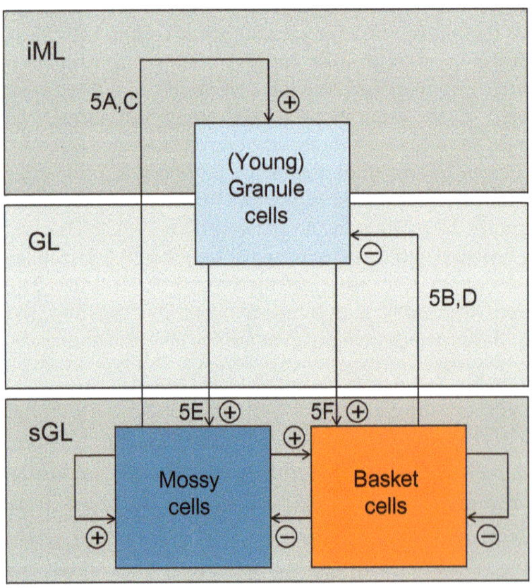

FIGURE 18.3 **Connectivity in the dentate gyrus (DG) of the model.** The model is composed of three populations of neurons: (1) excitatory granule cells, which are regularly added during the simulation to mimic neurogenesis; (2) excitatory mossy cells; and (3) inhibitory basket cells. Granule cells provide excitatory input to mossy and basket cells. In turn, mossy cells provide excitatory input to granule and basket cells, whereas basket cells provide inhibitory input to granule and mossy cells. Mossy and basket cells, but not granule cells, also form connections among themselves. With respect to the anatomy of the biological DG, mossy cells terminate on granule cells in the inner molecular layer (*iML*), basket cells predominantly form synapses on granule cells in the granular layer (*GL*), and granule cells target on mossy and basket cells in the subgranular layer (*sGL*). The numbers refer to the corresponding panels in Fig. 18.5.

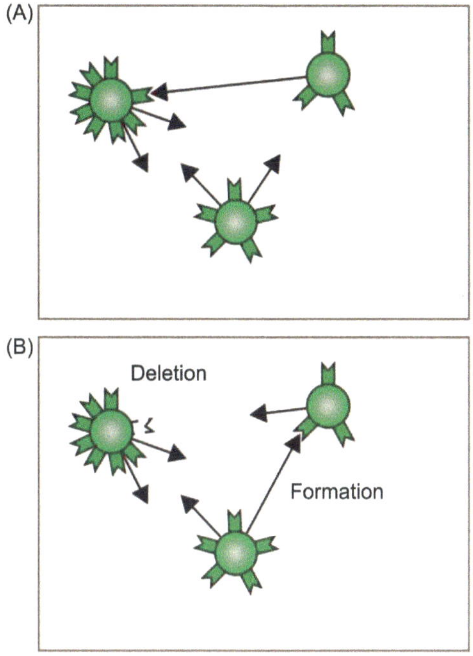

FIGURE 18.4 **Synaptic elements.** (A) In the model, we abstract away from detailed neuronal morphology and provide each neuron with a number of axonal elements (*lines with arrows* representing axon terminals) and dendritic elements (*green sockets* representing dendritic spines). A synapse consists of an axonal element that is bound to a dendritic element. In this example, two neurons are connected by a single synapse. (B) Synapse formation takes place by randomly combining vacant axonal and dendritic elements from different neurons. Synapse deletion occurs when a neuron removes a synaptic element (in this case, a dendritic element, indicated by the *broken contour line*) that is bound in a synapse. The counterpart (in this case, an axonal element) remains available for further synapse formation.

Each synapse consists of a single axonal and a single dendritic element (Fig. 18.4). An excitatory synapse consists of an excitatory axonal and an excitatory dendritic element. Likewise, an inhibitory synapse consists of an inhibitory axonal and an inhibitory dendritic element. Axonal and dendritic elements can be either free or bound in synapses. Free elements are available for synapse formation. Of the total numbers of axonal and dendritic elements—A_i, D_i^{exc}, and D_i^{inh}—there are a_i, d_i^{exc}, and d_i^{inh} free elements. The nonproliferating neurons are initialized with a set of synapses only, i.e., initially there are no free synaptic elements on these neurons.

Synapse strength is binary, i.e., it has a value of one when a synapse is present and zero when not. Multiple synapses can exist between neurons. Thus, the connection strength $C_{i,j}$ is a positive discrete number, for both inhibitory and excitatory connections, denoting the total number of synapses from presynaptic neuron j to postsynaptic neuron i. A neuron cannot connect to itself. As biological networks are not fully connected,[31] and most cells are connected by only a few synapses,[32] the network is initialized by randomly distributing $\frac{2}{3} \cdot N^0 \cdot N^0$ synapses onto the $N^0 \cdot N^0$ possible neuron combinations.

The numbers of axonal and dendritic elements change depending on the neuron's own level of electrical activity. Changes in the number of synaptic elements, and thus synaptic connectivity, occur on a slower time scale than changes in electrical activity.[9,33,34] Accordingly, we define two time scales in the model, a fast time scale t for activity changes and a slow morphogenetic time scale T for changes in connectivity, number of cells, and number of synaptic elements. Both time scales are discrete. The connectivity update $T \rightarrow T+1$ is done after every 1000 time steps in t. If we assume that spines and axonal branches remodel within a few days,[33,34] a time step in T roughly corresponds to 1 day in a biological network.

A simulation of the model consists of iterations through the following six steps:

Step 1: Calculating Neuronal Activity

The probability that neuron i fires at time t depends on its membrane potential X_i^t according to a standard sigmoidal function

$$F_i^t(X_i^t) = \frac{1}{1+e^{(\theta - X_i^t)/\beta}}, \qquad 1 \le i \le N \tag{18.1}$$

where F_i^t is the firing probability, θ represents the firing threshold, and β is the steepness of the sigmoidal function. The firing state of each neuron, z_i^t, which is either 1 or 0, indicates whether neuron i produced a spike at time t. The firing state is determined by the firing probability:

$$z_i^t = \begin{Bmatrix} 0 & \text{if} & F_i^t < r_i \\ 1 & \text{if} & F_i^t \ge r_i \end{Bmatrix}, \qquad 1 \le i \le N \tag{18.2}$$

where r_i is a random number drawn from the uniform distribution [0, 1]. Input from firing excitatory cells increases and input from firing inhibitory cells decreases the membrane potential of a postsynaptic neuron. Thus, the membrane potential of neuron i is given by

$$X_i^t = \sum_{j=1}^{NE} C_{i,j} \cdot z_j^{t-1} - \sum_{j=NE+1}^{N} C_{i,j} \cdot z_j^{t-1}, \qquad 1 \le i \le N \tag{18.3}$$

where z_i^0 is randomly set to 1 or 0 at the start of the simulation.

Step 2: Adding and Deleting Synaptic Elements and Deleting Synapses

As morphological changes are mediated by the intracellular Ca^{2+} concentration,[35–37] reflecting the long-term average of the neuron's firing activity,[38] the changes in the number of axonal and dendritic elements at morphogenetic time T are determined by the average s_i of the firing probability over the last 1000 time steps in t. Specifically, the changes in the number of axonal and dendritic elements are driven by the deviation Δs_i from a set point level that is arbitrarily set at 0.5 for all neurons:

$$\Delta A_i = v \cdot \Delta s_i \cdot A_i, \qquad 1 \le i \le N \tag{18.4}$$

$$\Delta D_i^{\text{exc}} = -v \cdot \Delta s_i \cdot D_i^{\text{exc}}, \qquad 1 \le i \le N \tag{18.5}$$

$$\Delta D_i^{\text{inh}} = v \cdot \Delta s_i \cdot D_i^{\text{inh}}, \qquad 1 \le i \le N \tag{18.6}$$

where v is the rate of change ($0 < v \ll 1$) and $\Delta s_i = s_i - 0.5$. Eq. (18.4) reflects that a neuron enhances its number of axonal elements with increased neuronal activity.[39,40] Eqs. (18.5) and (18.6) reflect that a neuron reduces the number of excitatory dendritic elements[41,42] and enhances the number inhibitory dendritic elements[43] when its average neuronal activity becomes too high, and that it does the opposite when its average activity becomes too low. If ΔA_i, ΔD_i^{exc}, or ΔD_i^{inh} is positive, new synaptic elements are added to neuron i. If they are negative, a corresponding number of synaptic elements are deleted from neuron i. Synaptic elements are randomly selected for deletion, regardless of whether they are bound in a synapse. If a deleted element was bound in a synapse, the synaptic connection is deleted as well, but the axonal or dendritic counterpart of the deleted element is kept and remains available for synapse formation.

The activity-dependent changes in the number of dendritic elements contribute to the homeostasis of neuronal activity. When a neuron's activity is below the set point level of 0.5, the increase in the number of free excitatory dendritic elements enhances the neuron's likelihood of making excitatory input connections. Likewise, the decrease in the number of inhibitory dendritic elements either reduces the neuron's likelihood of making inhibitory input connections or, when the deleted dendritic elements were bound in synapses, breaks up inhibitory connections. By contrast, when a neuron's activity is above the set point level, the increase in the number of free inhibitory dendritic elements enhances the neuron's likelihood of making inhibitory input connections. Likewise, the decrease in the number of excitatory dendritic elements either reduces the neuron's likelihood of making excitatory input connections or, when the deleted dendritic elements were bound in synapses, breaks up excitatory connections.

After the number of synaptic elements is changed, the total numbers of elements (A_i, D_i^{exc}, and D_i^{inh}) and the numbers of free elements (a_i, d_i^{exc}, and d_i^{inh}) are updated.

Step 3: Synapse Formation

For the formation of excitatory synapses, we consider the set of all a_j free axonal elements of all excitatory neurons j, $1 \leq j \leq NE$, and the set of all d_i^{exc} free dendritic elements of all neurons i, $1 \leq i \leq N$ (i.e., both excitatory and inhibitory neurons) and determine which set has the smaller number of elements. To make synaptic connections, we then fully distribute the smaller set of free elements randomly onto the larger set of free elements. Synapses can in principle be established between all types of neurons, but granule cells are excluded from making connections among each other, and neurons are not allowed to form synapses onto themselves.

Likewise, for the formation of inhibitory synapses, we consider the set of all a_j free axonal elements of all inhibitory neurons j, $NE+1 \leq j \leq N$, and the set of all d_i^{inh} free dendritic elements of all neurons i, $1 \leq i \leq N$ (i.e., both excitatory and inhibitory neurons) and determine which set has the smaller number of elements. To make synaptic connections, we fully distribute the smaller set of free elements randomly onto the larger set of free elements. Again, neurons are not allowed to form synapses onto themselves.

After synapse formation, the numbers of free synaptic elements —a_i, d_i^{exc}, and d_i^{inh}—are updated by the number of synaptic elements that got bound into synapses. Note that the total numbers of synaptic elements—A_i, D_i^{exc}, and D_i^{inh}—are not affected by synapse formation.

Step 4: Cell Deletion

As the survival of a cell depends on its level of electrical activity,[44–46] we assume that changes in the number of axonal and dendritic elements take place only in the range $0.35 \leq s_i \leq 0.85$, and that outside this range, a neuron will undergo apoptosis. At each morphogenetic time T, we check whether a neuron is inside or outside this range. Young neurons get a chance to become embedded into the network by excluding them from apoptosis for 30 time steps in T. If a neuron is deleted, all its synaptic elements are deleted as well. However, the synaptic elements of neurons that were connected to the deleted neuron remain. After cell deletion, the numbers of free synaptic elements on the remaining neurons are updated. Note that there is no change in the total number of synaptic elements on the remaining neurons.

Step 5: Adding New Cells

In three different sets of simulations, we add a new excitatory granule cell every 1, 5, or 10 time steps in T. These CP rates are denoted as CP 1:1, CP 1:5, and CP 1:10, respectively. As mentioned, there are no granule cells at the beginning of the simulation; all granule cells are added during the simulation. We consider CP 1:10 as a normal CP rate, and simulations with this rate represent NRC animals. Both CP 1:1 and CP 1:5 are considered as increased proliferation rates, and simulations with these rates correspond to IRC and IRC/MA animals, respectively. A new granule cell is endowed with a certain number of free axonal and dendritic elements. Each young granule cell i receives $a_i = \frac{2}{3} \cdot N^0$ free (excitatory) axonal elements and $\frac{2}{3} \cdot N^0$ free dendritic elements. Of the free dendritic elements, $d_i^{\mathrm{exc}} = \frac{NE^0}{N^0} \cdot \frac{2}{3} \cdot N^0$ are excitatory and $d_i^{\mathrm{inh}} = \frac{N^0 - NE^0}{N^0} \cdot \frac{2}{3} \cdot N^0$ are inhibitory. The young neurons do not carry any synapses initially. We also ran simulations in which young neurons did not have any synaptic elements initially. These simulations produced similar results, provided that the velocity v (see Eqs. (18.4)–(18.6)) was increased.

Step 6: Updating Connectivity

Before returning to the calculation of neuronal activity (iteration Step 1, above), we need to update the connectivity $C_{i,j}$. First, the number of excitatory neurons, NE, and the total number of neurons, N, are updated by the number of neurons that were added and deleted. Then, for each synapse between neurons j and i that lost either its axonal or dendritic element, the total number of synapses between these two neurons, $C_{i,j}$, is subtracted by one. Furthermore, all synapse numbers $C_{i,j}$ are set to zero if either neuron j or neuron i was deleted. Finally, for all neurons j that were able to bind one or more free axonal elements to free dendritic elements of neuron i, synapse number $C_{i,j}$ is updated accordingly.

5 RESULTS OF COMPUTATIONAL MODEL

In the model, the newly added granule cells, as they become integrated into the network, compete for free axonal and dendritic elements on the existing mossy cells (excitatory neurons) and basket cells (inhibitory neurons) to make synapses. The young granule cells, as well as the other neurons, adapt their number of synaptic elements so as to reach and maintain the set point level of electrical activity (homeostasis) (Eqs. (18.4)–(18.6)). A dynamical steady state is reached when all neurons have their average level of electrical activity around the set point.

The time course of the number of free axonal elements in the model corresponds to how the supply of excitatory and inhibitory axon terminals develops in the iML, GL, and sGL of the DG (Fig. 18.3). The free axonal elements in the model are either newly formed or arise from the breakdown of existing synapses. The number of free axonal elements in the model (both excitatory and inhibitory elements) that result from the breakdown of synapses can be taken to represent the LA count in the different layers of the DG in the experiment. LAs were counted on p90, which roughly corresponds to morphogenetic time $T = 90$ in the model. Simulations were carried out for different granule CP rates. Simulations with a normal CP rate (CP 1:10) compare with the NRC group in the experiment, whereas simulations with a strongly increased rate (CP 1:1) correspond to the IRC group. Simulation runs with an intermediate rate (CP 1:5) may represent the IRC/MA group, in which the high CP rate of IRC animals was (partly) suppressed by MA treatment.

5.1 Number of Free Excitatory and Inhibitory Axonal Elements

The total number of free excitatory axonal elements, provided by the mossy cells and comprising both newly formed axonal elements and axonal elements arising from the breakdown of synapses, showed a clear dependence on CP rate (Fig. 18.5A). For all CP rates, the total number of free excitatory axonal elements decreased with time, but free excitatory axonal elements were most quickly exhausted for the highest CP rate (CP 1:1). Thus, the higher the influx of new neurons, the fewer free excitatory axonal elements were available. Consequently, at morphogenetic time $T = 90$, the simulation with CP 1:10 had the highest and the simulation with CP 1:1 had the lowest number of free excitatory axonal elements, whereas the simulation with CP 1:5 showed an intermediate number of axonal elements.

A similar dependence on CP rate was observed for the total number of free inhibitory axonal elements, which are provided by the basket cells (Fig. 18.5B). Compared with the total number of free excitatory axonal elements, the total number of free inhibitory axonal elements showed an even more pronounced difference between high CP rate (CP 1:1) and low CP rate (CP 1:5 and CP 1:10). Whereas for CP 1:1 the number of free inhibitory axonal elements was already at a minimum around morphogenetic time $T = 25$, it took approximately 265 and 360 morphogenetic time steps before the number of elements was at its lowest for CP 1:5 and CP 1:10, respectively.

5.2 Number of Remodeling Excitatory and Inhibitory Axonal Elements

Of the total number of free axonal elements, we can determine how many free elements come from the breakdown of synapses. Synaptic rewiring in the network is driven by this pool of free axonal elements (called remodeling elements), and the sum of both excitatory and inhibitory remodeling axonal elements may be taken to represent the LA count in the experimental study. Although the changes in the number of remodeling axonal elements (Figs. 18.5C and 18.5D) were more erratic than those in the total number of free axonal elements (Figs. 18.5A and 18.5B), the number of remodeling axonal elements showed the same dependence on CP rate as the total number of free axonal elements. At $T = 90$ and before, the number of remodeling excitatory axonal elements provided by the mossy cells was almost always the highest for CP 1:10, followed by CP 1:5 and CP 1:1 (Fig. 18.5C). Like the total number of free inhibitory axonal elements, the number of remodeling inhibitory axonal elements, provided by the basket cells, was more quickly exhausted for high CP rates, with the simulation with CP 1:1 reaching its lowest number of remodeling axonal elements at around $T = 25$ and the simulation with CP 1:10 arriving at this level at around $T = 360$ (Fig. 18.5D).

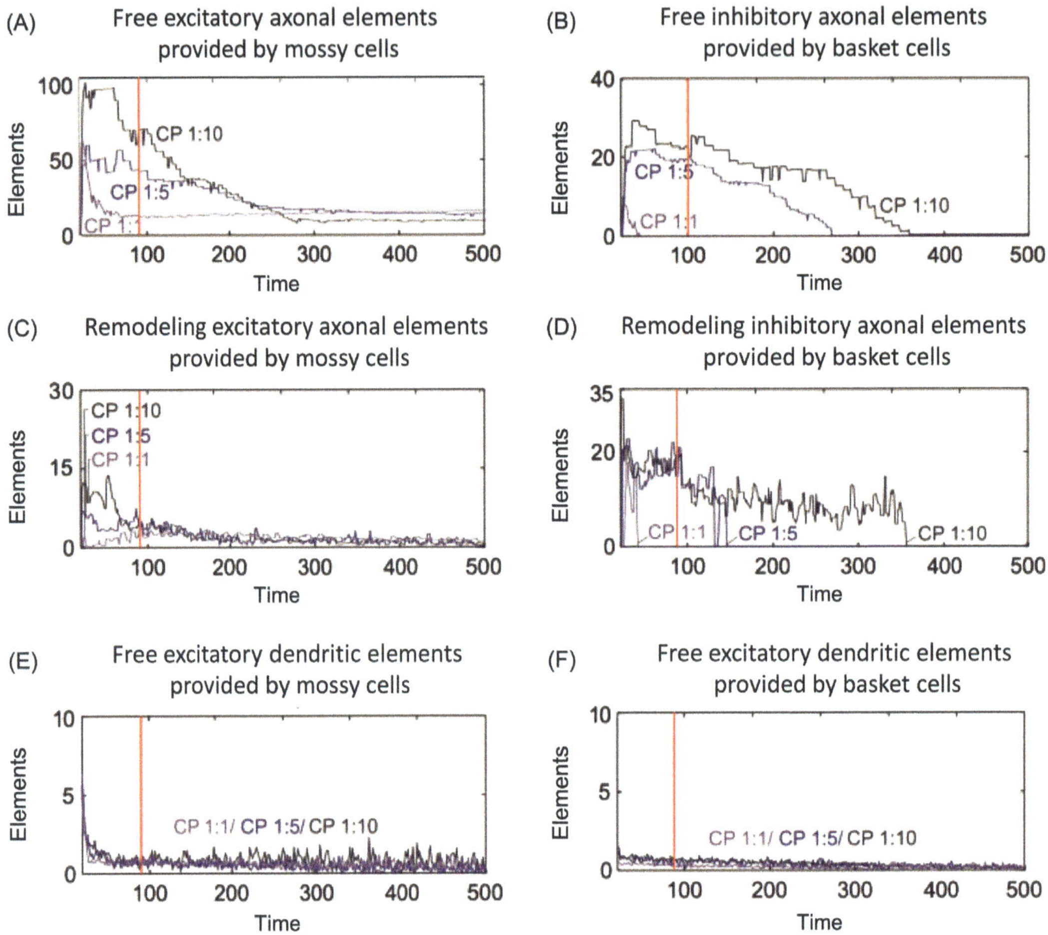

FIGURE 18.5 **Synaptic rewiring in the model.** Average number of synaptic elements per cell over morphogenetic time *T* for different cell proliferation (*CP*) rates. CP 1:1, CP 1:5, or CP 1:10 indicate that one granule cell is added every 1, 5, or 10 morphogenetic time steps, respectively. One morphogenetic time step corresponds to roughly 1 day in the development of the dentate gyrus (DG) of gerbils. Postnatal day 90, the day at which the DGs of the animals were stained, is marked by a *vertical red line*. (A) The total number of free excitatory axonal elements offered by mossy cells. This number comprises both free excitatory axonal elements that are newly formed and those that arise from the breakdown of existing synapses. (B) The total number of free inhibitory axonal elements, provided by basket cells. (C) The number of remodeling excitatory axonal elements offered by mossy cells. This is the number of free excitatory axonal elements that arise from the breakdown of existing synapses. (D) The number of remodeling inhibitory axonal elements, provided by basket cells. (E) The total number of free excitatory dendritic elements provided by mossy cells, which are target cells for young granule cells. (F). The total number of free excitatory dendritic elements provided by basket cells, which are target cells for young granule cells, too. The total numbers of free axonal elements (A, B) and the numbers of remodeling axonal elements (C, D) were dependent on CP rate, with higher values associated with lower CP rates, whereas the total numbers of free dendritic elements (E, F) were independent of CP rate. Source: *Reproduced, with permission, from Butz M, Teuchert-Noodt G, Grafen K, van Ooyen A. Inverse relationship between adult hippocampal cell proliferation and synaptic rewiring in the dentate gyrus. Hippocampus. 2008;18(9):879–898, John Wiley & Sons.*

5.3 Inverse Relationship Between CP Rate and Number of Free Axonal Elements

Taken together, the results revealed that during an extended middle period of the simulations, high CP rates were accompanied by a low number of free axonal elements (both total number of free elements and number of remodeling elements), and conversely that low CP rates were associated with a high number of free axonal elements. How did this inverse relationship between CP rate and number of free axonal elements arise in the model? When young granule cells enter the network, they make excitatory and inhibitory input synapses with free axonal elements (provided by mossy and basket cells) so as to reach and maintain their set point level of electrical activity (Eqs. (18.5) and (18.6)). As a result, the number of free axonal elements decreases. The higher the number of granule cells that enter the network per unit of time (i.e., CP rate), the higher the demand for free axonal elements and the stronger the decrease in number of free axonal elements. Because granule cells can make input synapses both with newly formed axonal elements and with remodeling axonal elements, both types of free axonal elements decrease over time.

When the activity of neurons exceeds the set point value, they will decrease their number of excitatory dendritic elements, both free dendritic elements and those bound in synapses (Eq. (18.5)). In the beginning of the simulation depicted in Fig. 18.5C, there was an increase in the number of remodeling excitatory axonal elements for CP 1:1, which arose from a breakup of excitatory synapses to compensate for the strong increase in network activity as a result of the high influx of excitatory granule cells.

5.4 Number of Free Dendritic Elements

Young granule cells make not only incoming connections, through the free excitatory and inhibitory axonal elements provided by the mossy and basket cells, respectively, but also outgoing connections terminating on the excitatory dendritic elements of mossy and basket cells. Both the total number of free excitatory dendritic elements on mossy cells (Fig. 18.5E) and the total number of free excitatory dendritic elements on basket cells (Fig. 18.5F) are much lower than the total number of free axonal elements (Figs. 18.5A and 18.5B). Moreover, unlike the total number of free axonal elements, they do not show a dependence on CP rate. Both the course over time of the number of free dendritic elements and the number of elements at any one time point are independent of CP rate. The reason for this is that free axonal elements on granule cells outnumber free dendritic elements on basket and mossy cells already for normal CP rates; therefore, increasing the CP rate will hardly enhance competition for dendritic elements. Thus, formation of input synapses of young granule cells is CP rate dependent, whereas the formation of output synapses are not affected by CP rate.

6 DISCUSSION

In the experimental study with gerbils, we found an inverse relationship between CP rate and synaptic rewiring in the DG. The higher the CP rate, the lower the amount of synaptic rewiring. A similar inverse relationship was observed in the neuronal network model between CP rate and the number of (remodeling) axonal elements. In the model, the need of neurons to reach and maintain a set point level of electrical activity (homeostasis) gives rise to reduced synaptic rewiring when CP rates are high. Young granule cells seek to establish input connections with free axonal elements so as to reach their set point level of electrical activity. Hence, the more granule cells that enter the network, the quicker the pool of free axonal elements, composed of both newly formed elements and elements arising from synapse breakdown (i.e., remodeling axonal elements), is exhausted.

6.1 Experimental Study

In all animal groups, the LA counts showed a characteristic distribution across the different dentate laminae, with the iML exhibiting the highest and the GL the lowest count (Fig. 18.2). What cell types and fibers may predominantly be affected by the synaptic rewiring indicated by these LA counts? Excitatory mossy cells and, to a lesser extent, inhibitory interneurons form an extended network of commissural and associative (C/A) fibers terminating on granule cells in the iML.[47,48] We hypothesize that C/A fibers together with septal afferents are the main source for synaptic rewiring in the iML. This hypothesis is supported by experimental findings showing that septal afferents and C/A fibers show pronounced reactive sprouting in response to perforant path lesions.[49,50]

In contrast to the iML, the oML shows a low LA count, comparable to the background LA count in the PC (Fig. 18.2). The dominating afferents terminating on granule cells in the oML are perforant path fibers, which thus may hardly undergo synaptic rewiring. Instead, perforant path fibers may increase in density to occupy young granule cells, a conjecture that is consistent with the observation that, unlike the iML, the oML grows thicker in the adult hippocampus.[51,52]

In the sGL, we expect that GABAergic interneurons predominantly contribute to the synaptic rewiring.[53] The low LA count in the GL (Fig. 18.2), which contains the cell bodies of mature granule cells, indicates that inhibitory basket cells, which form a dense plexus around granule cell somata, are hardly affected by synaptic rewiring.

How can young granule cells influence synaptic rewiring? Two-photon imaging studies have revealed that young granule cells extend dendritic filopodia toward nearby synapses so as to become occupied by axon terminals.[8,9] In addition, cell culture studies have shown that free postsynaptic receptor plates secrete neurotrophic factors to attract axon terminals for synapse formation.[54] Moreover, breaking of existing synapses has been observed[34,55] and correlates with the morphological development of dendritic spines.[56] Taken together, these experimental findings indicate that young granule cells may be able to induce synaptic rewiring.

What are the possible functional roles of synaptic rewiring in the DG? In general, synaptic rewiring adds an extra degree of freedom to neural plasticity[57] and enhances memory capacity.[58] In the DG in particular, neurogenesis and synaptic rewiring may be involved in pattern separation, memory resolution, and encoding temporal context,[59] as well as in clearance of memory traces.[60] An optimal balance between learning and forgetting may depend on neurogenesis-induced synaptic rewiring. As our modeling suggests, increased neurogenesis may not necessarily be beneficial to hippocampal function, as young granule cells bind to free axon terminals and hence reduce the capacity of the network to rewire. Moreover, in a previous modeling study, we found that a permanent increase in the rate of CP led to a diminished embedding and survival of young granule cells,[22] in agreement with experimental findings.[61]

The increased CP rates in IRC animals[17,18] are not in contradiction with other studies showing that enriched environments enhance CP rates.[62,63] In these studies, animals were housed in standard cages and only at a late stage in their development transferred to an enriched environment. In contrast, the animals that we used were housed their whole lives after weaning under either normal or isolated conditions, giving rise to a completely different animal model.

6.2 Modeling Study

Most traditional neural network models have a fixed connectivity structure, with plasticity merely arising from changes in the connection strength of existing synapses (synaptic plasticity). In contrast, our model implements structural plasticity, whereby unconnected neurons can become connected, connected neurons can become disconnected, and the number of synapses between connected neurons can change. The model is unique in the way it treats neurons as having separate axonal and dendritic elements that can combine to form synapses. The number of axonal and dendritic elements that a neuron has depends on its own level of electrical activity, in such a manner that a neuron strives to reach and maintain a desired level of electrical activity (homeostasis). As a result of the way synapses are modeled, synaptic rewiring is feasible, in the sense that when an axonal or dendritic element of a synapse has been deleted, its remaining counterpart can bind to a free synaptic element and hence form a new connection.

The model results were robust to changes in parameter values, in that the inverse relationship between CP rate and number of free axonal elements was not critically dependent on a particular choice of parameter values. The parameter setting influenced the quantitative outcomes of the simulations but not the qualitative results. Increasing the number of mossy cells increased the number of free excitatory axonal elements both in the beginning of the simulation and in the steady state. The same held true for the number of basket cells and the number of free inhibitory axonal elements. Simulations with $N_0 = 60$ or 100, instead of the default 30, had no impact on the qualitative outcomes of the model. Increasing the value of v, the rate at which the number of synaptic elements responds to changes in activity levels, accelerated connectivity changes in the network and meant that steady state was reached earlier. However, when v was too large, the network did not always reach steady state and activity homeostasis, because the network's response to adding new granule cells and deleting cells became too strong.

6.3 Future Experimental Studies

Future experimental studies may look at how CP rate affects synaptic rewiring in hippocampal areas CA3 or CA1. Based on our modeling study, we expect that CP rate-dependent synaptic rewiring as observed in the DG may not occur in the CA3 or CA1. In the model, the impact of CP rate on synaptic rewiring comes about by modulating the supply of axonal elements on the granule cells' presynaptic neurons, rather than by influencing the supply of dendritic elements on the granule cells' postsynaptic targets.

As mentioned earlier, our simulation results suggest that high CP rates could be detrimental to hippocampal function as a result of young granule cells exhausting the pool of free axon terminals and so reducing the capacity of the network to rewire. It would be interesting to investigate whether the advantages that adult-generated cells confer to cognition will diminish when CP rates are strongly increased by pharmacological or genetic approaches.

Further experimental studies are also needed to assess the extent to which acquiring and maintaining firing rate homeostasis guides the formation and breaking of connections as young granule cells become embedded in the DG. Without homeostatic mechanisms in place, insertion of new neurons into adult neural circuits may lead to a potentially adverse increase in excitation and hence impairment of circuit function.[14] Recent studies have lent support to a fundamental role of homeostasis in synapse formation by showing that under conditions of elevated CP, inhibitory innervation in the DG was enhanced, whereas under conditions of decreased CP, it was reduced.[64,65]

Finally, it would be important to determine whether the inverse relationship between CP rate and synaptic rewiring also occurs in the DG of other species than gerbils. In mice, no effect of deprived rearing was found on rates of synaptic turnover.[66] Furthermore, the effect of rearing conditions on synaptic remodeling may also be investigated in other brain areas than the hippocampus.[67]

6.4 Future Modeling Studies

In the model, we did not allow granule cells to form connections among themselves, in line with DG neuroanatomy.[68] Interestingly, when we did permit granule cells to form recurrent connections, the number of recurrent synapses remained rather low, provided CP rates were normal. This result arises as a consequence of competition between axonal elements on granule cells and those on mossy cells. However, when CP rates were high, the supply of axonal elements on mossy cells was not sufficient to fulfill the demand of granule cells for excitatory input; therefore, cells started to form recurrent connections. This reinforcement of recurrent circuitry may make the hippocampus prone to epilepsy.[69] It would therefore be interesting to investigate, both in the model and experimentally, whether a high CP rate would lead to a higher incidence of epileptic seizures.

Various model studies have investigated the influence of neurogenesis on learning and memory. One modeling study showed that adding new neurons to a network avoids catastrophic interference between sequentially stored patterns.[70] Other studies demonstrated that by regulating the rates of neurogenesis and apoptosis, a network can optimize storage and retrieval of novel information.[71,72] In yet another study, it was shown that neurogenesis and apoptosis lead to a more flexible network that can learn new information faster on top of existing memory content.[73] In these and similar models,[74–76] however, synapses are represented as single elements whose strengths can be modified by a learning rule.

Consequently, the formation of synapses that goes along with adding new cells may be viewed as simply increasing synaptic strengths that were zero before. Unlike our model, in which synapses are represented as consisting of separate axonal and dendritic elements, these models do not permit true synaptic rewiring, in the sense that a broken connection can become redirected to a new target. In our model, the capacity for synapse formation is dependent not only on de novo formation of synaptic elements but also on synapse deletion, which makes synaptic elements available that were previously bound into synapses. Thus, the full potential of neurogenesis to enhance learning and memory capabilities, which may depend on neurogenesis-induced synaptic rewiring, remains to be explored by future modeling studies.

As regards homeostatic regulation, we conjecture that the attempt of neurons to maintain their average electrical activity at a particular level is a main organizing principle of neuronal networks.[30,77,78] Homeostatic regulation may not only guide the formation of synaptic connectivity during development and govern the synaptic integration of newborn cells into the DG in adulthood but may also drive the compensatory structural changes after loss of input caused by lesions, stroke, or neurodegeneration. We have used minor variants of the model presented here to explain changes in visual cortex after focal retinal lesions,[78] to account for alterations in global network topology after deafferentation and focal stroke,[79] and to show what stimulation regime would most effectively promote network repair after cortical lesions.[80] We expect that future models will continue to play a crucial role in exploring how the striving of neurons for activity homeostasis may self-organize synaptic connectivity and neuronal properties.[81,82]

References

1. Eriksson PS, Perfilieva E, Björk-Eriksson T, et al. Neurogenesis in the adult human hippocampus. *Nat Med.* 1998;4(11):1313–1317.
2. Ming GL, Song H. Adult neurogenesis in the mammalian central nervous system. *Annu Rev Neurosci.* 2005;28:223–250.
3. Kempermann G, Wiskott L, Gage FH. Functional significance of adult neurogenesis. *Curr Opin Neurobiol.* 2004;14(2):186–191.
4. van Praag H, Schinder AF, Christie BR, Toni N, Palmer TD, Gage FH. Functional neurogenesis in the adult hippocampus. *Nature.* 2002;415(6875):1030–1034.
5. Altman J, Bayer SA. Migration and distribution of two populations of hippocampal granule cell precursors during the perinatal and postnatal periods. *J Comp Neurol.* 1990;301(3):365–381.
6. Markakis EA, Gage FH. Adult-generated neurons in the dentate gyrus send axonal projections to field CA3 and are surrounded by synaptic vesicles. *J Comp Neurol.* 1999;406(4):449–460.
7. Hastings NB, Gould E. Rapid extension of axons into the CA3 region by adult-generated granule cells. *J Comp Neurol.* 1999;413(1):146–154.
8. Zhao C, Teng EM, Summers RG, Ming GL, Gage FH. Distinct morphological stages of dentate granule neuron maturation in the adult mouse hippocampus. *J Neurosci.* 2006;26(1):3–11.
9. Toni N, Teng EM, Bushong EA, et al. Synapse formation on neurons born in the adult hippocampus. *Nat Neurosci.* 2007;10(6):727–734.
10. Cameron HA, Woolley CS, McEwen BS, Gould E. Differentiation of newly born neurons and glia in the dentate gyrus of the adult rat. *Neuroscience.* 1993;56(2):337–344.
11. Seki T, Arai Y. Highly polysialylated neural cell adhesion molecule (NCAM-H) is expressed by newly generated granule cells in the dentate gyrus of the adult rat. *J Neurosci.* 1993;13(6):2351–2358.
12. Linden R. The survival of developing neurons: a review of afferent control. *Neuroscience.* 1994;58(4):671–682.

13. Lehmann K, Butz M, Teuchert-Noodt G. Offer and demand: proliferation and survival of neurons in the dentate gyrus. *Eur J Neurosci*. 2005;21(12):3205–3216.
14. Meltzer LA, Yabaluri R, Deisseroth K. A role for circuit homeostasis in adult neurogenesis. *Trends Neurosci*. 2005;28(12):653–660.
15. Sierra A, Encinas JM, Deudero JJ, et al. Microglia shape adult hippocampal neurogenesis through apoptosis-coupled phagocytosis. *Cell Stem Cell*. 2010;7(4):483–495.
16. Butz M, Teuchert-Noodt G, Grafen K, van Ooyen A. Inverse relationship between adult hippocampal cell proliferation and synaptic rewiring in the dentate gyrus. *Hippocampus*. 2008;18(9):879–898.
17. Keller A, Bagorda F, Hildebrandt K, Teuchert-Noodt G. Effects of enriched and of restricted rearing on both neurogenesis and synaptogenesis in the hippocampal dentate gyrus of adult gerbils (*Meriones unguiculatus*). *Neurol Psychiatr Brain Res*. 2000;8:101–107.
18. Hildebrandt K, Teuchert-Noodt G, Dawirs RR. A single neonatal dose of methamphetamine suppresses dentate granule cell proliferation in adult gerbils which is restored to control values by acute doses of haloperidol. *J Neural Transm (Vienna)*. 1999;106(5–6):549–558.
19. Teuchert-Noodt G, Dawirs RR, Hildebrandt K. Adult treatment with methamphetamine transiently decreases dentate granule cell proliferation in the gerbil hippocampus. *J Neural Transm (Vienna)*. 2000;107(2):133–143.
20. Dawirs RR, Teuchert-Noodt G, Kacza J. Naturally occurring degrading events in axon terminals of the dentate gyrus and stratum lucidum in the spiny mouse (Acomys cahirinus) during maturation, adulthood and aging. *Dev Neurosci*. 1992;14(3):210–220.
21. Dawirs RR, Teuchert-Noodt G, Hildebrandt K, Fei F. Granule cell proliferation and axon terminal degradation in the dentate gyrus of gerbils (Meriones unguiculatus) during maturation, adulthood and aging. *J Neural Transm (Vienna)*. 2000;107(6):639–647.
22. Butz M, Lehmann K, Dammasch IE, Teuchert-Noodt G. A theoretical network model to analyse neurogenesis and synaptogenesis in the dentate gyrus. *Neural Netw*. 2006;19(10):1490–1505.
23. van Ooyen A, van Pelt J. Activity-dependent outgrowth of neurons and overshoot phenomena in developing neural networks. *J Theor Biol*. 1994;167:27–43.
24. van Ooyen A, van Pelt J, Corner MA. Implications of activity dependent neurite outgrowth for neuronal morphology and network development. *J Theor Biol*. 1995;172(1):63–82.
25. Winterfeld KT, Teuchert-Noodt G, Dawirs RR. Social environment alters both ontogeny of dopamine innervation of the medial prefrontal cortex and maturation of working memory in gerbils (*Meriones unguiculatus*). *J Neurosci Res*. 1998;52(2):201–209.
26. Gallyas F, Wolff JR, Böttcher H, Zaborszky L. A reliable and sensitive method to localize terminal degeneration and lysosomes in the central nervous system. *Stain Technol*. 1980;55(5):299–306.
27. Wolff JR, Leutgeb U, Holzgraefe M, Teuchert G. Synaptic remodelling during primary and reactive synaptogenesis. In: Rahmann H, ed. *Fundamentals of Memory Formation: Neuronal Plasticity and Brain Function*. Stuttgart, New York: Gustav Fischer Verlag; 1989:68–82.
28. Hayter AJ. The maximum familywise error rate of Fisher's least significant difference test. *J Am Stat Assoc*. 1986;81(396):1000–1004.
29. Tamhane AC. A comparison of procedures for multiple comparisons of means with unequal variances. *J Am Stat Assoc*. 1979;74(366):471–480.
30. van Ooyen A. Activity-dependent neural network development. *Network-Comp Neural*. 1994;5:401–423.
31. Stepanyants A, Hof PR, Chklovskii DB. Geometry and structural plasticity of synaptic connectivity. *Neuron*. 2002;34(2):275–288.
32. Peters A. Examining neocortical circuits: some background and facts. *J Neurocytol*. 2002;31(3-5):183–193.
33. Trachtenberg JT, Chen BE, Knott GW, et al. Long-term in vivo imaging of experience-dependent synaptic plasticity in adult cortex. *Nature*. 2002;420(6917):788–794.
34. De Paola V, Holtmaat A, Knott G, et al. Cell type-specific structural plasticity of axonal branches and boutons in the adult neocortex. *Neuron*. 2006;49(6):861–875.
35. Kater SB, Mattson MP, Cohan C, Connor J. Calcium regulation of the neuronal growth cone. *Trends Neurosci*. 1988;11(7):315–321.
36. Konur S, Ghosh A. Calcium signaling and the control of dendritic development. *Neuron*. 2005;46(3):401–405.

37. Lohmann C, Wong RO. Regulation of dendritic growth and plasticity by local and global calcium dynamics. *Cell Calcium*. 2005;37(5):403–409.
38. Ross WN. Changes in intracellular calcium during neuron activity. *Annu Rev Physiol*. 1989;51:491–506.
39. Rekart JL, Sandoval CJ, Routtenberg A. Learning-induced axonal remodeling: evolutionary divergence and conservation of two components of the mossy fiber system within Rodentia. *Neurobiol Learn Mem*. 2007;87 (2):225–235.
40. Zhong Y, Wu CF. Neuronal activity and adenylyl cyclase in environment-dependent plasticity of axonal outgrowth in Drosophila. *J Neurosci*. 2004;24(6):1439–1445.
41. Kirov SA, Harris KM. Dendrites are more spiny on mature hippocampal neurons when synapses are inactivated. *Nat Neurosci*. 1999;2(10):878–883.
42. Kirov SA, Goddard CA, Harris KM. Age-dependence in the homeostatic upregulation of hippocampal dendritic spine number during blocked synaptic transmission. *Neuropharmacology*. 2004;47(5):640–648.
43. Knott GW, Quairiaux C, Genoud C, Welker E. Formation of dendritic spines with GABAergic synapses induced by whisker stimulation in adult mice. *Neuron*. 2002;34(2):265–273.
44. Gallo V, Kingsbury A, Balázs R, Jørgensen OS. The role of depolarization in the survival and differentiation of cerebellar granule cells in culture. *J Neurosci*. 1987;7(7):2203–2213.
45. Monti B, Contestabile A. Blockade of the NMDA receptor increases developmental apoptotic elimination of granule neurons and activates caspases in the rat cerebellum. *Eur J Neurosci*. 2000;12(9):3117–3123.
46. Ono T, Sekino-Suzuki N, Kikkawa Y, Yonekawa H, Kawashima S. Alivin 1, a novel neuronal activity-dependent gene, inhibits apoptosis and promotes survival of cerebellar granule neurons. *J Neurosci*. 2003;23 (13):5887–5896.
47. Frotscher M, Heimrich B, Deller T, Nitsch R. Understanding the cortex through the hippocampus: lamina-specific connections of the rat hippocampal neurons. *J Anat*. 1995;187(Pt 3):539–545.
48. Frotscher M, Deller T, Heimrich B, Förster E, Haas C, Naumann T. Survival, regeneration and sprouting of central neurons: the rat septohippocampal projection as a model. *Ann Anat*. 1996;178(4):311–315.
49. Lynch G. Functional recovery after lesions of the nervous system. 3. Developmental processes in neural plasticity. The formation of new synaptic connections after brain damage and their possible role in recovery of function. *Neurosci Res Program Bull*. 1974;12(2):228–233.
50. Nadler JV, Cotman CW. Interactions between afferents to the dentate gyrus after entorhinal lesion during development: long-term regulation of choline acetyl-transferase activity. *Brain Res*. 1978;142(1):174–181.
51. Duffy CJ, Rakic P. Differentiation of granule cell dendrites in the dentate gyrus of the rhesus monkey: a quantitative Golgi study. *J Comp Neurol*. 1983;214(2):224–237.
52. Anthes DL, LeBoutillier JC, Petit TL. Structure and plasticity of newly formed adult synapses: a morphometric study in the rat hippocampus. *Brain Res*. 1993;626(1-2):50–62.
53. Ge S, Goh EL, Sailor KA, Kitabatake Y, Ming GL, Song H. GABA regulates synaptic integration of newly generated neurons in the adult brain. *Nature*. 2006;439(7076):589–593.
54. Purves D, Lichtman JW. *Principles of Neural Development*. Sunderland, MA: Sinauer Ass. Inc.; 1985.
55. Majewska AK, Newton JR, Sur M. Remodeling of synaptic structure in sensory cortical areas in vivo. *J Neurosci*. 2006;26(11):3021–3029.
56. Yuste R, Bonhoeffer T. Morphological changes in dendritic spines associated with long-term synaptic plasticity. *Annu Rev Neurosci*. 2001;24:1071–1089.
57. Mel BW. Have we been hebbing down the wrong path? *Neuron*. 2002;34(2):175–177.
58. Chklovskii DB, Mel BW, Svoboda K. Cortical rewiring and information storage. *Nature*. 2004;431 (7010):782–788.
59. Aimone JB, Li Y, Lee SW, Clemenson GD, Deng W, Gage FH. Regulation and function of adult neurogenesis: from genes to cognition. *Physiol Rev*. 2014;94(4):991–1026.
60. Feng R, Rampon C, Tang YP, et al. Deficient neurogenesis in forebrain-specific presenilin-1 knockout mice is associated with reduced clearance of hippocampal memory traces. *Neuron*. 2001;32(5):911–926.
61. Kitamura T, Sugiyama H. Running wheel exercises accelerate neuronal turnover in mouse dentate gyrus. *Neurosci Res*. 2006;56(1):45–52.
62. Lu L, Bao G, Chen H, et al. Modification of hippocampal neurogenesis and neuroplasticity by social environments. *Exp Neurol*. 2003;183(2):600–609.

63. Olson AK, Eadie BD, Ernst C, Christie BR. Environmental enrichment and voluntary exercise massively increase neurogenesis in the adult hippocampus via dissociable pathways. *Hippocampus*. 2006;16(3):250–260.
64. Bergami M, Masserdotti G, Temprana SG, et al. A critical period for experience-dependent remodeling of adult-born neuron connectivity. *Neuron*. 2015;85(4):710–717.
65. Singer BH, Gamelli AE, Fuller CL, Temme SJ, Parent JM, Murphy GG. Compensatory network changes in the dentate gyrus restore long-term potentiation following ablation of neurogenesis in young-adult mice. *Proc Natl Acad Sci USA*. 2011;108(13):5437–5442.
66. Schaefers AT, Grafen K, Teuchert-Noodt G, Winter Y. Synaptic remodeling in the dentate gyrus, CA3, CA1, subiculum, and entorhinal cortex of mice: effects of deprived rearing and voluntary running. *Neural Plast*. 2010;2010:870573.
67. Neufeld J, Teuchert-Noodt G, Grafen K, Winter Y, Witte AV. Synapse plasticity in motor, sensory, and limbo-prefrontal cortex areas as measured by degrading axon terminals in an environment model of gerbils (Meriones unguiculatus). *Neural Plast*. 2009;2009:281561.
68. Sutula T, He XX, Cavazos J, Scott G. Synaptic reorganization in the hippocampus induced by abnormal functional activity. *Science*. 1988;239(4844):1147–1150.
69. Dudek FE, Sutula TP. Epileptogenesis in the dentate gyrus: a critical perspective. *Prog Brain Res*. 2007;163:755–773.
70. Wiskott L, Rasch MJ, Kempermann G. A functional hypothesis for adult hippocampal neurogenesis: avoidance of catastrophic interference in the dentate gyrus. *Hippocampus*. 2006;16(3):329–343.
71. Chambers RA, Potenza MN, Hoffman RE, Miranker W. Simulated apoptosis/neurogenesis regulates learning and memory capabilities of adaptive neural networks. *Neuropsychopharmacology*. 2004;29(4):747–758.
72. Chambers RA, Conroy SK. Network modeling of adult neurogenesis: shifting rates of neuronal turnover optimally gears network learning according to novelty gradient. *J Cogn Neurosci*. 2007;19(1):1–12.
73. Crick C, Miranker W. Apoptosis, neurogenesis, and information content in Hebbian networks. *Biol Cybern*. 2006;94(1):9–19.
74. Appleby PA, Kempermann G, Wiskott L. The role of additive neurogenesis and synaptic plasticity in a hippocampal memory model with grid-cell like input. *PLoS Comput Biol*. 2011;7(1):e1001063.
75. Weisz VI, Argibay PF. Neurogenesis interferes with the retrieval of remote memories: forgetting in neurocomputational terms. *Cognition*. 2012;125(1):13–25.
76. Finnegan R, Becker S. Neurogenesis paradoxically decreases both pattern separation and memory interference. *Front Syst Neurosci*. 2015;9:136.
77. van Ooyen A. Using theoretical models to analyse neural development. *Nat Rev Neurosci*. 2011;12(6):311–326.
78. Butz M, van Ooyen A. A simple rule for dendritic spine and axonal bouton formation can account for cortical reorganization after focal retinal lesions. *PLoS Comput Biol*. 2013;9(10):e1003259.
79. Butz M, Steenbuck ID, van Ooyen A. Homeostatic structural plasticity can account for topology changes following deafferentation and focal stroke. *Front Neuroanat*. 2014;8:115.
80. Butz M, van Ooyen A, Wörgötter F. A model for cortical rewiring following deafferentation and focal stroke. *Front Comput Neurosci*. 2009;3:10.
81. Siegel M, Marder E, Abbott LF. Activity-dependent current distributions in model neurons. *Proc Natl Acad Sci USA*. 1994;91(24):11308–11312.
82. O'Leary T, van Rossum MC, Wyllie DJ. Homeostasis of intrinsic excitability in hippocampal neurones: dynamics and mechanism of the response to chronic depolarization. *J Physiol*. 2010;588(Pt 1):157–170.

CHAPTER

19

Modifications in Network Structure and Excitability May Drive Differential Activity-Dependent Integration of Granule Cells into Dentate Gyrus Circuits During Normal and Pathological Adult Neurogenesis

Quinton Skilling,[1], James Roach*,[1], Alison L. Althaus[1], Geoffrey G. Murphy[1,2], Leonard Sander[1] and Michal Zochowski[1]*

[1]University of Michigan, Ann Arbor, MI, United States

[2]University of Michigan Medical School, Ann Arbor, MI, United States

OUTLINE

*These authors contributed equally.

The Rewiring Brain
DOI: http://dx.doi.org/10.1016/B978-0-12-803784-3.00019-6

1 INTRODUCTION

For a long time, it was generally assumed that the dominant form of plasticity in the brain is the one connected with creation and modification of synaptic connections.[1,2] This was primarily due to the fact that it was universally accepted that after early development, the brain does not produce new neurons and thus undergoing changes in computation are mediated only through remodeling of the existing networks. Seminal papers of Altman[3,4] have shown that this assumption is not true and that there are places in the brain where adult neurogenesis does indeed take place. Of the regions known to exhibit adult neurogenesis, two areas have been extensively studied: the subventricular zone (SVZ) whose neurons migrate to the olfactory bulb and the subgranular zone (SGZ), whose neurons populate the dentate gyrus (DG) of hippocampal formation. This discovery opened completely new perspectives on how plasticity in these brain regions takes place during cognitive function, and also how this process can be altered during chronic pathologies such as epilepsy which specifically impacts the DG. While over the last half a century, there has been a lot of interest associated with understanding the differential roles of neurogenesis in brain function and pathology (see, e.g., Refs. [5–10] for review), its mechanistic role still remains to large extent unknown. Here we will concentrate on adult neurogenesis taking place in DG. First we will summarize basic neurobiological findings on integration of new cells into DG circuits in healthy and pathological brain and later we will discuss how this process can be regulated by pathological changes in the established circuits themselves and also by changes in intrinsic cellular excitability.

2 STRUCTURE AND INTEGRATION SEQUENCE INTO HEALTHY DG NETWORKS

The DG serves in large part as the gatekeeper to the hippocampal loop.[11,12] Namely its granule cells receive input from the entorhinal cortex (EC) and, on the other end, their axons project to CA3 pyramidal cells forming the Mossy fiber tract. Consequently, the axons originating in CA1/CA3 form Schaffer collateral pathway and project back to EC, closing the hippocampal computational loop. It is worth mentioning that in addition to this trisynaptic pathway there is a monosynaptic pathway that takes EC input directly to CA1/CA3 layers. Labeling studies have found that DG granule cells make on average synaptic contacts with around 12 CA3 pyramidal neurons, which they further interconnect with other pyramidal cells and inhibitory interneurons.[13] Within the DG circuit itself, in addition to excitatory granule cells, several types of inhibitory interneurons are present which provide control over granule cell activation. At the same time, hilar mossy cells provide distributed excitatory connections forming a feedback loop.[13]

Under normal circumstances the adult born neurons, that will become excitatory granule cells eventually, originate from the SGZ of DG[1,5,7,14] and go through a relatively rapid maturation sequence. Initially (during the first week after differentiation) the newly born cells receive tonic GABAergic input which, despite being inhibitory in mature neurons, acts as depolarizing current due to the membrane potential of newly born neurons being below the chloride reversal potential. The new cells then rapidly proceed to receive excitatory input connections from hilar cells, and at the same time start forming outgoing synapses on the CA3 pyramidal cells. The new GCs then proceed to form sparse excitatory connections with mature granule cells. This glutamatergic input together with back-propagating activation from CA3 pyramidal cells and hilar cell input constitutes the only early sources of excitatory input which is crucial for survival of these cells, as with time (around third week postdifferentiation) GABA-ergic signaling becomes inhibitory.[5] About a month after birth, the main glutamatergic input from EC becomes a dominating source of excitation. This suggests that during the initial stage of development signaling stemming from local DG circuit innervation by the new cells is crucial for their survival and future function.

The function of adult born neurons in the healthy hippocampal circuits is still being discussed. There is ample evidence that adult neurogenesis aids the creation of new memory representations and their storage. It has been hypothesized that the new cells act specifically as pattern separators, decorrelating the new input from existing memory traces.[12,15]

3 PATHOLOGICAL INTEGRATION OF NEWLY BORN NEURONS AND ITS FUNCTIONAL CORRELATES

Temporal lobe epilepsy (TLE) is the most common form of partial- or localization-related epilepsy.[16] Ictal activity in TLE typically originates in the hippocampal formation and adjacent structures. The chronic, spontaneous, recurrent seizures that typify TLE are associated with selective cell loss, predominantly in the hippocampal pyramidal cell layer and DG hilus. While there is notable cell loss and dispersion in the DG by comparison, the DG granule cell layer remains relatively intact. In DG, the observed hilar cell loss is thought to cause an increase in overall excitability as it was shown that the cells that are the most affected are the ones that drive activation of inhibitory interneurons.[16] Another source of pathological excitatory feedback loops is mossy fiber sprouting of granule cells that results in the formation of aberrant GC–GC connections that are generally not present in healthy DG circuits[17] and persistence of hilar basal dendrites on the GC population.[9,10] It is thus clear that epilepsy leads to a change in balance between excitation and inhibition within DG circuits as well as connectivity structure favoring pathological, namely, long-range excitation.

It has been found that some of these morphological changes are mediated by adult born neurons that are formed within a relatively short time window—in animal models only the granule cells that were 2–4 weeks old at the time of induced status epilepticus (SE) exhibited mossy fiber sprouting and resulting aberrant excitatory innervation within DG circuits.[18,19] Furthermore, the existence of basal hilar dendrites was again predominantly found on the adult born GCs that were formed around the time of SE.[10,20,21]

In addition to these pathological structural connectivity changes within DG circuits, it was observed that the number of adult-neurogenic cells increases dramatically during the pathology. Furthermore, their migratory patterns changed with many of them migrating to DG hilus or molecular layer instead of being confined to GC layer under normal conditions—so-called ectopic granule cells.[10,22] Finally, it has been observed that these ectopic granule cells exhibit modest but significant increases in excitability.[23]

The observed aberrant changes in integration of adult born neurons lead to the hypothesis that these cells may exacerbate the pathology rather than compensate for it and lead to additional cognitive deficits. This hypothesis was recently confirmed as it was found that the ablation of adult neurogenesis before induced seizure onset leads to a reduction of seizure frequency and alleviation of some of the observed cognitive deficits.[24]

While there is a number of specific mechanisms implicated as the ones causing aberrant integration of the new cells into DG circuits, their distinctive roles remain not clear.[10] Here we concentrate on changes in network-wide activity-dependent ques that may change integration patterns of the newborn cells even if the cells respond to these ques in the same way in normal and pathological conditions.

Below, we show in a reduced model that changes in the connectivity structure of the native network drive changes in network-wide activity patterns that in turn differentially mediate integration of the new cells. Details of this work can be found in Ref. [25]. In a later part of this chapter, we take a step back and sketch how experimentally observed intrinsic changes to neuronal excitability may also drive changes in network-wide neuronal firing patterns again eventually causing distinctive patterns of integration of newly born cells.

4 UNDERSTANDING NETWORK DRIVEN EFFECTS OF NEURONAL INCORPORATION DURING ADULT NEUROGENESIS

As we described in detail above, the adult neurogenesis occurs in the SGZ of DG of the hippocampus. Newly formed cells gravitate to the granular layer and attempt to incorporate into the existing network architecture. In the healthy adult brain, granule cells receive excitatory input from outside of the granular layer while simultaneously receiving inhibitory input and sending excitatory signals to nearby inhibitory cells, though few connections between local GCs are observed. We assume that as the new cells move into the granular zone, accepted spike-timing-dependent plasticity (STDP) rules[26] apply and the cells form incoming and outgoing connections preferentially with coactive neurons. One of the effects of status epilepticus is the change of excitatory/inhibitory balance of GC inputs toward increased excitation. As we show below, this results in new excitatory cells forming connections with an abnormally high level of other excitatory cells, further contributing to the high levels of synchronous activity. Thus the existing network architecture together with changing levels of inhibition has profound effects on the system dynamics which is compounded by the generation of new cells and their subsequent abnormal incorporation into the existing structural framework.

We focus on the spiking dynamics resulting from lowered inhibition and changing network topology and not directly on the biophysical environment leading to such dynamics. We employ the integrate-and-fire neuron model using 1200 neurons (1100 neurons in the case of low inhibition), where the membrane potential of a single neuron has the general form:

$$\frac{dV_k}{dt} = \left(I_{e/i} + E_k - \alpha_k V_k\right) + \sum_j \omega_{jk} A_{jk} I^j_{syn}(t + \tau_{delay})$$

with $I_{e/i}$ representing the intrinsic input to excitatory/inhibitory neurons (based on the class of k; $I_e = 0.73, I_i = 0.6$), $\boldsymbol{E_k}$ representing an external stimulus ($E_k = 0.4$ for cells receiving stimulus, $E_k = 0$ otherwise) and $\boldsymbol{\alpha_k}$ representing ionic leakage (pulled from a random distribution between 1 and 1.3 for each neuron). Neurons are placed randomly on a two-dimensional 40×40 grid, leaving empty nodes for new neurons to integrate into the network. When the membrane potential reaches firing threshold ($V_k \geq 1$), the neurons fire an action potential, its membrane potential is reset to zero and the cell enters a refractory period for 8 ms, whereby no input causes changes in membrane potential. Upon firing and after a short delay, postsynaptic targets receive a perturbation of the form:

$$I_{syn}(t) = \exp\left(-\frac{t}{3.0}\right) - \exp\left(-\frac{t}{0.3}\right)$$

which is modified by the weighted adjacency matrix $\omega_{jk}A_{jk}$. Here, for the native network (i.e., the one composed of mature cells) the weight is a fixed value based on the pairwise connectivity; $\omega_{e-e} = 0.2$, $\omega_{e-i} = 0.4$, $\omega_{i-e} = -0.4$, and $\omega_{i-i} = -0.7$.

The adjacency matrix is constructed based on the biological evidence of network structure in the context of healthy and epileptic brain states (Fig. 19.1). In the case of normal inhibition, 200 inhibitory cells each project onto up to 4 inhibitory cells and 100 random excitatory granule cells which, in turn, connect to local inhibitory cells. When modeling a pathological network (post-SE), we reduce the number of inhibitory neurons to 100. We introduce varied sparse excitatory-to-excitatory connectivity dependent on whether the networks have been ectopic for an extended period of time, essentially arranging as a small-world or random network corresponding to the healthy and epileptic states, respectively (Fig 19.1A). Using the Watts–Strogatz[27] formalism, small-world networks can be achieved by rewiring purely local connections to random connections with some probability; for a sufficiently small rewiring probability ($p < 0.05$), networks are predominantly local, whereas increasing the rewiring probability changes networks to be in the small-world regime ($0.05 < p < 0.2$) before becoming mostly random ($p \geq 0.7$). Increasing rewiring probability results in a larger fraction of long-range connections compared to low rewiring probability (Fig. 19.1B). For clarity, we focus on a small-world network ($p = 0.1$) to model healthy excitatory–excitatory connectivity in DG and a more random network ($p = 0.7$) to model pathological circuits. To represent input to the DG from the EC region, a persistent stimulus is applied to the network near the middle of the grid. Further, spontaneous activity is induced by manually causing a neuron to spike with probability 0.0003, corresponding to a 0.3-Hz firing activity.

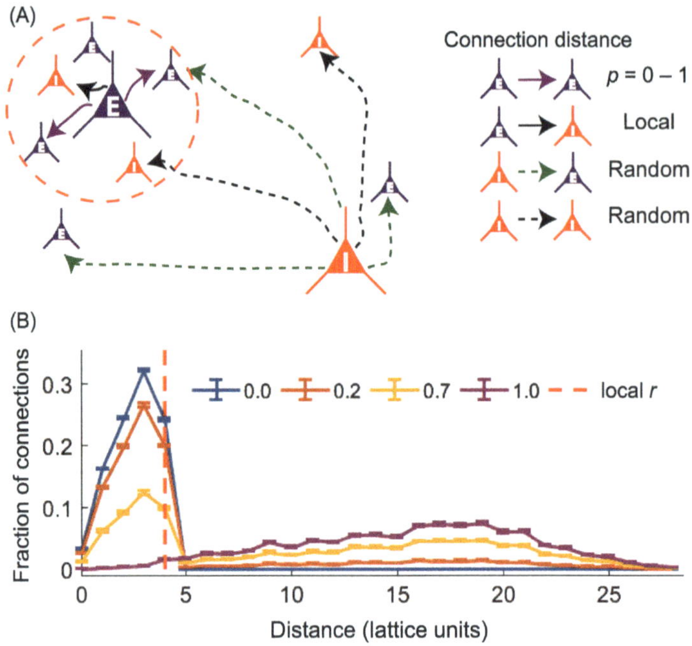

FIGURE 19.1 **Modeled network.** (A) Connectivity structure within the network. The span of the excitatory–excitatory connections varies between local and random. The excitatory to inhibitory connections are set to be local, while the connections originating from inhibitory neurons target random excitatory and inhibitory cells. (B) Relative fraction of local and long-distance connections as a function of the rewiring parameter. The total number of excitatory connections remains the same.

New granule cells are introduced to the existing networks every ~350 ms at random empty grid locations and sample the surrounding synaptic environment to form new incoming and outgoing connections. By observing the firing frequency f_i of nearby neurons, incoming connections are formed hierarchically through computing synaptic attachment probability

$$g_i = a\frac{f_i}{\max(f_i)} + (1-a)\gamma$$

The parameter $a \in [0, 1]$ controls the amount of stochasticity (white noise, γ) and has the effect of preferentially selecting large deviations in frequency for sufficiently high values (here, $a = 0.8$). Similarly, outgoing connections are selected randomly from the connections emanating from nearby cells. Connections are added until the connectivity degree is equal to the average degree of the existing network. In addition, periodically, the outgoing connections are analyzed for coincident spiking (defined as a postsynaptic cell firing within 10 ms after the presynaptic cell fires) and are pruned if this number is too low, whereby new connections are formed following the same rules as above. Newborn neurons fail to survive if their mean firing rate after 10 synaptic rewiring events is lower than a preset threshold and the cell is removed from the network.

Otherwise, the newly born GC matures and its connectivity stops to change. At the same time, random existing granule cell is randomly chosen and removed from the network to keep the network size constant.

In general, due to the set integration rules, the new cells formed through neurogenesis are more easily incorporated into the existing network if their connections are able to efficiently connect to high-frequency partners. Under normal inhibition, when the network is initially in a small-world architecture, surviving cells integrate into the network locally, near the site of persistent stimulation in few numbers relative to the size of the system. By contrast, in random networks, surviving cells are dispersed evenly in a random pattern across the entire network and in much greater numbers than for low rewiring (Fig 19.2A). This is due to the large number of long-range connections in random networks compared to small-world connectivity, with the neurons receiving persistent input able to cause coincident spiking far away from the stimulation site and thus integrate more easily far from the stimulation site. As a result, a time-correlated, global burst of activity shows greater synchronization in the random network compared to the small-world network (Fig 19.2B), in line with dynamics observed during the pathology.

Thus, under normal inhibition, local networks (low rewiring) are better at filtering the survivability of new neurons due to the low density of global coincident spiking events (Fig 19.2C, black points). This is not seen for increasingly random circuits as both survivability and synchrony increase globally, indicating that the observed increase in survivability (in terms of number and spatial extent) may further compound the morphological changes in the pathological DG circuits by increasing network synchrony. Furthermore, the spatial correlation between initial network activity and survival probability of the new cells, which corresponds to state-dependent integration of new cells needed to form new representations (Fig. 19.2D, black points), peaks for the small-world regime under normal inhibition and decays rapidly to zero correlation for increasing rewiring probability. At the same time, for low inhibition, survivability of the newly born GCs is dramatically increased, largely independent of network structure (i.e., rewiring parameter; Fig 19.2C, white dots), while the spatial correlation between network activity and cell survivability is significantly reduced especially for local circuit structures (Fig 19.2D, white dots). This indicates that both more randomized excitatory connectivity structure as well as decreased inhibition lead to greater survivability of newly born neurons as well as decreased specificity in integration patterns into the network.

To further understand the properties of activity-dependent integration of new cells into the networks under different conditions, we measured the mean distance of surviving newly born neurons from the stimulation site. Under normal conditions (normal inhibition, low rewiring), new neurons tend to concentrate closely around the stimulation site whereas diseased state circuits (high rewiring; low inhibition) support cell integration throughout the network (Fig 19.3A). Mechanistic explanation of this observation can be linked to number of synaptic rewiring events throughout the network. These events indirectly provide information about number of coincident firing events needed (or lack of thereof) for survivability of synapse and then the cell. For local network structures, the number of synaptic rewiring events dramatically increases away from the stimulation site. At the same time, they remain uniform for networks having random excitatory connectivity (Fig. 19.3B). This again indicates that the activation of the network and in particular

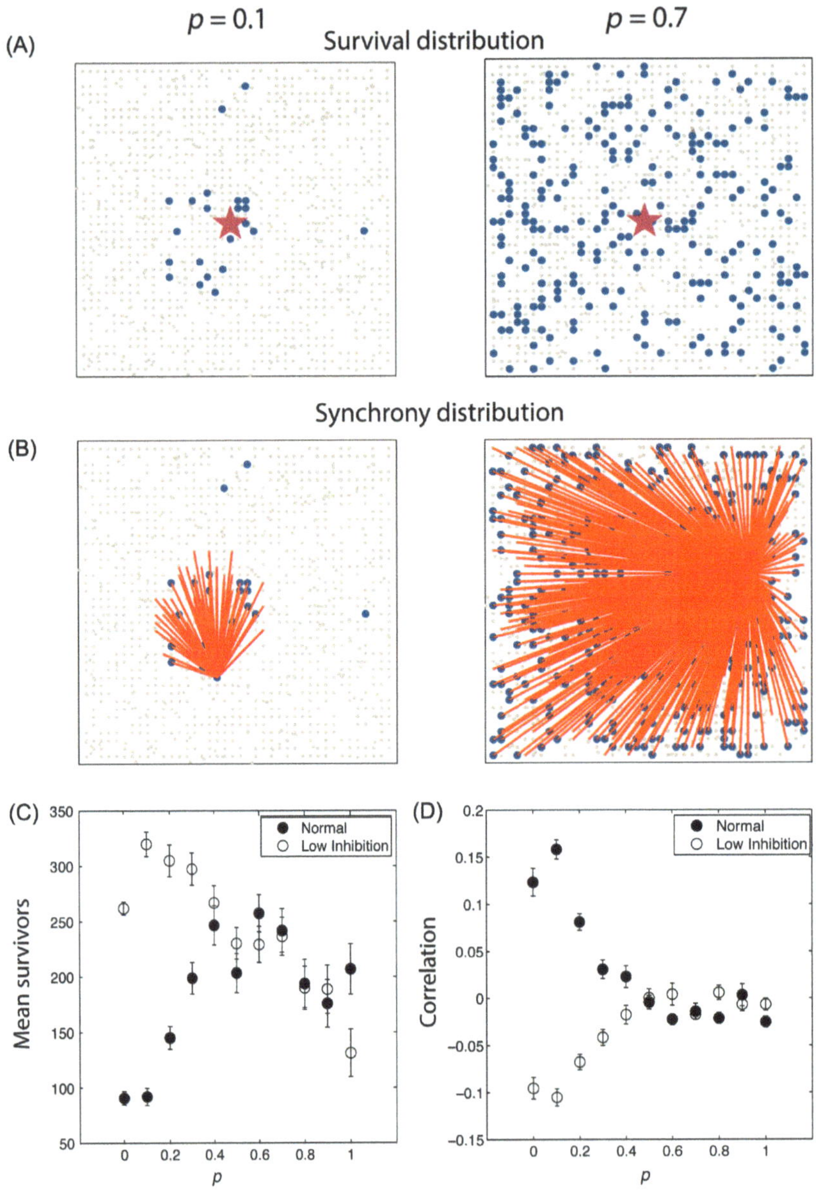

FIGURE 19.2 **Patterns of cell integration and synchrony for different network structures and inhibition levels.** (A) Example of distribution of surviving cells (*blue dots*) for small-world and random topology. The number of seeded cells was the same for both networks. (B) Synchrony patterns between the surviving cells. *Red lines* denote threshold value of synchrony between randomly chosen neuron and all other surviving cells. (C) Mean number of surviving neurons as a function of excitatory connectivity structure for networks with normal and low inhibition. (D) Spatial correlation between initial network activity and survival probability of the seeded cells as a function of excitatory connectivity structure for networks with normal and low inhibition.

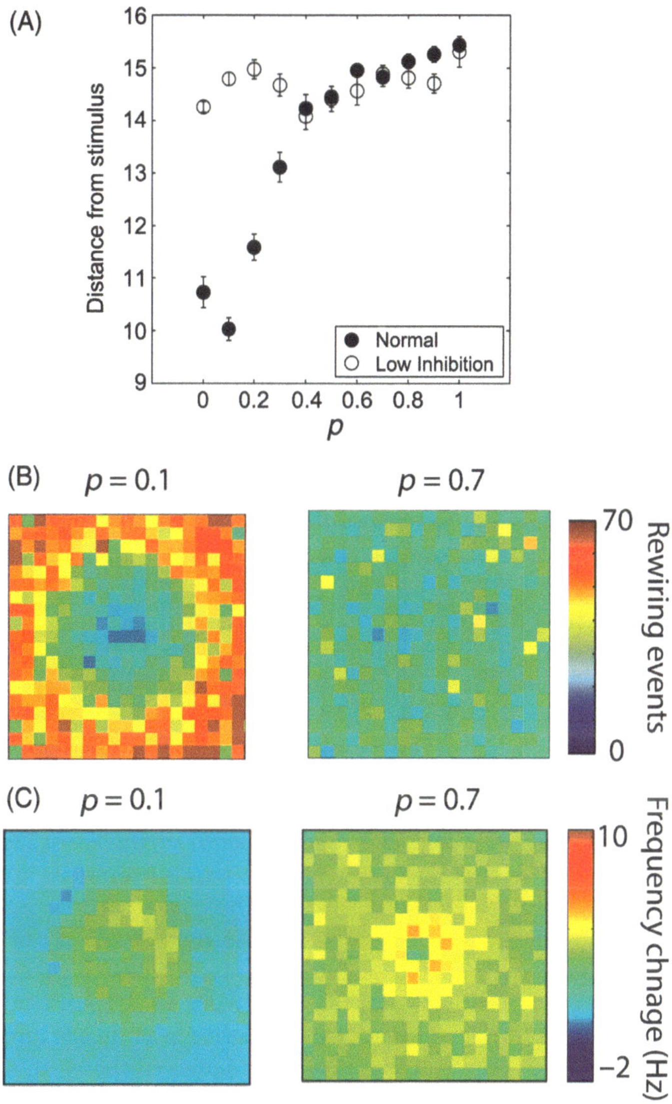

FIGURE 19.3 (A) Mean distance of surviving cells from stimulation site as a function of excitatory connectivity structure. For normal inhibition, the small-world connectivity provides highest localization of integration patterns. The effect subsides for networks with lower inhibition. (B) Number of synaptic rewiring events within a network for networks with local (left) and random (right) excitation. High number of rewiring events indicates low activation of a given synapse at the given location due to limited number of coincident firing events. (C) Functional effects of integration of new cells for networks with local (*left*) and random (*right*) excitatory connectivity structure: change of local firing frequency within the network from before to after neuronal integration.

the number of coincident events drops off rapidly for local connectivity while it remains spatially homogenous for random excitatory connectivity patterns.

How do the functional properties differentially change under those normal and pathological conditions? To answer this question, we measured overall change of network activity levels from before integration of the new cells into the network to after the integration process (Fig. 19.3C). For local network structures, the spiking frequency increases significantly near the stimulation site, but at the same time decreases away from it. This results in overall network activity staying at the same levels, and also creates discrete, heterogeneous regions of differentiated levels of activity. In contrast, a network with random excitatory connectivity structure exhibits a uniform increase in firing frequency throughout the network, for details, see Ref. [25].

5 CELLULAR CORRELATES OF INTEGRATION PATTERNS OF NEW CELLS INTO DG CIRCUITS

Above, we have focused on describing how changes in native network structure, due to pathology, may mediate changing patterns of incorporation of new cells in the network. We have shown that some of the changes observed experimentally are directly predicted by the model, without changing the cellular properties of the newly born cells, nor the incorporation rules that the cells follow.

However, there is seemingly contradictory evidence that the newborn granule cells under pathological conditions (post-SE) change their excitability properties. There are a large number of studies that have shown that new GCs are hyperexcitable,[23] and consequently fire in synchronized bursts. This can be at least partially explained by changes of the excitatory-to-inhibitory ratio of the synaptic inputs targeting these cells. However, there is another set of studies that shows a decrease of excitability of these cells. This has been observed in the post-SE born cells that exhibit healthy patterns of integration into DG circuits and therefore indicates that these excitability changes could be intrinsic to the cells rather than a result of their integration patterns into the networks.[28]

We have experimentally compared phase response curves (PRCs) of the healthy and ectopic granule cells (Fig. 19.4A). For those experiments, whole-cell patch recordings were made under current clamp conditions and the cells were held 10 mV below their resting membrane potential. A depolarizing sine wave was delivered (via the amplifier headstage) to elicit an action potential. Synaptic stimulation was achieved using an extracellular stimulating electrode placed in the perforant path (100 μs pulse width). The ectopic cells exhibit significantly stronger negative early phase response in comparison with healthy ones.

The PRCs are a measure of cellular membrane excitability properties.[29] The PRC characterizes the change in spike timing of a periodically firing neuron in response to brief, weak external stimulation. While clearly those are idealized conditions, we have shown earlier that PRCs remain relevant determinants of synchrony in extended networks. In general, they have been classified into two general categories: Type 1, which display only phase advances in response to excitatory stimuli, and Type 2, which respond with both phase advances and delays. Type 1 cells exhibit relatively poor propensity for synchronization under excitatory coupling, while Type II cells synchronize better.[30–32] Thus,

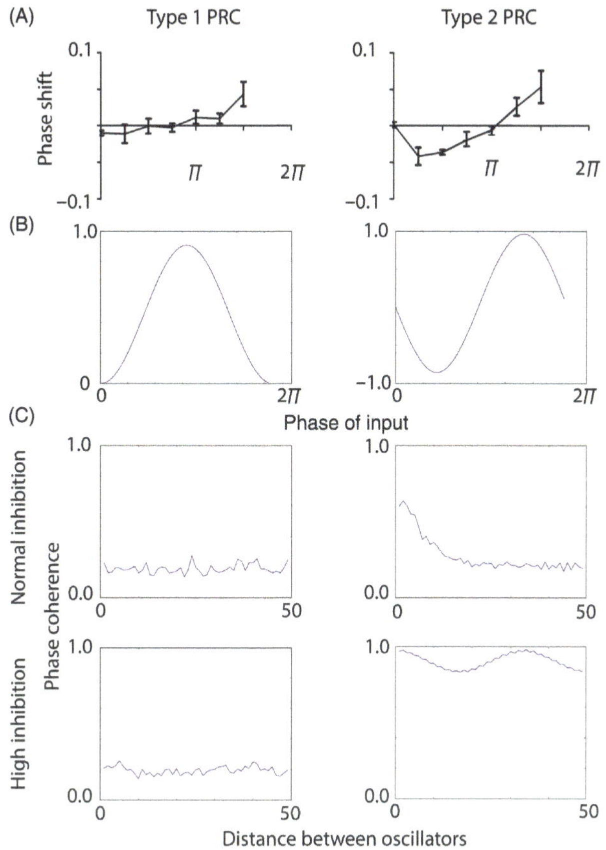

FIGURE 19.4 **Spatiotemporal activity patterns in a network for Winfree oscillators having Type 1 and Type 2 phase response curves (PRCs).** (A) Experimentally observed PRCs for the normal and ectopic granule cells in DG. (B) Shape of modeled Type 1 (*left*) and Type 2 (*right*) PRCs. (C) Coherence patterns as a function of spatial distance between the oscillators for networks with normal inhibition composed of oscillators with Type 1 and Type 2 PRCs. (D) Network-wide coherence patterns as a function of spatial distance between the oscillators for networks with high inhibition composed of oscillators with Type 1 and Type 2 PRCs. This is the case of wave formation as in Ref. [39].

broadly speaking, Type 2 neurons have reduced excitability but higher propensity for synchronization. We have shown extensively that networks having identical connectivity patterns but having cells endowed with different PRCs show very different spatiotemporal activity patterns.[33,34] We have further shown that changes in these patterns may lead to grossly different plastic changes in the networks. Namely, circuits that can change their connectivity patterns through generic[25] STDP rules will show overall potentiation when neuron exhibits Type 1 excitability (i.e., overall asynchronous dynamics), while identically structured circuits of Type 2 neurons will exhibit overall weakening of the connectivity.[35]

In order to show how PRCs could affect neurogenesis, we turn to a simplified system, the Winfree model of coupled oscillators,[36,37] which is closely related to the often-used

Kuramoto model.[38] In this scheme, spiking neurons are represented by oscillators which obey an equation of motion for their phases, θ_i:

$$\dot{\theta}_i = \omega_i + \sum_j \kappa_{i,j} P(\theta_j) R(\theta_i)$$

In this equation, $P(\theta_j)$ is the PRC for oscillator j. The coupling strength of the interaction oscillator j with oscillator i is $\kappa_{i,j}$, and R is the shape of the coupling pulse. In this work, we use strongly peaked pulse coupling, following[37]:

$$R(\theta) = a_n(1+\cos(\theta))^n$$

The constant a_n is chosen so that the area under the pulse is 2π. We take $n = 10$. For the PRC, we take a form that interpolates between Type 1 and Type 2, depending on a parameter, z:

$$P(\theta) = z\sin(\theta/2)^2 - (1-z)\sin(\theta)$$

For $z = 0$, we have Type 2 and for $z = 1$, Type 1.

The local coupling matrix, κ, is of similar shape as before—we investigate a simple case, local excitation, and long-range inhibition for N oscillators on a ring:

$$\kappa_{i,j} = \begin{cases} E, & r_{i,j} < X \\ -I, & X < r_{i,j} < 2X \end{cases}$$

Here X is the coupling range. We take $X = N/8$; E, I are the strengths of the excitatory and inhibitory coupling, respectively. It has been shown[39] in a Type 2 Kuramoto system that if inhibition dominates, the oscillators self-organize into network-wide waves of activity so that there is long-range correlation of phases even with short-range coupling. Of course, in the Type II case, if excitation dominates and the spread of natural frequencies is not too large, we can have global synchrony.[38]

A measure of the correlation is the pairwise phase coherence for oscillators k steps apart:

$$c_k = \left| \sum_t \frac{\exp(i[\theta_i - \theta_{i+k}])}{T} \right|$$

Here the sum is over T time points and we average over i. Fig 19.4 shows coherence patterns across the network as a function of distance between the oscillators for networks with normal and low inhibition, for purely Type 1 (Fig 19.4, left) or Type 2 PRCs. We observe that the networks having identical structures but different PRCs can have completely different spatiotemporal activity patterns ranging from completely incoherent to those with only short distance coherence to finally those with global coherence. Networks composed of oscillators with a Type 2 PRC tend to generate states with global coherence (or even synchrony, not shown). This indicates that ectopic cells that have a PRC skewed toward Type 2 may additionally promote synchrony in DG circuits. Also, for Type 2 networks which do not show global synchrony, there can still be strong short-range coherence. This could have interesting implications for patterns of neurogenesis.

6 DISCUSSION

Since the discovery that the brain is able to selectively generate new neurons throughout life, there has been extensive research focused on understanding mechanisms guiding the integration of the cells into the existing circuits and elucidating the functional roles of these neurons. The further discovery that adult neurogenesis results in significantly different patterns of cellular integration in healthy and pathological brains, and while in the normal brain it can play an important cognitive role, at the same time, it can exacerbate the pathology, has complicated this picture. Many lines of research investigated specific molecular and cellular changes in understanding this dramatic difference. Here we have shown that the change of incorporation patterns can be explained on a network level. Namely changes in the native DG network mediated by the pathology can change the distribution of activity-dependent factors in these circuit and consequently drive the newly born neurons to have altered incorporation patterns. We have shown that a simple network exhibiting such structure-dependent changes in activity-dependent pattern can explain a number of experimentally observed pathological changes.

We have also shown that simple changes in neuronal membrane excitability, as measured by PRCs, can also significantly change the network-wide integration patterns into DG circuits. This is due to the fact that so-called Type 2 PRCs promote the emergence of synchrony in the network.

While the presented models are clearly too simplified to provide answers about detailed mechanisms governing the specific changes of GC integration in healthy and pathological brains, they are intended to draw attention to the possible role of the network and its spatiotemporal patterning in explaining experimentally observed phenomena.

6.1 Future Experimental Studies

Future experiments will need to more carefully distinguish which pathological changes in the measured properties of GC are due to their direct synaptic input and thus extrinsic factors versus intrinsic ones determined by change in their receptor repertoire. This will allow to better address clinical interventions related to controlling epilepsy.

6.2 Future Modeling Studies

The next natural step in terms of modeling studies is to make more biophysically realistic models of the networks that include both variations on network structure and intrinsic neuronal excitability. This will allow to distinguish the differential roles of these two processes in the integration of newly born granule cells into DG circuits.

Acknowledgments

LS and MZ want to acknowledge contributions of Eshel Ben-Jacob and Casey Schneider-Mizell who took part in the work on original simulations. The work was supported by NSF PoLS grant 1058034 (LS and MZ) and NIH NIBIB R01EB018297 (QS, JR, MZ).

References

1. Hebb DO. *The Organization of Behavior: A Neuropsychological Theory*. New York: Wiley; 1949.
2. McNaughton BL, Douglas RM, Goddard GV. Synaptic enhancement in fascia dentata: cooperativity among coactive afferents. *Brain Res*. 1978;157:277–293.
3. Altman J. Autoradiographic investigation of cell proliferation in the brains of rats and cats. *Anat Record*. 1963;145:573–591.
4. Altman J. Postnatal growth and differentiation of the mammalian brain, with implications for a morphological theory of memory. In: Quarton G, Melnechuk T, Schmitt F, eds. *The Neurosciences: A Study Program*. New York, NY: Rockefeller University Press; 1967:723–743.
5. Carmen V, Van Praag H. Functional circuits of new neurons in the dentate gyrus. *Front Neural Circ*. 2013;7:1–15.
6. Peter J, Lisman J. Structure, function, and plasticity of hippocampal dentate gyrus microcircuits. *Front Neural Circ*. 2014;8:107.
7. Ming G, Song H. Adult neurogenesis in the mammalian brain: significant answers and significant questions. *Neuron*. 2011;70:687–702.
8. Yao B, Christian KM, He C, Jin P, Ming G, Song H. Epigenetic mechanisms in neurogenesis. *Nat Rev Neurosci*. 2016;17:537–549.
9. Jessberger S, Parent JM. Epilepsy and adult neurogenesis. *Cold Spring Harbor Perspect Biol*. 2015;7:a020677.
10 Parent JM, Kron MM. Neurogenesis and epilepsy. In: Noebels JL, Avoli M, Rogawski MA, et al., eds. *Jasper's Basic Mechanisms of the Epilepsies [Internet]*. 4th edition Bethesda, MD: National Center for Biotechnology Information (US); 2012.
11. Shepherd GM. *The Synaptic Organization of the Brain*. 5th edition New York, NY: Oxford University Press; 2004.
12. Yau S, Li A, So K. Involvement of adult hippocampal neurogenesis in learning and forgetting. *Neural Plast*. 2015;Article ID 717958.
13. Traub RD, Miles R. *Neuronal Networks of the Hippocampus*. New York, NY: Cambridge University Press; 2008.
14. Toni N, Sultan S. Synapse formation on adult-born hippocampal neurons. *Eur J Neurosci*. 2011;33:1062–1068.
15. Aimone JB. Computational modeling of adult neurogenesis. *Cold Spring Harbor Perspect Biol*. 2016;8:a018960.
16. Chang BS, Lowenstein DH. Epilepsy. *N Engl J Med*. 2003;349:1257–1266.
17. Okazaki MM, Evenson DA, Nadler JV. Hippocampal mossy fiber sprouting and synapse formation after status epilepticus in rats: visualization after retrograde transport of biocytin. *J Comp Neurol*. 1995;352(4):515–534.
18. Kron MM, Zhang H, Parent JM. The developmental stage of dentate granule cells dictates their contribution to seizure-induced plasticity. *J Neurosci*. 2010;30(6):2051–2059.
19. Jessberger S, Zhao C, Toni N, Clemenson Jr GD, Li Y, Gage FH. Seizure-associated, aberrant neurogenesis in adult rats characterized with retrovirus-mediated cell labeling. *J Neurosci*. 2007;27(35):9400–9407.
20. Ribak CE, Tran PH, Spigelman I, Okazaki MM, Nadler JV. Status epilepticus-induced hilar basal dendrites on rodent granule cells contribute to recurrent excitatory circuitry. *J Comp Neurol*. 2000;428(2):240–253.
21. Shapiro LA, Ribak CE. Newly born dentate granule neurons after pilocarpine-induced epilepsy have hilar basal dendrites with immature synapses. *Epilepsy Res*. 2006;69(1):53–66.
22. Parent JM, Elliott RC, Pleasure SJ, Barbaro NM, Lowenstein DH. Aberrant seizure-induced neurogenesis in experimental temporal lobe epilepsy. *Ann Neurol*. 2005;59:81–91.
23. Althaus AL, Sagher O, Parent JM, Murphy GG. Intrinsic neurophysiological properties of hilar ectopic and normotopic dentate granule cells in human temporal lobe epilepsy and a rat model. *J Neurophysiol*. 2015;113:1184–1194.
24. Cho K-O, Lybrand ZR, Ito N, Brulet R, Tafacory F, Zhang L, et al. Aberrant hippocampal neurogenesis contributes to epilepsy and associated cognitive decline. *Nat Comm*. 2015;6:6606.
25. Schneider-Mizell CM, Parent JM, Ben-Jacob E, Zochowski MR, Sander LM. From network structure to network reorganization: implications for adult neurogenesis. *Phys Biol*. 2010;7:046008.
26. Bi GQ, Poo MM. Synaptic modifications in cultured hippocampal neurons: dependence on spike timing, synaptic strength, and postsynaptic cell type. *J Neurosci*. 1998;18:10464–10472.
27. Watts DJ, Strogatz SH. Collective dynamics of "small-world" networks. *Nature*. 1998;393:440–442.

28. Jakubs K, Nanobashvili A, Bonde S, Ekdahl CT, Kokaia Z, Kokaia M, et al. Environment matters: synaptic properties of neurons born in the epileptic adult brain develop to reduce excitability. *Neuron.* 2006;52 (6):1047–1059.
29. Ermentrout B. Type I membranes, phase resetting curves, and synchrony. *Neural Comput.* 1996;8:979–1001.
30. Ermentrout B, Pascal M, Gutkin B. The effects of spike frequency adaptation and negative feedback on the synchronization of neural oscillators. *Neural Comput.* 2001;13:1285–1310.
31. Canavier C, Butera R, Dror R, Baxter D, Clark J, et al. Phase response characteristics of model neurons determine which patterns are expressed in a ring circuit model of gait generation. *Biol Cybern.* 1997;77:367–380.
32. Dror R, Canavier C, Butera R, Clark J, Byrne J. A mathematical criterion based on phase response curves for stability in a ring of coupled oscillators. *Biol Cybern.* 1999;80:11–23.
33. Bogaard A, Parent JM, Zochowski MR, Booth V. Interaction of cellular and network mechanisms in spatiotemporal pattern formation in neuronal networks. *J Neurosci.* 2009;29:1677–1687.
34. Fink CG, Booth V, Zochowski M. Cellularly-driven differences in network synchronization propensity are differentially modulated by firing frequency. *PLoS Comput Biol.* 2011;7:e1002062.
35. Fink CG, Murphy GG, Zochowski M, Booth V. A dynamical role for acetylcholine in synaptic renormalization. *PLoS Comput Biol.* 2013;9:e1002939.
36. Winfree AT. Biological rhythms and the behavior of populations of coupled oscillators. *J Theoret Biol.* 1967;16:15.
37. Ariaratnam JT, Strogatz SH. Phase diagram for the Winfree model of coupled nonlinear oscillators. *Phys Rev Lett.* 2001;86:4278.
38. Kuramoto Y. *Chemical Oscillations, Waves, and Turbulence.* Berlin: Springer; 1984.
39. Kazanci FG, Ermentrout B. Pattern formation in an array of oscillators with electrical and chemical coupling. *SIAM J Appl Math.* 2007;67:512.

CHAPTER

20

Computational Perspectives on Adult Neurogenesis

Kristofor D. Carlson, Fred Rothganger and James B. Aimone

Sandia National Laboratories, Albuquerque, NM, United States

OUTLINE

1 INTRODUCTION

Neurogenesis has been observed in a number of vertebrate and mammalian species across a variety of brain regions. Although neurogenesis is widespread throughout the vertebrate brains of reptiles and fish,[1] it is limited to a few brain structures in songbirds.[2] With the notable exception of cetaceans,[3] many mammals have been shown to have adult

The Rewiring Brain
DOI: http://dx.doi.org/10.1016/B978-0-12-803784-3.00020-2

neurogenesis in a few key brain regions such as the dentate gyrus (DG) region of the hippocampus and the olfactory bulb (OB).[4] Neurogenesis has been described in additional mammalian brain regions such as the neocortex[5,6] and the hypothalamus,[7] however, there is some debate as to the extent of neurogenesis in these regions. Adult neurogenesis in humans has been reliably observed in the DG[8,9] and recently, the striatum,[10] but appears to be absent in the OB.[11]

Over the past few decades, a combination of experimental paradigms has given neuroscientists a multiscale perspective of adult neurogenesis. Recent advances in immunohistological techniques that identify dividing cells with neuron-specific protein markers[12] and retroviral labeling[13] have allowed for the characterization of neurogenesis at the molecular and cellular scales. Other techniques have measured the overall levels of neurogenesis in the human DG by calculating neuron age using the presence of radioactive carbon in the atmosphere due to above ground nuclear testing.[9] At a much larger scope, behavioral studies have shown how extrinsic factors such as enriched environment, voluntary exercise, and diet positively impact levels of hippocampal neurogenesis.[14–17] Additional behavioral studies have shown that stress and aging can dramatically decrease neurogenesis levels,[18,19] demonstrating that extrinsic factors can have both positive and negative impacts on neurogenesis.

Although neurogenesis has been characterized with numerous experimental approaches at different scales, discerning its functional role remains a major challenge. This is partly because most behavioral studies of neurogenesis assess function within the context of the entire neural circuit (e.g., the hippocampus). Additionally, these experimental paradigms often influence multiple aspects of neural function, making it difficult to assign a specific functional role to adult neurogenesis in a given brain region. Even in instances where animal models with effective neurogenesis knockdown have been successfully developed, results concerning neurogenesis function have been inconclusive.[20]

The use of computational modeling can provide a number of benefits to help address the aforementioned challenges in understanding the role of adult neurogenesis in brain function. First computational models allow neuroscientists to generate "functional hypotheses" that describe how specific aspects of neurogenesis affect neural circuit function. Vague statements such as "neurogenesis plays a role in conjunctive encoding" or "neurogenesis knockout disrupts DG function," can be avoided by pointing to specific mechanisms in neural simulations responsible for changing functional behavior in the model neural circuit. Neural simulations can also be used to perform "virtual experiments" that would be impossible to achieve in an experimental setting, which allows for the exploration of large parameter ranges with precisely controlled model inputs. Finally, insights from simulations can be used to further develop theoretical models of brain function or design future experiments that can be used to verify how neurogenesis affects specific neural computations. As we will see in later discussions, neural models can provide quantitative insights into the advantage of introducing naive neurons into a neural system that can be used in the design of neuromorphic computing systems as well.

The remainder of this review will focus primarily on computational perspectives pertaining to adult neurogenesis, for a more general review on adult neurogenesis please see Aimone et al.[21] In the following section, we discuss a number of studies that examine the impact of new neurons on learning and memory with abstract neural network models. We then briefly discuss key aspects of neurogenesis found in the mammalian hippocampus before moving on to cover anatomically constrained models of neurogenesis. In these sections, we further restrict our discussion to hippocampal neurogenesis as it is the most studied neurogenic brain region and leave the topic of neurogenesis in the OB and striatum to more general neurogenesis reviews. We then briefly explore currently available simulation tools along with their limitations and describe new and upcoming simulation tools and frameworks. Finally we provide a summary, followed by a discussion of future experimental and computational research directions.

2 THE ADDITION OF NEW NEURONS INTO ANNS

Artificial neural networks (ANNs) are a family of mathematical models first proposed by McCulloch and Pitts[22] that are used to approximate unknown mathematical functions from the presentation of large and often imprecise data sets. ANNs consist of computational units which communicate with one another via connections composed of adaptable weights that change value depending on past inputs. The process of changing these connection weights based on past inputs is roughly analogous to biological learning processes observed in the brain and the algorithms that describe these weight changes are termed learning rules. The flexibility of these learning rules allows ANNs to learn complex input–output relationships resulting in their widespread use in a variety of applications including regression analysis, pattern classification, and blind source separation.

A major challenge to this class of neural models is the stability–plasticity dilemma, which states that quick and stable learning often leads to the forgetting of past information while very stable past memories often disrupt learning.[23] Inspired partly by this challenge, classification algorithms such as adaptive resonance theory (ART)[24] and a subset of radial basis functions (RBFs)[25] introduced naive neuronal units to describe inputs under certain conditions. The addition of naive neuronal units allowed these classification algorithms to effectively learn while preventing catastrophic forgetting. Interestingly, these algorithms utilized the addition of naive units before adult neurogenesis was widely accepted in the scientific community.

Multilayer perceptrons (MLPs) are a subset of feedforward ANN algorithms with nonlinear activation functions and at least one hidden layer between the input and output layers.[26,27] These algorithms often utilize supervised learning algorithms to learn the mapping between complex input data and output responses although unsupervised learning rules have also been used. Backpropagation of errors is a commonly used supervised learning algorithm that calculates a loss function in weight space and uses the minimization of this function to update the weight values.[28] The study of MLPs has become more important recently as they are a core ingredient of deep neural networks (DNNs), a set of popular machine learning algorithms used for speech and image recognition, natural language processing, and recommendation systems.[29–32]

Because MLP algorithms struggle with the stability–plasticity dilemma as well, a number of groups have investigated how the addition or removal of neurons from MLPs can improve (or degrade) learning and memory performance. Work done by one of the first of these groups, Chambers et al.[33] consisted of the implementation of neurogenesis and apoptosis in the hidden layer of a three-layer MLP model that utilized gradient decent backpropagation to learn alphabetic character representations. The network was first trained on the Roman alphabet and its recall was tested. Next, the network was trained on the Greek alphabet while its ability to recall character representations from both Greek and Roman alphabets was evaluated. Results showed that increasing the number of neurons undergoing turnover in the middle layer resulted in an increased network learning capacity for new information accompanied by an increased degradation of old memories. The effect of neurogenesis in this model illustrates the stability–plasticity dilemma quite well; new memories can be learned more effectively at the cost of degrading old memories.

In a follow-up study to Chambers et al., researchers found that the speed with which the network learned the first Roman alphabet affected the speed and accuracy with which the network learned the second Greek alphabet. Networks that learned the Roman alphabet quickly did so with less participation from the total number of middle layer neurons and instead had significantly fewer neurons participating in memory recall with weights at the maximum values.[34] Networks that learned the Roman alphabet quickly had memories that were less widely distributed across neurons resulting in longer training times to learn the Greek alphabet. Another follow-up study by Chambers et al.[35] found that the degree of contextual novelty in the new training set affected the memory recall benefits due to neuronal turnover. Specifically, networks with low turnover rates had the least degradation of previously learned memories and were best suited to learning relatively unnovel information over shorter time periods while networks with high turnover rates had the greatest memory degradation and were best for learning novel information over longer time periods.

Additional studies further examined the effect neurogenesis had on learning and recall of new and old information using a three-layer, feedforward MLP with unsupervised Hebbian learning. Deisseroth et al.[36] found that neurogenesis elicited larger memory degradation in older memories; however, newer memories were recalled with a higher fidelity than without neurogenesis despite the network remaining the same size. Another research group examined the relationship between homeostasis and neurogenesis with a similar simulation setup and found that networks that included some form of homeostasis with neurogenesis recalled memories with fewer errors.[37] Two forms of homeostasis were examined: a "turnover homeostasis" that required neuronal death for every neuronal birth and a "fixed activity-level homeostasis" present at downstream CA3 neurons. Although both homeostatic forms increased memory stability in the model, the "fixed activity-level homeostasis" resulted in the best performance and can be achieved in a biologically plausible fashion via a variety of biophysical mechanisms such as intrinsic plasticity[38] or synaptic scaling.[39] This implies that pairing neurogenesis with homeostatic mechanisms may allow networks to store new memories in a more rapid and stable manner.

Neurogenesis was also studied using a class of ANNs similar to MLPs called autoencoders. Autoencoders are similar to MLPs in that they have at least an input layer, a hidden layer, and an output layer but differ from MLPs by restricting the number of units in the output layer to be the same as in the input layer.[40] Instead of producing a target output response given input data as is true with MLPs, autoencoders attempt to reconstruct the input data in their output response using unsupervised learning algorithms. Wiskott et al.[41] tested the ability of a small three-layer linear autoencoder to learn two different input patterns (A & B) presented sequentially using three adaptation strategies: networks that adapted only to the first pattern, networks that adapted to both patterns, and networks with additive neurogenesis where the original neurons adapted only to pattern A while the newly added neurons adapted to pattern B. Results showed that additive neurogenesis significantly reduced the effect of catastrophic memory interference when compared with the two other nonneurogenesis adaptation strategies. In a follow-up study, Appleby and Wiskott[42] compared the memory performance of purely additive neurogenesis to neurogenesis with turnover and found additive neurogenesis was superior as it allowed the network to be responsive to changes in the statistics of the environment while preserving earlier memories. An additional follow-up study compared a network with an enhanced model of neurogenesis that included neuronal turnover and enhanced plasticity of new granule cells to networks that used solely additive neurogenesis, neurogenesis with turnover, or enhanced plasticity of young granule cells.[43] Results showed that the network with the enhanced neurogenesis model effectively learned spatially structured input patterns better than networks using any one of the three individual strategies alone.

3 BIOLOGICAL CONTEXT OF NEUROGENESIS

In the previous section, the addition of new neurons was shown to confer a number of functional advantages to a variety of ANN models. ANNs with neurogenesis were more adept at managing the stability–plasticity dilemma by striking a balance between learning new information and preserving old memories. In real biological systems, the utility of neurogenesis is strongly linked to the structure and function of the neural circuit in which it is active. To this end, neurogenesis has been incorporated into a number of anatomically constrained computational models of the DG and large portions of the hippocampal circuit. However, before we discuss computational models of adult neurogenesis in the DG and hippocampus, it is necessary to first describe the important properties of DG/hippocampus neuroanatomy as well as DG granule cell neurophysiology to place these models in the appropriate context.

3.1 The Hippocampus

The hippocampus is located in the medial temporal lobe and has been shown to influence spatial navigation[44] as well as the formation and consolidation of declarative memories.[45] Information from the cortex enters the hippocampus via the entorhinal cortex (EC), which acts as the primary interface between the cortex and the hippocampus.[46] The prevailing narrative of hippocampal function has been that of the trisynaptic loop: the EC provides a cortical representation of information, the DG performs pattern separation and generates a sparse representation, the cornu ammonus area 3 (CA3) acts as an autoassociator, and the cornu ammonus area 1 (CA1) performs temporal associations, eventually feeding back into the EC[47] (Fig. 20.1). However, it is important to understand that this narrative leaves out a number of details such as direct connections from the EC to the CA3 and CA1 and feedback connections from the CA3 back to the DG. Additional hippocampal structures such as the cornu ammonus area 2 (CA2) and the subiculum have

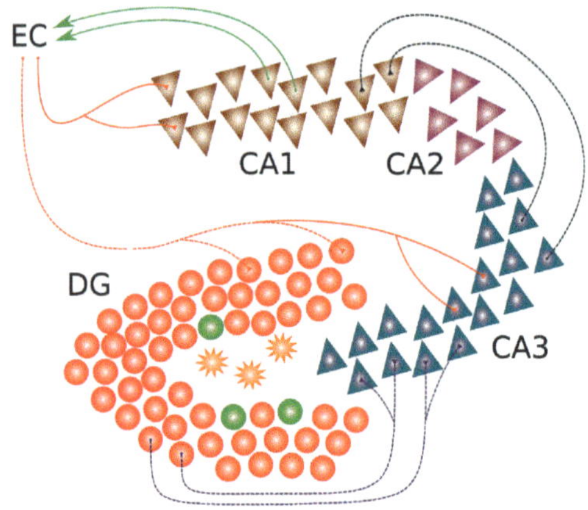

FIGURE 20.1 **The major information pathways in the hippocampus.** The *solid and dotted red lines* indicate the perforant pathway, the *dotted red, blue, and black lines* indicate the trisynaptic loop, the *dotted blue lines* indicate the mossy fibers, the *dotted black lines* indicate the Schaffer collaterals, and the *solid green arrows* indicate the cornu ammonus area 1 (CA1) connections feeding back onto the entorhinal cortex (EC). The green cells in the dentate gyrus (DG) indicate immature granule cells that have migrated to the DG granule cell layer from the subgranular zone (SGZ) while the starred cells in the DG indicate neural stem cells residing in the SGZ.

received little attention from researchers but will be need to be understood before a complete picture of hippocampal function is possible. We next discuss the structure in the hippocampus where adult neurogenesis is concentrated: the DG.

3.2 The Dentate Gyrus

The DG, thought to play a major role in episodic memory formation,[48] receives cortical input from the EC and projects to CA3 pyramidal cells via the mossy fiber pathway. The DG has been theorized to perform pattern separation by providing decorrelated input to the CA3 and pattern conjunction by combining EC object and spatial information into a single representation.[46,49,50] The DG has a number of anatomical and neurophysiological properties that support these theorized neural functions. The DG has roughly double the number of neurons than that of the EC,[45] which allows for the creation of a very sparse and decorrelated representation of the cortical information it receives. Additionally, the average firing rate of DG granule cells is low as they are tonically inhibited, further promoting temporal sparsity.[51] Finally, mossy fibers from DG granule cells form multiple synapses on CA3 pyramidal cells and can lead to the reliable activation of pyramidal neurons with relatively few DG granule cell action potentials.[52] This connectivity is conducive to reliably transmitting a sparse code from one brain region to another.

3.3 The Maturation of New Granule Cells

Granule cells possess unique neurophysiological properties that contribute to the overall DG circuit function. Granule cells arise from a progenitor population and are slowly integrated in the DG neural circuit, however, immature granule cells make up only a small fraction of the total granule cells in the DG at any one time.[13] There are a number of key features of granule cell maturation that play a role in DG circuit function. First, granule cells have neurophysiological properties that change gradually over time and go through critical survival and plasticity periods during their maturation.[53,54] Second, due in part to their distinct set of ion channels, enhanced plasticity, and reduced inhibition relative to mature granule cell neurons, young granule cells have been shown to have increased excitability.[53,55–57] Finally, only a small portion of young granule cells are integrated into the DG as many of them die before they reach full maturity.[58]

4 ANATOMICALLY CONSTRAINED COMPUTATIONAL MODELS OF NEUROGENESIS

A number of groups have investigated the role of neurogenesis in brain processing by constructing computational models of the full hippocampal circuit. One of the earliest computational models of neurogenesis was performed by Becker,[59] whose model included the main components of the hippocampus trisynaptic loop (EC → DG → CA3 → CA1). Simulations examined the effect of neurogenesis with neuronal turnover on the cued recall of unrelated items, pair associate items, and related items. To implement neuronal turnover, a fixed percentage of neurons were selected, their connections were deleted, and the deleted connections were randomly remapped. Findings indicated that neurogenesis provided a large benefit to those patterns that were related and had little effect for the two other cases. Due to the remapping of synaptic connections from the EC to the DG and from the DG to the CA3 that occurred during the neurogenesis process, similar items were assigned significantly different representations which resulted in a reduction of the interference between these items during recall. However, there was a slight negative effect incurred by neuronal turnover in the recall of related items as well, presumably because some information was lost during the neuronal turnover.

Becker et al.[60] performed a follow-up study where a similar training setup was used to examine the effect of neurogenesis using three test cases: no neurogenesis, unbiased neurogenesis, and neurogenesis biased toward newer neurons while also varying the rates of neurogenesis from 2% to 50%. Results showed that the elimination of neuronal turnover resulted in a steep forgetting curve during the retention phase. At high turnover rates, memory retention improved but without appreciable difference from the two neurogenesis cases. However, at low turnover rates, the preferential recruitment of new neurons allowed for better memory retention because the network was better able to utilize the relatively small number of new neurons.

A different set of studies used a computational model of the hippocampus to examine the effect of purely additive neurogenesis, a process where new neurons are added to the network without the removal of older neurons, on a number of memory-related tasks. In the first study, Weisz and Argibay[61] evaluated the performance of the model on the retrieval of a variable number of input patterns with the introduction of additive neurogenesis between the two training sets. They found neurogenesis significantly increased the ability of the network to recall recent memories (postneurogenesis memories) and also slightly impaired the recall of remote memories (preneurogenesis memories). Additionally, they found this effect was at least partially due to the fact that recent memories were represented by a statistically greater number of new neuron activity than remote memories and not merely due to the increase in the total number of neurons in the DG. In a follow-up study, results showed that while neurogenesis did significantly increase the retrieval of recent memories, it also slightly impaired the recall of remote memories due to interference from the new neuron activity.[62] This finding illustrates that, even without turnover of older neurons, neurogenesis can interfere with memories and may play an important role in forgetting.

As discussed in the previous section, the DG is theorized to perform specific computational functions, namely pattern separation and conjunctive encoding. While many anatomically constrained computational studies have modeled the role of neurogenesis in the hippocampal circuit, few models have examined the role of neurogenesis specifically on DG function. Aimone et al.[63] used a biologically detailed six-layer neural network model of the DG with realistic neuron number ratios, connection topologies, and immature granule cell electrophysiology to study the effect of neurogenesis on DG pattern separation. To examine the effect of neurogenesis on DG pattern separation, networks with and without neurogenesis were presented sets of patterns composed of two distinct inputs that represented spatial information from the medial entorhinal cortex (mEC) and contextual information from the lateral entorhinal cortex (lEC). The two networks were first trained on 400 patterns with neurogenesis processes occurring in both networks for every 10 pattern presentations. Afterwards, neurogenesis was disabled in one of the networks and a different set of patterns were presented to the networks, followed by the testing phase. Simulations showed immature neurons introduced a degree of similarity between patterns presented at the same time, a process the authors referred to as "pattern integration." The added similarity between patterns learned at the same time suggests that temporal information could be encoded in hippocampal memories giving rise to a previously hypothesized function of temporal conjunctive encoding,[64,65] a function that has been observed experimentally as well.[66]

Recent work by Finnegan and Becker[67] also examined how neurogenesis affects DG function by incorporating changes in excitability, connectivity, lateral inhibition, and synaptic plasticity that occur along the developmental trajectory of adult-born DG granule cells. The model consisted of a two-layer probabilistic generative ANN model used in deep learning applications, known as a Restricted Boltzmann Machine (RBM). RBMs are restricted in the sense that connections can only exist between the visible and hidden layers but not within them. The visible layer of the model represented input from the EC while the hidden layer represented the DG granule cells. Four different models of RBMs were trained sequentially on 11 groups of 90 patterns followed by a testing period where noisy versions of these patterns were presented after each group, to test proactive memory interference, and after all groups had been trained to test retroactive memory interference. The models were trained using a learning rule called contrastive divergence along with additional constraints on the sparseness of the code and connectivity of the models. The four different models trained were RBMs, sparse RBMs, RBMs with neurogenesis, and sparse RBMs with neurogenesis. Simulations showed that even though models with hyperexcitable young DG granule cells generated neural codes with more overlap (i.e., reduced pattern separation), they still outperformed models with sparser, less overlapping codes but lacking neurogenesis. For cases where highly overlapping patterns were learned within the same context, sparse connectivity played a key role in generating distinct memory traces. We next briefly discuss examining neurogenesis from a neurophysiological perspective through computational modeling of granule cells.

Granule cells have been well studied experimentally since before it was known that they were comprised of mixed population of young and mature neurons. Although there have been multiple high-biological fidelity computational models of granule cells that have examined dendritic integration,[68–70] there have been almost no computational studies of young granule cells and their unique neurophysiological properties. One notable exception includes work by Li et al.,[71] which focused on how maturing granule cell excitability changed with the gradual development of inhibition over time using slice physiology in mice and an abstract feedforward neural network model. The neural network model incorporated new neurons and kept the ratio of excitatory and inhibitory inputs constant to investigate the hypothesis that the gradual addition of new synapses may be sufficient to make young neurons significantly more active than mature granule cells. Results showed the reduced inhibition in young granule cells was sufficient to bias their activation to novel inputs when compared with mature granule cells. As a result of this activity bias, activity-dependent learning mechanisms occurred more frequently in younger granule cells during the presentation of novel stimulus inputs. These findings have been extended by recent slice experiments which found that low synaptic connectivity prevented immature granule cells from responding broadly to cortical activity, allowing them to contribute to a sparse and orthogonal DG representation.[72] This is an illustrative example of how experimental findings can be understood by investigating specific mechanisms, such as the gradual development of inhibition, using computational models.

Together, these computational studies suggest that the combination of immature and mature neurons in the DG provide a distinct computational function when compared to earlier frameworks only considering a homogeneous population of granule cells. A growing body of literature supports the idea that these two populations of neurons provide the DG a "mixed code," combining the ultrasparse code of the mature neurons with the more broadly active immature neuron population.[73–76] However, future computational research will be important in addressing one of the major open questions facing the neurogenesis community: how does this immature neuron population interact with the complex interneuron populations in the network and affect broader hippocampal coding? In particular, one perspective is that young neurons preferentially activate interneurons, thus increasing the sparsity of the overall DG population,[77,78] although there is evidence that new neurons preferentially influence the downstream CA3 network,[79] which could have interesting implications on increasing the information content of the overall DG population.[80]

We discussed a variety of anatomically constrained computational models of the hippocampus, the DG, and the DG granule cell that helped explain the utility of neurogenesis in neural systems. Due to the availability of extensive neural data and the existence of mature theories of hippocampus brain function, computational models have proven to be a useful tool for the study of neurogenesis in the hippocampus. We next discuss some of the challenges involved in performing computational neural simulations along with possible solutions in the form of new tools and methodologies.

5 CHALLENGES FACING NEUROGENESIS MODEL SIMULATION AND DESCRIPTION

Models of neurogenesis, as with developmental neural models in general, have unique requirements which distinguish them from the majority of other computational neuroscience models. In particular, they need to add and remove elements while the simulation is actively running. This adds significant complexity to the design and implementation of conventional numerical methods. For example, a system of ordinary differential equations (ODEs) may be mapped to the compute nodes on a cluster computer in a way that minimizes data movement between nodes. When new equations, and thus new couplings between variables, are added to the system, the optimal mapping changes. A simulator may simply ignore this problem or it may recompute the entire mapping. For a minor change, this remapping may be more computationally expensive than the communication overhead it saves. A third option is an incremental update to the mapping, perhaps assigning nodes for the new elements in a greedy manner based on communication costs. Alternatively, models may preallocate capacity (typically in system memory) for potential—yet currently unused—components, such as a neuron that is not yet born, which pushes the expense from processing and communication to memory.

Due to these complexities and the fairly small niche developmental models occupy among neuroscience models, support for neurogenesis modeling tends to receive limited support from the mainstream neural simulators. To our knowledge, NEST and Brian are the only general-purpose neural simulators to provide some form of structural plasticity as a standard feature.[81,82] Furthermore, Brian 2 has deprecated the DynamicConnectionMatrix in favor of a new Synapse class which does not yet (as of this writing) implement dynamic allocation.

Both the increased availability of experimental neural data and the decreased cost of supercomputing resources have led to the development of large-scale neural models. These models are important to the neuroscience community as they give researchers the ability to explore scale-dependent neural phenomena that may not occur in smaller scale neural models. Large-scale simulations that include structural plasticity are particularly challenging as these simulations must balance the need for dynamic addition and removal of computing elements (e.g., neurons and synapses) with run-time efficiency. This means the simulator must maintain a near optimal load balance and continue to minimize internode communication even as the model evolves. Small-scale neural simulations can support structural plasticity by allocating extra neuronal elements and synaptic connections that remain inactive until they are needed. However, this overallocation strategy is unsuitable for large-scale models. As a consequence of the current state of neural simulators, neurogenesis modelers are often forced to write custom code to support their simulation needs.

Another important aspect of computational modeling is the use of standard model interchange languages that allow for a simulation-independent representation of the computational model. Model interchange languages such as NeuroML are becoming more widespread and allow modelers to more easily adapt existing computational models and utilize a variety of simulator backends.[83] Despite including support for the active creation and destruction of elements that occurs, e.g., in reaction networks, explicit support for neurogenesis in model interchange languages such as NeuroML is limited. We next discuss

N2A: a neural modeling framework that can help meet some of the simulation and description challenges associated with building computational models of neurogenesis.[84]

6 N2A: A NEURAL MODELING FRAMEWORK WITH SUPPORT FOR STRUCTURAL PLASTICITY

We have developed a neural modeling language called Neurons to Algorithms (N2A). It proposes several constructs suitable for neurogenesis and developmental modeling in general. N2A is a model design platform that aims to facilitate the design and validation of neural models. N2A allows for the direct integration of information from neuroinformatics databases and from the direct contribution of neuroscience data from computational neural modelers. The N2A models are stored in an implementation-independent format that can be converted to run on a number of possible simulation backends (Fig. 20.2). We first give a brief summary of the language, and then describe how the constructs work.

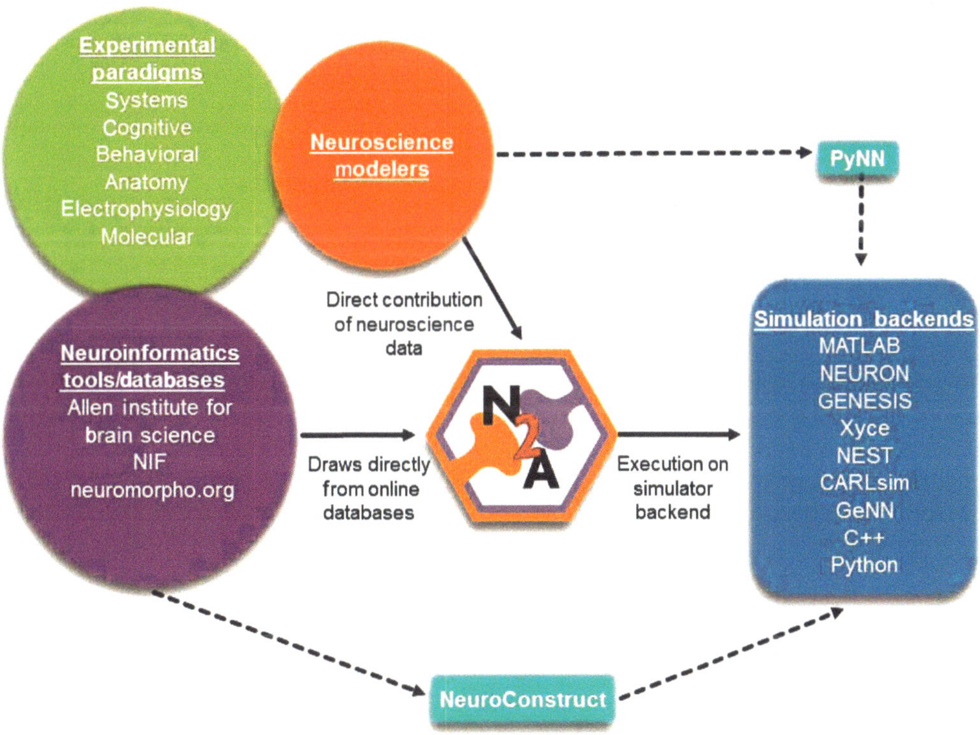

FIGURE 20.2 **Tools like NeuroConstruct draw from neuroinformatics databases while others, like PyNN, express neural models in an implementation-independent format.** N2A allows for both of these capabilities, making it a powerful computational neural modeling tool. N2A currently supports an internal (Java-based) backend, along with a C++ and a Xyce backend but integration with additional simulation backends is a long-term goal.

N2A represents each neural component, neuron or synapse, as a set of ODEs. Equation sets may inherit from other sets in an object-oriented manner, similar to languages such as C++ or Java. Equation sets may contain other sets as subcomponents. This is how full models are assembled for simulation. Equation sets include regular variables and reserved variables. The reserved variables begin with **$** and control aspects of the neural component outside its basic dynamics. All regular variables and most special variables may be expressed as derivatives of arbitrary order.

The attributes of a given component include **$n**, the quantity of instances to generate at runtime. As with all variables, **$n** may be changed at run-time, forcing a change in population size. Reducing **$n** forces instances to be killed at random. Increasing **$n** causes new instances to be created. The derivative **$n′** may be specified. In this case, new instances will come into being at a given rate.

When new neurons come into existence, it is necessary to determine new synaptic connections. A connection equation set may include the variable **$p**, the probability of connection between a given pair of instances. The equation for **$p** may be as complex as needed and include such values as the spatial location (**$xyz**) of each neuron involved in the synapse. The equation may also be a simple constant that represents the general probability that two populations will connect. Note that this approach is more expressive than connection-set algebra.[85]

While **$n** provides a crude form of structural dynamics in which populations grow or shrink at some rate, N2A allows much more targeted control. When **$p** is set to zero on a given neuron, the neuron dies. A population of neurons can be given a constant or regulated value of **$p**, causing them to die off randomly with a given probability.

Instances may split into multiple daughter instances and/or change to different equation sets, a process analogous to cell division and specialization. This allows control over the spatial location where new cells appear. In the case of neurogenesis, populations of stem cells may be distributed throughout the modeled system, with each new cell emerging close to its progenitor.

The special variable to control division and differentiation is **$type**. Suppose a part is of type A. If it contains the equation "**$type = A,B,C,**" the instance will continue to live, and two new parts will be added. The new parts of types B and C will each inherit any matching values from A. If the equation is "**$type = B,**" then the instance will be changed into a B, transferring any matching values. If the equation is "**$type = A,A,**" then the part does a symmetric split, as in cell proliferation. In order to further specialize, each new instance can *read* the value of **$type**, which returns the position of the new part in the list.

Currently, N2A supports three proprietary backends: Xyce, C++, and Internal. Xyce is a parallel version of the Spice circuit simulator used by electrical engineers, with added neural components.[86] The C++ backend generates code which may be compiled for example by the GNU compiler collection (GCC). The Internal simulator is a Java component included in the software which interprets models and is only suitable for small-scale testing and proof-of-concept simulations. N2A support for generating configurations for mainstream neural simulators is currently under development.

7 DISCUSSION

In the previous sections, we have seen that computational neural models of neurogenesis played a key role in describing the benefits of adding neurons to abstract neural networks and in uncovering the function of neurogenesis in the DG microcircuit. As we saw in abstract neural network simulations, neuronal birth and death provide a means for neural systems to strike a balance between learning novel information and minimizing degradation of older memories. Anatomically constrained computational models of neurogenesis in the DG have led to new insights into how neurogenesis affects pattern separation and temporal conjunctive encoding. These models were able to point to specific mechanisms, such as the increased excitability of immature granule cells, responsible for the functional behavior of the DG microcircuit. Findings from anatomically constrained simulations of the DG with neurogenesis have not only influenced thinking on the role of neurogenesis but also on the computational function of DG as well. Finally, we saw that there are practical challenges to overcome if neuroscientists want to run large-scale neural simulations that implement neurogenesis. However, recent advances in neural modeling frameworks, such as N2A, have already made progress in making it easier for modelers to design and construct large-scale neural models with descriptions for structural plasticity.

7.1 Future Experimental Studies

Results from anatomically constrained models of the DG and the hippocampus that examined the role of neurogenesis have suggested a number of experimental studies. Because immature granule cell neurophysiology in the DG has been theorized to play a large role in temporal conjunctive encoding,[64,75] further genetic and neurophysiological characterization of these cells could be performed to allow for more accurate simulation of neurogenesis in the DG. Experiments have shown that local environment and external factors, like stress[87] or exercise,[88] can influence neurogenesis. Further studies that clearly describe how particular external factors affect neurogenesis and turnover rates will allow computational modelers to perform simulations that study the behavioral effect of neurogenesis on an animal. Finally, additional experiments that examine the functional role of neurogenesis in the OB and the striatum may give us more insight into the general role of neurogenesis in brain function and explain why it is limited to only a few regions in the mammalian brain.

7.2 Future Modeling Studies

There are a number of interesting questions about the function of neurogenesis in neural systems where additional computational studies can be insightful. As mentioned earlier, work by Li et al.[71] showed that the number of synapses contacting immature granule cells may play a vital role in their excitability and significantly contributes to neurogenesis function. It seems likely that the number of synapses contacting immature granule cells may be scale dependent and thus may only occur in relatively large-scale network models.

While very large-scale models have been constructed to look at the DG, these approaches have not yet been directed toward learning and plasticity over long time scales, focusing instead on short-term dynamics in epilepsy,[89,90] although we have begun work along these lines.[91] Therefore, constructing large-scale neural models that explore neurogenesis function at a number of different scales may lead to additional insights. With a few exceptions,[63,65] there have been few computational studies that have examined the role neurogenesis plays in an animal's behavior. As additional experiments characterize the relationship between external factors and neurogenesis rates, computational models can incorporate these findings into simulations that link neurogenesis function to affect.

Acknowledgments

K.D.C., F.R., and J.B.A. are supported by Sandia National Laboratories' Laboratory Directed Research and Development (LDRD) program under the Hardware Acceleration of Adaptive Neural Algorithms Grand Challenge. Sandia National Laboratories is a multi-mission laboratory managed and operated by Sandia Corporation, a wholly owned subsidiary of Lockheed Martin Corporation, for the U.S. Department of Energy's National Nuclear Security Administration under contract DE-AC04-94AL85000.

References

1. Zupanc GK. A comparative approach towards the understanding of adult neurogenesis. *Brain Behav Evol*. 2002;58(5):246–249.
2. Nottebohm F. Why are some neurons replaced in adult brain? *J Neurosci*. 2002;22(3):624–628.
3. Patzke N, Spocter MA, Bertelsen MF, et al. In contrast to many other mammals, cetaceans have relatively small hippocampi that appear to lack adult neurogenesis. *Brain Struct Funct*. 2015;220(1):361–383.
4. Zhao C, Deng W, Gage FH. Mechanisms and functional implications of adult neurogenesis. *Cell*. 2008;132 (4):645–660.
5. Cameron HA, Gould E. Adult neurogenesis in the adult neocortex: small, sparse, but significant? *Biol Psychiat*. 2008;63(7):650–655.
6. Gould E. How widespread is adult neurogenesis in mammals? *Nat Rev Neurosci*. 2007;8(6):481–488.
7. Kokoeva MV, Yin H, Flier JS. Neurogenesis in the hypothalamus of adult mice: potential role in energy balance. *Science*. 2005;310(5748):679–683.
8. Eriksson PS, Perfilieva E, Björk-Eriksson T, et al. Neurogenesis in the adult human hippocampus. *Nat Med*. 1998;4(11):1313–1317.
9. Spalding KL, Bergmann O, Alkass K, et al. Dynamics of hippocampal neurogenesis in adult humans. *Cell*. 2013;153(6):1219–1227.
10. Ernst A, Alkass K, Bernard S, et al. Neurogenesis in the striatum of the adult human brain. *Cell*. 2014;156 (5):1072–1083.
11. Bergmann O, Liebl J, Bernard S, et al. The age of olfactory bulb neurons in humans. *Neuron*. 2012;74 (4):634–639.
12. Kuhn HG, Cooper-Kuhn CM. Bromodeoxyuridine and the detection of neurogenesis. *Curr Pharm Biotech*. 2007;8(3):127–131.
13. Zhao C, Teng EM, Summers RG, Ming GL, Gage FH. Distinct morphological stages of dentate granule neuron maturation in the adult mouse hippocampus. *J Neurosci*. 2006;26(1):3–11.
14. Goebbels S, Bode U, Pieper A, Funfschilling U, Schwab MH, Nave KA. Cre/loxP-mediated inactivation of the bHLH transcription factor gene NeuroD/BETA2. *Genesis*. 2005;42(4):247–252.
15. Morris AM, Churchwell JC, Kesner RP, Gilbert PE. Selective lesions of the dentate gyrus produce disruptions in place learning for adjacent spatial locations. *Neurobiol Learn Mem*. 2012;97(3):326–331.

16. Oh J, McCloskey MA, Blong CC, Bendickson L, Nilsen-Hamilton M, Sakaguchi DS. Astrocyte-derived interleukin-6 promotes specific neuronal differentiation of neural progenitor cells from adult hippocampus. *J Neurosci Res*. 2010;88(13):2798–2809.
17. Olah M, Ping G, De Haas AH, et al. Enhanced hippocampal neurogenesis in the absence of microglia T cell interaction and microglia activation in the murine running wheel model. *Glia*. 2009;57(10):1046–1061.
18. Gould E, Reeves AJ, Graziano MS, Gross CG. Neurogenesis in the neocortex of adult primates. *Science*. 1999;286(5439):548–552.
19. Maxwell IH, Maxwell F, Glode LM. Regulated expression of a diphtheria toxin A-chain gene transfected into human cells: possible strategy for inducing cancer cell suicide. *Cancer Res*. 1986;46(9):4660–4664.
20. Deng W, Aimone JB, Gage FH. New neurons and new memories: how does adult hippocampal neurogenesis affect learning and memory? *Nat Rev Neurosci*. 2010;11(5):339–350.
21. Aimone JB, Li Y, Lee SW, Clemenson GD, Deng W, Gage FH. Regulation and function of adult neurogenesis: from genes to cognition. *Physiol Rev*. 2014;94(4):991–1026.
22. McCulloch WS, Pitts W. A logical calculus of ideas immanent in nervous activity. *B Math Biophys*. 1943;5(4):115–133.
23. Carpenter GA, Grossberg S. ART 2: self-organization of stable category recognition codes for analog input patterns. *Appl Optic*. 1987;26(23):4919–4930.
24. Carpenter GA, Grossberg S. The ART of adaptive pattern recognition by a self-organizing neural network. *Computer*. 1988;21(3):77–88.
25. Yingwei L, Sundararajan N, Saratchandran P. A sequential learning scheme for function approximation using minimal radial basis function neural networks. *Neural Comput*. 1997;9(2):461–478.
26. Rosenblatt F. Principles of neurodynamics: perceptrons and the theory of brain mechanisms. No. VG-1196-G-8. Buffalo, NY: Cornell Aeronautical Lab Inc.; 1961.
27. Rumelhart DE, Hinton GE, Williams RJ. Learning internal representations by error propagation. No. ICS-8506. California Univ San Diego La Jolla Inst for Cognitive Science; 1985.
28. Rumelhart DE, Hinton GE, Williams RJ. Learning representations by back-propagating errors. *Cogn Model*. 1988;5(3):1.
29. Deng L, Hinton G, Kingsbury, B. New types of deep neural network learning for speech recognition and related applications: an overview. In: *2013 IEEE International Conference on Acoustics, Speech and Signal Processing (ICASSP)* 2013, 8599–8603.
30. Ciresan D, Meier U, Schmidhuber J. Multi-column deep neural networks for image classification. In: *2012 IEEE Conference on Computer Vision and Pattern Recognition (CVPR)*, June 2012, 3642–3649.
31. Socher R, Perelygin A, Wu J, et al. Recursive deep models for semantic compositionality over a sentiment treebank. In: *Proceedings of the Conference on Empirical Methods in Natural Language Processing (EMNLP)*. 2013;1631:1642.
32. Van den Oord A, Dieleman S, Schrauwen B. Deep content-based music recommendation. *Adv Neural Inf Process Syst*. 2013:2643–2651.
33. Chambers RA, Potenza MN, Hoffman RE, Miranker W. Simulated apoptosis/neurogenesis regulates learning and memory capabilities of adaptive neural networks. *Neuropsychopharmacology*. 2004;29(4):747–758.
34. Crick C, Miranker W. Apoptosis, neurogenesis, and information content in Hebbian networks. *Biol Cybern*. 2006;94(1):9–19.
35. Chambers RA, Conroy SK. Network modeling of adult neurogenesis: Shifting rates of neuronal turnover optimally gears network learning according to novelty gradient. *J Cogn Neurosci*. 2007;19(1):1–12.
36. Deisseroth K, Singla S, Toda H, Monje M, Palmer TD, Malenka RC. Excitation–neurogenesis coupling in adult neural stem/progenitor cells. *Neuron*. 2004;42(4):535–552.
37. Meltzer LA, Yabaluri R, Deisseroth K. A role for circuit homeostasis in adult neurogenesis. *Trends Neurosci*. 2005;28(12):653–660.
38. Desai NS. Homeostatic plasticity in the CNS: synaptic and intrinsic forms. *J Physiol Paris*. 2003;97(4):391–402.
39. Turrigiano GG, Leslie KR, Desai NS, Rutherford LC, Nelson SB. Activity-dependent scaling of quantal amplitude in neocortical neurons. *Nature*. 1998;391(6670):892–896.
40. Hinton GE, Zemel RS. Autoencoders, minimum description length, and Helmholtz free energy. *Adv Neur Inf Proc Syst*. 1994:3.
41. Wiskott L, Rasch MJ, Kempermann G. A functional hypothesis for adult hippocampal neurogenesis: avoidance of catastrophic interference in the dentate gyrus. *Hippocampus*. 2006;16(3):329–343.

42. Appleby PA, Wiskott L. Additive neurogenesis as a strategy for avoiding interference in a sparsely coding dentate gyrus. *Network*. 2009;20(3):137–161.
43. Appleby PA, Kempermann G, Wiskott L. The role of additive neurogenesis and synaptic plasticity in a hippocampal memory model with grid-cell like input. *PLoS Comput Biol*. 2011;7(1):e1001063.
44. O'Keefe J, Nadel L. *The Hippocampus as a Cognitive Map*. USA: Oxford University Press; 1978.
45. Squire LR. Memory and the hippocampus: a synthesis from findings with rats, monkeys, and humans. *Psychol Rev*. 1992;99(2):195–231.
46. Andersen P, Morris R, Amaral D, Bliss T, OKeefe J. *The Hippocampus Book*. Oxford University Press; 2006.
47. Rolls ET, Kesner RP. A computational theory of hippocampal function, and empirical tests of the theory. *Prog Neurobiol*. 2006;79(1):1–48.
48. Amaral D, Scharfman H, Lavenex P. The dentate gyrus: fundamental neuroanatomical organization (dentate gyrus for dummies). *Prog Brain Res*. 2007;163:3–22.
49. O'Reilly RC, McClelland JL. Hippocampal conjunctive encoding, storage, and recall: avoiding a trade-off. *Hippocampus*. 1994;4(6):661–682.
50. Kesner RP. A behavioral analysis of dentate gyrus function. *Prog Brain Res*. Dec. 2007;163:567–576.
51. Nusser Z, Mody I. Selective modulation of tonic and phasic inhibitions in dentate gyrus granule cells. *J Neurophysiol*. 2002;87(5):2624–2628.
52. McNaughton BL, Morris RG. Hippocampal synaptic enhancement and information storage within a distributed memory system. *Trends Neurosci*. 1987;10(10):408–415.
53. Tashiro A, Makino H, Gage FH. NMDA-receptor-mediated, cell-specific integration of new neurons in adult dentate gyrus. *Nature*. 2007;442(7105):929–933.
54. Ge S, Yang CH, Hsu KS, Ming GL, Song H. A critical period for enhanced synaptic plasticity in newly generated neurons of the adult brain. *Neuron*. 2007;54(4):559–566.
55. Ambrogini P, Lattanzi D, Ciuffoli S, et al. Morpho-functional characterization of neuronal cells at different stages of maturation in granule cell layer of adult rat dentate gyrus. *Brain Res*. 2004;1017(1):21–31.
56. Schmidt-Hieber C, Jonas P, Bischofberger J. Enhanced synaptic plasticity in newly generated granule cells of the adult hippocampus. *Nature*. 2004;429(6988):184–187.
57. Esposito MS, Piatti VC, Laplagne DA, et al. Neuronal differentiation in the adult hippocampus recapitulates embryonic development. *J Neurosci*. 2005;25(44):10074–10086.
58. Kempermann G, Gast D, Kronenberg G, Yamaguchi M, Gage FH. Early determination and long-term persistence of adult-generated new neurons in the hippocampus of mice. *Development*. 2003;130(2):391–399.
59. Becker S. A computational principle for hippocampal learning and neurogenesis. *Hippocampus*. 2005;15 (6):722–738.
60. Becker S, MacQueen G, Wojtowicz JM. Computational modeling and empirical studies of hippocampal neurogenesis-dependent memory: effects of interference, stress and depression. *Brain Res*. 2009;299:45–54.
61. Weisz VI, Argibay PF. A putative role for neurogenesis in neurocomputational terms: inferences from a hippcampal model. *Cognition*. 2009;112(2):229–240.
62. Weisz VI, Argibay PF. Neurogenesis interferes with the retrieval of remote memories: forgetting in neurocomputational terms. *Cognition*. 2012;125(1):13–25.
63. Aimone JB, Wiles J, Gage FH. Computational influence of adult neurogenesis in the encoding of time in new memories. *Neuron*. 2009;61(2):187–202.
64. Aimone JB, Wiles J, Gage FH. Potential role for adult neurogenesis in the encoding of time in new memories. *Nat Neurosci*. 2006;9(6):723–727.
65. Becker S, Wojtowicz JM. A model of hippocampal neurogenesis in memory and mood disorders. *Trends Cogn Sci*. 2007;11(2):70–76.
66. Kesner RP, Hui X, Sommer T, Wright C, Barrera VR, Fanselow MS. The role of postnatal neurogenesis in supporting remote memory and spatial metric processing. *Hippocampus*. 2014;24(12):1663–1671.
67. Finnegan R, Becker S. Neurogenesis paradoxically decreases both pattern separation and memory interference. *Front Syst Neurosci*. 2015:9.
68. Carnevale NT, Tsai KY, Claiborne BJ, Brown TH. Comparative electronic analysis of three classes of rat hippocampal neurons. *J Neurophysiol*. 1997;78(2):703–720.
69. Aradi I, Holmes WR. Role of multiple calcium and calcium-dependent conductances in regulation of hippocampal dentate granule cell excitability. *J Comput Neurosci*. 1999;6(3):215–235.

70. Schmidt-Hieber C, Jonas P, Bischofberger J. Subthreshold dendritic signal processing and coincidence detection in dentate gyrus granule cells. *J Neurosci*. 2007;27(31):8430–8441.
71. Li Y, Aimone JB, Xu X, Callaway EM, Gage FH. Development of GABAergic inputs controls the contribution of maturing neurons to the adult hippocampal network. *Proc Natl Acad Sci USA*. 2012;109(11):4290–4295.
72. Dieni CV, Panichi R, Aimone JB, Kuo CT, Wadiche JI, Overstreet-Wadiche L. Low excitatory innervation balances high intrinsic excitability of immature dentate neurons. *Nat Comm*. 2016:7.
73. Aimone JB, Deng W, Gage FH. Resolving new memories: a critical look at the dentate gyrus, adult neurogenesis, and pattern separation. *Neuron*. 2011;70(4):589–596.
74. Danielson NB, Kaifosh P, Zaremba JD, et al. Distinct contribution of adult-born hippocampal granule cells to context encoding. *Neuron*. 2016;90(1):101–112.
75. Rangel LM, Quinn LK, Chiba AA, Gage FH, Aimone JB. A hypothesis for temporal coding of young and mature granule cells. *Front Neurosci*. 2013;7:75.
76. Rangel LM, Alexander AS, Aimone JB, et al. Temporally selective contextual encoding in the dentate gyrus of the hippocampus. *Nat Comm*. 2014:5.
77. Sahay A, Wilson DA, Hen R. Pattern separation: a common function for new neurons in hippocampus and olfactory bulb. *Neuron*. 2011;70(4):582–588.
78. Ikrar T, Guo N, He K, et al. Adult neurogenesis modifies excitability of the dentate gyrus. *Front Neural Circuit*. 2013;7(1):204.
79. Temprana SG, Mongiat LA, Yang SM, et al. Delayed coupling to feedback inhibition during a critical period for the integration of adult-born granule cells. *Neuron*. 2015;85(1):116–130.
80. Vineyard CM, Verzi SJ, James CD, Aimone JB. Quantifying neural information content: a case study for the impact of hippocampal adult neurogenesis. In: *Proceedings of the International Joint Conference on Neural Networks 2016 (IJCNN)*, Vancouver, Canada; 2016.
81. Gewaltig MO, Diesmann M. NEST (NEural Simulation Tool). *Scholarpedia*. 2007;2(4):1430: Accessed 22.06.16.
82. Goodman D, Brette R. Brian: a simulator for spiking neural networks in Python. *Front Neuroinform*. 2008;2(5):254–263.
83. Gleeson P, Crook S, Cannon RC, et al. NeuroML: a language for describing data driven models of neurons and networks with a high degree of biological detail. *PLoS Comput Biol*. 2010;6(6):e1000815.
84. Rothganger F, Warrender CE, Trumbo D, Aimone JB. N2A: a computational tool for modeling from neurons to algorithms. *Front Neural Circuits*. 2014:8.
85. Djurfeldt M. The connection-set algebra—a novel formalism for the representation of connectivity structure in neuronal network models. *Neuroinformatics*. 2012;10(3):287–304.
86. Schiek RL, Warrender CE, Teeter C, et al. Simulating neural systems with Xyce. no. SAND2012-10628. Sandia National Laboratories; 2012.
87. Warner-Schmidt JL, Duman RS. Hippocampal neurogenesis: opposing effects of stress and antidepressant treatment. *Hippocampus*. 2006;16(3):239–249.
88. Van Praag H, Shubert T, Zhao C, Gage FH. Exercise enhances learning and hippocampal neurogenesis in aged mice. *J Neurosci*. 2005;25(38):8680–8685.
89. Hendrickson PJ, Gene JY, Song D, Berger TW. Interactions between inhibitory interneurons and excitatory associational circuitry in determining spatio-temporal dynamics of hippocampal dentate granule cells: a large-scale computational study. *Front Syst Neurosci*. 2015:9.
90. Schneider CJ, Bezaire M, Soltesz I. Toward a full-scale computational model of the rat dentate gyrus. *Front Neural Circuits*. 2012;6:83.
91. Vineyard CM, Verzi SJ, Caudell TP, Bernard ML, Aimone JB. Adult neurogenesis: implications on human and computational decision making. In: *International Conference on Augmented Cognition*. Springer, Berlin Heidelberg; 2013, pp. 531–540.

CHAPTER

21

Restricted Boltzmann Machine Models of Hippocampal Coding and Neurogenesis

Rory Finnegan[1], Mark Shaw[2] and Suzanna Becker[1]

[1]McMaster University, Hamilton, ON, Canada [2]University of Toronto, Toronto, ON, Canada

OUTLINE

1 INTRODUCTION

The hippocampus has intrigued researchers for many decades because of its vital role in the formation and retrieval of complex associative memories. Inspired by Marr's simple memory theory of neural coding in the archicortex,[1] a computational framework that included an initial coding stage followed by an associative retrieval stage, subsequent computational modelers have further fleshed out the contributions of different hippocampal subregions to these memory functions.[2–7] Under this widely accepted, traditional view of the hippocampus, the initial coding stage takes place in the dentate gyrus (DG),

The Rewiring Brain
DOI: http://dx.doi.org/10.1016/B978-0-12-803784-3.00021-4

followed by an associative retrieval stage in the CA3 and CA1 regions. In these models, the DG serves to generate less overlapping, more orthogonalized representations (pattern separation) through sparse coding, while the CA3, with its dense recurrent collateral connections, performs cued retrieval (pattern completion). The associative pathway from CA3 to CA1 in some models further contributes to pattern completion. The basic circuit of this traditional model is illustrated in Fig. 21.1.

The traditional view of the hippocampus as a memorization and cued retrieval device, as described earlier, has garnered considerable empirical support. Direct evidence from unit recordings of ongoing neuronal activity and immediate early gene (IEG) markers of recent neuronal activity supports a role for the DG in sparse coding and pattern separation.[9–13] Similarly, there is direct evidence of a role for the CA3 in cued recall and pattern completion.[14–16] However, more recent data suggest that this traditional view is too simplistic. Firstly, a large body of evidence suggests that the hippocampal "memory system" is as much involved in imagining and predicting the future as it is in remembering the past. For example, patients with medial temporal lobe (MTL) damage that includes the hippocampus not only show episodic and contextual memory deficits but also have great difficulty imagining future scenarios.[17–20] Events that they attempt to imagine, such as the past events that they attempt to remember, are notably lacking in episodic detail. Secondly, recent data call into question whether the MTL memory system is merely retrieving and remembering information or may also be involved in the ongoing classification and perception of stimuli. A prime example is the coding of space. In both rodents and humans, hippocampal "place cells" reflect the animal's current location in space and are critical for ongoing navigation.[21,22] The importance of the MTL for representing perceptual information also extends to nonspatial perceptual judgments. A growing body of evidence points to a central role for the hippocampus in representing higher order perceptual features and making complex perceptual decisions (for a review, see Lee et al.[23]), for example, discriminating faces and objects from novel stimuli, even when those faces and objects were not subsequently well remembered.[24]

The fact that the hippocampal system seems to be involved in perception, imagery, and predicting the future is inconsistent with the notion that its chief function is exact memorization for subsequent cued retrieval. Instead, the traditional view of the hippocampus (Fig. 21.1) must be rethought. An updated view is that the hippocampus operates integrally with the neocortex to encode, represent, perceive, and predict information. As a first step toward modeling the role of the hippocampus within this broader view, Kali and Dayan[25] proposed a hierarchical, probabilistic model of the cortex, with the hippocampus situated at the top of the hierarchy. The hippocampus and neocortex jointly form a multilayer model of the sensory input. Kali and Dayan's model provides a compelling alternative account of observations such as "sequence replay," the reactivation of recently experienced sequences of place representations during sleep.[26,27] The standard view is that hippocampal sequence replay serves to consolidate memories in the neocortex, making them independent of the hippocampus. In contrast, Kali and Dayan propose that event replay serves to maintain the correspondence between neocortical and hippocampal representations, as both may be evolving over time. This view fits within the broader framework of the brain as a probabilistic, generative, hierarchical model that includes the hippocampus.

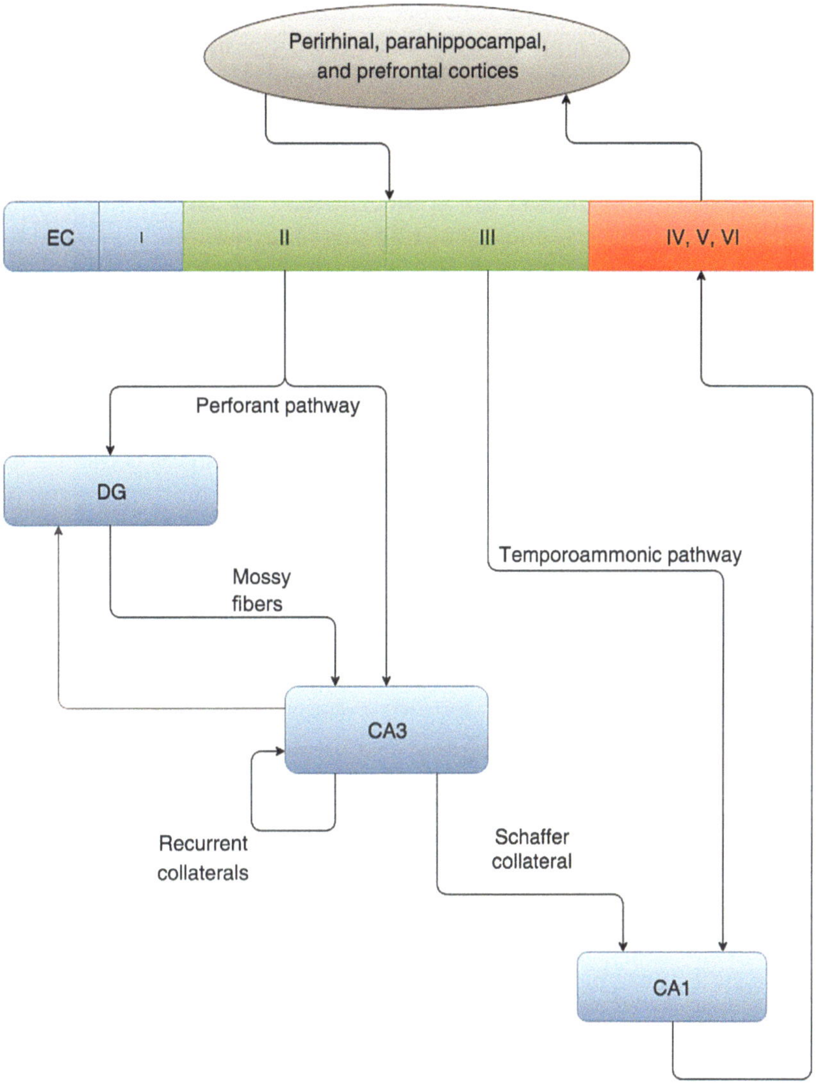

FIGURE 21.1 **Traditional view of the hippocampus as a memorization and cued retrieval device, with pattern separation in the dentate gyrus (*DG*) and pattern completion in the CA3.** The CA3 and CA1 regions receive both direct input from the entorhinal cortex (*EC*) via the perforant path, and indirect input through the trisynaptic circuit via the DG. The DG granule cells project to the CA3 field via mossy fiber synapses, which are few in number but are among the largest in the brain, such that only a few Mossy fiber synapses may be sufficient to activate a CA3 pyramidal cell.[8] It has therefore been suggested that these terminals act as "detonator synapses," so that during encoding, a sparse pattern of activation in the DG mandatorily causes a postsynaptic CA3 cell to fire.[2,3] However, during retrieval, the CA3 recurrent collaterals and CA3-to-CA1 Shaffer collaterals may dominate in driving CA3 and CA1 cells to perform associative recall.[3–5] Another prominent source of input to the hippocampus, frequently left out of the standard model, is from the subiculum.

Another key property of the hippocampus that is missing from the traditional model is the occurrence of ongoing neurogenesis throughout the life span. Adult neurogenesis in the hippocampus has profound implications for neural coding. Mature dentate granule cells (DGCs) fire very sparsely, as they are under tight control by inhibitory interneurons and might reasonably be assumed to contribute to pattern separation in the DG. However, the newly generated neurons have very different properties. By about 3–4 weeks of age, they have matured to the point that they are able to fire action potentials and contribute to neural coding, and yet at this stage they are much more plastic[28,29] and more highly excitable[30,31] than mature DGCs. Over the next several weeks, the young neurons mature and become progressively more like adult DGCs. How would this varying population of DGCs affect learning and memory? In this chapter, we consider how the idea of a probabilistic, generative model of encoding and prediction could be implemented within the hippocampal circuit using the Restricted Boltzmann Machine (RBM)[32] to simulate the learning process. Furthermore, we consider the contribution of neurogenesis to coding in this model. We then consider how the model could be extended to include more realistic inputs in the form of grid cells and boundary vector cells (BVCs) in the entorhinal cortex (EC) and subiculum/presubiculum. Simulated models with these more naturalistic inputs lead to the emergence of place cells when applying the RBM learning equations. Finally, we present new simulations of a more complete hippocampal model that includes multiple stacked RBMs, and we outline how this full hippocampal model should account for a wide range of neurobiological data including sequence coding and replay.

2 A RBM MODEL OF LEARNING AND NEUROGENESIS IN THE DG

The RBM is a type of artificial neural network that learns a set of weights so as to form a probabilistic, generative model of a data set.[32] The network consists of a set of reciprocally connected stochastic units, partitioned into visible and hidden layers of units, with no within-layer connections. The architecture of a one-layer RBM can be seen in the middle panel of Fig. 21.2. These bidirectional connections between the visible and hidden units form a bipartite graph. Weights in the network are updated based on the difference between the data-driven representations (and the probability distribution of the data set) and the model's expectation-driven representations. While exact calculation of these expectations is intractable, they can be approximated through brief Gibbs sampling.[32] We used one-step Gibbs sampling, and the contrastive divergence learning rule[32] in our RBM based model:

$$\Delta W_{ij} = \epsilon\left(\langle v_i h_j\rangle_{\text{data}} - \langle v_i h_j\rangle_{\text{recon}}\right) \tag{21.1}$$

where ΔW_{ij} is the update to the weight W_{ij} between the ith visible unit and jth hidden unit, calculated by taking the difference between the data-driven and reconstructed Hebbian terms (product of ith visible state and jth hidden state), multiplied by a learning rate ϵ. The input vector v_{data} is the data-driven visible state vector and h_{data} is the data-driven hidden state generated by clamping the visible units' states to v_{data} and sampling the hidden units' states according to Eq. 21.5. Thus, $(v_i h_j)_{\text{data}}$ represents the product of the ith visible and jth hidden activations in the data-driven phase. v_{recon} is a reconstruction of the input vector generated by clamping the states of the hidden units to the data-driven pattern h_{data} and sampling the states of the visible units according to Eq. 21.4. h_{recon} is then created in the same way as h_{data}, but by clamping the visible units' states to v_{recon}.

$$\Delta a_i = \epsilon\left(v_{i_{\text{data}}} - v_{i_{\text{recon}}}\right) \quad (21.2)$$

$$\Delta b_j = \epsilon\left(h_{j_{\text{data}}} - h_{j_{\text{recon}}}\right) \quad (21.3)$$

Units' binary states are updated probabilistically according to Eqs. 21.4 and 21.5 below:

$$p(v_i = 1/h) = \sigma\left(a_i + \textstyle\sum_j h_j w_{ij}\right) \quad (21.4)$$

$$p(h_j = 1/v) = \sigma\left(b_j + \textstyle\sum_i v_i w_{ij}\right) \quad (21.5)$$

where a_i and b_j represent biases, which provide a mechanism for shifting the output of the sigmoid activation function $\sigma(x) = 1/(1 + e^{-x})$. A unit's bias is similar to the negative of a unit's threshold in other neural network models. The weight updates presented in Eqs. 21.2 and 21.3 show that these bias weights are updated using the same positive and negative Hebbian terms used in updating W as shown in Eq. 21.1.

We can see in Eq. 21.1 that the positive Hebbian term $(v_i h_j)_{\text{data}}$ associates data-driven input and hidden state vectors, while the negative Hebbian term $(v_i h_j)_{\text{recon}}$ tries to "unlearn" the association between the corresponding reconstructed visible and hidden state vectors. Theoretically, the learning procedure should converge when the probability distribution of its internal reconstructions of the training patterns exactly match that of the corresponding data-driven states. In general, an RBM model's reconstructions of the training patterns are obtained by brief Gibbs sampling, alternatively sampling hidden and visible unit states that are nearby data-driven states using the model's bottom-up and top-down weights, respectively. Similar to a Hopfield network, the RBM utilizes a local and unsupervised learning rule, which also minimizes the free energy within the network. However, the presence of hidden units, along with the ability to stack RBMs to form deep networks, provides greater memory capacity. Furthermore, the ability to run the RBM in unclamped, top–down or generative mode may provide a way of simulating dreaming along with memory reconstruction and cued recall.

3 THE ROLE OF YOUNG DGCS IN MEMORY ENCODING

We recently proposed a novel RBM-based model of the DG, which incorporates the developmental trajectory of adult-born DGCs.[33] In this model, a single RBM represents the EC input and DGC layer with its visible and hidden units respectively. As the model DGCs undergo development, they become progressively less plastic, more sparse in their firing, and more densely connected to their entorhinal inputs. We demonstrate how these properties can explain the importance of adult-generated DGCs for memory across both short- and long-time scales.

In the model, the maturational trajectory of adult-born DGCs is loosely based on mouse data, for DGCs from the third week of maturation onward. It is at about 3–4 weeks of age that adult-born DGCs establish synaptic afferent and efferent connections and are able to fire action potentials.[34] As compared to more mature neurons, young DGCs have a higher input resistance, lower capacitance, lower activation threshold, and a slower membrane time constant. As a result, 3- to 4-week old DGCs can be described as being more excitable, while having a smaller and slower action potentials.[28,29] Moreover, the young neurons are more sparsely connected to their perforant pathway inputs from the EC relative to mature DGCs.[35] From weeks 5–8 the young neurons undergo a gradual decline in synaptic plasticity and are increasingly regulated by feedback inhibition.[36] By the eighth week, the physiological properties of adult-generated DGCs are largely indistinguishable from their mature counterparts.[35,36]

While several replacement and additive models of neurogenesis have looked at how new neurons affect learning,[37,38] few models have considered the full range of unique properties of adult hippocampal neurogenesis including the developmental trajectory of adult-generated neurons: changes in plasticity, connectivity, excitability, and survival versus apoptosis.

We model the neural trajectory of young DGCs by incorporating additional constraints into the learning equation, including a dynamic learning rate and separate penalty terms for sparse activation (or excitability) versus sparse connectivity. These dynamic constraints are calculated using a Gompertz function $g(t)$ (a roughly S-shaped curve)[39] to model each of the neural growth and decay parameters. At each simulated time step, the age of a hidden unit is increased, and its constraint parameters are updated as follows. A sparse connectivity constraint describes the level of interconnectedness between the visible and hidden layers. Young DGCs are randomly connected to only 30% of EC input units, whereas mature DGCs are fully connected to the input layer. The connectivity proportion varies smoothly from 30% to 100% in direct proportion to $g(t)$, the Gompertz function. Thus, young neurons are initially sparsely connected and transition to becoming fully connected to the input layer. Similarly, the learning rate, which can be thought of as a neuron's plasticity level, is defined as $1 - g(t)$ normalized to lie between 0.0025 and 0.1. Thus, the plasticity level decreases over time as the neuron matures. Finally, our sparse activation cost increases over time, in direct proportion to $g(t)$, scaled to lie between 0.0 (young DGCs) and 0.9 (mature DGCs). Thus, young neurons are unaffected by the sparseness constraint, whereas mature neurons are heavily constrained to have low firing probabilities, and therefore adopt sparse codes, simulating the increasingly tight regulation by feedback inhibition over DGCs as they mature. Given these variable properties, the learning rule can be redefined as

$$\Delta W_{ij} = \epsilon_j\left(\left(v_i h_j\right)_{\text{data}} - \left(v_i h_j\right)_{\text{recon}}\right) - \left(\lambda_j W_{ij}\right) - cost\left(q_j - p\right) \tag{21.6}$$

where the learning rate ϵ, weight decay λ, and sparsity cost term q are now each weighted by dynamically changing vectors of values rather than static hyperparameters, while p represents the current probability of hidden unit j activating.

Returning to our primary objective, what impact does the developmental trajectory of young DGCs have on hippocampal learning and memory? To investigate this, we designed a set of experiments to monitor proactive and retroactive memory interference over short- and long-time scales. Rodents with reduced neurogenesis show greater interference on challenging memory tasks, including *proactive interference* between previously learned information and newly learned information,[40] and *retroactive interference* of newly acquired memories on the retrieval of previously learned information,[41] when interfering items are learned in different sessions spaced over a relatively long time period of days to weeks. It is relatively easy to see why neurogenesis would reduce interference between similar items that are encountered several days apart, as there would be a substantial change in the population of young highly plastic DGCs when the second item is encountered. Interestingly, there is also evidence for a role of neurogenesis in resolving interference between similar items that are interleaved within a single learning/testing session.[42] It is more challenging to explain why neurogenesis would help in this situation, as the population of young DGCs would not be expected to change substantially over a time scale of minutes. We hypothesized that the very sparse, random connectivity of young neurons might be crucial for generating distinct representations of similar items encountered within the same session. This could allow different young DGCs to respond to different features of the input, although they are hyperexcitable and highly plastic.

Given the findings in the literature described above, we sought to test whether the model would exhibit proactive and/or retroactive interference, either within or across learning sessions, as a function of neurogenesis levels. Interference was created by training our models iteratively on highly overlapping sets of patterns with the expectation that new patterns would be more difficult to learn when similar patterns were learned previously (proactive interference), and distally learned patterns would be more difficult to retrieve when new similar patterns were subsequently learned (retroactive interference). We also examined whether this interference would manifest within a single learning session versus across a longer time scale of several days of simulated learning sessions.

In the same-session test conditions, the presence of a few young plastic neurons was expected to help with rapid encoding without creating interference between similar patterns. Likewise, for multisession testing, given that a new pool of young neurons would be available at each learning session, it was expected that the neurogenesis models would perform even better than same-session tests. While this initial set of simulations[33,43] focused on comparing the encoding performance of a single layer model with just the hippocampal DG layer, with and without neural turnover, later in this chapter we present new simulation results with a full multilayer hippocampal model.

The results of these experiments showed that the incorporation of neurogenesis led to improved encoding performance on both short- and long-time scales, in both the same and multisession tests, relative to a control model consisting of an RBM with a sparsely coded DG layer (but lacking neurogenesis). In the case of same session tests, the presence of a few young, more plastic, neurons likely provided the necessary advantage without introducing interference. Interestingly, the neurogenesis models had more overlap among hidden unit activations than the control sparse RBM network. Thus, although the RBM models were free to learn either a sparse code or a highly distributed code to represent the training patterns, our simulation results indicate that the neurogenesis models learned that it was advantageous to have less sparse activations. This may seem paradoxical, as sparse coding and pattern separation are widely believed to be the main mechanisms by which the DG avoids interference between overlapping items. Instead, our simulations suggest an additional mechanism for mitigating interference. The hyperactive young neurons generate highly distributed, overlapping codes for similar items, which allows them to learn distinct features despite being very sparsely connected to their inputs. At the same time, the neurogenesis models showed reduced proactive interference. The increase in accuracy on subsequently learned sets of overlapping patterns suggests that the neurogenesis models may be better at distinguishing novel and common elements to each group of patterns.

In the multisession tests, the incorporation of neural maturation and turnover provided less benefit to overall performance than expected. While the nonsparsely connected neurogenesis model did see about a 1% increase in performance over the same session tests, the addition of sparse connectivity imparted no further improvement; this model performed about the same as its nonsparsely connected counterpart. Thus, sparse connectivity was most important in protecting against interference between patterns learned within a single session, while neural turnover itself was sufficient to mitigate interference at longer time scales. These results suggest that sparse connectivity and neural turnover work in equilibrium with each other, and either one may be relatively more important, depending on the learning task demands.

4 SIMULATING THE EMERGENCE OF PLACE CELLS IN RBMS WITH MORE NATURALIST INPUTS

Our simulations described up to this point used highly simplified synthetic data sets of random patterns with varying degrees of overlap. An important next step in developing more realistic hippocampal models is to use more biologically realistic inputs. Since the discovery of place cells within the hippocampus by O'Keefe and Dovstrovsky,[44] their role in spatial memory and learning have been a focus of hippocampal research for decades. Located in the DG, CA3 and CA1 layers,[45] place cells fire in response to very specific spatial positions or place fields.[44] It is believed that these place cells help form a spatial map in hippocampal memory, but how are these place fields formed?

There are two major types of spatial inputs to place cells that have been relatively well characterized: grid cells and BVCs. Either or both of these inputs could drive the formation of place cells, as explored in the simulations reported in this section. Grid cells, first discovered by Hafting et al. in medial EC,[46] have subsequently also been found in pre- and parasubiculum,[47] all of which are prominent inputs to the hippocampus. Each grid cell has a tuning curve that exhibits a multimodal pattern of firing within an enclosed region of space, such that the cell fires in multiple regularly spaced locations. Remarkably, the locations of peak firing of a given grid cell correspond to the vertices of a hexagonal grid. An example of a simulated input grid cell's tuning curve can be seen in the left panel of Fig. 21.2. Thus, unlike place cells, grid cells respond to multiple locations.[46,48] It is believed that the diversity of the grid cell receptive fields, with varying spatial scales and phase offsets, can encode the spatial locations of an entire environment.[48] A second type of spatial feature that provides input to hippocampal place cells comes from BVCs, which are tuned to the distances and directions to environmental boundaries. For example, a particular BVC might respond best to a boundary along the west edge of an environment, whenever the rat is at short distance from that wall but respond equally well to all locations along

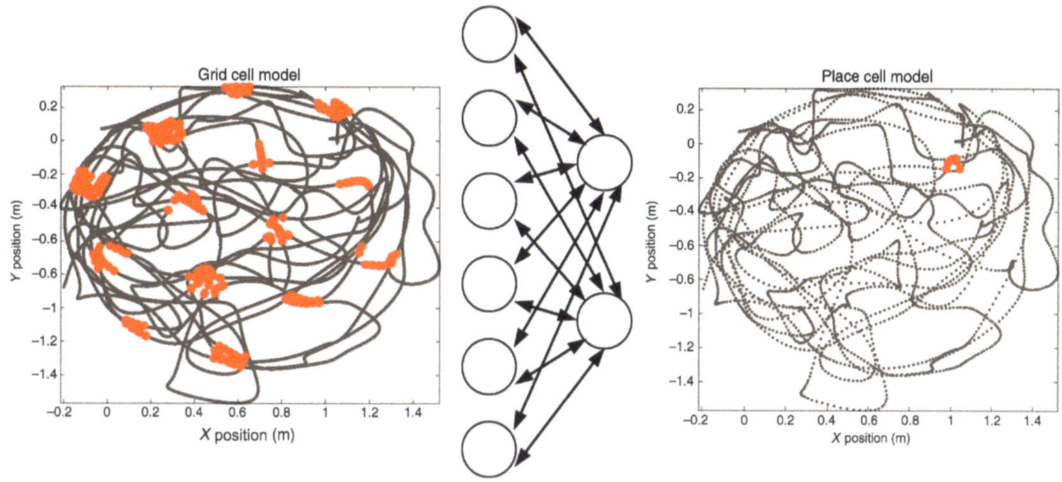

FIGURE 21.2 **Sample activation of a grid cell input and corresponding dentate gyrus place cell activation in our Restricted Boltzmann Machine (RBM) model.** *Left*: Example of firing pattern of a single grid cell used as input to the model. The red points show the grid cell activation, while the simulated rat's path is shown in gray. *Center*: Diagram of the RBM used to re-encode the grid cell firing patterns into place cell fields. The RBM model consisted of a set of visible and hidden units, fully interconnected, with no within-layer connections. *Right*: Sample firing pattern of a single place cell generated by the model's hidden units. Again, the red points show the place cell activation, while the simulated rat's path is shown in gray.

the wall. BVCs were first predicted by a place cell model developed by Burgess and colleagues,[49,50] and empirical evidence for BVCs was later discovered in the subiculum,[51] an important input to the hippocampus.

We have shown that an RBM that receives input from a visible layer consisting of simulated grid cells can generate spatially localized place-tuned neurons in the hidden layer of the RBM.[52] The grid cell input was generated by applying the oscillatory interference model of grid cell firing developed by Burgess and colleagues[53] and adapted by Hasselmo et al.[54] The data taken as input to the model are real data recorded by Hafting et al.[46] as a rat traversed a square environment. Hasselmo applied Burgess' model to this spatial trajectory data by using three head direction cells that respond in relation to the speed of the rat and its orientation. At the same time, their model incorporates an inherent oscillation function that together causes the formation of approximately realistic grid cells. These grid cell patterns are calculated along the same pathway of the rat recorded by Hafting et al. An example of a grid cell pattern imposed over the rat's pathway (gray lines) is shown in the left panel of Fig. 21.2. The activities of a population of 100 grid cells with different spatial scale and phase tuning formed the input to the RBM at each point in time. The dentate layer was also modeled by a population of 100 neurons. The expectation was that the RBM model would successfully be able to re-encode the simulated grid cell fields into place cell fields, which correspond to highly localized unimodal place fields within the source spatial trajectory data.

The results of this study were somewhat mixed. About 40% of DG cells formed small unimodal, localized place fields, such as the one shown in the right panel of Fig. 21.2.

These are typical of place tuning of DG cells. Another 25% formed bimodal place fields, with two different locations causing the cell to fire, and the remainder showed no specific tuning. For simulations with only 10 DG cells, in contrast, the model developed multimodal firing fields.

These simulations demonstrate that the RBM model was able to re-encode the grid cell fields into a place cell representation. However, to identify the expected unimodal place cell firing, the model needed to be trained for 4000 iterations with at least as many place cells as grid cell inputs.[52] Nonetheless, this model demonstrates how an RBM can be used to simulate a real-world situation based on biologically realistic input patterns, bridging the gap between theoretical and experimental research on hippocampal coding and memory.

While many have assumed that grid cells are the primary source of input driving the spatial selectivity of place-tuned hippocampal neurons, there is evidence that other factors influence place cell tuning, including the locations of environmental boundaries and context.[55,56] Moreover, it has been argued that BVCs, rather than grid cells, might be the major driving force for the development of place fields.[57] According to this alternative view, rather than grid cells and place cells forming a hierarchy of representations, they constitute complementary, interacting representations, with the grid cells maintaining one type of spatial representation driven by self-motion and the place cells maintaining another type of spatial representation driven by environmental features such as landmarks (BVCs), objects, and other contextual features. As originally demonstrated by Hartley et al.,[49] a place-selective tuning curve can be computed from a combination of appropriately tuned BVCs. We have further shown how this representation could be learned by an RBM with an input layer consisting of differently tuned BVCs.[58] The model was trained by "traversing" an enclosed environment, generating corresponding input patterns across the population of BVCs inputs. From this input, the model learned the mapping from BVCs to place-tuned neurons.[58]

Our simulations with realistic BVC or grid cell inputs lend further support to the notion that the hippocampus forms generative, probabilistic models of its input, and that the RBM can provide a biologically plausible mechanistic account of the learning process. However, all of the simulations reported up to this point have been with single layer networks. We next consider a multilayer stacked RBM model of the full hippocampal circuit.

5 THE EFFECT OF YOUNG DGCs ON LEARNING IN THE FULL HIPPOCAMPAL MODEL

The one-layer RBM models considered up to this point provide a good account of place tuning and neurogenesis in the DG, but in the absence of the CA3 and CA1 layers, they lack the capacity to model cued recall, sequence learning, and replay. As a continuation of our study on the role of young DGCs in hippocampal learning,[43] we now present a full multilayer hippocampal model to explore the role of young DGCs in cued recall performance. To model cued recall within the hippocampus, we must extend our model to include the associative CA3 layer, which, by the "traditional account" discussed in the Introduction section, performs associative retrieval and pattern completion. Another view of the role of the CA3, with its rich recurrent collateral interconnections, as well as its associative connections to area CA1, is that CA3 (possibly jointly with CA1) plays a crucial role in the formation of temporal associations, anticipatory coding, and sequence learning.[59–61] This is the view adopted here.

Interestingly, by adding another layer to the DG model, we can address an important flaw in the original DG model. Our existing DG model requires reciprocal connectivity between the input and output layers, whereas the known anatomy of the DG does not support this architecture: DGCs do not project back to the EC; however, this can be addressed by incorporating reciprocal connections between the CA3 and the DG,[60,62] and between the CA3 and the EC via a stacked RBM architecture. Specifically, by converting the DG layer into a Conditional Restricted Boltzmann Machine (CRBM), this layer maintains a set of bidirectional connections to the CA3, and only requires a set of conditional unidirectional weights from the EC. This full hippocampal circuit model is used to explore the functional impact of young versus mature DGCs on hippocampal learning, when investigating the performance changes on memory recall (pattern completion) and sequence replay tasks.

To incorporate the CA3 layer into the model, and model associative and temporal learning, a way of modeling the recurrent collaterals is required. While recurrent neural network learning procedures such as back propagation through time,[63–65] long- and short-term memory networks,[66] and liquid state machines[67] have proven effective for modeling sequential data,[68,69] they are not considered to be biologically plausible. We would like to retain the appealing features of the RBM, while extending the operation of the network to modeling sequential data. One such extension is the CRBM,[70] which offers a tractable approximation to a fully recurrent network. The CRBM extends the RBM by adding visible-to-visible and visible-to-hidden autoregressive weights from other (or conditional) visible inputs[70] but treats these extra inputs from the previous time step as fixed, rather than simulating a fully recurrent network. However, the CRBM is not limited to conditioning on these historical observations. By making the DG layer a CRBM with direct input from the EC and including bidirectional connectivity between the CA3 and DG, we can eliminate the need for the feedback projection from the DC to the EC. To simplify, we combine the CA3 and CA1 regions into a single recurrently connected layer, implemented as a CRBM with input from the EC and hidden-to-hidden layer temporal connections. This multilayer architecture is shown in Fig. 21.3. Our bidirectional weights between the CA and DG layers represent the mossy fiber inputs to the CA3 and back projections from the CA3 to the DG. Furthermore, we can use the autoregressive visible-to-hidden weights to represent the EC to DG connections. By doing so, the DG will learn patterns of activation in the CA3 by conditioning on the EC. This provides an RBM multilayer hippocampal model that correctly accounts for the directionality of the known connectivity within the hippocampal structure.

One advantage of multilayer models built upon the RBM is that they can be trained one layer at a time and stacked together. By greedily training a two-layer network and incorporating the developmental trajectory of newly generated neurons (as in our previous DG model) to the CRBM, we can build a full hippocampal model that supports both the recurrent connections in the CA3 and the unidirectional connections from the EC to the DG. We begin by training the CA3 layer on the EC input. We then lock the EC-to-CA3 layer weights and transform the EC input through these weights, producing inputs for training

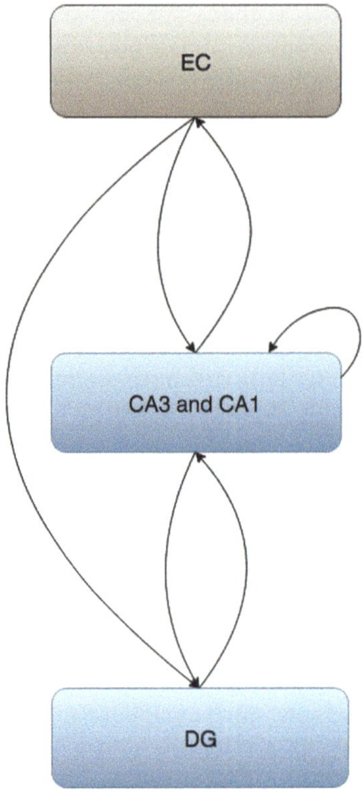

FIGURE 21.3 **A full hippocampal model consisting of stacked Conditional Restricted Boltzmann Machines.** A combined CA3 and CA1 layer is trained on entorhinal cortex (*EC*) input and conditioned on previous EC input. This CA3 and CA1 layer subsequently acts as the input to train the dentate gyrus (*DG*) layer via the bidirectional connections, representing the mossy fibers and CA3 back projections. In this architecture, the EC is treated as conditional input to the DG layer, replacing the bidirectional weights in our previous RBM-based DG models with unidirectional autoregressive weights. This architecture is based on the one proposed by Becker and Hinton (2007)[71] using Temporal Restricted Boltzmann Machines.

the DG layer. We proceed by training the DG layer on these CA3 outputs, conditioned on the initial EC patterns. While the CA3 layer supports the conditional input from previous time steps and will be included in future work, we did not model this property in the simulation experiments reported here. Once both layers of the network are trained, the cued recall performance can be tested by presenting a test pattern as the initial EC input, passing through both layers, sampling the DG hidden layer and reconstructing the pattern back through to the EC. This architecture is based on one proposed by Becker and Hinton[71] and is shown in Fig. 21.3.

To evaluate the role of young DGCs on learning in the full hippocampal model, the same experimental design from the previous study was reused in the current simulations. As before, we incorporated the same mechanisms for simulating the developmental trajectory of young DGCs. To review, we designed a set of experiments to monitor proactive and retroactive memory interference over short- and long-time scales. The model was trained for multiple sessions on a different group of patterns in each session, and there was interference across sessions and within sessions. This interference was created by training our models iteratively on highly similar patterns. The expectation was that new patterns would be more difficult to learn when similar patterns were learned previously (proactive interference), and distally learned patterns would be more difficult to retrieve when new similar patterns were subsequently learned (retroactive interference). We also examined whether this interference would manifest within a single learning session and/or across a longer time scale of several days of simulated learning sessions. The expectation was that the same improvements to encoding performance in the single layer model would be observed in the cued recall experiments with the full hippocampal model.

The results of these experiments are summarized in Fig. 21.4. As with the one-layer DG model, the multilayer hippocampal model showed reduced proactive and retroactive interference when neurogenesis was added, as shown in Fig. 21.4, panels A and C respectively. As with the one-layer model, the degree of DG hidden unit overlap versus input pattern overlap was higher for models with neurogenesis (a decrease in pattern separation), as the more active young DGCs are less selective in their firing patterns (Fig. 21.4, panel B). Across different simulation runs, the neurogenesis models achieved varying degrees of pattern separation (Fig. 21.4, panel B) but in all cases the DG overlap was much greater than that of the sparsely coded models lacking neurogenesis. Despite this reduced pattern separation, the models with neurogenesis usually outperformed those without (Fig. 21.4, panel D). Interestingly, the improved performance for the neurogenesis models appears to be magnified relative to the single EC-DG layer network from the earlier study.

6 DISCUSSION

In this chapter, we have presented several simulation studies that demonstrate how the RBM can be used to model the contribution of neurogenesis in the DG to learning and to model hippocampal learning more generally. We have demonstrated in both a single layer RBM, and a multilayer model of the full hippocampal circuit that neurogenesis contributes to interference reduction. We have also shown that the model can be extended to incorporate more biologically realistic grid cell and BVC inputs. The incorporation of these more realistic inputs led, in both cases, to the emergence of place cells.

Two novel insights emerge from our simulations of models with neurogenesis. First, neurogenesis contributes to interference reduction at both short- and long-time scales. At short-time scales, when there is interference between similar items encountered within a single learning session, neural turnover is less important than simply having a pool of highly plastic, highly active, and very sparsely connected neurons available for coding. There is very little turnover within the course of a single learning session. All else being equal, a set of highly plastic, highly active young neurons could potentially lead to even greater interference, by being equally activated for all items encountered within a session. However, in our simulations, the incorporation of sparse connectivity prevented this from happening, and in fact, sparse connectivity was critical in allowing the young neurons to generate distinct representations of overlapping items within a session. Over longer time scales of days to weeks, neurogenesis can reduce interference even without the sparseness constraint. This is because

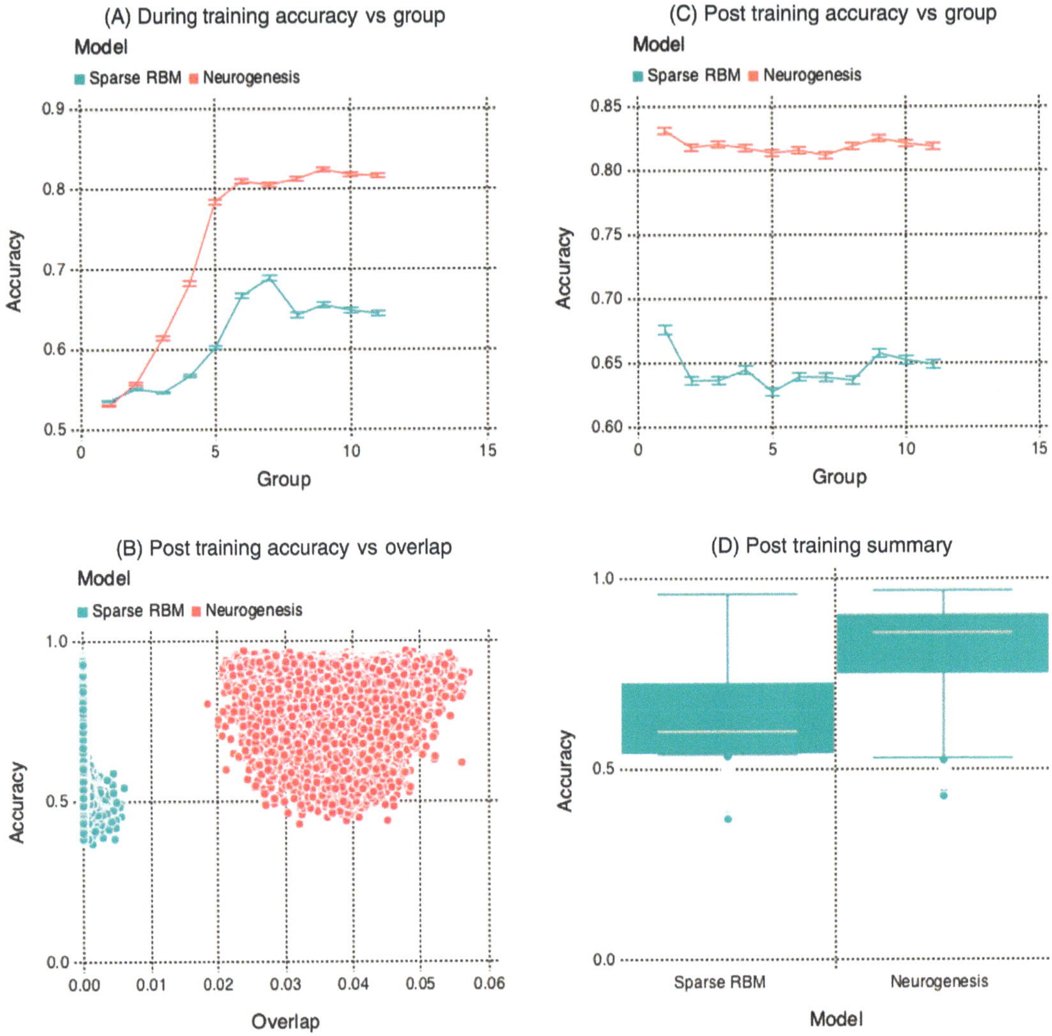

FIGURE 21.4 **Performance of the multilayer Restricted Boltzmann Machine (*RBM*) hippocampal models with and without neurogenesis on across-session cued recall tests.** The models were trained sequentially for 11 sessions, on 11 different groups of 90 patterns. Each group of patterns learned in a given session had high overlap with patterns learned in other sessions. The model was tested on noisy versions of these training patterns after each session to test proactive interference and after all sessions had completed to test retroactive interference. (A) Proactive interference for cued recall accuracies during training. (B) Retroactive interference for cued recall accuracies on each group of patterns. (C) The relationship between post-training recall accuracy and dentate gyrus hidden unit activation overlap. (D) The distribution of post-training accuracy over all sequentially learned groups of patterns.

of ongoing proliferation and maturation of the DGCs. Thus, from one session to the next, the pool of available hyperplastic and hyperexcitable young neurons will have evolved. This continuously evolving pool of young plastic neurons ensures that slightly different subsets of young neurons will be recruited for encoding different memories.

A second novel insight that emerges from our simulations of neurogenesis is that the young, highly active neurons can contribute to interference reduction, perhaps paradoxically, while at the same time decreasing pattern separation. Here, we use pattern separation in the manner in which it was originally intended, referring to the generation of more orthogonalized, or less overlapping, neural codes for overlapping input patterns. In the recent literature, the term pattern separation has also come into the widespread use to refer to almost any behavioral task that requires discrimination between similar items. This is rather problematic, as discussed in further detail by Becker,[72] as pattern separation (orthogonalization) and behavioral discrimination do not necessarily go hand-in-hand. As shown in our simulations, the addition of neurogenesis caused our models to generate more overlapping and yet distinct neural codes for similar items. Thus, orthogonalization (pattern separation, as originally conceived by computational modelers) is not the only possible mechanism for achieving interference reduction.

6.1 Future Experimental Studies

The widespread view that neurogenesis increases pattern separation has also been called into question by some empirical findings. For example, it is well established that neurogenesis levels decline with age.[73,74] If neurogenesis contributes to increased pattern separation, then pattern separation should also decline with age. However, direct assessment of neural activation patterns generated in similar contexts using IEG markers indicate that neural overlap declines with age (indicating stronger pattern separation) and this reduced overlap correlates with a decline in behavioral discrimination of similar contexts.[75] Thus, consistent with our simulation results, an age-related reduction in neurogenesis leads to greater pattern separation, and at the same time, greater interference between similar memories. Further empirical studies are required to tease apart the roles of the young versus older DG neurons in sparse coding and pattern separation, in encoding versus retrieval, in memory and interference reduction at short- (same session) versus long-time scales, and in representing novel information and fine details versus more general information.

While our simulations indicate that in the model, young DGCs contribute to reduced pattern separation within the DG as a whole, it has been suggested that the young neurons could indirectly contribute to pattern separation among the mature DGC population, by generating increased inhibitory feedback over mature DGCs.[76] It also should be noted that some, but not all, behavioral studies have found that neurogenesis knockdown leads to improved performance on high interference memory tasks.[77] These studies appear to further contradict the theory that neurogenesis increases sparse coding, and that this sparse coding reduces memory interference. Thus, further empirical research is needed to fully characterize the functional significance of neurogenesis for mitigating memory interference.

6.2 Future Modeling Studies

The simulations reported in this chapter demonstrated how an RBM model could be used to model place cell fields, neurogenesis, and the developmental trajectory of young

DGCs with a hippocampal circuit. We also showed how our one-layer RBM model can be extended to a stacked, multilayer RBM model of the full hippocampal circuit but have not yet demonstrated the potential of this multilayer model to simulate sequential behavior. Future developments of the model will combine the more realistic inputs, i.e., grid cells and BVCs, with the full multilayer model including temporal associative learning in the CA3 region. This model should learn to encode place sequences. Moreover, in the top-down or generative mode, the model provides a natural account of both forward and reverse sequence replay. A related model has been developed by Fox, Prescott, and colleagues based on particle filters.[78–80] Combining temporal sequence learning with neurogenesis, within a single model, should enable the encoding of distinct event memories for overlapping temporal sequences.

References

1. Marr D. Simple memory: a theory for archicortex. *Philos Trans R Soc B Biol Sci*. 1971;262(841):23–81.
2. McNaughton BL, Morris RGM. Hippocampal synaptic enhancement and information storage within a distributed memory system. *Trends Neurosci*. 1987;10(10):408–415.
3. Treves A, Rolls ET. Computational constraints suggest the need for two distinct input systems to the hippocampal CA3 network. *Hippocampus*. 1992;2:189–200.
4. Hasselmo ME, Schnell E. Laminar selectivity of the cholinergic suppression of synaptic transmission in rat hippocampal region CA1: computational modeling and brain slice physiology. *J Neurosci*. 1994;14(6):3898–3914.
5. Hasselmo ME, Wyble BP, Wallenstein GV. Encoding and retrieval of episodic memories: role of cholinergic and GABAergic modulation in the hippocampus. *Hippocampus*. 1996;6(6):693–708.
6. McClelland JL, McNaughton BL, O'Reilly RC. Why there are complementary learning systems in the hippocampus and neocortex: insights from the successes and failures of connectionist models of learning and memory. *Psychol Rev*. 1995;102(3):419–457.
7. O'Reilly RC, McClelland JL. Hippocampal conjunctive encoding, storage, and recall: avoiding a trade-off. *Hippocampus*. 1994;4(6):661–682.
8. Brown TH, Johnston D. Voltage-clamp analysis of mossy fiber synaptic input to hippocampal neurons. *J Neurophysiol*. 1983;50:487–507.
9. Jung MW, McNaughton BL. Spatial selectivity of unit activity in the hippocampal granular layer. *Hippocampus*. 1993;3(2):165–182.
10. Gothard KM, Hoffman KL, Battaglia FP, McNaughton BL. Dentate gyrus and CA1 ensemble activity during spatial reference frame shifts in the presence and absence of visual input. *J Neurosci*. 2001;21(18):7284–7292.
11. Chawla MK, Guzowski JF, Ramirez-Amaya V, et al. Sparse, environmentally selective expression of Arc RNA in the upper blade of the rodent fascia dentata by brief spatial experience. *Hippocampus*. 2005;15(5):579–586.
12. Leutgeb JK, Leutgeb S, Moser M-B, Moser EI. Pattern separation in the dentate gyrus and CA3 of the hippocampus. *Science*. 2007;315(5814):961–966.
13. Satvat E, Schmidt B, Argraves M, Marrone DF, Markus EJ. Changes in task demands alter the pattern of zif268 expression in the dentate gyrus. *J Neurosci*. 2011;31(19):7163–7167.
14. Nakazawa K, Quirk MC, Chitwood RA, et al. Requirement for hippocampal CA3 NMDA receptors in associative memory recall. *Science*. 2002;297(5579):211–218.
15. Neunuebel JP, Knierim JJ, Alme CB, et al. CA3 retrieves coherent representations from degraded input: direct evidence for CA3 pattern completion and dentate gyrus pattern separation. *Neuron*. 2014;81(2):416–427.
16. Wills TJ, Lever C, Cacucci F, Burgess N, O'Keefe J. Attractor dynamics in the hippocampal representation of the local environment. *Science*. 2005;308:873–876.
17. Tulving E. Memory and consciousness. *Can Psychol Can*. 1985;26(1):1–12.
18. Hassabis D, Kumaran D, Vann SD, Maguire EA. Patients with hippocampal amnesia cannot imagine new experiences. *Proc Natl Acad Sci USA*. 2007;104(5):1726–1731.

19. Rosenbaum RS, Gilboa A, Levine B, Winocur G, Moscovitch M. Amnesia as an impairment of detail generation and binding: evidence from personal, fictional, and semantic narratives in K.C. *Neuropsychologia*. 2009;47 (11):2181–2187.
20. Andelman F, Hoofien D, Goldberg I, Aizenstein O, Neufeld MY. Bilateral hippocampal lesion and a selective impairment of the ability for mental time travel. *Neurocase*. 2010;16(5):426–435.
21. O'Keefe J. Place units in the hippocampus of the freely moving rat. *Exp Neurol*. 1976;51:78–109.
22. Ekstrom AD, Kahana MJ, Caplan JB, et al. Cellular networks underlying human spatial navigation. *Nature*. 2003;425(6954):184–188.
23. Lee ACH, Yeung L-K, Barense MD. The hippocampus and visual perception. *Front Hum Neurosci*. 2012;6:91.
24. Barense MD, Henson RNA, Graham KS. Perception and conception: temporal lobe activity during complex discriminations of familiar and novel faces and objects. *J Cogn Neurosci*. 2011;23(10):3052–3067.
25. Káli S, Dayan P. Off-line replay maintains declarative memories in a model of hippocampal-neocortical interactions. *Nat Neurosci*. 2004;7(3):286–294.
26. Nádasdy Z, Hirase H, Czurkó A, Csicsvari J, Buzsáki G. Replay and time compression of recurring spike sequences in the hippocampus. *J Neurosci*. 1999;19(21):9497–9507.
27. Lee AK, Wilson MA. Memory of sequential experience in the hippocampus during slow wave sleep. *Neuron*. 2002;36(6):1183–1194.
28. Schmidt-Hieber C, Jonas P, Bischofberger J. Enhanced synaptic plasticity in newly generated granule cells of the adult hippocampus. *Nature*. 2004;429(6988):184–187.
29. Snyder JS, Kee N, Wojtowicz JM. Effects of adult neurogenesis on synaptic plasticity in the rat dentate gyrus. *J Neurophysiol*. 2001;85(6):2423–2431.
30. Mongiat LA, Espósito MS, Lombardi G, Schinder AF. Reliable activation of immature neurons in the adult hippocampus. *PLoS One*. 2009;4(4):e5320.
31. Marin-Burgin A, Mongiat LA, Pardi MB, Schinder AF. Unique processing during a period of high excitation/inhibition balance in adult-born neurons. *Science*. 2012;335:1238–1242.
32. Hinton GE. Training products of experts by minimizing contrastive divergence. *Neural Comput*. 2002;14 (8):1771–1800.
33. Finnegan R, Becker S. Neurogenesis paradoxically decreases both pattern separation and memory interference. *Front Syst Neurosci*. 2015;9:136.
34. Zhao C, Teng EM, Summers Jr RG, Ming G-L, Gage FH. Distinct morphological stages of dentate granule neuron maturation in the adult mouse hippocampus. *J Neurosci*. 2006;26(1):3–11.
35. Piatti VC, Ewell LA, Leutgeb JK. Neurogenesis in the dentate gyrus: carrying the message or dictating the tone. *Front Neurosci*. 2013;7:50.
36. Temprana SG, Mongiat LA, Yang SM, et al. Delayed coupling to feedback inhibition during a critical period for the integration of adult-born granule cells. *Neuron*. 2015;85(1):116–130.
37. Becker S. A computational principle for hippocampal learning and neurogenesis. *Hippocampus*. 2005;15 (6):722–738.
38. Weisz VI, Argibay PF. A putative role for neurogenesis in neuro-computational terms: inferences from a hippocampal model. *Cognition*. 2009;112(2):229–240.
39. Gompertz B. On the nature of the function expressive of the law of human mortality, and on a new method of determining the value of life contingencies. *Phil Trans Roy Soc London*. 1832;123:513–585.
40. Luu P, Sill O, Gao L, Becker S, Wojtowicz JM, Smith D. The role of adult hippocampal neurogenesis in reducing interference. *Behav Neurosci*. 2012;126(3):381–391.
41. Winocur G, Becker S, Luu P, Rosenzweig S, Wojtowicz JM. Adult hippocampal neurogenesis and memory interference. *Behav Brain Res*. 2012;227(2):464–469.
42. Clelland CD, Choi M, Romberg C, et al. A functional role for adult hippocampal neurogenesis in spatial pattern separation. *Science*. 2009;325(5937):210–213.
43. Finnegan R. Computational Modelling of Adult Hippocampal Neurogenesis. [master's thesis]. Hamilton, ON: McMaster University; 2016.
44. O'Keefe J, Dostrovsky J. The hippocampus as a spatial map: preliminary evidence from unit activity in the freely-moving rate. *Brain Res*. 1971;34:171–175.
45. Park E, Dvorak D, Fenton AA. Ensemble place codes in hippocampus: CA1, CA3, and dentate gyrus place cells have multiple place fields in large environments. *PLoS One*. 2011;6(7):e22349.

46. Hafting T, Fyhn M, Molden S, Moser MB, Moser EI. Microstructure of a spatial map in the entorhinal cortex. *Nature*. 2005;436(7052):1–8.
47. Boccara CN, Sargolini F, Thoresen VH, et al. Grid cells in pre- and parasubiculum. *Nat Neurosci*. 2010;13 (8):987–994.
48. Rolls ET, Stringer SM, Elliot T. Entorhinal cortex grid cells can map to hippocampal place cells by competitive learning. *Network*. 2006;17(4):447–465.
49. Hartley T, Burgess N, Lever C, Cacucci F, O'Keefe J. Modeling place fields in terms of the cortical inputs to the hippocampus. *Hippocampus*. 2000;10(4):369–379.
50. O'Keefe J, Burgess N. Geometric determinants of the place fields of hippocampal neurones. *Nature*. 1996;381:425–428.
51. Lever C, Burton S, Jeewajee A, O'Keefe J, Burgess N. Boundary vector cells in the subiculum of the hippocampal formation. *J Neurosci*. 2009;29(31):9771–9777.
52. Shaw M. *Modelling Place Cells Through Machine Learning*. [BSc thesis]. Hamilton, ON: McMaster University; 2012.
53. Burgess N, Barry C, O'Keefe J. An oscillatory interference model of grid cell firing. *Hippocampus*. 2007;17 (9):801–812.
54. Hasselmo ME, Giocomo LM, Zilli EA. Grid cell firing may arise from interference of theta frequency membrane potential oscillations in single neurons. *Hippocampus*. 2007;17(12):1252–1271.
55. Burgess N, Cacucci F, Lever C, O'Keefe J. Characterizing multiple independent behavioral correlates of cell firing in freely moving animals. *Hippocampus*. 2005;15(2):149–153.
56. Anderson MI, Jeffery KJ. Heterogeneous modulation of place cell firing by changes in context. *J Neurosci*. 2003;23:8827–8835.
57. Bush D, Barry C, Burgess N. What do grid cells contribute to place cell firing? *Trends Neurosci*. 2014;37 (3):136–145.
58. Byrne P, Becker S. A principle for learning egocentric-allocentric transformation. *Neural Comput*. 2008;20 (3):709–737.
59. Levy WB. A sequence predicting {CA3} is a flexible associator that learns and uses context to solve hippocampal-like tasks. *Hippocampus*. 1996;6:579–590.
60. Jensen O, Lisman JE. Hippocampal CA3 region predicts memory sequences: accounting for the phase precession of place cells. *Learn Mem*. 1996;3(2–3):279–287.
61. Mehta MR. From synaptic plasticity to spatial maps and sequence learning. *Hippocampus*. 2015;25(6):756–762.
62. Myers CE, Scharfman HE. Pattern separation in the dentate gyrus: a role for the CA3 backprojection. *Hippocampus*. 2011;21(11):1190–1215.
63. Williams RJ, Zipser DA. A learning algorithm for continually running fully recurrent neural networks. *Neural Comput*. 1989;1(2):270–280.
64. Pineda F.J. Generalization of back propagation to recurrent and higher order neural networks. In: *Proceedings of IEEE Conference on Neural Information Processing Systems*. Denver, CO; 1987.
65. Pearmutter BA. Learning state space trajectories in recurrent neural networks. *Neural Comput*. 1989;1 (2):263–269.
66. Hochreiter S, Schmidhuber J. Long short-term memory. *Neural Comput*. 1997;9(8):1735?1780.
67. Maass W, Markram H. On the computational power of circuits of spiking neurons. *J Comput Syst Sci*. 2004;69 (4):593–616.
68. Graves A. *Supervised Sequence Labelling with Recurrent Neural Networks*. Vol 385. Springer; 2012.
69. Schmidhuber J, Wierstra D, Gomez FJ. *Evolino: hybrid neuroevolution/optimal linear search for sequence learning. Edinburgh: Proceedings of the 19th International Joint Conference on Artificial Intelligence (IJCAI)*. San Francisco, CA: Morgan Kaufmann Publishers Inc., 2005:853–858.
70. Taylor G, Hinton G, Roweis S. Modeling human motion using binary latent variables. *NIPS*. 2007;19:1345–1352.
71. Becker S, Hinton GE. Caching and replay of place sequences in a Temporal Restricted Boltzmann Machine model of the hippocampus. *Society for Neuroscience*. 2007.
72. Becker S. Neurogenesis and pattern separation: time for a divorce. *Wiley Interdiscip Rev Cogn Sci*. 2016.
73. Seki T, Arai Y. Age-related production of new granule cells in the adult dentate gyrus. *NeuroReport*. 1995;6 (18):2479–2482.

74. Kuhn HG, Dickinson-Anson H, Gage FH. Neurogenesis in the dentate gyrus of the adult rat: age-related decrease of neuronal progenitor proliferation. *J Neurosci*. 1996;16(6):2027–2033.
75. Marrone DF, Adams AA, Satvat E. Increased pattern separation in the aged fascia dentata. *Neurobiol Aging*. 2011;32(12):2317.e23-2317.e32.
76. McAvoy K, Besnard A, Sahay A. Adult hippocampal neurogenesis and pattern separation in DG: a role for feedback inhibition in modulating sparseness to govern population-based coding. *Front Syst Neurosci*. 2015;9:120.
77. Hvoslef-Eide M, Oomen CA. Adult neurogenesis and pattern separation in rodents: A critical evaluation of data, tasks and interpretation. *Front Biol (Beijing)*. 2016;11:168.
78. Fox C, Prescott T. Hippocampus as unitary coherent particle filter. *Proc Int Jt Conf Neural Networks*. 2010.
79. Fox C, Prescott T. Learning in a unitary coherent hippocampus. *Lect Notes Comput Sci (including Subser Lect Notes Artif Intell Lect Notes Bioinformatics)*. 2010;6352(PART 1):388–394.
80. Boorman LW, Damianou AC, Martinez-Hernandez U, Prescott TJ. Extending a hippocampal model for navigation around a maze generated from real-world data. In: Conference on Biomimetic and Biohybrid Systems. Springer: 2015;441–452.

SECTION VI

STRUCTURAL PLASTICITY AND PATHOLOGY

CHAPTER

22

Modeling the Impact of Lesions in the Brain

Sol Lim[1,2], *Frances Hutchings*[1] *and Marcus Kaiser*[1]

[1]Newcastle University, Newcastle upon Tyne, United Kingdom [2]Indiana University, Bloomington, IN, United States

OUTLINE

1 INTRODUCTION

The brain shows dynamic activity and plasticity according to intrinsic and extrinsic stimuli constantly adapting and interacting with its surroundings. Neural systems also maintain robust functionality while undergoing massive internal changes during development and ageing. Moreover, the brain can be remarkably robust to physical damage. The robustness of the brain comes from its degeneracy[1–4]; multiple sets of structures can

The Rewiring Brain
DOI: http://dx.doi.org/10.1016/B978-0-12-803784-3.00022-6

encode the same function; thus if a set of structures fails to deliver the function, other sets of elements can provide alternatives that may secure comparable performance of the system. Significant loss of neural tissue can be compensated for in a relatively short time by large-scale adaptation of remaining brain parts when experienced during early development.[5,6] Patients with severe pathology in their brains may show little or no clinical manifestation of the disease,[7] and it was hypothesized that those patients may have had a greater "brain reserve", allowing them to have greater resilience against brain damage.[3,8] However, the removal of small amount of tissue (e.g., in Broca's area) can lead to a severe functional deficit and widespread disturbance.[9,10,12] These findings provide a somewhat contradictory picture of the robustness of the brain and suggest a number of questions. Can we systematically evaluate robustness of the brain functionality given this variability in the effects of brain lesions? Are severity and nature of effects of localized damage predictable? We delineate here how brain connectivity can be brought to answer these questions.

The functionality of any system is grounded in its structural properties. A brain can be represented as a graph (network) comprising sets of nodes and edges called *connectome*; microscopically neurons form synapses with dendrites and axons to transfer neural signals, and macroscopically areas of gray matter are interconnected to each other through white matter fiber bundles. Moreover, by representing a brain as a network, we can study brains from other species in the same framework using this graph-theoretical approach.[12,13] As the brain is also constrained by its embedding space and metabolic cost, both spatial characteristics and topological traits should be considered. For a synaptic network, as synapses are not necessarily reciprocal between two neurons, we consider a directed network. For a large-scale brain network using neuroimaging such as diffusion-weighted imaging or functional Magnetic Resonance Imaging (fMRI), we construct a brain network as an undirected network as information about directionality is difficult to derive. However, using invasive *postmortem* tract tracing after dye injection, we have information about the direction of fiber tracts in some cases. After representing the brain as a graph, we can investigate various topological, spatial, and dynamical properties of the brain network. This connectomics approach has led to the exploration of the structural properties of brain networks, such as topology, hodology, community structure (modularity), and hierarchical organization.[13–24] Recent advances in network science have extended those traditional structural descriptions,[25] allowing characterization of networks as small world,[26] random or scale-free.[27,28] Studies have shown that the brain network is a small-world network where brain areas are clustered with neighboring areas forming separate communities or *modules* for specialized functions, such as visual and auditory systems, while having short-cut pathways for efficient information flow.[13] In addition, the brain network is scale-free, where hub nodes, or highly connected nodes exist and they are densely connected to other hub nodes, making it resilient to random attacks[29] (Fig. 22.1). In addition to analyzing the brain network in terms of its topological properties, we can also explore network perturbation and failure by simulating dysfunctions of the network from brain diseases and injuries.[3]

Brain lesions caused by a stroke or traumatic brain injury can be modeled using connectome analysis. Stroke-affected or injured areas in the brain suffer from cell deaths and subsequent disconnections between neurons and between areas. We can simulate the effect of network perturbation by removing nodes (neurons or areas of the brain) or edges (synapses, dendrites, or axons) and we investigate how the lesions affect functional and

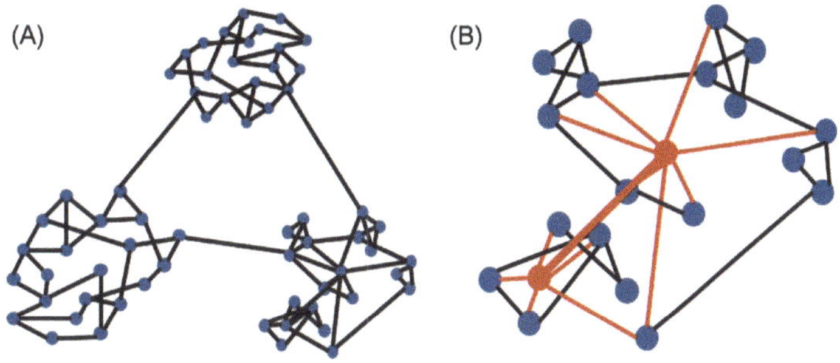

FIGURE 22.1 **Network organization with potentially crucial network components.** (A) Brain networks show a modular organization with many connections within modules and fewer connections between modules. These latter connections are bottlenecks for information transfer. Connections and nodes that are involved in linking modules are therefore potentially crucial for brain function. (B) Neural systems also show a rich-club organization where highly connected nodes or hubs (shown in *red*) are more strongly connected between each other than expected from the degree-preserving randomized benchmark network. This backbone of hubs, providing short cuts to reach other parts of the network, is functionally important.

structural integrity of the brain network and consequently how those would also influence the dynamic integrity of the network.

2 MODELS OF NETWORKS AND THEIR DYNAMICS

In the following sections we review the effects of lesions on networks and their dynamics. First, we go over virtual lesion impact on network measures by removing nodes and edges in the network. Next, we present how dynamics of the network is affected by virtually lesioning the network. Finally, we discuss how the brain network may adapt to brain damage during recovery.

2.1 Network Centrality

Hub nodes play an important role in integrating and distributing information of the network. Nodes with high betweenness and eigenvector centrality are used more often than others for shortest paths between nodes. Edges with high betweenness are used more often than other connections for shortest paths. When these nodes and edges with high centrality are removed or disrupted, the brain network can suffer serious problems in network efficiency. Edges with high-edge betweenness are chosen for targeted attack. Indeed, edge betweenness has been shown to highly correlate with structural network damage for cortical and other biological networks.[30] Removal of crucial nodes and edges of the brain network can lead to disrupted small-world topology and modular structure of the network. This can be measured by clustering coefficient, path length, global and local efficiency, and rich-club coefficient.[31–35]

The clustering coefficient shows how locally clustered a network is. It is the ratio of the number of existing edges between neighbors of a node i and the number of possible edges between all these neighbors. We considered neighboring nodes of node i to be all those nodes that have incoming or outgoing connections between them and node *i*. If a node *i* has k_i neighbors, then the number of all possible incoming and outgoing edges between the neighboring nodes is $k_i \times (k_i - 1)$. The coefficient itself is a local property of each node, and we define the average coefficient of all nodes to be the clustering coefficient of the graph.

Following Albert et al.,[27] we considered the average shortest path (ASP) or characteristic path length to characterize the network connectivity and integrity.[36] The ASP between any two nodes in the network is the average number of sequential connections that are necessary to link one node to another by the shortest possible route.[37] In case a network becomes disconnected in the process of removing edges or nodes and there is no path between two nodes, the pair of nodes is ignored. If no two connected points are left, the ASP is set to zero. Note that for networks with directed edges, the shortest path from node *i* to node *j* may not be the same as that from node *j* to node *i*.

2.2 Hierarchical Modular Network

The brain network consists of several modules where nodes are more interconnected within its module compared to other modules.[24,38] This community structure of the brain is formed in early development[39,40] and matures with age from "local to distributed" network, enabling the brain to have more efficient and integrated functionality while maintaining independent subsystems.[96–102] Moreover, this stable hierarchical community structure prevents fast spreading of the brain activity as seen for seizure spreading; instead such a hierarchy provides the substrate for sustained activity.[41] Therefore, perturbing the hierarchical modular structure of the brain can lead to serious damages to the brain.[3,42]

2.3 Dynamic Effects of Virtual Lesions

The network topology of the brain constrains the neural dynamics that arise from it, and the repercussions of disrupted brain connectivity in historical case studies and in computational models corroborate the importance of the network topology on neuronal dynamics.[43–45] In this context, neuronal dynamics refers to fluctuations in brain activity, both spontaneous and nonspontaneous (evoked by extrinsic stimuli). We can observe changes in network activity through computer simulations. The change before versus after removal of nodes or edges can indicate the influence of the removed network component. Effects can be measured both in terms of topological properties and dynamical features. Depending on the model, dynamical features might be spreading of activity, synchronization between nodes, transitions between different states, or oscillation properties of individual nodes. Such dynamical features can be observed not only in a computational model but also in experimental data ranging from fMRI and electroencephalogram (EEG) recordings to invasive measurements using *in vivo* or *in vitro* multielectrode array, electrocorticography (ECoG), and depth electrodes. Besides electrophysiological signals, indirect measures such as near-infrared, cognitive behavioral tests, and other techniques can be used.

Using computer models, lesion effects can be investigated systematically with low cost by inducing virtual lesions, targeting specific areas and connections.[44,46,47] Patients with brain injuries have different sizes of affected areas and differing etiologies, which complicate a systematic investigation of the lesion effect. Animal models can be used; however, they are costly, ethically controversial, and lesion experiments cannot be repeated many times, unlike computer simulations and models. Noninvasive brain stimulation techniques such as transcranial magnetic stimulation (TMS) can induce a temporary "virtual lesion" effect to observe immediate dynamic changes, but such experiments are also costly and the effects are short lived.[48,49]

In particular, lesion modeling studies using network analysis can be used to aid diagnosis and prognosis after the stroke or injury,[50] to predict the outcome of the surgery for epileptic patients,[51,52] and to help guide rehabilitation of stroke patients.[53]

In more abstract models, the impact of network topology on dynamics can be studied with a smaller parameter space. Simple oscillators coupled and connected according to human structural connectome data can simulate patterns of activity found in fMRI.[54,55] These oscillators represent the activity of a brain region as being either in or out of phase with other brain regions,[56] with regions interacting through phase modulations.[57] Functional networks can be constructed based on the statistical dependency of signals between brain regions in time domain (cross-correlation), frequency domain (coherence), or using relative phase differences (phase-locking value, entropy, etc.).[58,59] It is assumed that regions show similar activity patterns or "synchronized" patterns because they share information or communicate with each other,[60] and they are often directly linked through a fiber tract. Measures such as synchrony and metastability (the presence of transient periods of stability) have been used to quantify changes in dynamics when virtual lesions are applied to the model.[47,61] Abstract models such as these make it easy to analyze the dynamics of the network due to their simplicity; however, they are often criticized because they can be difficult to relate to actual biological systems.

More complex neural mass models (NMMs) estimating the dynamics of large populations of neurons have been used in place of simple oscillators and similarly coupled based on human connectome data.[62] While there are many variants, most NMMs use differential equations to estimate the expected mean firing rates over time of populations of different neuronal types to explain observed EEG and magnetoencephalography activity.[63] One unit of coupled neuron populations is often taken to represent a cortical column,[64] and NMMs assume that neurons of the same type in the same region tend to synchronize and share similar inputs.[65] These different populations (e.g., inhibitory interneurons, excitatory pyramidal cells, etc. [64,66]) are coupled together and impact each other's dynamics. They also receive input activity from other external connected cortical columns, and in turn send an output of internally generated activity to other units. NMMs therefore provide a clearer link with the underlying biology than simple oscillators. Both abstract and biologically more plausible models have been used to predict lesion effects on brain dynamics.[47,67,68]

2.4 Plasticity and Recovery From Virtual Lesions

The brain network's potential to recover from lesions using plasticity has been modeled mainly at the mesoscopic level of cortical maps rather than at the macroscopic whole brain connectome level. Lesions applied to cortical maps show recovery through reorganizing the maps[104,50,69] and behavioral changes after recovery can be investigated by testing the performance of the network with a comparison between pre- and postlesion.[50]

There have been a number of computational models on recovery process of the brain after damage, particularly with focal stroke damage. They have provided important insights for improving rehabilitation-training procedures. For instance, neuronal population models of sensorimotor cortex were trained to control a virtual arm,[70,71] and they were lesioned to investigate motor deficit recovery. These models can provide better ways to improve rehabilitation protocols,[71] in particular, about the importance of rehabilitation timing for optimal recovery.[70] Another mathematical model study made use of patient arm movements, and the encoding of them showed that bimanual movements were more effective than unimanual for motor recovery due to the increase in "preferred directions," inducing rotations in these directional preferences in damaged cortex.[53,72] In each case, neuronal populations of interest are simulated and then lesioned, with model implementations of plasticity taking various forms, by using different methods to determine how connections change over time such as reinforcement learning algorithms or optimization of a given gain function.

Other functional areas have been modeled; vision and language models can replicate the observable adverse effects lesions can cause—hemianopia and reading issues—and also the importance of lesion size and lateralization,[73] as well as showing that lateral interactions play a role in visual field recovery.[74] While many models focus on a particular functional area, some investigate the importance of lateralization of function in recovery. Models looking at both hemispheres and the corpus callosum have attempted to explain the use and importance of both sides of the brain in damage mitigation, finding the importance of preexisting lateralization and lesion size on recovery.[75] For more information, see detailed review articles.[76,77]

2.5 Further Applications of Lesion Modeling

Aberrant functional and structural connectivity has been shown in a variety of disorders including depression, epilepsy, schizophrenia, and Parkinson's disease (PD).[78–81] Models can provide a greater understanding of how network differences between health and disease can lead to knock-on alterations in dynamics. Endeavors in progress such as the Virtual Brain Project[82] aim to give a personalized simulation of large-scale brain dynamics that arise from an individual's connectome. Deviations in dynamics from the healthy norm may be detected, and predicting how lesioning may alter brain dynamics may be used to find a way to shift a disturbed network state of a patient closer to a healthier state.[82] At the mesoscopic level, software such as VERTEX[83,84] has been shown to replicate dynamics from simulations of connected neurons and has the potential to estimate effects of damage or drugs on local brain tissue dynamics.

On a disease-specific basis, more in-depth models of disease dynamics have provided already a useful modality for the study of network disruption and lesion effects. To illustrate with a recent example, we adapted a model of epilepsy to incorporate patient-specific connectivity and region surface area as derived from brain imaging. The activity of each node in the network was described by differential equations such that the activity could be either in a normal resting state or in a seizure-like state. Each node could transition into the seizure state with a probability dependent on its surface area and incoming connections with the network topology playing an important role in whether a node transitioned to a seizure state. In this model, simulated lesions were made to see whether removing them reduced the likelihood of seizures in areas that were most prone to them.[51] For other disorders where the network of the brain may exhibit abnormalities, similar models can be useful to evaluate how changes in network topology influence the disease spread and severity.

3 RESULTS

3.1 Important Nodes and Edges: Theoretical Studies

When observing the effects of virtual lesions within computer models, there are two strategies of removing components that are usually tested. Random removal picks nodes or edges randomly, whereas targeted (or attack) removal deletes components in order of their predicted importance for the network, starting with the most potentially important nodes first. For node removal, highly connected nodes or hubs are predicted to have a high impact as they could function as integrators or broadcasters (or both) of global network information. As we will see later, they are also implicated to play a role in many brain network diseases. Early theoretical studies based on cat and macaque connectome[36] showed that hub removal leads to a rapid disintegration of the network (Fig. 22.2). Random removal of nodes leads to a slow increase in average shortest path length before the network fragments into disconnected components, and the path length of all remaining paths between nodes goes down. For targeted removal, however, the loss of hubs leads to a rapid increase in path length followed by an early fragmentation, as seen through the path length decrease, when only 20% of the nodes are removed (Fig. 22.2A).

For edge removal, edge betweenness or the number of shortest paths that pass through an edge has been used as a predictor of importance. Again, targeting high-edge betweenness connections leads to a steeper increase and an earlier decrease in ASP, due to fragmentation, compared to random edge removal (Fig. 22.2B). Compared to node removal, the effects of edge removal are more gradual. However, one has to keep in mind that the targeted removal of nodes involves the removal of a large number of connections, whereas each step in the edge removal simulation plot only removes a single edge.

To investigate different effects of lesions depending on the topology of each network, we constructed benchmark networks with matched number of nodes and connections but with different topologies: rewired, random, scale-free and small-world (see Ref. [36] for more details). The same behavior to random and targeted removal appears for benchmark scale-free networks (Fig. 22.2C and D) but not for small-world or random benchmark networks (not shown). This indicated that the existence of hubs, irrespective of other organizational features, is sufficient to explain why brain networks are, on average, relatively robust to random removal but can react strongly to the removal of few components (hub nodes or edges with high-edge betweenness).

Other studies have investigated virtual lesions based on functional[85] and structural connectivity in humans.[44] Using resting-state fMRI, hub removal led to a more rapid increase in ASP and later disintegration of functional connectivity networks.[85] For structural

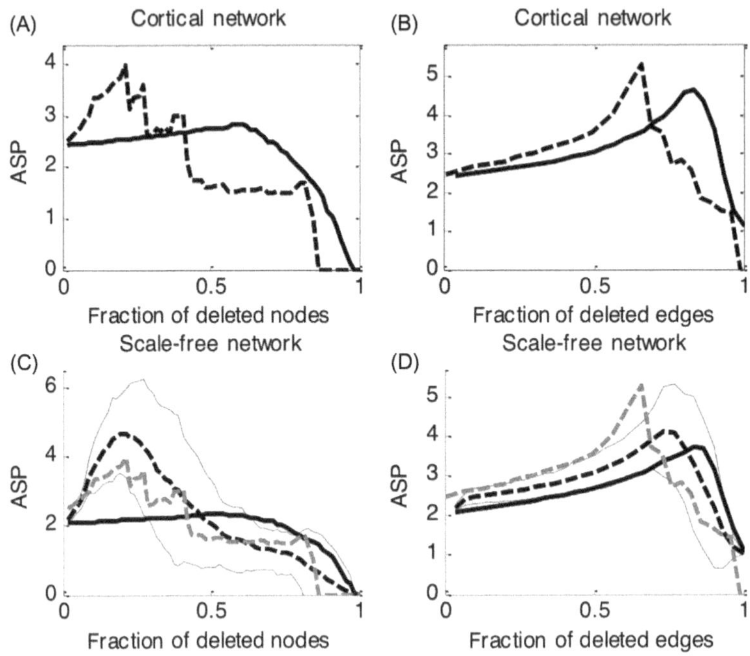

FIGURE 22.2 **Effect of node and edge removal on structural network features.** Cortical network of 66 brain regions of the macaque monkey shown in parts A and B; benchmark networks, generated to have similar properties to the real network and to be scale-free, shown in parts C and D. In all cases, *thick solid lines* represent a random removal of nodes or edges, whereas *dashed lines* represent a targeted approach attacking highly connected nodes (hubs) or edges with high-edge frequency (or edge betweenness), respectively. (A) The average shortest path (*ASP*), related to the characteristic path length[12], for the intact network (fraction = 0) and different stages of node removal up to removing all nodes of the network (fraction = 1). (B) The ASP for the intact network (fraction = 0) and different stages of edge removal up to removing all edges of the network (fraction = 1). (C and D) Node and edge removal for scale-free benchmark networks showing average ASP changes following targeted (*dashed thick lines*) and random (*solid thick lines*) node elimination; the thick gray dashed line represents the targeted removal in the real network (black dashed in parts A and B) for comparison. 95% confidence intervals (*thin gray lines*) are shown for the targeted removals in the benchmark networks. Source: *Kaiser M, Martin R, Andras P, Young MP. Simulation of robustness against lesions of cortical networks. Eur J Neurosci. 2007;25(10):3185–3192.*

connectivity, based on diffusion spectrum imaging, hub removal led to a steeper reduction in global efficiency compared to random removal. However, an even larger effect was observed for targeting nodes with high node betweenness centrality measuring how many shortest paths pass through a node. In addition, these potentially important nodes also have a significant effect on simulated network dynamics.[44]

While the importance of a node can be linked to the number of its connections, it is unclear what is special about an important edge. We generated benchmark networks with three distinct modules to observe the importance of inter- versus intramodular connections. We found that the removal of connections between modules led, on average, to a higher increase in ASP length than removal of connections within modules (Fig. 22.3). That means after removal of a connection, the shortest alternative pathway is considerably longer. At the same time, connections between modules also showed a higher edge

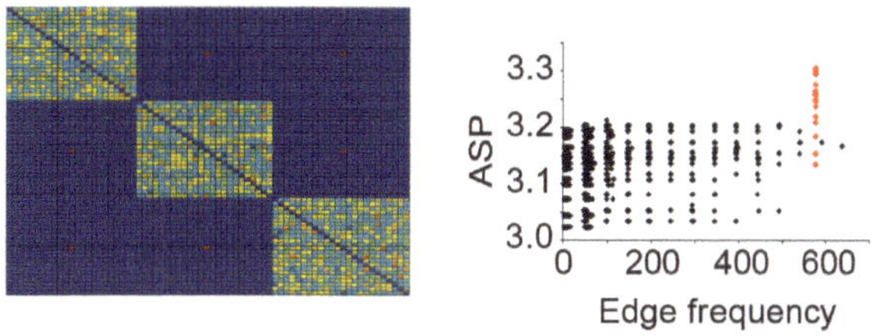

FIGURE 22.3 **Effect of edge removal on structural network features.** (*Left*) Connectivity matrix, averaged over 20 generated networks with three modules and fixed connections between modules. Color indicates edge betweenness, here called edge frequency, ranging from zero for absent connections (*blue*) to the maximum value (*red*). (*Right*) The average shortest path (*ASP*), comparable to the characteristic path length in other studies, after removing a single connection. Each data point represents the edge frequency of a connection and the ASP of the network after the removal of that connection. Connections between the three modules, highlighted in *red*, show both a higher edge frequency and, on average, a stronger effect as a result of removal. Source: *Kaiser M, Hilgetag CC. Edge vulnerability in neural and metabolic networks. Biol Cybern. 2004;90(5):311–317.*

betweenness than intramodular connections (Fig. 22.3). This explains why edge betweenness is a good predictor of the effect on ASP but also how the modular organization gives rise to connections that are much more important than others. Indeed, edge betweenness is a better predictor of importance with respect to changes in ASP than measures that observe the node degree of both nodes that are connected by an edge.[86–89]

In terms of brain dynamics a complementary picture can be seen. Vasa and colleagues,[90] by using anatomically coupled Kuramoto oscillators, found evidence that the extent of disruption to dynamics is related to the structural importance of nodes. Lesions caused most change to synchrony in the posterior regions, which are highly connected. Additionally, nodes with greater eigenvector centrality had a greater effect on global synchrony. Models suggest that focal lesions can have sweeping effects on neuronal dynamics dependent on the connectivity of the lesioned area, even reaching the contralateral hemisphere.[68]

In addition to revealing how and why neurological disorders can arise from a disrupted network, lesion models have been applied to aid in development of treatments for brain diseases. Deep brain stimulation has been tested for its efficacy in simulation models to treat PD, impeding abnormal synchronization of the network by inducing a temporary lesion.[91] Moreover, modeling coupled neuronal populations has been used to provide insight into the mechanisms behind network observations for schizophrenia.[92] Hutchings and colleagues[51] looked at simulations of virtual surgery as a treatment for temporal lobe epilepsy, using individual patient's connectome information to investigate seizure dynamics (Fig. 22.4). The effects of lesions on seizure dynamics were assessed by the impact on seizure likelihood. By severing nodes corresponding to areas that would be removed in corrective surgical procedures, the model could predict the effect of such lesioning on seizure likelihood. This can predict the prospective success of potential surgery in each individual, a step towards patient-specific surgery planning. The model showed that the conventional surgery procedure improved seizure likelihood in the majority of cases, although not always significantly. Additionally the model provisionally predicted alternative patient-specific surgery approaches, for further improving the seizure likelihood reduction. A similar study using ECoG data has also found promising results for surgery outcome prediction[52] supporting the viability of the approach. Similar models have been implemented to predict optimal stimulation sites, as an alternative to surgery for affecting seizure dynamics.[93]

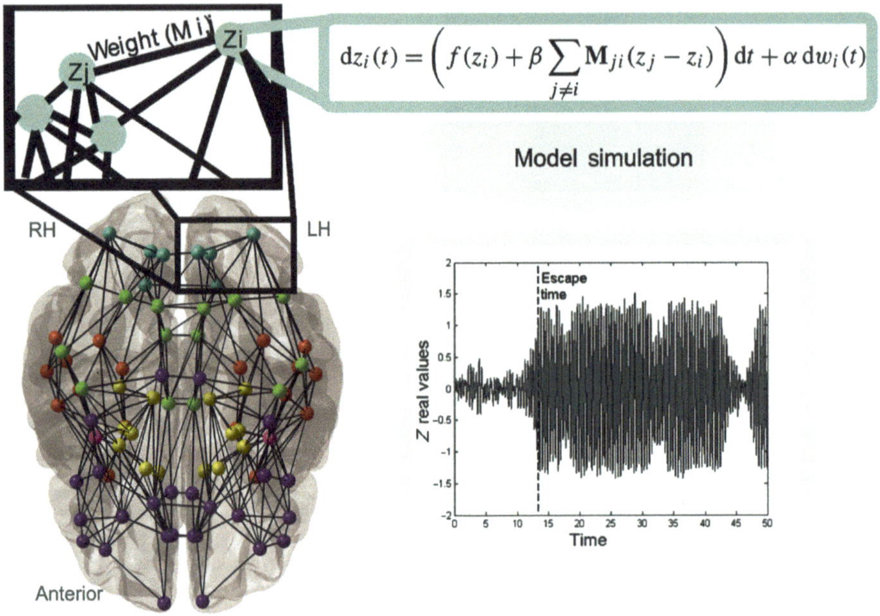

FIGURE 22.4 **Effect of virtual lesions on network dynamics.** The effect of removing nodes or edges on network dynamics can be studied through computer simulations. In this example of a computer simulation used for evaluating lesion effects, the image shows a schematic of how the activity of individual node is dependent on the network structure. The brain network is constructed from diffusion-weighted imaging (DWI) data, where nodes represent brain regions and the connections between them are defined by the number of streamlines using tractography based on DWI. Node colors show the lobes that the regions belong to. The activity of each node is determined by the equation (*blue box*), and the output activity (plotted in the *gray box*) depends on the node's current state and the activity of its neighbors. In this example, the model tests if and when the activity of a node (Z real values) resembles epileptic activity, and the "escape time" as marked on the activity plot is the time at which a node transitions to showing epileptiform activity. Virtual removal of connections and nodes can then be explored to find optimal strategies to reduce the likelihood of epileptic activity, resembling seizures in patients. Source: *Hutchings F, Han CE, Keller SS, Weber B, Taylor PN, Kaiser M. Predicting surgery targets in temporal lobe epilepsy through structural connectome based simulations. PLoS Comput Biol. 2015;11(12):e1004642.*

3.2 Important Nodes and Edges: Experimental Studies

We have seen that certain features of the brain are more important than others, which are kept and strengthened during maturation[94–102] (Fig. 22.5). For instance, long-distance connections and hub nodes formed in early development[103,104] (Fig. 22.6) are crucial to facilitate efficient information flow in the brain[23,105]; intermodular connections become strengthened during development, refining and integrating the network[96–98] (Fig. 22.2); and hub nodes themselves form a tight community called a rich-club responsible for major brain functions and failures.[106–109] Therefore, removing important topological and spatial features of the brain can be detrimental; the brain network becomes less stable and less efficient leading to dysfunctions of the brain, possibly giving rise to neurological disorders.

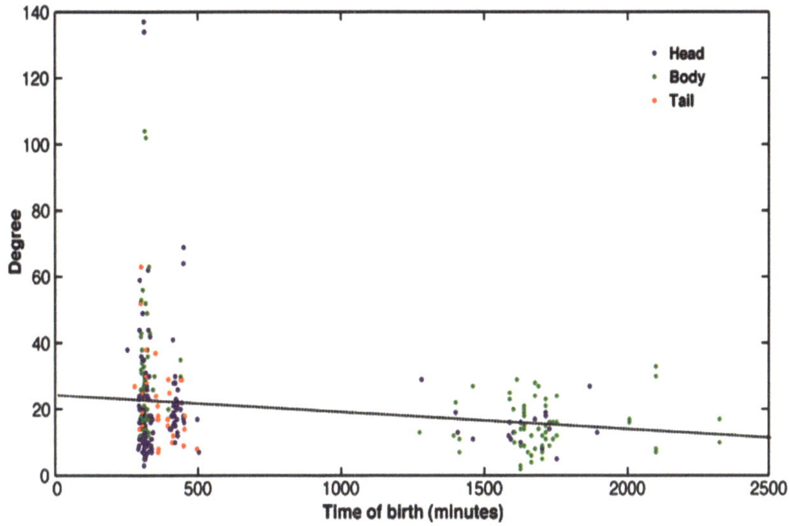

FIGURE 22.5 **Hubs arise early on during development.** For *Caenorhabditis elegans*, the birth time of a neuron is related to the number of connections (degree) that a node has at the adult stage. All nodes that are highly connected at the adult stage, with node degrees above 40, are born early on before the time of hatching (800 min). Source: *Varier S, Kaiser M Neural development features: spatio-temporal development of the Caenorhabditis elegans neuronal network. PLoS Comput Biol. 2011;7(1):e1001044*[146].

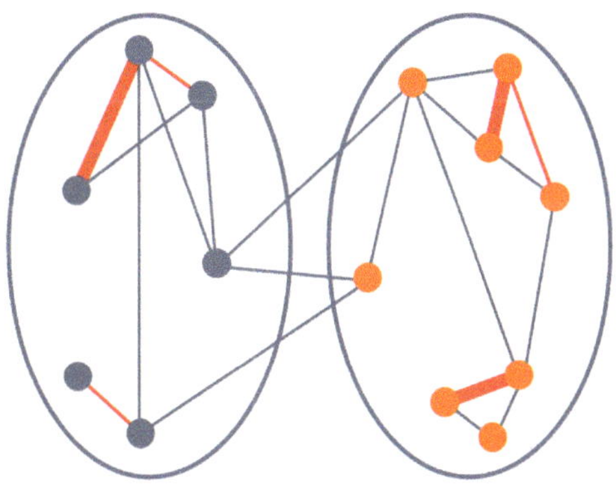

FIGURE 22.6 **Sparing of crucial connections during healthy brain development.** A schematic overview of reductions in structural connectivity strength between the ages of 4 and 40 years in humans. Pruning of connectivity (*red*) during development preferentially occurs for connections within modules, within hemispheres (*ellipses*), for short-distance connections and for strong fiber tracts. In other words, weaker long-distance connections that are crucial for functionally linking modules and hemispheres remain intact during development. Source: *Lim S, Han CE, Uhlhaas PJ, Kaiser M. Preferential detachment during human brain development: age- and sex-specific structural connectivity in diffusion tensor imaging (DTI) data. Cereb Cortex. 2015;25(6):1477–1489.*

4 DISCUSSION

4.1 Important Nodes and Edges: Brain Diseases

In line with the theoretical studies in the previous section, researchers have reported evidence of brain network disturbances in neurological disorders.[86] Studies have demonstrated that structural and functional hub nodes are adversely affected or lost in Alzheimer's disease (AD)[87] and PD.[88] The failure of the cognitive functions of patients has been associated with damaged hub nodes and subsequent failure of the rich-club members.[89,110] In addition, although they are less numerous than nonhub nodes, hub nodes seem to be more vulnerable to the influence of the aforementioned brain diseases.[111] For instance, stroke patients with less damaged hub nodes and rich-club members suffer less severity in aphasia.[112]

Damaged nodes and edges of the network lead to global disruptions in information flow and therefore also disturb small-world topology and modular organization of the brain, which may lead to schizophrenia,[110,113–115] AD,[116] epilepsy,[117–120] and autistic spectrum disorder (ASD)[121] to name just a few brain disorders.

With respect to disrupted long-distance connections, schizophrenic patients show reduced functional connectivity between the frontal and temporal lobe,[122] and ASD patients have shown deficiencies in long and intermodular connectivity and failure in pruning short-range connections in animal models.[121,123]

In relation to cognitive ability, changes in the network influence cognitive functions. A recent clinical study using an EEG correlation network in patients with AD suggests that increases in ASP lead to a reduced performance in memory tasks.[124] In this study, the ASP of the EEG synchronization network was higher in patients with AD compared to the control group. Furthermore, there was a negative correlation between the patients' ASP and their performance in a standard clinical memory test.

Dynamic simulations based on accurate network connectivity show a great potential to help to predict the surgery outcome for epilepsy patients and to guide surgery for better prognosis. Honey and Sporns,[47] using both Kuramoto oscillators and an NMM with structural connectome-based coupling, predicted that parietal and frontal region lesions seem to have the greatest potential for causing global network disruption. The comparison of the two models with the same connectivity also allows for highlighting differences—the results of both models were broadly similar, but there were differences in functional changes evoked by lesions, a warning that the intrinsic node dynamics do have a part to play in the global dynamics observed, and a model that is dependent only on network coupling is not sufficient for capturing the complexity of emergent brain dynamics. In summary, models representing healthy connectivity and dynamics can predict the consequences of lesioning different brain regions and can help uncover potential vulnerabilities. In addition, network models have explored the dynamic repercussions of lesions in the context of diseases, replicating abnormal networks and their dynamics. While lesions caused by brain diseases or injuries may disrupt the function and the structure of the brain, surgery—which can be thought of as selective and planned lesions—can restore some disrupted functions. Recent studies have shown clinical applications using surgery or neuromodulation in epilepsy and PD, which hold great promise for future medical applications in other brain network diseases.[51,52,125,126]

4.2 Future Experimental Studies

Further experimental validation will be necessary based on predictions from computational models such as VERTEX[87] and the surgery prediction and guidance models.[51,52] A larger sample size of patients and controls with higher temporal and spatial resolution imaging data from multiple modalities is crucial to validate and improve computational models. Combining different levels of data such as genetic information and imaging data is essential.[127] For dynamic lesion models more generally, stimulation-induced virtual lesions may prove to be an ethically viable alternative for validation in humans. TMS can transiently knock out activity in a target area,[49] and TMS effects on EEG rhythms can be replicated in models.[128] A thorough and systematic investigation and replication of TMS virtual lesion effects could be used to make a model that is capable of predicting changes in observed dynamics in patients following brain injury.

Once models can accurately predict dynamics resulting from lesioning, they can be applied to real-world situations, with the potential to greatly improve surgery outcomes in epilepsy and, in the case of the VERTEX simulator, with the longer term potential to better estimate effects of stimulation or pharmaceutical intervention on neural networks.

4.3 Future Modeling Studies

By using higher quality data and the processing power of computer modeling, studies can speed up the development of patient-specific optimal treatment for patients with various diseases, in surgery procedures or in determining the best drug and dosage to normalize divergent brain dynamics in a variety of disorders from PD to depression. The framework is beginning to be put in place for this application of computational modeling to practical medicine.[129]

Furthermore, incorporating plasticity to model brain dynamics may improve existing models to predict not only the immediate consequences of lesions but also the subsequent reorganization and adaptation of the brain network. There has been some progress in this area, but the scope was limited mainly to mesoscopic models for certain brain functions. A more comprehensive understanding of plasticity at the level of macroscopic brain connectivity is needed. We can, however, speculate as to what topological features may be beneficial for network reorganization after brain damage to recover. First, a network with a rich repertoire of alternative pathways is more likely to adapt well to a lesion, making it more resilient against functional failure by securing stable performance of the system. The degeneracy of the brain network or many-to-one mapping from structure to function allows the brain to make use of alternatives when certain areas or circuits of the brain are damaged.[7,130–132] People with presumably more alternative pathways connecting regions or bigger brain volume with possibly more neurons that can be recruited after brain damage are assumed to have higher cognitive or brain reserve.[3,7] Recently, this idea has been used to develop measures such as communicability,[133–135] maximum flow,[131] and resilience based on path ensembles[130] to quantify information flow in the network using all possible pathways from source to target as opposed to considering only the shortest path from previous studies. Second, the brain network can also be benefitted from de novo connections in an attempt to readjust or regain its original functionality.[5,136] As any kind of brain damage would deter efficient information flow in the network with longer APS, newly formed connections that may reduce the characteristic path length is likely to be beneficial. Finally, recuperating functions of important nodes and edges with high centralities in the network by neurostimulation such as TMS and transcranial direct current stimulation may facilitate the reorganization of the network.[137–140]

Lesion modeling combined with network analysis has revealed important aspects regarding brain networks in different species such as *Drosophila melanogaster*, *Caenorhabditis elegans*, and mammalian brain networks.[107] In doing so, most studies have used a single-layer network where nodes and edges are static and simple; however, complex networks in both biological and man-made systems intrinsically possess multilayer aspects, interacting with one another. Multilayer networks have multiple relationships between entities in the networks.[141] Considering multiple types of connectivity between different layers brings a wider range of dynamics into lesion modeling. For instance, interdependencies between layers can cause a cascade of failures of the networks.[142] Reis and colleagues[143] showed that correlated interlayer coupling influences vulnerability of multilayer networks. Therefore, future lesion modeling studies using a multilayer networks framework would improve our understanding of vulnerability or perturbation analysis of complex neural networks.

GLOSSARY OF NETWORK ANALYSIS TERMS

Centrality Measures quantify the relative "importance" of a node or an edge within a graph. There are various centrality measures depending on how to define the "importance" of a node or an edge. See Brandes and Erlebach, 2005.[144]

Betweenness centrality The relative frequency of all shortest paths that go through a node (node betweenness) or edge of the network (edge betweenness) out of all shortest paths.

Characteristic path length The characteristic path length L (also called "path length" or "average path length") is the average of the shortest path lengths of all pairs of nodes in the graph, which can also be defined by the median or the harmonic mean.

Clustering coefficient The clustering coefficient C_i of node i is the number of existing connections between its neighbors divided by all possible connections between neighbors. The clustering coefficient of a graph is the averaged clustering coefficients of all nodes in the graph. See Kaiser, 2008.[145]

Eigenvector centrality A centrality measure of a node as a function of the centralities of its neighbors—a node has a high eigenvector centrality if it is connected with nodes with high centralities.

Global efficiency The inverse of the harmonic mean of the shortest path lengths of all pair of nodes in the graph, which measures the efficiency of information transfer in parallel systems. Although inversely proportional to characteristic path length, this measure can be used for disconnected networks as opposed to characteristic path length.

Hierarchical organization A network with hierarchical relationship among nodes, e.g., a network of cortical areas based on the laminar origin and termination connection patterns.

Hodology The study of pathways and connections.

Hub A node with many neighbors in comparison with other nodes in the network if defined based on degree centrality.

Local efficiency Local efficiency of a node is the global efficiency of the subgraph without the node, measuring how neighbors are well connected, that is, how efficiently information would transfer in the absence of the node; local efficiency of a graph is the average local efficiency over all nodes in the graph.

Modular organization A network with multiple clusters or modules within which nodes are densely interconnected compared to between nodes in different modules.

Network integrity How "robust" a network is in terms of its connectedness and subsequent information transfer against removing nodes or edges.

Rich club coefficient A metric to quantify the extent to which hub nodes connect more to other hub nodes than expected from the degree-preserving randomized benchmark network.

Scale-free graph A graph with a power-law degree distribution.

Small-world graph A graph with the *clustering coefficient* much higher than in a benchmark random network, while the *characteristic path length* is comparable with the random network.

References

1. Tononi G, Sporns O, Edelman GM. Measures of degeneracy and redundancy in biological networks. *Proc Natl Acad Sci USA*. 1999;96(6):3257–3262.
2. Edelman GM, Gally JA. Degeneracy and complexity in biological systems. *Proc Natl Acad Sci USA*. 2001;98 (24):13763–13768
3. Sporns O. *Networks of the Brain*. Cambridge, MA: MIT press; 2011.
4. Marder E, Taylor AL. Multiple models to capture the variability in biological neurons and networks. *Nat Neurosci*. 2011;14(2):133–138.
5. Spear PD, Tong L, McCall MA. Functional influence of areas 17, 18 and 19 on lateral suprasylvian cortex in kittens and adult cats: implications for compensation following early visual cortex damage. *Brain Res*. 1988;447(1):79–91.
6. Stromswold K. The cognitive neuroscience of language acquisition. In: Gazzaniga M, ed. *The New Cognitive Neurosciences*. Cambridge, MA: MIT press; 2000:909–932.
7. Katzman R, Terry R, DeTeresa R, et al. Clinical, pathological, and neurochemical changes in dementia: a subgroup with preserved mental status and numerous neocortical plaques. *Ann Neurol*. 1988;23(2):138–144.
8. Stern Y. What is cognitive reserve? Theory and research application of the reserve concept. *J Int Neuropsychol Soc*. 2002;8(03):448–460.
9. Broca P. Perte de la parole, ramollissement chronique et destruction partielle du lobe antérieur gauche du cerveau. *Bull Soc Anthropol*. 1861;2(1):235–238.
10. Dronkers NF, Plaisant O, Iba-Zizen MT, Cabanis EA. Paul Broca's historic cases: high resolution MR imaging of the brains of Leborgne and Lelong. *Brain*. 2007;130(5):1432–1441.
11. Mesulam M-M. *Principles of Behavioral and Cognitive Neurology*. New York, NY: Oxford University Press; 2000.
12. Kaiser M. A tutorial in connectome analysis: topological and spatial features of brain networks. *NeuroImage*. 2011;57(3):892–907.
13. Sporns O, Chialvo DR, Kaiser M, Hilgetag CC. Organization, development and function of complex brain networks. *Trends Cogn Sci*. 2004;8(9):418–425.
14. Felleman DJ, van Essen DC. Distributed hierarchical processing in the primate cerebral cortex. *Cereb Cortex*. 1991;1:1–47.
15. Nicolelis MAL, Yu CH, Baccalá LA. Structural characterization of the neural circuit responsible for control of the cardiovascular functions in high vertebrates. *Comput Biol Med*. 1990;20(6):379–400.
16. Young MP. Objective analysis of the topological organization of the primate cortical visual system. *Nature*. 1992;358(6382):152–155.
17. Young MP, Scannell JW, Burns GA, Blakemore C. Analysis of connectivity: neural systems in the cerebral cortex. *Rev Neurosci*. 1994;5(3):227–250.
18. Hilgetag CC, O'Neill MA, Young MP. Indeterminate organization of the visual system. *Science*. 1996;271 (5250):776–777.
19. Hilgetag CC, Burns GAPC, O'Neill MA, Scannell JW, Young MP. Anatomical connectivity defines the organization of clusters of cortical areas in the macaque monkey and the cat. *Phil Trans R Soc Lond B*. 2000;355:91–110.
20. Sporns O, Tononi G, Edelman GM. Theoretical neuroanatomy: relating anatomical and functional connectivity in graphs and cortical connection matrices. *Cereb Cortex*. 2000;10:127–141.
21. Young MP. The architecture of visual cortex and inferential processes in vision. *Spat Vis*. 2000;13(2--3):137–146.

22. Petroni F, Panzeri S, Hilgetag CC, Koetter R, Young MP. Simultaneity of responses in a hierarchical visual network. *Neuroreport*. 2001;12:2753–2759.
23. Kaiser M, Hilgetag CC. Nonoptimal component placement, but short processing paths, due to long-distance projections in neural systems. *PLoS Computational Biology*. 2006;2(7):e95.
24. Meunier D, Lambiotte R, Bullmore ET. Modular and hierarchically modular organization of brain networks. *Front Neurosci*. 2010;4:200.
25. Strogatz SH. Exploring complex networks. *Nature*. 2001;410:268–276.
26. Watts DJ, Strogatz SH. Collective dynamics of 'small-world' networks. *Nature*. 1998;393:440–442.
27. Albert R, Jeong H, Barabási A-L. Error and attack tolerance of complex networks. *Nature*. 2000;406:378–382.
28. Barabási A-L, Albert R. Emergence of scaling in random networks. *Science*. 1999;286:509–512.
29. Barabasi AL. Scale-free networks: a decade and beyond. *Science*. 2009;325(5939):412–413.
30. Kaiser M, Hilgetag CC. Edge vulnerability in neural and metabolic networks. *Biol Cybern*. 2004;90(5):311–317.
31. Alstott J, Panzarasa P, Rubinov M, Bullmore ET, Vertes PE. A unifying framework for measuring weighted rich clubs. *Sci Rep*. 2014;4:7258.
32. Rubinov M, Sporns O. Complex network measures of brain connectivity: uses and interpretations. *NeuroImage*. 2010;52(3):1059–1069.
33. Rubinov M, Sporns O. Weight-conserving characterization of complex functional brain networks. *NeuroImage*. 2011;56(4):2068–2079.
34. Latora V, Marchiori M. Efficient behavior of small-world networks. *Phys Rev Lett*. 2001;87:198701.
35. Latora V, Marchiori M. Economic small-world behavior in weighted networks. *Eur Phys J B*. 2003;32 (2):249–263.
36. Kaiser M, Martin R, Andras P, Young MP. Simulation of robustness against lesions of cortical networks. *Eur J Neurosci*. 2007;25(10):3185–3192.
37. Diestel R. *Graph Theory*. New York: Springer; 1997.
38. Meunier D, Lambiotte R, Fornito A, Ersche KD, Bullmore ET. Hierarchical modularity in human brain functional networks. *Front Neuroinform*. 2009;3:37.
39. Fan Y, Shi F, Smith JK, Lin W, Gilmore JH, Shen D. Brain anatomical networks in early human brain development. *NeuroImage*. 2011;54(3):1862–1871.
40. Yap PT, Fan Y, Chen Y, Gilmore JH, Lin W, Shen D. Development trends of white matter connectivity in the first years of life. *PLoS One*. 2011;6(9):e24678.
41. Kaiser M, Hilgetag CC. Optimal hierarchical modular topologies for producing limited sustained activation of neural networks. *Front Neuroinform*. 2010;4:8.
42. Rubinov M, Bullmore E. Fledgling pathoconnectomics of psychiatric disorders. *Trends Cogn Sci*. 2013;17 (12):641–647.
43. Terry JR, Benjamin O, Richardson MP. Seizure generation: the role of nodes and networks. *Epilepsia*. 2012;53 (9):e166–e169.
44. Alstott J, Breakspear M, Hagmann P, Cammoun L, Sporns O. Modeling the impact of lesions in the human brain. *PLoS Comput Biol*. 2009;5(6):e1000408.
45. Ruff CC, Driver J, Bestmann S. Combining TMS and fMRI: from 'virtual lesions' to functional-network accounts of cognition. *Cortex*. 2009;45(9):1043–1049.
46. Deco G, McIntosh AR, Shen K, et al. Identification of optimal structural connectivity using functional connectivity and neural modeling. *J Neurosci*. 2014;34(23):7910–7916.
47. Honey CJ, Sporns O. Dynamical consequences of lesions in cortical networks. *Hum Brain Mapp*. 2008;29 (7):802–809.
48. Sack AT, Linden DEJ. Combining transcranial magnetic stimulation and functional imaging in cognitive brain research: possibilities and limitations. *Brain Res Rev*. 2003;43(1):41–56.
49. Pascual-Leone A. Transcranial magnetic stimulation: studying the brain–behaviour relationship by induction of 'virtual lesions'. *Philos Trans R Soc Lond B Biol Sci*. 1999;354(1387):1229–1238.
50. Goodall S, Reggia JA, Chen Y, Ruppin E, Whitney C. A computational model of acute focal cortical lesions. *Stroke*. 1997;28(1):101–109.
51. Hutchings F, Han CE, Keller SS, Weber B, Taylor PN, Kaiser M. Predicting surgery targets in temporal lobe epilepsy through structural connectome based simulations. *PLoS Comput Biol*. 2015;11(12): e1004642.

52. Sinha N, Dauwels J, Wang Y, Cash SS, Taylor PN. Predicting surgical outcome in intractable epilepsy using a computational model of seizure initiation. *BMC Neurosci.* 2015;16(Suppl 1):P230.
53. Takiyama K, Okada M. Recovery in stroke rehabilitation through the rotation of preferred directions induced by bimanual movements: a computational study. *PLoS One.* 2012;7(5):e37594.
54. Stam CJ, van Straaten ECW, Van Dellen E, et al. The relation between structural and functional connectivity patterns in complex brain networks. *Int J Psychophysiol.* 2016;103:149–160.
55. Honey CJ, Sporns O, Cammoun L, et al. Predicting human resting-state functional connectivity from structural connectivity. *Proc Natl Acad Sci USA.* 2009;106(6):2035–2040.
56. Breakspear M, Heitmann S, Daffertshofer A. Generative models of cortical oscillations: neurobiological implications of the Kuramoto model. *Front Hum Neurosci.* 2010;4(190):3389:10
57. Hoppensteadt FC, Izhikevich EM. *Weakly Connected Oscillators. Weakly Connected Neural Networks.* Springer; 1997:247–293.
58. Singer W. Neuronal synchrony: a versatile code for the definition of relations? *Neuron.* 1999;24(1):49–65.
59. Srinivasan R, Winter WR, Ding J, Nunez PL. EEG and MEG coherence: measures of functional connectivity at distinct spatial scales of neocortical dynamics, Elsevier. *J Neurosci Met.* 2007;166(1):41–52.
60. Bastos AM, Schoffelen J-M. A tutorial review of functional connectivity analysis methods and their interpretational pitfalls. *Front Sys Neurosci.* 2015:9.
61. Shanahan M. Metastable chimera states in community-structured oscillator networks. *Chaos.* 2010;20 (1):013108.
62. Sotero RC, Trujillo-Barreto NJ, Iturria-Medina Y, Carbonell F, Jimenez JC. Realistically coupled neural mass models can generate EEG rhythms. *Neur Comput.* 2007;19(2):478–512.
63. Spiegler A, Kiebel SJ, Atay FM, Knösche TR. Bifurcation analysis of neural mass models: impact of extrinsic inputs and dendritic time constants. *NeuroImage.* 2010;52(3):1041–1058.
64. Jansen BH, Rit VG. Electroencephalogram and visual evoked potential generation in a mathematical model of coupled cortical columns. *Biol Cybernet.* 1995;73(4):357–366.
65. Ursino M, Cona F, Zavaglia M. The generation of rhythms within a cortical region: analysis of a neural mass model. *NeuroImage.* 2010;52(3):1080–1094.
66. Wendling F, Bellanger J-J, Bartolomei F, Chauvel P. Relevance of nonlinear lumped-parameter models in the analysis of depth-EEG epileptic signals. *Biol Cybernet.* 2000;83(4):367–378.
67. Alstott J, Breakspear M, Hagmann P, Cammoun L, Sporns O. Modeling the impact of lesions in the human brain. *PLoS Comput Biol.* 2009;5(6):e1000408.
68. Fornito A, Zalesky A, Breakspear M. The connectomics of brain disorders. *Nat Rev Neurosci.* 2015;16 (3):159–172.
69. Armentrout SL, Reggia JA, Weinrich M. A neural model of cortical map reorganization following a focal lesion. *Artif Intell Med.* 1994;6(5):383–400.
70. Bains AS, Schweighofer N. Time-sensitive reorganization of the somatosensory cortex poststroke depends on interaction between Hebbian and homeoplasticity: a simulation study. *J Neurophysiol.* 2014;112 (12):3240–3250.
71. Han CE, Arbib MA, Schweighofer N. Stroke rehabilitation reaches a threshold. *PLoS Comput Biol.* 2008;4(8): e1000133.
72. Reinkensmeyer DJ, Guigon E, Maier MA. A computational model of use-dependent motor recovery following a stroke: optimizing corticospinal activations via reinforcement learning can explain residual capacity and other strength recovery dynamics. *Neural Netw.* 2012;29:60–69.
73. Reggia JA, Gittens SD, Chhabra J. Post-lesion lateralisation shifts in a computational model of single-word reading. *Laterality.* 2000;5(2):133–154.
74. Sabel BA, Kruse R, Wolf F, Guenther T. Local topographic influences on vision restoration hot spots after brain damage. *Restor Neurol Neurosci.* 2013;31(6):787–803.
75. Levitan S, Reggia JA. Interhemispheric effects on map organization following simulated cortical lesions. *Artif Intell Med.* 1999;17(1):59–85.
76. Casadio M, Tamagnone I, Summa S, Sanguineti V. Neuromotor recovery from stroke: computational models at central, functional, and muscle synergy level. *Front Comput Neurosci.* 2013;7:97.
77. Mikołajewska E, Mikołajewski D. Neuroplasticity in rehabilitation after central nervous system damages—computational models. *Adv Rehabil.* 2012;26(1):51–58.

78. Menon V. Large-scale brain networks and psychopathology: a unifying triple network model. *Trends Cogn Sci*. 2011;15(10):483–506.
79. Hammond C, Bergman H, Brown P. Pathological synchronization in Parkinson's disease: networks, models and treatments. *Trends Cogn Sci*. 2007;30(7):357–364.
80. Zhang Z, Liao W, Chen H, et al. Altered functional–structural coupling of large-scale brain networks in idiopathic generalized epilepsy. *Brain*. 2011;134(10):2912–2928.
81. Diessen E, Diederen SJH, Braun KPJ, Jansen FE, Stam CJ. Functional and structural brain networks in epilepsy: what have we learned? *Epilepsia*. 2013;54(11):1855–1865.
82. Jirsa V, Sporns O, Breakspear M, Deco G, McIntosh AR. Towards the virtual brain: network modeling of the intact and the damaged brain. *Arch Ital Biol*. 2010;148(3):189–205.
83. Tomsett RJ, Ainsworth M, Thiele A, et al. Modelling spatially realistic local field potentials in spiking neural networks using the VERTEX simulation tool. *BMC Neurosci*. 2014;15(Suppl 1):P130.
84. Tomsett RJ, Ainsworth M, Thiele A, et al. Virtual Electrode Recording Tool for EXtracellular potentials (VERTEX): comparing multi-electrode recordings from simulated and biological mammalian cortical tissue. *Brain Struct Func*. 2014:1–21.
85. Achard S, Salvador R, Whitcher B, Suckling J, Bullmore E. A resilient, low-frequency, small-world human brain functional network with highly connected association cortical hubs. *J Neurosci*. 2006;26(1):63–72.
86. Crossley NA, Mechelli A, Scott J, et al. The hubs of the human connectome are generally implicated in the anatomy of brain disorders. *Brain*. 2014.
87. de Haan W, Mott K, van Straaten EC, Scheltens P, Stam CJ. Activity dependent degeneration explains hub vulnerability in Alzheimer's disease. *PLoS Comput Biol*. 2012;8(8):e1002582.
88. Dubbelink KTO, Hillebrand A, Stoffers D, et al. Disrupted brain network topology in Parkinson's disease: a longitudinal magnetoencephalography study. *Brain*. 2014;137(1):197–207.
89. Daianu M, Dennis EL, Jahanshad N, Nir TM, Toga AW, Jack Jr. CR, et al. The Alzheimer's Disease Neuroimaging. Alzheimer's Disease Disrupts Rich Club Organization in Brain Connectivity Networks. *Proc IEEE Int Symp Biomed Imaging*. 2013;266–269.
90. Váša F, Shanahan M, Hellyer PJ, Scott G, Cabral J, Leech R. Effects of lesions on synchrony and metastability in cortical networks. *NeuroImage*. 2015;118:456–467.
91. Grill WM, Snyder AN, Miocinovic S. Deep brain stimulation creates an informational lesion of the stimulated nucleus. *NeuroReport*. 2004;15(7):1137–1140.
92. Cabral J, Fernandes H, Van Hartevelt T, James A, Kringelbach ML, Deco G. Disrupted connectivity in schizophrenia: modelling the impact of structural connectivity changes on the dynamics of spontaneous functional networks. *BMC Neurosci*. 2013;14(Suppl 1):P100.
93. Taylor PN, Thomas J, Sinha N, et al. Optimal control based seizure abatement using patient derived connectivity. *Front Neurosci*. 2015;9:202.
94. Wiener N. *Cybernetics*. Wiley, New York, 1948.
95. Sporns O. The human connectome: a complex network. *Ann NY Acad Sci*. 2011;1224:109–125.
96. Lim S, Han CE, Uhlhaas PJ, Kaiser M. Preferential detachment during human brain development: age- and sex-specific structural connectivity in diffusion tensor imaging (DTI) data. *Cereb Cortex*. 2015;25 (6):1477–1489.
97. Fair DA, Cohen AL, Power JD, et al. Functional brain networks develop from a "local to distributed" organization. *PLoS Comput Biol*. 2009;5(5):e1000381.
98. Hagmann P, Sporns O, Madan N, et al. White matter maturation reshapes structural connectivity in the late developing human brain. *Proc Natl Acad Sci USA*. 2010;107(44):19067–19072.
99. Supekar K, Musen M, Menon V. Development of large-scale functional brain networks in children. *PLoS Biol*. 2009;7(7):e1000157.
100. Supekar K, Uddin LQ, Prater K, Amin H, Greicius MD, Menon V. Development of functional and structural connectivity within the default mode network in young children. *NeuroImage*. 2010;52(1):290–301.
101. Uddin LQ, Supekar K, Menon V. Typical and atypical development of functional human brain networks: insights from resting-state FMRI. *Front Sys Neurosci*. 2010;4:21.
102. Uddin LQ, Supekar KS, Ryali S, Menon V. Dynamic reconfiguration of structural and functional connectivity across core neurocognitive brain networks with development. *J Neurosci*. 2011;31(50):18578–18589.

103. Kaiser M, Varier S. Evolution and development of brain networks: from *Caenorhabditis elegans* to *Homo sapiens*. *Network-Comp Neural*. 2011;22(1–4):143–147.
104. Varier S, Kaiser M, Forsyth R. Establishing, versus maintaining, brain function: a neuro-computational model of cortical reorganization after injury to the immature brain. *J Int Neuropsychol Soc*. 2011;17(06):1030–1038.
105. Kaiser M, Hilgetag C. Wiring principles, optimization. In: Jaeger D, Jung R, eds. *Encyclopedia of Computational Neuroscience*. New York: Springer; 2014:1–7.
106. Ball G, Aljabar P, Zebari S, et al. Rich-club organization of the newborn human brain. *Proc Natl Acad Sci USA*. 2014.
107. Kaiser M. Neuroanatomy: connectome connects fly and mammalian brain networks. *Curr Biol*. 2015;25(10): R416–R418.
108. Schroeter MS, Charlesworth P, Kitzbichler MG, Paulsen O, Bullmore ET. Emergence of rich-club topology and coordinated dynamics in development of hippocampal functional networks in vitro. *J Neurosci*. 2015;35 (14):5459–5470.
109. van den Heuvel MP, Sporns O. Rich-club organization of the human connectome. *J Neurosci*. 2011;31 (44):15775–15786.
110. van den Heuvel MP, Sporns O, Collin G, et al. Abnormal rich club organization and functional brain dynamics in schizophrenia. *JAMA Psychiatr*. 2013;70(8):783–792.
111. Yao Z, Zhang Y, Lin L, et al. Abnormal cortical networks in mild cognitive impairment and Alzheimer's disease. *PLoS Comput Biol*. 2010;6(11):e1001006.
112. Gleichgerrch E, Kocher M, Nesland T, Rorden C, Fridriksson J, Bonilha L. Preservation of structural brain network hubs is associated with less severe post-stroke aphasia. *Restor Neurol Neurosci*. 2015; (Preprint):1–11.
113. Wang Q, Su TP, Zhou Y, et al. Anatomical insights into disrupted small-world networks in schizophrenia. *NeuroImage*. 2011.
114. Alexander-Bloch AF, Gogtay N, Meunier D, et al. Disrupted modularity and local connectivity of brain functional networks in childhood-onset schizophrenia. *Front Sys Neurosci*. 2010;4.
115. Alexander-Bloch A, Lambiotte R, Roberts B, Giedd J, Gogtay N, Bullmore E. The discovery of population differences in network community structure: new methods and applications to brain functional networks in schizophrenia. *NeuroImage*. 2012;59:3889–3900.
116. de Haan W, van der Flier WM, Koene T, Smits LL, Scheltens P, Stam CJ. Disrupted modular brain dynamics reflect cognitive dysfunction in Alzheimer's disease. *NeuroImage*. 2012;59(4):3085–3093.
117. Bernhardt BC, Bonilha L, Gross DW. Network analysis for a network disorder: the emerging role of graph theory in the study of epilepsy. *Epilepsy Behav*. 2015;50:162–170.
118. Braakman HM, Vaessen MJ, Jansen JF, et al. Frontal lobe connectivity and cognitive impairment in pediatric frontal lobe epilepsy. *Epilepsia*. 2013;54(3):446–454.
119. Chiang S, Haneef Z. Graph theory findings in the pathophysiology of temporal lobe epilepsy. *Clin Neurophysiol*. 2014;125(7):1295–1305.
120. Chiang S, Stern JM, Engel Jr J, Haneef Z. Structural-functional coupling changes in temporal lobe epilepsy. *Brain Res*. 2015;1616:45–57.
121. Belmonte MK. Autism and abnormal development of brain connectivity. *J Neurosci*. 2004;24(42):9228–9231.
122. Lawrie SM, Buechel C, Whalley HC, Frith CD, Friston KJ, Johnstone EC. Reduced frontotemporal functional connectivity in schizophrenia associated with auditory hallucinations. *Biol Psychiatr*. 2002;51(12):1008–1011.
123. Anagnostou E, Taylor MJ. Review of neuroimaging in autism spectrum disorders: what have we learned and where we go from here. *Mol Autism*. 2011;2(1):1.
124. Stam CJ, Jones BF, Nolte G, Breakspear M, Scheltens P. Small-world networks and functional connectivity in Alzheimer's disease. *Cereb Cortex*. 2007;17(1):92–99.
125. Kerr CC, Van Albada SJ, Neymotin SA, Chadderdon GL, Robinson PA, Lytton WW. Cortical information flow in Parkinson's disease: a composite network/field model. *Front Comput Neurosci*. 2013;7:39.
126. Frankemolle AM, Wu J, Noecker AM, et al. Reversing cognitive–motor impairments in Parkinson's disease patients using a computational modelling approach to deep brain stimulation programming. *Brain*. 2010;133 (3):746–761.
127. Gaiteri C, Ding Y, French B, Tseng GC, Sibille E. Beyond modules and hubs: the potential of gene coexpression networks for investigating molecular mechanisms of complex brain disorders. *Genes Brain Behav*. 2014;13(1):13–24.

128. Cona F, Zavaglia M, Massimini M, Rosanova M, Ursino M. A neural mass model of interconnected regions simulates rhythm propagation observed via TMS-EEG. *NeuroImage*. 2011;57(3):1045–1058.
129. Wang Y, Hutchings F, Kaiser M. Computational modeling of neurostimulation in brain diseases. *Prog Brain Res*. 2015;222:191–228.
130. Avena-Koenigsberger A, Mišić B, Hawkins RXD, et al. Path ensembles and a tradeoff between communication efficiency and resilience in the human connectome. *Brain Struct Func*. 2016:1–16.
131. Yoo SW, Han CE, Shin JS, et al. A network flow-based analysis of cognitive reserve in normal ageing and Alzheimer's disease. *Sci Rep*. 2015;5:10057.
132. Price CJ, Friston KJ. Degeneracy and cognitive anatomy. *Trends Cogn Sci*. 2002;6(10):416–421.
133. Estrada E, Hatano N. Communicability in complex networks. *Phys Rev E*. 2008;77(3):036111.
134. Crofts JJ, Higham DJ. A weighted communicability measure applied to complex brain networks. *J R Soc Interface*. 2009;6(33):411–414.
135. Crofts JJ, Higham DJ, Bosnell R, et al. Network analysis detects changes in the contralesional hemisphere following stroke. *NeuroImage*. 2011;54(1):161–169.
136. Spear PD, Kalil RE, Tong L. Functional compensation in lateral suprasylvian visual area following neonatal visual cortex removal in cats. *J Neurophysiol*. 1980;43:851–869.
137. Berenyi A, Belluscio M, Mao D, Buzsaki G. Closed-loop control of epilepsy by transcranial electrical stimulation. *Science*. 2012;337(6095):735–737.
138. Bestmann S, de Berker AO, Bonaiuto J. Understanding the behavioural consequences of noninvasive brain stimulation. *Trend Cogn Sci*. 2015;19(1):13–20.
139. da Silva MC, Conti CL, Klauss J, et al. Behavioral effects of transcranial direct current stimulation (tDCS) induced dorsolateral prefrontal cortex plasticity in alcohol dependence. *J Physiol Paris*. 2013;107(6):493–502.
140. Silasi G, Murphy TH. Stroke and the connectome: how connectivity guides therapeutic intervention. *Neuron*. 2014;83(6):1354–1368.
141. Kivelä M, Arenas A, Barthelemy M, Gleeson JP, Moreno Y, Porter MA. Multilayer networks. *J Complex Networks*. 2014;2(3):203–271.
142. Bianconi G. Multilayer networks: dangerous liaisons? *Nat Phys*. 2014;10(10):712–714.
143. Reis SD, Hu Y, Babino A, et al. Avoiding catastrophic failure in correlated networks of networks. *Nat Phys*. 2014;10(10):762–767.
144. Brandes U, Erlebach T. *Network Analysis*. Heidelberg: Springer; 2005.
145. Kaiser M. Mean clustering coefficients: the role of isolated nodes and leafs on clustering measures for small-world networks. *New J Phys*. 2008;10:083042.
146. Varier S, Kaiser M. Neural development features: spatio-temporal development of the *Caenorhabditis elegans* neuronal network. *PLoS Comput Biol*. 2011;7(1):e1001044.

CHAPTER

23

Network Models of Epilepsy-Related Pathological Structural and Functional Alterations in the Dentate Gyrus

Ivan Raikov, Mark Plitt and Ivan Soltesz

Stanford University, Stanford, CA, United States

OUTLINE

1 INTRODUCTION

The dentate gyrus (DG) is considered to be the gateway to the hippocampus and forms part of the classic trisynaptic loop which connects the entorhinal cortex to the three major subfields of the hippocampus within the temporal lobe. The principal cells of the DG are the granule cells, which are relatively quiescent and receive a large portion of their excitatory input from layer II of the entorhinal cortex.[1,2] The hippocampal formation is of particular importance in epilepsy research, because spontaneous seizures arise from the

The Rewiring Brain
DOI: http://dx.doi.org/10.1016/B978-0-12-803784-3.00023-8

temporal lobe in certain common forms of medically intractable epilepsies, often associated with loss of cells and other histopathological changes (collectively referred to as hippocampal sclerosis) that have been observed throughout the hippocampal formation, including the DG.[3] In this chapter, we will primarily focus on temporal lobe epilepsy, which is the most common epilepsy syndrome in adults, with about a third of the patients suffering from seizures that cannot be adequately controlled with current anticonvulsant drugs.

Neuroimaging data can reveal a number of epilepsy-related structural and metabolic abnormalities in the mesial temporal structures.[4] Magnetic resonance imaging (MRI) has shown hippocampal atrophy and sclerosis, whereas positron emission tomography (PET) has shown functional changes, including a decrease in cerebral blood flow/glucose metabolism in the interictal (between seizures) period and a dramatic increase in the ictal (during seizures) period.[4]

Epilepsy can develop in the immature and adult brain after a brain-damaging insult through a number of molecular and cellular alterations that are collectively referred to as epileptogenesis. Sprouting of glutamatergic dentate granule cell axons (called mossy fibers) in the DG is a common form of axonal plasticity that most likely occurs during epileptogenesis before spontaneous seizures develop and has been shown to continue throughout the lifetime of the experimental epileptic animal.[5–9] It has been shown in experimental models that many epileptogenic insults trigger mossy fiber sprouting, including neurodegenerative disorders, traumatic brain injury (TBI), and stroke.[10]

The mechanisms underlying mossy fiber sprouting are the subject of active research, however, a number of studies have investigated the molecular signals that activate the pathways to initiate mossy fiber growth and synaptogenesis. Increased expression of GAP-43, a protein concentrated at growth cones and developing presynaptic terminals, has been proposed to promote mossy fiber sprouting.[11] Furthermore, it has been shown that the expression of brain-derived neurotrophic factor (BDNF) increases in the DG after seizures and that BDNF promotes granule cell hypertrophy and mossy fiber branching.[12–14] Koyama et al.[15] used hyperexcitation of ex vivo hippocampal cultures to show that the resulting upregulation of BDNF protein expression is necessary and sufficient to induce mossy fiber sprouting and the resultant network reorganization. Thus, while the specific signaling pathway has yet to be fully characterized, experimental evidence suggests that BDNF plays an important role in mossy fiber sprouting.

An unusual characteristic of the DG is that new granule cells are born throughout adulthood in both animals and humans, and epileptogenesis in the DG has been correlated with increased production of adult-born granule cells, as observed in rodent models of temporal lobe epilepsy.[16] Briefly, adult-born granule cells are produced by neural progenitor cells located below the granule cell layer. New cells in the normal brain are born and migrate into the granule cell layer throughout adulthood, and project axons (mossy fibers) into the dentate hilus (core) and dendrites into the molecular layer. Immature granule cells of up to several weeks old have a number of different functional properties compared to their mature counterparts and have much higher excitability. During epileptogenesis, an increased rate of granule cell neurogenesis is observed, and the newborn cells can develop in ectopic locations, such as the hilus, and show a number of structural abnormalities such as aberrant basal dendrites.[16–18] Furthermore, during epileptogenesis a number of cells in the DG are lost, including excitatory mossy cells, which receive convergent mossy fiber inputs and contact thousands of granule cells and interneurons, and some GABAergic interneurons. The prevalence of hilar ectopic dentate granule cells (referred to as "ectopic" because they are located outside of the granule cell layer), the degree of mossy fiber sprouting (aberrant branching of granule cell axons), and the extent of the loss of mossy cells all appear to increase with the frequency of seizures.[19] However, the causal relationship between these phenomena is an open question in epilepsy research.

From the perspective of network dynamics, the emergence of recurrent spontaneous seizures has been strongly correlated with high frequency oscillations (HFOs) greater than 200 Hz recorded in the DG, CA1, and CA3 subfields of the hippocampus, as well as in the subiculum and the entorhinal cortex.[20] These HFOs, called fast ripples (or pathological HFOs), are considered to be pathological events due to their tendency to precede seizure onset and to the greater number of fast-ripple generating sites in animals with higher rates of seizures.[20–22] Fast ripples are hypothesized to reflect hypersynchronous discharges of locally interconnected principal neurons capable of generating spontaneous seizures[23,24] and are considered as a biomarker for the epileptogenic region, defined as the area of brain necessary and sufficient for the generation of seizures.[22] In order to develop a more precise understanding of the functional consequences of these various forms of structural plasticity as they relate to the generation of pathological network dynamics, several groups have developed and analyzed detailed network models of the DG. In this chapter, we will review how these models are constructed and analyzed and the findings of several key computational investigations based on them.

2 MODEL

Several computational studies have focused on the functional consequences of structural alterations in the rat DG during epileptogenesis. Briefly, Santhakumar et al.[25] introduced a biophysically detailed model of the DG consisting of 500 granule cells,[26] 15 mossy, 6 basket, and 6 hilar cells with axonal projections to the perforant path termination zone, with intrinsic properties based on experimental data.[27,28] Neurons were distributed in a ring structure, and the spatial range of postsynaptic target cells was determined based on the axonal distribution pattern of the presynaptic cell types reported in the experimental literature.[29] Mossy fiber sprouting was modeled by adding synaptic connections from granule cells to the proximal dendrites of granule cells based on experimental data from filled and reconstructed granule cells in the experimental pilocarpine model of epilepsy, a model system with a particularly high density of mossy fiber sprouting.[6]

In two follow-up studies by Dyhrfjeld-Johnsen et al.[30] and Morgan and Soltesz,[31] the authors scaled up the dentate network model of Santhakumar et al. by including 50,000 granule cells and appropriate numbers of interneurons and synaptic connections, and used graph-theoretic measures to quantify the changes in network connectivity induced by mossy fiber sprouting and hilar cell loss. In the following sections, we will collectively review the approach taken by these models to represent structural plasticity and the insights gained from them.

2.1 DG Cell Types, Numbers, and Connections

The rat DG consists of over a million neurons of at least eight distinct cell types interconnected with over a billion synapses.[32] Eight types of dentate cells have been anatomically well described: granule cells (GCs), mossy cells, (MCs), basket cells (BCs), axo-axonic cells, hilar cells with axonal projections to the perforant path (HIPP cells), hilar cells with axonal projections to the commissural–associational pathway (HICAP cells), molecular layer cells with axonal projections to the perforant path (MOPP cells), and neurogliaform cells (NGFCs; a functionally distinct subclass of MOPP cells) (Fig. 23.1). Table 23.1 presents the connectivity matrix of the DG model constructed by Dyhrfjeld-Johnsen et al.,[30] consisting of the number of cells of each type, estimated from published data; the specific connectivity between cell types, which quantifies how many postsynaptic cells are innervated by each of the eight cell types.

Each pair of cells in the Dyhrfjeld-Johnsen et al. model were linked according to cell type-specific connection probabilities derived from the average number of projections from the pre- to postsynaptic neuronal class in the literature (Table 23.1). In general, in addition to the existence of connections between two particular cell types, the probability of connections from one particular cell A to a given cell B also depends on the extent of the axonal arbor of cell A and the relative distance between cells A and B. Therefore, Dyhrfjeld-Johnsen et al. further modified the cell type-specific connection probability by a factor obtained by the normalized Gaussian fits to the experimentally determined axonal distributions of the presynaptic cells and the relative positions of the pre- and postsynaptic neurons. Within these cell type-specific constraints, connections were made probabilistically on a neuron-to-neuron basis with a uniform synapse density along the axon (in agreement with in vivo data[33]), treating multiple synapses between two cells as a single connection and excluding autapses.

2.2 Structural Alterations in the Dentate During Epileptogenesis

As outlined in Section 1, several primary structural alterations occur in the adult DG during the process of epileptogenesis: mossy fiber sprouting, hilar cell loss (including mossy cells and hilar interneurons), and hilar ectopic adult-born granule cells. The first two phenomena, which we will refer to simply as sclerosis, have been represented in the DG models developed by Santhakumar et al.,[25] Dyhrfjeld-Johnsen et al.,[30] and Morgan and Soltesz.[31]

The distribution of sprouted mossy fiber to granule cell connections during sclerosis followed the same probability distribution as that of the healthy mossy fibers[29] and was added according to the same algorithm as the connections in the control DG network. Dyhrfjeld-Johnsen et al. modeled the progression of sclerosis by considering full (100%) sclerosis the state when all hilar cells are lost and the addition of a maximal number of previously nonexistent mossy fiber connections to other granule cells. As the densest, anatomically quantified sprouting reported in the literature from an experimental epilepsy model was an average of 275 extra mossy fiber connections per granule cell,[24]

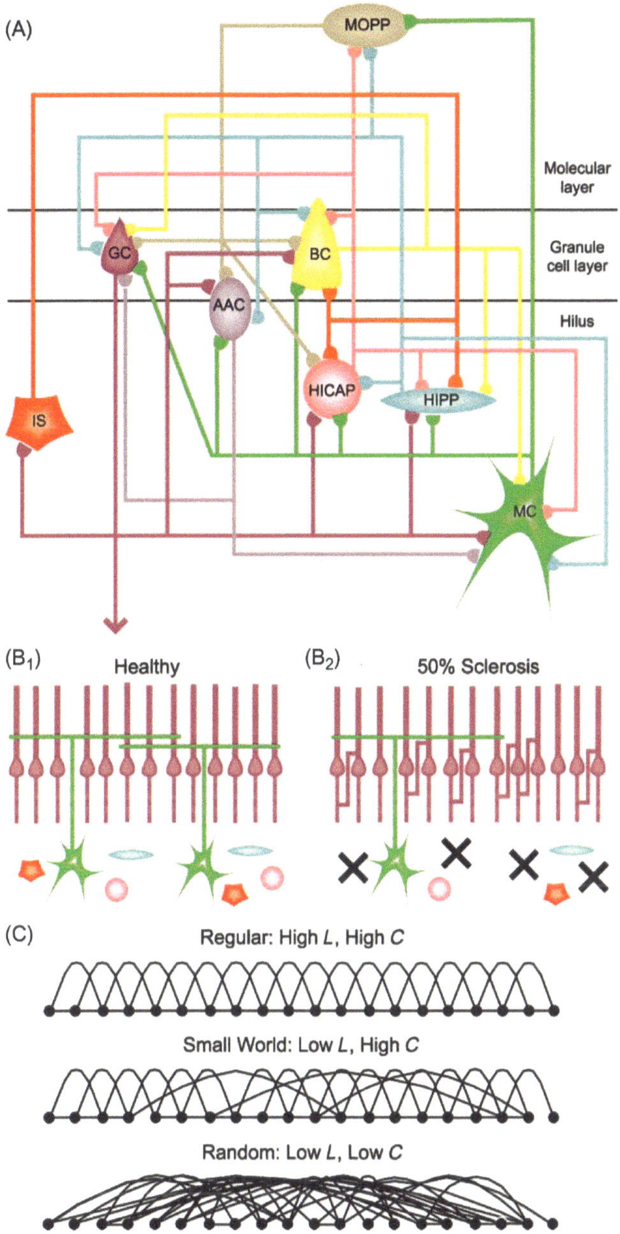

FIGURE 23.1 **Schematic of the basic circuitry of the dentate gyrus (DG) model constructed by Dyhrfjeld-Johnsen et al.[30] and the changes to the network during sclerosis.** Reproduced with permission from the authors. (A) Relational representation of the healthy DG illustrating the network connections between the eight major cell types: GC, granule cell; BC, basket cell; MC, mossy cell; AAC, axo-axonic cells; MOPP, molecular layer interneurons with axons in perforant path termination zone; HIPP, hilar interneurons with axons in perforant path termination zone; HICAP, hilar interneurons with axons in the commissural/associational pathway termination

TABLE 23.1 Connectivity Matrix of the Control Dentate Gyrus Model Constructed by Dyhrfjeld-Johnsen et al.[30]

	Granule cells	Mossy cells	Basket cells	Axoaxonic cells	MOPP cells	HIPP cells	HICAP cells	IS cells
Granule cells (1,000,000) [1–5]	N/A	**9.5** 7–12 [7]	**15** 10–20 [6–9]	**3** 1–5 [6,7,9]	N/A [6]	**110** 100–120 [4,10,11]	**40** 30–50 [4,7,10,11]	**20** 10–30 [7]
Mossy cells (30,000) [11]	**32,500** 30,000–35,000 [4,11–13]	**350** 200–500 [12,13]	**7.5** 5–10 [13]	**7.5** 5–10 [13]	**5** 5 [14]	**600** 600 [12,13]	**200** 200 [12,13]	N/A [15]
Basket cells (10,000) [16,17]	**1250** 1000–1500 [4,16–19]	**75** 50–100 [11,16,17,19]	**35** 20–50 [16,17,20,21]	N/A [18]	N/A [18]	**0.5** 0–1 [18]	N/A [18]	N/A [10,20]
Axo-axonic cells (2000) [4,22]	**3000** 2000–4000 [4,18,22]	**150** 100–200 [4,5,11,14,23]	N/A [5,18]	N/A [5,18]	N/A [5,18]	N/A [5,18]	N/A [5,18]	N/A [5,18,19]
MOPP cells (4000) [11,14]	**7500** 5000–10,000 [14]	N/A [14,24]	**40** 30–50 [14,25]	**1.5** 1–2 [14,26]	**7.5** 5–10 [14,25]	N/A [14,20,25]	7.5 5–10 [14,25]	N/A [14,15]
HIPP cells (12,000) [11]	**1550** 1500–1600 [4,11,20]	**35** 20–50 [4,11,12, 27,28]	**450** 400–500 [4,11,20]	**30** 20–40 [20,25]	**15** 10–20 [25]	N/A [14,20,25]	**15** 10–20 [25]	N/A [15,20]

(*Continued*)

zone; and IS, interneuron selective cells. Schematic shows the characteristic location of the various cell types within the three layers of the DG. Note, however, that this diagram does not indicate the topography of axonal connectivity (present in both the structural and functional dentate models) or the somatodendritic location of the synapses (incorporated in the functional network models). (B1) Schematic of the excitatory connectivity of the healthy DG is illustrated (only cell types in the hilus and granule cells are shown). Note that the granule cell axons (the mossy fibers) do not contact other granule cells in the healthy network. (B2) Schematic of the DG at 50% sclerosis shows the loss (indicated by the large X symbols) of half the population of all hilar cell types and the 50% sprouting of mossy fibers that results in abnormal connections between granule cells (note that, unlike in this simplified schematic, all granule cells formed sprouted contacts in the structural and functional models of sclerosis; thus progressive increase in sprouting was implemented by increasing the number of postsynaptic granule cells contacted by single sprouted mossy fibers; see Section 2). (C) Schematic of three basic network topologies: regular, small-world, and random. Nodes in a regular network are connected to their nearest neighbors, resulting in a high degree of local interconnectedness (high clustering coefficient *C*), and also requiring a large number of steps to reach other nodes in the network from a given starting point (high average path length *L*). Reconnection of even a few of the local connections in a regular network to distal nodes in a random manner results in the emergence of a small-world network, with a conserved high clustering coefficient (*C*) but a low average path length (*L*). In a random network, there is no spatial restriction on the connectivity of the individual nodes, resulting in a network with a low average path length *L* and also a low clustering coefficient *C*.

TABLE 23.1 (Continued)

	Granule cells	Mossy cells	Basket cells	Axoaxonic cells	MOPP cells	HIPP cells	HICAP cells	IS cells
HICAP cells	**700**	**35**	**175**	**N/A**	**15**	**50**	**50**	N/A
(3000)	700	30–40	150–200	[20]	10–20	50	50	
[5,29,30]	[4,11,20]	[20]	[4,11,20]		[14,20]	[20]	[20]	
IS cells	**N/A**	**N/A**	**7.5**	**N/A**	**N/A**	**7.5**	**7.5**	**450**
(3000)	[15]	[15]	5–10	[15]		5–10	5–10	100–800
[15,29,30]			[15,19]			[19]	[19]	[15]

Cell numbers and connectivity values were estimated by Dyhrfjeld-Johnsen et al. from the published data for granule cells, mossy cells, basket cells, axo-axonic cells, molecular layer interneurons with axonal projections to the perforant path (MOPP), hilar interneurons with axonal projections to the perforant path (HIPP), hilar interneurons with axonal projections to the commissural/associational pathway (HICAP), and interneuron specific cells (IS). Connectivity is given as number of connections to a postsynaptic population (rows) from a single presynaptic neuron (columns). The average number of connections used in the 1:1 structural model developed by Dyhrfjeld-Johnsen et al.[30] is given in bold. N/A, not applicable. References given in the table between square brackets correspond to:

[1] Gaarskjaer FB. *J Comp Neurol.* 1978;178:49–71.
[2] Boss BD, Peterson GM, Cowan WM. *Brain Res.* 1985;338:144–150.
[3] West MJ. *Prog Brain Res.* 1990;83:13–36.
[4] Patton P, Mcnaughton B. *Hippocampus.* 1995;5:245–286.
[5] Freund T, Buzsaki G. *Hippocampus.* 1996;6:347–470.
[6] Buckmaster P, Dudek F. *J Neurophysiol.* 1999;81:712–721.
[7] Acsady L, Kamondi A, Sik A, Freund T, Buzsaki G. *J Neurosci.* 1998;18:3386–3403.
[8] Geiger J, Lubke J, Roth A, Frotscher M., Jonas P. *Neuron.* 1997;18:1009–1023.
[9] Blasco-Ibanez J, Martinez-Guijarro F, Freund T. *Neuroreport.* 2000;11:3219–3225.
[10] Gulyas A, Miettinen R, Jacobowitz D, Freund T. *Neuroscience.* 1992;48:1–27.
[11] Buckmaster P, Jongen-Relo A. *J Neurosci.* 1999;19:9519–9529.
[12] Buckmaster PS, Wenzel HJ, Kunkel DD, Schwartzkroin PA. *J Comp Neurol.* 1996;366:271–292.
[13] Wenzel H, Buckmaster P, Anderson N, Wenzel M, Schwartzkroin P. *Hippocampus.* 1997;7:559–570.
[14] Han Z, Buhl E, Lorinczi Z, Somogyi P. *Eur J Neurosci.* 1993;5:395–410.
[15] Gulyas AI, Hajos N, Freund TF. *J Neurosci.* 1996;16:3397–3411.
[16] Babb TL, Pretorius JK, Kupfer WR, Brown WJ. *J Comp Neurol.* 1988;278:121–138.
[17] Woodson W, Nitecka L, Ben-ari Y. *J Comp Neurol.* 1989;280:254–271.
[18] Halasy K, Somogyi P. *Eur J Neurosci.* 1993;5:411–429.
[19] Acsady L, Katona I, Martinez-Guijarro F, Buzsaki G, Freund T. Unusual target selectivity of perisomatic inhibitory cells in the hilar region of the rat hippocampus. *J Neurosci.* 2000;20:6907–6919.
[20] Sik A, Penttonen M, Buzsaki G. *Eur J Neurosci.* 1997;9:573–588.
[21] Bartos M, Vida I, Frotscher M, Geiger JRP, Jonas P. *J Neurosci.* 2001;21:2687–2698.
[22] Li X, Somogyi P, Tepper J, Buzsaki G. *Exp Brain Res.* 1992;90:519–525.
[23] Ribak CE, Seress L, Amaral DG. *J Neurocytol.* 1985;14:835–857.
[24] Frotscher M, Seress L, Schwerdtfeger W, Buhl E. *J Comp Neurol.* 1991;312:145–163.
[25] Katona I, Acsady L, Freund T. *Neuroscience.* 1999;88:37–55.
[26] Soriano E, Nitsch R, Frotscher M. *J Comp Neurol.* 1990;293:1–25.
[27] Claiborne BJ, Amaral DG, Cowan WM. *J Comp Neurol.* 1990;302:206–219.
[28] Buckmaster P, Zhang G, Yamawaki R. *J Neurosci.* 2002;22:6650–6658.
[29] Nomura T, Fukuda T, Aika Y, Heizmann CW, Emson PC, Kobayashi T, Kosaka T. *Brain Res.* 1997a;764:197–204.
[30] Nomura T, Fukuda T, Aika Y, Heizmann CW, Emson PC, Kobayashi T, Kosaka T. *Brain Res.* 1997b:751:64–80.

Adapted with permission from the authors.

Dyhrfjeld-Johnsen et al. considered this number 100% sprouting. They then represented the severity of sclerosis as a percentage of the maximum: e.g., at 50% sclerosis, simulating epileptogenesis following a moderate head injury,[25] each granule cell would contact approximately 137 other GCs, and 50% of hilar cells would be lost (Fig. 23.1B).

In order to quantify and measure network structure and its alterations, Dyhrfjeld-Johnsen et al.,[30] and Morgan and Soltesz[31] applied graph theory to analyze the essential properties that define network topology. Two network measures were used by Watts and Strogatz[34] in their influential study on the structure of the nervous system of the worm *C. elegans*: the average path length, *L*, and the average clustering coefficient, *C*. The average path length is defined as the average number of steps required to reach any node from any other node in the network, and it is therefore a measure of how well connected the network is globally. The average clustering coefficient, on the other hand, is a measure of local connectivity. It is defined as the fraction of all possible connections between two cells that are actually formed. The average clustering coefficient for an entire network is simply the average of the clustering coefficients for each node in the network. Based on these two topological measures, networks can be divided into three classes: (1) regular, with a long path length and large clustering coefficient; (2) random, with a short path length and small clustering coefficient; and (3) small world, with a short path length and large clustering coefficient (Fig. 23.1C). Dyhrfjeld-Johnsen et al.[30] performed simulations of network configurations where sclerosis progresses in steps (e.g., 20% then 40%, and so on), quantified the resulting network structure by using the topological measures L and C, and characterized activity in the network by means of four different measures: (1) duration of granule cell firing (the time of the last granule cell action potential in the network minus the time of the first granule cell action potential); (2) average number of action potentials per granule cell; (3) time until activity of the most distant GCs from the stimulation point (i.e., latency to full network activity); and (4) synchrony of granule cell action potentials.[30]

3 RESULTS

3.1 The Role of Mossy Fiber Sprouting and Interneuron Loss in Determining Network Excitability

Computational studies have consistently found that even small amounts of mossy fiber sprouting can lead to network hyperexcitability. Santhakumar et al.[25] were among the first to show this phenomenon in a biophysically realistic network model.

Even a relatively small amount of sprouting 10%–15% of the amount of sprouting observed in the major experimental models of epilepsy[35] resulted in the spread of seizure-like activity to the adjacent model hippocampal lamella after focal stimulation of the perforant path. Their results also indicated that the spatially restricted, lamellar distribution of the sprouted mossy fiber contacts reported in studies performed in vivo was an important factor in sustaining seizure-like activity in the network. In contrast, removal of mossy cells resulted in decreased granule cell responses to perforant path activation.

These results indicate the crucial role of even moderate structural connectivity alterations underlying epileptogenesis. In the follow-up studies by Dyhrfjeld-Johnsen et al.[30] and Morgan and Soltesz,[31] the authors scaled up the dentate network model up to 50,000 granule cells. The simulation showed that the additional recurrent excitatory synapses created by sprouting led to long durations of activation in the dentate networks that otherwise remain relatively quiescent. Perhaps surprisingly, maximum excitability was seen at an intermediate level of pathology (40%–60% sclerosis), involving both sprouting and hilar cell loss, and at maximum levels of sclerosis (80%–100%), the network hyperexcitability declined. In the models with intermediate level of sclerosis, the pattern of activation was relatively spatially uniform, whereas in the maximum sclerosis configurations, the network developed activation patterns with distinct waves of activity that could collide and mutually annihilate, thus leading to an overall decrease in network activity.[30] Furthermore, although activity propagation was dependent on the presence of long-range excitatory mossy cells in the hilus, only relatively few mossy cells had to be present to sustain activity. These simulation results are consistent with histological results that indicate that some mossy cells frequently survive in the sclerotic hippocampus of temporal lobe epilepsy patients.[36]

In addition, these results are in agreement with experimental observations in epileptic patients. In surgically removed specimens from pharmacologically intractable human temporal lobe epilepsy patients,[37] cell counts showed that only approximately 80% of hilar cells were lost, even in patients with severe sclerosis.[36] This finding coincides perfectly with the point of maximal hyperexcitability found in the model network developed by Dyhrfjeld-Johnsen et al.,[30] when sclerosis was at 80%.

The model initially introduced by Santhakumar et al.[25] led to the development of a number of variants that were used to study network activity in epileptogenetic conditions. Tejada et al.[38] used a variant of the Santhakumar model to study the effects of the experimental pilocarpine model of medial temporal lobe epilepsy. Specifically, they tested how introducing pilocarpine-induced anatomical changes to granule cells (mossy fiber sprouting and dendritic spine loss) and how increasing the prevalence of adult-born granule cells in the model (increased neurogenesis is a known effect of this form of epilepsy) would impact the poststatus epilepticus network activity. The results suggested that the specific morphological alterations in granule cells may be responsible for both an increase (via mossy fiber sprouting or altered intrinsic excitability) or a decrease (caused by spine loss) of the activity in the network. The authors concluded that the potentiation provoked by mossy fiber sprouting and neurogenesis may be tempered by seizure-induced dendritic spine loss which reduces the activity of the network. These results indicate that it is the imbalance between structural alterations and seizure-induced neurogenesis that leads to the generation of epileptiform activity.

The connection between epileptic sclerosis in the dentate and pathological HFOs was explored in recent work by Kobayashi et al.[39] They developed another variant of the Morgan and Soltesz model[31] including 10,000 granule cells, 300 mossy cells, 100 basket cells, and 120 hilar interneurons. They modeled various degrees of structural change including loss of a fraction of mossy cells and hilar cells as well as connectivity between granule cells induced by mossy fiber sprouting. The effect of sprouting on cell connectivity was evaluated by arranging two sets of simulations with different probability of connectivity (sparse and dense), where the number of postsynaptic granule cells contacted by sprouted mossy fibers differed by a factor of 2. Kobayashi et al. discovered that HFOs with peak frequencies at around 250 Hz could be produced in the model with 100% interneuron cell loss and dense sprouting connectivity, supporting the hypothesis of Dzhala and Staley[40] that synchronous onset of action potential (AP) bursts mediated by recurrent excitatory synaptic transmission is important for the generation of abnormal fast ripples. In the model variants with no or partial interneuron cell loss, only a small granule cell population discharge was evoked by the stimulation.

3.2 Network Architectures That Induce Hyperexcitability

The simulations by Dyhrfjeld-Johnsen et al.[30] and Morgan and Soltesz[31] were complemented with formal graph analyses. These analyses showed that the normal dentate network has the characteristics of a small-world network,[34] that is to say, a network with dense local connections (i.e., a high clustering coefficient) and few long-distance connections (i.e., a short characteristic path length).

This finding means that any cell is reachable from any of the other 1 million cells in the rat dentate network by a chain of less than three synaptic connections, and cells are highly connected locally, resulting in fast local computations and the ability to efficiently relay signals to distant parts of the network. While the analysis of the healthy DG provided useful and novel information about its structure, perhaps the most interesting and counterintuitive structural information comes from the analysis of networks undergoing sclerosis. Dyhrfjeld-Johnsen et al.[30] found that as the DG undergoes pathological structural changes, the features that define its small-world characteristics are enhanced, and despite the loss of a large proportion of highly connected hilar mossy cells, the mossy fiber sprouting was able to compensate for the hilar cell loss. The graph analysis suggested that scale-free network topologies, a topology in which the logarithm of the probability of a cell making a given number of connections is a decreasing linear function of the logarithm of the number of connections, greatly enhanced network excitability, but the presence of highly interconnected nodes (network hubs) with four to seven times as many connections as the average granule cell in the model was sufficient to maintain hyperexcitability and distribute seizure activity extensively even without a scale-free topology.[31] At higher levels of sclerosis, however, the effects of mossy fiber sprouting were not sufficient to compensate for the loss of the long-range projecting mossy cells, which ultimately lead to the transition of the dentate network to a more regular structure which, in turn, caused the decline in hyperexcitability.[30] These results underscore the importance of network topology analysis of epileptic circuits and highlight the important role played by even a few surviving long-distance projections in maintaining global hyperexcitability in the epileptic DG.

The above forms of structural plasticity and changes to network excitability have implications for information processing in the DG as well. The DG is classically thought to perform pattern separation, which is the process of making overlapping patterns of neural activity more distinct.[41] Scharfman and Myers[42] recently proposed that the axons of dentate granule cells, the mossy fibers, have structural and functional properties that impart output variance that facilitates pattern separation. They have shown via computational modeling that if a subset of granule cells becomes more excitable, mossy fibers axons will show decreased variance and consequently a reduction in the pattern separation capability. Furthermore, they showed data of abnormal granule cells that have been identified in animal models of disease where DG-dependent behaviors are impaired. Together these results provide quantitative insight into the disruption of cognitive function in the epileptic DG where a subpopulation of hyperexcitable granule cells, such as hilar ectopic granule cells, may become dominant.

4 DISCUSSION

Computational models of structural plasticity in the DG have increasingly included more biological detail. As layers of biological complexity are added to these models, more interesting insights have been drawn from them. These studies have brought to light the specific structural plasticity mechanisms that are crucial for determining network excitability. The increase in recurrent granule cell connectivity increases the excitability of the network as a whole. In addition, the loss of specific interneurons is a strong determinant of overall functional connectivity, and particularly the presence of hub cells is important for maintaining excitability. Furthermore, insights from applications of graph theory to these modeling studies have shown that only specific network topologies are capable of maintaining hypersynchronous activity.

The application of graph theory to models of epilepsy in the work of Dyhrfjeld-Johnsen et al.[30] and Morgan and Soltesz[31] coincided with a larger trend of utilizing graph analysis to understand brain networks. Recent advances in the field of connectomics[43–45] have brought a wide array of techniques to neuroscience, allowing researchers to draw structure from the large number of detailed connections between elements of neural circuits. These graph measures are also being applied to neurological disorders, partially redefining the disorders as forms of pathological neural circuit rewiring.[46] Graphs constructed from brain connectivity data (structural or functional) can be used to detect functional integration and segregation, quantify centrality of individual brain regions or pathways, and characterize patterns of local anatomical circuitry and resilience of networks to insult. Such connectivity data can come from animals or from humans, using structural imaging techniques such as diffusion tensor imaging (DTI) or functional imaging such as fMRI.[47,48]

Graph theory may also improve localization of temporal lobe epilepsy, leading to more precise resection of the epileptogenic cortex and sparing of normal tissue. We refer the reader to a recent review by Haneef and Chiang[49] that surveys specific clinical scenarios in which the graph theory analyses could be used for improved seizure focus localization. In terms of experimental work, Feldt Muldoon et al.[50] applied network clustering analysis to data from slices of the epileptic DG of mice and showed that epileptic networks are comprised of multiple functional clusters of spatially localized neurons and that large-scale synchronization events are composed of the concurrent activation of temporally and spatially variable combinations of these clusters. This raises the possibility that pathological and normal network dynamics could be distinguished based on the stability of their spatial components and indicates that even though individual neurons may exhibit highly variable dynamics, the topology of the network can induce pathological synchronization.

Graph-theoretical analysis has been successfully used to reveal progressive structural and functional topological changes and characterize the role of individual regions and nodes in epileptic disorders such as idiopathic generalized epilepsy (IGE), which is characterized by widespread generalized spike-and-waves.[51] Although the conventional viewpoint for IGE has been that the entire brain might be homogeneously involved, Zhang et al.[52] refined this view by applying graph-theoretical analysis to reveal that patients with IGE characterized by tonic clonic seizures had lost optimal topological organization in specific functional and structural connectivity networks. Their analysis quantified significant increases in nodal topological characteristics in a number of cortical and subcortical regions, further supporting the hypothesis that regions involved in epileptogenesis display abnormal hub properties in network analysis. Graph-theoretical analysis has been applied to ictal electrophysiology data as well, such as in the study by Ponten et al.,[53] in which the authors applied graph theory methods to EEG from seizures, and compared them with interictal periods. They found that EEG networks show increased clustering and path lengths during absence seizures, highlighting the significant alterations of functional brain connectivity during a seizure. These and other studies provide important tools for determining abnormalities in the global characteristics of structural and functional brain networks and identifying the specific regions where such abnormalities first arise.

Graph models and graph measures of brain connectivity and network architecture will likely have an increasing clinical relevance, as the normal range and pathological deviations of network metrics are determined in large populations of patients over long time periods.[54,55] Consequently, the computational modeling of epilepsy should benefit greatly from the application of connectomics and network analysis techniques.

While analysis of graph models of normal and epileptic brain networks has produced detailed quantitative measures of altered connectivity structure, the question of understanding the dynamical mechanisms of epileptiform activity still remains. Coupled with the rapid rise in knowledge about the structural and functional properties of hippocampal networks, detailed biophysical modeling provides an approach to understanding the dynamics that emerge in large-scale brain networks under normal and abnormal conditions. The stage is set for full-scale network models that closely integrate detailed structural and connectivity data with biophysically realistic neuron models and provide answers to the fundamental questions related to the dynamical and structural changes associated with epilepsy.

Through detailed literature review and data mining, recent efforts resulted in the most detailed to-date quantitative estimates of the cellular and synaptic constituents of the CA1 region of the rat hippocampus.[56] In addition to deriving the numbers of each neuron type, the authors have calculated the interneuron convergence onto pyramidal cells and compared it with the known input synapses on pyramidal cells. This type of quantitative knowledge base forms the basis for currently on-going, data-driven, large-scale hippocampus computational modeling efforts. Such knowledge bases (see also Wheeler et al.[57]) will be critically important for the construction of more accurate, biologically constrained computer models of the hippocampus, and these knowledge bases can also incorporate the substantial data becoming available from experimental research in epilepsy.

While dendritic morphology has been shown to have a dramatic impact on neuronal function,[58] the inherent variability in dendritic morphology between cells of the same type has so far been overlooked in computational modeling. To address this shortcoming of biophysically detailed models, Schneider et al. used the structural context of the hippocampal neural tissue in which granule cell dendritic trees exist to drive their generation in silico and to construct a 1:1 scale, biophysically detailed computational model of the

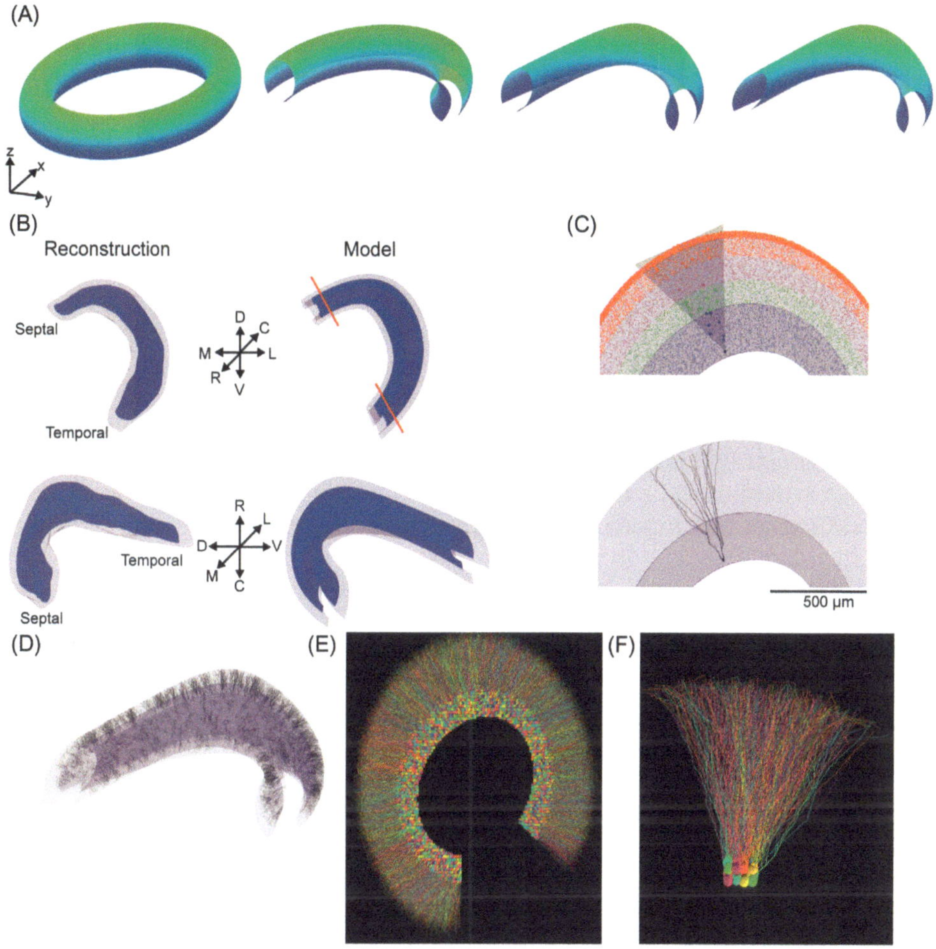

FIGURE 23.2 **Context-dependent generation of dentate gyrus (DG) granule cell dendritic morphologies**; adapted from Schneider et al.[59] (published under the Creative Commons Attribution (CC BY) license). (A) The outer boundary for the granule cell layer was created by cutting a piece out of an elliptical torus (*left*) into a partial structure (*middle left*), then adding a deflection in the *z* direction (*middle right*), and finally creating unequal septal and temporal ends (*right*). (B) Comparison between the experimental reconstruction and model structure. The rostrocaudal (R-C), dorsoventral (D-V), and mediolateral (M-L) axes are shown. (C) Target points lying within an elliptical cone are selected from the pool of available target points, and a subset of these target points is selected to generate the tree. (D) Visualization of the boundary surfaces of the DG model structure (GCL and ML) and 1000 synthetic dendritic trees. (E) Rendering of the complete morphologies for somata of all granule cells in a 20-μm transverse slice from the center of the model DG. (F) Rendering of 48 granule cells from the crest of the slice in (E).

DG in the rat (Fig. 23.2; Schneider et al.[59,60]). The morphology of each neuron in the entire population of DG granule cells in the rat was synthesized by growing dendritic trees within their characteristic dendritic fields bounded by a realistic reconstructed surface.[61] This methodology enables morphological statistics to be linked to large-scale neuroanatomical features and provides an important basis for detailed modeling of the morphological alterations of neurite structures that are often associated with recurrent seizures in experimental models of epilepsy, yet are poorly understood.[62]

4.1 Future Modeling Studies

Connecting together dynamic models of hippocampal subfields according to the appropriate anatomical connectivity structure will be an important step toward a more comprehensive mechanistic explanation of epileptic events in this brain area. Full-scale data-driven computational models can be explored for emergent properties in a way that is not feasible in experimental work, and they provide a methodology to elucidate the possible underlying mechanisms that disrupt network balance and allow seizures to emerge. The disturbances of interactions between individual neurons and local circuits may vary between individuals, but it is likely that these different effects act in combination to cause hyperexcitability via a limited number of convergent pathways. Full-scale computational models that encompass the rich repertoire of oscillatory behaviors observed in vivo might provide an important tool in making the causal connections between local circuit mechanisms and large-scale abnormal oscillations.

To model the electrophysiological activity of single neurons, as well as the detailed synaptic connections within neuronal networks, it is necessary to have precise reconstructions and generative models of neuronal morphology. The methodology for linking of macroscopic and microscopic neuroanatomy developed by Schneider et al.[59] provides a framework that can be expanded upon with additional cell types and axons, and by importing the synthetic morphologies in a simulator such as NEURON, provides an avenue to link neuroanatomical features with electrophysiological function. In addition, the context-driven generation methodology can be applied to creating realistic axonal morphologies connectivity and comparing them directly to the growing base of connectomics data. Once detailed dendritic and axonal morphologies are established, precise and realistic network connectivity and synaptic locations will be unambiguously determined by the positions and distances between neurites in the neuroanatomical context, thus allowing more detailed investigations of the relationship between structure and network function. Furthermore, the pathological changes of hippocampal sclerosis have been found in clinical studies to be variable along the longitudinal axis[63] and the probabilistic approximations utilized by Dyhrfjeld-Johnsen et al.[30] and Morgan and Soltesz[31] may not most ideally represent the progression of the disease. When applied to modeling axonal sprouting in epilepsy, the context-driven generation methodology could produce pathological connectivity patterns that accurately represent the diverse focal and diffuse patterns of sclerosis identified in the hippocampus and allow for more nuanced measurements of the change in network activity.

A further topic of modeling investigation is the role of immature adult-born granule cells during the development of epilepsy. The function of these new neurons under normal and pathological conditions has been the subject of numerous recent studies,[64] and Scharfman and Myers[42] recently reported that an increase of the proportion of excitable granule cells would lead to disruption of pattern separation cognitive function.

Further modeling studies would be required to establish the precise effects on network function of the abnormal structural integration of adult-born ectopic granule cells.

4.2 Future Experimental Studies

There are a number of antiepileptic drugs that suppress seizures, but the mechanisms of action of most are not fully understood,[65] particularly in a network context where different cell types can have different responses to drugs.[66] An important application for computational modeling would be to explore the predicted effects and hypothesized pharmacological mechanisms of treatment, to design new treatments that are focused on the mechanisms of epileptogenesis as opposed to merely stopping seizures, or to explore nonpharmacological treatments. Regarding the latter, for example, neurostimulation is an increasingly used treatment that reduces seizures and avoids many adverse effects of medications, but it is unclear which are the optimal brain sites and structures to stimulate, as well as which is the most effective stimulation protocol.[67] Similarly, closed-loop interventions[68] such as responsive neurostimulation provide additional options for seizure control while reducing side effects. The success of such approaches critically depends on fast and accurate seizure detection.[69] Full-scale biophysical computational models provide the means to cheaply test closed-loop approaches under an extensive range of conditions and oscillatory regimes. In particular, Krook-Magnuson et al.[71,72] and Paz et al.[70] have shown that the ability to target appropriate local and long-range connections to the seizure focus provides a powerful way to control seizures, especially in cases where the seizure focus is in an inaccessible area of the brain. Krook-Magnuson et al.[71] showed that targeting hippocampal neurons contralateral to the seizure focus was as effective at suppressing seizures as intervening at the focus itself. Even more surprisingly, Krook-Magnuson et al.[72] showed that optogenetic modulation of the cerebellum, a structure separated by a long distance, could also control seizures originating in the hippocampus. These studies suggest that a promising area of research is to identify and target for intervention hub cells that can control brain activity on a large scale because of their extensive long-distance connections. Dyhrfjeld-Johnsen et al.[30] identified mossy cells in the DG as potentially important candidates for hub cell intervention, as one mossy cell projects onto more than 30,000 granule cells and thus a modulation of just a few mossy cells might be sufficient to control seizures while minimally interfering with the function of the rest of the neuronal population. More concretely, a population of widely projecting early-born GABAergic cells, referred to as hub cells, appear to meet the criteria for hub cells and were shown to influence network synchrony in the developing hippocampus.[73–75] Large-scale biophysically detailed network models can be used to obtain precise quantitative measurements on the spatial distribution and number of hub cells that would be necessary and sufficient for seizure control.

Although hippocampal sclerosis and mossy fiber sprouting have been implicated in temporal lobe epilepsy, the network mechanisms of propagating epileptic activity remain unclear, particularly in cases where a large proportion of pyramidal cells in the CA3 region (which receives input from the DG mossy fibers) and the CA1 region (which receives input from CA3) are lost during epileptogenesis. The recent results of Haussler et al.[76] suggest that the activity of the DG could be transferred to the CA2 region, and a possible candidate for the propagation of epileptic activity to the cortex is the connection from CA2 to the entorhinal cortex which has been recently described.[77,78] Therefore, experimental studies of the changes in connectivity within hippocampal regions and between the hippocampal formation and the surrounding cortical regions will be necessary to determine the extent and mechanisms of large-scale temporal lobe epilepsy-associated pathological restructuring.

Acknowledgments

This chapter was supported by the NIH (NS35915 to I.S.), NSF (PRAC1614622 to I.S.), by NSF (I.R. individual award), and by the Stanford Neuroscience Graduate Program (M.P.).

References

1. Steward O. Topographic organization of the projections from the entorhinal area to the hippocampal formation of the rat. *J Comp Neurol*. 1976;167(3):285–314.
2. Varga C, Lee S, Soltesz I. Target-selective GABAergic control of entorhinal cortex output. *Nat Neurosci*. 2010;13(7):822–824.
3. Margerison J, Corsellis J. Epilepsy and the temporal lobes. a clinical, electroencephalographic and neuropathological study of the brain in epilepsy, with particular reference to the temporal lobes. *Brain*. 1966;89 (3):499–530.
4. Ben-Ari Y, Dudek FE. Primary and secondary mechanisms of epileptogenesis in the temporal lobe: there is a before and an after. *Epilepsy Currents*. 2010;10(5):118–125.
5. Dudek FE, Sutula TP. Epileptogenesis in the dentate gyrus: a critical perspective. *Prog Brain Res*. 2007;163:755–773.
6. Nadler JV. The recurrent mossy fiber pathway of the epileptic brain. *Neurochem Res*. 2003;28:1649–1658.
7. Buckmaster P, Zhang G, Yamawaki R. Axon sprouting in a model of temporal lobe epilepsy creates a predominantly excitatory feedback circuit. *J Neurosci*. 2002;22:6650–6658.
8. Nissinen J, Lukasiuk K, Pitkanen A. Is mossy fiber sprouting present at the time of the first spontaneous seizures in rat experimental temporal lobe epilepsy? *Hippocampus*. 2001;11:299–310.
9. Mikkonen M, Soininen H, Kalvianen R, et al. Remodeling of neuronal circuitries in human temporal lobe epilepsy: increased expression of highly polysialylated neural cell adhesion molecule in the hippocampus and the entorhinal cortex. *Ann Neurol*. 1998;44:923–934.
10. Pitkanen A, Kharatishvili I, Karhunen H, et al. Epileptogenesis in experimental models. *Epilepsia*. 2007;48 (Suppl. 2):13–20.
11. Cantallops I, Routtenberg A. Rapid induction by kainic acid of both axonal growth and f1/gap-43 protein in the adult rat hippocampal granule cells. *J Comp Neurol*. 1996;366:303–319.
12. Danzer S, Crooks K, Lo D, McNamara J. Increased expression of brain-derived neurotrophic factor induces formation of basal dendrites and axonal branching in dentate granule cells in hippocampal explant cultures. *J Neurosci*. 2002;22:9754–9763.
13. Guilhem D, Dreyfus P, Makiura Y, Suzuki F, Onteniente B. Short increase of bdnf messenger rna triggers kainic acid-induced neuronal hypertrophy in adult mice. *Neuroscience*. 1996;72:923–931.
14. Suzuki F, Junier M-P, Guilhem D, Sorensen J-C, Onteniente B. Morphogenetic effect of kainate on adult hippocampal neurons associated with a prolonged expression of brain-derived neurotrophic factor. *Neuroscience*. 1995;64:665–674.
15. Koyama R, Yamada M, Fujisawa S, et al. Brain-derived neurotrophic factor induces hyperexcitable reentrant circuits in the dentate gyrus. *J Neurosci*. 2004;24(33):7215–7224.
16. Parent JM, Yu TW, Leibowitz RT, et al. Dentate granule cell neurogenesis is increased by seizures and contributes to aberrant network reorganization in the adult rat hippocampus. *J Neurosci*. 1997;17(10):3727–3738.
17. Pierce J, Punsoni M, McCloskey D, Scharfman H. Mossy cell axon synaptic contacts on ectopic granule cells that are born following pilocarpine-induced seizures. *Neurosci Lett*. 2007;422(2):136–140.

18. Parent JM. Adult neurogenesis in the intact and epileptic dentate gyrus. *Prog Brain Res*. 2007;163:529–540.
19. Hester MS, Danzer SC. Accumulation of abnormal adult-generated hippocampal granule cells predicts seizure frequency and severity. *J Neurosci*. 2013;33(21):8926–8936.
20. Jacobs J, Staba R, Asano E, et al. High-frequency oscillations (HFOs) in clinical epilepsy. *Prog Neurobiol*. 2012;98(3):302–315.
21. Engel Jr J, Bragin A, Staba R, Mody I. High-frequency oscillations: what is normal and what is not? *Epilepsia*. 2009;50(4):598–604.
22. Bragin A, Wilson CL, Engel J. Spatial stability over time of brain areas generating fast ripples in the epileptic rat. *Epilepsia*. 2003;44(9):1233–1237.
23. Jirsch JD, Urrestarazu E, LeVan P, et al. High-frequency oscillations during human focal seizures. *Brain*. 2006;129(6):1593–1608.
24. Staba RJ, Wilson CL, Bragin A, Fried I, Engel Jr J. Quantitative analysis of high-frequency oscillations (80–500 Hz) recorded in human epileptic hippocampus and entorhinal cortex. *J Neurophysiol*. 2002;88(2):1743–1752.
25. Santhakumar V, Aradi I, Soltesz I. Role of mossy fiber sprouting and mossy cell loss in hyperexcitability: a network model of the dentate gyrus incorporating cell types and axonal topography. *J Neurophysiol*. 2005;93(1):437–453.
26. Aradi I, Holmes WR. Role of multiple calcium and calcium-dependent conductances in regulation of hippocampal dentate granule cell excitability. *J Comput Neurosci*. 1999;6(3):215–235.
27. Lubke J, Frotscher M, Spruston N. Specialized electrophysiological properties of anatomically identified neurons in the hilar region of the rat fascia dentata. *J Neurophysiol*. 1998;79:1518–1534.
28. Staley KJ, Otis TS, Mody I. Membrane properties of dentate gyrus granule cells: comparison of sharp microelectrode and whole-cell recordings. *J Neurophysiol*. 1992;67:1346–1358.
29. Buckmaster PS, Wenzel HJ, Kunkel DD, Schwartzkroin PA. Axon arbors and synaptic connections of hippocampal mossy cells in the rat in vivo. *J Comp Neurol*. 1996;366:271–292.
30. Dyhrfjeld-Johnsen J, Santhakumar V, Morgan RJ, et al. Topological determinants of epileptogenesis in large-scale structural and functional models of the dentate gyrus derived from experimental data. *J Neurophysiol*. 2007;97(2):1566–1587.
31. Morgan RJ, Soltesz I. Nonrandom connectivity of the epileptic dentate gyrus predicts a major role for neuronal hubs in seizures. *Proc Natl Acad Sci USA*. 2008;105(16):6179–6184.
32. Patton P, McNaughton B. Connection matrix of the hippocampal-formation 1. The dentate gyrus. *Hippocampus*. 1995;5:245–286.
33. Sik A, Penttonen M, Buzsaki G. Interneurons in the hippocampal dentate gyrus: an in vivo intracellular study. *Eur J Neurosci*. 1997;9(3):573–588.
34. Watts DJ, Strogatz SH. Collective dynamics of small-world networks. *Nature*. 1998;393(6684):440–442.
35. Curia G, Longo D, Biagini G, Jones R, Avolia M. The pilocarpine model of temporal lobe epilepsy. *J Neurosci Methods*. 2008;172(2–4):143–157.
36. Bluemcke I, Suter B, Behle K, et al. Loss of hilar mossy cells in ammon's horn sclerosis. *Epilepsia*. 2000;41(Suppl. 6):S174–S180.
37. Gabriel S, Njunting M, Pomper J, et al. Stimulus and potassium-induced epileptiform activity in the human dentate gyrus from patients with and without hippocampal sclerosis. *J Neurosci*. 2004;24:10416–10430.
38. Tejada J, Garcia-Cairasco N, Roque AC. Combined role of seizure-induced dendritic morphology alterations and spine loss in newborn granule cells with mossy fiber sprouting on the hyperexcitability of a computer model of the dentate gyrus. *PLoS Comput Biol*. 2014;10:e1003601.
39. Kobayashi K, Akiyama T, Ohmori I, Yoshinaga H, Gotman J. Action potentials contribute to epileptic high-frequency oscillations recorded with electrodes remote from neurons. *Clin Neurophysiol*. 2015;126:873–881.
40. Dzhala VI, Staley KJ. Mechanisms of fast ripples in the hippocampus. *J Neurosci*. 2004;24(40):8896–8906.
41. O'Reilly R, McClelland J. Hippocampal conjunctive encoding, storage, and recall: avoiding a trade-off. *Hippocampus*. 1994;4(6):661–682.
42. Scharfman HE, Myers CE. Corruption of the dentate gyrus by "dominant" granule cells: implications for dentate gyrus function in health and disease. *Neurobiol Learn Mem*. 2015;129:69–82.
43. Oh S, Harris J, Ng L, et al. A mesoscale connectome of the mouse brain. *Nature*. 2014;508(7495):207–214.

44. Bargmann C, Marder E. From the connectome to brain function. *Nat Methods*. 2013;10:483–490.
45. Denk W, Briggman K, Helmstaedter M. Structural neurobiology: missing link to a mechanistic understanding of neural computation. *Nat Rev Neurosci*. 2012;13:351–358.
46. Sporns O. *Discovering the Human Connectome*. MIT Press; 2012.
47. Engel Jr J, Thompson PM, Stern JM, et al. Connectomics and epilepsy. *Curr Opin Neurol*. 2013;26(2):186–194.
48. Taylor PN, Kaiser M, Dauwels J. Structural connectivity based whole brain modelling in epilepsy. *J Neurosci Methods*. 2014;246:51–57.
49. Haneef Z, Chiang S. Clinical correlates of graph theory findings in temporal lobe epilepsy. *Seizure*. 2014;23:809–818.
50. Feldt Muldoon S, Soltesz I, Cossart R. Spatially clustered neuronal assemblies comprise the microstructure of synchrony in chronically epileptic networks. *Proc Natl Acad Sci USA*. 2013;110(9):3567–3572.
51. Engel J. A proposed diagnostic scheme for people with epileptic seizures and with epilepsy: report of the ILAE task force on classification and terminology. *Epilepsia*. 2001;42(6):796–803.
52. Zhang Z, Liao W, Chen H, et al. Altered functional-structural coupling of large-scale brain networks in idiopathic generalized epilepsy. *Brain*. 2011;134(10):2912–2928.
53. Ponten SC, Douw L, Bartolomei F, Reijneveld JC, Stam CJ. Indications for network regularization during absence seizures: weighted and unweighted graph theoretical analyses. *Exp Neurol*. 2009;217(1):197–204.
54. Daianu M, Jahanshad N, Nir TM, et al. Breakdown of brain connectivity between normal aging and Alzheimer's disease: a structural k-core network analysis. *Brain Connect*. 2013;3(4):407–422.
55. Dennis EL, Jahanshad N, McMahon KL, et al. Development of brain structural connectivity between ages 12 and 30: a 4-Tesla diffusion imaging study in 439 adolescents and adults. *NeuroImage*. 2013;64:671–684.
56. Bezaire MJ, Soltesz I. Quantitative assessment of CA1 local circuits: knowledge base for interneuron-pyramidal cell connectivity. *Hippocampus*. 2013;23(9):751–785.
57. Wheeler DW, White CM, Rees CL, et al. Hippocampome.org: a knowledge base of neuron types in the rodent hippocampus. *Elife*. 2015;4:e09960.
58. Poirazi P. Information processing in single cells and small networks: insights from compartmental models. In: Maroulis G, Simos TE, editors, *International Conference on Computational Methods in Sciences and Engineering 2008 (ICCMSE 2008)*, volume 1108, pp. 158–167, 2009.
59. Schneider CJ, Cuntz H, Soltesz I. Linking macroscopic with microscopic neuroanatomy using synthetic neuronal populations. *PLoS Comput Biol*. 2014;10:e1003921.
60. Schneider CJ, Bezaire M, Soltesz I. Toward a full-scale computational model of the rat dentate gyrus. *Front Neural Circuits*. 2012;6(83).
61. Ropireddy D, Bachus SE, Ascoli GA. Non-homogeneous stereological properties of the rat hippocampus from high-resolution 3D serial reconstruction of thin histological sections. *Neuroscience*. 2012;205:91–111.
62. Furtado M, Castro O, Del Vecchio F, de Oliveira J, Garcia-Cairasco N. Study of spontaneous recurrent seizures and morphological alterations after status epilepticus induced by intrahippocampal injection of pilocarpine. *Epilepsy Behav*. 2011;20(2):257–266.
63. Thom M, Liagkouras I, Martinian L, et al. Variability of sclerosis along the longitudinal hippocampal axis in epilepsy: a post mortem study. *Epilepsy Res*. 2012;102:45–59.
64. Song J, Christian K, Ming G, Song H. Modification of hippocampal circuitry by adult neurogenesis. *Dev Neurobiol*. 2012;72:1032–2043.
65. Duncan JS, Sander JW, Sisodiya SM, Walker MC. Adult epilepsy. *Lancet*. 2006;367(9516):1087–1100.
66. Pothmann L, Mueller C, Averkin R, et al. Function of inhibitory micronetworks is spared by Na^+ channel-acting anticonvulsant drugs. *J Neurosci*. 2014;34(29):9720–9735.
67. Fisher RS, Velasco AL. Electrical brain stimulation for epilepsy. *Nat Rev Neurol*. 2014;10:261–270.
68. Krook-Magnuson E, Gelinas JN, Soltesz I, Buzsaki G. Neuroelectronics and biooptics: closed-loop technologies in neurological disorders. *JAMA Neurol*. 2015;72(7):823–829.
69. Nagaraj V, Lee ST, Krook-Magnuson E, et al. Future of seizure prediction and intervention: closing the loop. *J Clin Neurophysiol*. 2015;32(3):194–206.
70. Paz JT, Davidson TJ, Frechette ES, et al. Closed-loop optogenetic control of thalamus as a tool for interrupting seizures after cortical injury. *Nat Neurosci*. 2013;16(1):64–70.
71. Krook-Magnuson E, Armstrong C, Oijala M, Soltesz I. On-demand optogenetic control of spontaneous seizures in temporal lobe epilepsy. *Nat Commun*. 2013;4(1376).

72. Krook-Magnuson E, Szabo GG, Armstrong C, Oijala M, Soltesz I. Cerebellar directed optogenetic intervention inhibits spontaneous hippocampal seizures in a mouse model of temporal lobe epilepsy. *eNeuro*. 2014;1(1).
73. Bonifazi P, Goldin M, Picardo M, et al. GABAergic hub neurons orchestrate synchrony in developing hippocampal networks. *Science*. 2009;326(5958):1419–1424.
74. Picardo M, Guigue P, Bonifazi P, et al. Pioneer GABA cells comprise a subpopulation of hub neurons in the developing hippocampus. *Neuron*. 2011;71(4):695–709.
75. Cossart R. Operational hub cells: a morpho-physiologically diverse class of GABAergic neurons united by a common function. *Curr Opin Neurobiol*. 2014;26:51–56.
76. Haussler U, Rinas K, Kilias A, Egert U, Haas C. Mossy fiber sprouting and pyramidal cell dispersion in the hippocampal CA2 region in a mouse model of temporal lobe epilepsy. *Hippocampus*. 2016;26(5):577–588.
77. Hitti F, Siegelbaum S. The hippocampal CA2 region is essential for social memory. *Nature*. 2014;508 (7494):88–92.
78. Rowland D, Weible A, Wickersham I, et al. Transgenically targeted rabies virus demonstrates a major monosynaptic projection from hippocampal area ca2 to medial entorhinal layer II neurons. *J Neurosci*. 2013;33:14889–14898.

CHAPTER

24

Computational Models of Stroke Recovery

Vittorio Sanguineti
University of Genoa, Genova, Italy

OUTLINE

1 INTRODUCTION

In the nervous system, a cerebrovascular accident (stroke) elicits a complex series of reorganization processes at molecular, cellular, neural population, behavioral (sensorimotor and cognitive), and social interaction levels, with temporal scales that range from hours to months to years.[1–4] Alterations occur well beyond the actual lesion, including a

The Rewiring Brain
DOI: http://dx.doi.org/10.1016/B978-0-12-803784-3.00024-X

low-activity "penumbra" region in the surrounding areas and interhemispheric unbalance due to a decreased activity in the ipsilesional side.[5]

Animal models and human studies suggest that functional recovery is mediated by use-dependent reorganization of the preserved neural circuitry. A key to neuromotor recovery and the basis of many neurorehabilitation interventions is movement associated with a task[1,2] and with volitional effort.[6–8] Repeated exercise based on voluntary goal-directed movements produces alterations in neuronal excitability,[9] leading to changes in neural circuitry, with a process resembling that occurring in the developing brain.

Redundancy in the musculoskeletal system plays a key role in recovery. As the motor system has more degrees of freedom than task variables, the same functional behavior may be achieved with different movements. The pressure toward regaining functional independence may lead to the development of compensatory strategies that, even when adequate for carrying out activities of daily life, may be stereotypical or energetically inefficient so that they may ultimately prevent true recovery.[10,11] For instance, an excess use of the nonparetic limb can have a negative influence on the process of cortical reorganization[12] by further reinforcing the imbalance between the two hemispheres.

In summary, neuromotor recovery through exercise is the end result of a complex interplay between activity-dependent reorganization of the brain areas close to the lesion, the recruitment of new neural pathways, and the development of novel motor strategies.

Understanding the complex set of interactions between neural structures that underlie voluntary movements is of paramount importance to understand neuromotor impairments and to identify treatments that facilitate recovery.[13]

A deeper understanding of the functional and physiological mechanisms underlying recovery would likely have a strong impact on approaches to neuromotor rehabilitation. Computational motor control and, more in general, computational models may greatly contribute to this understanding.[14] There is a specific need for "improved models of human motor recovery to provide a more rational framework for designing robotic therapy control strategies."[15] However, while musculoskeletal models have a long history in the personalization of treatment of movement disorders,[16] computational models of neuromotor recovery through exercise are still in their infancy.

In this chapter we review the state-of-the-art of computational models for neuromotor recovery, with a focus on functional behaviors. We then relate these models with the mechanisms of synaptic and structural plasticity, and discuss their implications for treatment. We finally suggest directions for future research.

2 MODELS OF NEUROMOTOR RECOVERY

There have been several attempts to model the time course of recovery, either spontaneous or facilitated by some form of treatment, e.g., neuromodulation, electrical stimulation, or assistance by a robot. In this chapter we will specifically focus on models of activity-dependent recovery.

Existing models of recovery cover different levels of description, ranging from cortical or subcortical lesions, to muscle control, to functional behavior in the context of a specific task.

Models of neuromotor recovery at the level of cortical or corticospinal (CS) circuitry[17–19] address how focal cortical lesions elicit neural reorganization phenomena, and the way these lesions affect motor behavior. Models of muscle control focus on characterizing the impairment in individual subjects in terms of altered muscle synergies.[20] Other models address the "functional" level of description, related to the ability to complete a specific task and to how it changes over time.[21,22] Only few models focus on how physical interaction affects voluntary control[23] and how such voluntary control changes with exercise .[24]

Even fewer models encompass multiple levels of description,[25–27] by addressing the mechanisms of cortical reorganization in the context of voluntary motor activity in a skill learning scenario. In this case, the emphasis is on how voluntary movements promote recovery through cortical and subcortical reorganization.

In the following sections we review a number of computational models of neuromotor recovery, at central and functional level respectively, which have recently appeared in the literature. For each model, we provide a general description. We then discuss their main findings or predictions, their limitations, their relations, and the implications for rehabilitation.

2.1 Models of Recovery at Central Level: Focal Cortical Lesions and Activity-Dependent Reorganization

Several models of neuromotor recovery explicitly address the mechanisms of cortical reorganization following a focal lesion. The work of Goodall et al.[17] uses a computational model of the sensorimotor control of arm movements,[28] incorporating a model of both the somatosensory and the motor cortex, to investigate the reorganization processes that occur immediately after a focal cortical lesion. The model assumes "Mexican-hat" lateral cortical connections, a competitive activation dynamics, and a Hebbian plasticity mechanism for incoming cortical connections. Lesions are modeled by setting the activation levels of selected units to zero and by eliminating connections to and from those units.

The main prediction of this model is a two-phase reorganization process. Immediately after the lesion, low activity is observed in the areas surrounding the lesion. A second phase is characterized by a gradual increase of the size of this area and by a general reorganization of the intact cortical regions. Both effects are mediated by activity-dependent synaptic changes. The low-activity perilesional area and its expansion over time are due to lack of activation. The model also predicts that a small uniform excitatory perilesional input may favor the participation of this area in the reorganization process.

The same group also studied the short- and long-term changes in lateralization that occur after a focal lesion and investigated the possible contribution to recovery of the intact hemisphere and of interhemispheric communication.[29,30] Cortical reorganization was modeled in terms of synaptic changes (self-organization) of a topological map.

Butz et al.[19] specifically addressed the mechanisms of activity-dependent synaptic rewiring that occur immediately after a focal cortical lesion. Synapse formation is accounted for by models of axonal and dendritic elements. On the presynaptic side, activity is assumed to promote axonal outgrowth. On the postsynaptic side, each neuron is assumed to change its input connectivity in an homeostatic manner, with the goal of keeping the firing probability within a specified range. In this framework, rewiring after a lesion can be seen as a form of compensation, driven by the need to regain a stable (homeostatic) regime.

The main prediction of this model is that neural populations that are already in homeostatic conditions (e.g., in adult individuals) are much less likely to compensate for lesions than networks that are still under development. An additional prediction is that external stimulation may promote axonal outgrowth, thus accelerating rewiring. However, prolonged stimulation may induce a saturation effect; hence, it can be detrimental to recovery or, anyway, less effective than paused stimulation. In terms of rehabilitation, these findings suggest that training with pauses in between may be more effective than continuous intensive training without pauses. The model is discussed in more detail elsewhere in this book.

Varier et al.[31] used the model proposed by Reggia and colleagues[28] to examine the effects of focal and distributed lesions at various stages of development. In partial contrast with Butz et al.,[19] this model predicts that mature systems are relatively more robust to lesions than systems that are still under development. This apparent discrepancy may be a consequence of the different assumptions on structural plasticity—highly modifiable[19] versus hardwired.[28]

Reinkensmeyer et al.[18] developed a model of cortical damage and its consequences on arm reaching movements. Different from the previous approaches, this model does not address intracortical connectivity and its reorganization. Based on experiments on non-human primates,[32] neurons in the motor cortex are assumed to collectively encode the initial direction of the movement (population vector coding). Specifically, each neuron's firing rate is assumed to be a function (truncated cosine) of the difference between the actual direction and the "preferred direction" for that neuron plus a noise term, whose standard deviation is proportional to the deterministic part of the cell response (signal-dependent noise). The overall encoded direction is the sum of the preferred directions of each individual neuron, weighted by their activity levels.

Cortical lesions were simulated by eliminating a fraction of the neurons (cell death)—hence resulting in underrepresented or nonrepresented preferred directions. Movement performance was measured in terms of the discrepancy between intended and encoded movement direction. They specifically looked at the variability of directional error within the same intended direction and across directions, and how these quantities are affected by cell death. They found that both types of error are inversely correlated with the fraction of surviving cells. In a number of experimental studies with stroke survivors, they found that the same indicators exhibit similar relationships with the subjects' clinical impairment score.[33–35]

This study addresses the problem of how cortical damage results in impaired movements. As such, it is a model of impairment, which does not explicitly address the mechanisms of recovery.

More recently, cortical dynamics has been investigated in terms of complex network theory. The patterns of functional connectivity between cortical areas have been shown to follow a "small world" topology (prevalence of short-distance over long-distance connections). Stroke results in an alteration of connectivity at a whole network level. After stroke, the overall connectivity shifts from a small world pattern toward a more random topology. Li et al.[36] used a statistical model of the evolution of connectivity to simulate the process degeneration from normal to pathological connectivity. As before, this model addresses impairment and not the recovery mechanisms. However, it is interesting in that the simulations allow to characterize the effect of stroke on the whole-brain topology of individual subjects.

Two related studies[26,37] focus on how reorganization of the preserved CS pathways and the recruitment of new ones underlie the recovery of the ability to generate force.

The proposed model—inspired by single-cell recordings of the neural correlates of wrist force generation in primates—specifically addresses the "residual capacity" phenomenon, i.e., the empirical evidence that additional motor training may still improve the movement capacities even years after a stroke. Specifically, the model is based on the notion that experience of movement "practice" induces a reoptimization in the recruitment of the intact CS connections to motor neurons (MNs). A pool of CS cells is assumed to receive a movement command as input. The activity of CS cells summates at the level of the MN pools in the spinal cord. Wrist force (aimed toward either flexion or extension) is determined by the difference between the activities of the "flexor" and "extensor" MN pools. This can be considered as the simplest muscle synergy. The synaptic weights of the CS–MN connections are assumed to be fixed, whereas the weights of the input connections to the CS cells are learned through a reinforcement mechanism, in which the experienced attempt to move the limb represents the reward signal that guides the refinement of activation.

Furthermore, the model assumes that if CS connections originating from the primary motor cortex (M1) are lacking, CS connections from the supplementary motor area (SMA) may be recruited as well.

Model simulations predict that the size of the residual CS population determines the maximum strength an individual stroke patient can achieve. This is consistent with observations based on transcranial magnetic stimulation (TMS) and magnetic resonance diffusion tensor imaging[38] suggesting a strong correlation between white matter integrity and maximum grip force. In addition, the model predicts that the same dose of exercise is more effective when administered to subacute subjects (as compared to chronic) and to less impaired subjects (as compared to more severe). Furthermore, severe M1 lesions are predicted to induce an increase in SMA activity. These predictions have been confirmed experimentally.[39,40]

This model has not been directly used in rehabilitation but provides some hints on how to make rehabilitation more effective. The simulations show that recovery would be facilitated if noise level were decreased or if noise were signal-dependent. Another prediction is that inhibition of the stronger connections, e.g., the residual connections originating from M1, would facilitate the recruitment of alternative pathways. The latter prediction is somehow consistent with, and actually provides a possible interpretation for, the empirical observation[41] that repetitive TMS of the intact cortical hemisphere facilitates recovery.

This model only makes qualitative predictions and cannot describe the behavior of one specific subject. Furthermore, apart from the M1-SMA shift of activity, the model does not directly address the compensation issue and does not provide new insight about the interplay between neural reorganization and task re-learning.

Although the models are with different emphases (synaptic vs structural plasticity; local vs whole-brain topology), all the above approaches focus on the neural mechanisms of use-dependent reorganization. Impairment is only modeled qualitatively, and as such these models cannot immediately be related to the behavior of a specific subject, if not in qualitative terms. A notable exception is represented by Li's brain network evolution model,[36] which provides a characterization of the impairment of individual subjects, on the basis of functional brain imaging data (fMRI and possibly electroencephalography or magnetoencephalography).

2.2 Models at Functional Level

2.2.1 *Temporal Evolution of Performance Over Training Time*

Variants of the power law of practice have often been used to describe the trial-by-trial evolution of motor performance during exercise, both within and between training sessions (the "functional" level of description).

In a widely used approach to robot-assisted rehabilitation,[42,43] subjects are allowed a specific time interval to complete the task without assistance. After that, the robot completes the task through a high-stiffness position controller. In this way, the amount of robot intervention (fraction of the movement operated by the robot) automatically adjusts to subject's performance (more active performance, less robot intervention). However, to prevent movements from being totally passive, at least initially, this technique requires at least a minimum residual amount of voluntary control. In a systematic analysis of various performance indicators during robot-assisted rehabilitation based on this protocol, Colombo et al.[21,22] used an exponential model to account for their temporal evolution.

The main finding of these studies was that some movement features, namely force control and movement smoothness, improved more quickly than parameters that relate to the fine-tuning of the movement, such as speed. Specifically, speed improvement exhibited a much longer (two to three times) time constant than force control and smoothness. Furthermore, in several subjects the path curvature exhibited a nonmonotonic time course, with an initial increase until a peak, followed by a steady decrease.

These findings shed some light on the process of functional recovery. Subjects first explore the action space with the primary objective of exploiting their residual abilities to achieve the goal; then they undergo effort optimization to make the movement more efficient. The distinctive behavior of path curvature is a consequence of this two-phase process: the increase relates to the exploration phase, whereas the decay denotes the beginning of the effort optimization phase. The final step is the fine-tuning of movement performance, denoted by the increase in movement speed.[44] These processes likely start at the same time and run in parallel but have different time constants (faster the former, slower the latter).

A different but related view of recovery is based on the notion that each movement is built by combining multiple submovements and the empirical observation[45] that neuromotor recovery is characterized by a gradual decrease in the number of submovements, which results in improved smoothness. These mechanisms suggest that training-mediated recovery shares common features with motor skill learning.

In a subsequent study, Colombo et al.[44] developed a control algorithm (Progressive Task Regulation, PTR) that automatically adjusts various aspects of the task (amplitude of the movement, number of subtasks, and assistance modality) according to the evolution of the different performance indicators. After each training epoch, performance is measured (moving average over the last three epochs) and the difficulty level of the task is modified according to a set of rules. Depending on the fraction of movement completed without assistance of the robot, exercise difficulty can switch to more challenging assistance modalities. The observed evolution of the different performance indicators was incorporated into the threshold values used to determine when the switch occurs. The decisions of switching task difficulty based on the PTR algorithm were found to be highly consistent with those based on subjective therapists' assessment.

This model only focuses on performance time series and only provides information on the temporal evolution of the recovery process. However, it does not attempt to describe the mechanisms underlying recovery. Specifically, the model describes the evolution of performance in nonassisted trials. The same modeling framework can address situations in which the degree of assistance is constant or changes slowly and monotonically but cannot distinguish between the "robot" and the actual "voluntary" contributions in situations when they act together.

Recently, the same group proposed a dynamical model of recovery,[46] attempting to reproduce the mechanisms underlying the trial-by-trial evolution of performance. The model derives from a computational framework that was originally developed by Smith et al.[47] in the context of associative learning in animal studies. In this model, an internal state variable denoting "motor improvement" is modeled as a random walk. This reflects the assumption that motor improvement builds up as the effect of many different factors. As an addition to the original model, at a given trial the logarithms of a number of observable performance indicators are assumed to be proportional to the amount of motor improvement. The model was found to reproduce the variety of time constants of the recovery observed in the different performance indicators. Although interesting, this model is nothing more than a noisy version of the exponential model. In particular, the random walk assumption says little on what determines recovery.

2.2.2 *The Role of Physical Assistance*

A few modeling studies have addressed the issue of how robot assistance affects recovery. Emken et al.[18] and Reinkensmeyer[23,48] focus on adaptive changes in gait movements in the presence of assistive forces. The study specifically addresses the trial-by-trial evolution of performance during adaptation to an assistive force field and suggests that adaptation can be explained as an optimization process, which accounts for a combination of motor error and effort.

Similar to force field adaptation experiments,[49–52] the adaptation process is modeled as a linear autoregressive model. A controller receives the desired trajectory, the performance error, and the motor command (muscle force) on the previous trial as inputs. The motor command on the next trial is the controller output. The controller consists of two modules: (1) an inverse model of limb dynamics, which transforms a desired trajectory into an appropriate "feedforward" motor command and (2) a corrective module, which adapts the motor command to changes in the dynamics of the environment. The controller output (muscle force) is applied to the body, whose movements are disturbed by an external perturbation.

Adaptation within the corrective module relies on the notion that while dealing with assistive forces, the motor system behaves as a "greedy" optimizer, so that these forces are quickly incorporated into the motor plan to minimize the effort while maintaining the required performance level (e.g., a small error). A force estimator predicts the disturbance on the next trial, and an optimizing controller generates the next motor command by minimizing a cost function on a trial-by-trial basis. This optimization framework results in a control law with two terms: a corrective term, which increments the motor command in proportion to the previous error, in the direction that reduces error, and a decay term, which gradually reduces the motor command—this latter term captures the "slacking" mechanism.

Emken et al.[23] also suggested that slacking may have adverse effects on recovery. As a consequence, in robot-assisted rehabilitation assistance should be kept to a minimum (assist-as-needed). Furthermore, it has to be decreased—manually or automatically—as performance improves. A variety of techniques has been proposed to adaptively regulate the magnitude of assistive force as a function of the observed outcome. In Casadio et al.[53] the therapist manually selected the assistance level to keep it to the minimum level that evoked a functional response needed for accomplishing the task. In Vergaro et al.,[54] a linear controller continuously and automatically regulated the assistive force provided by the robot, depending on online performance measures. Similar mechanisms have been proposed in the context of the upper limb, the performance-based progressive robot-assisted therapy used by the MIT-Manus robot,[55] and the lower limb, the patient-cooperative training modality used by the Lokomat system.[56,57]

Using a computational model of human-robot load sharing, Reinkensmeyer et al.[58] suggested that to achieve assistance as needed, the robot controller should possess a slacking mechanism that resembles that observed in humans. Wolbrecht et al.[59] designed a more complex adaptive control scheme, based on task performance that automatically negotiates an error-reducing and an effort-enhancing component. The controller needs an explicit model of the subject's arm and its neural control. In this specific study, this model took the form of a radial basis function neural network, which was built by experience.

The "greedy optimizer" model has been highly influential to rehabilitation but does not really address the recovery mechanisms. In fact, it has been derived from studies on healthy subjects, and its implications for recovery are largely speculative. The slacking hypothesis has never been directly tested in clinical rehabilitation trials.

In a recent study, Casadio et al.[24] developed a linear dynamical model to describe the trial-by-trial evolution of the motor performance of chronic stroke survivors who underwent a rehabilitation protocol based on a robot-assisted arm extension task.

The model is based on the notion that in robot-assisted exercise the robot device and the subject cooperate toward a common goal—a form of shared control. Specifically, the model assumes that task performance is a function of a voluntary, human-generated command (taken as the model's state variable) and of a robot-generated assistive force (taken as one of the model's inputs; Fig. 24.1, top right).

As regards the dynamics of the actual recovery process, the model assumes that the amount of voluntary control on the next trial is the sum of three components: (1) a

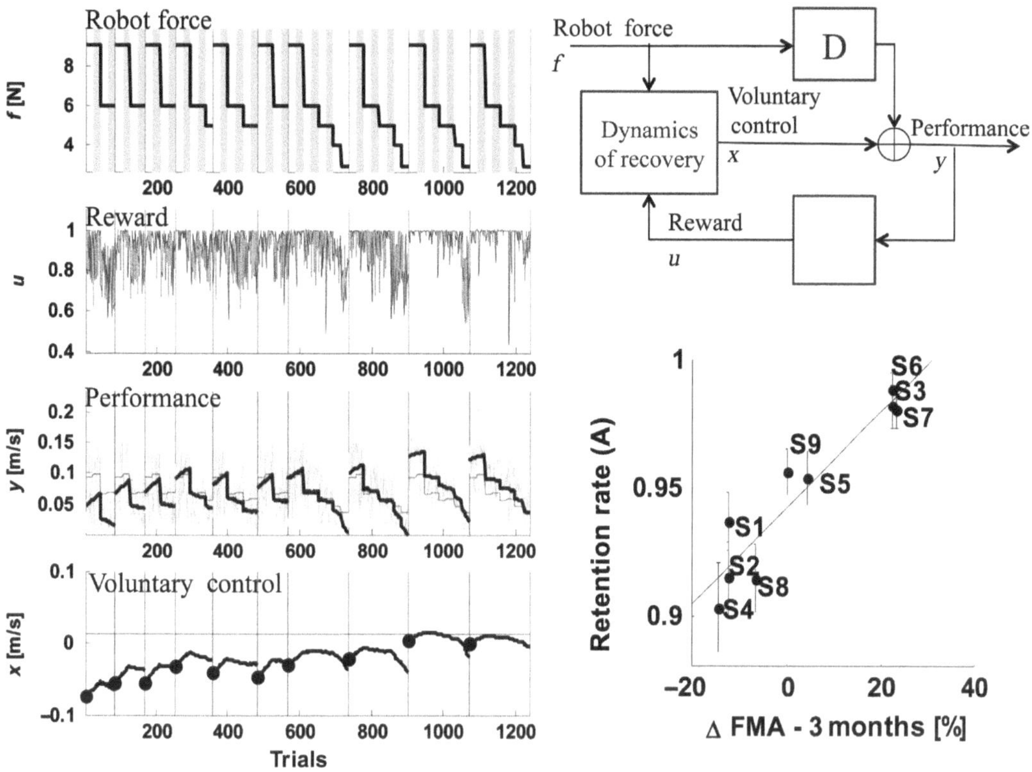

FIGURE 24.1 **State-space model of stroke recovery through robot assistance.** *Left*: Fitting performance. *Right*: Model schematic (*top*) and the main result that the model's rate of retention parameter predicts the long-term outcome—change in the Fugl-Meyer Assessment (*FMA*) in the 3 months after the end of the treatment (*bottom*). Source: *Modified from Casadio M, Sanguineti V. Learning, retention and slacking: a model of the dynamics of recovery in robot therapy. IEEE T Neur Sys Reh. 2012;20(3):286–296.*

"memory" or "retention" term—a fraction of the current amount of voluntary control; (2) a "learning" component, proportional to a "driving" signal—a function of movement performance that can be interpreted as a reward, and (3) an assistance-related component—proportional to the magnitude of the assistive force.

A distinctive feature of this model is that it posits separate mechanisms for "retention" and for the effect of assistance, i.e., the actual "slacking." These terms have often been used interchangeably; see Ref. [60] for a review that specifically covers slacking models.

The model was used to analyze the trial-by-trial time series of performance in nine chronic stroke survivors, who underwent a 10-session training protocol (Fig. 24.1, left).

The estimates of the model parameters for each subject suggested that recovery is determined by a complex interplay of memory (retention), performance, and slacking. One specific finding was that in severely impaired subjects, recovery is greater when the driving (reward) signal is greater; hence, recovery improves when the performance—not the motor error—is greater. Another finding, which somewhat confirmed the observation of Emken et al.,[23] was that a greater assistive force has a negative impact on recovery (slacking). However, only a few subjects—the least impaired—exhibited a significant "slacking" effect. The single most important finding was that the retention rate (memory decay) parameter accurately predicts the long-term outcome of the rehabilitation trial (see Fig. 24.1, bottom right). This finding is consistent with[25] the hypothesis that recovery must reach a threshold to self-sustain implies a build-up mechanism that integrates the effect of repeated motor activities. High retention is an essential prerequisite of this mechanism.

This model suggests that mechanisms of recovery may differ in different subjects (Fig. 24.2).

Specifically, some subjects turned out to be little sensitive to assistive force, whereas their performance was much more affected by the presence or absence of vision of the hand; see Fig. 24.2 (left). Other subjects exhibited a large forgetting rate, so that the inter-trial improvement in performance did not build up into a massive improvement; see Fig. 24.2 (right). Intersubject differences call for an adaptive regulation of assistance, in which peculiarities of the individual subjects are to be taken into account. Squeri et al.[61] designed an adaptive Bayesian regulator that adjusts the magnitude of assistive force (or

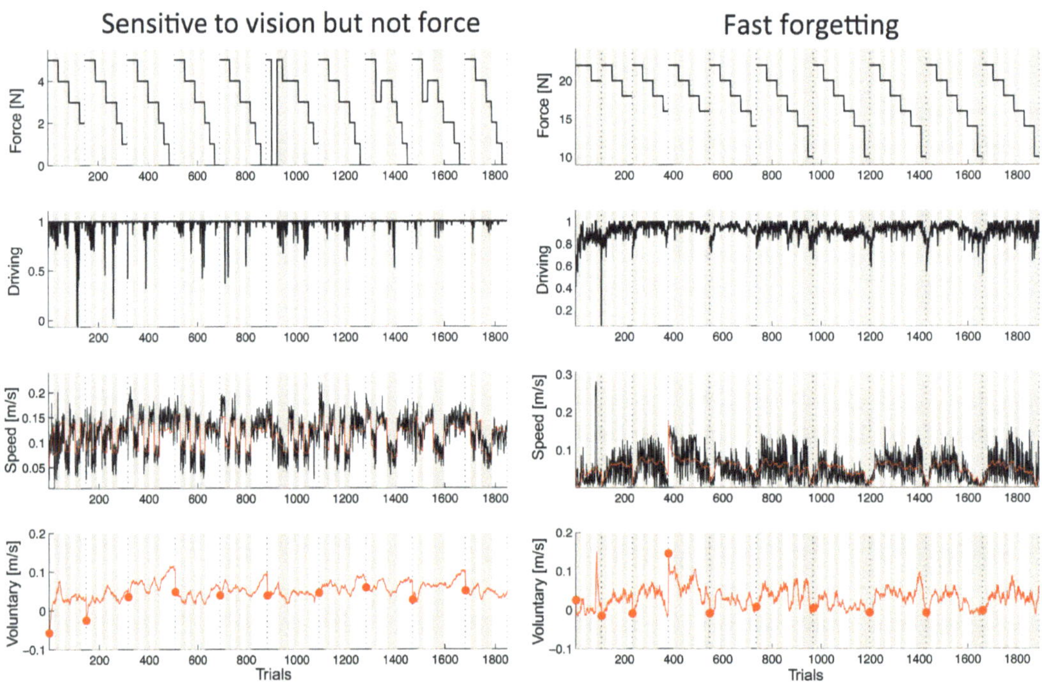

FIGURE 24.2 **Fitting examples from the state-space model.** *Left*: Performance and model fitting in a subject who is little sensitive to assistive force and is more sensitive to the presence/absence of arm vision (*white and shaded areas*). *Right*: Performance and model fitting in a fast-forgetting (low retention) subject, who is unable to build-up performance from trial to trial. Source: *Data (Right) from Casadio M, Sanguineti V. Learning, retention and slacking: a model of the dynamics of recovery in robot therapy. IEEE T Neur Sys Reh. 2012;20(3):286–296.*

other task parameter) to keep the average performance around a target magnitude (performance clamp). In this way, as performance improved, the controller automatically reduced the amount of assistance. Squeri et al.[62] used this model to analyze the temporal evolution of the subjects' voluntary control in a task that involved submovements in different directions. One single controller was used to adaptively regulate the degree of assistance in all submovements. The model suggested that the dynamics of recovery is direction-dependent, mostly due to the different sensitivity to assistance exhibited by the arm moving in different directions. These results suggest that modulation of assistance would better be made separately for each direction. More in general, they point at the need for therapy personalization.

While this model attempts to distinguish between recovery and slacking (i.e., adaptation to assistance), it remains purely descriptive and does not explicitly address the underlying mechanisms.

2.3 Multilevel Models

Other models focus on cortical reorganization in the context of a specific motor task. Han et al.[25] look at how lesions in cortical motor areas affect the mechanisms of arm selection to achieve a goal (reaching a target) and how impairment evolves through spontaneous arm use. The model accounts for motor cortical dynamics (both hemispheres) and action selection (Fig. 24.3).

The cortex is modeled as in Reinkensmeyer et al.,[18] with the addition of a Hebbian learning mechanism to account for cortical reorganization. In addition, the process of deciding which arm to use is modeled as a form of reinforcement learning.

More specifically, the preferred directions for each neuron are assumed to adapt as a function of activity. Adaptation has two aims: (1) shifting the actual encoded direction closer to the desired direction (supervised component) and (2) shifting the preferred directions of the individual neurons toward the desired direction (self-organizing component).

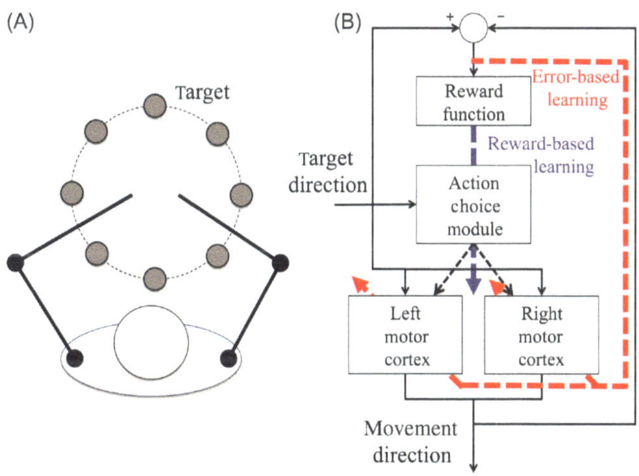

FIGURE 24.3 **Model of spontaneous recovery after stroke.** The model focuses on reaching movements (A), for which subjects spontaneously select what arm to use. (B) Includes a model of the motor cortex (both hemispheres) and an action selection module. Source: *Modified from Han CE, Arbib MA, Schweighofer N. Stroke rehabilitation reaches a threshold. PLoS Comput Biol. 2008;4(8):e1000133.*

The action selection module accounts for the process of selecting the hand that will actually make the movement. A model of the action-value mapping, based on radial basis functions, generates the expected reward as a function of the direction of the actual movement. The hand whose expected reward is maximal is selected to execute the movement. This module is driven by a reinforcement learning mechanism. After every movement, a reward signal is provided, defined as the sum of two terms, respectively reflecting (1) how close the cortex's encoded direction is to the desired movement direction and (2) the fact that the left hand is more likely chosen for leftward movements, whereas the right hand is more likely selected for rightward movements. The action-value model is updated to minimize the difference between actual and expected reward.

As in Reinkensmeyer et al.,[18] the effect of a stroke was modeled by eliminating part of the neurons within one hemisphere's motor cortex. As a result, the impaired side is initially unlikely to be selected for movements on that side, and lack of use makes its selection even less likely. In contrast, its forced use induces reorganization, so that the intact portion of that hemisphere gradually shifts its preferred directions toward those that were once covered by the impaired portion of the cortex. In summary, the model addresses the mechanisms of interaction between activity-driven cortical reorganization and functional compensation, i.e., the change in the motor strategy (in this case, from the impaired to intact arm) that is driven by the need to preserve functional performance (e.g., a high reward).

Takiyama et al.[27] used a similar model, with emphasis on bimanual training. Their main prediction is that bimanual training facilitates the adaptation of the preferred directions of the unimpaired neurons in the ipsilesional cortical hemisphere.

Han et al.'s[25] model predicts that recovery will self-sustain if the amount of spontaneous use of the impaired arm reaches a certain threshold. If this is not the case, the impaired arm will be less likely to be selected, and recovery (if any) will gradually wash out. The model makes an important qualitative prediction—an activity threshold is a necessary condition for recovery to self-sustain. This can help explaining the mechanisms of action of rehabilitation strategies that rely on forced use of the impaired arm. As a matter of fact, observations from a rehabilitation trial based on constraint-induced movement therapy were found to be consistent with the "threshold" notion.[63] Aiming to achieve an "activity threshold" rather than providing a fixed amount of training, can be seen as a form of personalization of the therapy. Studies in this direction are currently under way, one first step being to quantify the amount of arm nonuse in activities of daily living.[64]

This model is important because it is a first attempt to address the interplay between cortical reorganization and the development of compensatory strategies.

3 TOWARD A COMPUTATIONAL REHABILITATION

Computational models are widely used to investigate the mechanisms underlying motor skill learning and sensorimotor adaptation. A similar approach might potentially contribute to our understanding of the mechanisms of recovery after a stroke at "functional" level and to the design of optimal, individualized rehabilitation strategies.

3.1 Do Computational Models Provide Testable Hypotheses on the Mechanisms of Recovery?

It has been suggested[65,66] that neuromotor recovery shares at least some features with motor learning. The class of models focusing on the recovery of functions highlights this "motor learning" component. To acquire (or reacquire) a motor skill, an individual must understand how to achieve the movement goal and, more in general, learn how to get high rewards, while minimizing the necessary muscle effort. Some of the models discussed in the previous sections unveil specific aspects of this process. The observation[21] that different aspects of motor performance exhibit a different temporal evolution, which may not be monotonic, reflects the notion that recovery is a complex multifactorial process, in which maximization of performance is only one of the components. Generalization from simpler to more complex movements[45] is another crucial aspect of recovery.

Some models[25,26] describe neuromotor recovery within a reinforcement learning paradigm. Casadio et al.[24] suggest that in severe chronic stroke survivors, improvements in voluntary control are determined by performance and not error—another indication that recovery has much in common with motor skill learning.

Other models focus on the "central" level and describe cortical reorganization in terms of Hebbian and/or self-organization principles. These models highlight a number of physiological mechanisms of recovery. The model of Goodall et al.[17] predicts an initial increase of the size of perilesional "silent" areas immediately after the lesion, followed by overall reorganization. Another specific prediction is that tonic stimulation of the lesioned side would limit the size of perilesional "silent" areas. A similar effect is obtained by limiting the activity of the contralateral hemisphere. Butz et al.[19] focuses on axonal outgrowth driven by the push toward restoring the homeostatic interhemispheric balance. Its main prediction is that stimulation may facilitate axonal outgrowth but, because of a saturation effect, paused stimulation is more effective than sustained stimulation.

Still other models address reorganization mechanisms either at the level of the musculoskeletal system or at the level of CS circuitry. Reinkensmeyer et al.'s model[26] includes a basic form of reorganization, based on the recruitment of CS pathways that originate from cortical areas (e.g., the SMA) which in the intact brain are normally not used because they are less efficient, and therefore more "costly," in contributing to force generation.

In Han et al.'s model,[25] the activity of the impaired limb induces a reorganization of the ipsilesional cortical areas, which in turn makes this same hand more likely to be selected for movement. Its main prediction is that for recovery to self-sustain, activity of the impaired hand must reach a threshold. The hand selection model[25] suggests that if there is too much emphasis on performing the task but there is too little pressure toward reorganization—for instance, if the affected arm is not likely to be selected and hence the impaired hemisphere is not active enough to undergo reorganization—recovery of the paretic side functions will not self-sustain and will possibly wash out. One testable prediction is that techniques that facilitate reorganization independent of motor learning, e.g., by increasing cortical excitability, as with transcranial direct current stimulation[67] might lower the "threshold" activity level that allows recovery to self-sustain. Similarly, Takiyama's model[27] predicts that the interhemispheric activation induced by bimanual exercise facilitates cortical reorganization on the ipsilesional side.

A complementary view of the recovery process is provided by the attempts to characterize muscle synergies in a quantitative way.[20,68–70] These descriptions—not covered here—do not directly address the recovery mechanisms but provide a window into the way muscle groups are recruited in different phases of the recovery process and allow to distinguish whether recovery occurs through structural changes in muscle synergies or as a tuning of existing synergies.

Only few models address the facilitatory role of assistive forces. In principle, assistive forces enable achieving the same motor performance with less voluntary contribution,[23] but how this mechanism plays a role in the recovery process has not been extensively investigated. The "slacking" hypothesis predicts that reduced voluntary commands have a detrimental effect on recovery. A recent study confirmed this prediction by demonstrating that subjects undergoing passive training exhibit less recovery than subjects that are actively involved in the exercise through an electromyography-driven robot.[71] Casadio et al.[24] found that this effect may not be equally important in all subjects.

3.2 Can Computational Models Predict the Recovery on a Single Patient Basis?

By "predicting the recovery," we mean estimating the future evolution—either spontaneous or induced by treatment—of the impairment and/or the functional performance of a specific subject, measured in terms of clinical scales and task-specific performance indicators. A number of mathematical models[72–77] have been proposed to predict the recovery outcome within different time scales. These models use different types of information, such as the initial degree of impairment and the nature, size, or location of the lesion. In contrast, computational models explicitly account for the mechanisms by which recovery takes place, namely use-dependent neural reorganization and motor learning.

Most computational models focus on general principles but do not address the dynamics of the recovery process of a specific individual. All models of the central level account for brain areas, CS circuitry, and neural plasticity mechanisms in a way that cannot be immediately associated to empirical observations on individual subjects. However, some comparison of these models with empirical data is still possible. For instance, in Reinkensmeyer et al.,[18] the lesion is modeled as a reduction in the number of available direction-tuned cells in the motor map. The size of the lesion can be related to the degree of impairment, a correlation that has been observed empirically. In principle it would be possible to personalize the model in terms of location and size of the lesion or fraction of intact CS pathways (e.g., from imaging data), but the general focus is on quantities (e.g., the activity and the changes of the spatial tuning individual cortical columns) that are hard to associate to empirical measures.

Other models allow inferring how recovery would take place in a specific individual. These include all functional models. Specifically, Colombo et al.[21] predict the time constant of the recovery process by fitting the time course of the performance data during the rehabilitation treatment. Casadio et al.[24] predict the long-term retention of the recovery by using a state-space model and by looking at the memory decay of the learning process. These models directly refer to observable quantities, so that it is relatively easy to identify subject-specific model parameters. Model parameters capture the modality with which one subject undergoes recovery. For instance, a model may allow estimating the subject's rate of spontaneous recovery or may provide information on his/her peculiar response to mechanical perturbations (mechanical impedance).

3.3 Do Models Allow Designing Patient-Specific "Optimal" Therapy?

Computational models may enable designing patient-specific therapy aimed at maximizing speed and amount of recovery of that patient. Patient-specific models provide a description of his/her status and characteristics, e.g., impairment, articular, and muscular synergies; residual movements; and force generation abilities. They also provide a better understanding of what determines the recovery of that patient. This information can be used to define specific goals for treatment and to assess its efficacy in terms of progress of the subject's status (motor strategies and functional behavior) toward the treatment goals.

Colombo et al.[44] designed and tested a controller (PTR) that automatically selects exercise parameters (amplitude of the movement, number of submovements, and assistance modality) based on their previous work on modeling the evolution of different performance indicators.[21]

Reinkensmeyer et al.[58] proposed a robot controller based on the assist-as-needed principle, with a slacking mechanism similar to the one observed in humans. In several other applications, the magnitude of assistive force is adaptively regulated as a function of the observed outcome, in both the upper limb[54,55] and the lower limb.[56,57]

Bayesian regulation of robot-generated assistance[61] is a direct derivation of the model of recovery proposed by Casadio et al.[24]

The success of the above-mentioned approaches suggests that an even greater advantage would come from the much more ambitious goal of designing treatment modalities that directly rely on patient-specific recovery models. In principle, if a model allows predicting evolution and final outcome of a specific rehabilitation intervention, it should be possible to use this model as a basis for an optimal planning of the intervention. This may include the timing of the individual exercise sessions, the specific exercises to be administered, the trial-by-trial regulation of the degree of assistance, and the online planning of assistive or resistive forces (or other forms of stimulation, such as functional electrical stimulation (FES)).

Only few examples of model-based robot controllers for treatment have been proposed so far.[59,78] One major difficulty is that incorporating recovery models in the robot controller requires to achieve a dual goal: the controller should select the goal and the difficulty level of the exercise based on the subject's state as predicted by the model and, at the same time, the model should be continuously adjusted, on the basis of the observed subject and robot performance while the treatment proceeds. Therefore, the resulting treatment protocol has to be a trade-off between exploitation (of the model) and exploration (of the treatment control space to keep the model up to date).

4 DIRECTIONS FOR FUTURE RESEARCH

Computational approaches to the study of neuromotor recovery after stroke are innovative and promising but still in their infancy. They may shed new light on the recovery process, and along these lines there is still a lot to understand, to discover, and to integrate in the clinical practice.

4.1 Future Modeling Studies

The models reviewed in this chapter capture important and complementary aspects of the recovery process at central, muscular, and functional level.

With few exceptions, the focus is on one single level of description. Neuromotor recovery is facilitated by exercise and is mediated by neural reorganization at cortical and subcortical levels, whose physiological substrate is synaptic plasticity and rewiring through axonal outgrowth. On the one hand, the neural plasticity mechanisms constrain the way the recovery proceeds. For instance, different mechanisms, e.g., synaptic and structural plasticity, may be reflected, at functional level, in dynamic processes at different time scales. In fact, in sensorimotor adaptation, some peculiar aspects of the adaptation process have been explained in terms of concurrent multiscale processes.[79] On the other hand, sensorimotor behaviors determine the signals (sensory, motor, reward, etc.) that determine the patterns of neural activity and ultimately drive neural plasticity, thus inducing specific activity-dependent synaptic changes. Therefore, behavioral mechanisms and neural reorganization cannot be simply treated as different levels of description. Rather, both aspects need to be accounted for when modeling the neuromotor recovery process. The importance of such interplay is exemplified by the prediction that the possibility that recovery will self-sustain can be inferred by evaluating if the amount of spontaneous use of the impaired arm reaches a certain threshold.[25]

Convergence of future modeling studies toward multiple levels of description may provide a more comprehensive representation of the different aspects of the recovery process and of their interconnections. Multilevel models are being successfully used in other related fields such as musculoskeletal disorders (see Fregly et al.[16] for review) and are of great importance both for a more comprehensive understanding of the mechanisms related to the observed phenomenon, neuromotor recovery in our case, and for planning the most appropriate intervention.

One important issue is how to incorporate in these models the specific features of one patient (nature, size, and location of the lesion, and type and degree of impairment). One main difficulty is our limited understanding of the physiological mechanisms underlying synaptic plasticity and neural reorganization. These phenomena are hard to monitor and quantify experimentally, especially in patients. Modeling approaches relying on quantities that can be observed with noninvasive procedures, such as fMRI and EEG would facilitate progress in this respect.

Novel modeling approaches may also be beneficial to the study of the functional level of description. The computational approaches used for describing the neural control of movements and motor learning might find insightful application in the study of the recovery process. Optimization is a particularly viable concept to describe neuromotor recovery. Given the available sensory information and the constraints that derive from the actual impairment (sensory, motor), it may suggest how the next voluntary command is selected. Bayesian inference, optimal control, and reinforcement learning may play a role here.

4.2 Future Experimental Studies

The need to incorporate multiple levels of description to improve the models' ability to predict recovery on an individual subject basis also highlights the need for new experimental studies. Specifically, experimental studies are needed in which noninvasive recordings of both neural activity and sensorimotor performance are carried out during the recovery process, particularly to unveil how the latter is affected by exercise or other types of intervention, e.g., noninvasive neuromodulation. In particular, the roles of synaptic plasticity and the structural changes in neural connectivity should be related to improved functions, due to either compensation or true recovery.

Specific aspects of recovery that would deserve attention in experimental studies include the "slacking" phenomenon, whose impact on neuromotor recovery has never been investigated systematically[23]; the neural correlates of the reward signals; patient motivation during exercise; and their impact on neural reorganization and structural plasticity.

5 CONCLUSIONS

Computational models have been successfully applied to the study of motor learning and adaptation, providing important insights with respect to how brain controls movement and reacts to the environment or task variables' changes. Their application to the rehabilitation field is fairly new, and the early attempts[80] suggest that they will lead to a deeper understanding of the mechanisms underlying neuromotor recovery.

Recent and promising routes toward the restoration of normal functions include the interaction with virtual environments, physical assistance provided by haptic robots, neuromodulation through noninvasive stimulation of the brain cortex, and combinations of the above. Multilevel information on neuromotor performance incorporating computational models would help clinicians and neuroscientists in identifying the abnormal neural activity responsible for impaired movements and to design more effective and highly personalized forms of treatment. Even more importantly, models might be directly incorporated into technological solutions, thus constituting the basis for personalized therapy.

Acknowledgments

This work was partly supported by the EU Grant FP7-ICT- 271724 (HUMOUR), by grants from the Italian Ministry of Research (PRIN 2009, PRIN 2015), and by the COST Action TD1006 (European Network on Robotics for NeuroRehabilitation).

References

1. Nudo RJ. Postinfarct cortical plasticity and behavioral recovery. *Stroke*. 2007;38(2 Suppl):840–845.
2. Nudo RJ. Mechanisms for recovery of motor function following cortical damage. *Curr Opin Neurobiol*. 2006;16(6):638–644.
3. Barbay S, Plautz EJ, Friel KM, et al. Behavioral and neurophysiological effects of delayed training following a small ischemic infarct in primary motor cortex of squirrel monkeys. *Exp Brain Res*. 2006;169(1):106–116.
4. Schaechter JD. Motor rehabilitation and brain plasticity after hemiparetic stroke. *Prog Neurobiol*. 2004;73(1):61–72.
5. Hummel FC, Cohen LG. Non-invasive brain stimulation: a new strategy to improve neurorehabilitation after stroke? *Lancet Neurol*. 2006;5(8):708–712.
6. Blennerhassett J, Dite W. Additional task-related practice improves mobility and upper limb function early after stroke: a randomised controlled trial. *Aust J Physiother*. 2004;50(4):219–224.
7. Higgins J, Salbach NM, Wood-Dauphinee S, Richards CL, Cote R, Mayo NE. The effect of a task-oriented intervention on arm function in people with stroke: a randomized controlled trial. *Clin Rehabil*. 2006;20(4):296–310.
8. Timmermans AA, Spooren AI, Kingma H, Seelen HA. Influence of task-oriented training content on skilled arm-hand performance in stroke: a systematic review. *Neurorehabil Neural Repair*. 2010;24(9):858–870.
9. Ward NS, Cohen LG. Mechanisms underlying recovery of motor function after stroke. *Arch Neurol*. 2004;61(12):1844–1848.
10. Cirstea MC, Levin MF. Compensatory strategies for reaching in stroke. *Brain*. 2000;123(Pt 5):940–953.
11. Levin MF. Should stereotypic movement synergies in hemiparetic patients be considered adaptive? *Behav Brain Sci*. 1996;19(1):79–80.
12. Avanzino L, Bassolino M, Pozzo T, Bove M. Use-dependent hemispheric balance. *J Neurosci*. 2011;31(9):3423–3428.
13. Casadio M, Tamagnone I, Summa S, Sanguineti V. Neuromotor recovery from stroke: computational models at central, functional, and muscle synergy level. *Front Comput Neurosci*. 2013;7:97.
14. Huang VS, Krakauer JW. Robotic neurorehabilitation: a computational motor learning perspective. *J Neuroeng Rehabil*. 2009;6:5.
15. Marchal-Crespo L, Reinkensmeyer DJ. Review of control strategies for robotic movement training after neurologic injury. *J Neuroeng Rehabil*. 2009;6:20.
16. Fregly BJ, Boninger ML, Reinkensmeyer DJ. Personalized neuromusculoskeletal modeling to improve treatment of mobility impairments: a perspective from European research sites. *J Neuroeng Rehabil*. 2012;9:18.
17. Goodall S, Reggia JA, Chen Y, Ruppin E, Whitney C. A computational model of acute focal cortical lesions. *Stroke*. 1997;28(1):101–109.
18. Reinkensmeyer DJ, Iobbi MG, Kahn LE, Kamper DG, Takahashi CD. Modeling reaching impairment after stroke using a population vector model of movement control that incorporates neural firing-rate variability. *Neural Comput*. 2003;15(11):2619–2642.
19. Butz M, van Ooyen A, Worgotter F. A model for cortical rewiring following deafferentation and focal stroke. *Front Comput Neurosci*. 2009;3:10.
20. Cheung VC, Turolla A, Agostini M, et al. Muscle synergy patterns as physiological markers of motor cortical damage. *Proc Natl Acad Sci USA*. 2012;109(36):14652–14656.
21. Colombo R, Sterpi I, Mazzone A, Delconte C, Minuco G, Pisano F. Measuring changes of movement dynamics during robot-aided neurorehabilitation of stroke patients. *IEEE Trans Neural Syst Rehabil Eng*. 2010;18(1):75–85.
22. Colombo R, Pisano F, Micera S, et al. Assessing mechanisms of recovery during robot-aided neurorehabilitation of the upper limb. *Neurorehabil Neural Repair*. 2008;22(1):50–63.
23. Emken JL, Benitez R, Sideris A, Bobrow JE, Reinkensmeyer DJ. Motor adaptation as a greedy optimization of error and effort. *J Neurophysiol*. 2007;97(6):3997–4006.
24. Casadio M, Sanguineti V. Learning, retention and slacking: a model of the dynamics of recovery in robot therapy. *IEEE Trans Neural Syst Rehabil Engg*. 2012;20(3):286–296.
25. Han CE, Arbib MA, Schweighofer N. Stroke rehabilitation reaches a threshold. *PLoS Comput Biol*. 2008;4(8):e1000133.

26. Reinkensmeyer DJ, Guigon E, Maier MA. A computational model of use-dependent motor recovery following a stroke: optimizing corticospinal activations via reinforcement learning can explain residual capacity and other strength recovery dynamics. *Neural Netw.* 2012;29–30:60–69.
27. Takiyama K, Okada M. Recovery in stroke rehabilitation through the rotation of preferred directions induced by bimanual movements: a computational study. *PLoS One.* 2012;7(5):e37594.
28. Chen Y, Reggia JA. Alignment of coexisting cortical maps in a motor control model. *Neural Comput.* 1996;8 (4):731–755.
29. Shkuro Y, Glezer M, Reggia JA. Interhemispheric effects of simulated lesions in a neural model of single-word reading. *Brain Lang.* 2000;72(3):343–374.
30. Reggia JA, Gittens SD, Chhabra J. Post-lesion lateralisation shifts in a computational model of single-word reading. *Laterality.* 2000;5(2):133–154.
31. Varier S, Kaiser M, Forsyth R. Establishing, versus maintaining, brain function: a neuro-computational model of cortical reorganization after injury to the immature brain. *J Int Neuropsychol Soc.* 2011;17(6):1030–1038.
32. Georgopoulos AP, Kalaska JF, Caminiti R, Massey JT. On the relations between the direction of two-dimensional arm movements and cell discharge in primate motor cortex. *J Neurosci.* 1982;2(11):1527–1537.
33. Takahashi CD, Reinkensmeyer DJ. Hemiparetic stroke impairs anticipatory control of arm movement. *Exp Brain Res.* 2003;149(2):131–140.
34. Reinkensmeyer DJ, McKenna Cole A, Kahn LE, Kamper DG. Directional control of reaching is preserved following mild/moderate stroke and stochastically constrained following severe stroke. *Exp Brain Res.* 2002;143 (4):525–530.
35. Kamper DG, McKenna-Cole AN, Kahn LE, Reinkensmeyer DJ. Alterations in reaching after stroke and their relation to movement direction and impairment severity. *Arch Phys Med Rehabil.* 2002;83(5):702–707.
36. Li W, Huang Y, Li Y, Chen X. Brain network evolution after stroke based on computational experiments. *PLoS One.* 2013;8(12):e82845.
37. Reinkensmeyer DJ, Maier MA, Guigon E, et al. Do robotic and non-robotic arm movement training drive motor recovery after stroke by a common neural mechanism? experimental evidence and a computational model. *Conf Proc IEEE Eng Med Biol Soc.* 2009;2009:2439–2441.
38. Stinear CM, Barber PA, Smale PR, Coxon JP, Fleming MK, Byblow WD. Functional potential in chronic stroke patients depends on corticospinal tract integrity. *Brain.* 2007;130(Pt 1):170–180.
39. Feydy A, Carlier R, Roby-Brami A, et al. Longitudinal study of motor recovery after stroke: recruitment and focusing of brain activation. *Stroke.* 2002;33(6):1610–1617.
40. Ward NS, Newton JM, Swayne OB, et al. The relationship between brain activity and peak grip force is modulated by corticospinal system integrity after subcortical stroke. *Eur J Neurosci.* 2007;25(6):1865–1873.
41. Bolognini N, Pascual-Leone A, Fregni F. Using non-invasive brain stimulation to augment motor training-induced plasticity. *J Neuroeng Rehabil.* 2009;6:8.
42. Krebs HI, Hogan N, Aisen ML, Volpe BT. Robot-aided neurorehabilitation. *IEEE Trans Rehabil Eng.* 1998;6 (1):75–87.
43. Colombo R, Pisano F, Mazzone A, et al. Design strategies to improve patient motivation during robot-aided rehabilitation. *J Neuroeng Rehabil.* 2007;4:3.
44. Colombo R, Sterpi I, Mazzone A, Delconte C, Pisano F. Taking a lesson from patients' recovery strategies to optimize training during robot-aided rehabilitation. *IEEE Trans Neural Syst Rehabil Eng.* 2012;20(3):276–285.
45. Dipietro L, Krebs HI, Fasoli SE, Volpe BT, Hogan N. Submovement changes characterize generalization of motor recovery after stroke. *Cortex.* 2009;45(3):318–324.
46. Panarese A, Colombo R, Sterpi I, Pisano F, Micera S. Tracking motor improvement at the subtask level during robot-aided neurorehabilitation of stroke patients. *Neurorehabil Neural Repair.* 2012;26(7):822–833.
47. Smith AC, Frank LM, Wirth S, et al. Dynamic analysis of learning in behavioral experiments. *J Neurosci.* 2004;24(2):447–461.
48. Reinkensmeyer DJ. How to retrain movement after neurologic injury: a computational rationale for incorporating robot (or therapist) assistance. In: *25th Annual International Conference of the IEEE Engineering in Medicine and Biology Society;* September 17–21, 2003; Cancun, Mexico.
49. Donchin O, Francis JT, Shadmehr R. Quantifying generalization from trial-by-trial behavior of adaptive systems that learn with basic functions: theory and experiments in human motor control. *J Neurosci.* 2003;23 (27):9032–9045.

50. Scheidt RA, Dingwell JB, Mussa-Ivaldi FA. Learning to move amid uncertainty. *J Neurophysiol.* 2001;86 (2):971–985.
51. Thoroughman KA, Shadmehr R. Learning of action through adaptive combination of motor primitives. *Nature.* 2000;407(6805):742–747.
52. Cheng S, Sabes PN. Modeling sensorimotor learning with linear dynamical systems. *Neural Comput.* 2006;18 (4):760–793.
53. Casadio M, Morasso P, Sanguineti V, Giannoni P. Minimally assistive robot training for proprioception enhancement. *Exp Brain Res.* 2009;194(2):219–231.
54. Vergaro E, Casadio M, Squeri V, Giannoni P, Morasso P, Sanguineti V. Self-adaptive robot training of stroke survivors for continuous tracking movements. *J Neuroeng Rehabil.* 2010;7(1):13.
55. Krebs HI, Palazzolo JJ, Dipietro L, Volpe BT, Hogan N. Rehabilitation robotics: performance-based progressive robot-assisted therapy. *Auton Robot.* 2003;15(1):7–20.
56. Mihelj M, Nef T, Riener R. A novel paradigm for patient-cooperative control of upper-limb rehabilitation robots. *Adv Robotics.* 2007;21(8):843–867.
57. Riener R, Lunenburger L, Jezernik S, Anderschitz M, Colombo G, Dietz V. Patient-cooperative strategies for robot-aided treadmill training: first experimental results. *IEEE Trans Neural Syst Rehabil Eng.* 2005;13 (3):380–394.
58. Reinkensmeyer DJ, Wolbrecht E, Bobrow J. A computational model of human-robot load sharing during robot-assisted arm movement training after stroke. *Conf Proc IEEE Eng Med Biol Soc.* 2007;2007:4019–4023.
59. Wolbrecht ET, Chan V, Reinkensmeyer DJ, Bobrow JE. Optimizing compliant, model-based robotic assistance to promote neurorehabilitation. *IEEE Trans Neural Syst Rehabil Eng.* 2008;16(3):286–297.
60. Reinkensmeyer DJ, Akoner O, Ferris DP, Gordon KE. Slacking by the human motor system: computational models and implications for robotic orthoses. *Conf Proc IEEE Eng Med Biol Soc.* 2009;2009:2129–2132.
61. Squeri V, Basteris A, Sanguineti V. Adaptive regulation of assistance 'as needed' in robot-assisted motor skill learning and neurorehabilitation. In: *IEEE 12th International Conference on Rehabilitation Robotics (ICORR 2011). Zurich, Switzerland,* 2011.
62. Squeri V, Sterpi I, Basteris A, et al. Robot therapy for severely impaired stroke survivors: toward a concurrent regulation of task difficulty and degree of assistance. In: *4th IEEE RAS/EMBS International Conference on Biomedical Robotics and Biomechatronics (BioRob2012);* Rome, Italy, 2012; 1026–1031.
63. Schweighofer N, Han CE, Wolf SL, Arbib MA, Winstein CJ. A functional threshold for long-term use of hand and arm function can be determined: predictions from a computational model and supporting data from the Extremity Constraint-Induced Therapy Evaluation (EXCITE) Trial. *Phys Ther.* 2009;89(12):1327–1336.
64. Han CE, Kim S, Chen S, et al. Quantifying arm nonuse in individuals poststroke. *Neurorehabil Neural Repair.* 2013;27(5):439–447.
65. Krakauer JW. Motor learning: its relevance to stroke recovery and neurorehabilitation. *Curr Opin Neurol.* 2006;19(1):84–90.
66. Dipietro L, Krebs HI, Volpe BT, et al. Learning, not adaptation, characterizes stroke motor recovery: evidence from kinematic changes induced by robot-assisted therapy in trained and untrained task in the same workspace. *IEEE Trans Neural Syst Rehabil Eng.* 2012;20(1):48–57.
67. Reis J, Schambra HM, Cohen LG, et al. Noninvasive cortical stimulation enhances motor skill acquisition over multiple days through an effect on consolidation. *Proc Natl Acad Sci USA.* 2009;106(5):1590–1595.
68. Tresch MC, Cheung VC, d'Avella A. Matrix factorization algorithms for the identification of muscle synergies: evaluation on simulated and experimental data sets. *J Neurophysiol.* 2006;95(4):2199–2212.
69. Ting LH, Macpherson JM. A limited set of muscle synergies for force control during a postural task. *J Neurophysiol.* 2005;93(1):609–613.
70. Raghavan P, Santello M, Gordon AM, Krakauer JW. Compensatory motor control after stroke: an alternative joint strategy for object-dependent shaping of hand posture. *J Neurophysiol.* 2010;103(6):3034–3043.
71. Hu XL, Tong KY, Song R, Zheng XJ, Leung WW. A comparison between electromyography-driven robot and passive motion device on wrist rehabilitation for chronic stroke. *Neurorehabil Neural Repair.* 2009;23 (8):837–846.
72. Oczkowski WJ, Barreca S. Neural network modeling accurately predicts the functional outcome of stroke survivors with moderate disabilities. *Arch Phys Med Rehabil.* 1997;78(4):340–345.

73. Stineman MG, Maislin G, Fiedler RC, Granger CV. A prediction model for functional recovery in stroke. *Stroke*. 1997;28(3):550–556.
74. Lofgren B, Gustafson Y, Nyberg L. Cross-validation of a model predicting discharge home after stroke rehabilitation. Validating stroke discharge predictors. *Cerebrovasc Dis*. 2000;10(2):118–125.
75. Mirbagheri MM, Niu X, Varoqui D. Prediction of stroke motor recovery using reflex stiffness measures at one month. *IEEE Trans Neural Syst Rehabil Eng*. 2012;20(6):762–770.
76. Chaudhuri G, Harvey RF, Sulton LD, Lambert RW. Computerized tomography head scans as predictors of functional outcome of stroke patients. *Arch Phys Med Rehabil*. 1988;69(7):496–498.
77. Saeki S, Ogata H, Hachisuka K, Okubo T, Takahashi K, Hoshuyama T. Association between location of the lesion and discharge status of ADL in first stroke patients. *Arch Phys Med Rehabil*. 1994;75(8):858–860.
78. Reinkensmeyer DJ, Wolbrecht ET, Chan V, Chou C, Cramer SC, Bobrow JE. Comparison of three-dimensional, assist-as-needed robotic arm/hand movement training provided with Pneu-WREX to conventional tabletop therapy after chronic stroke. *Am J Phys Med Rehabil*. 2012;91(11 Suppl 3):S232–S241.
79. Smith MA, Ghazizadeh A, Shadmehr R. Interacting adaptive processes with different timescales underlie short-term motor learning. *PLoS Biol*. 2006;4(6):e179.
80. Reinkensmeyer DJ, Burdet E, Casadio M, et al. Computational neurorehabilitation: modeling plasticity and learning to predict recovery. *J Neuroeng Rehabil*. 2016;13(1):42.

CHAPTER

25

Neural Plasticity in Human Brain Connectivity: The Effects of Deep Brain Stimulation

Tim J. van Hartevelt[1,2], *Henrique M. Fernandes*[1,2], *Angus B.A. Stevner*[1,2], *Gustavo Deco*[3] and *Morten L. Kringelbach*[1,2]

[1]University of Oxford, Oxford, United Kingdom [2]Aarhus University, Aarhus, Denmark
[3]Pompeu Fabra University, Barcelona, Spain

OUTLINE

1 INTRODUCTION

It has been almost a century since one of the most important pioneers in neuroscience, Ramon y Cajal,[1] stated unequivocally that nerve paths in the adult central nervous system are fixed and cannot be regenerated once damaged. This view dominated the field for nearly half a century until the first studies showing neuronal regeneration came out in the 1960s and 1970s.[2,3] Since the discovery of regeneration and structural plasticity in the adult brain,

The Rewiring Brain
DOI: http://dx.doi.org/10.1016/B978-0-12-803784-3.00025-1

this topic has been of increasing interest to a large number of researchers. In the past decades, it has also become clear that structural plasticity and learning are closely related to each other. Whereas plasticity allows for new skills and knowledge to be stored, the act of learning can in fact stimulate this process. Much work has been done recently showing both functional and structural changes including gray and white matter connectivity changes.[4–6]

Especially today, in an ever-aging society, regeneration research is of great importance as the number of people suffering from degenerative disorders (e.g., Parkinson's disease (PD) and Alzheimer's disease) increases. The most common treatment for PD consists of pharmaceutical interventions such as L-DOPA. Although these drugs show great efficacy, they do not stimulate regeneration of affected brain areas. An alternative treatment method, although mostly used as a last resort intervention, is deep brain stimulation (DBS). DBS is a neurosurgical procedure that is increasingly used to alleviate the symptoms of a number of otherwise intractable disorders including PD, essential tremor, dystonia, and chronic pain.[7,8] DBS for PD has become well established since the 1990s with two main surgical targets, namely the subthalamic nucleus (STN) and the internal globus pallidus (GPi).[9–12] DBS treatment in PD has been shown to alleviate motor symptoms, as well as affect a number of non-motor PD symptoms. For example, direct stimulation of the STN in PD patients can also improve their sense of smell.[13] In this chapter, we will explore how DBS targeting, such as the STN, can have widespread effects reaching far beyond the first-degree neural connections from the area of stimulation.

Especially in the past 5 years, more insights into the underlying neural mechanisms of DBS have been gained through animal as well as human neuroimaging studies. So far, we know that the effects of DBS must be closely linked to at least three factors: (1) the stimulation parameters such as frequency, amplitude, pulse width, and duration; (2) the physiological properties of the neural tissue (which may be dependent on disease state); and (3) the interactions between the electrode and the surrounding neural tissue and specific anatomy of the targeted region.[7,8] The third factor includes the extended brain-wide connectivity pattern from the electrode target. This connectivity pattern is called the specific structural "fingerprint" of connectivity and is an important factor for the efficacy of DBS.[14]

The next question then becomes how DBS affects the (whole-brain) connectivity networks. Before attempting to answer this question, it is important to be aware of the issues and limitations related to investigating this problem. To investigate structural brain networks in humans, it is necessary to use noninvasive neuroimaging techniques such as magnetic resonance imaging (MRI). One immediately obvious problem here is that patients with DBS are generally not suitable for this type of investigation. Submitting a patient with DBS to an MRI scan comes with a considerable risk. When placed inside the magnetic field of the scanner, the implanted electrodes can heat up or even short-circuit and cause brain damage. Although there are ways to safely scan patients with DBS in an MRI scanner,[15,16] it is often still difficult to obtain approval from research ethics committees. Apart from the technical and ethical difficulties, the DBS electrodes are known to cause artifacts leading to signal dropout (Fig. 25.1). Although this naturally limits the quality of the data, the remaining data are of great importance as they offer the first insights into the underlying mechanisms of DBS. Due to this limitation, there are only a few studies so far that have investigated DBS using neuroimaging in humans with some of these studies looking at functional connectivity. Although this chapter focuses on structural

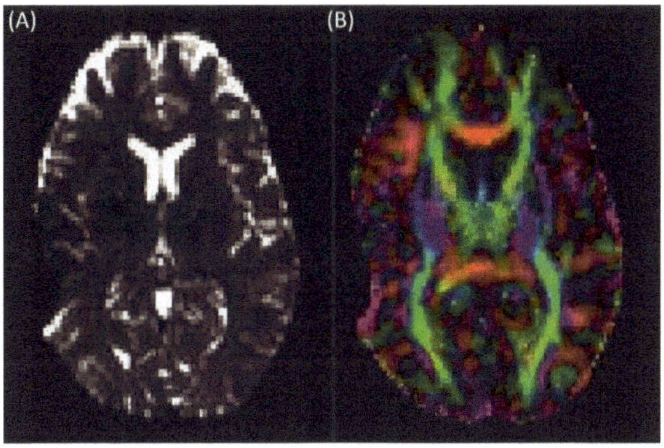

FIGURE 25.1 **Deep brain stimulation (DBS) electrode artifact from lead.** The artifact from the external lead of the DBS electrodes is visible in the form of dropout in left hemisphere in (A) the b_0-weighted image (i.e., a scan with no diffusion gradient) and (B) the diffusion tensor image with the colors representing the direction of the white matter fibers. The dropout is recognizable as a lack of signal on the left side of the brain, which in this case shows as a "gap." Source: *van Hartevelt TJ, Cabral J, Deco G, et al. Neural plasticity in human brain connectivity: the effects of long term deep brain stimulation of the subthalamic nucleus in Parkinson's disease. PLoS One. 2014;9(1):e86496.*

plasticity, these functional studies are highly relevant and do provide a significant amount of new information on a very young topic. As such, this chapter will include *structural connectivity*, which is a measure of white matter fibers connecting different regions of the brain creating a network. This chapter will also include a measure of *functional connectivity* in which connections between brain areas are expressed as a correlation matrix. This correlation is based on the similarity of activity between separate brain areas and can be measured using functional MRI (fMRI) and magnetoencephalography (MEG).

2 DEEP BRAIN STIMULATION

A deep brain stimulator consists of an electrode that is implanted in the brain and is connected to an implantable pulse generator through the stimulator lead (Fig. 25.2). The electrodes are implanted through a small hole that is drilled in the skull. This procedure is generally performed under local anesthesia using a stereotactic frame applied to the patient's skull.

As a neurosurgical procedure, DBS is increasingly being used to alleviate the symptoms of PD, as well as a number of otherwise intractable disorders including essential tremor, dystonia, and chronic pain.[8,17] DBS for PD has become well established since the 1990s with two main surgical targets, namely the STN and the GPi.[9,10]

How the efficacy of DBS for a multitude of disorders is reached is however more complicated. Crucial factors for effective symptom alleviation in patients are the selected DBS parameters, which include the stimulus amplitude, duration, and frequency band. These parameters can be set externally and can be easily changed. The efficacy of the treatment is dependent on a set of parameters that may vary with the disorder being treated, as well as

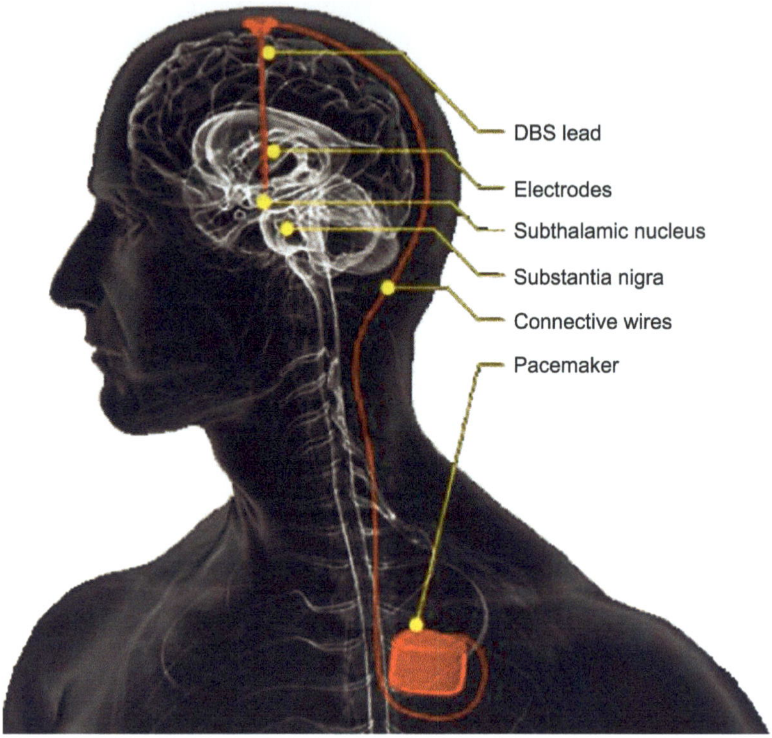

FIGURE 25.2 **Deep brain stimulation (*DBS*).** The DBS electrode, lead, and pulse generator are shown in this illustration. Also shown are the locations of the subthalamic nucleus and the substantia nigra.

the neurophysiological properties of the targeted brain area.[8] Although current systems can easily be changed externally, they do not take into account the continuous neural feedback from a patient. The changes made to the settings are mostly based on a patient's behavioral state and symptom alleviation. Finally, the parameters that are used for effective symptom alleviation generally lead to the stimulation of a volume of neural tissue larger than the target area alone. It is believed that this may be an explanation for certain side effects.[8]

To make matters more complicated, the parameters for clinical symptom alleviation depend also on the targeted brain tissue and the physiological properties of the target region. This is of course different for specific disorders and could even change with disease state, as is the case with neurodegenerative disorders such as PD.

3 DEEP BRAIN STIMULATION–INDUCED FUNCTIONAL CONNECTIVITY CHANGES

Functional connectivity changes following DBS across the whole human brain were first explored using MEG in patients with DBS for chronic pain. Specific functional changes were found in brain activity associated with pain relief such as the caudal and rostral anterior cingulate cortex (ACC).[7] A more recent study investigating the long-term changes in functional connectivity in patients with chronic pain undergoing DBS treatment showed that suppression of activity in the caudal ACC differed 12 months after surgery compared to only 1 week after surgery.[18] This difference in activity suppression after 1 year of continuous DBS is indicative of underlying structural changes attributed to the DBS treatment itself. It should be noted here that this study shows changes in ACC activity while the DBS target was also in the ACC. So although the findings are suggestive of underlying structural changes, these would most likely involve short-range connections and do not provide much information on whole-brain changes.

Although MEG studies benefit from a high temporal resolution, the spatial resolution is still inferior to that of MRI. However, until recently, submitting patients receiving DBS treatment to an MRI scan was deemed unsafe due to the implanted electrodes. In 2012, Kahan and colleagues conducted an fMRI study to investigate the effects of DBS in PD patients on functional connectivity and in the process presented a safe protocol for conducting MRI studies in patients receiving DBS.[19] In this study, the effect of DBS was investigated using dynamic causal modeling (DCM),[20] which is a Bayesian framework that aims to explain how observed blood oxygenation level dependent (BOLD) responses are generated by estimating the effective connectivity between specified regions of interest (ROI). This effective connectivity in turn can point toward a causal mechanism and includes directionality of the connection between brain regions. Here, the STN was one of the main ROIs as it is the target for DBS in PD patients. The effects of DBS are shown here to affect functional connectivity beyond short-range connections. Therapeutic STN DBS affects longer-range connections and specifically alters the cortico-thalamic coupling. Turning DBS on reverses the inhibitory effects of the insula on the thalamus during movement. There are other areas involved in this process than simply the insula and the STN. In other words, STN DBS affects the patterns of functional connectivity of cortical and basal ganglia areas toward successfully alleviating motor symptoms.[19]

In a follow-up study, the effect of STN DBS on the resting motor network in PD was investigated, again using DCM. Again, it was found that STN DBS causes activity changes (both increases and decreases in different areas) through a set of regions in the basal ganglia.[21]

It has become evident that DBS changes activity and functional connectivity in short- and long-range connections, either directly or through a network of cortical and subcortical regions. Combining the change in activity suppression after 12 months of DBS (indicative of underlying structural changes) with its effects on long-range functional connectivity raises the question whether these long-range connections show a similar long-term change. In other words, does DBS induce structural plasticity over time and over long-range connections?

4 DEEP BRAIN STIMULATION–INDUCED STRUCTURAL CONNECTIVITY CHANGES

Based on these functional imaging studies aiming to elucidate the effects of DBS, and thereby contributing to the knowledge of the underlying mechanisms of DBS, it is possible to speculate and hypothesize about structural changes in the brain as a result of DBS. The first evidence for structural changes following DBS has come from an animal study using the 6-OHDA rat model of PD showing that, alongside alleviation of motor symptoms, prolonged high-frequency STN stimulation has a neurorestorative action.[22] Specifically, STN stimulation increases new cell survival in the subventricular zone and in the striatum. This study clearly shows that STN stimulation promotes structural neural plasticity in a rat model and in two specific areas.

As mentioned before, the step to testing this in humans is difficult due to the inherent problems of scanning patients with implanted DBS electrodes, explaining the small number of studies (which are all case studies). One such case study by van Hartevelt and colleagues shows a way of using the valuable and rare MRI data from a patient with DBS.[23] As can be seen from Fig. 25.1, the artifact resulting from the externalized lead renders one hemisphere nearly useless in terms of analysis, especially when investigating the long-range connections and looking at the whole-brain network. It is however perfectly possible to investigate only one hemisphere.[24] We will come back to this point after describing the methods of the study by van Hartevelt and colleagues in more detail.

To investigate the long-term effects of DBS on structural connectivity, preoperative and 5-month postoperative diffusion tensor imaging (DTI) data from a PD patient were collected. DTI data allow for comprehensive tracking of the white matter fibers across the entire brain and create a whole-brain network (or in this case one hemisphere). This is done by parcellating the brain into a predefined number of ROIs. In this study, the Automated Anatomical Labeling (AAL) template was used resulting in a total of 45 ROIs (using both hemispheres would result in a total of 90 ROIs).[25]

The structural network can be quantified and characterized using advanced graph theoretical measures (Box 25.1). To use these measures, a network matrix has to be created based on the tractography data. In short, using a parcellated brain in combination with the DTI data allows for tracking the white matter pathways connecting one region to any other region of the brain. The full connectivity matrix will thus be a 90×90 matrix containing the number of fibers connecting every region to all the other 89 regions in the AAL (see Fig. 25.3 for an analysis pipeline). In addition to these 90 AAL regions, the STN was added to the template, as it is not originally part of the AAL extending the matrix to a total of 92 areas (90 areas from the AAL plus the bilateral STN). As mentioned before, there is a significant signal loss in the left hemisphere as a result of the externalized lead. Due to the left hemisphere being removed from further structural connectivity analysis, the reported findings are solely based on the right hemisphere consisting of 46 regions (including the STN). For the full details of the methods, please see the article by van Hartevelt and colleagues.[23]

In the study mentioned here, the authors investigated the long-term structural connectivity changes after 5 months of continuous DBS in a PD patient. The first analysis shows that there are no major differences between the first DTI scan and the 5-month follow-up scan in terms of global changes using the global graph theory measures explained in Box 25.1. Considering the findings of Kahan and colleagues that DBS leads to both increases and decreases in functional connectivity,[21] we might not be surprised of the fact that there are no global differences, as these local changes may cancel each other out at a global scale. In other words, there is no global change up or down in terms of connectivity

BOX 25.1

GRAPH THEORETICAL MEASURES

Connection density. The connection density is the number of edges in a network divided by the maximum number of possible edges [$(N^2 - N)/2$, where N is the number of nodes in the network].

Degree. The degree of a node is equivalent to the number of edges connected to that node.

Global efficiency. The global efficiency, E_g, of a network C reflects how efficiently information can be exchanged over that network. The global efficiency is calculated as the average inverse shortest path length, d_{ij}, and is inversely related to the characteristic path length.

$$E_g(C) = \frac{1}{N(N-1)} \sum_{i \neq j \in C} \frac{1}{d_{ij}} \quad (25.1)$$

Nodal efficiency. Nodal efficiency, E_n, is used to calculate how efficiently a node is connected to all other nodes in the network. Nodes with high efficiency can reach all other nodes in the network taking, on average, fewer edges. The nodal efficiency is calculated as follows:

$$E_n(i) = \frac{1}{N-1} \sum_{j \in C} \frac{1}{d_{ij}} \quad (25.2)$$

Clustering coefficient. The clustering coefficient (Cl) is the fraction of triangles around a node and is equivalent to the fraction of a node's neighbors that are neighbors of each other.

$$Cl(C) = \frac{1}{|V'|} \sum_{v \in V'} \frac{\delta_v}{\tau_v}, \quad (25.3)$$

where δ_v is the number of connected triangles, τ_v is the number of connected triples, and V' is the set of nodes with degree >2.

Small-worldness. A network C is considered to be small-world ($\sigma > 1$), if the average shortest path length is small (which is equivalent to having high efficiency E) and the clustering coefficient Cl is high, when compared to an equivalent random graph R. σ is calculated as

$$\sigma = \frac{Cl(C)/Cl(R)}{E_g(R)/E_g(C)}. \quad (25.4)$$

Hierarchy. The hierarchy coefficient β is the (positive) exponent of the power–law relationship between the clustering C_i and the degree d_i of the nodes in the network such that $C \sim d^{-\beta}$.[26,27] β was estimated using the least-squares nonlinear fitting function from Matlab. The higher the hierarchy coefficient, the more network hubs, defined so for having a large degree, have low clustering, meaning that they are more connected to nodes poorly connected to each other.

metrics, but rather a change in balance. Therefore, the hypothesis becomes that specific regions and connections will show long-term changes.

Indeed in this case study, it is clearly shown that only a subset of the 46 brain regions (AAL template + STN) show changes after long-term STN DBS. Specifically, only 17 regions exhibited changes in a local structural graph measure (*nodal efficiency*, see Box 25.1) when comparing the 5-month postoperative DTI data with the preoperative DTI data. It should be noted that this study employed a threshold of more than 20% increase or decrease

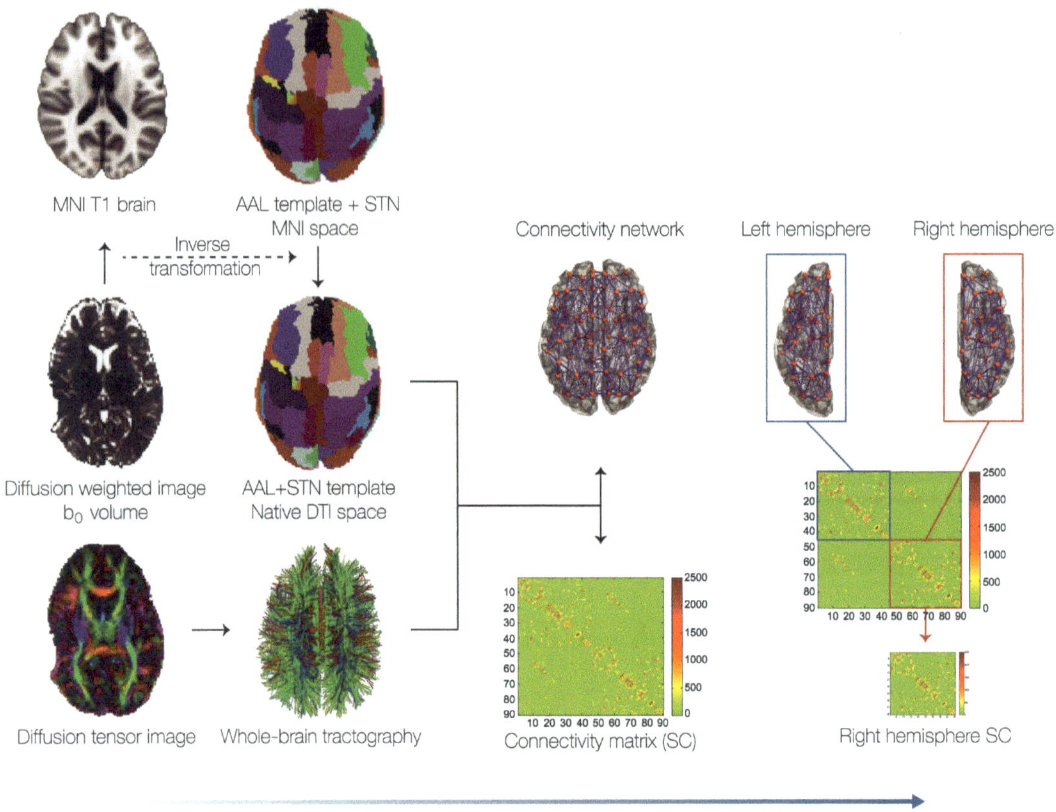

FIGURE 25.3 **Pipeline for generating the connectivity matrices.** The whole-brain connectivity matrix between the 92 brain regions using the methods described in Ref. [23]. Due to the electrode artifact in the left hemisphere (see bottom left and Fig. 25.1), the analyses on structural connectivity with deep brain stimulation (DBS) were only conducted on the right hemisphere connectivity matrices. The pipeline consists of preprocessing the diffusion tensor imaging (*DTI*) data using the FSL toolbox.[43] After the initial steps, including eddy current correction and motion correction, the brain is parcellated into 92 regions using the Automated Anatomical Labeling (*AAL*) template plus the bilateral subthalamic nucleus (*STN*). Then the connectivity matrix is calculated using the fiber tracking in FSL from each AAL region to all of the other regions in the AAL creating a 92 × 92 connectivity matrix. Source: *Taken from Van Hartevelt TJ, Cabral JRB, Møller A, et al. Evidence from a rare case-study for Hebbian-like changes in structural connectivity induced by long-term deep brain stimulation. Front Behav Neurosci. 2015;9:167.*

between post-DBS and pre-DBS measures due to the nature of the case study. Most of these areas exhibit an increased nodal efficiency after long-term DBS. The findings are summarized in Table 25.1 and Fig. 25.4, revealing which brain regions exhibit structural changes over time. These increases in nodal efficiency could possibly be explained by an increase in the number of connections from these nodes effectively decreasing the number of steps (or connections) needed between two regions in order to communicate with each other. However, this measure is also sensitive to the strength of the connections between regions. A study investigating the effects of STN DBS in a rat model of PD has shown that high-frequency stimulation did not affect proliferation of cells. However, it did result in an

TABLE 25.1 Regions of Interest From the Right Hemisphere (AAL + STN)

AAL region	Left	Right	Preoperative nodal efficiency	Postoperative nodal efficiency	Difference (%)
Precentral gyrus	1	47	154.83	111.39	**− 28.05**
Superior frontal gyrus, dorsolateral	2	48	142.49	163.69	14.87
Superior frontal gyrus, orbital part	3	49	97.46	128.07	**31.41**
Middle frontal gyrus	4	50	111.20	124.39	11.86
Middle frontal gyrus, orbital part	5	51	65.53	75.46	15.14
Inferior frontal gyrus, opercular part	6	52	145.84	126.38	− 13.34
Inferior frontal gyrus, triangular part	7	53	91.04	94.65	3.97
Inferior frontal gyrus, orbital part	8	54	95.03	123.60	**30.06**
Rolandic operculum	9	55	224.32	227.93	1.60
Supplementary motor area	10	56	123.85	102.34	− 17.36
Olfactory cortex	11	57	113.32	161.20	**42.26**
Superior frontal gyrus, medial	12	58	111.14	110.65	− 0.43
Superior frontal gyrus, medial orbital	13	59	126.44	138.98	9.92
Gyrus rectus	14	60	109.46	154.57	**41.21**
Insula	15	61	108.30	126.86	17.13
Anterior cingulate and paracingulate gyri	16	62	165.09	194.93	18.08
Middle cingulate and paracingulate gyri	17	63	169.76	174.84	3.00
Posterior cingulate gyrus	18	64	129.20	162.06	**25.43**
Hippocampus	19	65	112.79	166.70	**47.80**
Parahippocampal gyrus	20	66	110.91	119.84	8.05
Amygdala	21	67	113.45	156.20	**37.68**
Calcarine fissure	22	68	120.80	178.42	**47.70**
Cuneus	23	69	200.12	207.92	3.90
Lingual gyrus	24	70	91.98	135.11	**46.90**
Superior occipital gyrus	25	71	139.42	173.18	**24.22**
Middle occipital gyrus	26	72	136.09	130.53	− 4.08
Inferior occipital gyrus	27	73	109.82	124.72	13.57
Fusiform gyrus	28	74	149.29	157.77	5.68
Postcentral gyrus	29	75	128.16	109.54	− 14.53
Superior parietal gyrus	30	76	118.85	114.12	− 3.98
Inferior parietal gyri	31	77	193.01	160.54	− 16.82

(Continued)

TABLE 25.1 (Continued)

AAL region	Left	Right	Preoperative nodal efficiency	Postoperative nodal efficiency	Difference (%)
Supramarginal gyrus	32	78	135.67	163.88	**20.79**
Angular gyrus	33	79	131.60	136.13	3.44
Precuneus	34	80	136.77	162.83	19.05
Paracentral lobule	35	81	124.05	126.07	1.63
Caudate nucleus	36	82	99.06	105.85	6.85
Putamen (lenticular nucleus)	37	83	102.70	110.43	7.52
Pallidum (lenticular nucleus)	38	84	113.83	127.50	12.00
Thalamus	39	85	81.77	114.11	**39.54**
Heschl's gyrus	40	86	222.10	279.70	**25.93**
Superior temporal gyrus	41	87	154.29	174.56	13.13
Temporal pole: superior temporal gyrus	42	88	133.97	138.50	3.38
Middle temporal gyrus	43	89	93.18	92.25	−1.00
Temporal pole: middle temporal gyrus	44	90	108.33	136.92	**26.38**
Inferior temporal gyrus	45	91	81.71	97.19	18.95
Subthalamic nucleus	46	92	96.72	164.80	**70.38**

The regions identified to have more than 20% difference between preoperative and postoperative nodal efficiency are shown in bold. The numbers in the column corresponding to the right hemisphere correspond to the numbering used in Fig. 25.4.
From van Hartevelt TJ, Cabral J, Deco G, et al. Neural plasticity in human brain connectivity: the effects of long term deep brain stimulation of the subthalamic nucleus in Parkinson's disease. PLoS One. 2014;9(1):e86496.

increased survival of cells. In other words, although the amount of new cells is unchanged, due to the increased cell survival, DBS can lead to a net increase of neurons later on.[22]

One main finding in this study is the decrease in nodal efficiency in the precentral sensory motor area, while there were increases in the subthalamic nucleus, the thalamus, and the supramarginal gyrus. Earlier studies (e.g., Kahan and colleagues[21]) have clearly shown that these areas are implicated in thalamocortical loops. Therefore, it is remarkable to find corresponding structural network changes in these areas indicating neural plasticity after prolonged STN DBS.

Additional increases in nodal efficiency were found in prefrontal/limbic regions, which include regions of the olfactory cortex, amygdala, hippocampus, temporal pole, and orbitofrontal cortex. Most of these regions are known to be involved in olfactory processing.[28,29] The importance of finding significant differences in these olfactory areas lies in the fact that olfactory dysfunction is one of the most common non-motor symptoms in PD.[30,31] These changes in structural connectivity in olfactory areas after long-term STN DBS might point toward changed olfactory functioning after long-term DBS. This corresponds to the significant increase in odor discrimination in PD patients with DBS *on* compared to *off* stimulation.[13]

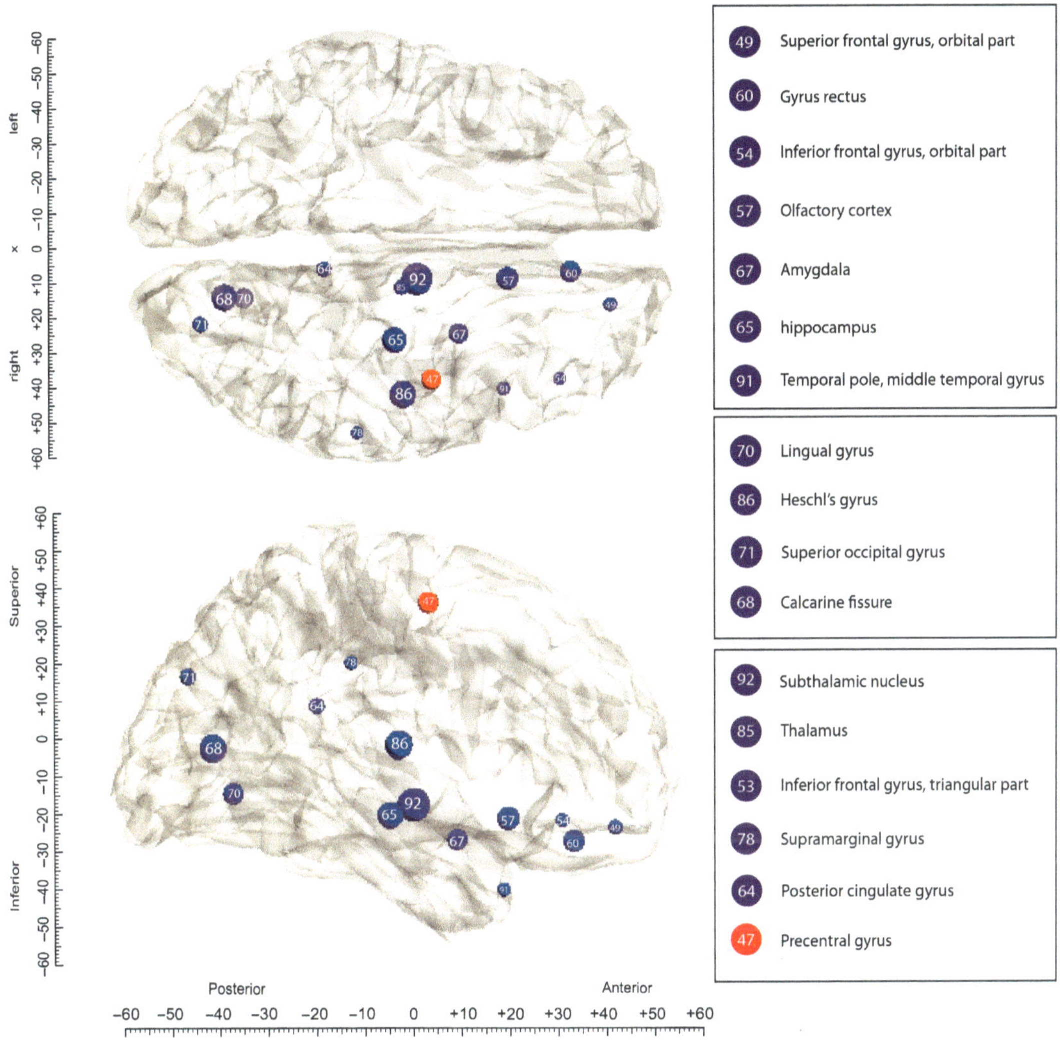

FIGURE 25.4 **Nodal efficiency changes between pre-deep brain stimulation (DBS) and post-DBS structural networks.** The regions exhibiting a difference in nodal efficiency between pre-DBS and post-DBS measures are plotted on three-dimensional renderings of the human brain as seen from above (*top*) and from the side (*bottom*). While the size of the spheres indicates the magnitude (size) of the difference, the color indicates an increase (*blue*) or decrease (*red*) in nodal efficiency after DBS. The number inside each sphere corresponds to the Automatic Anatomical Labeling index reported in Table 25.1. Source*: van Hartevelt TJ, Cabral J, Deco G, et al. Neural plasticity in human brain connectivity: the effects of long term deep brain stimulation of the subthalamic nucleus in Parkinson's disease. PLoS One. 2014;9(1):e86496.*

Changes in structural connectivity in the form of higher nodal efficiency were also found in sensory regions including visual and auditory cortices. Specifically changes in the calcarine fissure and surrounding cortex, the superior occipital gyrus, the lingual gyrus, and Heschl's gyrus were found after long-term DBS. This of course could be linked to a normalization of sensory systems with DBS given that visual and auditory hallucinations and other sensory impairments are common in PD.[32] Perhaps more convincing is the finding that DBS is known to modulate auditory processing in advanced PD.[33]

So it has now been shown that structural changes occur after long-term STN DBS in areas that were specifically shown to exhibit changes in functional connectivity following STN DBS.[21] This case study also expanded on those findings by identifying more remote brain areas also affected by long-term STN DBS. All these findings add pieces to solving the puzzle of finding the exact underlying mechanisms of how DBS works. The next question then becomes how STN DBS induces long-term structural changes in remote areas. In a follow-up study to the case study above, van Hartevelt and colleagues used computational modeling as a means to answer this question.[34] Using a dynamic mean field model (see Box 25.2 for details on this model), it is possible to simulate functional connectivity based on structural connectivity. First, the differences in structural connectivity between the pre-DBS and post-DBS structural networks are calculated. Then the model is applied and the pre-DBS functional connectivity is generated. The simulated post-DBS functional connectivity is calculated by perturbing the model inputs on the pre-DBS data so that the differences in functional connectivity (post-DBS − pre-DBS) maximally correlate with the differences previously calculated in structural connectivity. Using this computational modeling approach allows for specific tuning of network inputs from selected brain regions. In short, the connections from the region of the DBS electrode, i.e., the STN, can be modulated in the model.

This follow-up study shows that the modeled functional connectivity derived from the post-DBS data can be best reproduced by selectively changing the inputs from the STN to the thalamus, putamen, caudate, and the STN itself. Particularly interesting here is the finding that the connectivity strength (or coupling weight) between the STN and the thalamus is most sensitive to changes in terms of predicting long-term changes. Fig. 25.5 clearly shows the result correlation between the modeled post-DBS functional connectivity and the predicted modeled functional connectivity based on model perturbations for these four main connections from the STN. The results from this study suggest that DBS-induced changes arise in a fashion akin to the Hebbian-learning principle on a macroscale, that is, constant stimulation of specific connections leads to structural changes in those connections. The important role of the thalamus in STN DBS is shown both in this study[34] as well as is in the previously found structural nodal efficiency changes[23] and the functional connectivity analysis by Kahan and colleagues.[19] As the stimulation parameters were fixed to maximally alleviate symptoms, it is currently unknown how different settings such as an increase or decrease in stimulation frequency would affect long-term structural changes. Especially as different stimulation parameters can have an excitatory or inhibitory effect, it would be plausible that these would result in different long-term structural changes.

5 IMPACT OF STRUCTURAL CONNECTIVITY CHANGES ON BRAIN DYNAMICS

To gain insight into the impact of the local structural changes (as shown in Figs. 25.4 and 25.5) induced by DBS in terms of whole-brain functional connectivity, the computational modeling again plays an important role. The dynamic mean field model described

BOX 25.2

DYNAMIC MEAN FIELD MODEL

In short, the dynamic mean field approximation reduces a spiking network model of a local cortical area, including the whole dynamics of excitatory and inhibitory populations of spiking neurons interconnected by AMPA, GABA, and NMDA receptors and their respective equations,[37] to a single one-dimensional equation.[38,39] Thus, the global brain dynamics, represented as a network of interconnected local networks, can be simply and consistently described by the following set of coupled differential equations:

$$\frac{dS_i(t)}{dt} = -\frac{S_i}{\tau_S} + (1 - S_i)\gamma H(x_i) + \sigma v_i(t), \tag{25.5}$$

$$H(x_i) = \frac{ax_i - b}{1 - \exp(-d(ax_i - b))}, \tag{25.6}$$

$$x_i = wJ_N S_i + GJ_N \sum_j C_{ij} S_j + I_0, \tag{25.7}$$

where $H(x_i)$ and S_i denote the population firing rate and the average synaptic gating variable (i.e., the average fraction of open channels of all three receptor types) at the local cortical area i (from 1 to $N = 45$ areas in our case), respectively. Here x_i is the total input current in area i. $w = 0.9$ is the local excitatory recurrence and C_{ij} corresponds to the coupling weight between the areas i and j. Note that C_{ij} is estimated from the structural connectivity, i.e., in proportion to the number of white-matter tracts detected between areas i and j, and therefore this parameter is changed between, for example, pre-DBS and post-DBS. G is the global coupling weight that scales all the couplings from C uniformly. Parameter values for the input–output function (Eq. (25.6)) are $a = 270$ VnC, $b = 108$ Hz, and $d = 0.154$ s. The kinetic parameters are $\gamma = 0.641/1000$ (the factor 1000 is for expressing everything in ms), and $\tau_s = 100$ ms. The synaptic couplings are $J_N = 0.2609$ nA and the overall effective external input is $I_0 = 0.3$ nA. In Eq. (25.5), v_i is uncorrelated standard Gaussian noise and the noise amplitude at each node is $\sigma = 0.001$ nA.

The Balloon–Windkessel hemodynamic model can be used to transform the mean field activity into (simulated) BOLD signal, which describes the transduction of neuronal activity into BOLD signal.[20] The level of synaptic activity in each brain region represented by the synaptic variable S_i is used to compute the BOLD signal estimation in that specific brain region (as in Refs. [23,38,40]). The simulated BOLD signal is then downsampled at 2 s to have the same temporal resolution typically used in fMRI. The simulated functional connectivity between all brain areas is obtained by computing the temporal correlation matrix of the simulated BOLD signals.

in Box 25.2 can be used again to simulate the spontaneous dynamics of the brain based on the empirical structural input. Going one step further than simply applying graph theory, the modeled functional connectivity matrices can be used to calculate other global functional connectivity properties, such as measures of segregation and integration or the

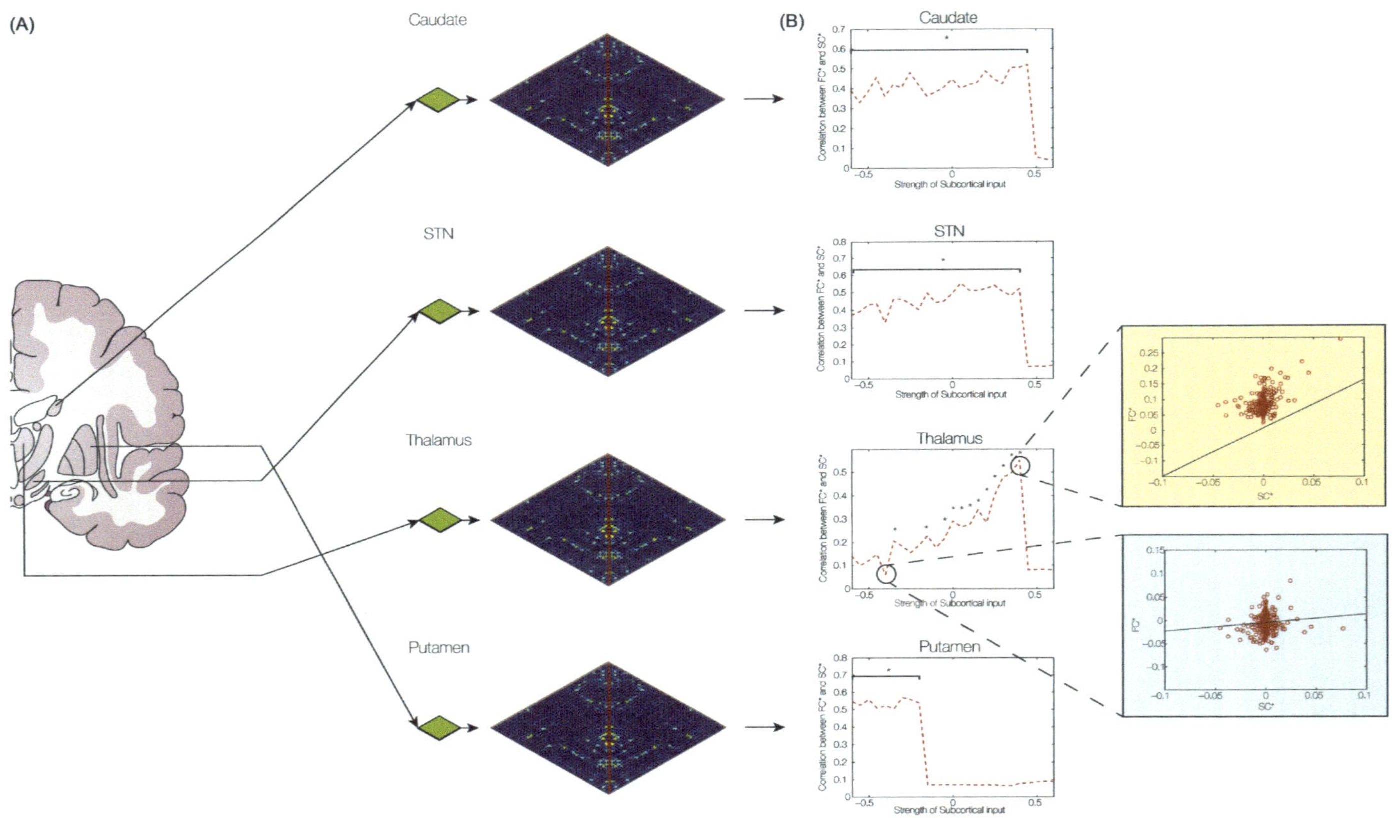

FIGURE 25.5 **Determining the contribution of subcortical connections of the subthalamic nucleus (*STN*).** (A) The optimum model fit was calculated and each of the parameters of the model (see Box 25.2) was varied for the connections from the STN to the putamen, caudate, thalamus and STN itself to estimate their contributions to the changes in functional connectivity caused by deep brain stimulation (DBS). The resulting simulated functional connectivity matrices (*dark blue*) are shown for each of the connections from the STN. Compared to no DBS, the optimum fit was highly significant ($P < 10^{-7}$), showing that including the DBS input of the STN into the model accurately predicts the simulated functional connectivity. (B) Clearly demonstrates the importance of the changes in connectivity weights to the thalamus and the putamen with the consequences of the best fit in thalamic connectivity shown in top right (*light yellow*) and poorest fit shown in bottom right (*light blue*) (with a star indicating significance levels of $P < 0.05$). In other words, changing the coupling strength between the STN and the thalamus from −0.5 to 0.5 significantly improves the correlation between the observed difference in structural connectivity between pre- and post-DBS and the simulated difference in functional connectivity between pre- and post-DBS. Furthermore, it is shown that the optimal coupling strength between the STN and the thalamus is around 0.4. Source: *van Hartevelt TJ, Cabral JRB, Møller A, et al. Evidence from a rare case-study for Hebbian-like changes in structural connectivity induced by long-term deep brain stimulation. Front Behav Neurosci. 2015;9:167.*

global bifurcation point, which represents an abrupt qualitative change in the system's dynamics when one or more parameters pass through critical values. Although there are models such as the Hopf model, which uses local bifurcation, the dynamic mean field model calculates the global bifurcation point as the critical point where the whole-brain dynamics become unstable. The bifurcation analysis uses rigorous simulations and is based on finding the fixed bifurcation point of a network, that is, the point at which the global dynamics become unstable, as a function of the model parameters. Long-lasting simulations are run until convergence is reach to find the maximum stable point on the edge of the bifurcation and each simulation starts for 1000 different random initial conditions. The *segregation* and *integration* measures, two fundamental aspects of brain organization, refer to the segregation of brain areas (that differ in terms of local functional specialization) and their global integration during perception and behavior.[35,36] In Fig. 25.6 it can be seen that long-term DBS restores the bifurcation point back to a healthy regime compared to the pre-DBS bifurcation point.

In other words, the structural changes due to long-term DBS are highly correlated with a change of the bifurcation point back to a healthy regime. Furthermore, as can be seen in Fig. 25.7, DBS-induced structural changes also bring the integration and segregation measures toward a healthy regime, again not yet to the level of a healthy control group though. Taken together, these results indicate that prolonged DBS has restorative effects in the structural connectome and furthermore affects the functional connectivity at a global level.

6 DISCUSSION

Neural plasticity has only been known and investigated since the 1960s before which it was believed that nerve paths in the adult central nervous system were fixed. The field of DBS is even younger than that and DBS was not established as a proper treatment until the 1990s. Advanced tools for investigating possible neural plasticity in humans undergoing DBS treatment have only become available in the past 5 years with safe MRI protocols. Nonetheless, despite the short amount of time, there are already a handful of studies of which the results all point unequivocally in one direction. DBS has a restorative action on both structural and functional connectivity networks. This has been assessed using methods ranging from MEG to fMRI and DTI. The first MEG study showed an immediate effect of DBS on nearby brain regions,[41] followed by the first longitudinal study also showing prolonged changes in those same brain regions.[18] Interestingly, this follow-up study showed that the long-term effects were significantly different from the short-term changes, indicative of underlying (structural) connectivity changes.

A few fMRI studies subsequently showed, using DCM, that DBS leads to functional activity changes in brain regions far away from the stimulation site.[19,21] Among the brain regions identified by these studies were the insula and thalamus, indicating changes in the cortico-thalamic loop. The evidence so far indicates that DBS results in functional changes in both nearby and remote brain regions through the connectivity of the DBS target region and this long-term difference indicates underlying structural changes.

In recent DTI studies, the first evidence emerged for these structural changes in humans after long-term DBS.[23,34] Previous to this finding, an animal model of PD already revealed

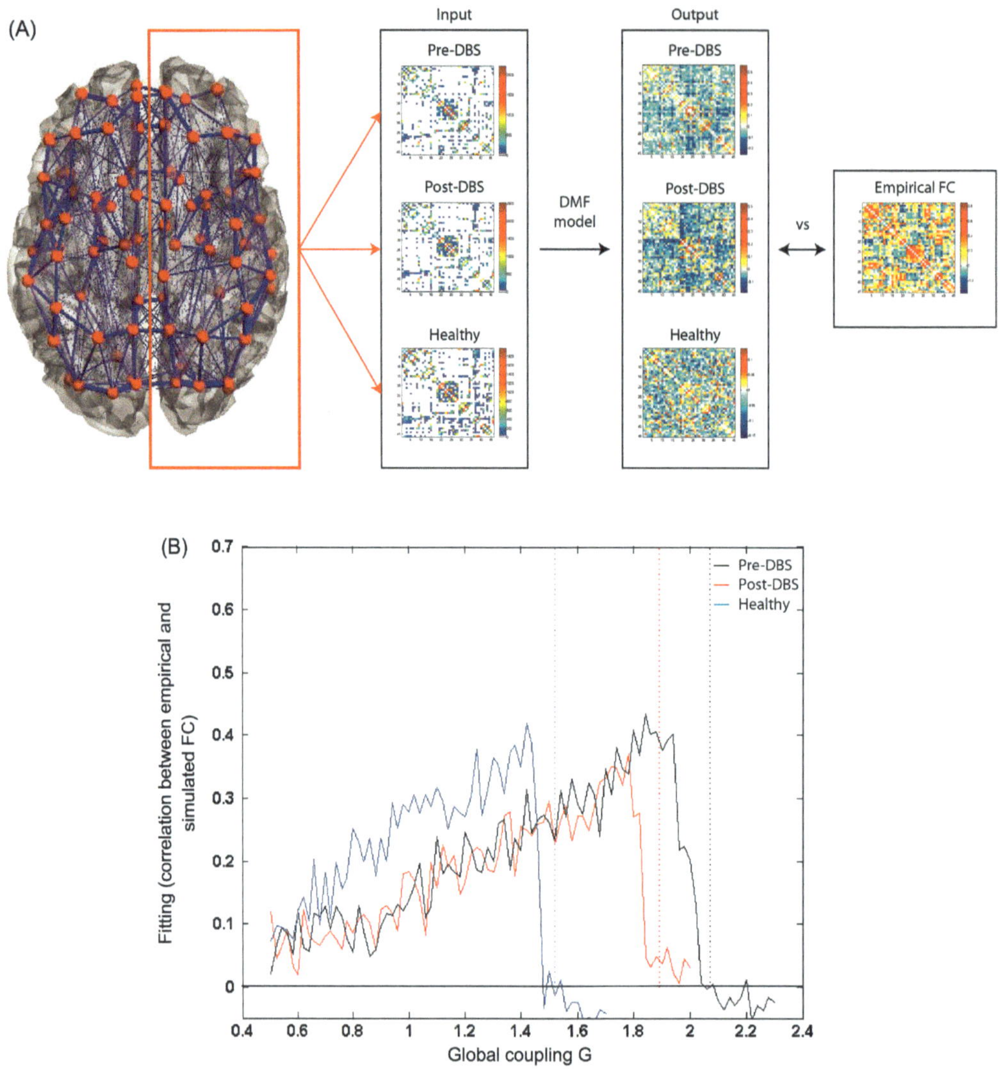

FIGURE 25.6 The impact of deep brain stimulation (*DBS*)-induced structural changes on resting-state functional connectivity. (A) Schematic overview of how the simulated functional connectivity matrices are obtained from the structural connectivity using the dynamic mean field model (*DMF*). The simulated functional connectivity (*FC*) matrices are here compared to an empirical resting state FC matrix. (B) *Solid lines* indicate the goodness of fit of simulated functional connectivity matrices with an empirical healthy functional connectivity matrix as a function of the global coupling weight (G). The simulated functional connectivity has been obtained with the structural connectivity matrices from pre-DBS (*black*), post-DBS (*red*), and healthy control (*blue*) data. *Vertical dashed lines* indicate the corresponding bifurcation points, above which the dynamics becomes unstable. Source: *van Hartevelt TJ, Cabral J, Deco G, et al. Neural plasticity in human brain connectivity: the effects of long term deep brain stimulation of the subthalamic nucleus in Parkinson's disease. PLoS One. 2014;9(1):e86496.*

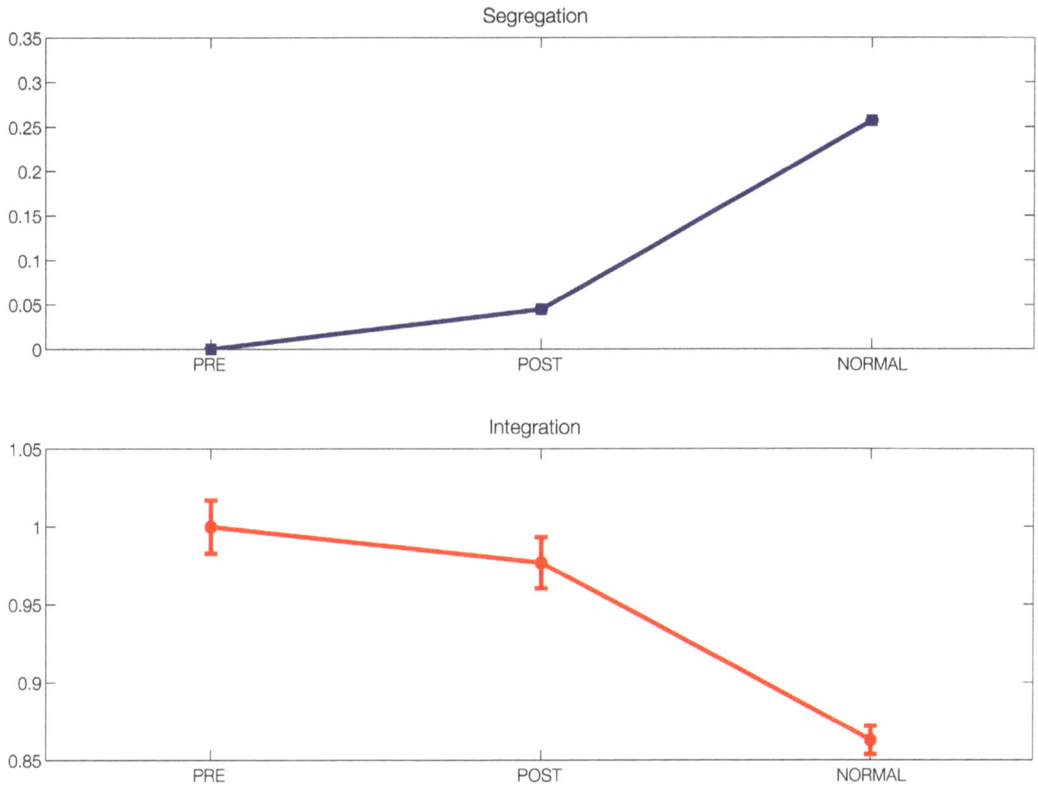

FIGURE 25.7 **Segregation and integration.** Segregation and integration measures in the simulated functional connectivity networks pre-deep brain stimulation (DBS) and post-DBS compared to those found in the simulated functional network obtained using the connectivity from a healthy population. Both measures improved after 5 months of continuous DBS, albeit not quite to the level found in a normal population. Source: *van Hartevelt TJ, Cabral JRB, Møller A, et al. Evidence from a rare case-study for Hebbian-like changes in structural connectivity induced by long-term deep brain stimulation. Front Behav Neurosci. 2015;9:167.*

the neurorestorative action of DBS of the STN in rats.[22] However, this animal model focused on two specific regions linked to neurogenesis. The recent studies using DTI to investigate connectivity changes after long-term DBS were not limited in terms of ROIs. DBS was shown to affect remote brain regions after STN DBS for 5 months. These local structural changes in turn revealed that global functional changes move back to a healthy regime (e.g., the shift in bifurcation point), corroborating findings from the previous fMRI studies by Kahan and colleagues.[19,21] In studies where MEG has been acquired in patients with STN DBS, these remote areas such as auditory cortices in the temporal lobe and frontal cortices have shown high coherence with the STN activity indicating strong functional connectivity between the STN and cortical areas.[11,33,42]

While all current evidence (albeit limited) points toward a remarkable long-term effect of DBS on structural and functional connectivity, the main purpose of DBS treatment remains symptom alleviation and not rewiring the brain (yet). Using these recent findings in combination with computational modeling might allow us to predict connectivity changes for different DBS targets other than the STN. This does require a significant number of neuroimaging studies on larger scales and in different patient groups to be performed. The feasibility of conducting such studies has been showcased as part of this chapter.

It should be pointed out that although functional and structural changes are found after DBS, these functional changes might only be apparent when stimulation of the STN is continuous. That is, these structural changes may not be fully utilized when stimulation is turned off, thus not showing any long-term effects of symptom changes while off stimulation. Kahan and colleagues have demonstrated this very clearly by comparing patients *on* and *off* DBS stimulation showing immediate changes when turning the DBS either *on* or *off* between fMRI acquisition.[19,21]

6.1 Future Experimental Studies

An interesting future direction comes from mapping the direct connections of a DBS target and finding which connections are necessary for a positive clinical outcome. Fernandes and colleagues have recently performed such a study for patients suffering from chronic pain receiving DBS in the ACC.[14] This method identified a fingerprint of structural connectivity from the activated tissue around the electrodes for both successful and unsuccessful DBS cases. In this study, only pre-DBS data were used and a postoperative computed tomographic scan was acquired to localize electrodes. Those locations were then used in the pre-DBS DTI data to map the connectivity from the electrode site and create the structural fingerprint. Combining such methods with whole-brain computational modeling and findings from the studies mentioned in this chapter could possibly lead to a new way of predicting the effectiveness and long-term effects of DBS. Making a small leap, this could be used to identify new surgical targets for disorders not (yet) treated with DBS for which affected brain areas are well known.

6.2 Future Modeling Studies

Using computational modeling with a structural network as initial input, it would be possible to predict the effect of DBS on the whole-brain network. It would of course be of great importance to go beyond using a global model and employ, for example, a Hopf bifurcation model to investigate and simulate the effects of DBS on separate brain regions by looking at the local bifurcations. In other words, how does the stability and corresponding bifurcation of any given region change as a result of DBS.

This process could of course also be approached from the other direction. If we know which regions we need to change for optimal symptom alleviation with minimal side effects, it might be possible to find a central hub that would be ideal as a DBS target. Using these different types of computational models thus not only would be able to predict changes in the whole-brain network after DBS but might also be able to identify target regions for DBS to generate changes in the brain regions that are needed for symptom alleviation.

In conclusion, it has become clear that long-term DBS leads to significant changes in local functional and structural connectivity as well as global functional dynamics. DBS helps to rebalance the brain network, not only over the short-term but also in terms of structural neuroplasticity. Furthermore, structural changes were shown to arise in a fashion similar to Hebbian learning. This could indicate that continuous stimulation of damaged fiber tracts might even have a restorative action possibly opening up a new avenue of treatment for certain types of brain damage such as stroke or mild traumatic brain injury.

References

1. Ramon y Cajal S, May RM. *Degeneration & regeneration of the nervous system*. London: Oxford University Press, Humphrey Milford; 1928.
2. Raisman G. Neuronal plasticity in the septal nuclei of the adult rat. *Brain Res*. 1969;14(1):25–48.
3. Bjorklund A, Katzman R, Stenevi U, West KA. Development and growth of axonal sprouts from noradrenaline and 5-hydroxytryptamine neurones in the rat spinal cord. *Brain Res*. 1971;31(1):21–33.
4. Sampaio-Baptista C, Filippini N, Stagg CJ, Near J, Scholz J, Johansen-Berg H. Changes in functional connectivity and GABA levels with long-term motor learning. *Neuroimage*. 2015;106:15–20.
5. Sampaio-Baptista C, Khrapitchev AA, Foxley S, et al. Motor skill learning induces changes in white matter microstructure and myelination. *J Neurosci*. 2013;33(50):19499–19503.
6. Sampaio-Baptista C, Scholz J, Jenkinson M, et al. Gray matter volume is associated with rate of subsequent skill learning after a long term training intervention. *Neuroimage*. 2014;96:158–166.
7. Kringelbach ML, Jenkinson N, Owen SLF, Aziz TZ. Translational principles of deep brain stimulation. *Nat Rev Neurosci*. 2007;8:623–635.
8. Kringelbach ML, Green AL, Owen SL, Schweder PM, Aziz TZ. Sing the mind electric—principles of deep brain stimulation. *Eur J Neurosci*. 2010;32(7):1070–1079.
9. Aziz TZ, Peggs D, Sambrook MA, Crossman AR. Lesion of the subthalamic nucleus for the alleviation of 1-methyl-4-phenyl-1,2,3,6-tetrahydropyridine (MPTP)-induced parkinsonism in the primate. *Mov Disord*. 1991;6(4): 288–292.
10. Odekerken VJ, van Laar T, Staal MJ, et al. Subthalamic nucleus versus globus pallidus bilateral deep brain stimulation for advanced Parkinson's disease (NSTAPS study): a randomised controlled trial. *Lancet Neurol*. 2012;12(1):37–44.
11. Litvak V, Jha A, Eusebio A, et al. Resting oscillatory cortico-subthalamic connectivity in patients with Parkinson's disease. *Brain*. 2011;134(Pt 2):359–374.
12. Benabid AL, Chabardes S, Mitrofanis J, Pollak P. Deep brain stimulation of the subthalamic nucleus for the treatment of Parkinson's disease. *Lancet Neurol*. 2009;8(1):67–81.
13. Hummel T, Jahnke U, Sommer U, Reichmann H, Muller A. Olfactory function in patients with idiopathic Parkinson's disease: effects of deep brain stimulation in the subthalamic nucleus. *J Neural Transm*. 2005;112(5): 669–676.
14. Fernandes HM, Van Hartevelt TJ, Boccard SGJ, et al. Novel fingerprinting method characterises the necessary and sufficient structural connectivity from deep brain stimulation electrodes for a successful outcome. *N J Phys*. 2015;17(1):015001.
15. Carmichael DW, Pinto S, Limousin-Dowsey P, et al. Functional MRI with active, fully implanted, deep brain stimulation systems: safety and experimental confounds. *Neuroimage*. 2007;37(2):508–517.
16. Kahan J, Papadaki A, White M, et al. The safety of using body-transmit MRI in patients with implanted deep brain stimulation devices. *PLoS One*. 2015;10(6):e0129077.
17. Kringelbach ML, Jenkinson N, Owen SL, Aziz TZ. Translational principles of deep brain stimulation. *Nat Rev Neurosci*. 2007;8(8):623–635.
18. Mohseni HR, Smith PP, Parsons CE, et al. MEG can map short and long-term changes in brain activity following deep brain stimulation for chronic pain. *PLoS One*. 2012;7(6):e37993.
19. Kahan J, Mancini L, Urner M, et al. Therapeutic subthalamic nucleus deep brain stimulation reverses cortico-thalamic coupling during voluntary movements in Parkinson's disease. *PloS One*. 2012;7(12):e50270.

20. Friston KJ, Harrison L, Penny W. Dynamic causal modelling. *Neuroimage*. 2003;19(4):1273–1302.
21. Kahan J, Urner M, Moran R, et al. Resting state functional MRI in Parkinson's disease: the impact of deep brain stimulation on 'effective' connectivity. *Brain*. 2014;137(Pt 4):1130–1144.
22. Khaindrava V, Salin P, Melon C, Ugrumov M, Kerkerian-Le-Goff L, Daszuta A. High frequency stimulation of the subthalamic nucleus impacts adult neurogenesis in a rat model of Parkinson's disease. *Neurobiol Dis*. 2011;42(3):284–291.
23. van Hartevelt TJ, Cabral J, Deco G, et al. Neural plasticity in human brain connectivity: the effects of long term deep brain stimulation of the subthalamic nucleus in Parkinson's disease. *PLoS One*. 2014;9(1):e86496.
24. Iturria-Medina Y, Perez Fernandez A, Morris DM, et al. Brain hemispheric structural efficiency and interconnectivity rightward asymmetry in human and nonhuman primates. *Cereb Cortex*. 2011;21(1):56–67.
25. Tzourio-Mazoyer N, Landeau B, Papathanassiou D, et al. Automated anatomical labeling of activations in SPM using a macroscopic anatomical parcellation of the MNI MRI single-subject brain. *Neuroimage*. 2002;15(1): 273–289.
26. Ravasz E, Barabasi AL. Hierarchical organization in complex networks. *Phys Rev E Stat Nonlin Soft Matter Phys*. 2003;67(2 Pt 2):026112.
27. Bassett DS, Bullmore E, Verchinski BA, Mattay VS, Weinberger DR, Meyer-Lindenberg A. Hierarchical organization of human cortical networks in health and schizophrenia. *J Neurosci*. 2008;28(37):9239–9248.
28. Rolls ET, Kringelbach ML, de Araujo IET. Different representations of pleasant and unpleasant odours in the human brain. *European J Neurosci*. 2003;18(3):695–703.
29. van Hartevelt TJ, Kringelbach ML. The olfactory system. In: Juergen KM, George P, eds. *The Human Nervous System*. Third Edition San Diego, CA: Academic Press; 2012:1219–1238.
30. Haehner A, Boesveldt S, Berendse HW, et al. Prevalence of smell loss in Parkinson's disease—a multicenter study. *Parkinsonism Relat Disord*. 2009;15(7):490–494.
31. Hawkes CH, Shephard BC, Daniel SE. Olfactory dysfunction in Parkinson's disease. *J Neurol Neurosurg Psychiatry*. 1997;62(5):436–446.
32. Shine JM, Halliday GH, Carlos M, Naismith SL, Lewis SJ. Investigating visual misperceptions in Parkinson's disease: a novel behavioral paradigm. *Mov Disord*. 2012;27(4):500–505.
33. Airaksinen K, Makela JP, Taulu S, et al. Effects of DBS on auditory and somatosensory processing in Parkinson's disease. *Hum Brain Mapp*. 2011;32(7):1091–1099.
34. Van Hartevelt TJ, Cabral JRB, Møller A, et al. Evidence from a rare case-study for Hebbian-like changes in structural connectivity induced by long-term deep brain stimulation. *Front Behav Neurosci*. 2015;9:167.
35. Tononi G, Sporns O, Edelman GM. A measure for brain complexity: relating functional segregation and integration in the nervous system. *Proc Natl Acad Sci USA*. 1994;91(11):5033–5037.
36. Deco G, Tononi G, Boly M, Kringelbach ML. Rethinking segregation and integration: contributions of whole-brain modelling. *Nat Rev Neurosci*. 2015;16(7):430–439.
37. Wong KF, Wang XJ. A recurrent network mechanism of time integration in perceptual decisions. *J Neurosci*. 2006;26(4):1314–1328.
38. Deco G, Jirsa VK. Ongoing cortical activity at rest: criticality, multistability, and ghost attractors. *J Neurosci*. 2012;32(10):3366–3375.
39. Deco G, Ponce-Alvarez A, Mantini D, Romani GL, Hagmann P, Corbetta M. Resting-state functional connectivity emerges from structurally and dynamically shaped slow linear fluctuations. *J Neurosci*. 2013;33(27):11239–11252.
40. Cabral J, Fernandes HM, Van Hartevelt TJ, James AC, Kringelbach ML, Deco G. Structural connectivity in schizophrenia and its impact on the dynamics of spontaneous functional networks. *Chaos*. 2013;23(4):046111.
41. Kringelbach ML, Jenkinson N, Green AL, et al. Deep brain stimulation for chronic pain investigated with magnetoencephalography. *Neuroreport*. 2007;18(3):223–228.
42. Hirschmann J, Özkurt TE, Butz M, et al. Distinct oscillatory STN-cortical loops revealed by simultaneous MEG and local field potential recordings in patients with Parkinson's disease. *Neuroimage*. 2011;55(3):1159–1168.
43. Jenkinson M, Beckmann CF, Behrens TE, Woolrich MW, Smith SM. FSL. *Neuroimage*. 2012;62(2):782–790.

Index

Note: Page numbers followed by "*f*" and "*t*" refer to figures and tables, respectively.

B

E

M

Q

R

X

Y

Z

FSC
www.fsc.org
MIX
Papier | Fördert
gute Waldnutzung
FSC® C083411

Kolibri 360 GmbH
Ferdinand-Jühlke-Straße 7
99095 Erfurt, Deutschland
produktsicherheit@kolibri360.de